Ergebnisse der Mathematik und ihrer Grenzgebiete

Volume 69

3. Folge

A Series of Modern Surveys in Mathematics

For further volumes:
www.springer.com/series/728

Gaëtan Chenevier • Jean Lannes

Automorphic Forms and Even Unimodular Lattices

Kneser Neighbors of Niemeier Lattices

Translated by Reinie Erné

Gaëtan Chenevier
CNRS, Institut de Mathématique d'Orsay
Université Paris-Sud
Orsay, France

Jean Lannes
Institut de Mathématiques de Jussieu
Université Paris Diderot
Paris, France

ISSN 0071-1136 ISSN 2197-5655 (electronic)
Ergebnisse der Mathematik und ihrer Grenzgebiete. 3. Folge / A Series of Modern Surveys in Mathematics
ISBN 978-3-319-95890-3 ISBN 978-3-319-95891-0 (eBook)
https://doi.org/10.1007/978-3-319-95891-0

Library of Congress Control Number: 2018963583

Mathematics Subject Classification (2010): 11E, 11F, 11H, 11M, 11R, 11E12, 11F27, 11F46, 11F55, 11F60, 11F66, 11F80, 11H06, 11H55, 11R39

This Springer imprint is published by the registered company Springer Nature Switzerland AG
The registered company address is: Gewerbestrasse 11, 6330 Cham, Switzerland

Preface

Automorphic forms are functions defined on *adele groups*, derived from harmonic analysis, whose theory forms a far-reaching generalization of that of modular forms. Langlands' famous *functoriality conjecture* predicts unexpected connections between automorphic forms associated with quite different groups. Recent advances confirm part of these general conjectures, as well as their refinements by Arthur, for the classical groups. The technicality of the proofs is formidable, but, in contrast, the statements are fascinating due to their extreme beauty, their wide range of applications, and to some extent their simplicity. Our aim in this book is to reconsider several problems of classical origin, from number theory and the theory of quadratic forms, in light of these recent results.

A special case, in which the Langlands conjectures nevertheless conserve all their flavor while being freed from numerous difficulties present in general, is that where one restricts oneself to studying automorphic forms that are *unramified at all primes*. These forms are also called *level* 1 automorphic forms. When one deals with classical or Siegel modular forms, historic examples of automorphic forms if there ever were any, this assumption means that one considers only forms that are modular for the groups $\mathrm{SL}_2(\mathbb{Z})$ or $\mathrm{Sp}_{2g}(\mathbb{Z})$, and not for general congruence subgroups.

The interest of the case of level 1 automorphic forms does not lie uniquely in the simplifications it provides; it is also very appealing for the number-theorist because of the mix of scarcity and elegance of the examples (here too, think of modular forms for $\mathrm{SL}_2(\mathbb{Z})$). Moreover, these forms are linked, sometimes very directly, sometimes much less so, and sometimes only conjecturally, to objects of algebraic geometry (varieties, stacks) that are both proper and smooth over the ring $\mathbb{Z}$ of integers, and even to *motives* over the rational numbers with everywhere good reduction, objects that are as fascinating as they are mysterious.

In this work, we aim to study the conjectures of Arthur and Langlands in the context of level 1 automorphic forms, to give precise formulations of the statements arising from the work of Arthur in this framework, and to illustrate the latter through examples that are more specific but particularly spicy. We will also compare Arthur's results with those derived from more classical constructions, namely *theta series*, which put numerous examples within reach. Some of these constructions turn out to be even richer, as we discovered, when they are combined with the *triality principle*. Let us emphasize that we wish to work, if possible, with groups of high rank, as they best reveal the richness of the general phenomena, and to move away from the classical examples provided by "small" groups such as GL_2, which have already been the subject of an extensive literature.

Our illustrations will mainly concern the theory of quadratic forms over $\mathbb{Z}$ that are nondegenerate and positive definite, in other words, the theory of *even (integral)* Euclidean lattices whose determinant is 1 or 2. This condition on the determinant means exactly that the associated projective quadric is smooth over $\mathbb{Z}$, in which case the associated special orthogonal group is smooth (and even reductive) over $\mathbb{Z}$. In the dimensions (less than or equal to 25) for which these objects are classified, the concrete problem we are going to address is the determination, for each prime number p, of the number of *p-neighborhoods* in Kneser's sense, between the classes of such objects. We will call this the *p-neighbor problem*.

The p-neighbor problem allows for a quite elementary approach: this is the point of view that we chose to follow in the introduction (Chap. 1), and also in the organization of this book, where it will serve as a connecting thread. This will also make it possible to begin by exposing the rich and fascinating history of the subject, and to highlight some simple but striking statements that are consequences of our results (the dimension 16 case, the determination of the *p-neighborhood graphs* in dimension 24, the affirmation of the Nebe–Venkov conjecture on the linear combinations of higher-genus theta series of Niemeier lattices. . .). However, we think it is helpful to explain our original motivation beforehand, which was to test Arthur's results in a context that is both concrete and of high dimension, a motivation that we will not emphasize in the beginning of the introduction.

In the remainder of this preface, we will explain the place of the p-neighbor problem in the general landscape of Langlands' conjectures, or even motives, as well as the line of thought that led us to this problem. We hope that this enlightenment (or darkening depending on the viewpoint!) will arouse the interest of the readers who are maybe less sensitive to the appeal of the theory of Euclidean lattices. In any case, this passage will be inevitable in order to understand the ideas of the solution of the p-neighbor problem we propose, which uses, in a crucial way, the aforementioned recent developments. This apparent disproportion between the sophistication of methods and the elementary aspect of the p-neighbor problem is one of the charms of the latter.

The remainder of the preface will be organized as follows. First, we return, in a more precise way, to the notion of level 1 automorphic forms (studied in Chap. 4). After having discussed a few examples, we briefly present Langlands' conjectures, emphasizing a statement that we call the *Arthur–Langlands conjecture* (Chaps. 6 and 8). We explain how Langlands and Arthur motivate this conjecture by means of a certain hypothetical group, the *Langlands group of* $\mathbb{Z}$, which we denote by $\mathbf{L}_{\mathbb{Z}}$. When one specializes the statements to *algebraic* automorphic forms, the Langlands group $\mathbf{L}_{\mathbb{Z}}$ can, to a large extent, be replaced by the absolute Galois group of $\mathbb{Q}$. We will then be in a position to provide the enlightenment we promised above and also a glimpse at some of the problems still to be solved once Arthur's results have been "put into the machine."

Automorphic Forms of Level 1

Let us fix an algebraic group (scheme) G defined and *reductive* over the ring $\mathbb{Z}$ of integers. This means that G is connected, smooth over $\mathbb{Z}$, and that its reduction modulo p is reductive over $\mathbb{Z}/p\mathbb{Z}$ for each prime p. The most important examples are GL_n and the famous Chevalley groups, or the groups that are isogenous to them such as PGL_n, but other examples will also play an important role further on.

The adele group $G(\mathbb{A})$ is a locally compact topological group in a natural way, it is the *restricted product* of the real Lie group $G(\mathbb{R})$ and of the p-adic Lie groups $G(\mathbb{Q}_p)$ over all primes p; the subgroup $G(\mathbb{Q})$ is discrete in $G(\mathbb{A})$. We denote by Z the neutral component of the center of $G(\mathbb{R})$ (so Z equals 1 if G is semisimple). The homogeneous space $G(\mathbb{Q})\backslash G(\mathbb{A})/Z$ is endowed with a finite $G(\mathbb{A})$-invariant Borel measure. A central question is to describe the Hilbert space $\mathrm{L}^2(G(\mathbb{Q})\backslash G(\mathbb{A})/Z)$ of *square integrable automorphic forms of* G, viewed as a unitary representation of $G(\mathbb{A})$ for the right translations. In accordance with our objectives, we limit ourselves to considering the subspace

$$\mathcal{A}^2(G) = \mathrm{L}^2(G(\mathbb{Q})\backslash G(\mathbb{A})/Z \cdot G(\widehat{\mathbb{Z}}))$$

of automorphic forms *of level* 1, which is nothing but the subspace of $G(\widehat{\mathbb{Z}})$-invariants of $\mathrm{L}^2(G(\mathbb{Q})\backslash G(\mathbb{A})/Z)$. This is a Hilbert space equipped with a natural unitary action of the group $G(\mathbb{R})$ and, for each prime p, with an action of the convolution ring

$$\mathrm{H}_p(G) = \mathbb{Z}[G(\mathbb{Z}_p)\backslash G(\mathbb{Q}_p)/G(\mathbb{Z}_p)] \; ,$$

whose elements are the *Hecke operators* at p; all these actions commute pairwise. The aim is to describe $\mathcal{A}^2(G)$ endowed with the commuting actions of the group $G(\mathbb{R})$ and of the commutative ring with unit $\mathrm{H}(G) := \otimes_p \mathrm{H}_p(G)$.

Denote by $\Pi(G)$ the set of isomorphism classes of objects of the form $\pi_\infty \otimes \pi_f$, with π_∞ an irreducible unitary representation of $G(\mathbb{R})$ and π_f a 1-dimensional complex representation of the ring $\mathrm{H}(G)$. Such a π_f may equally be viewed as a collection of ring homomorphisms[1] $\pi_p \colon \mathrm{H}_p(G) \to \mathbb{C}$; we also talk about *systems of eigenvalues of Hecke operators*. Moreover, denote by $\mathrm{m}(\pi)$ the *multiplicity* of π as a subrepresentation of $\mathcal{A}^2(G)$; it is finite according to Harish-Chandra. A *level 1 discrete automorphic representation* of G (from here on, "level 1" will always be dropped from the terminology) is an element π of $\Pi(G)$ with $\mathrm{m}(\pi) \neq 0$. Finally, denote by $\Pi_{\mathrm{disc}}(G) \subset \Pi(G)$ the subset of these representations. For general reasons,

---

[1] Here, we do not follow the tradition according to which π_p rather denotes the isomorphism class of the (irreducible) $\mathbb{C}[G(\mathbb{Q}_p)]$-submodule of $\mathrm{L}^2(G(\mathbb{Q})\backslash G(\mathbb{A})/Z)$ generated by an arbitrary nonzero element of π. The difference is, however, artificial, as it is a well-known consequence of the commutativity of $\mathrm{H}_p(G)$ that the two definitions contain exactly the same information.

we may write

$$\mathcal{A}^2(G) = \mathcal{A}^2_{\rm disc}(G) \overset{\perp}{\oplus} \mathcal{A}^2_{\rm cont}(G) \quad \text{with} \quad \mathcal{A}^2_{\rm disc}(G) \simeq \overset{\perp}{\bigoplus_{\pi \in \Pi_{\rm disc}(G)}} \mathrm{m}(\pi)\, \pi\,. \quad (1)$$

The space $\mathcal{A}^2_{\rm disc}(G)$ contains the subspace $\mathcal{A}^2_{\rm cusp}(G)$ of *cusp forms*, whose definition is a natural generalization of that of a cuspidal modular form. We denote by $\Pi_{\rm cusp}(G) \subset \Pi_{\rm disc}(G)$ the subset of elements appearing in $\mathcal{A}^2_{\rm cusp}(G)$. The description of the subsets $\Pi_{\rm cusp}(G) \subset \Pi_{\rm disc}(G)$ of $\Pi(G)$, and of the multiplicities $\mathrm{m}(\pi)$ above, is the heart of the problem. Indeed, we know since Langlands how to describe the *continuous* part $\mathcal{A}^2_{\rm cont}(G)$ in terms of the $\mathcal{A}^2_{\rm cusp}(L)$, where L runs through the Levi subgroups of all the proper parabolic subgroups of G defined over $\mathbb{Z}$. We will not be interested in $\mathcal{A}^2_{\rm cont}(G)$ in this book.

Two Examples

The representations π in $\Pi_{\rm disc}(G)$ have very different concrete manifestations depending on the nature of their *Archimedean component* π_∞. If U is an arbitrary irreducible unitary representation of $G(\mathbb{R})$, and if we set $\mathcal{A}_U(G) := \mathrm{Hom}_{G(\mathbb{R})}(U, \mathcal{A}^2(G))$, then we have

$$\mathcal{A}_U(G) = \mathrm{Hom}_{G(\mathbb{R})}(U, \mathcal{A}^2_{\rm disc}(G)) \simeq \bigoplus_{\{\pi\, \in\, \Pi_{\rm disc}(G)\,|\, \pi_\infty \simeq U\}} \mathrm{m}(\pi)\, \pi_f\,.$$

This is an $\mathrm{H}(G)$-module in an obvious way, and a finite-dimensional complex vector space according to Harish-Chandra. It is equivalent to describe the whole of $\Pi_{\rm disc}(G)$ or the $\mathrm{H}(G)$-modules $\mathcal{A}_U(G)$ when U runs through the unitary dual of $G(\mathbb{R})$.

In order to illustrate these notions, it is instructive to specify them in the special case of the group $G = \mathrm{PGL}_2$.[2] If U is a *discrete series* representation, say with lowest weight the (even) integer $k > 0$, then $\mathcal{A}_U(G)$ can be naturally identified with the space of cusp forms of weight k for $\mathrm{SL}_2(\mathbb{Z})$ endowed with the action of the standard Hecke operators on the latter. If $U := U_s$ is a *principal or complementary* series, parametrized in the usual way by an element $s \in i\mathbb{R} \cup [0,1[$, then $\mathcal{A}_{U_s}(G)$ can be identified with the Hecke-module of cuspidal *Maass forms* with eigenvalue $(1-s^2)/4$ for the action of the Laplace operator on the Poincaré upper half-plane. Contrary to the previous case, these spaces are very mysterious: Selberg has proved[3] $\mathcal{A}_{U_s}(G) = 0$ for $s > 0$, but we do not know any exact value of s such that $\mathcal{A}_{U_s}(G)$ is nonzero, or whether the latter can be of dimension greater than 1. Finally, according to Bargmann, the unique remaining unitary representation of $\mathrm{PGL}_2(\mathbb{R})$ is the trivial representation 1, and we obviously have $\dim \mathcal{A}_1(G) = 1$ (consider the constant functions).

---

[2] Following our definitions, we have a canonical isomorphism $\mathcal{A}^2(\mathrm{PGL}_n) \overset{\sim}{\to} \mathcal{A}^2(\mathrm{GL}_n)$.

[3] This is an Archimedean analog of Ramanujan's conjecture, still open for general congruence subgroups.

Let us now discuss the example that will be of great importance in this book. Let $n \geq 1$ be an integer and $\mathbb{R}^n$ the standard Euclidean space of dimension n. It turns out that the (compact) special orthogonal group of $\mathbb{R}^n$ is of the form $G(\mathbb{R})$ with G reductive over $\mathbb{Z}$ if and only if the integer n is congruent to -1, 0, or $+1$ modulo 8. Let us describe such a G under the assumption $n \equiv 0 \bmod 8$. It is well known that in this case, $\mathbb{R}^n$ has *even unimodular lattices*. Such a lattice L is naturally endowed with an integral quadratic form, positive definite and nondegenerate over $\mathbb{Z}$. The associated orthogonal group (scheme) O_L is smooth over $\mathbb{Z}$, and its neutral component SO_L is semisimple over $\mathbb{Z}$, with real points $\mathrm{SO}(\mathbb{R}^n)$.

We denote by $\mathcal{L}_n$ the set of even unimodular lattices in $\mathbb{R}^n$. Any two elements of $\mathcal{L}_n$ are in the same *genus*, that is, are isometric over $\mathbb{Z}_p$ for every prime p (hence over the rationals as well, according to Hasse and Minkowski). This implies, first, that the space $\mathcal{A}^2(\mathrm{SO}_L)$ depends in a nonessential way on the choice of the lattice L. In order to fix ideas, in this book, we will focus on the group $\mathrm{SO}_n := \mathrm{SO}_{\mathrm{E}_n}$, where E_n denotes the *standard* even unimodular lattice generated by $\frac{1}{2}(1,\ldots,1)$ and the n-tuples of integers $(x_1,\ldots,x_n)$ with $\sum_i x_i$ even. Another consequence is that we have a natural identification

$$\mathcal{L}_n \xrightarrow{\sim} \mathrm{SO}_n(\mathbb{Q})\backslash\mathrm{SO}_n(\mathbb{A})/\mathrm{SO}_n(\widehat{\mathbb{Z}})$$

that is compatible with the obvious actions of $\mathrm{SO}_n(\mathbb{R})$ on both sides. If 1 denotes the trivial representation of $G(\mathbb{R})$, and if $\widetilde{\mathrm{X}}_n = \mathrm{SO}(\mathbb{R}^n)\backslash\mathcal{L}_n$ denotes the finite set of proper isometry classes of elements in $\mathcal{L}_n$, we therefore have natural isomorphisms

$$\mathcal{A}_1(\mathrm{SO}_n) \simeq \{f : \widetilde{\mathrm{X}}_n \to \mathbb{C}\} \simeq \bigoplus_{\{\pi \in \Pi_{\mathrm{disc}}(\mathrm{SO}_n)\,;\,\pi_\infty \simeq 1\}} \mathrm{m}(\pi)\ \pi_f\ .$$

The vector space $\mathbb{C}[\widetilde{\mathrm{X}}_n]$, dual of $\mathcal{A}_1(\mathrm{SO}_n)$, is therefore an $\mathrm{H}(\mathrm{SO}_n)$-module in a natural way. For instance, if p is a fixed prime, it is an exercise to see that the endomorphism of $\mathbb{C}[\widetilde{\mathrm{X}}_n]$ mapping the class of a lattice to the sum of the classes of its p-neighbors is induced by an element of $\mathrm{H}_p(\mathrm{SO}_n)$, which we denote by T_p. The determination of this endomorphism is exactly the problem considered at the beginning of the introduction.[4] Let us add that the spaces $\mathcal{A}_U(\mathrm{SO}_n)$, with U arbitrary (but necessarily finite-dimensional), have similar interpretations as spaces of $\mathrm{SO}_n(\mathbb{R})$-equivariant functions $\mathcal{L}_n \to U^*$; many such spaces will play a role in this book.

Langlands' Functoriality Principle

Let us describe, rather briefly, Langlands' general conjectures in the case of level 1 automorphic forms. A starting point is the notion of *dual* group, introduced by

[4] Actually, we will mostly consider the analogous, only slightly simpler, problem in which SO_n is replaced by $\mathrm{O}_n := \mathrm{O}_{\mathrm{E}_n}$, whose only flaw is that it does not quite fit the conventions adopted here because O_n is not connected, but this slight difference is inessential.

Langlands. If G is reductive over $\mathbb{Z}$, its dual in the sense of Langlands is simply "the" complex linear algebraic reductive group, denoted by $\widehat{G}$, whose *based root datum* is *dual* (or *inverse*) to that of $G_{\mathbb{C}}$:

$G_{\mathbb{C}}$	GL_n	PGL_n	Sp_{2g}	PGSp_{2g}	SO_{2n+1}	SO_{2n}	PGSO_{2n}
$\widehat{G}$	GL_n	SL_n	SO_{2g+1}	Spin_{2g+1}	Sp_{2n}	SO_{2n}	Spin_{2n}

This group first allows Langlands to parametrize the elements of $\Pi(G)$. He observes that the *Satake isomorphism* provides a canonical bijection, for each prime p, between the set of ring homomorphisms $\mathrm{H}_p(G) \to \mathbb{C}$ and the set of semisimple conjugacy classes in $\widehat{G}(\mathbb{C})$. In a similar way, he interprets the *infinitesimal character* (in the sense of Harish-Chandra) of a unitary representation of $G(\mathbb{R})$ as a semisimple conjugacy class in the Lie algebra of $\widehat{G}$. Finally, with each element π of $\Pi(G)$ is associated a collection of conjugacy classes

$$\mathrm{c}(\pi) = (\mathrm{c}_\infty(\pi), \mathrm{c}_2(\pi), \mathrm{c}_3(\pi), \cdots)$$

that uniquely determines π_p for each prime p, as well as the infinitesimal character of π_∞, which only leaves finitely many possibilities for π_∞. These parametrizations, recalled in Chap. 6, have some very concrete aspects. For example, we will see that for π in $\Pi(\mathrm{SO}_n)$, we have the relation

$$\pi_p(\mathrm{T}_p) \quad = \quad p^{n/2-1}\,\mathrm{trace}\,\mathrm{c}_p(\pi)\ . \tag{2}$$

Let G and G' be two reductive groups over $\mathbb{Z}$, and consider a morphism of algebraic groups $r\colon \widehat{G} \to \widehat{G'}$. Langlands' *functoriality principle* predicts, for each *constituent* π of $\mathcal{A}^2(G)$, the existence of a constituent π' of $\mathcal{A}^2(G')$ that *corresponds* to π, in the sense that we have an equality of conjugacy classes $r(\mathrm{c}_v(\pi)) = \mathrm{c}_v(\pi')$ for each v in the set $\mathrm{V} := \{\infty, 2, 3, 5, \ldots\}$ of all places of $\mathbb{Q}$. It is only a principle, rather than a conjecture, as it is not quite accurate as stated, even if we give a reasonable sense to the term "constituent." In what follows, we propose to make the statement of the functoriality principle precise in the important case $G' = \mathrm{GL}_n$, in which r is nothing but an n-dimensional representation of the algebraic group $\widehat{G}$. We will later refer to this statement as the *Arthur–Langlands conjecture*.

The Langlands Group of $\mathbb{Z}$

Langlands observed that the formulation of his conjectures is enlightened if one assumes the existence of a certain group, which we will denote here by[5] $\mathbf{L}_{\mathbb{Z}}$, whose

[5] To be completely honest, Langlands considers a group that applies to all automorphic forms, rather than to level 1 forms only, of which our $\mathbf{L}_{\mathbb{Z}}$ would merely be a quotient [139, Sect. 2]. Moreover, following Arthur in [9, Sect. 8], we adopt Kottwitz's point of view [129, Sect. 12] on the

$\widehat{G}$-valued representations parametrize the automorphic representations of G in an appropriate sense. We may think of this group as being an extension of the absolute Galois group of $\mathbb{Z}$ (... trivial according to Minkowski!). For our needs in this preface, we only assume that $\mathbf{L}_{\mathbb{Z}}$ is a compact Hausdorff topological group (hence an inverse limit of compact Lie groups) satisfying the axioms denoted by (L1), (L2), and (L3) that we introduce below.

For every prime p, $\mathbf{L}_{\mathbb{Z}}$ *is endowed with a conjugacy class* Frob_p. *Moreover, the complex pro-Lie-algebra of* $\mathbf{L}_{\mathbb{Z}}$ *is endowed with a semisimple conjugacy class* Frob_∞. (L1)

Let G be reductive over $\mathbb{Z}$. Following Arthur and Langlands, we denote by $\Psi(G)$ the set of $\widehat{G}(\mathbb{C})$-conjugacy classes of continuous group homomorphisms

$$\psi\colon \mathbf{L}_{\mathbb{Z}} \times \mathrm{SL}_2(\mathbb{C}) \longrightarrow \widehat{G}(\mathbb{C}) \tag{3}$$

that are polynomial on the $\mathrm{SL}_2(\mathbb{C})$-factor. Such a ψ is called *discrete* if the centralizer C_ψ of $\mathrm{Im}\ \psi$ in $\widehat{G}(\mathbb{C})$ is finite modulo the center $\mathrm{Z}(\widehat{G})$ of $\widehat{G}(\mathbb{C})$. For example, if G is GL_n, in which case we also have $\widehat{G} = \mathrm{GL}_n$ and ψ is nothing but an n-dimensional representation of $\mathbf{L}_{\mathbb{Z}} \times \mathrm{SL}_2(\mathbb{C})$, then ψ is discrete if and only if it is an irreducible representation. We denote by $\Psi_{\mathrm{disc}}(G) \subset \Psi(G)$ the subset of classes of discrete morphisms.

In parallel with what has been done for $\Pi(G)$, Arthur and Langlands associate with each ψ in $\Psi(G)$ a collection of conjugacy classes $\mathrm{c}(\psi) = (\mathrm{c}_v(\psi))_{v \in \mathrm{V}}$ defined by $\mathrm{c}_\infty(\psi) = \psi(\mathrm{Frob}_\infty, \mathrm{e}_\infty)$ and $\mathrm{c}_p(\psi) = \psi(\mathrm{Frob}_p, \mathrm{e}_p)$, where the e_v are the elements of $\mathfrak{sl}_2(\mathbb{C})$ for $v = \infty$ and of $\mathrm{SL}_2(\mathbb{C})$ for $v = p$, defined by

$$\mathrm{e}_\infty = \begin{bmatrix} -1/2 & 0 \\ 0 & 1/2 \end{bmatrix} \quad \text{and} \quad \mathrm{e}_p = \begin{bmatrix} p^{-1/2} & 0 \\ 0 & p^{1/2} \end{bmatrix}.$$

For every integer $n \geq 1$, *there is a unique bijection*

$$\Pi_{\mathrm{disc}}(\mathrm{GL}_n) \xrightarrow{\sim} \Psi_{\mathrm{disc}}(\mathrm{GL}_n)\ , \quad \pi \mapsto \psi_\pi \tag{L2}$$

such that we have $\mathrm{c}(\pi) = \mathrm{c}(\psi_\pi)$ *for all* $\pi \in \Pi_{\mathrm{disc}}(\mathrm{GL}_n)$. *Moreover,* ψ_π *is trivial on* $\mathrm{SL}_2(\mathbb{C})$ *if and only if we have* $\pi \in \Pi_{\mathrm{cusp}}(\mathrm{GL}_n)$.

This axiom, together with the compactness of $\mathbf{L}_{\mathbb{Z}}$, implies that for any π in $\Pi_{\mathrm{cusp}}(\mathrm{GL}_n)$ and for any prime p, the eigenvalues of the conjugacy class $\mathrm{c}_p(\pi)$ all have absolute value 1: this is the so-called *generalized Ramanujan conjecture*. It also shows[6] $|\mathbf{L}_{\mathbb{Z}}^{\mathrm{ab}}| = \dim \mathcal{A}(\mathrm{GL}_1) = 1$.

Langlands group, which amounts to viewing it as a topological group rather than a pro-algebraic one as Langlands does. See [11] for another reference on the Langlands group.

[6] We will prove that (L2) also implies that $\mathbf{L}_{\mathbb{Z}}$ is connected; see Proposition 9.3.4.

For every G reductive over $\mathbb{Z}$, there exists a decomposition

$$\mathcal{A}_{\mathrm{disc}}(G) = \bigoplus^{\perp}_{\psi \in \Psi_{\mathrm{disc}}(G)} \mathcal{A}_{\psi}(G) , \tag{L3}$$

stable under $G(\mathbb{R})$ and $\mathrm{H}(G)$ and satisfying the following property: if $\pi \in \Pi(G)$ appears in $\mathcal{A}_{\psi}(G)$, then we have $\mathrm{c}(\pi) = \mathrm{c}(\psi)$.

In particular, if a representation $\pi \in \Pi(G)$ appears in a summand $\mathcal{A}_{\psi}(G)$ as in (L3), then π satisfies the Ramanujan conjecture (in the sense that for every prime p, $\mathrm{c}_p(\pi)$ is the conjugacy class of a "compact element" of $\widehat{G}(\mathbb{C})$) if and only if ψ is trivial on $1 \times \mathrm{SL}_2(\mathbb{C})$. It is Arthur's idea that the failure of Ramanujan's conjecture may, in general, be entirely explained by the presence of $\mathrm{SL}_2(\mathbb{C})$ in the definition of $\Psi(G)$ (formula (3)).

Arthur and Langlands strengthen the axiom (L3) by adding a converse statement, called the *multiplicity formula*, whose formulation, however, requires the introduction of more technical ingredients. Let us simply say that if $\psi \in \Psi_{\mathrm{disc}}(G)$ and $\pi \in \Pi(G)$ satisfy $\mathrm{c}(\pi) = \mathrm{c}(\psi)$, this formula expresses the multiplicity of π in the subspace $\mathcal{A}_{\psi}(G)$ as the scalar product of two "explicit"[7] characters of the finite group $\mathrm{C}_{\psi}/\mathrm{Z}(\widehat{G})$.

The Arthur–Langlands Conjecture

Let us go back to the statement of the Arthur–Langlands conjecture alluded to above. In order to state it, let us first assume the existence of a compact group $\mathbf{L}_{\mathbb{Z}}$ satisfying the axioms (L1), (L2), and (L3). Let G be reductive over $\mathbb{Z}$, π in $\Pi_{\mathrm{disc}}(G)$, and $r\colon \widehat{G} \to \mathrm{GL}_n$ a representation. Let $\psi \in \Psi_{\mathrm{disc}}(G)$ be such that π appears in $\mathcal{A}_{\psi}(G)$; such a ψ exists by Axiom (L3). The decomposition into irreducibles of the representation $r \circ \psi$ of the direct product $\mathbf{L}_{\mathbb{Z}} \times \mathrm{SL}_2(\mathbb{C})$ can be written as $\oplus_i r_i \otimes \mathrm{Sym}^{d_i - 1}\mathbb{C}^2$ for some irreducible representations r_i of dimension n_i of $\mathbf{L}_{\mathbb{Z}}$ and certain integers $d_i \geq 1$. By Axiom (L2), we have $r_i \simeq \psi_{\pi_i}$ for a unique π_i in $\Pi_{\mathrm{cusp}}(\mathrm{GL}_{n_i})$. In particular, for every $v \in \mathrm{V}$, we have the identity between conjugacy classes

$$r(\mathrm{c}_v(\pi)) = \bigoplus_i \mathrm{c}_v(\pi_i) \otimes \mathrm{Sym}^{d_i - 1}(e_v) \tag{4}$$

(the reader will have no trouble deciphering the meaning of the right-hand side of this equality).

[7] The definition of these characters is very delicate. One of them is a group homomorphism $\mathrm{C}_{\psi}/\mathrm{Z}(\widehat{G}) \to \mathbb{C}^{\times}$ defined by Arthur in [9, p. 55] with the help of the ε*-factors* of certain L-functions associated with ψ. The other one depends on the definition of a certain finite subset of irreducible unitary representations of $G(\mathbb{R})$ associated with ψ, denoted by $\Pi_{\infty}(\psi)$, nowadays usually called an *Arthur packet* [9, Sect. 4]. This character is nonzero if and only if π_{∞} belongs to $\Pi_{\infty}(\psi)$. In the important special case $\mathrm{C}_{\psi} = \mathrm{Z}(\widehat{G})$, the multiplicity of π in $\mathcal{A}_{\psi}(G)$ is thus nonzero if and only if we have $\pi_{\infty} \in \Pi_{\infty}(\psi)$.

As a consequence of this analysis, we have shown that the existence of a compact group $\mathbf{L}_{\mathbb{Z}}$ satisfying the axioms (L1), (L2) and (L3) implies the following statement, whose formulation does not involve $\mathbf{L}_{\mathbb{Z}}$: *For every reductive group G over $\mathbb{Z}$, every π in $\Pi_{\rm disc}(G)$, and every representation $r\colon \widehat{G} \to \mathrm{GL}_n$, there exists a unique collection of triples (d_i, n_i, π_i), with $d_i, n_i \geq 1$ integers satisfying $n = \sum_i d_i n_i$ and π_i a representation in $\Pi_{\rm cusp}(\mathrm{GL}_{n_i})$, such that Equality* (4) *holds.* This is the precise form of the Arthur–Langlands conjecture that had been promised.

In his work mentioned earlier, Arthur proved the following special cases of this conjecture: $G_{\mathbb{Q}}$ is either the symplectic group Sp_{2g} of a symplectic space over $\mathbb{Q}$ of dimension $2g$, or the special orthogonal group of a quadratic space of dimension $2n$ or $2n+1$ over $\mathbb{Q}$ that has a totally isotropic subspace of dimension n, $\pi \in \Pi_{\rm disc}(G)$ is arbitrary, and r is the natural representation of $\widehat{G}$, called the *standard representation*, whose dimension is $2g+1$, $2n$, and $2n$, respectively. For such groups, Arthur also proves a version of the multiplicity formula to which we alluded during the discussion of Axiom (L3). We will state more precise forms of Arthur's results in Chap. 8. However, let us stress that we will not say anything about Arthur's proofs; they go far beyond the scope of this work.

Galois Representations and Motives

The group $\mathbf{L}_{\mathbb{Z}}$ is subject to several other conjectures. A most tempting one is that it satisfies the *Sato–Tate property*: the Frob_p are equidistributed in the set of conjugacy classes of $\mathbf{L}_{\mathbb{Z}}$, endowed with its invariant probability measure.[8] In this section, we will instead discuss the conjectural relation between $\mathbf{L}_{\mathbb{Z}}$, Grothendieck motives, and Galois representations.

These links will only concern the quotient of $\mathbf{L}_{\mathbb{Z}}$, whose irreducible representations parametrize, in the sense of Axiom (L2), the representations π in $\Pi_{\rm cusp}(\mathrm{GL}_n)$ that are *algebraic*. Here, this adjective means that if we denote by λ_i the eigenvalues of the conjugacy class $\mathrm{c}_\infty(\pi) \subset \mathrm{M}_n(\mathbb{C})$, we have $\lambda_i - \lambda_j \in \mathbb{Z}$ for all i, j. We then denote by $\mathrm{w}(\pi)$ the maximum of the differences $\lambda_i - \lambda_j$, and call it the *motivic weight* of π.

Denote by $\overline{\mathbb{Q}} \subset \mathbb{C}$ the subfield of algebraic numbers. Fix a prime ℓ, an algebraic closure $\overline{\mathbb{Q}}_\ell$ of the field of ℓ-adic numbers, and an embedding $\iota\colon \overline{\mathbb{Q}} \to \overline{\mathbb{Q}}_\ell$. Thanks to the works of a number of mathematicians (including Clozel, Deligne, Fontaine, Grothendieck, Langlands, Mazur, Serre, Shimura, Taniyama, Tate, Weil ...), one conjectures the existence of a natural bijection $\pi \mapsto \rho_{\pi,\iota}$ between the set of algebraic π in $\Pi_{\rm cusp}(\mathrm{GL}_n)$ and the set of isomorphism classes of irreducible continuous representations $\mathrm{Gal}(\overline{\mathbb{Q}}/\mathbb{Q}) \to \mathrm{GL}_n(\overline{\mathbb{Q}}_\ell)$ that are *unramified* at each prime $p \neq \ell$ and *crystalline at ℓ* in the sense of Fontaine, with lowest *Hodge–Tate weight* 0. In particular, one requires this bijection to satisfy the equality[9]

[8] Given the connectedness of $\mathbf{L}_{\mathbb{Z}}$, it would be easy to see, for instance, that this property implies the usual Sato–Tate conjecture for modular forms for $\mathrm{SL}_2(\mathbb{Z})$.

[9] This equality makes sense because we also conjecture that we have $\det(t - \mathrm{c}_p(\pi)) \in \overline{\mathbb{Q}}[t]$ if π is algebraic.

$$\det(t - \rho_{\pi,\iota}(\mathrm{Frob}_p)) = \iota(\det(t - p^{\mathrm{w}(\pi)/2}\mathrm{c}_p(\pi)))$$

for each prime $p \neq \ell$, which determines it uniquely.

This conjecture may readily be seen as an "algebraic" analog of Axiom (L2). Many difficult and important special cases of it are known. According to Fontaine and Mazur, one expects that the Galois representations above are exactly those appearing in the ℓ-adic realizations of pure motives over $\mathbb{Q}$ with *everywhere good reduction*.

Conclusion

Let G be reductive over $\mathbb{Z}$ and r a representation of $\widehat{G}$. As we have seen, the Arthur–Langlands conjecture predicts that for every π in $\Pi_{\mathrm{disc}}(G)$, the collection of conjugacy classes $r(\mathrm{c}(\pi))$ can be expressed in a very precise way in terms of *building blocks* that are elements π_i of $\Pi_{\mathrm{cusp}}(\mathrm{GL}_{n_i})$ and integers d_i with $\dim r = \sum_i n_i d_i$. Here are some questions that arise naturally: Assuming that a representation π in $\Pi_{\mathrm{disc}}(G)$ is given, for instance such that π_f appears concretely in a specific $\mathcal{A}_U(G)$, can we determine the associated representations π_i and integers d_i? Is it easier to determine them rather than π itself?

A first obstacle we encounter when trying to illustrate these questions is to have at our disposal examples of groups G and of irreducible unitary representations U of $G(\mathbb{R})$ for which we know how to determine whether $\mathcal{A}_U(G)$ is nonzero, or even better determine its dimension. When U is a discrete series representation, this is an accessible but notoriously difficult problem: for example, when we have $G = \mathrm{Sp}_{2g}$, it contains the question of determining[10] the dimension of spaces of Siegel modular cusp forms for $\mathrm{Sp}_{2g}(\mathbb{Z})$. When U is not in the discrete series, it seems hopeless to obtain a formula for $\dim \mathcal{A}_U(G)$, as is shown by the example $\mathrm{G} = \mathrm{PGL}_2$.

The special case where $G(\mathbb{R})$ is compact, for which all the irreducible representations are in the discrete series, has the peculiar feature that the question of determining $\dim \mathcal{A}_U(G)$ is significantly more elementary. We will give many such examples with $G = \mathrm{SO}_n$. The case $G = \mathrm{SO}_{24}$ is especially interesting from this point of view, as it is one of the groups of highest rank for which $\dim \mathcal{A}_U(G)$ can be computed for at least *one* U (and with $\mathcal{A}_U(G) \neq 0$). Indeed, we have $\dim \mathcal{A}_1(G) = |\widetilde{\mathrm{X}}_{24}|$, and this cardinality is 25 because the Leech lattice is the only one, among the 24 Niemeier lattices, not to admit any improper isometry. We are forced to ask ourselves the following question.

Question 1. Let r be the standard representation of $\widehat{\mathrm{SO}_{24}}$ and π in $\Pi_{\mathrm{disc}}(\mathrm{SO}_{24})$ with $\pi_\infty = 1$; can we determine the collection of representations π_i and the integers d_i corresponding to π and r according to the Arthur–Langlands conjecture?[11]

[10] This determination is classical for $g = 1$, due to Igusa (in the scalar-valued case) and Tsushima (in the general case) for $g = 2$, to Tsuyumine for $g = 3$ (again in the scalar-valued case), and has been solved only very recently by Taïbi in general for $g \leq 7$.

[11] Observe that Arthur's results do not immediately apply here because SO_n is not (quasi-)split over $\mathbb{Q}$. Nevertheless, we will prove that the Arthur–Langlands conjecture is satisfied when π and

This is the question at the origin of this work. Formulas (4) and (2) show that a positive answer to this question gives decisive information about the p-neighbor problem in dimension 24.

Before saying more about Question 1, let us add that the π_i that appear in its statement are not arbitrary: they are algebraic. More generally, if G is reductive over $\mathbb{Z}$ and if π is in $\Pi_{\mathrm{disc}}(G)$ with π_∞ a discrete series representation, then the eigenvalues of $\mathrm{c}_\infty(\pi)$ in the adjoint representation of Lie $\widehat{G}$ are in $\mathbb{Z}$ (Harish-Chandra); it follows that if r is an arbitrary representation of $\widehat{G}$, then the representations π_i associated with π and r by the Arthur–Langlands conjecture are necessarily algebraic. As a consequence, those π are related to motives and Galois representations, which makes them even more interesting. Those links are deep. We will show, for example, that Arthur's multiplicity formula suggests that if π in $\Pi_{\mathrm{cusp}}(\mathrm{GL}_{8k})$ is algebraic, isomorphic to its *dual*, and if the eigenvalues of $\mathrm{c}(\pi_\infty)$ are distinct integers, then there exists a π' in $\Pi_{\mathrm{disc}}(\mathrm{SO}_{8k})$ satisfying $r(\mathrm{c}(\pi')) = \mathrm{c}(\pi)$. These unexpected relations between Galois representations and even unimodular lattices clearly show the interest of studying $\Pi_{\mathrm{disc}}(\mathrm{SO}_n)$ for the number-theorist.

Let us return to Question 1. An obstacle we immediately faced, at least when we started working on this question, is that very few results were known about $\Pi_{\mathrm{cusp}}(\mathrm{GL}_n)$ with $n > 2$, even if we restrict ourselves to algebraic representations.[12] For instance, assuming that there exists a representation π in $\Pi_{\mathrm{disc}}(\mathrm{SO}_{24})$ satisfying $\pi_\infty = 1$ and such that one of the associated π_i is in $\Pi_{\mathrm{cusp}}(\mathrm{GL}_{n_i})$ with n_i big, it is very likely that we would never be able to say anything interesting either about this π, or about the p-neighbor problem in dimension 24. Note that we always have $n_i \leq 24$, but also $\mathrm{w}(\pi_i) \leq 22$, as can be seen by considering $\mathrm{c}_\infty(\pi)$.

One of our main results will be the proof in Sect. 9.3 of a classification of the automorphic representations π in $\Pi_{\mathrm{cusp}}(\mathrm{GL}_n)$, with $n \geq 1$ arbitrary, that are algebraic of motivic weight $\mathrm{w}(\pi) \leq 22$. We will see that there are only 11 such representations and that they all appear (as π_i) in the answer to Question 1. We furthermore have $n \leq 4$ in all cases, with exactly four of the representations in $\Pi_{\mathrm{cusp}}(\mathrm{GL}_4)$. These four, which actually *come* from certain vector-valued Siegel cusp form of genus 2, will play an important role in this book.

The scope of the classification above is broader: for arbitrary G, the Arthur–Langlands conjecture suggests that every representation π in $\Pi_{\mathrm{disc}}(G)$ with π_∞ in the discrete series and such that $\mathrm{c}_\infty(\pi)$ is "small enough," is *built* from the 11 automorphic representations mentioned above. For example, we will see how to use this approach to determine the dimension of the space of Siegel modular cusp forms of weight at most 12 for $\mathrm{Sp}_{2g}(\mathbb{Z})$.

It seems reasonable to end this preface here, and to leave to the reader the pleasure of immersing themselves in the actual introduction of the book.

---

r are as in the statement of Question 1, by applying Arthur's results to Sp_{2g} and using some theta series arguments.

[12] The situation is very different by now, thanks to the works [55] and [195]. Note that although these works were published before the present book, they were actually entirely motivated by it. Many important questions remain; for example, we do not know the number of algebraic π in $\Pi_{\mathrm{cusp}}(\mathrm{GL}_3)$ with a given Archimedean component π_∞.

Acknowledgements The authors wish to thank the Centre de Mathématiques de l'École Polytechnique, whose structures have allowed the improbable collaboration between a number-theorist and a topologist fan of quadratic forms. They warmly thank Colette Moeglin, David Renard, and Olivier Taïbi for useful discussions, as well as Richard Borcherds and Thomas Mégarbané for their sagacious remarks. They are also very grateful to Reinie Erné for the care she took in translating their text from the French and for the patience she displayed in the face of their whims. Finally, the authors thank, respectively, Valeria and Martine for their support throughout this project, which sometimes seemed endless.

Orsay, France Gaëtan Chenevier
Paris, France Jean Lannes

Contents

Chapter 1
Introduction

1.1 Even Unimodular Lattices

Let $n \geq 1$ be an integer, and consider the Euclidean space $\mathbb{R}^n$ endowed with the standard inner product $(x_i) \cdot (y_i) = \sum_i x_i y_i$. An even unimodular lattice of rank n is a lattice $L \subset \mathbb{R}^n$ with covolume 1 such that $x \cdot x$ is an even integer for all x in L. The set $\mathcal{L}_n$ of these lattices is endowed with an action of the orthogonal group $\mathrm{O}(\mathbb{R}^n)$; we denote the set of isometry classes of even unimodular lattices of rank n by

$$\mathrm{X}_n \ := \ \mathrm{O}(\mathbb{R}^n)\backslash\mathcal{L}_n \ .$$

To each L in $\mathcal{L}_n$, there corresponds a quadratic form

$$\mathrm{q}_L \colon L \to \mathbb{Z} \ , \ \ x \mapsto \frac{x \cdot x}{2} \ ,$$

whose associated bilinear form $x \cdot y$ has determinant 1. The map $L \mapsto \mathrm{q}_L$ then induces a bijection between X_n and the set of isomorphism classes of positive definite quadratic forms over $\mathbb{Z}$ of rank n and determinant 1.

As is well known, the set X_n is finite. It is nonempty if and only if $n \equiv 0 \bmod 8$. A standard example of an element of $\mathcal{L}_n$ is the lattice

$$\mathrm{E}_n \ := \ \mathrm{D}_n \ + \ \mathbb{Z}e \ ,$$

where $\mathrm{D}_n = \{(x_i) \in \mathbb{Z}^n \,;\, \sum_i x_i \equiv 0 \bmod 2\}$, $e = \frac{1}{2}(1, 1, \ldots, 1)$, and $n \equiv 0 \bmod 8$. Let us explain this notation. With each element L of $\mathcal{L}_n$ is associated a root system

$$\mathrm{R}(L) \ := \ \{x \in L\,; \ \ x \cdot x = 2\}$$

of rank at most n. Its irreducible components are of type $\mathbf{A}_n$, $\mathbf{D}_n$, $\mathbf{E}_6$, $\mathbf{E}_7$, or $\mathbf{E}_8$; we will say that it is "of type ADE." The root system $\mathrm{R}(\mathrm{E}_8)$ is, for example, of type $\mathbf{E}_8$ and generates the lattice E_8 over $\mathbb{Z}$. For $n > 8$, the root system $\mathrm{R}(\mathrm{E}_n)$ is of type $\mathbf{D}_n$ and generates D_n.

G. Chenevier, J. Lannes, *Automorphic Forms and Even Unimodular Lattices*, Ergebnisse der Mathematik und ihrer Grenzgebiete. 3. Folge / A Series of Modern Surveys in Mathematics 69, https://doi.org/10.1007/978-3-319-95891-0_1

The set X_n has only been determined in dimension $n \leq 24$. Mordell and Witt proved, respectively,

$$X_8 = \{E_8\} \text{ and } X_{16} = \{E_{16}, E_8 \oplus E_8\} .$$

The two lattices E_{16} and $E_8 \oplus E_8$ will play an important role in this book. They are both easy and difficult to distinguish from each other: their root systems are different, but they represent each integer exactly the same number of times. This last, well-known, property leads, for example, to Milnor's isospectral tori.

The elements of X_{24} were classified by Niemeier [158], who proved, in particular, $|X_{24}| = 24$. Before saying more about these lattices, let us mention that for $n \geq 32$, the Minkowski–Siegel–Smith mass formula shows that the size of X_n explodes. For example, we have $|X_{32}| > 8 \cdot 10^6$ [177]; in fact, X_{32} even has more than 10^9 elements, as shown by King [118].

An element of $\mathcal{L}_{24}$ is called a *Niemeier lattice*; the most famous one is the Leech lattice. Up to isometry, it is the only element L of $\mathcal{L}_{24}$ with $R(L) = \emptyset$ (Conway). A remarkable fact is that if L is a Niemeier lattice that is not isomorphic to the Leech lattice, then $R(L)$ has rank 24 and all its irreducible components have the same Coxeter number. A simple proof of this was given by Venkov [201]. The miracle is then that the map $L \mapsto R(L)$ induces a bijection between $X_{24} - \{\text{Leech}\}$ and the set of isomorphism classes of root systems R of rank 24 and type ADE with components all having the same Coxeter number $h(R)$. The proof is a rather tedious case-by-case verification.

Table 1.1 The 23 roots systems of type ADE and rank 24 with components all having the same Coxeter number

R	$\mathbf{D}_{24}$	$\mathbf{D}_{16}\mathbf{E}_8$	$3\mathbf{E}_8$	$\mathbf{A}_{24}$	$2\mathbf{D}_{12}$	$\mathbf{A}_{17}\,\mathbf{E}_7$	$\mathbf{D}_{10}\,2\mathbf{E}_7$	$\mathbf{A}_{15}\,\mathbf{D}_9$
$h(R)$	46	30	30	25	22	18	18	16
R	$3\mathbf{D}_8$	$2\mathbf{A}_{12}$	$\mathbf{A}_{11}\,\mathbf{D}_7\,\mathbf{E}_6$	$4\mathbf{E}_6$	$2\mathbf{A}_9\mathbf{D}_6$	$4\mathbf{D}_6$	$3\mathbf{A}_8$	$2\mathbf{A}_7\,2\mathbf{D}_5$
$h(R)$	14	13	12	12	10	10	9	8
R	$4\mathbf{A}_6$	$4\mathbf{A}_5\,\mathbf{D}_4$	$6\mathbf{D}_4$	$6\mathbf{A}_4$	$8\mathbf{A}_3$	$12\mathbf{A}_2$	$24\mathbf{A}_1$	
$h(R)$	7	6	6	5	4	3	2	

The results mentioned in this section are discussed in Chap. 2, which is mainly devoted to recalling classical results. We first develop prerequisites from bilinear and quadratic algebra necessary to understand the constructions of the quadratic forms to which we have alluded above, as well as others we will need. In particular, we recall Venkov's theory and explain the construction of certain Niemeier lattices. We also take the opportunity to recall some basic facts on classical group schemes over $\mathbb{Z}$ that will be used later. Appendix B contains, among other things, a variant of the results of Chap. 2: in it, we study the even lattices in $\mathbb{R}^n$ of determinant 2 as well as the corresponding theory of quadratic forms over $\mathbb{Z}$ (in odd dimensions).

1.2 Kneser Neighbors

Let p be a prime. The notion of p-neighbors was introduced by M. Kneser; it can be viewed as a tool for constructing a set of even unimodular lattices from a given lattice and the prime p. In Chap. 3, we study several variations on this notion and give many examples.

Kneser defines two lattices L and M in $\mathcal{L}_n$ to be *p-neighbors* if $L \cap M$ has index p in L (and therefore in M). It is easy to construct all p-neighbors of a given lattice L. Indeed, with any isotropic line ℓ in $L \otimes \mathbb{F}_p$, say generated by an element x of L satisfying $\mathrm{q}_L(x) \equiv 0 \bmod p^2$, we can associate the even unimodular lattice[1]

$$\mathrm{vois}_p(L;\ell) := H + \mathbb{Z}\,\frac{x}{p}\,,$$

where $H = \{y \in L\,;\, x\cdot y \equiv 0 \bmod p\}$ (the lattice above depends only on the choice of ℓ). The map $\ell \mapsto \mathrm{vois}_p(L;\ell)$ induces a bijection between $\mathrm{C}_L(\mathbb{F}_p)$ and the set of p-neighbors of L, where C_L denotes the projective (and smooth) quadric over $\mathbb{Z}$ defined by $\mathrm{q}_L = 0$. This quadric turns out to be hyperbolic over $\mathbb{F}_p$ for every prime p, so the number of p-neighbors of L is exactly $|\mathrm{C}_L(\mathbb{F}_p)| = 1+p+p^2+\cdots+p^{n-2}+p^{n/2-1}$, which we will denote by $\mathrm{c}_n(p)$.

Consider, for example, the element $\rho = (0,1,2,\ldots,23)$ of E_{24}. It generates an isotropic line in $\mathrm{E}_{24} \otimes \mathbb{F}_{47}$ because of the congruence $\sum_{i=1}^{23} i^2 \equiv 0 \bmod 47$. It is not very difficult to verify that $\mathrm{vois}_{47}(\mathrm{E}_{24};\rho)$ does not have any roots, so that we have an isometry

$$\mathrm{vois}_{47}(\mathrm{E}_{24};\rho) \simeq \mathrm{Leech}\,.$$

This particularly simple construction of the Leech lattice is attributed to Thompson in [68]; we will return to it later. It illustrates the saying that many constructions of lattices are special cases of constructions of neighbors.

Returning to the general setting, for every L in $\mathcal{L}_n$, we have a partition of the quadric $\mathrm{C}_L(\mathbb{F}_p)$ given by the isometry class of the associated p-neighbor. One of the aims of this book is to study this partition in dimension $n \leq 24$. For example, can we determine the number $\mathrm{N}_p(L,M)$ of p-neighbors of L isometric to a given $M \in \mathcal{L}_n$? The first interesting case is, of course, that of dimension $n = 16$. To state the result, we introduce the linear map $\mathrm{T}_p : \mathbb{Z}[\mathrm{X}_n] \to \mathbb{Z}[\mathrm{X}_n]$ defined by $\mathrm{T}_p\,[L] = \sum [M]$, where the sum is taken over the p-neighbors M of L.

Theorem A. *In the basis* $\mathrm{E}_8 \oplus \mathrm{E}_8,\ \mathrm{E}_{16}$, *the matrix of* T_p *is*

$$\mathrm{c}_{16}(p)\begin{bmatrix}1&0\\0&1\end{bmatrix} + (1+p+p^2+p^3)\,\frac{1+p^{11}-\tau(p)}{691}\begin{bmatrix}-405&286\\405&-286\end{bmatrix},$$

where τ is Ramanujan's function defined by $q\prod_{m\geq 1}(1-q^m)^{24} = \sum_{n\geq 1}\tau(n)q^n$.

[1] The notation *vois* comes from the French word *voisin* for *neighbor*.

For example, this theorem asserts that for every prime p, we have $\mathrm{N}_p(\mathrm{E}_8 \oplus \mathrm{E}_8, \mathrm{E}_{16}) = (405/691)(1+p^{11}-\tau(p))(p^4-1)/(p-1)$. This theorem is probably known to specialists, but we have not been able to find it stated this way in the literature. We will give several proofs of it further on. In view of the theory of theta series and modular forms for $\mathrm{SL}_2(\mathbb{Z})$, the presence of $\tau(n)$ in the statement seems, at first sight, rather classical. For example, if we set $\mathrm{r}_L(n) = |\{x \in L\,;\, x \cdot x = 2n\}|$, then we can easily show that we have $\mathrm{r}_{\mathrm{Leech}}(p) = (65520/691)(1+p^{11}-\tau(p))$ for every prime p, a formula that resembles that of the theorem. Nevertheless, the presence of the term $\tau(p)(p^4-1)(p-1)^{-1}$ in the formula for $\mathrm{N}_p(\mathrm{E}_8 \oplus \mathrm{E}_8, \mathrm{E}_{16})$ given above appears to be much more subtle; it will turn out to be equivalent to a nontrivial case of the Arthur–Langlands functoriality conjecture.[2]

Our main theorem is similar to Theorem A but concerns Niemeier lattices. We can state it in the same style as Theorem A, namely as an explicit formula for the matrix of T_p on $\mathbb{Z}[\mathrm{X}_{24}]$, but the result is very hard to digest. This explicit formula involves rational coefficients with such large denominators that it appears quite exceptional that $\mathrm{N}_p(L, M)$ is an integer! We will state a more conceptual (and equivalent) version of our result in Sect. 1.4 (Theorem E). A remarkable feature is that the statement involves all cuspidal modular forms of weight $k \leq 22$ for the group $\mathrm{SL}_2(\mathbb{Z})$, as well as four vector-valued Siegel modular forms for $\mathrm{Sp}_4(\mathbb{Z})$. Let us already discuss a number of consequences concerning the Niemeier lattices that follow from our formulas.

Consider the graph $\mathrm{K}_n(p)$ with set of vertices X_n, where the classes of two nonisomorphic lattices L and M are joined by an edge if and only if $\mathrm{N}_p(L, M) \neq 0$. Kneser proved that $\mathrm{K}_n(p)$ is connected for all n and p, as a consequence of his famous strong approximation theorem. This nice result shows that we can, theoretically, reconstruct X_n from the lattice E_n and a prime p. Niemeier used this to compute X_{24} using 2-neighborhoods.

The graph $\mathrm{K}_{16}(p)$ is the connected graph with 2 vertices (Kneser). This is, of course, compatible with the bound $|\tau(p)| < 2p^{11/2}$ (Deligne–Ramanujan) and the formula for $\mathrm{N}_p(\mathrm{E}_8 \oplus \mathrm{E}_8, \mathrm{E}_{16})$ given by Theorem A. On the other hand, the graph $\mathrm{K}_{24}(2)$, determined by Borcherds [68], is not at all trivial. It has diameter 5 and the Wikipedia page http://en.wikipedia.org/wiki/Niemeier_lattice gives a nice representation of it, also due to Borcherds. Our results allow us, for example, to determine $\mathrm{K}_{24}(p)$ for every prime p (Sect. 10.2).[3]

Theorem B. (i) *Let L be a Niemeier lattice with roots. Then L is a p-neighbor of the Leech lattice if and only if $p \geq \mathrm{h}(\mathrm{R}(L))$.*
(ii) *The graph $\mathrm{K}_{24}(p)$ is complete if and only if $p \geq 47$.*

---

[2] The comparison of Theorem A with the formula for $\mathrm{r}_{\mathrm{Leech}}(p)$ given above leads to the "purely quadratic" relation $\mathrm{N}_p(\mathrm{E}_8 \oplus \mathrm{E}_8, \mathrm{E}_{16}) = (9/1456) \cdot \mathrm{r}_{\mathrm{Leech}}(p) \cdot (p^4-1)(p-1)^{-1}$, which we do not know how to prove directly.

[3] A list of these graphs can be found at http://gaetan.chenevier.perso.math.cnrs.fr/niemeier/niemeier.html.

Statement (i) of this theorem concerns the constructions of the Leech lattice as a p-neighbor of a Niemeier lattice with roots. For example, on the Borcherds graph $\mathrm{K}_{24}(2)$, we see that the distance from the Leech lattice to E_{24} is 5 and that the Leech lattice is linked only to the lattice with root system $24\mathbf{A}_1$. (The latter is the Niemeier lattice with roots that is the most delicate to construct, as it needs the Golay code; see Sect. 2.3.) This last property is, in fact, quite easy to understand: if the Leech lattice is a 2-neighbor of a Niemeier lattice L (with roots), then L has an index 2 subgroup without any roots. In particular, $\mathrm{R}(L)$ has the property that the sum of two roots is not a root, so that its irreducible components are of rank 1, which implies that $\mathrm{R}(L) = 24\mathbf{A}_1$. Of the root systems in Table 1.1, this is also the only one with Coxeter number 2, in accordance with statement (i).

The most elementary part of Theorem B, which is proved in Sect. 3.4 and generalizes the observation above, consists in verifying that $p \geq \mathrm{h}(\mathrm{R}(L))$ if the Leech lattice is a p-neighbor of L. This is a formal analog of a result of Kostant [128] asserting that the minimal order of a regular element of finite order in a connected, compact, adjoint Lie group coincides with the Coxeter number of its root system. The proof of the other statements, on the other hand, requires Theorem E as well as a number of Ramanujan-type inequalities. It will be completed only in Chap. 10 (Sects. 10.2 and 10.3).

In Chap. 3, we also study the limit cases of assertion (i) of Theorem B (with direct arguments, that is, without using Theorem E). For this, we carry out a detailed analysis of the elements c of $\mathrm{C}_L(\mathbb{F}_p)$ satisfying $\mathrm{vois}_p(L; c) \simeq$ Leech, where L is a Niemeier lattice with nonempty root system $R = \mathrm{R}(L)$. For the relevance of the statements, we need to study, more generally, the d-neighbors of L, where $d \geq 1$ is an integer that need not be prime (Sect. 3.1). We prove that if ρ is a Weyl vector of R and we set $h = \mathrm{h}(R)$, then we have isometries (Theorem 3.4.2.10)

$$\mathrm{vois}_h(L; \rho) \simeq \mathrm{vois}_{h+1}(L; \rho) \simeq \text{Leech} . \tag{1.2.1}$$

This is well defined because $\rho \in L$ (Borcherds) and $\mathrm{q}_L(\rho) = h(h+1)$ (Venkov). This statement contains, for example, the observation by Thompson mentioned earlier. In fact, these 23 (or 46) constructions of the Leech lattice are none other than the famous *holy constructions* of Conway and Sloane [67]. We, however, give a new proof of the isometries (1.2.1) using the theory of neighbors and show the identities

$$\mathrm{N}_h(L, \text{Leech}) = \frac{|W|}{\varphi(h)g} \quad \text{and} \quad \mathrm{N}_{h+1}(L, \text{Leech}) = \frac{|W|}{\varphi(h+1)} , \tag{1.2.2}$$

where W denotes the Weyl group of R and g^2 its index of connection in the sense of Bourbaki. We conclude Chap. 3 with an analysis of $\mathrm{vois}_2(L; \rho)$ inspired by results of Borcherds (Fig. 3.1).

1.3 Theta Series and Siegel Modular Forms

Let us return to the determination of the operator T_p on $\mathbb{Z}[\mathrm{X}_n]$. We begin with a few simple observations. The T_p commute and are self-adjoint for a suitable inner product on $\mathbb{R}[\mathrm{X}_n]$ [156] (Sect. 3.2). We must therefore determine a basis of common eigenvectors of the T_p, as well as the corresponding sets of eigenvalues. There is only one obvious stable line, generated by $\sum_{L \in \mathrm{X}_n} [L]/|\mathrm{O}(L)|$, on which the operator T_p has "trivial" eigenvalue $\mathrm{c}_n(p)$.

As hinted at in the preface, we are in fact dealing with a disguised problem belonging to the spectral theory of automorphic forms. Indeed, if $G = \mathrm{O}_n$ denotes the orthogonal group scheme over $\mathbb{Z}$ defined by the quadratic form $\mathrm{q}_{\mathrm{E}_n}$ and $\mathbb{A}$ the adele ring of $\mathbb{Q}$, then arguments from genus theory lead to an isomorphism of $G(\mathbb{R})$-sets $\mathcal{L}_n \simeq G(\mathbb{Q})\backslash G(\mathbb{A})/G(\widehat{\mathbb{Z}})$ (Sects. 2.2 and 4.1). Consequently, the dual of $\mathbb{R}[\mathrm{X}_n]$ can be identified with the space of real-valued functions on $G(\mathbb{Q})\backslash G(\mathbb{A})$ that are invariant under the right action of $G(\mathbb{R}) \times G(\widehat{\mathbb{Z}})$ by translation. In this description, the operator T_p is induced by a specific element of the ring $\mathrm{H}(G)$ of Hecke operators of G.

These classical observations are recalled in Chap. 4. Although we are mainly interested in the automorphic forms for the $\mathbb{Z}$-group O_n, our statements and proofs will require the introduction of several variants (automorphic forms for SO_n, PGO_n and PGSO_n), as well as modular forms for $\mathrm{SL}_2(\mathbb{Z})$, vector-valued Siegel modular forms for $\mathrm{Sp}_{2g}(\mathbb{Z})$, and even, through Arthur's results, automorphic forms for PGL_n. Therefore, from the beginning, we need to adopt a sufficiently general point of view embracing all these objects (Sect. 4.3). In Sects. 4.1 and 4.2, the reader can find an elementary exposition on Hecke operators. The emphasis is on the examples provided by the classical groups and their variants (Hecke, Satake, Shimura); these lead to a wider overview of p-neighbors and their generalizations. Sections 4.4 and 4.5 are devoted to recalling some properties of automorphic forms for O_n and Siegel modular forms. Let us emphasize that this chapter is intended for nonspecialists and does not pretend to any originality.

One approach to studying the $\mathrm{H}(\mathrm{O}_n)$-module $\mathbb{Z}[\mathrm{X}_n]$ is to examine the Siegel theta series $\vartheta_g(L)$ of each genus $g \geq 1$ of the elements L of $\mathcal{L}_n$. For every $n \equiv 0 \bmod 8$ and $g \geq 1$, they allow us to define a linear map

$$\vartheta_g : \mathbb{C}[\mathrm{X}_n] \to \mathrm{M}_{n/2}(\mathrm{Sp}_{2g}(\mathbb{Z}))\ , \quad [L] \mapsto \vartheta_g(L)\ ,$$

where $\mathrm{M}_k(\mathrm{Sp}_{2g}(\mathbb{Z}))$ denotes the space of Siegel modular forms of weight $k \in \mathbb{Z}$ for $\mathrm{Sp}_{2g}(\mathbb{Z})$ (Sect. 5.1). The relevance of this map for our problem comes from the generalized Eichler commutation relations; they assert that ϑ_g intertwines each element of $\mathrm{H}(\mathrm{O}_n)$ with an "explicit" element of $\mathrm{H}(\mathrm{Sp}_{2g})$ (Eichler, Freitag, Yoshida, Andrianov, Sect. 5.1). The map ϑ_g is trivially injective for $g \geq n$. It seems, however, quite difficult to determine the structure of the $\mathrm{H}(\mathrm{Sp}_{2g})$-module $\mathrm{M}_k(\mathrm{Sp}_{2g}(\mathbb{Z}))$, especially for large g. Nevertheless, in Chap. 9, we develop a strategy that allows us to solve new cases of this problem. Our strategy relies, among other things, on results of Arthur [13].

The map ϑ_g has been widely studied. Its kernel, which shrinks when g increases, describes the linear relations between the theta series of genus g of the elements of $\mathcal{L}_n$, and determining its image is an example of Eichler's famous *basis problem*. More precisely, ϑ_g induces an injective map

$$\mathrm{Ker}\,\vartheta_{g-1}/\mathrm{Ker}\,\vartheta_g \longrightarrow \mathrm{S}_{n/2}(\mathrm{Sp}_{2g}(\mathbb{Z}))\,, \tag{1.3.1}$$

where $\mathrm{S}_k(\mathrm{Sp}_{2g}(\mathbb{Z})) \subset \mathrm{M}_k(\mathrm{Sp}_{2g}(\mathbb{Z}))$ denotes the subspace of cusp forms (see Sect. 5.1 or footnote 4 below for the convention on ϑ_0), and Eichler asks whether this map is surjective. An important result of Böcherer [27] gives a necessary and sufficient condition for an eigenform for $\mathrm{H}(\mathrm{Sp}_{2g})$ to be in the image of (1.3.1), in terms of the vanishing of an associated L-function at the integer $n/2-g$ (Sect. 7.2).

The Case $n = 16$

The case $n = 16$ is the subject of a famous story, recalled in Sect. 5.2. Indeed, a classical result of Witt and Igusa asserts that we have

$$\vartheta_g(\mathrm{E}_8 \oplus \mathrm{E}_8) = \vartheta_g(\mathrm{E}_{16}) \quad \text{if} \;\; g \leq 3\,. \tag{1.3.2}$$

These remarkable identities mean that $\mathrm{E}_8 \oplus \mathrm{E}_8$ and E_{16} represent each positive integral quadratic form of rank at most 3 exactly the same number of times. This is well known in genus $g=1$, as a consequence of the vanishing $\mathrm{S}_8(\mathrm{SL}_2(\mathbb{Z}))=0$ (and leads to the isospectral tori of Milnor mentioned earlier). This, incidentally, shows[4] that "the" nontrivial eigenvector of $\mathbb{Z}[\mathrm{X}_{16}]$ is $[\mathrm{E}_{16}]-[\mathrm{E}_8 \oplus \mathrm{E}_8]$. The difficulty in genera 2 and 3 is that the vanishing of $\mathrm{S}_8(\mathrm{Sp}_{2g}(\mathbb{Z}))$, though still true, is more difficult to prove. In Appendix A, we will give another proof of the identities (1.3.2) which does not rely on any such vanishing results (that ingenious proof is due to Kneser).

The form $J = \vartheta_4(\mathrm{E}_8 \oplus \mathrm{E}_8) - \vartheta_4(\mathrm{E}_{16})$, which is nothing but the famous *Schottky form*, is easily shown to be nonzero. By results of Poor and Yuen [167], we even know that it generates $\mathrm{S}_8(\mathrm{Sp}_8(\mathbb{Z}))$. Theorem A then follows from the resolution by Ikeda [108] of the Duke–Imamoḡlu conjecture [40]. Indeed, when applied to Jacobi's modular form Δ in $\mathrm{S}_{12}(\mathrm{SL}_2(\mathbb{Z}))$, Ikeda's theorem shows the existence of a nonzero Siegel modular form in $\mathrm{S}_8(\mathrm{Sp}_8(\mathbb{Z}))$ that is an eigenform for $\mathrm{H}(\mathrm{Sp}_8)$, with Hecke eigenvalues explicitly determined by the $\tau(p)$. Ikeda's proof is quite difficult; one of the main contributions of this book to Theorem A is to give another, very different, proof of Ikeda's result in this specific case.

Our main result is the following. For any map $f\colon \mathcal{L}_n \to \mathbb{C}$, we define $\mathrm{T}_p(f)\colon \mathcal{L}_n \to \mathbb{C}$ by setting $\mathrm{T}_p(f)(L) = \sum_M f(M)$ for every $L \in \mathcal{L}_n$, where the sum is taken over the p-neighbors M of L. If $1 \leq g \leq n/2$, we denote by $\mathrm{H}_{d,g}(\mathbb{R}^n)$ the space of polynomials $(\mathbb{R}^n)^g \to \mathbb{C}$ that are harmonic for the Euclidean

---

[4] This assertion can be proved much more directly. Indeed, if ϑ_0 denotes the linear map $\mathbb{C}[\mathrm{X}_n] \to \mathbb{C}$ that sends the class of any element of $\mathcal{L}_n$ to 1, then we have $\vartheta_0 \circ \mathrm{T}_p = \mathrm{c}_n(p)\,\vartheta_0$, so that $\mathrm{Ker}\,\vartheta_0$ is stable by T_p.

Laplace operator on $(\mathbb{R}^n)^g$ and satisfy $P \circ \gamma = (\det\gamma)^d P$ for all $\gamma \in \mathrm{GL}_g(\mathbb{C})$ (Sect. 5.4). This space is endowed with a linear action of $\mathrm{O}(\mathbb{R}^n)$.

Theorem C. *Let $q + \sum_{n \geq 2} a_n q^n$ be a modular form of weight k for $\mathrm{SL}_2(\mathbb{Z})$ that is an eigenform for the Hecke operators, and let $d = k/2 - 2$. There exists a map $f\colon \mathcal{L}_8 \to \mathbb{C}$ such that*

(i) *for every prime p, we have* $\mathrm{T}_p(f) = p^{-d}\,(p^4 - 1)(p-1)^{-1}\,a_p\,f$;
(ii) *under the action of* $\mathrm{O}(\mathbb{R}^8)$, *the function f generates a representation isomorphic to* $\mathrm{H}_{d,4}(\mathbb{R}^8)$.

Section 5.4 is mainly devoted to proving a specific case of this theorem when $k = 12$, which leads to a complete and relatively elementary proof of Theorem A. The general case will be addressed and made more precise in Sect. 7.2.

Let us sketch the proof. We begin by realizing the initial modular form as a theta series $\sum_{x \in \mathrm{E}_8} P(x)\, q^{x \cdot x/2}$, where $P\colon \mathbb{R}^8 \to \mathbb{C}$ is a suitable harmonic polynomial. In the case of Δ, we verify that any nonzero harmonic polynomial of degree 8 which is invariant under the Weyl group $\mathrm{W}(\mathbf{E}_8)$ does the trick, and in general we invoke a result of Waldspurger [205]. This construction defines a subspace of the functions $\mathcal{L}_8 \to \mathbb{C}$ with the following two properties: First, they are eigenvectors for the Hecke operators in $\mathrm{H}(\mathrm{O}_8)$, with eigenvalues related to the a_p by the Eichler commutation relations. Second, they generate a representation isomorphic to $\mathrm{H}_{8,1}(\mathbb{R}^8)$ under the action of $\mathrm{O}(\mathbb{R}^8)$. The main idea then consists in applying, at the source, an automorphism of $\mathcal{L}_8$ of order 3 arising from triality. Such an automorphism is constructed from a structure of Coxeter octonions on the lattice E_8 and from an isomorphism $\mathcal{L}_8 \simeq G(\mathbb{Q})\backslash G(\mathbb{A})/G(\widehat{\mathbb{Z}})$, where $G = \mathrm{PGSO}_8$. The resulting functions satisfy the conditions of the theorem: we refer to Sect. 5.4 for the details.

Condition (ii) of Theorem C implies that the function f generates a Siegel theta series of genus 4 (with "pluriharmonic" coefficients). When nonzero, this theta series is a substitute for the Ikeda lift of genus 4 of the initial modular form. We show that it is nonzero when $k = 12$; Theorem A easily follows.

Finally, let us mention that we will prove the vanishing of $\mathrm{S}_8(\mathrm{Sp}_{2g}(\mathbb{Z}))$ for all $g \neq 4$ and $g \leq 8$ further on (Theorem 9.5.9). For $g = 5, 6$, it had already been obtained by Poor and Yuen [168] by different methods. Consequently, the map $\vartheta_g\colon \mathbb{C}[\mathrm{X}_{16}] \to \mathrm{M}_8(\mathrm{Sp}_{2g}(\mathbb{Z}))$ is surjective for every genus $1 \leq g \leq 8$.

The Case $n = 24$

This case is the subject of remarkable work by Erokhin [80], Borcherds–Freitag–Weissauer [31], and Nebe–Venkov [156] (Sect. 5.3). Erokhin showed that $\mathrm{Ker}\,\vartheta_{12} = 0$, and the three authors of [31] proved that $\mathrm{Ker}\,\vartheta_{11}$ has dimension 1. In [156], Nebe and Venkov undertook a detailed study of the entire filtration of $\mathbb{Z}[\mathrm{X}_{24}]$ given by the sequence of the $\mathrm{Ker}\,\vartheta_g$ for $g \geq 1$. Their starting point is an explicit expression for the operator T_2 on $\mathbb{Z}[\mathrm{X}_{24}]$, which they deduce from results of Borcherds (Sect. 3.3). They observe that the eigenvalues of T_2 are distinct integers, which allows them to give an explicit basis of $\mathbb{Q}[\mathrm{X}_{24}]$ consisting of common eigenvectors for all of the T_p. They also state a conjecture on the dimension of the image of the map (1.3.1) for every integer $1 \leq g \leq 12$, which they prove in many, but not all, cases. We establish their conjecture and even show that the Eichler basis problem admits a positive solution in dimension $n = 24$ for every genus $1 \leq g \leq 12$ (Theorem 9.5.2 and Corollary 9.5.6).

Theorem D. *The map* $\vartheta_g\colon \mathbb{C}[\mathrm{X}_{24}] \to \mathrm{M}_{12}(\mathrm{Sp}_{2g}(\mathbb{Z}))$ *is surjective and induces an isomorphism* $\mathrm{Ker}\,\vartheta_{g-1}/\mathrm{Ker}\,\vartheta_g \overset{\sim}{\to} \mathrm{S}_{12}(\mathrm{Sp}_{2g}(\mathbb{Z}))$ *for every integer* $g \leq 12$. *The dimension of* $\mathrm{S}_{12}(\mathrm{Sp}_{2g}(\mathbb{Z}))$ *for* $g \leq 12$ *is given by the following table:*

g	1	2	3	4	5	6	7	8	9	10	11	12
$\dim \mathrm{S}_{12}(\mathrm{Sp}_{2g}(\mathbb{Z}))$	1	1	1	2	2	3	3	4	2	2	1	1

We will sketch the proof in Sect. 1.6; the most difficult part is the first assertion. The theorem leads to a complete description of the filtration $(\mathrm{Ker}\,\vartheta_g)_{g\geq 1}$ on $\mathbb{Z}[\mathrm{X}_{24}]$. We note that the Eichler basis problem has a negative answer in dimension $n = 32$ and genus $g = 14$, as stated in Corollary 7.3.5.

1.4 Automorphic Forms for the Classical Groups

Siegel modular forms, as well as the automorphic forms for O_n, can be studied from the perspective of recent work by Arthur [13], published in 2013. However, in order to state the results, we first need to recall some basic features of the Langlands point of view on the theory of automorphic forms [135, 33], which we gather in Chap. 5. The main aspects of this point of view have already been touched upon in the preface. We briefly recall it.

Let G be a semisimple group scheme[5] over $\mathbb{Z}$. We denote by $\Pi_{\mathrm{disc}}(G)$ the set of topologically irreducible subrepresentations of the space of square-integrable functions on $G(\mathbb{Q})\backslash G(\mathbb{A})/G(\widehat{\mathbb{Z}})$ for the natural actions of $G(\mathbb{R})$ and of the commutative ring $\mathrm{H}(G)$ of Hecke operators of G (Sect. 4.3). The Satake isomorphism associates with each $\pi \in \Pi_{\mathrm{disc}}(G)$ and each prime p a semisimple conjugacy class $\mathrm{c}_p(\pi)$ in $\widehat{G}(\mathbb{C})$, where $\widehat{G}$ denotes the complex semisimple algebraic group that is dual to $G_{\mathbb{C}}$ in the sense of Langlands (Sects. 6.1 and 6.2). In Sect. 6.2.8, we make this enlightening point of view on the eigenvalues of the Hecke operators due to Langlands explicit in the case of the classical groups and Hecke operators we are interested in; we follow Gross' article [97]. Furthermore, we recall how the Harish-Chandra isomorphism

[5] The discussion that follows does not apply verbatim to certain nonconnected group schemes that naturally occur here, such as O_n or PGO_n. We will, when necessary, indicate any modifications needed to include them, but in this introduction we will ignore this detail.

allows us to view the infinitesimal character of the Archimedean component π_∞ of π as a semisimple conjugacy class $\mathrm{c}_\infty(\pi)$ in the Lie algebra of $\widehat{G}$ (Sect. 6.3).

As explained in the preface, a central and structuring conjecture, initially due to Langlands in the "tempered case" and extended by Arthur to the general case [9], asserts that these collections of conjugacy classes can all be expressed in terms of similar data relative to the elements of $\Pi_{\mathrm{disc}}(\mathrm{PGL}_m)$ for $m \geq 1$. This conjecture is discussed in Sect. 6.4.4. Let us state it another, particularly direct, way, using L-functions. Fix $\pi \in \Pi_{\mathrm{disc}}(G)$ and an algebraic representation $r\colon \widehat{G}(\mathbb{C}) \to \mathrm{SL}_n(\mathbb{C})$. According to Langlands, the Euler product

$$\mathrm{L}(s,\pi,r) = \prod_p \det\bigl(1 - p^{-s}\, r(\mathrm{c}_p(\pi))\bigr)^{-1}$$

is absolutely convergent for every complex number s with sufficiently large real part. When G is the $\mathbb{Z}$-group PGL_m and r is the tautological representation of $\widehat{G} = \mathrm{SL}_m$, we simply write $\mathrm{L}(s,\pi)$ for $\mathrm{L}(s,\pi,r)$. The Arthur–Langlands conjecture for the pair (π, r) predicts the existence of an integer $k \geq 1$ and, for $i = 1, \ldots, k$, a representation[6] $\pi_i \in \Pi_{\mathrm{cusp}}(\mathrm{PGL}_{n_i})$ and an integer $d_i \geq 1$ such that we have the equality

$$\mathrm{L}(s,\pi,r) = \prod_{i=1}^{k} \prod_{j=0}^{d_i-1} \mathrm{L}\Bigl(s + j - \frac{d_i - 1}{2}, \pi_i\Bigr)\,. \tag{1.4.1}$$

By slight abuse of language, we call the collection of conjugacy classes $(r(\mathrm{c}_v(\pi)))$ the *Langlands parameter* of the pair (π, r); we denote it by $\psi(\pi, r)$. When the equality (1.4.1) holds, we will symbolically[7] denote it by

$$\psi(\pi,r) = \oplus_{i=1}^{k} \pi_i[d_i]\,.$$

If G is a *classical group over* $\mathbb{Z}$ (Sects. 6.4.7 and 8.1), its dual $\widehat{G}$ is a complex classical group (that is, special orthogonal or symplectic). In particular, $\widehat{G}$ comes with a "tautological" representation called the *standard representation*, denoted by St. An important result proved by Arthur in [13] asserts that the Arthur–Langlands conjecture is true for (π, St) for all π in $\Pi_{\mathrm{disc}}(G)$ if G is either Sp_{2g} or a split special orthogonal $\mathbb{Z}$-group.

In Chap. 7, we illustrate this theory by giving many examples of specific cases of the Arthur–Langlands conjecture, concerning automorphic forms for SO_n or Siegel modular forms for $\mathrm{Sp}_{2g}(\mathbb{Z})$. They do not rely on the work of Arthur, but rather on more classical constructions of theta series. We recall Rallis' point of view on the Eichler commutation relations (Sect. 7.1), as well as important results of Böcherer [27] and Ikeda [108]. We prove Theorem C and give other applications of the triality

[6] As is customary, we denote by $\Pi_{\mathrm{cusp}}(G) \subset \Pi_{\mathrm{disc}}(G)$ the subset of representations occurring in the subspace of cusp forms [92] (Sect. 4.3).

[7] Strictly speaking, our notation includes the corresponding natural identity at the Archimedean place (Sect. 6.4.4). We also denote the summand $\pi_i[d_i]$ simply by $[d_i]$ (resp. π_i) when $\pi_i = 1$ (resp. $d_i = 1$). These conventions are used in Table 1.2.

to the construction of certain elements of $\Pi_{\rm disc}(\mathrm{SO}_8)$ (Sect. 7.2). One ingredient for the proofs is a slight refinement of the Rallis identities to the pair $(\mathrm{PGO}_n, \mathrm{PGSp}_{2g})$ (Sect. 7.1.4). In the end, our analysis recovers sufficiently many constructions to allow us, for example, to determine $\psi(\pi, \mathrm{St})$ for 13 of the "first" 16 representations π in $\Pi_{\rm disc}(\mathrm{SO}_8)$ (Sect. 7.4).

We can now state the analog of Theorem A for X_{24}; we refer to Sect. 10.1 for a statement of this theorem in terms of representations of $\mathrm{Gal}(\overline{\mathbb{Q}}/\mathbb{Q})$, in the spirit of what we announced in [53]. We need to introduce some additional notation:

- The representation Δ_w for $w \in \{11, 15, 17, 19, 21\}$ denotes the element of $\Pi_{\rm cusp}(\mathrm{PGL}_2)$ generated by the 1-dimensional vector space $\mathrm{S}_{w+1}(\mathrm{SL}_2(\mathbb{Z}))$ of cusp forms of weight $w+1$ for $\mathrm{SL}_2(\mathbb{Z})$.
- The representation $\mathrm{Sym}^2\Delta_w$ is the Gelbart–Jacquet symmetric square of Δ_w [90]. This is the unique element of $\Pi_{\rm cusp}(\mathrm{PGL}_3)$ satisfying the equality $\mathrm{c}_v(\mathrm{Sym}^2\Delta_w) = \mathrm{Sym}^2\,\mathrm{c}_v(\Delta_w)$ for all places v of $\mathbb{Q}$.
- If (w, v) is one of the four ordered pairs $(19, 7)$, $(21, 5)$, $(21, 9)$, and $(21, 13)$, then the representation $\Delta_{w,v}$ is an element of $\Pi_{\rm cusp}(\mathrm{PGL}_4)$, defined and studied in Sect. 9.1. Its infinitesimal character $\mathrm{c}_\infty(\Delta_{w,v})$, which, by definition, is the conjugacy class of a semisimple element in $\mathrm{M}_4(\mathbb{C})$, has eigenvalues[8] $\pm w/2$ and $\pm v/2$; further on, we will even see that this property uniquely characterizes $\Delta_{w,v}$.

Theorem E. *The parameters $\psi(\pi, \mathrm{St})$ of the 24 elements π of $\Pi_{\rm disc}(\mathrm{O}_{24})$ generated by the eigenfunctions $\mathrm{X}_{24} \to \mathbb{C}$ for $\mathrm{H}(\mathrm{O}_{24})$ are those in Table 1.2 below.*

Table 1.2 The standard parameters of the $\pi \in \Pi_{\rm disc}(\mathrm{O}_{24})$ such that $\pi_\infty = 1$

$[23] \oplus [1]$	$\mathrm{Sym}^2\Delta_{11} \oplus \Delta_{17}[4] \oplus \Delta_{11}[2] \oplus [9]$
$\mathrm{Sym}^2\Delta_{11} \oplus [21]$	$\mathrm{Sym}^2\Delta_{11} \oplus \Delta_{15}[6] \oplus [9]$
$\Delta_{21}[2] \oplus [1] \oplus [19]$	$\Delta_{15}[8] \oplus [1] \oplus [7]$
$\mathrm{Sym}^2\Delta_{11} \oplus \Delta_{19}[2] \oplus [17]$	$\Delta_{21}[2] \oplus \Delta_{17}[2] \oplus \Delta_{11}[4] \oplus [1] \oplus [7]$
$\Delta_{21}[2] \oplus \Delta_{17}[2] \oplus [1] \oplus [15]$	$\Delta_{19}[4] \oplus \Delta_{11}[4] \oplus [1] \oplus [7]$
$\Delta_{19}[4] \oplus [1] \oplus [15]$	$\Delta_{21,9}[2] \oplus \Delta_{15}[4] \oplus [1] \oplus [7]$
$\mathrm{Sym}^2\Delta_{11} \oplus \Delta_{19}[2] \oplus \Delta_{15}[2] \oplus [13]$	$\mathrm{Sym}^2\Delta_{11} \oplus \Delta_{19}[2] \oplus \Delta_{11}[6] \oplus [5]$
$\mathrm{Sym}^2\Delta_{11} \oplus \Delta_{17}[4] \oplus [13]$	$\mathrm{Sym}^2\Delta_{11} \oplus \Delta_{19,7}[2] \oplus \Delta_{15}[2] \oplus \Delta_{11}[2] \oplus [5]$
$\Delta_{17}[6] \oplus [1] \oplus [11]$	$\Delta_{21}[2] \oplus \Delta_{11}[8] \oplus [1] \oplus [3]$
$\Delta_{21}[2] \oplus \Delta_{15}[4] \oplus [1] \oplus [11]$	$\Delta_{21,5}[2] \oplus \Delta_{17}[2] \oplus \Delta_{11}[4] \oplus [1] \oplus [3]$
$\Delta_{21,13}[2] \oplus \Delta_{17}[2] \oplus [1] \oplus [11]$	$\mathrm{Sym}^2\Delta_{11} \oplus \Delta_{11}[10] \oplus [1]$
$\mathrm{Sym}^2\Delta_{11} \oplus \Delta_{19}[2] \oplus \Delta_{15}[2] \oplus \Delta_{11}[2] \oplus [9]$	$\Delta_{11}[12]$

[8] Likewise, $\mathrm{c}_\infty(\Delta_w)$ has eigenvalues $\pm w/2$.

Let us emphasize that in his remarkable work [109], Ikeda was able to determine 20 of the 24 parameters given in the table, namely those that do not contain any representation of the form $\Delta_{w,v}$.

Given the importance of the role played by the $\Delta_{w,v}$ in this book, we should say a bit more about their origin. Let (j, k) be one of the four ordered pairs $(6, 8)$, $(4, 10)$, $(8, 8)$, and $(12, 6)$. A dimension formula due to R. Tsushima shows that the space of vector-valued cuspidal Siegel modular forms for $\mathrm{Sp}_4(\mathbb{Z})$ with coefficients in $\mathrm{Sym}^j \otimes \det^k$ has dimension 1 [199]. We will give an explicit generator of this space using a construction of theta series with "pluriharmonic" coefficients based on the lattice E_8. If $\pi_{j,k}$ denotes the element of $\Pi_{\mathrm{cusp}}(\mathrm{PGSp}_4)$ generated by this form (which is necessarily an eigenform), then we have the relation $\psi(\pi_{j,k}, \mathrm{St}) = \Delta_{w,v}$ with $(w, v) = (2j + k - 3, j + 1)$. Note that PGSp_4 is isomorphic to the split classical $\mathbb{Z}$-group $\mathrm{SO}_{3,2}$, whose dual is the group Sp_4 over $\mathbb{C}$, so that Arthur's theory applies to $(\pi_{j,k}, \mathrm{St})$.

We will prove Theorem E in Sect. 9.4.3, using a method we will describe in Sect. 1.6. However, we will first give two other *conditional* proofs in Sects. 9.2.10 and 9.2.11. These proofs, obtained by applying Arthur's *multiplicity formula* [13], will ultimately be the most natural, but at present they depend on certain statements refining Arthur's that, though expected, are not yet available.

In Chap. 8, we therefore return to the general results of Arthur [13], which we will specify in the case of classical groups over $\mathbb{Z}$ and their "everywhere unramified" automorphic forms. Such an analysis has already been partially carried out in [55, Sect. 3]; we develop and complete it. The bulk of Chap. 8 is devoted to explaining the famous multiplicity formula mentioned above. This formula gives a necessary and sufficient condition for a given collection of (π_i, d_i) to "come" from a $\pi \in \Pi_{\mathrm{disc}}(G)$, in the sense that $\psi(\pi, \mathrm{St}) = \oplus_{i=1}^{k} \pi_i[d_i]$, where π moreover has a prescribed Archimedean component π_∞ (Sect. 8.3). We limit ourselves to the case where π_∞ is a discrete series representation of $G(\mathbb{R})$ and make explicit the parametrization of the latter by Shelstad (Sect. 8.4). It is this parametrization that plays a key role in Arthur's formula. The version of this formula that we give has, at present, been proved only if G is split over $\mathbb{Z}$ and all integers d_i are equal to 1. We will, however, discuss the general case, indicating the conjectures on which specific cases depend (Sect. 8.4.21), because it greatly clarifies the specific constructions studied in this book. In particular, we give explicit formulas in the cases of Siegel modular forms for $\mathrm{Sp}_{2g}(\mathbb{Z})$ and automorphic forms for SO_n (Sect. 8.5). We verify that they are compatible with the results of Chap. 7 and the results of Böcherer on the image of the map (1.3.1) (Sect. 8.6). As promised, we finally show, in Sect. 9.2, that these formulas lead to a rather miraculous, but simple, conditional proof of Theorem E.

1.5 Algebraic Automorphic Representations of Small Weight

Let $n \geq 1$ be an integer. We call a representation $\pi \in \Pi_{\rm cusp}(\mathrm{PGL}_n)$ *algebraic* if the eigenvalues λ_i of $\mathrm{c}_\infty(\pi)$ satisfy $\lambda_i \in \frac{1}{2}\mathbb{Z}$ and $\lambda_i - \lambda_j \in \mathbb{Z}$ for all i, j (Sect. 8.2.6). The greatest difference between two eigenvalues of $\mathrm{c}_\infty(\pi)$ is then called the *motivic weight* of π; it is a nonnegative integer, denoted by $\mathrm{w}(\pi)$. As we saw in the preface,[9] these algebraic cuspidal automorphic representations are interesting in their own right, because they are exactly those that are related to the ℓ-adic "geometric" Galois representations through the yoga of Fontaine–Mazur and Langlands (Sect. 8.2.16). We are interested in these representations for a slightly different reason, as explained by the following observation.

Let G be a semisimple group scheme over $\mathbb{Z}$, let $\pi \in \Pi_{\rm disc}(G)$ be such that π_∞ has the same infinitesimal character as a finite-dimensional algebraic representation V of $G(\mathbb{C})$, and let $r\colon \widehat{G}(\mathbb{C}) \to \mathrm{SL}_n(\mathbb{C})$ be an algebraic representation. Suppose $\psi(\pi, r) = \oplus_{i=1}^k \pi_i[d_i]$, following Arthur and Langlands. The representations π_i are then algebraic, with motivic weight bounded in terms of the highest weights of V and r (Sect. 8.2). For example, if $G = \mathrm{Sp}_{2g}$ and $\pi \in \Pi_{\rm cusp}(\mathrm{Sp}_{2g})$ is generated by a Siegel modular eigenform of weight k for $\mathrm{Sp}_{2g}(\mathbb{Z})$ (with, say, $k > g$, but this condition can be weakened), then we can write $\psi(\pi, \mathrm{St}) = \oplus_{i=1}^k \pi_i[d_i]$ thanks to Arthur, where the π_i are algebraic of motivic weight at most $2k - 2$. An important ingredient for our proofs is the following classification statement, which is also of independent interest. We prove it in Sect. 9.3.

Theorem F. *Let $n \geq 1$, and let $\pi \in \Pi_{\rm cusp}(\mathrm{PGL}_n)$ be algebraic of motivic weight at most 22. Then π is one of the following 11 representations:*

$$1\,,\ \Delta_{11}\,,\ \Delta_{15}\,,\ \Delta_{17}\,,\ \Delta_{19}\,,\ \Delta_{19,7}\,,\ \Delta_{21}\,,\ \Delta_{21,5}\,,\ \Delta_{21,9}\,,\ \Delta_{21,13}\,,\ \mathrm{Sym}^2\Delta_{11}\,.$$

In motivic weight strictly less than 11, this theorem states that we have $n = 1$ and that π is the trivial representation, a result already known to Mestre and Serre (in a somewhat different language; see [144, Sect. III, Remarque 1]). In this specific case, it gives, among other things, an "automorphic" analog of the classical Minkowski theorem asserting that every number field other than $\mathbb{Q}$ contains at least one ramified prime (the case $\mathrm{w}(\pi) = 0$), and also of Shafarevich's conjecture, proved independently by Abrashkin and Fontaine, according to which there are no abelian varieties over $\mathbb{Z}$ (the case $\mathrm{w}(\pi) = 1$). As far as we know, the result of Theorem F is already new in the specific case $\mathrm{w}(\pi) = 11$. Let us emphasize that we make no assumptions on the integer n and that the theorem implies that $n \leq 4$.

[9] The definition of algebraic given in the preface, which seems more restrictive, is in fact equivalent to this one: see Remark 8.2.14. The motivic weight $\mathrm{w}(\pi)$ is also twice the greatest eigenvalue of $\mathrm{c}_\infty(\pi)$.

Our proof of this theorem, in the spirit of the work of Stark, Odlyzko, and Serre on lower bounds for the discriminants of number fields, relies on an analog in the setting of automorphic L-functions of the *explicit formulas* of Riemann and Weil in prime number theory. This analog was developed by Mestre [144] and applied by Fermigier to the standard L-functions $\mathrm{L}(s,\pi)$ for $\pi \in \Pi_{\mathrm{cusp}}(\mathrm{PGL}_n)$ to show the nonexistence of certain elements π [84]. We apply it, more generally, to the "Rankin–Selberg L-function" of an arbitrary pair $\{\pi,\pi'\}$ of cuspidal automorphic representations of PGL_n and $\mathrm{PGL}_{n'}$ (Jacquet, Piatetski-Shapiro, Shalika).

In the specific case where π' is the dual of π, this method has already been succesfully applied by Miller [147]; however, our study contains some new results that deserve to be mentioned. First of all, we discovered that certain real-valued symmetric bilinear forms on the Grothendieck ring K_∞ of the Weil group of $\mathbb{R}$ that occur naturally in the statements of the explicit formulas are positive definite on sufficiently large subgroups of K_∞. It is this phenomenon that is responsible for the finiteness of the list given in Theorem F. Moreover, we establish simple, but efficient, criteria to prevent the simultaneous existence of π and π' (for example involving only π_∞ and π'_∞). We refer to Sect. 9.3 for precise statements.

1.6 Proofs of Theorems D and E

Let us sketch the proof of Theorem E (Sect. 9.4.3). Let $\pi \in \Pi_{\mathrm{disc}}(\mathrm{O}_{24})$ be such that π_∞ is the trivial representation. We first claim that (π,St) satisfies the Arthur–Langlands conjecture. Indeed, in all but one case, the results of Erokhin and Borcherds–Freitag–Weissauer recalled in Sect. 1.3 show that π admits a "ϑ-correspondent" π' in $\Pi_{\mathrm{cusp}}(\mathrm{Sp}_{2g})$ that is generated by a Siegel modular form of weight 12 and genus $g \leq 11$ for $\mathrm{Sp}_{2g}(\mathbb{Z})$ (Sect. 7.1). The claim follows from Arthur's theorem applied to π' and from the point of view of Rallis on the Eichler relations. The exceptional π, already determined by Ikeda [108], satisfies $\psi(\pi,\mathrm{St}) = \Delta_{11}[12]$, hence the Arthur–Langlands conjecture as well. Next, a simple combinatorial argument relying only on Theorem F shows that there are at most 24 possibilities for $\psi(\pi,\mathrm{St})$, namely those given in Table 1.2. On the other hand, there are at least 24 possibilities for $\psi(\pi,\mathrm{St})$, because by Nebe and Venkov, the operator T_2 has distinct eigenvalues on $\mathbb{C}[\mathrm{X}_{24}]$. This concludes the proof.

This method allows us to study, more generally, the elements of the group $\Pi_{\mathrm{cusp}}(\mathrm{Sp}_{2g})$ generated by a Siegel modular form of weight $k \leq 12$ for the group $\mathrm{Sp}_{2g}(\mathbb{Z})$ (Sect. 9.5). Theorem D is the result of this study in the specific case $k = 12$. We find 23 Siegel modular forms for $\mathrm{Sp}_{2g}(\mathbb{Z})$ that are eigenforms for $\mathrm{H}(\mathrm{Sp}_{2g})$ and have weight 12 and genus $g \leq 12$. We give their standard parameters in Table C.1. In the case of forms of weight $k \leq 11$, we prove the following theorem, which generalizes results of [77] and [168] (Theorem 9.5.9).

Theorem G. *Let $g \geq 1$ and k be integers with $g \leq k$.*

(i) *If $k \leq 10$, then $\mathrm{S}_k(\mathrm{Sp}_{2g}(\mathbb{Z})) = 0$ unless (k, g) is one of*

$$(8,4)\,, \qquad (10,2)\,, \qquad (10,4)\,, \qquad (10,6)\,, \qquad (10,8)\,,$$

in which case $\mathrm{S}_k(\mathrm{Sp}_{2g}(\mathbb{Z}))$ has dimension 1. The standard parameters of the five elements of $\Pi_{\mathrm{disc}}(\mathrm{Sp}_{2g})$ generated by these spaces are, respectively,

$$\Delta_{11}[4]\oplus[1]\,, \quad \Delta_{17}[2]\oplus[1]\,, \quad \Delta_{15}[4]\oplus[1]\,, \quad \Delta_{17}[2]\oplus\Delta_{11}[4]\oplus[1]\,, \quad \text{and} \quad \Delta_{11}[8]\oplus[1]\,.$$

(ii) *If $k = 11$ and $g \neq 6$, then $\mathrm{S}_k(\mathrm{Sp}_{2g}(\mathbb{Z})) = 0$.*

Let us point out certain difficulties in the proofs of Theorems D and G that do not appear in that of Theorem E. Let π be an element of $\Pi_{\mathrm{cusp}}(\mathrm{Sp}_{2g})$ generated by a Siegel modular form of weight $k \leq 12$ and genus $g < 12+k$. Theorem F implies that $\psi(\pi, \mathrm{St})$ belongs to an explicit finite list of possibilities. In contrast to the situation of Theorem E, certain elements of this list should not actually occur, as shown by reviewing the multiplicity formula. We bypass the use of this formula by turning to results of Böcherer [27] and Ikeda [108, 110], as well as to various constructions of theta series. We also expect the vanishing of $\mathrm{S}_{11}(\mathrm{Sp}_{12}(\mathbb{Z}))$ but cannot give an unconditional proof. The cases where $g = k$ are more delicate (we do not even know how to write Arthur's multiplicity formula explicitly in this case). We exclude them in an ad hoc fashion by using the work of S. Mizumoto [150] on the poles of the L-function $\mathrm{L}(s, \pi, \mathrm{St})$ (Sect. 8.7).

1.7 A Few Applications

By Theorem E, the original problem of determining the numbers $\mathrm{N}_p(L, M)$ for L and M in X_{24} and p prime becomes equivalent to that of determining the eigenvalues of the Hecke operators in $\mathrm{H}(\mathrm{PGSp}_4)$ acting on the four genus 2 vector-valued Siegel modular eigenforms mentioned in Sect. 1.4. In Sect. 10.3, we give a method we discovered to compute these eigenvalues, using the analysis of the p-neighbors of the Leech lattice carried out in Sect. 3.4.

Let (j, k) be one of the four ordered pairs considered in Sect. 1.4, namely $(6, 8)$, $(4, 10)$, $(8, 8)$, and $(12, 6)$. Denote by (w, v) the corresponding ordered pair $(2j + k - 3, j + 1)$. If q is an integer of the form p^m with p prime, we set

$$\tau_{j,k}(q) \;=\; q^{w/2}\, \mathrm{trace}\; \mathrm{c}_p(\Delta_{w,v})^m \;;$$

this complex number is in fact in $\mathbb{Z}$.

Theorem H. *Let (j, k) be one of the four ordered pairs $(6, 8), (4, 10), (8, 8)$, and $(12, 6)$. The integers $\tau_{j,k}(p)$ with p prime and at most 113, and the integers $\tau_{j,k}(p^2)$ with p prime and at most 29, are given by Tables C.3 and C.4, respectively.*

These results confirm and extend the prior computations by Faber and Van der Geer [83], [89, Sect. 25] for $p \leq 37$ by completely different methods. Our computation allows us to determine the exact value of $\mathrm{N}_p(L, M)$ for all L and M in X_{24} and all primes $p \leq 113$.

Theorem F shows that the computation of $\tau_{j,k}(q)$ may be less futile than it seems. Indeed, in view of the Langlands conjecture, this theorem suggests a parallel classification, which still needs proving on the "ℓ-adic side," of the effective pure motives over $\mathbb{Q}$ with everywhere good reduction and motivic weight at most 22. For example, it imposes a remarkable conjectural constraint on the Hasse–Weil zeta function of the Deligne–Mumford stack $\overline{\mathcal{M}_{g,n}}$ classifying the stable curves of genus g endowed with n marked points, with $g \geq 2$, $n \geq 0$, and $3g - 3 + n \leq 22$. We should be able to express the zeta function uniquely in terms of the $\tau_{j,k}(q)$ and the coefficients of the normalized cusp forms of weight at most 22 for $\mathrm{SL}_2(\mathbb{Z})$. This confirms certain results (resp. conjectures) of Bergström, Faber, and Van der Geer [83, 82, 21] when $g = 2$ (resp. $g = 3$).

In Sect. 10.4, we use Theorem E to prove congruences satisfied by the integers $\tau_{j,k}(p)$ with p prime. We obtain these congruences by studying the eigenvectors of T_2 in the natural basis of $\mathbb{Z}[\mathrm{X}_{24}]$ and using arguments involving Galois representations. Among other things, we prove the congruence conjectured by Harder in [100].

Theorem I (Harder Conjecture). *For every prime p, we have the congruence*

$$\tau_{4,10}(p) \equiv \tau_{22}(p) + p^{13} + p^8 \bmod 41 ,$$

where $\tau_{22}(p)$ denotes the pth coefficient of the normalized cusp form of weight 22 for the group $\mathrm{SL}_2(\mathbb{Z})$.

Finally, let us return to the proof of Theorem E sketched in Sect. 1.6. It relies on the equality $|\mathrm{X}_{24}| = 24$, a consequence of Niemeier's classification. However, in Sect. 9.6, we explain how the combination of the ideas mentioned above and Arthur's multiplicity formula (including Conjectures 8.4.22 and 8.4.25 stated in Chap. 8) allows us to bypass the use of this equality and even recover it "without any computations involving Euclidean lattices." Even better, we recover not only that there are exactly 24 Niemeier lattices up to isometry, but also that only one of them does not have an isometry of determinant -1.

Is it reasonable to hope that we can give a sharp estimate for the cardinality of X_{32} through such a method? The question remains open, but the example of dimension 24 shows that this approach, dear to the first author, is less absurd than it may seem. A necessary ingredient for this project is the knowledge of (say, "self-dual, regular") algebraic representations in $\Pi_{\mathrm{cusp}}(\mathrm{PGL}_n)$ of motivic weight at most 30; progress in this direction has been achieved in [55] and [195].

*

* *

To conclude this introduction, let us say a few words on the use of results of Arthur in this book. These results rely on an impressive collection of difficult works, some of which appeared only shortly before the completion of this book (see [13, 152, 207], and the discussion in Sect. 8.1). That is why, in the main part of this book, we felt it useful to indicate with a star* the statements that depend on the results of Arthur's book [13]. In this introduction, this concerns the proofs of Theorems B, D, E, F,[10] G, H, and I. On the other hand, let us mention that, contrary to what we stated in [53] and [51], our proofs no longer depend on the results announced of Chap. 9 of [13] on inner forms, nor on the conjectural properties of the Arthur packets of the type studied by Adams and Johnson.

[10] For Theorem F, we in fact prove a variant that is nearly as strong without using Arthur's theory; see Theorem 9.3.2.

Chapter 2
Bilinear and Quadratic Algebra

2.1 Basic Concepts in the Theory of Bilinear and Quadratic Forms

Let A be a commutative ring with unit.

A b-*module* over A is a projective A-module L of finite type endowed with a symmetric bilinear form that is nondegenerate, that is, such that the induced homomorphism $L \to \mathrm{Hom}_A(L, A)$ is an isomorphism; when A is a field, we obviously replace "b-module" by "b-*vector space*." Most of the time, we denote the symmetric bilinear form $L \times L \to A$ by $(x, y) \mapsto x.y$. In this book, the rings A we encounter are principal ideal domains or fields, so that in the definition we just gave, we could have replaced "projective A-module of finite type" by "free A-module of finite dimension."

Let S be a symmetric $n \times n$ matrix with coefficients in A. We use the notation $\langle S \rangle$ to denote the A-module A^n endowed with the bilinear form whose matrix in the canonical basis is S; it is clear that $\langle S \rangle$ is a b-module if and only if we have $\det S \in A^\times$. If S is a diagonal matrix with diagonal entries $a_1, a_2, \ldots, a_n$, then $\langle S \rangle$ is denoted by $\langle a_1, a_2, \ldots, a_n \rangle$.

Let us now assume that A is a Dedekind domain; we denote its field of fractions by K. (In fact, given the rings we have in mind, namely $\mathbb{Z}$ and $\mathbb{Z}_p$, we could replace "Dedekind domain" by "principal ideal domain.")

Let V be a finite-dimensional K-vector space. A *lattice* in V (with respect to A) is a sub-A-module L of V that generates V as a K-vector space and is of finite type over A; it is a projective A-module of rank $\dim_K V$.

Let V be a b-vector space over K, and let L be a lattice in V. The sub-A-module of V consisting of the elements y such that $x.y$ belongs to A for all x in L is a lattice in V that we call the *dual* of L and denote by $L^\sharp$. We call the lattice L *integral* if $x.y$ belongs to A for all x and y in L, in other words, if we have $L \subset L^\sharp$.

G. Chenevier, J. Lannes, *Automorphic Forms and Even Unimodular Lattices*, Ergebnisse der Mathematik und ihrer Grenzgebiete. 3. Folge / A Series of Modern Surveys in Mathematics 69, https://doi.org/10.1007/978-3-319-95891-0_2

Let L be an integral lattice in V. Consider the quotient $L^\sharp/L$. The following hold:

- The quotient $L^\sharp/L$ is a torsion A-module of finite type.
- The symmetric bilinear form associated with V induces a nondegenerate symmetric bilinear "form" on $L^\sharp/L$ with values in K/A. Here, "nondegenerate" means that the homomorphism $L^\sharp/L \to \mathrm{Hom}_A(L^\sharp/L, K/A)$ induced by the form is an isomorphism.

We call this type of object an e-*module* over A; the letter "e" is for "enlacement", the French word for "linking." An e-module over A is thus a torsion A-module of finite type endowed with a nondegenerate bilinear form with values in K/A (called the *linking form*). We call the e-module $L^\sharp/L$ the *residue* of L. Of course, studying the quotient $L^\sharp/L$ is far from new, but our terminology is not the classical one; for example, $L^\sharp/L$ is called the "dual quotient group" or "glue group" in [68] and the "cokernel of L" in [17]; we will often denote it by $\mathrm{res}\, L$.

Whenever possible, we will extend to e-modules the notation and terminology used for modules endowed with a symmetric bilinear form with values in A. Here are a few examples, where C is an e-module:

- The symmetric bilinear form $C \times C \to K/A$ is generally denoted by $(x, y) \mapsto x.y$.
- We call a submodule I of C *isotropic* if we have $x.y = 0$ for all elements x and y of I, that is, if we have $I \subset I^\perp$, where $I^\perp$ denotes the orthogonal complement of I.
- We call a submodule I of C a *Lagrangian* if we have $I = I^\perp$.

A $\widetilde{\mathrm{b}}$-*module* over A (which we still assume to be a Dedekind domain) is a projective A-module L of finite type endowed with a symmetric bilinear form such that the induced homomorphism $L \to \mathrm{Hom}_A(L, A)$ is injective (or, equivalently, such that the induced bilinear form on $K \otimes_A L$ is nondegenerate). An integral lattice in a b-vector space over K is the prototype of such an object. Conversely, every $\widetilde{\mathrm{b}}$-module L over A can be viewed as an integral lattice in the b-vector space $K \otimes_A L$. A $\widetilde{\mathrm{b}}$-module L therefore has a residue $\mathrm{res}\, L$ that is an e-module; the A-module underlying $\mathrm{res}\, L$ can be identified with the cokernel of the injection $L \hookrightarrow \mathrm{Hom}_A(L, A)$.

By replacing symmetric bilinear forms by alternating bilinear forms in the previous definitions, we obtain, mutatis mutandis, the definitions of, respectively, a-*module*, ae-*module*, $\widetilde{\mathrm{a}}$-*module*, and *residue* of an $\widetilde{\mathrm{a}}$-module. For example, an ae-module over A is a torsion A-module C of finite type endowed with an alternating bilinear form $C \times C \to K/A$ such that the induced homomorphism $C \to \mathrm{Hom}_A(C, K/A)$ is an isomorphism. Note that an a-module over A always has even rank and that the same holds for an $\widetilde{\mathrm{a}}$-module if A is a Dedekind domain.

Recall that a map $f\colon M \to N$ between two A-modules is called *quadratic* if it satisfies the following two properties:

- We have $f(ax) = a^2 f(x)$ for every a in A and every x in M.
- The map $M \times M \to N$ defined by $(x, y) \mapsto f(x+y) - f(x) - f(y)$ is bilinear.

In the case $N = A$, we say that f is a quadratic form on M.

By replacing symmetric bilinear forms by quadratic forms, we obtain, mutatis mutandis, the definitions of, respectively, q-*module*, qe-*module*, $\widetilde{\mathrm{q}}$-*module*, and *residue* of a $\widetilde{\mathrm{q}}$-module. For example, a q-module over A is a projective A-module L of finite type endowed with a quadratic form $\mathrm{q}\colon L \to A$ such that the symmetric bilinear form

$$L \times L \to A\,, \quad (x, y) \mapsto \mathrm{q}(x+y) - \mathrm{q}(x) - \mathrm{q}(y)$$

is nondegenerate (we call it the *associated* symmetric bilinear form). A qe-module over A is a torsion A-module of finite type endowed with a nondegenerate quadratic form with values in K/A (the *quadratic linking form*); the *residue* of a $\widetilde{\mathrm{q}}$-module is now a qe-module. A submodule I of a qe-module is *isotropic* if we have $\mathrm{q}(I) = 0$ (a condition that implies $I \subset I^{\perp}$); it is a *Lagrangian* if we have $\mathrm{q}(I) = 0$ in addition to the condition $I = I^{\perp}$. A lattice L in a q-vector space is *integral* if $\mathrm{q}(x)$ belongs to A for every x in L (a condition that implies that $x.y$ belongs to A for all x and y in L).

If 2 is not a zero divisor, then a q-module over A is just an *even* b-module, that is, a b-module L such that $x.x$ is divisible by 2 for every x in L; in this case, the quadratic form is determined by the equality $x.x = 2\mathrm{q}(x)$. If 2 is invertible in A, the notions of q-module and b-module coincide; even in this case, remember that quadratic forms and associated symmetric bilinear forms are linked by the equality we just mentioned.

The following proposition is obvious..., which does not prevent it from being quite useful.

Proposition 2.1.1. *Let V be a* b*-vector space (resp. an* a*-vector space, resp. a* q*-vector space) over K and L an integral lattice in V. Let $\gamma\colon L^{\sharp} \to \operatorname{res} L$ be the homomorphism obtained by passing to the quotient.*

(a) *Let I be a submodule of* $\operatorname{res} L$. *The following conditions are equivalent:*

  (i) *The submodule I is isotropic.*
  (ii) *The lattice $\gamma^{-1}(I)$ is integral.*

(b) *The map $I \mapsto \gamma^{-1}(I)$ is an inclusion-preserving bijection from the set of isotropic submodules of* $\operatorname{res} L$ *to the set of integral lattices of V containing L (and therefore contained in $L^{\sharp}$).*

(c) *Let I be an isotropic submodule of* $\operatorname{res} L$. *The symmetric bilinear form $I^{\perp}/I \times I^{\perp}/I \to K/A$ (resp. the alternating bilinear form $I^{\perp}/I \times I^{\perp}/I \to K/A$, resp. the quadratic form $I^{\perp}/I \to K/A$) induced by the corresponding form associated with* $\operatorname{res} L$ *gives $I^{\perp}/I$ the structure of an* e*-module (resp.* ae*-module, resp.* qe*-module) that can be identified with the residue of the integral lattice $\gamma^{-1}(I)$.*

Hyperbolic Functors

Let A be a commutative ring with unit, and let L be a projective A-module of finite type. Then $\operatorname{Hom}_A(L, A)$ is also a projective A-module of finite type, which we denote by L^* (and call the *dual* of L). The map

$$L \oplus L^* \to A\,, \quad (x, \xi) \mapsto \langle x, \xi \rangle$$

is a nondegenerate quadratic form that gives the projective A-module of finite type $L \oplus L^*$ the structure of a q-module. This q-module is denoted by $\mathrm{H}(L)$ and called the *hyperbolic* q-module over L.

The associated symmetric bilinear form is the map

$$((x, \xi), (y, \eta)) \mapsto \langle x, \eta \rangle + \langle y, \xi \rangle\;;$$

$L \oplus L^*$ endowed with this form is called the *hyperbolic* b-module over L and is also denoted by $\mathrm{H}(L)$.

Likewise, $L \oplus L^*$ endowed with the alternating bilinear form

$$((x, \xi), (y, \eta)) \mapsto \langle x, \eta \rangle - \langle y, \xi \rangle$$

is an a-module over A, called the *hyperbolic* a-module over L and again denoted by $\mathrm{H}(L)$.

Let H be a q-module (resp. a-module) over A; recall that in this context, a *Lagrangian* of H is a direct summand L with $L = L^{\perp}$ and $\mathrm{q}(L) = 0$ (resp. $L = L^{\perp}$).

Proposition 2.1.2. *Let H be a* q*-module (resp.* b*-module,* a*-module) and L a Lagrangian of H. The inclusion of L in H extends to an isomorphism of* q*-modules (resp.* a*-modules)* $\mathrm{H}(L) \simeq H$.

Proof. We prove the "quadratic version" of the statement; we will shamelessly follow the proof of [16, Proposition 2.1.5], which treats the "alternating version" implicitly. Let $i\colon L \to H$ be the inclusion of L in H and $p\colon H \to L^*$ the homomorphism (of A-modules) that is the composition of the isomorphism $H \to H^*$ induced by the bilinear form and the homomorphism i^*. Since L is a Lagrangian, the sequence of A-modules

$$0 \longrightarrow L \xrightarrow{\;i\;} H \xrightarrow{\;p\;} L^* \longrightarrow 0$$

is exact. We need to show that there exists an A-linear section $s\colon L^* \to H$ of p satisfying $\mathrm{q}(s(\xi)) = 0$ for every ξ in L^*. Let Σ be the set of A-linear sections of p. The set Σ is nonempty because L^* is projective; moreover, Σ has the canonical structure of an affine space under $\mathrm{Hom}_A(L^*, L)$. We identify the latter with the A-module of bilinear forms on L^* and denote it by $\mathcal{B}_{L^*}$. Let $\mathcal{Q}_{L^*}$ be the A-module of quadratic forms on L^* and $\gamma\colon \Sigma \to \mathcal{Q}_{L^*}$ the map that sends a section s to the quadratic form $\xi \mapsto \mathrm{q}(s(\xi))$. Let u be an element of $\mathrm{Hom}_A(L^*, L)$. We have

$$\gamma(s + u) = \gamma(s) + \widetilde{\gamma}(u)\,,$$

where $\widetilde{\gamma}$ denotes the map $\mathcal{B}_{L^*} \to \mathcal{Q}_{L^*}$ that sends a bilinear form u to the quadratic form $\xi \mapsto u(\xi, \xi)$. The fact that $\gamma^{-1}(0)$ is nonempty now follows from the surjectivity of $\widetilde{\gamma}$. This is clear when L^* is free; the general case follows by introducing an A-module M such that the direct sum $L^* \oplus M$ is free of finite dimension. □

Remark. By considering the b-vector space $\langle \left[\begin{smallmatrix} 0 & 1 \\ 1 & 1 \end{smallmatrix}\right] \rangle$ over $\mathbb{F}_2$, we see that the statement analogous to Proposition 2.1.2 for b-modules does not hold in general.

A q-module (resp. b-module, a-module) is called *hyperbolic* if it is isomorphic to an $\mathrm{H}(L)$ for some projective A-module L of finite type. Proposition 2.1.2 says, in particular, that a q-module is hyperbolic if and only if it has a Lagrangian.

Now, let A be a Dedekind domain (with field of fractions K) and I a torsion A-module of finite type. Then $\mathrm{Hom}_A(I, K/A)$ is also a torsion A-module of finite type; we denote it by $I^\vee$ (and call it the *dual* of I). The map

$$I \oplus I^\vee \to K/A\,, \quad (x, \xi) \mapsto \langle x, \xi \rangle$$

is a nondegenerate quadratic form that gives the torsion A-module of finite type $I \oplus I^\vee$ the structure of a qe-module. This qe-module is denoted by $\mathrm{H}(I)$ and called the *hyperbolic* qe-module over I.

Let H be a qe-module over A. Recall that in this context, a Lagrangian of H is a submodule I with $I = I^\perp$ and $\mathrm{q}(I) = 0$. The proof of the following proposition is left to the reader (compare part (b) with part (b) of [16, Proposition 2.1.5]).

Proposition-Definition 2.1.3. *Let H be a* qe-*module, and let I and J be two Lagrangians of H. We say that I and J are* transverse *(or that J is* transverse to *I) if $I \cap J = 0$.*

(a) *Let I and J be two transverse Lagrangians of H. Then the linking form of H induces an isomorphism $J \cong I^\vee$ and the composition*

$$\mathrm{H}(I) = I \oplus I^\vee \to I \oplus J \to H$$

is an isomorphism of qe-*modules (the first arrow is the direct sum of the identity on I and the inverse of the automorphism $J \cong I^\vee$).*

(b) *Let I be a Lagrangian of H and $\mathcal{T}_I$ the (possibly empty) set of Lagrangians transverse to I. Then $\mathcal{T}_I$ admits the canonical structure of an affine space under the A-module $(\Lambda^2(I^\vee))^\vee$ (that is, the A-module consisting of the alternating bilinear maps $I^\vee \times I^\vee \to K/A$).*

A qe-module is called *hyperbolic* if it is isomorphic to an $\mathrm{H}(I)$ for some A-module I of finite type. Proposition 2.1.3 (a) says, in particular, that a qe-module is hyperbolic if and only if it has two transverse Lagrangians.

The Tensor Products of Forms

Let A be a commutative ring, and let L_1 and L_2 be two b-modules over A. The homomorphism of A-modules $L_1 \otimes_A L_2 \to (L_1 \otimes_A L_2)^* \cong L_1^* \otimes_A L_2^*$ that is the tensor product of the structural homomorphisms $L_1 \to L_1^*$ and $L_2 \to L_2^*$ turns

$L_1 \otimes_A L_2$ into a b-module that we call the *tensor product* of the b-modules L_1 and L_2. The symmetric bilinear form of $L_1 \otimes_A L_2$ is characterized by the fact that $(x_1 \otimes_A x_2).(y_1 \otimes_A y_2) = (x_1.y_1)(x_2.y_2)$ for all x_1, y_1 in L_1 and all x_2, y_2 in L_2.

Mutatis mutandis, we define:

- the tensor product of a b-module and a q-module, which is a q-module;
- the tensor product of a b-module and an e-module, which is an e-module;
- the tensor product of a b-module and a qe-module, which is a qe-module;
- the tensor product of a q-module and an e-module, which is a qe-module.

For example, if L_1 is a b-module and L_2 is a q-module, then the quadratic form of $L_1 \otimes_A L_2$ is characterized by the fact that $\mathrm{q}(x_1 \otimes_A x_2) = (x_1.x_1)\,\mathrm{q}(x_2)$ for every x_1 in L_1 and every x_2 in L_2.

The Discriminant of a q-Module of Even Constant Rank; the Dickson–Dieudonné Determinant

Let A be a commutative ring and L a q-module over A of even constant rank, which we assume to be nonzero. Let $\Delta(L)$ be the center of the even part, denoted by $\mathrm{C}^+(L)$, of the Clifford algebra of L (see, for example, [56, Chap. III]).

- The commutative A-algebra $\Delta(L)$ is a "double cover" of A, that is, an étale A-algebra and a projective A-module of rank 2 [71, Exp. XII, Proposition 1.5] that we should view as the *discriminant* of L. In the case $L = \mathrm{H}(P)$ with P a projective A-module of constant rank, this cover is trivial (and even trivialized): $\Delta(L) = A \times A$.
- An automorphism α of the q-module L induces an automorphism $\Delta(\alpha)$ of the A-algebra $\Delta(L)$. If we identify the automorphism group of the A-algebra $\Delta(L)$ with $\mathbb{Z}/2(A)$, then $\Delta(\alpha)$ can be identified with an element of $\mathbb{Z}/2(A)$ that we call the *Dickson–Dieudonné determinant* of α; we denote it by $\widetilde{\det}\,\alpha$. Let us recall the definition of the group functor $A \mapsto \mathbb{Z}/2(A)$ that just appeared surreptitiously above: $\mathbb{Z}/2(A)$ is the set of elements x of A satisfying $x^2 = x$, endowed with the group law $(x, y) \mapsto x + y - 2xy$.

These two statements are a predictable "globalization" of well-known results in the case where A is a field. The subtle case is that where A is a field of characteristic 2; see [74], [38, Sect. 9, Exercice 9].

Let L be a b-module over A of constant rank n. The *determinant* of L is the b-module $\Lambda^n L$ (this is a projective A-module of rank 1 endowed with the symmetric bilinear form induced by that on L); we denote it by $\det L$. When L is free, the isomorphism class of $\det L$ is identified with an element of $A^\times/A^{\times 2}$. This element is the class in $A^\times/A^{\times 2}$ of the determinant of the Gram matrix $[e_i.e_j]$, for any basis $(e_1, e_2, \ldots, e_n)$ of L; this class is often also denoted by $\det L$.

Let L be a q-module over A of even constant rank $2n$ with $n \geq 1$. The relation between the discriminant of L and the determinant of the underlying b-module is given by Proposition 2.1.4 below, whose proof Pierre Deligne kindly provided us with. To state this result, we must introduce some notation. We denote by $\mathrm{D}(L)$ the cokernel of the unit $\eta\colon A \to \Delta(L)$. It is a projective A-module of rank 1 (this follows, for example, by a faithfully flat descent argument [SGA $4\frac{1}{2}$, I, Proposition 4.2] because we have $\Delta(L) \otimes_A \Delta(L) \cong \Delta(L) \times \Delta(L)$, and it is endowed with a canonical nondegenerate symmetric bilinear form that we denote by θ. We can define θ, for example, as induced by the symmetric bilinear form $\Delta(L) \times \Delta(L) \to A$ given by $(x, y) \mapsto \mathrm{tr}_{\Delta(L)/A}((x - \bar{x})y)$ (where $\bar{x}$ denotes the "conjugate" of x). In other words, $\mathrm{D}(L)$ has the natural structure of a b-module of rank 1 over A. Finally, we denote by $(-1)^n \det L$ the b-module of rank 1 over A obtained by multiplying the symmetric bilinear form associated with $\det L$ by $(-1)^n$. Here is the result announced above.

Proposition 2.1.4. *The two* b*-modules* $\mathrm{D}(L)$ *and* $(-1)^n \det L$ *of rank* 1 *over* A *are naturally isomorphic.*

Classical Groups

The main object of this last part of the section is to fix the notation and terminology we will be using in this book with regards to the orthogonal and symplectic groups (and their variants). Convenient references on this subject are [63, 72, 75, 114].

Let A be a commutative ring with unit.

– Let L be a projective A-module of finite type. We denote the automorphism group of L by $\mathrm{GL}(L)$. The functor $R \mapsto \mathrm{GL}(R \otimes_A L)$, defined on the category of commutative A-algebras and with values in the category of groups, is an A-group scheme that we denote by GL_L. Note that if the rank of L is 1, then $\mathrm{GL}(L)$ and GL_L can be identified, respectively, with the group $A^\times$ and the A-group scheme $\mathbb{G}_\mathrm{m}$ (if G is a $\mathbb{Z}$-group scheme, we also denote by G the A-group scheme obtained after base change). If L has constant rank n, then we denote the kernel of the "determinant" homomorphism $\det\colon \mathrm{GL}_L \to \mathrm{GL}_{\Lambda^n L} = \mathbb{G}_\mathrm{m}$ by SL_L. We denote by PGL_L the A-group scheme defined as the functor that sends a commutative A-algebra R to the group of automorphisms of the R-algebra $\mathrm{End}_R(R \otimes_A L)$; we can also view PGL_L as the quotient A-group scheme $\mathrm{GL}_L/\mathbb{G}_\mathrm{m}$. Of course, in the case $A = \mathbb{Z}$ and $L = A^n$, we replace the notation $\mathrm{GL}_{\mathbb{Z}^n}$, $\mathrm{SL}_{\mathbb{Z}^n}$, and $\mathrm{PGL}_{\mathbb{Z}^n}$, by GL_n, SL_n, and PGL_n, respectively.

– Let L be a q-module (resp. b-module) over A. We say that an endomorphism α of the underlying A-module is *orthogonal* if $\mathrm{q}(\alpha(x)) = \mathrm{q}(x)$ for all x in L (resp. $\alpha(x).\alpha(y) = x.y$ for all x and y in L). The orthogonal endomorphisms form a group for the composition (so the orthogonal endomorphisms are in fact automorphisms) that we call the *orthogonal group* of L and denote by $\mathrm{O}(L)$. The functor $R \mapsto \mathrm{O}(R \otimes_A L)$, defined on the category of commutative A-algebras and with values in the category of groups, is an A-group scheme that we denote by O_L.

In fact, in this book, quadratic forms play a more important role than symmetric bilinear forms. One reason for this distinction is the statement below (which does not hold in all generality for b-modules).

Proposition 2.1.5. *For every* q*-module L over a commutative ring A, the A-group scheme* O_L *is smooth over A.*

Proof. Since the property we want to verify is local for the Zariski topology, we may assume that L is free, say $L = A^n$ for some integer n, and will do so from now on. The quadratic form becomes

$$(x_1, x_2, \ldots, x_n) \mapsto \sum_{i,j} q_{i,j}\, x_i x_j\ ,$$

where $[q_{i,j}] := Q$ denotes an $n \times n$ matrix, defined up to the addition of an alternating matrix (an *alternating matrix* is a skew-symmetric matrix A with zeros on the diagonal; we may also define an alternating matrix as an antisymmetrization). An $n \times n$ matrix M with coefficients in an A-algebra R belongs to $\mathrm{O}_L(R) := \mathrm{O}(R \otimes_A L)$ if and only if the matrix ${}^{\mathrm{t}}MQM - Q$ is alternating. The proposition follows from the fact that the equations resulting from this description ($n(n+1)/2$ polynomials in n^2 variables with coefficients in A) satisfy the Jacobian criterion for smoothness. □

We now suppose that L has even constant rank $2n$. The map that sends an orthogonal automorphism α of a q-module L to its Dickson–Dieudonné determinant $\widetilde{\det}\,\alpha$ (see earlier on) induces a homomorphism of A-group schemes, which we denote by $\widetilde{\det}\colon \mathrm{O}_L \to \mathbb{Z}/2$. Proposition 2.1.4 implies that this homomorphism lifts the homomorphism $\det\colon \mathrm{O}_L \to \mu_2$, in other words, that the diagram

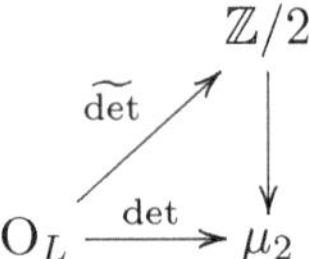

is commutative (let us recall the definition of the vertical homomorphism: let R be a commutative A-algebra and $\mathbb{Z}/2\,(R)$ the set of elements x of R satisfying $x^2 = x$, endowed with the group law $(x, y) \mapsto x + y - 2xy$; then $\mathbb{Z}/2\,(R) \to \mu_2(R)$ is the homomorphism that sends x to $1 - 2x$). To show this implication, consider the following commutative diagram with exact lines:

$$\begin{array}{ccccccccc}
0 & \longrightarrow & A & \xrightarrow{\ \eta\ } & \Delta(L) & \longrightarrow & \Lambda^{2n} L & \longrightarrow & 0 \\
 & & \big\downarrow{\scriptstyle \mathrm{id}} & & \big\downarrow{\scriptstyle \Delta(\alpha)} & & \big\downarrow{\scriptstyle \Lambda^{2n}\alpha} & & \\
0 & \longrightarrow & A & \xrightarrow{\ \eta\ } & \Delta(L) & \longrightarrow & \Lambda^{2n} L & \longrightarrow & 0\,.
\end{array}$$

We denote the A-group scheme that is the kernel of $\widetilde{\det}$ by SO_L (of course, the group $\mathrm{SO}_L(A)$ is simply denoted by $\mathrm{SO}(L)$).

Let L be a q-module over A; we denote by $\mathrm{GO}(L)$ the subgroup of the product $\mathrm{GL}(L) \times A^\times$ consisting of the pairs (α, ν) such that $\mathrm{q}(\alpha(x)) = \nu\,\mathrm{q}(x)$ for every x in L. The A-group scheme GO_L is defined by proceeding as previously; it can be viewed as a subgroup of $\mathrm{GL}_L \times \mathbb{G}_\mathrm{m}$. We leave it to the reader to verify that the restriction of the projection $\mathrm{GL}_L \times \mathbb{G}_\mathrm{m} \to \mathrm{GL}_L$ to GO_L is a closed embedding and that GO_L can therefore also be viewed as a closed subgroup of GL_L. We denote by $\nu\colon \mathrm{GO}_L \to \mathbb{G}_\mathrm{m}$ the homomorphism obtained by restricting the projection $\mathrm{GL}_L \times \mathbb{G}_\mathrm{m} \to \mathbb{G}_\mathrm{m}$ to GO_L. The group $\mathrm{GO}(L)$, introduced above, is called the *orthogonal similitude group* of L, while the element ν of $A^\times$ is called the *similitude factor* of α. We denote the quotient A-group scheme $\mathrm{GO}_L/\mathbb{G}_\mathrm{m}$ by PGO_L.

Once again, we assume that L has rank $2n$. Let (α, ν) be an element of $\mathrm{GO}(L)$; note that the element $d = \nu^{-n}\det\alpha$ of $A^\times$ satisfies $d^2 = 1$. Let $\mathrm{d}\colon \mathrm{GO}(L) \to \mu_2(A)$ be the group homomorphism defined by $(\alpha,\nu) \mapsto d$. We also denote by $\mathrm{d}\colon \mathrm{GO}_L \to \mu_2$ the associated A-group scheme homomorphism; note that d extends the homomorphism $\det\colon \mathrm{O}_L \to \mu_2$. Proposition 2.1.4 implies, as before, that d lifts to a homomorphism $\widetilde{\mathrm{d}}\colon \mathrm{GO}_L \to \mathbb{Z}/2$ that extends the homomorphism $\widetilde{\det}\colon \mathrm{O}_L \to \mathbb{Z}/2$. Let us be a bit more precise. Recall that we identify the automorphism group of $\Delta(L)$ with $\mathbb{Z}/2(A)$; through this identification, $\widetilde{\mathrm{d}}(\alpha,\nu)$ corresponds to the composition

$$\Delta(L) \xrightarrow{\Delta(\alpha)} \Delta(\nu L) \xrightarrow{[\nu]} \Delta(L)\,,$$

where νL denotes the q-module obtained by multiplying the quadratic form on L by ν and $[\nu]$ denotes the isomorphism induced by the isomorphism $[\nu]\colon \mathrm{C}^+(\nu L) \to \mathrm{C}^+(L)$ introduced in [71, Exp. XII, Sect. 1.3]. The A-group scheme GSO_L is defined as the kernel of $\widetilde{\mathrm{d}}$. By passing to the quotient, $\widetilde{\mathrm{d}}$ induces a homomorphism $\mathrm{PGO}_L \to \mathbb{Z}/2$; the A-group scheme PGSO_L is defined as the kernel of this induced homomorphism. We can also define PGSO_L as the quotient $\mathrm{GSO}_L/\mathbb{G}_\mathrm{m}$.

- Finally, let L be an a-module over A. We denote the automorphism group of L by $\mathrm{Sp}(L)$. The A-group schemes Sp_L, GSp_L, and PGSp_L and the homomorphism $\nu\colon \mathrm{GSp}_L \to \mathbb{G}_\mathrm{m}$ are defined mutatis mutandis. If L has constant rank $2n$, the theory of the Pfaffian shows that the homomorphism $\det\colon \mathrm{GSp}_L \to \mathbb{G}_\mathrm{m}$ (induced by the homomorphism $\det\colon \mathrm{GL}_L \to \mathbb{G}_\mathrm{m}$) coincides with ν^n (a relation that implies, in particular, that Sp_L is a subgroup of SL_L). Let $n \geq 1$ be an integer; in the case where A is $\mathbb{Z}$ and L is the hyperbolic a-module $\mathrm{H}(\mathbb{Z}^n)$, we replace the notation $\mathrm{Sp}_{\mathrm{H}(\mathbb{Z}^n)}$, $\mathrm{GSp}_{\mathrm{H}(\mathbb{Z}^n)}$, and $\mathrm{PGSp}_{\mathrm{H}(\mathbb{Z}^n)}$ by Sp_{2n}, GSp_{2n}, and PGSp_{2n}, respectively. Recall that we can identify Sp_2 with SL_2.

All A-group schemes introduced above are affine and of finite presentation over A; for short, we will call such A-group schemes *A-groups*.

Let us conclude this section on the classical groups with one last remark. Let $PG = G/\mathbb{G}_\mathrm{m}$ be one of the "projective" A-group schemes we just defined; in Chap. 4, we will only need to consider the group $PG(A)$ for rings A with $\mathrm{Pic}(A) = 0$, so that the canonical injection $G(A)/A^\times \to PG(A)$ will be an isomorphism.

2.2 On b-Modules and q-Modules over $\mathbb{Z}$

In this section, we recall some very classical results from the theory of b-modules and q-modules over $\mathbb{Z}$ (see, for example, [177, Chap. V], [148, Chap. II]).

To organize these results, which we number 1, 2, and 3, we need the concept of Witt ring; see, for example, [148, Chap. I, Sect. 7]. Let us recall the definition. Let A be a commutative ring. The set of isomorphism classes of b-modules over A, which we denote by $\mathcal{B}(A)$, is a commutative monoid for the orthogonal sum. We denote by $\mathrm{W}(A)$ the quotient monoid $\mathcal{B}(A)/\mathcal{N}(A)$, where $\mathcal{N}(A)$ is the submonoid generated by the isomorphism classes of the split b-modules (a b-module is called *split* if it has a *Lagrangian*, that is, a direct summand that is its own orthogonal complement). The monoid $\mathrm{W}(A)$ is a group, and the tensor product of b-modules gives it the structure of a commutative ring. The abelian group $\mathrm{WQ}(A)$ is defined mutatis mutandis in terms of q-modules over A; see, for example, [148, App. 1] (recall that in this case, the split q-modules are in fact hyperbolic; see Proposition 2.1.2). The group $\mathrm{WQ}(A)$ has the natural structure of a $\mathrm{W}(A)$-module.

Let us turn to the case $A = \mathbb{Z}$.

(1) The first result we wish to recall is the determination of $\mathrm{W}(\mathbb{Z})$: the canonical homomorphism $\mathrm{W}(\mathbb{Z}) \to \mathrm{W}(\mathbb{R})$ is an isomorphism. There are two ways to state this:

- The "signature" homomorphism, which we denote by $\tau\colon \mathrm{W}(\mathbb{Z}) \to \mathbb{Z}$, is an isomorphism. Let us explain what we mean by *signature*. Let E be a b-vector space over $\mathbb{R}$; such an E is isomorphic to a b-vector space of the form

$$\langle +1, +1, \ldots, +1, -1, -1, \ldots, -1 \rangle \,,$$

  and the signature of E, which we denote by $\tau(E)$, is the difference $n_+ - n_-$ between the number of terms $+1$ and the number of terms -1 in this form. It is clear that the homomorphism $\tau\colon \mathrm{W}(\mathbb{R}) \to \mathbb{Z}$ is an isomorphism.
- The "unit" homomorphism, which we denote by $\eta\colon \mathbb{Z} \to \mathrm{W}(\mathbb{Z})$, is an isomorphism. For a nice proof of this result that does not use the Hasse–Minkowski theorem, see [148, Chap. IV, Sect. 2].

Scholium 2.2.1. *Let L_1 and L_2 be two* b*-modules over $\mathbb{Z}$. The following conditions are equivalent:*

(i) *The two* b*-vector spaces $\mathbb{Q} \otimes_{\mathbb{Z}} L_1$ and $\mathbb{Q} \otimes_{\mathbb{Z}} L_2$ over $\mathbb{Q}$ are isomorphic.*
(ii) *The two* b*-vector spaces $\mathbb{R} \otimes_{\mathbb{Z}} L_1$ and $\mathbb{R} \otimes_{\mathbb{Z}} L_2$ over $\mathbb{R}$ are isomorphic.*

(2) The second result we wish to mention concerns the theory of "Wu vectors." Let L be a b-module over $\mathbb{Z}$; since $\mathbb{F}_2 \otimes_{\mathbb{Z}} L$ is a b-vector space over $\mathbb{F}_2$, there exists an element u of L, well-defined modulo $2L$, such that

$$x.x \equiv u.x \pmod 2$$

for every x in L. We call u a *Wu vector* (it is also called a "characteristic vector"; the term "Wu vector" refers to classes defined by Wen-Tsün Wu in the cohomology modulo 2 of compact manifolds [214]). Note that the reduction modulo 8 of the integer

$u.u$ is independent of the choice of u and that the map $L \mapsto u.u$ induces a homomorphism of commutative rings with units, which we denote by $\sigma\colon \mathrm{W}(\mathbb{Z}) \to \mathbb{Z}/8$. Also note that in the above, we can replace $\mathbb{Z}$ by $\mathbb{Z}_2$ and that the homomorphism σ factors through $\mathrm{W}(\mathbb{Z}_2)$. Finally, the reduction modulo 2 of σ factors through a homomorphism $\mathrm{W}(\mathbb{F}_2) \to \mathbb{Z}/2$ that coincides with the isomorphism "dimension modulo 2."

In view of what we stated earlier concerning $\mathrm{W}(\mathbb{Z})$, the homomorphism σ can be identified with the reduction modulo 8 from $\mathbb{Z}$ to $\mathbb{Z}/8$.

Scholium 2.2.2. (a) *Let L be a* b*-module over $\mathbb{Z}$, and let u be a Wu vector of L. Then we have the congruence*

$$\tau(L) \equiv u.u \pmod 8 .$$

(b) *The signature of a* q*-module over $\mathbb{Z}$ is divisible by* 8.

(3) The last result we wish to mention is more technical. Let L be an *odd* (that is, not even) b-module over $\mathbb{Z}$. Let M be the submodule of index 2 of L consisting of the vectors x satisfying $x.x \equiv 0 \bmod 2$. The map $M \to \mathbb{Z}$ defined by $x \mapsto x.x/2$ turns M into a $\widetilde{\mathrm{q}}$-module whose residue we will determine further on.

Consider the exact sequence

$$0 \to L/M \to M^\sharp/M \to M^\sharp/L \to 0 .$$

Let u be a Wu vector of L and v an element of L with $v.v \equiv 1 \bmod 2$ (or, equivalently, $u.v \equiv 1 \bmod 2$). The quotients L/M and $M^\sharp/L$ are cyclic groups of order 2, generated, respectively, by the classes of v and $u/2$ (the quotient L/M is in fact a Lagrangian of the e-module underlying the qe-module $\operatorname{res} M$ and $M^\sharp/L$ is canonically isomorphic to the dual of this Lagrangian). The exact sequence above can be split if and only if u belongs to M, that is, if we have $u.u \equiv 0 \bmod 2$ or, equivalently, in view of result 2 mentioned above, if the dimension of L is even. We therefore distinguish between two cases according to the parity of this dimension:

- In the case $\dim L \equiv 1 \bmod 2$, the residue $\operatorname{res} M$ is isomorphic to $\mathbb{Z}/4$ and is generated by the class of $u/2$; in $\mathbb{Q}/\mathbb{Z}$, we have the equality

$$\mathrm{q}\Big(x\frac{u}{2}\Big) = \frac{\tau(L)}{8}\, x^2$$

  for all x in $\mathbb{Z}$.
- In the case $\dim L \equiv 0 \bmod 2$, the residue $\operatorname{res} M$ is isomorphic to the sum $\mathbb{Z}/2 \oplus \mathbb{Z}/2$ and is generated by the classes of $u/2$ and v or, equivalently, the classes of $u/2$ and $u/2 - v$; in $\mathbb{Q}/\mathbb{Z}$, we have the equality

$$\mathrm{q}\Big(x\frac{u}{2} + y\Big(\frac{u}{2} - v\Big)\Big) = \frac{\tau(L)}{8}\,(x^2+y^2) + \Big(\frac{\tau(L)}{4} + \frac{1}{2}\Big)\, xy$$

  for all x and y in $\mathbb{Z}$.

We can see that the equation $\mathrm{q}(\iota) = 0$ with $\iota \in \operatorname{res} M - \{0\}$ has no solution for $\tau(L) \not\equiv 0 \bmod 8$ and that for $\tau(L) \equiv 0 \bmod 8$, it has exactly two, namely $u/2$ and $u/2 - v$. In summary, we have obtained the following result.

Scholium 2.2.3. *Let L be an odd* b*-module over $\mathbb{Z}$ and M the submodule of index* 2 *of L consisting of the vectors x satisfying $x.x \equiv 0 \bmod 2$ (M is therefore a* $\widetilde{\mathrm{q}}$*-module over $\mathbb{Z}$). The following conditions are equivalent:*

(i) *The* qe*-module* $\operatorname{res} M$ *is isomorphic to* $\mathrm{H}(\mathbb{Z}/2)$.
(ii) *The signature of L is divisible by* 8.

The Partition of the Wu Vectors into Two Classes

Let L be an odd b-module over $\mathbb{Z}$ and M the submodule of index 2 of L consisting of the vectors x satisfying $x.x \equiv 0 \bmod 2$.

Let $\mathrm{Wu}(L)$ be the set of Wu vectors of L. The action of L on $\mathrm{Wu}(L)$ defined by $(u, x) \mapsto u + 2x$ (for u in $\mathrm{Wu}(L)$ and x in L) is free and transitive. The action of M on $\mathrm{Wu}(L)$ induced by restriction has exactly two orbits (those of u and $u - 2v$, where u denotes an arbitrary Wu vector of L and v is an element of L with $v.v \equiv 1 \bmod 2$); we say that two Wu vectors in the same orbit are *equivalent*. Let u_1 and u_2 be two nonequivalent Wu vectors; we can paraphrase the discussion before Scholium 2.2.3 as follows:

- The classes of $u_1/2$ and $u_2/2$ generate the abelian group $\operatorname{res} M$.
- If the dimension of L is odd, these classes are of order 4 and opposite.
- If the dimension of L is even, these classes are of order 2 and form a basis of the $\mathbb{Z}/2$-vector space $\operatorname{res} M$.
- We have the following equalities in $\mathbb{Q}/\mathbb{Z}$:

$$\mathrm{q}\Big(\frac{u_1}{2}\Big) = \frac{\tau(L)}{8}\,, \quad \mathrm{q}\Big(\frac{u_2}{2}\Big) = \frac{\tau(L)}{8}\,, \quad \mathrm{q}\Big(\frac{u_1}{2} + \frac{u_2}{2}\Big) = \frac{1 + \dim L}{2}\,;$$

  note that in both cases, these equalities determine the quadratic linking form $\mathrm{q}\colon \operatorname{res} M \to \mathbb{Q}/\mathbb{Z}$.

Let us give an illustration of the above. Let n be a positive integer; we consider the "Euclidean" symmetric bilinear form

$$\mathbb{Z}^n \times \mathbb{Z}^n \to \mathbb{Z}\,, \quad \big((x_1, x_2, \ldots, x_n), (y_1, y_2, \ldots, y_n)\big) \mapsto \sum_{i=1}^{n} x_i y_i\,.$$

Endowed with this form, $\mathbb{Z}^n$ is an odd b-module over $\mathbb{Z}$, which we denote by I_n. We denote by D_n the submodule of I_n of index 2 consisting of the vectors x satisfying $x.x \equiv 0 \bmod 2$, that is, $\sum_{i=1}^{n} x_i \equiv 0 \bmod 2$ (this notation consciously evokes the theory of root systems; we will return to this topic in Sect. 2.3). Let $(\varepsilon_1, \varepsilon_2, \ldots, \varepsilon_n)$ be the canonical basis of I_n; note that the vectors $u_1 := \varepsilon_1 + \varepsilon_2 + \ldots + \varepsilon_n$ and

$u_2 := -\varepsilon_1 + \varepsilon_2 + \ldots + \varepsilon_n$ are nonequivalent Wu vectors of I_n. If n is divisible by 8, we can describe the qe-module $\operatorname{res} \mathrm{D}_n$ explicitly as follows:

- As a $\mathbb{Z}$-module, the residue $\operatorname{res} \mathrm{D}_n$ is a $\mathbb{Z}/2$-vector space of dimension 2 with basis given by the classes of the vectors $\iota_1 := u_1/2$ and $\iota_2 := u_2/2$.
- The quadratic linking form of $\operatorname{res} \mathrm{D}_n$ is determined by $\mathrm{q}(\iota_1) = 0$, $\mathrm{q}(\iota_2) = 0$, and $\iota_1.\iota_2 = 1/2$.

Items (b) and (c) of Proposition 2.1.1 show that the lattice in $\mathbb{Q} \otimes_{\mathbb{Z}} \mathrm{I}_n$ generated by D_n and $\frac{1}{2}(\varepsilon_1 + \varepsilon_2 + \ldots + \varepsilon_n)$ is a q-module over $\mathbb{Z}$, which we denote by E_8 for $n = 8$ and by E_n or D_n^+ for $n \geq 16$ (and $n \equiv 0 \bmod 8$).

Scholium 2.2.4. *The composition*

$$\mathrm{WQ}(\mathbb{Z}) \xrightarrow{\text{forget}} \mathrm{W}(\mathbb{Z}) \xrightarrow{\ \tau\ } \mathbb{Z}$$

induces an isomorphism (of $\mathrm{W}(\mathbb{Z})$*-modules) from* $\mathrm{WQ}(\mathbb{Z})$ *to the ideal* $8\mathbb{Z}$*; the group* $\mathrm{WQ}(\mathbb{Z})$ *is infinite, cyclic, and generated by the class of* E_8.

Proof. It suffices to observe that the forgetful map $\mathrm{WQ}(\mathbb{Z}) \to \mathrm{W}(\mathbb{Z})$ is injective. □

The Genus of a q-Module over $\mathbb{Z}$

This heading refers to item (b) of the following statement.

Scholium 2.2.5. *Let L be a* q*-module over* $\mathbb{Z}$ *and p a prime.*

(a) *The* q*-vector space* $\mathbb{F}_p \otimes_{\mathbb{Z}} L$ *is hyperbolic.*
(b) *The* q*-module* $\mathbb{Z}_p \otimes_{\mathbb{Z}} L$ *is hyperbolic.*

Proof. The first statement implies the second; indeed, two q-modules L_1 and L_2 over $\mathbb{Z}_p$ are isomorphic if and only if the q-vector spaces $\mathbb{F}_p \otimes_{\mathbb{Z}_p} L_1$ and $\mathbb{F}_p \otimes_{\mathbb{Z}_p} L_2$ are isomorphic. To prove statement (a), it suffices to show that the natural homomorphism $\mathrm{WQ}(\mathbb{Z}) \to \mathrm{WQ}(\mathbb{F}_p)$ is trivial. The easy case where p is odd is left to the reader. For $p = 2$, we can use the following arguments:

- The Arf invariant $\mathrm{Arf}\colon \mathrm{WQ}(\mathbb{F}_2) \to \mathrm{H}^1_{\text{ét}}(\mathbb{F}_2; \mathbb{Z}/2) \cong \mathbb{Z}/2$ is an isomorphism.
- The group $\mathrm{H}^1_{\text{ét}}(\mathbb{Z}; \mathbb{Z}/2)$ is trivial.

The argument above can be replaced by the following, more prosaic, one:

- The homomorphism $\mathrm{WQ}(\mathbb{Z}) \to \mathrm{WQ}(\mathbb{F}_2)$ factors through $\mathrm{WQ}(\mathbb{Z}_2)$ and the determinant of a q-module L over $\mathbb{Z}_2$ is equal to the class of the element $(-1)^{(\dim L)/2}\,(-3)^{\mathrm{Arf}(\mathbb{F}_2 \otimes_{\mathbb{Z}_2} L)}$ in $\mathbb{Z}_2^\times/\mathbb{Z}_2^{\times 2}$ (to see this, note, for example, that such an L decomposes into an orthogonal sum of q-modules of dimension 2, each endowed with a basis (e, f) with $e.f = 1$). □

We can also deduce the results of Scholium 2.2.5 from the following proposition, whose proof is left to the reader.

Proposition 2.2.6. *Let p be a prime, and let L_1 and L_2 be two* q*-modules over $\mathbb{Z}_p$. The following conditions are equivalent:*

(i) *L_1 and L_2 are isomorphic.*
(ii) *The two* q*-vector spaces $\mathbb{F}_p \otimes_{\mathbb{Z}} L_1$ and $\mathbb{F}_p \otimes_{\mathbb{Z}} L_2$ over $\mathbb{F}_p$ are isomorphic.*
(iii) *L_1 and L_2 have the same dimension and determinant.*
(iv) *The two* q*-vector spaces $\mathbb{Q}_p \otimes_{\mathbb{Z}} L_1$ and $\mathbb{Q}_p \otimes_{\mathbb{Z}} L_2$ over $\mathbb{Q}_p$ are isomorphic.*

For the sake of future reference, we add the following statements to the ones we have recalled so far; they refine the results of Scholium 2.2.1.

Theorem 2.2.7. *Let L_1 and L_2 be two* q*-modules over $\mathbb{Z}$. Assume that the* q*-vector space $\mathbb{R} \otimes_{\mathbb{Z}} L_1$ over $\mathbb{R}$ is indefinite; then the following conditions are equivalent:*

(i) *L_1 and L_2 are isomorphic.*
(ii) *The two* q*-vector spaces $\mathbb{R} \otimes_{\mathbb{Z}} L_1$ and $\mathbb{R} \otimes_{\mathbb{Z}} L_2$ over $\mathbb{R}$ are isomorphic.*

Theorem 2.2.8. *Let L_1 and L_2 be two* q*-modules over $\mathbb{Z}$, and let p be a prime. The following conditions are equivalent:*

(i) *The two* q*-modules $\mathbb{Z}[1/p] \otimes_{\mathbb{Z}} L_1$ and $\mathbb{Z}[1/p] \otimes_{\mathbb{Z}} L_2$ over $\mathbb{Z}[1/p]$ are isomorphic.*
(ii) *The two* q*-vector spaces $\mathbb{R} \otimes_{\mathbb{Z}} L_1$ and $\mathbb{R} \otimes_{\mathbb{Z}} L_2$ over $\mathbb{R}$ are isomorphic.*

These two theorems can be proved using the strong approximation theorem for spin groups [162]. We can also prove Theorem 2.2.7 as follows: begin by noting that Scholium 2.2.1 implies that an indefinite b-module over $\mathbb{Z}$ represents zero and conclude as Serre does in [177, Chap. V, Sect. 3].

The Classical Terminology in the Positive Definite Case

Let V be a Euclidean space, that is, an $\mathbb{R}$-vector space of finite dimension endowed with a positive definite symmetric bilinear form (called an inner product). In this context, a *lattice* in V is a cocompact discrete subgroup. Such a lattice L is called *integral* if the inner product $x.y$ is an integer for all x and y in L. Endowed with the symmetric bilinear form induced by the inner product, L is a $\widetilde{\mathrm{b}}$-module over $\mathbb{Z}$ that is positive definite (in other words, the b-vector space $\mathbb{R} \otimes_{\mathbb{Z}} L$ is positive definite). Conversely, a positive definite $\widetilde{\mathrm{b}}$-module L over $\mathbb{Z}$ is an integral lattice in the Euclidean space $\mathbb{R} \otimes_{\mathbb{Z}} L$. An integral lattice L in V is called a *unimodular lattice* if it has covolume 1 (the word "integral" is implicit in this case), in other words, if the $\widetilde{\mathrm{b}}$-module L is in fact a b-module. An integral lattice L in V is called *even* if $x.x$ is even for all x in L, that is, if the $\widetilde{\mathrm{b}}$-module L is in fact a $\widetilde{\mathrm{q}}$-module. A positive definite q-module L over $\mathbb{Z}$ is therefore an even unimodular lattice in the Euclidean space $\mathbb{R} \otimes_{\mathbb{Z}} L$.

In this book, we will more frequently use the classical terminology of even unimodular lattice (resp. unimodular lattice, even integral lattice, integral lattice) than the terminology of positive definite q-module (resp. b-module, $\widetilde{\mathrm{q}}$-module, $\widetilde{\mathrm{b}}$-module) over $\mathbb{Z}$. (In fact, the terminology of q-module, b-module, ..., is seldom used by anyone else than the second author of this book!)

2.3 Root Systems and Even Unimodular Lattices

With the exception of the Leech lattice, the even unimodular lattices that occur in dimensions 8, 16, and 24 are all constructed from certain root systems using a process we will now describe.

In fact, the root systems in question are certain direct sums of root systems of type $\mathbf{A}_l$, $\mathbf{D}_l$, $\mathbf{E}_6$, $\mathbf{E}_7$, and $\mathbf{E}_8$; we will say that such direct sums are root systems of *type ADE* (a more common terminology, justified by considering the Dynkin diagram, is *simply laced*). It is clear that among the irreducible root systems, those of type ADE are characterized by the property that all roots have the same length. The reader can easily verify that as a definition of root systems of type ADE, we can also take the following ad hoc one, which is a variant of [39, Chap. VI.1, Définition 1].

Definition 2.3.1. Let V be a Euclidean space and R a subset of V consisting of vectors α satisfying $\alpha.\alpha = 2$. We say that R is a *root system of type ADE* in V if the following conditions are satisfied:

(I) The subset R is finite and generates V.
(II) For all α in R, the orthogonal reflection $x \mapsto x - (\alpha.x)\,\alpha$ in V (which we denote by s_α) leaves R unchanged.
(III) For all α and β in R, the inner product $\alpha.\beta$ is an integer.

Here, we encounter the Weyl group $\mathrm{W}(R)$ generated by the s_α as a subgroup of the orthogonal group of V; the same holds for the (finite) group $\mathrm{A}(R)$ consisting of the automorphisms of the $\mathbb{R}$-vector space V that leave R unchanged (this follows, for example, from [39, Chap. VI, Sect. 1, Propositions 3 and 7]).

Recall that when developing the general theory of a root system R in a $\mathbb{R}$-vector space V, one shows that $\mathrm{s}_\alpha(x)$ can be written uniquely as $x - \langle x, \alpha^\vee \rangle\,\alpha$ with $\alpha^\vee \in V^*$ and that the $\alpha^\vee$ form a root system in V^*. The latter is denoted by $R^\vee$ and called the *dual* of R. In the setting of Definition 2.3.1, if we identify V and V^* via the inner product, the map $\alpha \mapsto \alpha^\vee$ is the identity and the roots systems R and $R^\vee$ coincide.

Let $R \subset V$ be a root system of type ADE; the lattice in V generated by R, which we denote by $\mathrm{Q}(R)$ and call the *root lattice*, is an even integral lattice, and R can be identified with the subset of $\mathrm{Q}(R)$ consisting of the elements α satisfying $\alpha.\alpha = 2$ (this last property is not, a priori, an immediate consequence of Definition 2.3.1; to see that it holds, note that it does so for the root systems $\mathbf{A}_l$, $\mathbf{D}_l$, $\mathbf{E}_6$, $\mathbf{E}_7$, and $\mathbf{E}_8$). The group $\mathrm{A}(R)$ can be identified with the automorphism group of $\mathrm{Q}(R)$, where we view the latter as a $\widetilde{\mathrm{q}}$-module over $\mathbb{Z}$. The notation $\mathrm{Q}(R)$ is that of Bourbaki [39]. Conversely, let L be an integral lattice (in other words, a positive definite $\widetilde{\mathrm{b}}$-module over $\mathbb{Z}$), let $\mathrm{R}(L)$ be the (finite) subset of L consisting of the elements α satisfying

$\alpha.\alpha = 2$, which we call the *roots* of L, and let $\mathrm{V}(L)$ be the subspace of $\mathbb{R} \otimes_{\mathbb{Z}} L$ generated by $\mathrm{R}(L)$; then $\mathrm{R}(L)$ is a root system of type ADE in $\mathrm{V}(L)$ (take $\Lambda = \{2\}$ at the beginning of item 4 of [39, Chap. VI, Sect. 4]). In summary, the classification of the root systems of type ADE coincides with that of the (even) integral lattices generated by their roots.

The process that allows us to obtain an even unimodular lattice from certain root systems of type ADE is simply a particular case of the general process provided by Proposition 2.1.1. Let R be a root system of type ADE. Suppose that the qe-module $\operatorname{res}\mathrm{Q}(R)$ has a Lagrangian I. Then the inverse image of I under the canonical map $\mathrm{Q}(R)^\sharp \to \mathrm{Q}(R)^\sharp/\mathrm{Q}(R) =: \operatorname{res}\mathrm{Q}(R)$ is a positive definite q-module (see parts (b) and (c) of Proposition 2.1.1) over $\mathbb{Z}$, in other words, an even unimodular lattice. The lattice $\mathrm{Q}(R)^\sharp$ is the *weight lattice* of the root system R; it is denoted by $\mathrm{P}(R)$ in [39].

Example. Let $n \geq 1$ be an integer; we endow the $\mathbb{R}$-vector space $\mathbb{R}^n$ with its canonical Euclidean structure and denote its canonical basis by $(\varepsilon_1, \varepsilon_2, \ldots, \varepsilon_n)$.

Once again, consider the even integral lattice $\mathrm{D}_n \subset \mathrm{I}_n := \mathbb{Z}^n \subset \mathbb{R}^n$ introduced in Sect. 2.2. The root system $\mathbf{D}_n$, for $n \geq 3$, is defined by the equality $\mathbf{D}_n := \mathrm{R}(\mathrm{D}_n)$; the set $\mathrm{R}(\mathrm{D}_n)$ generates D_n for $n \geq 2$.

Let us recall what we saw in Sect. 2.2:

- The qe-module $\operatorname{res}\mathrm{D}_n$ has a Lagrangian if and only if n is divisible by 8, which we will assume to be true from now on.
- In this case, we have an isomorphism of qe-modules $\operatorname{res}\mathrm{D}_n \cong \mathrm{H}(\mathbb{Z}/2)$, and the two Lagrangians of $\operatorname{res}\mathrm{D}_n$ are generated by the classes of the vectors $\frac{1}{2}(\varepsilon_1+\varepsilon_2+\ldots+\varepsilon_n)$ and $\frac{1}{2}(-\varepsilon_1+\varepsilon_2+\ldots+\varepsilon_n)$. Note that these two vectors are interchanged by the involutive automorphism $(x_1, x_2, \ldots x_n) \mapsto (-x_1, x_2, \ldots x_n)$ of D_n.

We have denoted by E_n the lattice generated by D_n and the vector $\frac{1}{2}(\varepsilon_1 + \varepsilon_2 + \ldots + \varepsilon_n)$; the lattice E_n is the simplest example of an even unimodular lattice obtained by the process described above (for more sophisticated examples, see the classification of even unimodular lattices of dimension 24 mentioned further on).

The root system $\mathbf{E}_8$ is defined by the equality $\mathbf{E}_8 := \mathrm{R}(\mathrm{E}_8)$; the set $\mathrm{R}(\mathrm{E}_8)$ generates E_8; in other words, we have $\mathrm{E}_8 = \mathrm{Q}(\mathbf{E}_8)$. For $n \geq 16$ (and $n \equiv 0 \bmod 8$), we have $\mathrm{R}(\mathrm{E}_n) = \mathrm{R}(\mathrm{D}_n) = \mathbf{D}_n$; the subgroup generated by $\mathrm{R}(\mathrm{E}_n)$ has index 2 in E_n. The even unimodular lattice E_n is also denoted by D_n^+ (at least for $n \geq 16$); this notation is justified by the fact that the automorphism group of D_n acts transitively on the set consisting of the two Lagrangians of $\operatorname{res}\mathrm{D}_n$.

After describing, above, the root systems $\mathbf{D}_n$ for $n \geq 3$ and $\mathbf{E}_8$, as well as their root lattices, we provide hereafter, for the sake of completeness, similar information about the remaining irreducible root systems of type ADE. The definitions of the even lattice A_n and of the root system $\mathbf{A}_n := \mathrm{R}(\mathrm{A}_n)$, for $n \geq 1$, may be found in the second example after Scholium 2.3.15; again, we have $\mathrm{A}_n = \mathrm{Q}(\mathbf{A}_n)$. We denote by E_7 (resp. E_6) the orthogonal complement in E_8 of the vector $\varepsilon_7 + \varepsilon_8$ (resp. the vectors $\varepsilon_6 - \varepsilon_7$ and $\varepsilon_7 + \varepsilon_8$). The root system $\mathbf{E}_l$, for $l = 7, 6$, is defined by $\mathbf{E}_l := \mathrm{R}(\mathrm{E}_l)$, and we have $\mathrm{E}_l = \mathrm{Q}(\mathbf{E}_l)$.

THE CLASSIFICATION OF THE EVEN UNIMODULAR LATTICES IN DIMENSIONS 8, 16, AND 24

The classification of the even unimodular lattices is due to Louis J. Mordell in dimension 8, to Ernst Witt in dimension 16, and to Hans-Volker Niemeier in dimension 24 [158]. We will now recall the ingenious strategy developed by Boris Venkov [201] for recovering Niemeier's classification. His strategy also works in dimensions 8 and 16. Indeed, the initial idea of Venkov is to consider theta series "with harmonic coefficients of degree 2" and observe that these series are identically zero because all cusp forms of weight $14 = 24/2 + 2$ for $\mathrm{SL}_2(\mathbb{Z})$ are zero. For the arguments he then uses, it is no longer crucial that the dimension of the lattice be 24, and every cusp form of weight $n/2 + 2$ for $\mathrm{SL}_2(\mathbb{Z})$ is also zero for $n = 8, 16$.

By considering the "coefficient of $e^{2\imath\pi\tau}$" in the theta series mentioned above, Venkov obtains the following identity.

Proposition 2.3.2. *Let L be an even unimodular lattice of dimension $n =$ 8, 16, or 24. Then we have the identity*

$$\sum_{\alpha\in\mathrm{R}(L)} (\alpha.x)^2 \quad = \quad \frac{2\,|\mathrm{R}(L)|}{n} \; x.x$$

for every x in the Euclidean space $\mathbb{R}\otimes_{\mathbb{Z}} L$ (where $|-|$ denotes the cardinality of a finite set).

Venkov then deduces the following result.

Proposition-Definition 2.3.3. *Let L be an even unimodular lattice of dimension $n =$ 8, 16, or 24. If the set $\mathrm{R}(L)$ of roots of L (recall that these are the elements α of L satisfying $\alpha.\alpha = 2$) is nonempty, then it satisfies the following properties:*

(a) *The set $\mathrm{R}(L)$ is a root system (of type ADE) of rank n (in $\mathbb{R}\otimes_{\mathbb{Z}} L$); in other words, $\mathrm{R}(L)$ generates the $\mathbb{R}$-vector space $\mathbb{R}\otimes_{\mathbb{Z}} L$.*
(b) *All irreducible components of the root system $\mathrm{R}(L)$ have the same Coxeter number, which we call the* Coxeter number *of L and denote by $\mathrm{h}(L)$; we will say that such a root system is* equi-Coxeter.
(c) *We have $|\mathrm{R}(L)| = n\,\mathrm{h}(L)$.*

Remarks.

- Venkov shows properties (b) and (c) simultaneously using Proposition 2.3.2 and [39, Chap. VI, Sect. 1, Proposition 32]. We can also deduce property (c) from property (b) by using the relation $|R| = n\,h$ that links the number of roots, rank, and Coxeter number for every reduced irreducible root system (see [104, 3.18], [39, Chap. VI, Sect. 1, Exercice 20]).
- Since the Coxeter number of the direct sum of two root systems is the least common multiple of their Coxeter numbers (recall that the Coxeter number of a root system is defined as the order of a Coxeter element), $\mathrm{h}(L)$ is also the Coxeter number of the root system $\mathrm{R}(L)$.

Scholium 2.3.4. *Let L be an even unimodular lattice of dimension 24 with $\mathrm{R}(L) \neq \emptyset$. Then we have the identity*

$$\sum_{\alpha \in \mathrm{R}(L)} (\alpha.x)^2 \;=\; 2\,\mathrm{h}(L)\; x.x$$

for every x in the Euclidean space $\mathbb{R} \otimes_{\mathbb{Z}} L$.

Corollary 2.3.5. *Every even unimodular lattice of dimension 8 is isomorphic to* E_8.

Proof. Let L be such a lattice. The theta series of L, which is modular of weight 4 for $\mathrm{SL}_2(\mathbb{Z})$, is necessarily equal to the normalized Eisenstein series $\mathbb{E}_4$ (this unusual notation is due to the overuse of the letter "E"). It follows that we have $|\mathrm{R}(L)| = 240$ and $\mathrm{h}(L) = 30$. The latter implies $\mathrm{R}(L) \simeq \mathbf{E}_8$. Since the lattice $\mathrm{E}_8 = \mathrm{Q}(\mathbf{E}_8)$ is unimodular, we indeed have $L \simeq \mathrm{E}_8$. □

Corollary 2.3.6. *Every even unimodular lattice of dimension 16 is isomorphic to either D_{16}^{+} or $\mathrm{E}_8 \oplus \mathrm{E}_8$ (and these two lattices are not isomorphic).*

Proof. Let L be such a lattice. The theta series of L is necessarily equal to the normalized Eisenstein series $\mathbb{E}_8 = \mathbb{E}_4^2$. Consequently, we have $|\mathrm{R}(L)| = 480$ and $\mathrm{h}(L) = 30$. The latter implies that we have either $\mathrm{R}(L) \simeq \mathbf{D}_{16}$ or $\mathrm{R}(L) \simeq \mathbf{E}_8 \coprod \mathbf{E}_8$ (where $\coprod$ denotes the *direct sum of root systems*).

In the case $\mathrm{R}(L) \simeq \mathbf{E}_8 \coprod \mathbf{E}_8$, we conclude as before that we have $L \simeq \mathrm{E}_8 \oplus \mathrm{E}_8$.

In the case $\mathrm{R}(L) \simeq \mathbf{D}_{16}$, the lattice D_{16} appears, up to an isomorphism, as a sublattice of L. We may therefore assume that in $\mathbb{Q} \otimes_{\mathbb{Z}} L$, we have the inclusions $\mathrm{D}_{16} \subset L \subset \mathrm{D}_{16}^{\sharp}$. Then L/D_{16} is a Lagrangian of the qe-module $\mathrm{res}\, \mathrm{D}_{16}$ and we have $L \simeq \mathrm{D}_{16}^{+}$. □

Let us now turn to more serious matters, namely determining the isomorphism classes of even unimodular lattices of dimension 24.

Let L be an even unimodular lattice of dimension 24 with $\mathrm{R}(L) \neq \emptyset$. Properties (a) and (b) of Proposition 2.3.3 tell us that $\mathrm{R}(L)$ is a root system of type ADE, of rank 24 and equi-Coxeter. Venkov begins by listing the isomorphism classes of such root systems explicitly. This list has 23 elements $\mathbf{R}_1, \mathbf{R}_2, \ldots, \mathbf{R}_{23}$. For the complete list, we refer to the second column of [68, Chap. 16, Table 16.1] (our Table 1.1); the reader will notice that we use the integers $1, 2, \ldots, 23$ to index the elements, rather than the Greek letters $\alpha, \beta, \ldots, \psi$ used by Conway and Sloane. Here are some examples:

$$\begin{gathered}
\mathbf{R}_1 = \mathbf{D}_{24}\ ,\ \mathbf{R}_2 = \mathbf{D}_{16} \coprod \mathbf{E}_8\ ,\ \mathbf{R}_3 = \mathbf{E}_8 \coprod \mathbf{E}_8 \coprod \mathbf{E}_8\ , \\
\mathbf{R}_4 = \mathbf{A}_{24}\ ,\ \mathbf{R}_5 = \mathbf{D}_{12} \coprod \mathbf{D}_{12}\ ,\ \mathbf{R}_6 = \mathbf{A}_{17} \coprod \mathbf{E}_7\ , \\
\mathbf{R}_7 = \mathbf{D}_{10} \coprod \mathbf{E}_7 \coprod \mathbf{E}_7\ ,\ \mathbf{R}_{23} = \mathbf{A}_1 \coprod \mathbf{A}_1 \coprod \cdots \coprod \mathbf{A}_1\ ,
\end{gathered}$$

where the last direct sum consists of 24 irreducible components equal to $\mathbf{A}_1$. Before continuing our description of Venkov's arguments, we need to make a few observations and recall some results. These can be found below, numbered 1, 2, and 3.

(1) Let M be a $\widetilde{\mathrm{q}}$-module over $\mathbb{Z}$. If M is positive definite, then, in addition to its structure of qe-module, the quotient $\operatorname{res} M := M^\sharp/M$ has a structure that we will now describe. Let ξ be an element of $\operatorname{res} M$ and $\gamma\colon M^\sharp \to \operatorname{res} M$ the canonical surjection. We define a map $\mathrm{qm}\colon \operatorname{res} M \to \mathbb{Q} \cap [0,\infty[$ by setting

$$\mathrm{qm}(\xi) = \inf_{x\in\gamma^{-1}(\xi)} \mathrm{q}(x)\ .$$

This map clearly makes the following diagram commutative:

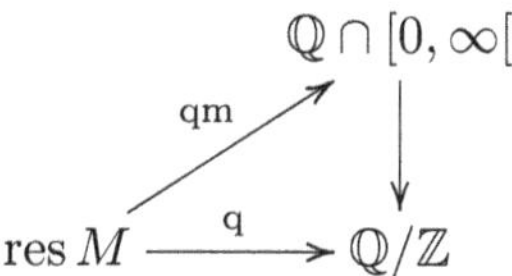

(the vertical arrow is the restriction of the reduction modulo $\mathbb{Z}$). We call $\operatorname{res} M$ endowed with this additional structure a *Venkov* qe-*module*.

Example. Let $n > 0$ be an integer divisible by 8, and let $M = \mathrm{D}_n$. We saw in Sect. 2.2 that $\operatorname{res} \mathrm{D}_n$ is isomorphic to $\mathbb{Z}/2 \oplus \mathbb{Z}/2$ endowed with the quadratic linking form defined by $\mathrm{q}(\bar{0},\bar{0}) = 0$, $\mathrm{q}(\bar{1},\bar{0}) = 0$, $\mathrm{q}(\bar{0},\bar{1}) = 0$, and $\mathrm{q}(\bar{1},\bar{1}) = 1/2$. The map qm, in turn, is given by $\mathrm{qm}(\bar{0},\bar{0}) = 0$, $\mathrm{qm}(\bar{1},\bar{0}) = n/8$, $\mathrm{qm}(\bar{0},\bar{1}) = n/8$, and $\mathrm{qm}(\bar{1},\bar{1}) = 1/2$.

The following proposition is obvious.

Proposition 2.3.7. *Let M be an even integral lattice. Let I be a submodule of $\operatorname{res} M$ with $I \subset I^\perp$, and let L be the associated integral lattice. Then the following conditions are equivalent:*

(i) $\mathrm{R}(L) = \mathrm{R}(M)$.
(ii) *We have* $\mathrm{qm}(\xi) > 1$ *for every* ξ *in* $I - \{0\}$.

Example. We return to the previous example. The proposition shows that we have $\mathrm{R}(\mathrm{E}_n) = \mathrm{R}(\mathrm{D}_n)$ for $n \geq 16$ and $\mathrm{R}(\mathrm{E}_8) \supsetneq \mathrm{R}(\mathrm{D}_8)$.

(2) Let $R \subset V$ be a root system of type ADE. Let us recall how to determine the map $\mathrm{qm}\colon \operatorname{res} \mathrm{Q}(R) \to \mathbb{Q} \cap [0,\infty[$. We may clearly assume that R is irreducible, and will do so from now on.

We fix a chamber C of R; we denote the corresponding basis of R and highest root by $(\alpha_1,\alpha_2,\ldots,\alpha_l)$ and $\widetilde{\alpha}$, respectively. Recall that C is the subset of V consisting of the elements ξ satisfying $\alpha_i.\xi > 0$ for $i = 1,2,\ldots,l$. Also recall that we have $\widetilde{\alpha} = n_1\alpha_1 + n_2\alpha_2 + \cdots + n_l\alpha_l$ with $n_i \in \mathbb{N} - \{0\}$ for $i = 1,2,\ldots,l$. Let J be the subset of $\{1,2,\ldots,l\}$ consisting of the indices i for which $n_i = 1$. Let $\varpi_1,\varpi_2,\ldots,\varpi_l$ be the *fundamental weights*, that is, the elements of V defined by $\alpha_i.\varpi_j = \delta_{i,j}$ (where $\delta_{i,j}$ is the Kronecker delta). Then $(\varpi_1,\varpi_2,\ldots,\varpi_l)$ is a basis of the $\mathbb{Z}$-module $\mathrm{Q}(R)^\sharp$ and C is the open cone of V generated by this basis.

Proposition 2.3.8. *Let R be an irreducible root system of type ADE, endowed with a chamber C. Let $\gamma\colon \mathrm{Q}(R)^\sharp \to \operatorname{res}\mathrm{Q}(R)$ be the canonical map and $\Pi(R)$ the subset $\{0\}\coprod\{\varpi_j\,;\, j\in \mathrm{J}\}$ of $\mathrm{Q}(R)^\sharp$.*

(a) *The restriction of γ to $\Pi(R)$ is a bijection.*
(b) *For every ξ in $\Pi(R)$, we have $\mathrm{qm}(\gamma(\xi)) = \mathrm{q}(\xi)$.*

Proposition 2.3.8 allows us to determine the map $\mathrm{qm}\colon \operatorname{res}\mathrm{Q}(R) \to \mathbb{Q}\cap[0,\infty[$. A reference for statement (a) is [39, Chap. VI, Sect. 2, Corollary of Proposition 6]; statement (b) is implicit in [201]. For the sake of the reader, we include, in a smaller font, a proof of these two statements; it is based on Proposition 2.3.9 below (for which we refer to [39, Chap. V, Sect. 3, Théorème 2]), which is a fundamental result on the action of the affine Weyl group on V.

Before stating the proposition, let us recall the definition of the *alcove* associated with the chamber C (we denote it by Alc hereafter and in Chap. 3):

$$Alc \;:=\; \{\,\xi\,;\,\xi\in V,\ \alpha_i.\xi > 0 \text{ for } i=1,2,\ldots,l \text{ and } \widetilde{\alpha}.\xi < 1\,\}\,;$$

the closure $\overline{Alc}$ of Alc in V is, of course, defined by the nonstrict inequalities $\alpha_i.\xi \geq 0$ and $\widetilde{\alpha}.\xi \leq 1$.

Proposition 2.3.9. *Every orbit of the canonical action of the semidirect product $\mathrm{W}(R)\ltimes \mathrm{Q}(R)$ (the affine Weyl group) on V meets $\overline{Alc}$ at a single point.*

Proof of Proposition 2.3.8.

(a) By definition, $\Pi(R)$ is contained in $\overline{Alc}$ because for every ξ in $\Pi(R)$, the inner products $\widetilde{\alpha}.\xi$ and $\alpha_i.\xi$ for $i=1,2,\ldots,l$ are 0 or 1. In fact, it is not difficult to see that we have $\Pi(R) = \mathrm{Q}(R)^\sharp\cap\overline{Alc}$: Let ξ be an element of $\mathrm{Q}(R)^\sharp$. We write it as $\xi=\sum_{i=1}^{l}(\alpha_i.\xi)\,\varpi_i$; if ξ is in $\overline{C}$, then the integers $\alpha_i.\xi$ are nonnegative, and if ξ is nonzero and we also have $\widetilde{\alpha}.\xi\leq 1$, then ξ must be one of the ϖ_j with j in J.

In view of what we just wrote, statement (a) is a consequence of Proposition 2.3.9:

– The "uniqueness" part of Proposition 2.3.9 shows that the restriction of γ to $\Pi(R)$ is injective.

– The "existence" part shows that it is also surjective. Indeed, let ξ be an element of $\mathrm{Q}(R)^\sharp$; then there exist η in $\overline{Alc}$, w in $\mathrm{W}(R)$, and x in $\mathrm{Q}(R)$ such that we have $\xi = w\eta + x$. Consequently, we have $\eta\in\Pi(R)$, and therefore $\gamma(\xi)=\gamma(w\eta)$, and finally $\gamma(\xi)=\gamma(\eta)$ because the action of $\mathrm{W}(R)$ on $\operatorname{res}\mathrm{Q}(R)$, induced by that of $\mathrm{W}(R)$ on $\mathrm{Q}(R)$, is trivial [39, Chap. VI, Sect. 1, Proposition 27].

(b) This statement is a consequence of the implication $(v)\Rightarrow(i)$ of Proposition 2.3.10 below. □

Proposition 2.3.10. *Let $R\subset V$ be an irreducible root system of type ADE. Let ξ be an element of V. The following conditions are equivalent:*

(i) *We have $\mathrm{q}(\xi)\leq\mathrm{q}(\xi+x)$ for every x in $\mathrm{Q}(R)$.*
(ii) *We have $\mathrm{q}(\xi)\leq\mathrm{q}(\xi+\alpha)$ for every α in R.*
(iii) *We have $\alpha.\xi\leq 1$ for every α in R.*
(iv) *We have $|\alpha.\xi|\leq 1$ for every α in R.*
(v) *There exists an element w of $\mathrm{W}(R)$ such that $w\xi$ is in $\overline{Alc}$ (we assume that R is endowed with a chamber C).*

Proof of (i) $\Rightarrow$ (ii) $\Rightarrow$ (iii) $\Longleftrightarrow$ (iv). These implications are trivial (for (ii) $\Rightarrow$ (iii) $\Longleftrightarrow$ (iv), note that if α is a root, then so is $-\alpha$). □

Proof of (iv) $\iff$ (v). Let

$$\Psi \ := \ \{\,\xi\,;\,\xi \in V,\ \ \alpha.\xi \le 1 \ \text{ for all } \alpha \ \text{ in } \ R\,\}\ ;$$

we must show that we have $\Psi = \bigcup_{w \in \mathrm{W}(R)} w\,\overline{Alc}$.

Given an element ξ of V, there exists an element w of $\mathrm{W}(R)$ with $w^{-1}\xi \in \overline{C}$. If we, moreover, assume $\xi \in \Psi$, then we have $(w\widetilde{\alpha}).\xi = \widetilde{\alpha}.(w\xi) \le 1$ and therefore $w^{-1}\xi \in \overline{Alc}$; this proves the inclusion $\Psi \subset \bigcup_{w \in \mathrm{W}(R)} w\,\overline{Alc}$.

Let α be a positive root (for the chamber C); since α can be written as $\sum_{i=1}^{l} \nu_i \alpha_i$ with $0 \le \nu_i \le n_i$, we have $0 \le \alpha.\xi \le \widetilde{\alpha}.\xi$ for every ξ in $\overline{C}$. The inclusion $\overline{Alc} \subset \Psi$ follows; consequently, we have $\bigcup_{w \in \mathrm{W}(R)} w\,\overline{Alc} \subset \Psi$ because Ψ is invariant under the action of $\mathrm{W}(R)$. □

Proof of (iii) $\Rightarrow$ (i). Let Γ be an arbitrary lattice in the Euclidean space V. Set

$$\Phi_\Gamma \ := \ \{\,\xi\,;\,\xi \in V,\ \ \mathrm{q}(\xi) \le \mathrm{q}(\xi - x) \ \text{ for all } \ x \ \text{ in } \ \Gamma\,\}\ ;$$

that is, Φ_Γ is the subset of V consisting of the points whose distance to the origin is less than or equal to the distance to any point of Γ. The study of these subsets has, of course, a very rich history. By definition, Φ_Γ is the intersection of the semispaces consisting of the q with $x.\xi \le \mathrm{q}(x)$, where x runs through Γ; one easily sees that Φ_Γ is compact and is the intersection of a finite subfamily of these semispaces. The subset Φ_Γ has the following three properties:

(1) The set Φ_Γ is the closure of its interior in V.
(2) The translates $\Phi_\Gamma + x$, where x runs through Γ, cover V.
(3) The translates $\mathring{\Phi}_\Gamma + x$ for x in Γ are pairwise disjoint (where $\mathring{\Phi}_\Gamma$ denotes the interior of Φ_Γ).

Lemma 2.3.11. *Let* Φ *(resp.* Φ'*) be a subset of* V *with properties* (1), (2), *and* (3) *(resp.* (1) *and* (3)*). If* $\Phi \subset \Phi'$, *then* $\Phi = \Phi'$.

Proof. Suppose $\Phi' \not\subset \Phi$. In this case, we also have $\mathring{\Phi}' \not\subset \Phi$ because Φ is closed and Φ' is the closure of its interior. Let ξ be an element of V with $\xi \in \mathring{\Phi}'$ and $\xi \notin \Phi$. Since Φ has property (2), there exists an x in Γ such that $\xi + x \in \Phi$ and a fortiori $\xi + x \in \Phi'$. We therefore have $\mathring{\Phi}' \cap (\Phi' + x) \neq \emptyset$; by a general topological argument analogous to the one we just gave, we also have $\mathring{\Phi}' \cap (\mathring{\Phi}' + x) \neq \emptyset$. Since Φ has property (3), we have $x = 0$. This gives a contradiction. □

Proof of Proposition 2.3.10 (iii) $\Rightarrow$ (i), *Continued.* We can use Lemma 2.3.11 to prove the implication (iii) $\Rightarrow$ (i). Set $\Phi := \Phi_{\mathrm{Q}(R)}$; then Φ is the subset of V consisting of the ξ satisfying condition (i). We must show that we have $\Phi = \Psi$. It is clear that Ψ has property (1); to conclude, it therefore suffices to show that it also has property (3). Let x be an element of $\mathrm{Q}(R)$ with $\mathring{\Psi} \cap (\mathring{\Psi} + x) \neq \emptyset$; let ξ be an element of this intersection. The inequalities $-1 < \alpha.\xi < 1$ and $-1 < \alpha.(\xi + x) < 1$ imply $-2 < \alpha.x < 2$; since $\alpha.x$ is an integer, we also have $-1 \le \alpha.x \le 1$, that is, $x \in \Psi$. Proposition 2.3.12 below then says that x is zero. □

Proposition 2.3.12. *We have* $\Psi \cap \mathrm{Q}(R) = \{0\}$.

Proof. In view of the equality $\Psi = \bigcup_{w \in \mathrm{W}(R)} w\,\overline{Alc}$, it suffices to show that we have $\overline{Alc} \cap \mathrm{Q}(R) = \{0\}$. This equality follows from Proposition 2.3.9. □

Remark (On the Terminology). The elements ξ of $\mathrm{Q}(R)^\sharp$ satisfying $\mathrm{q}(\xi) \le \mathrm{q}(\xi + x)$ for every x in $\mathrm{Q}(R)$ are called *minuscule weights*; however, this term is often reserved for the fundamental weights ϖ_j for $j \in J$ considered earlier (see, for example, [193]).

(3) Let R be a root system of type ADE. The action of $\mathrm{A}(R)$ on $\mathrm{Q}(R)$ induces an action of $\mathrm{A}(R)$ on the Venkov qe-module $\operatorname{res}\mathrm{Q}(R)$. The restriction of the action of $\mathrm{A}(R)$ on $\operatorname{res}\mathrm{Q}(R)$ to $\mathrm{W}(R)$ is trivial (recall the reference: [39, Chap. VI, Sect. 1, Proposition 27]); since $\mathrm{W}(R)$ is a normal subgroup of $\mathrm{A}(R)$ [39, Chap. I. Sect. 1, Proposition 16], there is a canonical (in fact faithful [39, Chap. VI, Sect. 4, Exercice 7]) action of the quotient group $\mathrm{G}(R) := \mathrm{A}(R)/\mathrm{W}(R)$ on $\operatorname{res}\mathrm{Q}(R)$.

Let us now return to the classification of even unimodular lattices of dimension 24 with nonempty set of roots. To complete this classification, Venkov verifies the following (miraculous) statement case by case.

Proposition 2.3.13. *Let R be a root system of type ADE, of rank 24 and equi-Coxeter.*

(a) *The Venkov* qe-*module* $\operatorname{res}\mathrm{Q}(R)$ *has a Lagrangian I with* $\mathrm{qm}(\xi) > 1$ *for every ξ in $I - \{0\}$.*
(b) *Such a Lagrangian is unique up to the action of* $\mathrm{G}(R)$.

Corollary 2.3.14. *The map $L \mapsto \mathrm{R}(L)$ induces a bijection from the set of isomorphism classes of even unimodular lattices of dimension 24 with nonempty root system $\mathrm{R}(L)$ onto the set of isomorphism classes of equi-Coxeter root systems of type ADE and rank 24.*

Scholium 2.3.15. *Let L be an even unimodular lattice of dimension 24 with $\mathrm{R}(L) \neq \emptyset$. Then the following hold:*

(a) *The action of the Weyl group $\mathrm{W}(\mathrm{R}(L))$ on $\mathrm{R}(L)$ extends to an (orthogonal) action on L, so that we can identify $\mathrm{W}(\mathrm{R}(L))$ with a subgroup of the orthogonal group $\mathrm{O}(L)$.*
(b) *The group $\mathrm{W}(\mathrm{R}(L))$ is a normal subgroup of $\mathrm{O}(L)$, and the quotient $\mathrm{O}(L)/\mathrm{W}(\mathrm{R}(L))$ is canonically isomorphic to the subgroup of $\mathrm{G}(R)$ that stabilizes the Lagrangian $L/\mathrm{Q}(\mathrm{R}(L))$ of the* qe-*module* $\operatorname{res}\mathrm{Q}(R)$ *(this Lagrangian is one of those considered in Proposition* 2.3.13*).*

(Part (b) of this observation follows from the fact that the canonical action of $\mathrm{W}(\mathrm{R}(L))$ on $\operatorname{res}\mathrm{Q}(R)$ is trivial.)

Examples.
(1) $\mathrm{R}(L) \cong \mathbf{D}_{24}$, $\mathrm{R}(L) \cong \mathbf{D}_{16} \coprod \mathbf{E}_8$, and $\mathrm{R}(L) \cong \mathbf{E}_8 \coprod \mathbf{E}_8 \coprod \mathbf{E}_8$

It follows from previous results that in these three cases, we have, respectively, $L \cong \mathrm{D}_{24}^+$, $L \cong \mathrm{D}_{16}^+ \oplus \mathrm{E}_8$, and $L \cong \mathrm{E}_8 \oplus \mathrm{E}_8 \oplus \mathrm{E}_8$.

(2) $\mathrm{R}(L) \cong \mathbf{A}_{24}$

Let $n \geq 1$ be an integer; we endow the $\mathbb{R}$-vector space $\mathbb{R}^{n+1}$ with its canonical Euclidean structure and denote its canonical basis by $(\varepsilon_1, \varepsilon_2, \ldots, \varepsilon_{n+1})$. We denote by A_n the submodule of $\mathrm{I}_{n+1} := \mathbb{Z}^{n+1} \subset \mathbb{R}^{n+1}$ consisting of the $(n+1)$-tuples $(x_1, x_2, \ldots, x_{n+1})$ with $x_1 + x_2 + \ldots + x_{n+1} = 0$; A_n is an even integral lattice (in the hyperplane of $\mathbb{R}^{n+1}$ consisting of the $(n+1)$-tuples $(\xi_1, \xi_2, \ldots, \xi_{n+1})$

with $\xi_1 + \xi_2 + \ldots + \xi_{n+1} = 0$). The root system $\mathbf{A}_n$ is defined by the equality $\mathbf{A}_n := \mathrm{R}(\mathrm{A}_n)$ (so that we also have $\mathrm{A}_n = \mathrm{Q}(\mathbf{A}_n)$).

The underlying abelian group of the Venkov qe-module $\mathrm{res}\,\mathrm{A}_n$ is cyclic of order $n+1$, generated by the class of the orthogonal projection of ε_1 onto the hyperplane $\sum_{i=1}^{n+1} \xi_i = 0$, which we denote by ϖ. The quadratic linking form $\mathrm{q}\colon \mathbb{Z}/(n+1) \to \mathbb{Q}/\mathbb{Z}$, defined by transport of structure, is given by

$$\mathrm{q}(\bar{k}) = k^2\,\mathrm{q}(\varpi) = \frac{n\,k^2}{2(n+1)}\,.$$

The map $\mathrm{qm}\colon \mathbb{Z}/(n+1) \to \mathbb{Q} \cap [0,\infty[$, in turn, is given by

$$\mathrm{qm}(\bar{k}) = \frac{k\,(n+1-k)}{2(n+1)} \quad \text{for } 0 \le k \le n$$

(in the case of $\mathbf{A}_n$, all fundamental weights are minuscule).

It is clear that $\mathrm{res}\,\mathrm{A}_n$ has a Lagrangian, in the bilinear sense, if and only if the integer $n+1$ is a square, that is, $n+1 = r^2$; the Lagrangian is the submodule generated by $r\,\varpi$ and is then unique.

Assume $n+1 = r^2$ and denote by I the submodule generated by $r\,\varpi$; note that I is a Lagrangian in the quadratic sense if and only if n is even, that is, r is odd, and that in this case, I satisfies condition (ii) of Proposition 2.3.7 if and only $r \ge 5$.

Let A_{24}^+ be the lattice in $\mathbb{Q} \otimes_{\mathbb{Z}} \mathrm{A}_{24}$ generated by A_{24} and $5\,\varpi$. The discussion above shows that A_{24}^+ is an even unimodular lattice of dimension 24 with $\mathrm{R}(\mathrm{A}_{24}^+) \cong \mathbf{A}_{24}$ and that this property characterizes A_{24}^+, up to isomorphism, among the even unimodular lattices of dimension 24.

(3) $\mathrm{R}(L) \cong \mathbf{D}_{12} \coprod \mathbf{D}_{12}$

The Venkov qe-module $\mathrm{res}\,\mathrm{D}_{12}$ is isomorphic to $\mathbb{Z}/2 \oplus \mathbb{Z}/2$ endowed with the maps q and qm defined, respectively, by

$$\mathrm{q}(\bar{0},\bar{0}) = 0\,,\ \mathrm{q}(\bar{1},\bar{0}) = \tfrac{1}{2}\,,\ \mathrm{q}(\bar{0},\bar{1}) = \tfrac{1}{2}\,,\ \text{and } \mathrm{q}(\bar{1},\bar{1}) = \tfrac{1}{2}\,;$$
$$\mathrm{qm}(\bar{0},\bar{0}) = 0\,,\ \mathrm{qm}(\bar{1},\bar{0}) = \tfrac{3}{2}\,,\ \mathrm{qm}(\bar{0},\bar{1}) = \tfrac{3}{2}\,,\ \text{and } \mathrm{qm}(\bar{1},\bar{1}) = \tfrac{1}{2}\,.$$

The Venkov qe-module $\mathrm{res}(\mathrm{D}_{12} \oplus \mathrm{D}_{12})$ is isomorphic to $\mathrm{res}\,\mathrm{D}_{12} \oplus \mathrm{res}\,\mathrm{D}_{12}$. The Lagrangians of $\mathrm{res}(\mathrm{D}_{12} \oplus \mathrm{D}_{12})$ are the graphs of the permutations ϕ of $\mathrm{res}\,\mathrm{D}_{12}$ that preserve 0 (such a permutation is linear and preserves the quadratic linking form). The graph of ϕ satisfies condition (ii) of Proposition 2.3.7 if and only if we have $\phi(\mathrm{qm}^{-1}(1/2)) \neq \mathrm{qm}^{-1}(1/2)$; there are four of these ϕ, and we see that the group $\mathrm{G}(\mathbf{D}_{12} \coprod \mathbf{D}_{12})$ (which is isomorphic to $\mathfrak{S}_2 \wr \mathrm{G}(\mathbf{D}_{12}) = \mathfrak{S}_2 \wr \mathfrak{S}_2$) indeed acts transitively on the set consisting of the four corresponding graphs.

(4) $\mathrm{R}(L) \cong \mathbf{A}_{17} \coprod \mathbf{E}_7$

The Venkov qe-module $\mathrm{res}(\mathrm{A}_{17} \oplus \mathrm{E}_7)$ is isomorphic to $\mathrm{res}\,\mathrm{A}_{17} \oplus \mathrm{res}\,\mathrm{E}_7$. The structure of the Venkov qe-module $\mathrm{res}\,\mathrm{A}_n$ has already been given explicitly for every n in the example (2) above. The qe-module $\mathrm{res}\,\mathrm{E}_7$, in turn, is isomorphic to $\mathbb{Z}/2$ endowed with the quadratic linking form defined by $\mathrm{q}(\bar{1}) = -1/4$ (see Proposition B.2.2 (d)); we easily verify that we have $\mathrm{qm}(\bar{1}) = 3/4$. We see that the unique Lagrangian of the qe-module $\mathrm{res}(\mathrm{A}_{17} \oplus \mathrm{E}_7)$ indeed satisfies condition (ii) of Proposition 2.3.7.

(5) $\mathrm{R}(L) \cong \mathbf{D}_{10} \coprod \mathbf{E}_7 \coprod \mathbf{E}_7$

The structure of the Venkov qe-module $\mathrm{res}(\mathrm{D}_{10} \oplus \mathrm{E}_7 \oplus \mathrm{E}_7)$ is determined by that of $\mathrm{res}\,\mathrm{E}_7$, which we made explicit above, and that of $\mathrm{res}\,\mathrm{D}_{10}$. The latter is as follows: $\mathrm{res}\,\mathrm{D}_{10}$ is isomorphic to $\mathbb{Z}/2 \oplus \mathbb{Z}/2$ endowed with the map qm defined by $\mathrm{qm}(\bar{0}, \bar{0}) = 0$, $\mathrm{qm}(\bar{1}, \bar{0}) = 5/4$, $\mathrm{qm}(\bar{0}, \bar{1}) = 5/4$, and $\mathrm{qm}(\bar{1}, \bar{1}) = 1/2$.

The qe-module $\mathrm{res}(\mathrm{D}_{10} \oplus \mathrm{E}_7 \oplus \mathrm{E}_7)$ has two Lagrangians, which we denote by I_1 and I_2; they are the graphs of the two isomorphisms of qe-modules from $\mathrm{res}\,\mathrm{D}_{10}$ to $\langle -1 \rangle \otimes \mathrm{res}(\mathrm{E}_7 \oplus \mathrm{E}_7)$. We easily verify that I_1 and I_2 both satisfy condition (ii) of Proposition 2.3.7.

The group $G := \mathrm{G}(\mathbf{D}_{10} \coprod \mathbf{E}_7 \coprod \mathbf{E}_7)$ can be identified with $\mathbb{Z}/2 \times \mathbb{Z}/2$ because it is isomorphic to the product of the groups $\mathrm{G}(\mathbf{D}_{10})$ and $\mathrm{G}(\mathbf{E}_7 \coprod \mathbf{E}_7)$, which are both cyclic of order 2. We easily verify that G acts transitively on the set $\{I_1, I_2\}$; also note that the diagonal $\mathbb{Z}/2$ acts trivially (we will use this observation at the end of Sect. B.2).

(6) $\mathrm{R}(L) \cong 24\,\mathbf{A}_1$ (the right-hand side denotes the direct sum of 24 copies of the root system $\mathbf{A}_1$)

The Venkov qe-module $\mathrm{res}\,\mathrm{A}_1$ is isomorphic to $\mathbb{Z}/2$ endowed with the map qm defined by $\mathrm{qm}(x) = \lambda(x)/4$, where $\lambda \colon \mathbb{Z}/2 \to \mathbb{N}$ is the map defined by $\lambda(\bar{0}) = 0$ and $\lambda(\bar{1}) = 1$. It follows that the Venkov qe-module $\mathrm{res}(\mathrm{A}_1^{\oplus 24})$ is isomorphic to $(\mathbb{Z}/2)^{24}$ endowed with the map qm defined by

$$\mathrm{qm}(x_1, x_2, \ldots, x_{24}) \;=\; \frac{1}{4} \sum_{i=1}^{24} \lambda(x_i) \;=:\; \frac{1}{4}\,\mathrm{wt}(x_1, x_2, \ldots, x_{24})\,.$$

The group $\mathrm{G}(24\,\mathbf{A}_1)$ is isomorphic to $\mathfrak{S}_{24} \wr \mathrm{G}(\mathbf{A}_1) = \mathfrak{S}_{24}$, and its action on $\mathrm{res}(\mathrm{A}_1^{\oplus 24})$ can be identified with the obvious action of $\mathfrak{S}_{24}$ on $(\mathbb{Z}/2)^{24}$.

A Lagrangian of $\mathrm{res}(\mathrm{A}_1^{\oplus 24})$ can be identified with a linear subspace I of $(\mathbb{Z}/2)^{24}$ with $\dim I = 12$ and $\mathrm{wt}(x) \equiv 0 \bmod 4$ for every x in I. Such an I is called an *even self-dual binary code* of length 24. Up to isomorphism (that is, modulo the action of $\mathfrak{S}_{24}$), there exist nine even self-dual binary codes of length 24 [164].

The Lagrangian I moreover satisfies condition (ii) of Proposition 2.3.7 if and only if $\mathrm{wt}(x) \geq 8$ for every x in $I - \{0\}$. Again, there exists an even self-dual binary code of length 24 satisfying this property, and, up to isomorphism, there is only one, namely the Golay code [163].

In 1964, John Leech described in [141] a remarkable even unimodular lattice of dimension 24, denoted by Λ_{24} in [68] (Leech does not introduce a notation for the lattice in question!), with $x.x \geq 4$ for all nonzero x in Λ_{24}; of course, Λ_{24} is called the *Leech lattice*. In 1969, John H. Conway proved the following result [66].

Theorem 2.3.16. *Every even unimodular lattice of dimension* 24 *with empty set of roots is isomorphic to the Leech lattice.*

This theorem allows the completion of Corollary 2.3.14. Finally, one obtains the following result.

Theorem 2.3.17. *The map* $L \mapsto \mathrm{R}(L)$ *induces a bijection from the set of isomorphism classes of even unimodular lattices of dimension* 24 *to the set of* 24 *elements consisting of the empty set and the isomorphism classes of equi-Coxeter root systems of type ADE and rank* 24.

The construction of Λ_{24} by Leech (elegantly presented in Appendix 5 of [148]) involves in a crucial way the Golay code mentioned above when constructing a Niemeier lattice, say L_{23}, with $\mathrm{R}(\mathrm{L}_{23}) = 24\mathbf{A}_1$. In fact, the contruction by Leech is equivalent to one of the "holy constructions" in [68]; see item (c) in Theorem 3.4.2.10, which shows, in particular, the Leech lattice as a suitable 2-neighbor of L_{23}. The notion of neighbor, due to Martin Kneser, is the theme of Chap. 3.

In this book, the Leech lattice will denoted by Leech (most often) or L_{24}; its isomorphism class will be denoted in the same way.

Chapter 3
Kneser Neighbors

3.1 Variations on the Notion of Kneser Neighbors

The notion of 2-neighbor unimodular lattices was introduced by Martin Kneser in [122]. In that paper, Kneser uses this notion to describe an algorithm to classify unimodular lattices (the completeness is essentially due to Theorem 2.2.8) and applies this algorithm to explicitly list the isomorphism classes of unimodular lattices of dimension at most 15 (see also [159, Sect. 106F]).

We begin our variations with two very general observations (we state them for an arbitrary Dedekind domain R, but the applications we have in mind are $R = \mathbb{Z}$ and $R = \mathbb{Z}_p$).

Proposition 3.1.1. *Let R be a Dedekind domain and K its field of fractions. Let V be a finite-dimensional* q*-vector space over K. Let L_1 and L_2 be two self-dual integral lattices in V (in particular, L_1 and L_2 are* q*-modules, and the lattice $L_1 \cap L_2$ is a* $\widetilde{\mathrm{q}}$*-module, over R).*

We set $I_1 = L_1/(L_1 \cap L_2)$ and $I_2 = L_2/(L_1 \cap L_2)$.

(a) *The submodules I_1 and I_2 are two transverse Lagrangians of the* qe*-module* $\mathrm{res}(L_1 \cap L_2)$*. The linking form of this* qe*-module induces an isomorphism ι from I_2 to $I_1^\vee$, and the composition*

$$\mathrm{H}(I_1) = I_1 \oplus I_1^\vee \xrightarrow{\mathrm{id} \oplus \iota^{-1}} I_1 \oplus I_2 \longrightarrow \mathrm{res}(L_1 \cap L_2)$$

is an isomorphism of qe*-modules over R (recall that the notation $\mathrm{H}(I_1)$ denotes the hyperbolic* qe*-module over the torsion R-module of finite type I_1).*

(b) *Let r be the minimal number of generators of the R-module I_1; we have the inequality*

$$2\,r \;\leq\; \dim_K V\,.$$

G. Chenevier, J. Lannes, *Automorphic Forms and Even Unimodular Lattices*, Ergebnisse der Mathematik und ihrer Grenzgebiete. 3. Folge / A Series of Modern Surveys in Mathematics 69, https://doi.org/10.1007/978-3-319-95891-0_3

Proof. The first part of statement (a) is obvious; the second corresponds to part (a) of Definition-Proposition 2.1.3. Let us turn to statement (b). Let I be a torsion R-module of finite type. We denote the minimal number of generators of I by $\mathrm{r}(I)$; this number can be seen, for example, as the maximal integer k such that the exterior power $\Lambda^k I$ is nonzero. Since the dual $I^\vee$ is (noncanonically) isomorphic to I (to see this, note that R is "locally principal"), we have $\mathrm{r}(I) = \mathrm{r}(I^\vee)$. Consequently, by part (a), we have $\mathrm{r}(\mathrm{res}(L_1 \cap L_2)) = 2\,\mathrm{r}(I_1)$. Since $\mathrm{res}(L_1 \cap L_2)$ is a quotient of the lattice $(L_1 \cap L_2)^\sharp$, we have $\mathrm{r}(\mathrm{res}(L_1 \cap L_2)) \leq \dim_K V$. □

Let us specialize to $R = \mathbb{Z}$.

Scholium-Definition 3.1.2. *Let V be a* q*-vector space over $\mathbb{Q}$; let L_1 and L_2 be two self-dual integral lattices in V (in particular, the lattices L_1 and L_2 are* q*-modules, and the lattice $L_1 \cap L_2$ is a* $\widetilde{\mathrm{q}}$*-module, over $\mathbb{Z}$).*

Let A be a finite abelian group; the following conditions are equivalent:

(i) *The quotient $L_1/(L_1 \cap L_2)$ is isomorphic to A.*
(ii) *The quotient $L_2/(L_1 \cap L_2)$ is isomorphic to A.*

*If these conditions are satisfied, we say that L_1 and L_2 are A-*neighbors *(or that L_2 is an A-*neighbor *of L_1).*

d-Neighbors, Asymmetric Point of View

What makes the notion of A-neighbors special when A is a cyclic group is the statement below, which can be view as a corollary of Proposition-Definition 2.1.3 (b).

Proposition 3.1.3. *Let A be a finite cyclic group; then $A^\vee$ is the unique Lagrangian of the* qe*-module* $\mathrm{H}(A)$ *transverse to the Lagrangian A (in the sense of Proposition-Definition* 2.1.3).

Fix a q-module L over $\mathbb{Z}$ and an integer $d \geq 2$, and consider the set of $\mathbb{Z}/d$-neighbors of L in $\mathbb{Q} \otimes_{\mathbb{Z}} L$ (which is a q-vector space over $\mathbb{Q}$). To lighten the notation, we shorten $\mathbb{Z}/d$-neighbor to *d-neighbor* (and when we write p-neighbor, p is assumed to be prime).

In this context, a d-neighbor of L is an integral lattice L' in $\mathbb{Q} \otimes_{\mathbb{Z}} L$ with $L'^\sharp = L'$ and $L/(L \cap L') \simeq \mathbb{Z}/d$. Set $M = L \cap L'$. By the above, the following hold:

- The lattice dL' is contained in M.
- The image of the composition $dL' \subset M \subset L \to L/dL$ is an isotropic line in L/dL, which we denote by c, endowed with its structure of q-module over $\mathbb{Z}/d$. Let us explain what we mean by an *isotropic line* in L/dL: the c introduced above is a free submodule of dimension 1 of the $\mathbb{Z}/d$-module L/dL such that the restriction of the quadratic form $\mathrm{q}\colon L/dL \to \mathbb{Z}/d$ to c is zero (c is necessarily a direct summand because $\mathbb{Z}/d$ is an injective $\mathbb{Z}/d$-module).

- The lattice M is the inverse image of $c^\perp$ under the homomorphism $L \to L/dL$; here $c^\perp$ denotes the submodule of the $\mathbb{Z}/d$-module L/dL orthogonal to the line c.
- The lattice L' is the inverse image of the unique Lagrangian transverse to the Lagrangian L/M under the homomorphism $M^\sharp \to \operatorname{res} M$.

The above shows that the map $L' \mapsto c$ is injective, while the proposition below shows that it is also surjective.

Proposition 3.1.4. *Let c be an isotropic line in L/dL, and let M be the submodule of L defined as the inverse image of $c^\perp$ under the homomorphism $L \to L/dL$. Then the* qe*-module* $\operatorname{res} M$ *is isomorphic to* $\mathrm{H}(\mathbb{Z}/d)$ *and the quotient L/M is one of its Lagrangians. Moreover, the inverse image of the unique Lagrangian transverse to L/M under the homomorphism $M^\sharp \to \operatorname{res} M$ is a d-neighbor L' of L with $L \cap L' = M$.*

Before proving the proposition, let us introduce some terminology and notation that we will use in the remainder of this chapter.

Let L be a free, finite-dimensional $\mathbb{Z}$-module; we call an element u of L indivisible or *primitive* if it is nonzero and the quotient $L/\mathbb{Z}u$ has no torsion. Every nonzero element u of L can be written uniquely as $\mathrm{c}(u)v$ with v primitive and $\mathrm{c}(u)$ in $\mathbb{N} - \{0\}$. Let $d \geq 2$ be an integer; we say that an element u of L is *d-primitive* if it is nonzero and d is relatively prime to $\mathrm{c}(u)$. In other words, u is d-primitive if the submodule of L/dL generated by the class of u is a free $\mathbb{Z}/d$-module of dimension 1. As we already observed, such a submodule is necessarily a direct summand, so that a d-primitive element u of L defines an element of $\mathrm{P}_L(\mathbb{Z}/d)$, where P_L denotes the scheme whose R-points, for any commutative ring R, are the direct summands of rank 1 of the free R-module $R \otimes_{\mathbb{Z}} L$ (in other words, P_L is the projective space of L). We denote this element by $[u]$.

Proof of Proposition 3.1.4. It clearly suffices to verify that $\operatorname{res} M$ is isomorphic to $\mathrm{H}(\mathbb{Z}/d)$. Let u be a d-primitive element of L whose class modulo d generates the line c; since this line is isotropic, we have $\mathrm{q}(u) \equiv 0 \bmod d$. Let v be an element of L with $u.v \equiv 1 \bmod d$. Note that v and u/d belong to $M^\sharp$ and that $\operatorname{res} \mathrm{M}$ is a free $\mathbb{Z}/d$-module of dimension 2 with basis consisting of these two elements. We set $w = u/d - (\mathrm{q}(u)/d)v$. We observe that, in $\mathbb{Q}/\mathbb{Z}$, we have the equalities $\mathrm{q}(v) = 0$, $\mathrm{q}(w) = 0$, and $v.w = 1/d$. □

All things considered, we can conclude that the lattice L' is fully determined by c. Let us introduce some notation that highlights this dependence.

We denote the set of isotropic lines in L/dL by $\mathrm{C}_L(\mathbb{Z}/d)$. We justify this notation as follows: Let $\mathrm{C}_L \subset \mathrm{P}_L$ be the (projective) quadric defined by the quadratic form associated with L. Note, incidentally, that since the quadratic form is nondegenerate, C_L is smooth over $\mathbb{Z}$ (by the projective version of the Jacobian criterion for smoothness). It is clear that the $\mathbb{Z}/d$-points of C_L are the isotropic lines in L/dL. Let c be an element of $\mathrm{C}_L(\mathbb{Z}/d)$. We denote the lattices M and L' introduced in Proposition 3.1.4 by $\mathrm{M}_d(L; c)$ and $\mathrm{vois}_d(L; c)$, respectively (again, when we write

$\mathrm{vois}_p(L;c)$, p is assumed to be prime). Finally, we denote the set of d-neighbors of L in $\mathbb{Q}\otimes_{\mathbb{Z}} L$ by $\mathrm{Vois}_d(L)$. With this notation, we have the following statement.

Proposition 3.1.5. *The map*

$$\mathrm{C}_L(\mathbb{Z}/d) \to \mathrm{Vois}_d(L)\ , \quad c \mapsto \mathrm{vois}_d(L;c)$$

is a bijection.

Let u be a d-primitive element of L with $\mathrm{q}(u) \equiv 0 \bmod d$; we will also denote the lattices $\mathrm{M}_d(L;[u])$ and $\mathrm{vois}_d(L;[u])$ by $\mathrm{M}_d(L;u)$ and $\mathrm{vois}_d(L;u)$, respectively. For further reference, let us give the algorithm $u \rightsquigarrow \mathrm{vois}_d(L;u)$ provided by the proof of Proposition 3.1.4 explicitly.

Let v be an element of L with $u.v \equiv 1 \bmod d$; then $\mathrm{vois}_d(L;u)$ is the lattice in $\mathbb{Q}\otimes_{\mathbb{Z}} L$ generated by

$$\mathrm{M}_d(L;u) := \{\, x\,;\, x \in L\,,\ u.x \equiv 0 \bmod d\,\} \quad \text{and} \quad \frac{u-\mathrm{q}(u)v}{d}\ .$$

Set $\widetilde{u} = u - \mathrm{q}(u)v$; note that we have $\widetilde{u} \equiv u \bmod d$, in other words, $[\widetilde{u}] = [u]$ in $\mathrm{C}_L(\mathbb{Z}/d)$, and $\mathrm{q}(\widetilde{u}) \equiv 0 \bmod d^2$. This observation leads, in an obvious way, to an alternative presentation of the algorithm: Given u, we determine an element $\widetilde{u}$ of L with $\widetilde{u} \equiv u \bmod d$ and $\mathrm{q}(\widetilde{u}) \equiv 0 \bmod d^2$. The lattice $\mathrm{vois}_d(L;u)$ of u is then the lattice in $\mathbb{Q}\otimes_{\mathbb{Z}} L$ generated by $\mathrm{M}_p(L;u)$ and $\widetilde{u}/d$.

d-Neighbors, Abstract Point of View

Let L_1 and L_2 be two q-modules over $\mathbb{Z}$ of the same dimension n. Unsurprisingly, we say that L_2 is a d-neighbor of L_1 if, as a q-module, L_2 is isomorphic to a d-neighbor of L_1 in $\mathbb{Q}\otimes_{\mathbb{Z}} L_1$. In view of the above, if L_2 is a d-neighbor of L_1, then L_1 is a d-neighbor of L_2; we therefore also say that L_1 and L_2 are d-neighbors. To avoid confusion, we will sometimes call the notion we just introduced the "abstract" neighborhood, while calling the notion introduced in Scholium-Definition 3.1.2 the "concrete" neighborhood. If the q-vector space $\mathbb{R}\otimes_{\mathbb{Z}} L_1$ is indefinite, then L_2 is a d-neighbor of L_1 if and only if the two q-vector spaces $\mathbb{R}\otimes_{\mathbb{Z}} L_1$ and $\mathbb{R}\otimes_{\mathbb{Z}} L_2$ are isomorphic (Theorem 2.2.7): the relation of being abstract d-neighbors is not very interesting in this case! Therefore, from now on, we assume that L_1 and L_2 are positive definite (which implies that n is divisible by 8). As agreed on before, we abandon the term "positive definite q-module over $\mathbb{Z}$" and use "even unimodular lattice" instead.

Let L_1 and L_2 be two even unimodular lattices. We denote by $\widetilde{\mathrm{Vois}}_d(L_1,L_2)$ the set of isomorphisms of q-vector spaces $\phi\colon \mathbb{Q}\otimes_{\mathbb{Z}} L_2 \to \mathbb{Q}\otimes_{\mathbb{Z}} L_1$ with $L_1/(L_1\cap\phi(L_2))$ cyclic of order d. The set $\widetilde{\mathrm{Vois}}_d(L_1,L_2)$ is finite; by definition, it is nonempty if and only if L_1 and L_2 are d-neighbors. Moreover, $\widetilde{\mathrm{Vois}}_d(L_1,L_2)$ is endowed with a free left action of the orthogonal group $\mathrm{O}(L_1)$ and a free right action of the orthogonal group $\mathrm{O}(L_2)$; these actions commute. We denote by $\mathrm{Vois}_d(L_1,L_2)$ the subset of

$\mathrm{Vois}_d(L_1)$ consisting of the d-neighbors of L_1 in $\mathbb{Q} \otimes_{\mathbb{Z}} L_1$ that are isomorphic, as q-modules, to L_2. The set $\mathrm{Vois}_d(L_1, L_2)$ is canonically endowed with a left action of $\mathrm{O}(L_1)$. Again by definition, the map $\widetilde{\mathrm{Vois}}_d(L_1, L_2) \to \mathrm{Vois}_d(L_1, L_2)$ given by $\phi \mapsto \phi(L_2)$ induces an $\mathrm{O}(L_1)$-equivariant bijection

$$\widetilde{\mathrm{Vois}}_d(L_1, L_2)/\mathrm{O}(L_2) \;\cong\; \mathrm{Vois}_d(L_1, L_2)\,.$$

We denote the cardinality of the set $\mathrm{Vois}_d(L_1, L_2)$ by $\mathrm{N}_d(L_1, L_2)$. When we write $\mathrm{N}_p(-,-)$, p is assumed to be prime.

Take note: the introduction of this notation is not insignificant. The study of these cardinalities in dimensions 16 and 24 for d prime is the main subject of this book (see Chap. 1)!

We denote by $[-]$ the isomorphism class of an even unimodular lattice. The number $\mathrm{N}_d(L_1, L_2)$ clearly depends only on $[L_1]$ and $[L_2]$; consequently, we will also denote this integer by $\mathrm{N}_d([L_1], [L_2])$.

Lemma 3.1.6. *The map $\phi \mapsto \phi^{-1}$ induces a bijection from $\widetilde{\mathrm{Vois}}_d(L_1, L_2)$ to $\widetilde{\mathrm{Vois}}_d(L_2, L_1)$.*

Proof. This follows from the "symmetric point of view" on d-neighbors, that is, from Proposition 3.1.1. □

Scholium 3.1.7. *We have the relation*

$$\frac{1}{|\mathrm{O}(L_1)|}\ \mathrm{N}_d(L_1, L_2) \;=\; \frac{1}{|\mathrm{O}(L_2)|}\ \mathrm{N}_d(L_2, L_1)\,,$$

where $|-|$ denotes the cardinality of a finite set.

Proof. We have $|\widetilde{\mathrm{Vois}}_d(L_1, L_2)| = \mathrm{N}_d(L_1, L_2)\,|\mathrm{O}(L_2)|$. □

Scholium 3.1.7 can be made more precise, giving Scholium 3.1.8 below. For an illustration of the latter, see Sects. 3.3.1 and 3.3.2.

By the above, set-theoretically, the quotient $\mathrm{O}(L_1)\backslash\widetilde{\mathrm{Vois}}_d(L_1, L_2)/\mathrm{O}(L_2)$ is in canonical bijection with the quotient $\mathrm{O}(L_1)\backslash\mathrm{Vois}_d(L_1, L_2)$ and with the quotient $\mathrm{O}(L_2)\backslash\mathrm{Vois}_d(L_2, L_1)$, so that we have a canonical bijection between the latter two.

Scholium 3.1.8. *Let Ω_1 be an $\mathrm{O}(L_1)$-orbit in $\mathrm{Vois}_d(L_1, L_2)$ and Ω_2 an $\mathrm{O}(L_2)$-orbit in $\mathrm{Vois}_d(L_2, L_1)$ that correspond through the canonical bijection*

$$\mathrm{O}(L_1)\backslash\mathrm{Vois}_d(L_1, L_2) \;\cong\; \mathrm{O}(L_2)\backslash\mathrm{Vois}_d(L_2, L_1)\,;$$

then we have

$$\frac{|\Omega_1|}{|\mathrm{O}(L_1)|} \;=\; \frac{|\Omega_2|}{|\mathrm{O}(L_2)|}\,.$$

A more direct way to obtain the equality above is by using the concrete notion of d-neighborhood.

Proposition 3.1.9. *Let V be a positive definite* q*-vector space over $\mathbb{Q}$. Assume that V contains two even unimodular lattices L_1 and L_2 and that these lattices are d-neighbors in V. Let Ω_1 be the $\mathrm{O}(L_1)$-orbit of L_2 and Ω_2 the $\mathrm{O}(L_2)$-orbit of L_1. We then have the following relations:*

$$\frac{|\Omega_1|}{|\mathrm{O}(L_1)|} = \frac{|\Omega_2|}{|\mathrm{O}(L_2)|} = \frac{1}{|\mathrm{O}(L_1) \cap \mathrm{O}(L_2)|},$$

where $\mathrm{O}(L_1)$ and $\mathrm{O}(L_2)$ are identified with subgroups of $\mathrm{O}(V)$ and the intersection is taken in $\mathrm{O}(V)$.

Proof. The stabilizers of L_2 for the action of $\mathrm{O}(L_1)$ and of L_1 for the action of $\mathrm{O}(L_2)$ can both be identified with $\mathrm{O}(L_1) \cap \mathrm{O}(L_2)$. □

2-Neighbors, the Point of View of Borcherds [68, Chap. 17]

We begin with two remarks concerning Proposition 3.1.9. We use its notation and consider the lattice $L_1 \cap L_2$.

- This lattice is a $\widetilde{\mathrm{q}}$-module whose residue is endowed with an ordered pair of Lagrangians that are both cyclic of order d and transverse to each other, namely $\omega := (L_1/(L_1 \cap L_2), L_2/(L_1 \cap L_2))$. The group $\mathrm{O}(L_1) \cap \mathrm{O}(L_2)$ can be identified with the subgroup of $\mathrm{O}(L_1 \cap L_2)$ consisting of the elements that preserve ω; we denote this subgroup by $\mathrm{O}(L_1 \cap L_2; \omega)$. The equalities in Proposition 3.1.9 can therefore also be written as follows:

$$|\Omega_1| = \frac{|\mathrm{O}(L_1)|}{|\mathrm{O}(L_1 \cap L_2; \omega)|}, \quad |\Omega_2| = \frac{|\mathrm{O}(L_2)|}{|\mathrm{O}(L_1 \cap L_2; \omega)|}.$$

- When d is prime, the unordered pair of Lagrangians underlying ω is uniquely determined in terms of $L_1 \cap L_2$. This implies that $\mathrm{O}(L_1 \cap L_2; \omega)$ has index at most 2 in $\mathrm{O}(L_1 \cap L_2)$. In particular, we have $\mathrm{O}(L_1 \cap L_2; \omega) = \mathrm{O}(L_1 \cap L_2)$ if L_1 and L_2 are not isomorphic.

Having made these remarks, we could continue studying d-neighbors for arbitrary $d \geq 2$, but to simplify the exposition, we will treat only p-neighbors with p prime. In fact, the case we have in mind is $p = 2$.

Let $n > 0$ be an integer divisible by 8; recall that X_n is the finite set of isomorphism classes of even unimodular lattices of dimension n. For p prime, we introduce three other finite sets:

- $\mathrm{Y}_n(p)$ is the set of isomorphism classes of ordered pairs (L_1, L_2) with L_1 an even unimodular lattice of dimension n and L_2 a p-neighbor of L_1 in $\mathbb{Q} \otimes_{\mathbb{Z}} L_1$.
- $\mathrm{B}_n(p)$ is the set of isomorphism classes of the $\widetilde{\mathrm{q}}$-modules M over $\mathbb{Z}$ with $\dim M = n$, $\mathbb{R} \otimes_{\mathbb{Z}} M > 0$, and $\mathrm{res}\, M \simeq \mathrm{H}(\mathbb{Z}/p)$.

– $\widetilde{\mathrm{B}}_n(p)$ is the set of isomorphism classes of pairs $(M;\omega)$ with M as above and ω a bijection from the set of Lagrangians of $\operatorname{res} M$ to the set $\{1,2\}$. By definition, $\widetilde{\mathrm{B}}_n(p)$ is endowed with a left action of the symmetric group $\mathfrak{S}_2$; the quotient $\mathfrak{S}_2\backslash\widetilde{\mathrm{B}}_n(p)$ can be identified with $\mathrm{B}_n(p)$.

We have done everything to ensure that the sets $\mathrm{Y}_n(p)$ and $\widetilde{\mathrm{B}}_n(p)$ are in canonical bijection. For $(M;\omega)$ as above, we denote by $\mathrm{d}_i(M;\omega)$ for $i = 1,2$ the inverse image of $\omega^{-1}(i)$ under the surjection $M^\sharp \to \operatorname{res} M$; $\mathrm{d}_1(M;\omega)$ and $\mathrm{d}_2(M;\omega)$ are even unimodular lattices (of dimension n) that are p-neighbors in $\mathbb{Q}\otimes_{\mathbb{Z}} M$. By passing to isomorphism classes, we obtain two maps from $\widetilde{\mathrm{B}}_n(p)$ to X_n that we also denote by d_1 and d_2.

We have now introduced the notation necessary to state the following proposition.

Proposition 3.1.10. *Let p be a prime, and let x_1 and x_2 be two elements of X_n. We have*

$$\mathrm{N}_p(x_1,x_2) \quad = \sum_{\beta\,\in\,\mathrm{d}_1^{-1}(x_1)\cap\mathrm{d}_2^{-1}(x_2)} \frac{|\mathrm{O}(x_1)|}{|\mathrm{O}(\beta)|}$$

(we leave it to the reader to decode the notation $|\mathrm{O}(x_1)|$ and $|\mathrm{O}(\beta)|$).

Proposition 3.1.10 admits a slightly more concrete avatar for $p = 2$ because the choice of a $\widetilde{\mathrm{q}}$-module M over $\mathbb{Z}$ with $\dim M = n$, $\mathbb{R}\otimes_{\mathbb{Z}} M > 0$, and $\operatorname{res} M \simeq \mathrm{H}(\mathbb{Z}/2)$ is equivalent to the choice of an odd unimodular lattice L with $\dim L = n$. Let us explain why (following Borcherds).

– With an M as above, we associate the lattice L that is the inverse image under the surjection $M^\sharp \to \operatorname{res} M$ of the line in the $\mathbb{Z}/2$-vector space $\operatorname{res} M$ that is not isotropic in the quadratic sense; this line is isotropic in the bilinear sense, so that L is an odd unimodular lattice.

– With an odd unimodular lattice L of dimension n, we associate the submodule M consisting of the elements x with $x.x \equiv 0 \bmod 2$ (see Scholium 2.2.3).

Moreover, the set consisting of the two Lagrangians of $\operatorname{res} M$ is in natural bijection with the set consisting of the two classes of Wu vectors of L (see the discussion following Scholium 2.2.3).

We therefore need to introduce the following notation:

– B_n is the finite set of isomorphism classes of odd unimodular lattices L of dimension n.

– $\widetilde{\mathrm{B}}_n$ is the finite set of isomorphism classes of pairs $(L;\omega)$ with L an odd unimodular lattice of dimension n and ω a bijection from the set consisting of the two classes of Wu vectors of L to $\{1,2\}$. By definition, $\widetilde{\mathrm{B}}_n$ is endowed with a left action of the symmetric group $\mathfrak{S}_2$; the quotient $\mathfrak{S}_2\backslash\widetilde{\mathrm{B}}_n$ can be identified with B_n.

– For $i = 1,2$, let L_i be the even unimodular lattice in $\mathbb{Q}\otimes_{\mathbb{Z}} L$ generated by the submodule of L consisting of the elements x with $x.x \equiv 0 \bmod 2$ and the vector $u_i/2$, where u_i denotes a representation of the class $\omega^{-1}(i)$. We use the notation

d_i again, this time for the map $\mathrm{d}_i \colon \widetilde{\mathrm{B}}_n \to \mathrm{X}_n$ that sends the isomorphism class of $(L;\omega)$ to the isomorphism class of L_i.

By construction, the unimodular lattices L and L_i, for $i = 1, 2$, are 2-neighbors "in the bilinear sense," that is, $L \cap L_i$ has index 2 in L and L_i. The reader can verify that among the unimodular lattices in $\mathbb{Q} \otimes_{\mathbb{Z}} L$ that are 2-neighbors of L, the lattices L_1 and L_2 are characterized by the fact that they are even. We will say that L_1 and L_2 are the *even 2-neighbors* of L.

We can now finally state the avatar of Proposition 3.1.10 for $p = 2$ mentioned earlier. The statement seems identical; the difference is in the meaning of the notation. The maps d_1 and d_2 are now the maps from $\widetilde{\mathrm{B}}_n$ to X_n introduced above, β belongs to $\widetilde{\mathrm{B}}_n$, and if β is represented by an odd unimodular lattice L endowed with a bijection ω from the set of its two classes of Wu vectors to $\{1, 2\}$, then $\mathrm{O}(\beta)$ is the subgroup of $\mathrm{O}(L)$ that preserves ω.

Proposition 3.1.11. *Let x_1 and x_2 be two elements of X_n. We have*

$$\mathrm{N}_2(x_1, x_2) \quad = \sum_{\beta \,\in\, \mathrm{d}_1^{-1}(x_1) \cap \mathrm{d}_2^{-1}(x_2)} \frac{|\mathrm{O}(x_1)|}{|\mathrm{O}(\beta)|} \,.$$

GRAPHS OF p-NEIGHBORS

Let $n > 0$ be an integer divisible by 8 and p a prime. The *graph of p-neighbors* $\mathrm{K}_n(p)$ is defined as follows: The set of vertices is X_n, the set of classes of even unimodular lattices of dimension n. The edges are the subsets $\{[L_1], [L_2]\}$ of X_n with L_1 and L_2 p-neighbors (recall that $[L]$ is the isomorphism class of an even unimodular lattice L).

Theorem 3.1.12 (M. Kneser). *For every n and p, the graph $\mathrm{K}_n(p)$ is connected.*

Proof. Let L and M be two even unimodular lattices of the same dimension; we must show that there exists a finite sequence of even unimodular lattices

$$L = L_0\,,\ L_1\,,\ L_2\,,\ \ldots\,,\ L_{m-1}\,,\ L_m = M$$

with L_k and L_{k+1} (abstract) p-neighbors for $0 \leq k \leq m-1$.

Theorem 2.2.8 shows that there exists an isomorphism of q-modules $\phi \colon \mathbb{Z}[1/p] \otimes_{\mathbb{Z}} L \to \mathbb{Z}[1/p] \otimes_{\mathbb{Z}} M$. After replacing M by $\phi^{-1}(M)$ if necessary, we may assume $M \subset \mathbb{Z}[1/p] \otimes_{\mathbb{Z}} L \subset \mathbb{Q} \otimes_{\mathbb{Z}} L$; set $V = \mathbb{Q} \otimes_{\mathbb{Z}} L = \mathbb{Q} \otimes_{\mathbb{Z}} M$. Recall part (a) of Proposition 3.1.1. We view L and M as lattices in V and set $N = L \cap M$. We have $N^\sharp = L + M$ and $\operatorname{res} N = L/N \oplus M/N$ (as an abelian group). Set $I = L/N$ and $J = M/N$; the pairing $I \times J \to \mathbb{Q}/\mathbb{Z}$ induced by the linking form of $\operatorname{res} N$ is nondegenerate. Consequently, J and $\operatorname{res} N$ are canonically isomorphic to the Pontryagin dual $I^\vee$ and the hyperbolic qe-module $\mathrm{H}(I)$, respectively. Since we have $M \subset \mathbb{Z}[1/p] \otimes_{\mathbb{Z}} L$, the finite abelian group I is a p-group. Let

$$I = I_0 \supset I_1 \supset I_2 \supset \ldots \supset I_{m-1} \supset I_m = 0$$

be a finite decreasing sequence of subgroups of I with $I_k/I_{k+1} \simeq \mathbb{Z}/p$ for $0 \leq k \leq m-1$. Let

$$0 = J_0 \subset J_1 \subset I_2 \subset \ldots \subset J_{m-1} \subset J_m = J$$

be the "orthogonal" sequence of subgroups of J. We set $K_k = I_k \oplus J_k$. Note that K_k is a Lagrangian of res N. Let L_k be the inverse image of K_k under the canonical homomorphism $N^\sharp \to \operatorname{res} N$. By construction,

$$L = L_0\,,\ L_1\,,\ L_2\,,\ \ldots\,,\ L_{m-1}\,,\ L_m = M$$

is a sequence of even unimodular lattices with L_k and L_{k+1} (concrete) p-neighbors for $0 \leq k \leq m-1$. □

MISCELLANIES

In this section, we gather four more or less technical statements (Propositions 3.1.13, 3.1.14, and 3.1.17 and Corollary 3.1.16) concerning the notion of d-neighborhood; we will use them further on.

The least technical of the four is Propositions 3.1.13. Its proof, which is an illustration of Proposition 2.1.1, is left to the reader.

Proposition 3.1.13. *Let L be a q-module over $\mathbb{Z}$ and $d \geq 2$ an integer. Let u be a d-primitive element of L with $\mathrm{q}(u) \equiv 0 \bmod d^2$. Assume $d = d_1 d_2$ with $d_1 \geq 2$ and $d_2 \geq 2$. Then u/d_1 is a d_2-primitive element of* $\mathrm{vois}_{d_1}(L;u)$ *and in $\mathbb{Q} \otimes_{\mathbb{Z}} L$, we have the equality*

$$\mathrm{vois}_d(L;u) \quad = \quad \mathrm{vois}_{d_2}\Big(\mathrm{vois}_{d_1}(L;u); \frac{u}{d_1}\Big)\,.$$

Proposition 3.1.14. *Let L be a q-module over $\mathbb{Z}$ and $d \geq 2$ an integer. Let u be a d-primitive element of L with $\mathrm{q}(u) = d$. Let s_u be the orthogonal reflection of $\mathbb{Q} \otimes_{\mathbb{Z}} L$ with respect to the hyperplane $u^\perp$.*

(a) *In $\mathbb{Q} \otimes_{\mathbb{Z}} L$, we have the equality*

$$\mathrm{vois}_d(L;u) \quad = \quad \mathrm{s}_u(L)\,;$$

in particular, $\mathrm{vois}_d(L;u)$ *is isomorphic to L.*

(b) *Suppose $d = d_1 d_2$ with $d_1 \geq 2$ and $d_2 \geq 2$. Then u is d_i-primitive for $i = 1, 2$, and in $\mathbb{Q} \otimes_{\mathbb{Z}} L$, we have the equality*

$$\mathrm{vois}_{d_2}(L;u) \quad = \quad \mathrm{s}_u(\mathrm{vois}_{d_1}(L;u))\,;$$

in particular, $\mathrm{vois}_{d_1}(L;u)$ *and* $\mathrm{vois}_{d_2}(L;u)$ *are isomorphic.*

Proof. The map s_u is given by

$$\mathrm{s}_u(x) \;=\; x - \frac{u.x}{d}\, u$$

for all x in $\mathbb{Q} \otimes_{\mathbb{Z}} L$. This expression shows that s_u induces an automorphism of $\mathrm{M}_d(L;u)$ as a $\widetilde{\mathrm{q}}$-module over $\mathbb{Z}$ and therefore an automorphism of $\operatorname{res}\mathrm{M}_d(L;u)$ as a qe-module over $\mathbb{Z}$. Let I and J be the two transverse Lagrangians of $\operatorname{res}\mathrm{M}_d(L;u)$ that correspond to the lattices L and $\mathrm{vois}_d(L;u)$, respectively, through the map in Proposition 2.1.1 (b). Let v be an element of L with $u.v \equiv 1 \bmod d$; recall that I and J are generated by the classes of v and $(u - \mathrm{q}(u)\,v)/d$, respectively, in $\operatorname{res}\mathrm{M}_d(L;u)$. We thus have $\mathrm{s}_u(I) = J$; this equality implies part (a) of the proposition.

Through the bijection in Proposition 2.1.1 (b), the lattices $\mathrm{vois}_{d_1}(L;u)$ and $\mathrm{vois}_{d_2}(L;u)$ correspond to the Lagrangians $d_1 I \oplus d_2 J$ and $d_2 I \oplus d_1 J$, respectively, of $\operatorname{res}\mathrm{M}_d(L;u)$. By the above, we have $\mathrm{s}_u(d_1 I \oplus d_2 J) = d_2 I \oplus d_1 J$; this equality implies part (b) of the proposition. □

Remark. If we agree that L is the only 1-neighbor of L in $\mathbb{Q} \otimes_{\mathbb{Z}} L$, then part (a) of Proposition 3.1.4 is a special case of part (b).

Proposition 3.1.15. *Let L be a* b*-module over $\mathbb{Z}$. Assume that there exists an element e of L with $e.e = 1$ (and therefore that L is odd). Then the orthogonal reflection of L with respect to the hyperplane $e^{\perp}$ interchanges the two classes of Wu vectors of L.*

Proof. Let s_e be the reflection in question, and let u be a Wu vector of L; the equality $\mathrm{s}_e(u) = u - 2(e.u)e$ and the congruence $e.u \equiv e.e \bmod 2$ show that the Wu vectors u and $\mathrm{s}_e(u)$ are not equivalent. □

Corollary 3.1.16. *Let L be an odd unimodular lattice of dimension divisible by 8. Assume that there exists an element e of L with $e.e = 1$. Then the orthogonal reflection of $\mathbb{Q} \otimes_{\mathbb{Z}} L$ with respect to the hyperplane $e^{\perp}$ interchanges the two even unimodular lattices that are 2-neighbors of L.*

The following proposition, whose proof is obvious, shows that the special case $d = 2$ of part (a) of Proposition 3.1.14 and Corollary 3.1.16 are closely related.

Proposition 3.1.17. *Let L be an even unimodular lattice and u an element of L with $\mathrm{q}(u) = 2$ (this equality implies that u is 2-primitive). Let B be the odd lattice in $\mathbb{Q} \otimes_{\mathbb{Z}} L$ whose even 2-neighbors are L and* $\mathrm{vois}_2(L;u)$ *(see the end of the discussion "2-neighbors, the point of view of Borcherds"). Let e be the element $u/2$ of $\mathbb{Q} \otimes_{\mathbb{Z}} L$. Then:*

- *We have $e.e = 1$.*
- *The lattice B is generated, in $\mathbb{Q} \otimes_{\mathbb{Z}} L$, by* $\mathrm{M}_2(L;u)$ *and e.*
- *The orthogonal reflections s_u and s_e of $\mathbb{Q} \otimes_{\mathbb{Z}} L$ coincide.*

3.2 Hecke Operators Associated with the Notion of Neighborhood

Let $n > 0$ be an integer divisible by 8. Recall that X_n is the finite set of isomorphism classes of unimodular lattices of dimension n. We denote by $\mathbb{Z}[\mathrm{X}_n]$ the free $\mathbb{Z}$-module generated by the set X_n.

Let A be a finite abelian group and L an even unimodular lattice of dimension n. We denote the finite set consisting of the A-neighbors of L in $\mathbb{Q} \otimes_{\mathbb{Z}} L$ by $\mathrm{Vois}_A(L)$ (the notation $\mathrm{Vois}_d(-)$ introduced in the previous section is an abbreviation of $\mathrm{Vois}_{\mathbb{Z}/d}(-)$).

The Hecke operator T_A is the endomorphism of $\mathbb{Z}[\mathrm{X}_n]$ defined by

$$\mathrm{T}_A[L] \;:=\; \sum_{L' \in \mathrm{Vois}_A(L)} [L']$$

for every even unimodular lattice L of dimension n.

Remarks.

– Let $\mathrm{r}(A)$ be the minimal number of generators of the abelian group A. Part (b) of Proposition 3.1.1 shows that T_A is zero if we have $2\,\mathrm{r}(A) > n$. We could therefore assume $2\,\mathrm{r}(A) \leq n$ in the definition above.

– Let A and B be two finite abelian groups. It is not very difficult to see that the Hecke operators T_A and T_B commute if the cardinalities of A and B are relatively prime. In fact, T_A and T_B commute for all A and B; we will prove this and greatly generalize it in Chap. 4.

Let $d \geq 2$ be an integer; we shorten the notation $\mathrm{T}_{\mathbb{Z}/d}$ to T_d (and when we write T_p, p is assumed to be prime). By the definition of the integers $\mathrm{N}_d(x, y)$, we have

$$\mathrm{T}_d\, x \;=\; \sum_{y \in \mathrm{X}_n} \mathrm{N}_d(x, y)\, y$$

for every x in X_n. In other words, if we view T_d as an $\mathrm{X}_n \times \mathrm{X}_n$ matrix, then its entry with index (y, x) is $\mathrm{N}_d(x, y)$.

Proposition 3.1.5 shows that we also have

$$\mathrm{T}_d[L] \;=\; \sum_{c \in \mathrm{C}_L(\mathbb{Z}/d)} [\mathrm{Vois}_d(L; c)]$$

for every even unimodular lattice L.

We state Proposition 3.2.2 below for future reference. It is implied by Proposition-Definition 3.2.1 and Proposition 3.1.5; the former is essentially a consequence of Scholium 2.2.5.

Proposition-Definition 3.2.1. *Let L be an even unimodular lattice, and let $d \geq 2$ be an integer. Then the* q*-module $\mathbb{Z}/d \otimes_{\mathbb{Z}} L$ is hyperbolic. In particular, the cardinality of the quadric $\mathrm{C}_L(\mathbb{Z}/d)$ depends only on the dimension of L, which we denote by n.*

We denote the cardinality in question by $\mathrm{c}_n(d)$. *We have*

$$\mathrm{c}_n(p) \;=\; \sum_{m=0}^{n-2} p^m \;+\; p^{n/2-1}$$

for every prime p.

Remarks (Continued). The computation of $\mathrm{c}_n(d)$ for any d easily follows from the computation for d prime and the fact that the quadrics C_L are smooth over $\mathbb{Z}$.

Proposition 3.2.2. *Let* $d \geq 2$ *be an integer. We have*

$$\sum_{y \in \mathrm{X}_n} \mathrm{N}_d(x,y) \;=\; \mathrm{c}_n(d)$$

for every x *in* X_n.

Remarks (Continued and Concluded).

- Scholium 3.1.7 can also be stated as follows.

Proposition 3.2.3. *Let* $d \geq 2$ *be an integer. The endomorphism* T_d *of* $\mathbb{Z}[\mathrm{X}_n]$ *is self-adjoint for the inner product* $(-|-)$ *defined by*

$$(x|y) \;=\; |\mathrm{O}(x)|\;\delta_{x,y}$$

for x *and* y *in* X_n, *where* $\delta_{x,y}$ *denotes the Kronecker delta.*

Again, this statement will be greatly generalized in Chap. 4.

- Let $\epsilon : \mathbb{Z}[\mathrm{X}_n] \to \mathbb{Z}$ be the homomorphism of $\mathbb{Z}$-modules given by $\epsilon(x) = 1$ for every x in X_n. Proposition 3.2.2 says that we have $\epsilon \circ \mathrm{T}_d = \mathrm{c}_n(d)\,\epsilon$, in other words, that ϵ is an eigenvector of the endomorphism T_d^* of $(\mathbb{Z}[\mathrm{X}_n])^*$ with eigenvalue $\mathrm{c}_n(d)$ (where T_d^* replaces the notation $\mathrm{T}_d^{\mathrm{t}}$ used elsewhere). This observation and Proposition 3.2.3 lead to the following statement.

Proposition 3.2.4. *Let* $d \geq 2$ *be an integer. The element*

$$\sum_{x \in \mathrm{X}_n} \frac{1}{|\mathrm{O}(x)|}\; x$$

of $\mathbb{Q} \otimes_{\mathbb{Z}} \mathbb{Z}[\mathrm{X}_n]$ *is an eigenvector of* T_d *with eigenvalue* $\mathrm{c}_n(d)$.

- Proposition 3.2.3 implies that T_d is diagonalizable, at least after extension of scalars to $\mathbb{R}$. In fact, for $n = 8, 16, 24$, the eigenvalues of T_d are integers. This is trivial for $n = 8$ because we have $|\mathrm{X}_8| = 1$; it is nearly as trivial for $n = 16$ because we have $|\mathrm{X}_{16}| = 2$ and we already know one integral eigenvalue, namely $\mathrm{c}_{16}(d)$. The case $n = 24$ requires more effort. We will explain in Sect. 3.3.3 how Gabriele Nebe and Boris Venkov determined T_2 from the work of Borcherds; thanks to the program PARI, we know that the roots of the characteristic polynomial of T_2 are integral and simple. Since T_d commutes with T_2 for every d, the eigenvectors of T_2 are also eigenvectors of T_d and the eigenvalues of T_d are integers.

One motivation of this book is the study of the arithmetic properties of these eigenvalues for d prime.

3.3 Examples

3.3.1 Determination of T_2 *for* $n = 16$

It is well known that the canonical homomorphism $\mathrm{O}(\mathrm{E}_8) \to \mathrm{O}(\mathbb{F}_2 \otimes_{\mathbb{Z}} \mathrm{E}_8)$ induces an isomorphism $\mathrm{O}(\mathrm{E}_8)/\{\pm 1\} \cong \mathrm{O}(\mathbb{F}_2 \otimes_{\mathbb{Z}} \mathrm{E}_8)$ (see, for example, [39, Chap. VI, Sect. 4, Exercice 1]). It follows that the action of $\mathrm{O}(\mathrm{E}_8)$ partitions $\mathbb{F}_2 \otimes_{\mathbb{Z}} \mathrm{E}_8 - \{0\}$ into two orbits, namely $\mathrm{q}^{-1}(0)$ and $\mathrm{q}^{-1}(1)$; these orbits have, respectively, 135 and 120 elements.

Consider the lattices $\mathrm{E}_8 \oplus \mathrm{E}_8$ and E_{16}, which we embed into $\mathbb{Q}^{16}$ in the usual way. We denote the canonical basis of $\mathbb{Q}^{16}$ by $(\varepsilon_1, \varepsilon_2, \ldots, \varepsilon_{16})$. The group $\mathrm{O}(\mathrm{E}_8 \oplus \mathrm{E}_8)$ is clearly canonically isomorphic to the wreath product $\mathfrak{S}_2 \wr \mathrm{O}(\mathrm{E}_8)$. By the above, the action of $\mathrm{O}(\mathrm{E}_8 \oplus \mathrm{E}_8)$ partitions the quadric $\mathrm{C}_{\mathrm{E}_8 \oplus \mathrm{E}_8}(\mathbb{F}_2)$ into three orbits:

- the orbit of the point $[2\varepsilon_1]$, with $2 \cdot 135 = 270$ elements (the vector $2\varepsilon_1$ belongs to the lattice $\mathrm{E}_8 \oplus \mathrm{E}_8$ and satisfies $\mathrm{q}(2\varepsilon_1) = 2$; we use $[2\varepsilon_1]$ to denote its class in $\mathrm{C}_{\mathrm{E}_8 \oplus \mathrm{E}_8}(\mathbb{F}_2)$);
- the orbit of the point $[\varepsilon_1 + \varepsilon_2 + \varepsilon_9 + \varepsilon_{10}]$, with $120^2 = 14400$ elements (we again have $\mathrm{q}(\varepsilon_1 + \varepsilon_2 + \varepsilon_9 + \varepsilon_{10}) = 2$);
- the orbit of the point $[2\varepsilon_1 + 2\varepsilon_9]$, with $135^2 = 18225$ elements (this time, we have $\mathrm{q}(2\varepsilon_1 + 2\varepsilon_9) = 4$).

By part (a) of Proposition 3.1.14, the lattice $\mathrm{vois}_2(\mathrm{E}_8 \oplus \mathrm{E}_8; c)$ is isomorphic (as a q-module) to $\mathrm{E}_8 \oplus \mathrm{E}_8$ for every c in one of the first two orbits.

Since the graph of the 2-neighbors is connected (Theorem 3.1.12), we must have $\mathrm{N}_2(\mathrm{E}_8 \oplus \mathrm{E}_8, \mathrm{E}_{16}) = 18225$. This equality determines the Hecke operator $\mathrm{T}_2 \colon \mathbb{Z}[\mathrm{X}_{16}] \to \mathbb{Z}[\mathrm{X}_{16}]$, in view of Scholium 3.1.7 and Proposition 3.2.2. Its matrix in the basis $(\mathrm{E}_{16}, \mathrm{E}_8 \oplus \mathrm{E}_8)$, which we also denote by T_2, is

$$\mathrm{T}_2 = \begin{bmatrix} 20025 & 18225 \\ 12870 & 14670 \end{bmatrix}.$$

We can, in fact, easily verify that we have $\mathrm{vois}_2(\mathrm{E}_8 \oplus \mathrm{E}_8; [2\varepsilon_1 + 2\varepsilon_9]) = \mathrm{E}_{16}$ (which implies $\mathrm{N}_2(\mathrm{E}_8 \oplus \mathrm{E}_8, \mathrm{E}_{16}) = 18225$). Indeed, the lattice $\mathrm{M}_2(\mathrm{E}_8 \oplus \mathrm{E}_8; [2\varepsilon_1 + 2\varepsilon_9])$ is generated by $\mathrm{D}_8 \oplus \mathrm{D}_8$ and $\frac{1}{2}\sum_{i=1}^{16} \varepsilon_i$, where the first (resp. second) D_8 is the orthogonal complement modulo 2 of the vector of $2\varepsilon_1$ (resp. $2\varepsilon_9$) in the first (resp. second) E_8. Since we have $\mathrm{q}(2\varepsilon_1+2\varepsilon_9)=4$, the lattice $\mathrm{vois}_2(\mathrm{E}_8\oplus\mathrm{E}_8,[2\varepsilon_1+2\varepsilon_9])$ is generated by $\mathrm{M}_2(\mathrm{E}_8\oplus\mathrm{E}_8,[2\varepsilon_1+2\varepsilon_9])$ and $\varepsilon_1+\varepsilon_9$ (recall the algorithm $u \rightsquigarrow \mathrm{vois}_d(L;u)$). But the lattice generated by $\mathrm{D}_8\oplus\mathrm{D}_8$ and $\varepsilon_1+\varepsilon_9$ is D_{16}, so that the lattice $\mathrm{vois}_2(\mathrm{E}_8\oplus\mathrm{E}_8,[2\varepsilon_1+2\varepsilon_9])$ coincides with the lattice generated by D_{16} and $\frac{1}{2}\sum_{i=1}^{16}\varepsilon_i$, that is, E_{16}.

Variant. To illustrate Scholium 3.1.8 and, in doing so, reassure ourselves, we now consider the 2-neighbors of E_{16}.

The group $\mathrm{O}(\mathrm{E}_{16})$ can be identified with the subgroup of $\mathrm{O}(\mathrm{I}_{16})$ consisting of the elements that preserve the class of the Wu vector $\sum_{i=1}^{16}\varepsilon_i$ (in the sense of the discussion following Scholium 2.2.3). We therefore have a canonical isomorphism $\mathrm{O}(\mathrm{E}_{16})\cong\mathfrak{S}_{16}\ltimes(\{\pm1\}^{16})^0$, where $(\{\pm1\}^{16})^0$ is the subgroup of $\{\pm1\}^{16}$ consisting of the 16-tuples $(\eta_1,\eta_2,\ldots,\eta_{16})$ with $\eta_1\eta_2\ldots\eta_{16}=1$.

The action of $\mathrm{O}(\mathrm{E}_{16})$ partitions the quadric $\mathrm{C}_{\mathrm{E}_{16}}(\mathbb{F}_2)$ into four orbits:

- the orbit of the point $[2\varepsilon_1]$, with a single element (the vector $2\varepsilon_1$ belongs to the lattice E_{16} and satisfies $\mathrm{q}(2\varepsilon_1)=2$);
- the orbit of the point $[\sum_{i=1}^{4}\varepsilon_i]$, with $2\binom{16}{4}=3640$ elements (note that we have $\mathrm{q}(\sum_{i=1}^{4}\varepsilon_i)=2$);
- the orbit of the point $[\sum_{i=1}^{8}\varepsilon_i]$, with $\binom{16}{8}=12870$ elements (note that we have $\mathrm{q}(\sum_{i=1}^{8}\varepsilon_i)=4$);
- the orbit of the point $[\frac{1}{2}\sum_{i=1}^{16}\varepsilon_i]$, with $2^{14}=16384$ elements (note that we have $\mathrm{q}(\frac{1}{2}\sum_{i=1}^{16}\varepsilon_i)=2$).

Again by part (a) of Proposition 3.1.14, the lattice $\mathrm{vois}_2(\mathrm{E}_{16};c)$ is isomorphic (as a q-module) to E_{16} for every c that is not in the third orbit.

We conclude as before: since the graph of the 2-neighbors is connected, we must have $\mathrm{N}_2(\mathrm{E}_{16},\mathrm{E}_8\oplus\mathrm{E}_8)=12870$.

3.3.2 Determination of T_3 *for* $n=16$

As before, we embed E_{16} into $\mathbb{Q}^{16}$ in the usual way. We see that the action of $\mathrm{O}(\mathrm{E}_{16})$ partitions the quadric $\mathrm{C}_{\mathrm{E}_{16}}(\mathbb{F}_3)$ into five orbits, namely the orbits of the classes of the following vectors in E_{16} (that, in fact, belong to D_{16}):

$$u_1=2\varepsilon_1+\varepsilon_2+\varepsilon_3\ ,$$
$$u_2=\varepsilon_1+\varepsilon_2+\ldots+\varepsilon_6\ ,$$
$$u_3=2\varepsilon_1+\varepsilon_2+\varepsilon_3+\ldots+\varepsilon_9\ ,$$
$$u_4=\varepsilon_1+\varepsilon_2+\ldots+\varepsilon_{12}\ ,$$
$$u_5=2\varepsilon_1+\varepsilon_2+\varepsilon_3+\ldots+\varepsilon_{15}\ .$$

Note that we have $(\mathrm{q}(u_i))_{i=1,2,\ldots,5}=(3,3,6,6,9)$. By part (a) of Proposition 3.1.14, the lattice $\mathrm{vois}_3(\mathrm{E}_{16},[u_i])$ is isomorphic (as a q-module) to E_{16} for

$i = 1, 2$. The cardinality of the orbit of $[u_i]$ is $\binom{16}{3i} 2^{3i-1}$. This number is not divisible by 286 for $i = 4, 5$; by Scholium 3.1.8, the lattice $\mathrm{vois}_3(\mathrm{E}_{16}; [u_i])$ is also isomorphic (as a q-module) to E_{16} for $i = 4, 5$. Indeed, we have

$$\frac{|\mathrm{O}(\mathrm{E}_8 \oplus \mathrm{E}_8)|}{|\mathrm{O}(\mathrm{E}_{16})|} = \frac{405}{286}$$

with 405 and 286 relatively prime; Scholium 3.1.8 shows that if the lattice $\mathrm{vois}_3(\mathrm{E}_{16}; [u_i])$ is isomorphic to $\mathrm{E}_8 \oplus \mathrm{E}_8$, then the number of elements of the orbit of $[u_i]$ is divisible by 286.

Since the graph of the 3-neighbors is connected, the lattice $\mathrm{vois}_3(\mathrm{E}_{16}; [u_3])$ must be isomorphic (as a q-module) to $\mathrm{E}_8 \oplus \mathrm{E}_8$ (this is confirmed by the program `PARI`) and we have $\mathrm{N}_3(\mathrm{E}_{16}, \mathrm{E}_8 \oplus \mathrm{E}_8) = \binom{16}{9} 2^8 = 2928640$. In view of Scholium 3.1.7 and Proposition 3.2.2, this equality determines the Hecke operator $\mathrm{T}_3 \colon \mathbb{Z}[\mathrm{X}_{16}] \to \mathbb{Z}[\mathrm{X}_{16}]$. Its matrix in the basis $(\mathrm{E}_{16}, \mathrm{E}_8 \oplus \mathrm{E}_8)$, which we also denote by T_3, is

$$\mathrm{T}_3 = \begin{bmatrix} 4248000 & 4147200 \\ 2928640 & 3029440 \end{bmatrix}.$$

3.3.3 Determination of T_2 *for* $n = 24$ *(Following Nebe–Venkov [156])*

Let $n > 0$ be an integer divisible by 8.

In Sect. 3.1 we explained, following Borcherds, why the set $\mathrm{Y}_n(2)$ of isomorphism classes of ordered pairs (L_1, L_2) with L_1 an even unimodular lattice of dimension n and L_2 a 2-neighbor of L_1 in $\mathbb{Q} \otimes_{\mathbb{Z}} L_1$ is in natural bijection with the set $\widetilde{\mathrm{B}}_n$ of isomorphism classes of pairs $(L; \omega)$ with L an odd unimodular lattice of dimension n and ω a bijection from the set consisting of the two classes of the Wu vectors of L to $\{1, 2\}$. Recall that we use B_n to denote the set of isomorphism classes of odd unimodular lattices L of dimension n. We, moreover, denote by B_n^1 the subset of B_n consisting of the classes $[L]$ where L represents 1 (in other words, such that there exists an e in L with $e.e = 1$) and by B_n^2 its complement $\mathrm{B}_n - \mathrm{B}_n^1$. In [29], Borcherds uses the bijection $\mathrm{Y}_{24}(2) \cong \widetilde{\mathrm{B}}_{24}$ to determine B_{24}. In [68, Chap. 17], he lists the 156 elements b of B_{24}^2 explicitly and for each of these b, he gives sufficient information to determine $|\mathrm{O}(b)|$. He limits himself to B_{24}^2 because a lattice L that represents 1 is isomorphic to an orthogonal sum $\mathrm{I}_1 \oplus L'$ and the unimodular lattices that do not represent 1 and have dimension strictly less than 23 have already been listed (see [68, Chap. 16, Table 16.7], B_{24}^1 has 117 elements). Following Nebe and Venkov, we note that if an odd unimodular lattice L of dimension n represents 1, then the two even unimodular lattices that are 2-neighbors of L are isomorphic by Corollary 3.1.16. Let $\mathrm{B}_n^{2,0}$ be the subset of B_n^2 consisting of the isomorphism classes of the odd unimodular lattices L of dimension n such that the two even unimodular lattices L_1 and L_2 that are 2-neighbors of L are not isomorphic. Let e be the map from

$\mathrm{B}_n^{2,0}$ to the set of unordered pairs of elements of X_n that sends $[L]$ to $\{[L_1],[L_2]\}$. Proposition 3.1.11 specializes as follows.

Proposition 3.3.3.1. *Let x_1 and x_2 be two distinct elements of X_n. We have*

$$\mathrm{N}_2(x_1,x_2) \;=\; \sum_{b\,\in\,\mathrm{e}^{-1}(\{x_1,x_2\})} \frac{|\mathrm{O}(x_1)|}{|\mathrm{O}(b)|}\,.$$

Nebe and Venkov determine T_2 using the statement above and Borcherds' table, taking into account Proposition 3.2.2. Note that our conventions lead to the matrix $(24,24)$ of [156, page 59] being the transpose of our T_2.

Remarks.

- For $n=8$, Proposition 3.1.11 gives the relation

$$\frac{|\mathrm{O}(\mathrm{E}_8)|}{|\mathrm{O}(\mathrm{I}_8)|} \;=\; \frac{\mathrm{c}_8(2)}{2}\,.$$

More generally, for every n divisible by 8, Proposition 3.1.11 leads to the following relation between mass formulas:

$$\sum_{b\in\mathrm{B}_n}\frac{1}{|\mathrm{O}(b)|} \;=\; \frac{\mathrm{c}_n(2)}{2}\sum_{x\in\mathrm{X}_n}\frac{1}{|\mathrm{O}(x)|} \qquad (*)$$

(see [68, Chap. 16, Sect. 2]). Let us explain why. Proposition 3.1.11 says that we have

$$\mathrm{N}_2(x,y)\,\frac{1}{|\mathrm{O}(x)|} \;=\; \sum_{\mathrm{d}_1(\beta)=x\;,\;\mathrm{d}_2(\beta)=y}\frac{1}{|\mathrm{O}(\beta)|}$$

for all x and y in X_n; taking the sum over y and then over x, we obtain

$$\sum_{\beta\in\widetilde{\mathrm{B}}_n}\frac{1}{|\mathrm{O}(\beta)|} \;=\; \mathrm{c}_n(2)\sum_{x\in\mathrm{X}_n}\frac{1}{|\mathrm{O}(x)|}\,. \qquad (**)$$

Let $\mathrm{p}\colon \widetilde{\mathrm{B}}_n \to \mathrm{B}_n$ be the obvious map; the equality

$$\sum_{\beta\in\mathrm{p}^{-1}(b)}\frac{1}{|\mathrm{O}(\beta)|} \;=\; \frac{2}{|\mathrm{O}(b)|}$$

shows that the relations (*) and (**) are equivalent.

- Our analysis, in Sect. 3.3.1, of the 2-neighborhoods between even unimodular lattices of dimension 16 leads to the statement below, which we will use in Appendix A. As in Sect. 3.3.1, we identify, in the usual way, the lattices E_{16}, $\mathrm{E}_8\oplus\mathrm{E}_8$, and $\mathrm{D}_8\oplus\mathrm{D}_8$ with lattices in $\mathbb{Q}^{16}$; we denote the canonical basis of $\mathbb{Q}^{16}$ by $(\varepsilon_1,\varepsilon_2,\ldots,\varepsilon_{16})$.

Scholium-Definition 3.3.3.2. *The lattice in* $\mathbb{Q}^{16}$ *generated by*

$$\mathrm{D}_8 \oplus \mathrm{D}_8\,, \quad \frac{1}{2}\sum_{i=1}^{16} \varepsilon_i\,, \quad -\varepsilon_1 - \varepsilon_9 + \frac{1}{2}\sum_{i=1}^{8} \varepsilon_i$$

is an odd unimodular lattice, which we denote by Bor_{16}*; its even* 2*-neighbors are* E_{16} *and* $\mathrm{E}_8 \oplus \mathrm{E}_8$. *Up to isomorphism, the lattice* Bor_{16} *is the only odd unimodular lattice of dimension* 16 *that does not represent* 1. *The root system* $\mathrm{R}(\mathrm{Bor}_{16})$ *is isomorphic to* $\mathbf{D}_8 \coprod \mathbf{D}_8$.

Proof. Let B be an odd unimodular lattice of dimension 16 with isomorphic even 2-neighbors; the analysis mentioned above and Proposition 3.1.17 show that B represents 1. Now, let B be an odd unimodular lattice of dimension 16 that does not represent 1; by the above, one of the even 2-neighbors of B is isomorphic to E_{16} and the other is isomorphic to $\mathrm{E}_8 \oplus \mathrm{E}_8$. Let M be the submodule of B consisting of the elements x with $x.x$ even; by our analysis, M is isomorphic (as a $\widetilde{\mathrm{q}}$-module) both to the submodule of E_{16} that is the orthogonal complement modulo 2 of $\sum_{i=1}^{8} \varepsilon_i$ and to the submodule of $\mathrm{E}_8 \oplus \mathrm{E}_8$ that is the orthogonal complement modulo 2 of $2\varepsilon_1 + 2\varepsilon_9$. Seen as lattices in $\mathbb{Q}^{16}$, these two orthogonal complements modulo 2 coincide with the lattice generated by $\mathrm{D}_8 \oplus \mathrm{D}_8$ and $\frac{1}{2}\sum_{i=1}^{16} \varepsilon_i$, which we denote by M_{16}. Indeed, M_{16} has index 2 in both E_{16} and $\mathrm{E}_8 \oplus \mathrm{E}_8$, and we have $(\sum_{i=1}^{8} \varepsilon_i).x \equiv 0 \bmod 2$ (resp. $(2\varepsilon_1 + 2\varepsilon_9).x \equiv 0 \bmod 2$) for every x in M_{16}. Incidentally, this shows that we have $\mathrm{M}_{16} = \mathrm{E}_{16} \cap (\mathrm{E}_8 \oplus \mathrm{E}_8)$. Let ξ be the element $-\varepsilon_1 - \varepsilon_9 + \frac{1}{2}\sum_{i=1}^{8} \varepsilon_i$ of $\mathbb{Q}^{16}$; we easily see that it belongs to $\mathrm{M}_{16}^{\sharp}$ and that we have $\xi.\xi = 3$. It follows that ξ generates the "nonquadratically isotropic" line of $\mathrm{res}\, \mathrm{M}_{16}$, so that Bor_{16} is the odd unimodular lattice corresponding to M_{16} by Borcherds' theory. According to this theory, $M \simeq \mathrm{M}_{16}$ implies $B \simeq \mathrm{Bor}_{16}$ (note that Bor_{16} does not represent 1 because its even 2-neighbors are not isomorphic). The last assertion of the observation can be proved using Proposition 2.3.7 (we have $\mathrm{R}(\mathrm{M}_{16}) = \mathrm{R}(\mathrm{E}_{16}) \cap \mathrm{R}(\mathrm{E}_8 \oplus \mathrm{E}_8) \simeq \mathbf{D}_8 \coprod \mathbf{D}_8$ and the image of the class of ξ in $\mathrm{res}\, \mathrm{M}_{16}$ by the function qm must be $3/2$ because we cannot have $x.x = 1$ for x in Bor_{16}). □

3.4 d-Neighborhoods Between a Niemeier Lattice with Roots and the Leech Lattice

The justification for this section is the following.

Let $d \geq 2$ be an integer; by determining the integers $\mathrm{N}_d(L, \mathrm{Leech})$ for every Niemeier lattice with roots L, we also obtain the Hecke operator T_d (in the expression $\mathrm{N}_d(L, \mathrm{Leech})$, "Leech" is an abbreviation for "Leech lattice" that we will often use).

Let us explain why. In Sect. 3.3, we saw that, thanks to Nebe–Venkov, we know the Hecke operator $\mathrm{T}_2 \colon \mathbb{Z}[\mathrm{X}_{24}] \to \mathbb{Z}[\mathrm{X}_{24}]$ explicitly; we easily verify (thanks, PARI) that the elements $\mathrm{T}_2^k[\mathrm{Leech}]$ for $0 \leq k \leq 23$ are linearly independent. Since the

Hecke operators T_d and T_2 commute, by determining $\mathrm{T}_d\,[\mathrm{Leech}]$, we also obtain T_d. In view of Scholium 3.1.7 and Proposition 3.2.2, determining $\mathrm{T}_d\,[\mathrm{Leech}]$ is, in turn, equivalent to determining the integers $\mathrm{N}_d(x, [\mathrm{Leech}])$ for all x in $\mathrm{X}_{24} - \{[\mathrm{Leech}]\}$.

3.4.1 Necessary Conditions for a Niemeier Lattice with Roots to Have a d-Neighbor with No Roots

Proposition 3.4.1.1. *Let L be a Niemeier lattice with roots and $d \geq 2$ an integer. Let $\mathrm{h}(L)$ be the Coxeter number of L (see Proposition-Definition 2.3.3). If L has a d-neighbor with no roots, then the following inequality holds:*

$$d \;\geq\; \mathrm{h}(L)\;.$$

Proof. Suppose that there exists an element c of $\mathrm{C}_L(\mathbb{Z}/d)$ such that we have $\mathrm{R}(\mathrm{vois}_d(L;c)) = \emptyset$. A fortiori, we then have $\mathrm{R}(\mathrm{M}_d(L;c)) = \emptyset$ (recall that $\mathrm{M}_d(L;c)$ is the intersection in $\mathbb{Q} \otimes_{\mathbb{Z}} L$ of the lattices L and $\mathrm{vois}_d(L;c)$) or, equivalently, $\mathrm{R}(L) \cap \mathrm{M}_d(L;c) = \emptyset$. Let u be an element of L that represents c. The condition $\mathrm{R}(\mathrm{vois}_d(L;c)) = \emptyset$ therefore implies

$$\alpha.u \;\not\equiv\; 0 \quad (\mathrm{mod}\ d) \quad \text{for every } \alpha \text{ in } \mathrm{R}(L)\;.$$

We consequently obtain the inequality $d \geq \mathrm{h}(L)$ by applying Proposition 3.4.1.2 below with R an irreducible component of $\mathrm{R}(L)$ and f the linear form $x \mapsto x.u$. □

Proposition 3.4.1.2. *Let V be a finite-dimensional $\mathbb{R}$-vector space and $R \subset V$ an irreducible and reduced root system; let h be the Coxeter number of R. Let $f : V \to \mathbb{R}$ be a linear form whose restriction to R takes on integer values, and let $d \geq 2$ be an integer with $d < h$. Then there exists a root α in R such that we have*

$$f(\alpha) \equiv 0 \quad (\mathrm{mod}\ d)\;.$$

Proof. We fix a chamber C of the root system R; we denote the corresponding basis of R and highest root by $\{\alpha_1, \alpha_2, \ldots, \alpha_l\}$ and $\widetilde{\alpha}$, respectively. Recall that we have $\widetilde{\alpha} = n_1\alpha_1 + n_2\alpha_2 + \cdots + n_l\alpha_l$ with $n_i \in \mathbb{N} - \{0\}$ for $i = 1, 2, \ldots, l$ and

$$n_1 + n_2 + \cdots + n_l = h - 1 \tag{mh}$$

[39, Chap. VI, Sect. 1, Proposition 31] (here, mh stands for "maximal height"). Finally, we denote by $C^\vee$ the chamber of the dual root system $R^\vee$ determined by C (see [39, Chap. VI, Sect. 1, n°5]) and by Alc the alcove of V^* with $Alc \subset C^\vee$ and $0 \in \overline{Alc}$ [39, Chap. VI, Sect. 2, Proposition 4]. Thus, Alc (resp. $\overline{Alc}$) is the open (resp. closed) subset of V^* consisting of the elements ϕ satisfying the inequalities $\langle \alpha_i, \phi \rangle > 0$ (resp. $\langle \alpha_i, \phi \rangle \geq 0$) for $i = 1, 2, \ldots, l$ and $\langle \widetilde{\alpha}, \phi \rangle < 1$ (resp. $\langle \widetilde{\alpha}, \phi \rangle \leq 1$).

Let ϕ be an element of V^*. Since $\overline{Alc}$ is a fundamental domain for the action of the affine Weyl group on V^* (see, for example, [39, Chap. VI, Sect. 2, n°1 et n°2]), there exist an element w of the Weyl group of R and an element θ of the lattice $\mathrm{Q}(R^\vee)$ of V^* such that we have the inequalities

$$\langle w\alpha_i\,,\phi-\theta\rangle \geq 0 \quad \text{for } i=1,2,\ldots,l \quad \text{and} \quad \langle w\widetilde{\alpha}\,,\phi-\theta\rangle \leq 1\,.$$

We obtain a proof of the proposition by taking $\phi=(1/d)f$: there exist w and θ as above such that we have

$$f(w\alpha_i)-d\theta(w\alpha_i)\geq 0 \quad \text{and} \quad f(w\widetilde{\alpha})-d\theta(w\widetilde{\alpha})\leq d\,.$$

Note that the $w\alpha_i$ and $w\widetilde{\alpha}$ are roots and that the $f(w\alpha_i)$ and $\theta(w\alpha_i)$, and $f(w\widetilde{\alpha})$ and $\theta(w\widetilde{\alpha})$, are integers; we set $x_i=f(w\alpha_i)-d\theta(w\alpha_i)$ and $y=d-(f(w\widetilde{\alpha})-d\theta(w\widetilde{\alpha}))$. We then have $x_i\geq 0$, $y\geq 0$, and

$$n_1x_1+n_2x_2+\cdots+n_lx_l+y=d\,.$$

In view of the equality (mh), one of the integers x_1, x_2, ..., x_l, y must be zero, which proves the proposition. □

Remark. In fact, we use Proposition 3.4.1.2 only for irreducible root systems of type ADE, that is, for $R=\mathbf{A}_n$ $(n\geq 1)$, $R=\mathbf{D}_n$ $(n\geq 3)$, $R=\mathbf{E}_6$, $R=\mathbf{E}_7$, and $R=\mathbf{E}_8$. There exist elementary proofs of Proposition 3.4.1.2 in the first two cases. We treat the second one below; the treatment of the first case is similar (and in fact simpler).

We endow $\mathbb{R}^n$ with its canonical Euclidean structure; recall that we have $\mathbf{D}_n=\mathrm{R}(\mathrm{D}_n)$, where D_n is the submodule of $\mathbb{Z}^n$ consisting of the n-tuples $(x_1,x_2,\ldots,x_n)$ with $\sum_{i=1}^n x_i$ even. Let $f:\mathbb{R}^n\to\mathbb{R}$ be a linear form that takes on integer values on $\mathbf{D}_n$ or, equivalently, on D_n. Let $\varepsilon_1,\varepsilon_2,\ldots,\varepsilon_n$ be the canonical basis of $\mathbb{R}^n$. Since $2\varepsilon_i$ belongs to D_n, $f(2\varepsilon_i)$ is an integer; since $\varepsilon_i-\varepsilon_j$ belongs to D_n, the parity of this integer does not depend on the choice of the subscript i. We set $\lambda=0$ if $f(2\varepsilon_i)$ is even and $\lambda=1$ if $f(2\varepsilon_i)$ is odd; we set $\nu_i=f(\epsilon_i)-\lambda/2$ (the ν_i are thus integers). If we have $f(\alpha)\not\equiv 0 \pmod{d}$ for all α in $\mathbf{D}_n$, then the map from $\{1,2,\ldots,n\}$ to $\mathbb{Z}/d$ that sends i to the class of ν_i modulo d induces an injection from $\{1,2,\ldots,n\}$ to the quotient of $\mathbb{Z}/d$ by the involution $t\mapsto -t-\lambda$ (recall that $\mathbf{D}_n$ consists of the elements $\pm\varepsilon_i\pm\varepsilon_j$ for $i\neq j$). Since the cardinality of the quotient in question is bounded above by $d/2+1$, we have $n\leq d/2+1$ or, equivalently, $d\geq 2n-2$.

Notation Related to Niemeier Lattices with Roots

Since all statements in the remainder of Sect. 3.4 concern Niemeier lattices with roots, we first recall and complete the notation we use for these lattices.

Let L be a Niemeier lattice with roots:

- $R=\mathrm{R}(L)$ denotes the set of roots of L.
- We set $V=\mathbb{R}\otimes_{\mathbb{Z}}L$; this is a Euclidean space of dimension 24 and $R\subset V$ is an equi-Coxeter root system of type ADE of rank 24.

- $W = \mathrm{W}(L)$ denotes the Weyl group of the root system R. Recall that W can be identified with a subgroup of the orthogonal group $\mathrm{O}(L)$ (item (a) of Scholium 2.3.15).
- $Q = \mathrm{Q}(R) \subset L$ is the lattice in V generated by R.
- $Q^\sharp = (\mathrm{Q}(R))^\sharp \supset L$ is the dual lattice of Q.
- $f = \mathrm{f}(L)$ denotes the index of Q in $Q^\sharp$ (which Bourbaki calls the *index of connection* of the root system R).
- $g = \mathrm{g}(L)$ denotes the index of Q in L (since L/Q is a Lagrangian of the qe-module $Q^\sharp/Q$, we have $f = g^2$).

We denote by $R_1, R_2, \ldots, R_c$ the irreducible components of R:

- The number of irreducible components of R is therefore $c = \mathrm{c}(L)$.
- All these irreducible components have the same Coxeter number, namely $h = \mathrm{h}(L)$.

We choose a chamber C of R:

- $B \subset R$ denotes the basis of R corresponding to this choice.
- $R_+ \subset R$ is the set of positive roots for the order relation on V defined by C.
- $H \subset R_+ \subset R$ denotes the set of maximal elements of R for the order in question (the set H has c elements; more precisely, we have $H \cap R_k = \{\widetilde{\alpha}_k\}$, where $\widetilde{\alpha}_k$ denotes the highest root of R_k).

We denote by Alc the alcove of V in C that contains 0; Alc (resp. $\overline{Alc}$) is the open (resp. closed) subset of V consisting of the elements x satisfying the inequalities $\alpha.x > 0$ (resp. $\alpha.x \geq 0$) for $\alpha \in B$ and $\alpha.x < 1$ (resp. $\alpha.x \leq 1$) for $\alpha \in H$.

Finally, we set $\Pi := Q^\sharp \cap \overline{Alc}$. The subset Π of $Q^\sharp$ can be identified with the product of sets $\prod_{k=1}^{c} \Pi(R_i)$ (we introduced the notation $\Pi(-)$ in Proposition 2.3.8; the equality $\Pi(S) = \mathrm{Q}(S)^\sharp \cap \overline{Alc}$ for S an irreducible root system of type ADE was established in the proof of part (a) of the same proposition).

The proof of Proposition 3.4.1.2 leads to the following observation.

Scholium 3.4.1.3. *Let L be a Niemeier lattice with roots. Let ξ be an element of $Q^\sharp$ and $d \geq 1$ an integer. Then there exist $w \in W$ and $x \in Q$ such that the element $\eta := w\xi + dx$ of $Q^\sharp$ belongs to $d\,\overline{Alc}$, in other words, such that we have the inequalities $\alpha.\eta \geq 0$ for $\alpha \in B$ and $\alpha.\eta \leq d$ for $\alpha \in H$. Moreover, if we have $\alpha.\xi \not\equiv 0 \bmod d$ for every α in R, then the pair (w, x) is uniquely determined in terms of ξ.*

We say that an element x of V is *regular* if we have $\alpha.x \neq 0$ for every α in R (in other words, if x is in a chamber). Let $d \geq 1$ be an integer; we say that x is *d-regular* if we have $\alpha.x \notin d\mathbb{Z}$ for every α in R (in other words, if $(1/d)x$ is in an alcove). Let $d \geq 2$ be an integer; an element of $\mathrm{P}_L(\mathbb{Z}/d)$ is called *regular* if it is represented by a d-regular element u of L, that is, by an element satisfying $\alpha.u \not\equiv 0 \bmod d$ for every α in R (this condition does not depend on the choice of u). We denote the subset of $\mathrm{P}_L(\mathbb{Z}/d)$ consisting of such elements by $\mathrm{P}_L^{\mathrm{reg}}(\mathbb{Z}/d)$. Finally, we set $\mathrm{C}_L^{\mathrm{reg}}(\mathbb{Z}/d) := \mathrm{C}_L(\mathbb{Z}/d) \cap \mathrm{P}_L^{\mathrm{reg}}(\mathbb{Z}/d)$. The proof of Proposition 3.4.1.1 that we gave amounts in fact to verify the following more precise statement (where item (a) is obvious).

Scholium 3.4.1.4. *Let L be a Niemeier lattice with roots and $d \geq 2$ an integer.*

(a) *Let c be an element of $\mathrm{C}_L(\mathbb{Z}/d)$; if the lattice $\mathrm{vois}_d(L;c)$ has no roots, then c belongs to $\mathrm{C}_L^{\mathrm{reg}}(\mathbb{Z}/d)$.*
(b) *If the set $\mathrm{P}_L^{\mathrm{reg}}(\mathbb{Z}/d)$ is nonempty, then we have the inequality $d \geq \mathrm{h}(L)$.*

3.4.2 On the h-Neighborhoods and $(h+1)$-Neighborhoods Between a Niemeier Lattice with Roots and Coxeter Number h and the Leech Lattice

Let L be a Niemeier lattice with roots and Coxeter number h. In Sect. 3.4.1, we saw that a necessary condition for L to have the Leech lattice as a d-neighbor is the inequality $d \geq h$. Below we show, in particular, that this inequality is optimal; this is intimately linked to the *holy constructions* of the Leech lattice due to Conway and Sloane [67]. We also determine the integer $\mathrm{N}_d(L, \mathrm{Leech})$ for $d = h, h+1$; we will use this computation in Sect. 10.3.

We begin by recalling the definition of a Weyl vector of a Niemeier lattice and gathering some of the properties of these vectors that we will need.

Weyl Vectors

Let L be a Niemeier lattice with roots and $C \subset V$ a chamber of the root system R. Let ρ be the half-sum of the positive roots (for the order relation on V defined by C):

$$2\rho = \sum_{\alpha \in R_+} \alpha$$

(this equality shows that ρ belongs to $\frac{1}{2}L$; Proposition 3.4.2.1 below states that ρ in fact belongs to L). We call ρ a *Weyl vector* of the root system R or of the lattice L. Let α be a root of R; then α belongs to B if and only if we have $\rho.\alpha = 1$ (see [39, Chap. VI, Sect. 1, Proposition 29]). This observation shows that the map $C \mapsto \rho$ is bijective. It follows that the action of W on the set of Weyl vectors is simply transitive.

Proposition 3.4.2.1 (Borcherds). *Let L be a Niemeier lattice with roots and ρ a Weyl vector of L. Then ρ belongs to L.*

Proof. Before recalling the argument given by Borcherds in [29, 30], we state several results that will be useful further on.

Proposition 3.4.2.2. *Let L be a Niemeier lattice with roots.*

(a) *We have*

$$h\, x.y \;=\; \sum_{\alpha\in R_+} (\alpha.x)(\alpha.y)$$

for all x and y in V.

(b) *Let ρ be a Weyl vector of L; we have*

$$h\,\mathrm{q}(x) - \rho.x \;=\; \sum_{\alpha\in R_+} \frac{(\alpha.x)^2 - \alpha.x}{2}$$

for every x in V.

Proof. Part (a) is equivalent to Scholium 2.3.4. Part (b) (due to Borcherds) follows from part (a) and the definition of ρ. □

Statement (a) of Proposition 3.4.2.2 implies the following result.

Corollary 3.4.2.3. *Let L be a Niemeier lattice with roots. Then the quotient $Q^\sharp/Q$ is annihilated by h.*

Proof. Indeed, part (a) of Proposition 3.4.2.2 shows that if ξ and η are two elements of $Q^\sharp$, then $h\,\xi.\eta$ is integral and therefore $h\,\xi$ belongs to Q. □

Let ρ be a Weyl vector of L. Since the canonical action of W on $Q^\sharp/Q$ is trivial, the image of ρ in $Q^\sharp/Q$ does not depend on the choice of this Weyl vector. Here is another explanation of this phenomenon. We use the qe-module structure of $Q^\sharp/Q =: \mathrm{res}\, Q$. By Corollary 3.4.2.3, the map $Q^\sharp/Q \to \mathbb{Q}/\mathbb{Z}$ defined by $\xi \mapsto h\,\mathrm{q}(\xi)$ is linear (and with values in $\frac{1}{2}\mathbb{Z}/\mathbb{Z}$). Hence there exists an element σ of $Q^\sharp/Q$, uniquely determined (and annihilated by 2), such that we have $h\,\mathrm{q}(\xi) = \sigma.\xi$ for every ξ in $Q^\sharp/Q$. Part (b) of Proposition 3.4.2.2 shows that σ is the class of ρ. Indeed, that statement implies that $h\,\mathrm{q}(\xi) - \rho.\xi$ is integral for every ξ in $Q^\sharp$. We have obtained the following result.

Proposition 3.4.2.4. *Let L be a Niemeier lattice with roots and ρ a Weyl vector of L. Then the image of ρ in $Q^\sharp/Q$, which we denote by $\overline{\rho}$, is characterized by the property*

$$h\,\mathrm{q}(\xi) \;=\; \overline{\rho}.\xi$$

for every ξ in $Q^\sharp/Q$.

Proof of Proposition 3.4.2.1, *Continued.* Expressed using the formalism introduced above, Borcherds' argument is the following: Let I be a Lagrangian of the qe-module $Q^\sharp/Q$. The equality $\mathrm{q}(I) = 0$ implies $\overline{\rho} \in I^\perp = I$; by taking $I = L/Q$, we obtain $\rho \in L$. □

Proposition 3.4.2.5 (Venkov). *Let L be a Niemeier lattice with roots and ρ a Weyl vector of L. Then we have*

$$\mathrm{q}(\rho) \;=\; h(h+1)\,.$$

Proof. It suffices, for example, to observe that we have $\rho.\rho = (n/12)h(h+1)$ for every irreducible root system of type ADE of rank n and to invoke part (b) of Proposition-Definition 2.3.3. □

Proposition 3.4.2.6. *Let L be a Niemeier lattice with roots. For every element x of V, we have the inequality*

$$\inf_{\alpha \in R} (\alpha.x)^2 \leq \frac{\mathrm{q}(x)}{h(h+1)} .$$

Moreover, equality holds if and only if there exist a Weyl vector ρ of L and a real number $\lambda \geq 0$ such that we have $x = \lambda\rho$.

Proof. Let $C \subset V$ be a chamber of R such that x belongs to $\overline{C}$. Let B be the basis of R and ρ the Weyl vector associated with C. Let $\{\varpi_\alpha\}_{\alpha \in B}$ be the dual basis of B (with respect to the inner product). Consider the equality

$$x.x = \sum_{(\alpha,\beta) \in B \times B} (\alpha.x)(\beta.x)\ \varpi_\alpha.\varpi_\beta .$$

By using that we have $\alpha.x \geq 0$ and $\beta.x \geq 0$ (by definition of C), $\varpi_\alpha.\varpi_\beta \geq 0$ (see, for example, [39, Chap. VI, Sect. 1, Théorème 2, Remarque 2]), and $\rho.\rho = \sum_{(\alpha,\beta)} \varpi_\alpha.\varpi_\beta$ (a specialization of the equation above), we deduce the inequality

$$x.x \geq \rho.\rho \inf_{\alpha \in B} (\alpha.x)^2 . \qquad (*)$$

Set $\lambda = \inf_{\alpha \in B} \alpha.x$; inequality (*) can be refined to

$$x.x \geq \rho.\rho \inf_{\alpha \in B} (\alpha.x)^2 + \sum_{\alpha \in B} \left((\alpha.x)^2 - \lambda^2\right) \varpi_\alpha.\varpi_\alpha .$$

This shows that if equality holds in (*), we have $\alpha.x = \lambda$ for every α in B and therefore $x = \lambda\rho$. □

Scholium 3.4.2.7. *Let L be a Niemeier lattice with roots. For every element ξ of $\mathrm{Q}(R)^\sharp$ that is regular, that is, satisfies $\alpha.\xi \neq 0$ for every α in R, we have the inequality*

$$\mathrm{q}(\xi) \geq h(h+1) .$$

Moreover, equality holds if and only if ξ is a Weyl vector of L.

Part (a) of Proposition 3.4.2.8 below implies, in particular, that part (b) of Scholium 3.4.1.4 is "optimal."

Proposition 3.4.2.8. *Let L be a Niemeier lattice with roots and ρ a Weyl vector of L.*

(a) *The Weyl vector ρ is a primitive (a fortiori h-primitive) and h-regular element of L.*
(b) *Let ξ be an element of $Q^\sharp$. The following conditions are equivalent:*

(i) *The element* ξ *is* h*-regular.*
(ii) *There exist an element* w *of* W *and an element* x *of* Q *such that we have* $\xi = w\rho + hx$.

Moreover, if these conditions hold, then the pair (w, x) *from condition* (ii) *is uniquely determined in terms of* ξ.

Proof of Part (a). The equality $\alpha.\rho = 1$ for $\alpha \in B$ shows that ρ is primitive. Before proving that ρ is h-regular, let us recall the definition and some properties of the *height function*, which we denote by $\mathrm{H}\colon R_+ \to \mathbb{N} - \{0\}$. Let β be an element of R_+. We can write β as $\sum_{\alpha \in B} n_\alpha \alpha$ with n_α in $\mathbb{N}$ [39, Chap. VI, Sect. 1, Théorème 3]; we set $\mathrm{H}(\beta) := \sum_{\alpha \in B} n_\alpha$. The function H has the following properties (the notation H below is one in the list preceding Scholium 3.4.1.3):

- $\mathrm{H}(\beta) = \beta.\rho$;
- $\mathrm{H}(\beta) \geq 1$, and $\mathrm{H}(\beta) = 1 \iff \beta \in B$;
- $\mathrm{H}(\beta) \leq h - 1$, and $\mathrm{H}(\beta) = h - 1 \iff \beta \in H$.

The last property follows from [39, Chap. VI, Sect. 1, Proposition 31] (we have already invoked this reference in the proof of Proposition 3.4.1.2) and the very definition of the subset $H \subset R_+$. The equality $R = R_+ \coprod -R_+$ and the inequalities $1 \leq \beta.\rho \leq h - 1$ for every β in R_+ show that we have $\alpha.\rho \not\equiv 0 \bmod h$ for every α in R. □

Proof of Part (b). The implication (ii) ⇒ (i) follows from the fact that ρ is h-regular. Let us prove (i) ⇒ (ii). In view of Scholium 3.4.1.3, we may assume $\xi \in h\,\overline{Alc}$, that is,

- $\alpha.\xi \geq 0$ for every α in B;
- $\widetilde{\alpha}.\xi \leq h$ for every $\widetilde{\alpha}$ in H.

The first inequality shows that if ξ is h-regular, then we have $\alpha.\xi \geq 1$ for every α in B (by definition, $\beta.\xi$ is in $\mathbb{Z}$ for every β in R) or, equivalently, $\alpha.(\xi - \rho) \geq 0$ for every α in B. Likewise, the second inequality shows that we have $\widetilde{\alpha}.\xi \leq h - 1$ for every $\widetilde{\alpha}$ in H or, equivalently, $\widetilde{\alpha}.(\xi - \rho) \leq 0$ for every $\widetilde{\alpha}$ in H. But an element η of V that satisfies $\alpha.\eta \geq 0$ for every α in B and $\widetilde{\alpha}.\eta \leq 0$ for every $\widetilde{\alpha}$ in H is zero (for example because it belongs to $\epsilon\,\overline{Alc}$ for every $\epsilon > 0$). We therefore have $\xi = \rho$. The last part of statement (b), on the uniqueness of the pair (w, x) given by condition (ii), follows from the proof we just gave. □

We now arrive at the last statement concerning Weyl vectors of Niemeier lattices that we wish to highlight; parts (b) and (c) are again due to Borcherds.

Proposition 3.4.2.9. *Let* L *be a Niemeier lattice with roots,* ρ *a Weyl vector of* L*, and* ξ *an element of* $Q^\sharp$.

(a) *The element* $\rho - h\xi$ *of* $Q^\sharp$ *belongs to* L *and is* h*-regular.*
(b) *We have the inequality*

$$\mathrm{q}(\rho - h\xi) \;\geq\; \mathrm{q}(\rho) \;=\; h(h+1)\,.$$

(c) *The following conditions are equivalent:*

(i) *Equality holds in part* (b).
(ii) *The vector* ξ *belongs to* Π.
(iii) *The difference* $\rho - h\xi$ *is a Weyl vector of* L.

Proof. The element $\rho - h\xi$ belongs to L by Proposition 3.4.2.1 and Corollary 3.4.2.3. It is h-regular by part (a) of Proposition 3.4.2.8. It is a fortiori regular, so that the inequality of statement (b) can be seen as a consequence of Scholium 3.4.2.7. However, Borcherds' argument [29, 30], which uses part (b) of Proposition 3.4.2.2, is more effective for treating the case of equality. Indeed, we have $\mathrm{q}(\rho - h\xi) - \mathrm{q}(\rho) = h(h\mathrm{q}(\xi) - \rho.\xi)$. Since we have $t^2 - t \geq 0$ for every t in $\mathbb{Z}$, the right-hand side of the equality in the statement in question, with $x = \xi$, is nonnegative and is zero if and only if we have $\alpha.\xi \in \{0, 1\}$ for every α in R_+. This last property characterizes the elements of Π (see the proof of part (a) of Proposition 2.3.8). This proves the equivalence (i) $\Longleftrightarrow$ (ii) of part (c). The equivalence (i) $\Longleftrightarrow$ (iii) follows from Scholium 3.4.2.7 (the equality case). □

Holy Constructions

We now arrive at the main statement of Sect. 3.4.2. This statement deserves the name "Theorem" because of part (c), which is implicit in [67], at least as far as the lattice $\mathrm{vois}_h(L;\rho)$ is concerned.

Theorem 3.4.2.10. *Let* L *be a Niemeier lattice with roots and* ρ *a Weyl vector of* L. *We denote by* s_ρ *the orthogonal reflection of* $\mathbb{Q} \otimes_{\mathbb{Z}} L$ *with respect to the hyperplane orthogonal to* ρ.

(a) *The class of* ρ *in the projective space* $\mathrm{P}_L(\mathbb{Z}/h)$ *(resp.* $\mathrm{P}_L(\mathbb{Z}/(h+1))$*) belongs to the quadric* $\mathrm{C}_L(\mathbb{Z}/h)$ *(resp.* $\mathrm{C}_L(\mathbb{Z}/(h+1))$*).*
(b) *The lattices* $\mathrm{vois}_h(L;\rho)$ *and* $\mathrm{vois}_{h+1}(L;\rho)$ *are interchanged by the reflection* s_ρ*:*

$$\mathrm{vois}_{h+1}(L;\rho) = \mathrm{s}_\rho(\mathrm{vois}_h(L;\rho)) .$$

(c) *The lattices* $\mathrm{vois}_h(L;\rho)$ *and* $\mathrm{vois}_{h+1}(L;\rho)$ *have no roots:*

$$\mathrm{vois}_h(L;\rho) \simeq \mathrm{Leech} , \quad \mathrm{vois}_{h+1}(L;\rho) \simeq \mathrm{Leech} .$$

(d) *We have*

$$\mathrm{N}_h(L, \mathrm{Leech}) = \frac{|W|}{\varphi(h)\,g} , \quad \mathrm{N}_{h+1}(L, \mathrm{Leech}) = \frac{|W|}{\varphi(h+1)}$$

(where $\varphi(-)$ *above denotes the Euler totient function of a positive integer and* $|-|$ *denotes the cardinality of a finite set; recall that* g *denotes the index of* Q *in* L *or, equivalently, the square root of the index of connection* f *of* R*).*

Proof of Parts (a) *and* (b). Part (a) follows from Proposition 3.4.2.5, and part (b) is a particular case of part (b) of Proposition 3.1.14. □

Proof of Part (c). In view of part (b), it suffices to prove that $\mathrm{vois}_h(L;\rho)$ or $\mathrm{vois}_{h+1}(L;\rho)$ has no roots. We propose three proofs.

(1) The first proof is very prosaic. The program PARI allows us to compute, without difficulty, the *minimum* $\mathrm{m}(\Lambda)$ of an integral lattice Λ of dimension 24, namely the integer $\inf_{x\in\Lambda-\{0\}} x.x$. We thus verify that we have $\mathrm{m}(\mathrm{vois}_h(L;\rho)) = 4$ for the 23 Niemeier lattices with roots.

(2) The second proof consists in identifying the lattice $\mathrm{vois}_h(L;\rho)$ with the construction of the Leech lattice given by Conway and Sloane in [67], where, as mentioned before, this identification is implicit. The construction in question, which Conway and Sloane call a *holy construction*, is recalled below.

Conway and Sloane first associate two finite subsets of L with $(L;\rho)$:

- The first, which we denote by F, is the disjoint union $B\coprod -H$; recall that the elements of H are the highest roots $\widetilde{\alpha}_1, \widetilde{\alpha}_2, \ldots, \widetilde{\alpha}_c$ of the irreducible components $R_1, R_2, \ldots, R_c$ of R. To alleviate the notation, we denote the elements of B by $\alpha_1, \alpha_2, \ldots, \alpha_{24}$ and set $\alpha_{24+i} = -\widetilde{\alpha}_i$; the elements of F (which are all roots of R) correspond to the vertices of the *extended Dynkin graph* of R.
- The second subset, which we denote by G, consists of the minuscule weights of R that are in $L\cap\overline{C}$ (for the definition of minuscule weights, see the remark following Proposition 2.3.12). We therefore have $G = L\cap\Pi$, and the canonical map from G to L/Q is a bijection. We denote the elements of G by $\mu_0, \mu_1, \ldots, \mu_{g-1}$, where $\mu_0 = 0$.

Conway and Sloane then consider the lattice $\mathrm{HC}(L;\rho)$ in $\mathbb{Q}\otimes_{\mathbb{Z}} L$ consisting of the elements of the form

$$\sum_{i=1}^{24+c} m_i\alpha_i + \sum_{j=0}^{g-1} n_j\left(\frac{\rho}{h} - \mu_j\right),$$

where the m_i and n_j denote integers satisfying $\sum_i m_i + \sum_j n_j = 0$.

The following proposition shows that the lattice $\mathrm{HC}(L;\rho)$ can be described in terms of Kneser neighborhoods.

Proposition 3.4.2.11. *Let L be a Niemeier lattice with roots and ρ a Weyl vector of L. The lattice* $\mathrm{HC}(L;\rho)$ *of Conway and Sloane coincides with the lattice* $\mathrm{vois}_h(L;\rho)$.

Proof. This is rather a verification than a proof, which is why we use a smaller font. We have seen (see the discussion following Proposition 3.1.5) that the lattice $\mathrm{vois}_h(L;\rho)$ is the sub-$\mathbb{Z}$-module of $\mathbb{Q}\otimes_{\mathbb{Z}} L$ generated by M and $\widetilde{\rho}/h$, where M denotes the kernel of the homomorphism from L to $\mathbb{Z}/h$ that sends an element x of L to the class mod h of the integer $\rho.x$ and $\widetilde{\rho}$ denotes an element of L with $\widetilde{\rho} \equiv \rho \bmod h$ and $\mathrm{q}(\widetilde{\rho}) \equiv 0 \bmod h^2$. We can take $\widetilde{\rho} = \rho - h\alpha_1$, because we have $\mathrm{q}(\rho - h\alpha_1) = h(h+1) - h + h^2 = 2h^2$.

Having recalled the above, we observe that the $\alpha_1 - \alpha_i$ belong to M; indeed, we have $\rho.\alpha_i = 1$ for $i \leq 24$ and $\rho.\alpha_i = 1 - h$ for $i > 24$. Moreover, the μ_j belong to M; indeed, part (c) of

Proposition 3.4.2.9 shows that we have $\mathrm{q}(\rho - h\mu_j) = \mathrm{q}(\rho)$, an equality that is equivalent to $\rho.\mu_j = h\,\mathrm{q}(\mu_j)$.

Now, let x be an element of $\mathbb{Q} \otimes_{\mathbb{Z}} L$ of the form considered by Conway and Sloane. The observations above show that x can also be written as $y + (\sum_j n_j)\widetilde{\rho}/h$ with y in M; in other words, in $\mathbb{Q} \otimes_{\mathbb{Z}} L$, we have the inclusion $\mathrm{HC}(L;\rho) \subset \mathrm{vois}_h(L;\rho)$. The opposite inclusion $\mathrm{vois}_h(L;\rho) \subset \mathrm{HC}(L;\rho)$ follows from the fact that L is generated by $\mathrm{Q}(R)$ and the μ_j. Let us give a few more details. From the fact in question, we deduce that M is generated by the μ_j, the $\alpha_1 - \alpha_i$, and $h\alpha_1$. We deduce that the μ_j belong to $\mathrm{HC}(L;\rho)$ by considering the equality $\mu_j = (\rho/h - \mu_0) - (\rho/h - \mu_j)$. It is clear that the same holds for the $\alpha_1 - \alpha_i$. Suppose $\alpha_1 \in R_1$; we deduce that $h\alpha_1$ belongs to $\mathrm{HC}(L;\rho)$ by observing that we have $\widetilde{\alpha}_1 = \sum_{\beta \in B \cap R_1} m_\beta\, \beta$, where the m_β are integers with $\sum_{\beta \in B \cap R_1} m_\beta = h - 1$. Finally, we can deduce that $\widetilde{\rho}/h$ belongs to $\mathrm{HC}(L;\rho)$ by writing $\widetilde{\rho}/h = -\alpha_1 + (\rho/h - \mu_0)$. □

In [67], Conway and Sloane state that they have verified, case by case, that $\mathrm{HC}(L;\rho)$ has no roots... so that the proof of part (c) of Theorem 3.4.2.10 using [67] greatly resembles the proof we first gave. We must, however, note that shortly after the publication of [67], Borcherds [29, 30] discovered a uniform proof in terms of "Lorentzian" lattices.

(3) The third proof of part (c) that we propose systematically uses the theory of Kneser neighbors. It is a proof by contradiction: we begin by supposing $\mathrm{R}(\mathrm{vois}_h(L;\rho)) \neq \emptyset$.

Let us begin with an ad hoc statement concerning the general theory of d-neighbors.

Lemma 3.4.2.12. *Let L be a* q*-module over $\mathbb{Z}$. Let $d \geq 2$ be an integer and u a d-primitive element of L with $\mathrm{q}(u) \equiv 0 \bmod d$. Let x be an element of $\mathrm{vois}_d(L;u)$. Then $\mathrm{q}(u)x$ belongs to $\mathrm{M}_d(L;u)$ and in L, we have the following congruence:*

$$\mathrm{q}(u)x \;\equiv\; (u.x)\,u \mod d$$

(note that u belongs to $\mathrm{M}_d(L;u) \subset \mathrm{vois}_d(L;u)$, so that $u.x$ is integral and $(u.x)\,u$ also belongs to $\mathrm{M}_d(L;u)$).

Proof. We set $M = \mathrm{M}_d(L;u)$ and $L' = \mathrm{vois}_d(L;u)$. The element $\mathrm{q}(u)x$ belongs to M because the quotient L'/M is cyclic of order d and $\mathrm{q}(u)$ is divisible by d. Let v be an element of L with $u.v \equiv 1 \bmod d$; since L' is generated by M and $(u - \mathrm{q}(u)v)/d$, it suffices to verify the congruence of the lemma for $x \in M$ and for $x = (u - \mathrm{q}(u)v)/d$. In the first case, the two elements are divisible by d; the second case is obvious. □

Scholium 3.4.2.13. *Let L be a Niemeier lattice with roots and ρ a Weyl vector of L. Let x be an element of $\mathrm{vois}_h(L;\rho)$. Then $\rho.x$ is integral, hx and $(h+1)\mathrm{s}_\rho(x)$ belong to $\mathrm{M}_h(L;\rho)$ and $\mathrm{M}_{h+1}(L;\rho)$, respectively, and in L we have the following congruences:*

$$hx \;\equiv\; (\rho.x)\,\rho \mod h\,, \quad (h+1)\mathrm{s}_\rho(x) \;\equiv\; (\rho.x)\,\rho \mod (h+1)\,.$$

Proof. This follows from Lemma 3.4.2.12 and the equalities $\mathrm{q}(\rho) = h(h+1)$, $\mathrm{vois}_{h+1}(L;\rho) = \mathrm{s}_\rho(\mathrm{vois}_h(L;\rho))$, and $\rho.\mathrm{s}_\rho(x) = -\rho.x$. □

Let us now analyze the constraints that the condition $\mathrm{R}(\mathrm{vois}_h(L;\rho)) \neq \emptyset$ imposes.

Proposition 3.4.2.14. *Let L be a Niemeier lattice with roots, ρ a Weyl vector of L, and α' a root of the lattice* $\mathrm{vois}_h(L;\rho)$.

(a.1) *The integer $\rho.\alpha'$ is not divisible by h.*
(a.2) *The integers $\rho.\alpha'$ and h are not relatively prime.*
(b.1) *The integer $\rho.\alpha'$ is not divisible by $h+1$.*
(b.2) *The integers $\rho.\alpha'$ and $h+1$ are not relatively prime.*

Proof. We set $L' = \mathrm{vois}_h(L;\rho)$ and $M = \mathrm{M}_h(L;\rho)$; recall that we have $M = L \cap L'$.

(a.1). Scholium 3.4.2.13 shows that if $\rho.\alpha'$ is divisible by h, then α' belongs to L, which is impossible because we have $\mathrm{R}(M) = \emptyset$.

(a.2). Scholium 3.4.2.13 shows that if the integer $\rho.\alpha'$ is relatively prime to h, then $h\alpha'$ is an h-primitive element of L and the class of this element in the quadric $\mathrm{C}_L(\mathbb{Z}/h)$ is equal to that of ρ. This is impossible because of the equality $\mathrm{q}(h\alpha') = h^2$ and Scholium 3.4.2.7.

(b.1) and (b.2). First, note that if α' is a root of L', then $\mathrm{s}_\rho(\alpha')$ is a root of $\mathrm{s}_\rho(L') = \mathrm{vois}_{h+1}(L;\rho)$ and that we have $\rho.\mathrm{s}_\rho(\alpha') = -\rho.\alpha'$. We then proceed as before. For the proof of part (b.2), we use Propositions 3.4.2.15 and 3.4.2.16 and Corollary 3.4.2.17 below. Proposition 3.4.2.15 is the counterpart of Proposition 3.4.2.8, Proposition 3.4.2.16 is similar to statement (b) of Proposition 3.4.2.9, and Corollary 3.4.2.17 is similar to Scholium 3.4.2.7. □

Proposition 3.4.2.15. *Let L be a Niemeier lattice with roots and ρ a Weyl vector of L.*

(a) *The Weyl vector ρ is $(h+1)$-regular.*
(b) *Let ξ be an element of $Q^\sharp$. The following conditions are equivalent:*

  (i) *The element ξ is $(h+1)$-regular.*
  (ii) *There exist an element w of W, an element ϖ of Π, and an element x of Q such that we have $\xi = w(\rho+\varpi) + (h+1)x$.*

  Moreover, if these conditions hold, then the triple (w,ϖ,x) from condition (ii) *is uniquely determined in terms of ξ.*

(c) *Let u be an element of L. The following conditions are equivalent:*

  (i) *The element u is $(h+1)$-regular.*
  (ii) *There exist an element w of W and an element x of L such that we have $u = w\rho + (h+1)x$.*

  Moreover, if these conditions hold, then the pair (w,x) from condition (ii) *is uniquely determined in terms of u.*

Proof of Parts (a) *and* (b). These are variants of parts (a) and (b) of Proposition 3.4.2.8. Let us prove, for example, the implication (i) $\Rightarrow$ (ii) of part (b). In view of Scholium 3.4.1.3, we may assume $\xi \in (h+1)\overline{Alc}$ (again, the last part of statement (b), on the uniqueness of the triple (w, ϖ, x), will be a consequence of our proof). If ξ is $(h+1)$-regular, then we now have the inequalities $\alpha.(\xi - \rho) \geq 0$ for every α in B and $\widetilde{\alpha}.(\xi - \rho) \leq 1$ for every $\widetilde{\alpha}$ in H. We therefore have $\xi - \rho \in Q^\sharp \cap \overline{Alc} =: \Pi$. □

Proof of Part (c). The implication (ii) ⇒ (i) follows from the fact that ρ is $(h+1)$-regular. The implication (i) ⇒ (ii) follows from the implication (i) ⇒ (ii) of part (b). Indeed, if u is $(h+1)$-regular, then it can be written uniquely as $u = w_0(\rho + \varpi) + (h+1)x_0$ with $(w_0, \varpi, x_0) \in W \times \Pi \times Q$. Since u is in L, the same holds for ϖ, that is, we have $\varpi \in \Pi \cap L$. We write $u = w_0(\rho - h\varpi) + (h+1)(w_0\varpi + x_0)$. By part (c) of Proposition 3.4.2.9, $\rho - h\varpi$ is a Weyl vector of L; hence there exists an element w_1 of W, uniquely determined in terms of ϖ, such that we have $\rho - h\varpi = w_1\rho$. By setting $w = w_0 w_1$ and $x = w_0\varpi + x_0$, we indeed have $u = w\rho + (h+1)x$. The uniqueness of the pair (w, x) follows from the fact that the canonical map $\Pi \cap L \to L/Q$ is a bijection. □

Proposition 3.4.2.16. *Let x be a nonzero element of L; then we have the inequality*

$$\mathrm{q}(\rho - (h+1)x) \;\geq\; (h+1)(h+2)\,.$$

Proof. We adapt the argument of Borcherds used in the proof of Proposition 3.4.2.9. Let x be an element of L; we see that we now have

$$\mathrm{q}(\rho - (h+1)x) \;=\; (h+1)\left(h + \mathrm{q}(x) + \sum_{\alpha \in R_+} \frac{(\alpha.x)^2 - \alpha.x}{2}\right).$$

To estimate the right-hand side of this equality, we distinguish between two cases:

(1) There exists a positive root α with $\alpha.x \notin \{0, 1\}$. In this case, we have $(\alpha.x)^2 - \alpha.x \geq 2$ and therefore $\mathrm{q}(x) + \sum_{\alpha \in R_+} ((\alpha.x)^2 - \alpha.x)/2 \geq 2$.

(2) We have $\alpha.x \in \{0, 1\}$ for every α in R_+. If x is nonzero, then we have $\mathrm{q}(x) \geq 2$. Indeed, x cannot be a root because x or $-x$ would then be a positive root, say β, with $\beta.x \notin \{0, 1\}$. □

Corollary 3.4.2.17. *Let u be an element of L. If u is $(h+1)$-primitive and we have $\mathrm{q}(u) = (h+1)^2$, then u is not $(h+1)$-regular.*

Proof. By part (c) of Proposition 3.4.2.15, if such a u is $(h+1)$-regular, there exist an element w of the Weyl group of R and an element x of L such that we have $u = w\rho + (h+1)x$. Propositions 3.4.2.5 and 3.4.2.16 show that we then have either $\mathrm{q}(u) = h(h+1)$ or $\mathrm{q}(u) \geq (h+1)(h+2)$. □

Proposition 3.4.2.14 shows that the lattice $\mathrm{vois}_h(L; \rho)$ has no roots if h or $h+1$ is prime. This is the case for 19 of the Niemeier lattices with no roots. The four that resist correspond to $h = 25, 14, 9, 8$. To overcome this problem, we refine the previous argument.

Proposition 3.4.2.18. *Let L be a Niemeier lattice with roots. We assume that the lattice* $\mathrm{vois}_h(L;\rho)$ *also has roots, and we denote its Coxeter number by h'. There exists an integer ν satisfying the following conditions:*

(1) $\nu > 0$;
(2) $\nu^2 \leq \big(h(h+1)\big)\big(h'(h'+1)\big)^{-1}$;
(3) $\gcd(\nu, h) \neq 1$ *and* $\gcd(\nu, h+1) \neq 1$.

Proof. Set $L' = \mathrm{vois}_h(L;\rho)$. We apply Proposition 3.4.2.6 to the lattice L' by taking the element ρ for x: there exists a root α' of L' such that we have $(\rho.\alpha')^2 \leq \big(h(h+1)\big)\big(h'(h'+1)\big)^{-1}$. Set $\nu = |\rho.\alpha'|$. By the above, ν is an integer that satisfies the three given conditions. □

We denote by $\mathrm{S}(h,h')$ the subset of $\mathbb{Z}$ consisting of the integers ν satisfying the three conditions of Proposition 3.4.2.18. We clearly have $\mathrm{S}(h,h'_1) \subset \mathrm{S}(h,h'_2)$ for $h'_1 \geq h'_2$. Note that $\mathrm{S}(h,2)$ is empty for $h \neq 25$. The lattice $\mathrm{vois}_h(L;\rho)$ therefore has no roots for $h \neq 25$.

We are left with the case $h = 25$. We have $\mathrm{S}(25,2) = \{10\}$ and $\mathrm{S}(25,3) = \emptyset$. By the second equality, we have $h' = 2$, and therefore $\mathrm{R}(L') = 24\,\mathbf{A}_1$. Consider the equality (Scholium 2.3.4 and Proposition 3.4.2.5)

$$\sum_{\beta \in R'_+} (\rho.\beta)^2 \;=\; 2600\;,$$

where R'_+ denotes the set of 24 elements consisting of the positive roots of L' for some choice of a chamber. This equality shows that we cannot have $|\rho.\beta| = 10$ for every β in R'_+. Proposition 3.4.2.14 shows that there exists a β_1 in R'_+ with $|\rho.\beta_1| \geq 20$. Consequently, there exists a β_2 in R'_+ with $|\rho.\beta_2| \leq 9$ (note that we have $23 \times 10^2 > 2600 - 20^2$). This contradicts Proposition 3.4.2.14.

This contradiction completes our third proof of statement (c) of Theorem 3.4.2.10. □

Proof of Part (d) *of Theorem* 3.4.2.10. Proposition 3.4.2.8 shows that the class of ρ in $\mathrm{P}_L(\mathbb{Z}/h)$ belongs to $\mathrm{P}_L^{\mathrm{reg}}(\mathbb{Z}/h)$ and that the action of W on this set is transitive. Since the class of ρ belongs to $\mathrm{C}_L^{\mathrm{reg}}(\mathbb{Z}/h)$, we see that we have $\mathrm{C}_L^{\mathrm{reg}}(\mathbb{Z}/h) = \mathrm{P}_L^{\mathrm{reg}}(\mathbb{Z}/h)$. Likewise, Proposition 3.4.2.15 shows that the class of ρ in $\mathrm{P}_L(\mathbb{Z}/(h+1))$ belongs to $\mathrm{P}_L^{\mathrm{reg}}(\mathbb{Z}/(h+1))$, that the action of W on this set is transitive (by part (c)), and that we have $\mathrm{C}_L^{\mathrm{reg}}(\mathbb{Z}/(h+1)) = \mathrm{P}_L^{\mathrm{reg}}(\mathbb{Z}/(h+1))$. Since a necessary condition for $\mathrm{vois}_d(L;c)$ to be isomorphic to the Leech lattice is that c belongs to $\mathrm{C}_L^{\mathrm{reg}}(\mathbb{Z}/d)$ (Scholium 3.4.1.4 (a)), the proof of part (d) of Theorem 3.4.2.10 consists in verifying that the stabilizer of the class of ρ in $\mathrm{P}_L(\mathbb{Z}/h)$ (resp. $\mathrm{P}_L(\mathbb{Z}/(h+1))$) for the action of W has $\phi(h)g$ (resp. $\phi(h+1)$) elements. This follows from Propositions 3.4.2.19 and 3.4.2.20 below. □

Proposition 3.4.2.19. *Let L be a Niemeier lattice with roots and ρ a Weyl vector of L. Then the stabilizer of the class of ρ in $\mathrm{P}_L(\mathbb{Z}/h)$ for the action of W is a canonical extension of $(\mathbb{Z}/h)^\times$ by L/Q.*

Proof. Let $S \subset W$ be the stabilizer in question and w an element of S. By definition, we have $w\rho = \lambda\rho + hx$ with λ in $\mathbb{Z}$ relatively prime to h and x in L. We see that the class $\overline{\lambda}$ of λ in $(\mathbb{Z}/h)^{\times}$ depends only on w and that the map $w \mapsto \overline{\lambda}$ is a group homomorphism; we denote it by $\pi\colon S \to (\mathbb{Z}/h)^{\times}$. The implication (i) $\Rightarrow$ (ii) of Proposition 3.4.2.8 (b) shows that π is surjective because $\lambda\rho$ is h-regular for every λ in $\mathbb{Z}$ relatively prime to h. Next, we consider the subgroup $\ker\pi$. Let w be in $\ker\pi$; we have $w\rho = \rho + hx$ with x in L. The map that sends w to the class of x in L/Q is a group homomorphism, which we denote by $\iota\colon\ \ker\pi \to L/Q$ (we use that the action of W on $Q^{\sharp}/Q$ is trivial). The same argument as above shows that ι is surjective. The uniqueness in Proposition 3.4.2.8 (b) shows that ι is injective. □

Remarks.

- By Corollary 3.4.2.3, the $\mathbb{Z}$-module L/Q is a $\mathbb{Z}/h$-module, so that we have a natural action of $(\mathbb{Z}/h)^{\times}$ on L/Q. This action coincides with that defined by the extension in Proposition 3.4.2.19. If ρ belongs to Q (which is not always the case; see Proposition 3.4.2.4), then the proof of Proposition 3.4.2.19 shows, implicitly, that S is canonically isomorphic to the semi-direct product $L/Q \rtimes (\mathbb{Z}/h)^{\times}$. We can, in fact, verify that the extension in question is always trivial.
- By construction, the homomorphism ι that appears in the proof of Proposition 3.4.2.19 factors through a set-theoretic map from $\ker\pi$ to L. This shows that we have a set-theoretic section of the homomorphism $L \to L/Q$ (which depends only on the choice of ρ). The image of this section is $\Pi \cap L$, and this second remark is intimately linked to the beginning of the proof we gave of item (a) of Proposition 2.3.8.

Proposition 3.4.2.20. *Let L be a Niemeier lattice with roots and ρ a Weyl vector of L. Then the stabilizer of the class of ρ in $\mathrm{P}_L(\mathbb{Z}/(h+1))$ for the action of W is canonically isomorphic to $(\mathbb{Z}/(h+1))^{\times}$.*

Proof. This proof is analogous to that of Proposition 3.4.2.19, where Proposition 3.4.2.15 replaces Proposition 3.4.2.8. It is in fact simpler. This time, the homomorphism $\pi : S \to (\mathbb{Z}/(h+1))^{\times}$ is an isomorphism. The reason for this simplification is the following: in Proposition 3.4.2.8 (b), the element x belongs to Q, whereas in Proposition 3.4.2.15 (c), it belongs to L. □

3.4.3 On the Stabilizers for the Action of W on $\mathrm{P}_L^{\mathrm{reg}}(\mathbb{Z}/d)$, for L a Niemeier Lattice with Roots

In this subsection, we give an upper bound for the size of the stabilizers for the action of W on $\mathrm{P}_L^{\mathrm{reg}}(\mathbb{Z}/d)$, for L a Niemeier lattice with roots. Our motivation is Scholium-Definition 3.4.3.3, which will prove to be useful in Sect. 10.3.1.

Let L be a Niemeier lattice with roots and $d \geq 2$ an integer.

From here on, we assume that d is relatively prime to the index g of Q in L. In this case, the canonical homomorphism $Q/dQ \to L/dL$ is an isomorphism. We introduce, mutatis mutandis, the notation P_Q, C_Q (this time, this scheme is only smooth over $\mathbb{Z}[1/g]$), $\mathrm{P}_Q(\mathbb{Z}/d)$, $\mathrm{P}_Q(\mathbb{Z}/d)$, $\mathrm{P}_Q^{\mathrm{reg}}(\mathbb{Z}/d)$, and $\mathrm{C}_Q^{\mathrm{reg}}(\mathbb{Z}/d)$, as well as the corresponding terminology. It is clear that the canonical bijections $\mathrm{P}_Q(\mathbb{Z}/d) \cong \mathrm{P}_L(\mathbb{Z}/d)$, $\mathrm{C}_Q(\mathbb{Z}/d) \cong \mathrm{C}_L(\mathbb{Z}/d)$, $\mathrm{P}_Q^{\mathrm{reg}}(\mathbb{Z}/d) \cong \mathrm{P}_L^{\mathrm{reg}}(\mathbb{Z}/d)$, and $\mathrm{C}_Q^{\mathrm{reg}}(\mathbb{Z}/d) \cong \mathrm{C}_L^{\mathrm{reg}}(\mathbb{Z}/d)$ are W-equivariant.

Proposition 3.4.3.1. *Let L be a Niemeier lattice with roots and $d \geq 2$ an integer relatively prime to the index of Q in L. Let S be the stabilizer of an element of $\mathrm{P}_L^{\mathrm{reg}}(\mathbb{Z}/d)$ for the action of W.*

(a) *The group S is canonically isomorphic to a subgroup of $(\mathbb{Z}/d)^\times$.*
(b) *If d is prime, then the action of S on R (induced by that of W) is free.*

Proof of Part (a). In view of what we wrote earlier, we may replace $\mathrm{P}_L(\mathbb{Z}/d)$ by $\mathrm{P}_Q(\mathbb{Z}/d)$. Let u be a d-primitive element of Q, let $S \subset W$ be the stabilizer of the class of u in $\mathrm{P}_Q(\mathbb{Z}/d)$, and let w be an element of S. We proceed as in the proof of Proposition 3.4.2.19. By definition, we have $wu = \lambda u + dx$ with λ in $\mathbb{Z}$ relatively prime to d and x in Q. The class $\overline{\lambda}$ of λ in $(\mathbb{Z}/d)^\times$ depends only on w, and the map $\pi\colon S \to (\mathbb{Z}/d)^\times$ defined by $w \mapsto \overline{\lambda}$ is a group homomorphism. Using Scholium 3.4.1.3, we easily verify that π is injective if u is d-regular. □

Proof of Part (b). The equality $wu = \lambda u + dx$ implies $\alpha.(wu) \equiv \lambda(\alpha.u) \bmod d$ for every α in R or, equivalently, $(w^{-1}\alpha).u \equiv \lambda(\alpha.u) \bmod d$. If we have $w\alpha = \alpha$ (and therefore $w^{-1}\alpha = \alpha$), then we have $(\lambda - 1)(\alpha.u) \equiv 0 \bmod d$ or, equivalently, $(\alpha.u)(\pi(w) - 1) = 0$ in $\mathbb{Z}/d$. If u is d-regular and d is prime, we obtain $\pi(w) = 1$. Since π is injective when u is d-regular, we indeed have the implication $w\alpha = \alpha \Rightarrow w = \mathrm{id}$. □

Remark. Proposition 3.4.2.20 is an illustration of Proposition 3.4.3.1 (a); Proposition 3.4.2.19 shows that the hypothesis on d is necessary.

Corollary-Definition 3.4.3.2. *Let L be a Niemeier lattice with roots and p a prime; we denote by $\mathrm{D}_p(L)$ the gcd of the integers $p-1$, $24h$, and $|W|$. If p does not divide the index of Q in L, then the stabilizer of an element of $\mathrm{P}_L^{\mathrm{reg}}(\mathbb{F}_p)$ for the action of W is canonically isomorphic to a subgroup of the group $\mu_{\mathrm{D}_p(L)}(\mathbb{F}_p)$ (which is cyclic of order $\mathrm{D}_p(L)$).*

Proof. Let $S \subset W$ be one of these stabilizers. By Proposition 3.4.3.1 (a), S can be identified with a subgroup of $\mathbb{F}_p^\times$. By Proposition 3.4.3.1 (b), the cardinality of S divides the cardinality of R, namely $24h$. □

Remark. We see that $24\mathrm{h}(L)$ divides $|\mathrm{W}(L)|$ except in the case $\mathrm{R}(L) = 24\mathbf{A}_1$, where the gcd of the integers $24\mathrm{h}(L)$ and $|\mathrm{W}(L)|$ is 16. We therefore have $\gcd(p-1, 24\mathrm{h}(L), |\mathrm{W}(L)|) = \gcd(p-1, 24\mathrm{h}(L))$ in all other cases.

Scholium-Definition 3.4.3.3. *Let L be a Niemeier lattice with roots and p a prime. We denote by* $\mathrm{pas}\,(L;p)$ *the integer defined by*

$$\mathrm{pas}\,(L;p) \;:=\; \frac{|\mathrm{W}(L)|}{\gcd\,(p-1, 24\mathrm{h}(L), |\mathrm{W}(L)|)} \,.$$

If p does not divide the index of Q in L, then $\mathrm{N}_p(L,\mathrm{Leech})$ *is divisible by* $\mathrm{pas}\,(L;p)$. *In this case, we denote by* $\mathrm{n}_p(L)$ *the integer defined by the equality*

$$\mathrm{N}_p(L,\mathrm{Leech}) \;=\; \mathrm{n}_p(L)\;\mathrm{pas}\,(L;p)\,.$$

The notation pas comes from the French word "pas" that refers, in this context, to the common difference in an arithmetic sequence.

Proof. The integer $\mathrm{N}_p(L,\mathrm{Leech})$ is the sum of the cardinalities of the W-orbits of the points c of $\mathrm{C}_L(\mathbb{F}_p)$ with $\mathrm{vois}_p(L;c)\simeq$ Leech; these points belong to $\mathrm{P}_L^{\mathrm{reg}}(\mathbb{F}_p)$ by Scholium 3.4.1.4 (a). □

Remark. The integer $\mathrm{pas}\,(L;p)$ is the product of the integer $\mathrm{pas}_1(L)$ (which does not depend on p) and the integer $\mathrm{pas}_2\,(L;p)$ defined, respectively, by

$$\mathrm{pas}_1(L) = \frac{|\mathrm{W}(L)|}{\gcd\,(24\mathrm{h}(L), |\mathrm{W}(L)|)}\,, \quad \mathrm{pas}_2(L;p) = \frac{24\,\mathrm{h}(L)}{\gcd\,(p-1, 24\mathrm{h}(L))}\,.$$

EXAMPLES

Let us illustrate the above by considering the Niemeier lattice A_{24}^+ associated with the root system $\mathbf{A}_{24}$ (see the second example following Scholium 2.3.15) and the prime numbers 29 and 31. The choice of this illustration is deliberate: we will use the computation of the integers $\mathrm{N}_{29}(\mathrm{A}_{24}^+,\mathrm{Leech})$ and $\mathrm{N}_{31}(\mathrm{A}_{24}^+,\mathrm{Leech})$ in Sect. 10.3.1.

Recall that by construction, we have $\mathrm{Q}(\mathbf{A}_{24}) = \mathrm{A}_{24}$, where A_{24} is the sub-$\mathbb{Z}$-module of $\mathbb{Z}^{25}$ consisting of the 25-tuples $(x_1, x_2, \ldots, x_{25})$ with $\sum_i x_i = 0$, endowed with the even bilinear form induced by the Euclidean structure of $\mathbb{R}^{25}$. The Weyl group W can be identified with the symmetric group $\mathfrak{S}_{25}$; its action on A_{24} is the obvious one. It follows that the $\mathbb{F}_p$-vector space $\mathbb{F}_p\otimes_{\mathbb{Z}}\mathrm{A}_{24}$ can be identified with the linear subspace of $\mathbb{F}_p^{25}$ consisting of the elements $(x_1, x_2, \ldots, x_{25})$ with $\sum_i x_i = 0$; the induced action of $\mathfrak{S}_{25}$ is again the obvious one.

We denote by $\widetilde{\mathrm{C}}_{\mathrm{A}_{24}}(\mathbb{F}_p)$ and $\widetilde{\mathrm{C}}^{\mathrm{reg}}_{\mathrm{A}_{24}}(\mathbb{F}_p)$, respectively, the inverse images of $\mathrm{C}_{\mathrm{A}_{24}}(\mathbb{F}_p)$ and $\mathrm{C}^{\mathrm{reg}}_{\mathrm{A}_{24}}(\mathbb{F}_p)$ in $\mathbb{F}_p\otimes_{\mathbb{Z}}\mathrm{A}_{24}-\{0\}$. For $p\neq 2$, we can identify $\widetilde{\mathrm{C}}_{\mathrm{A}_{24}}(\mathbb{F}_p)$, as a $(\mathfrak{S}_{25}\times\mathbb{F}_p^{\times})$-set, with the subset of $\mathbb{F}_p^{25}-\{(0,0,\ldots,0)\}$ consisting of the elements $(x_1, x_2, \ldots, x_{25})$ satisfying

$$x_1 + x_2 + \ldots + x_{25} = 0 \quad\text{and}\quad x_1^2 + x_2^2 + \ldots + x_{25}^2 = 0\,. \qquad (*)$$

By definition, the 25-tuple $(x_1, x_2, \ldots, x_{25})$ is a set-theoretic map from $\{1, 2, \ldots, 25\}$ to $\mathbb{F}_p$; we denote it by x. The map x belongs to $\widetilde{\mathrm{C}}^{\mathrm{reg}}_{\mathrm{A}_{24}}(\mathbb{F}_p)$ if and only if it is injective. Indeed, if we view the root system $\mathbf{A}_{24}$, as usual, as a subset of the Euclidean space $\mathbb{R}^{25}$ endowed with its canonical basis $\{\varepsilon_1, \varepsilon_2, \ldots, \varepsilon_{25}\}$ [39, Planche I], then the roots are the $\varepsilon_i - \varepsilon_j$ with $i \neq j$. Note, incidentally, that this observation allows us to deduce, at little cost, that if the set $\mathrm{P}^{\mathrm{reg}}_{\mathrm{A}^+_{24}}(\mathbb{F}_p)$ is nonempty, we necessarily have $p \geq 25$, at least for $p \neq 5$.

We now arrive at the two examples we had in mind.

(1) *The lattices* $\mathrm{vois}_{29}(\mathrm{A}^+_{24}; c)$ *for* c *in* $\mathrm{C}^{\mathrm{reg}}_{\mathrm{A}^+_{24}}(\mathbb{F}_{29})$

Let $x\colon \{1, 2, \ldots, 25\} \to \mathbb{F}_{29}$ be an injective map, and let $\{y_1, y_2, y_3, y_4\} \in \mathbb{F}_{29}$ be the complement of the image of x. Since we have $\sum_{z \in \mathbb{F}_{29}} z = 0$ and $\sum_{z \in \mathbb{F}_{29}} z^2 = 0$, the map x satisfies (*) if and only the four elements in the complement satisfy

$$y_1 + y_2 + y_3 + y_4 = 0 \quad \text{and} \quad y_1^2 + y_2^2 + y_3^2 + y_4^2 = 0\ .$$

We denote by $\mathcal{C}$ (resp. $\widetilde{\mathcal{C}}$) the subscheme of $\mathbf{P}^3$ (resp. $\mathbf{A}^4 - \{0\}$), say over $\mathbb{F}_{29}$, defined by the equations above; $\mathcal{C}$ is clearly isomorphic to $\mathbf{P}^1$. Finally, we denote by $\mathcal{C}^{\mathrm{reg}}$ (resp. $\widetilde{\mathcal{C}}^{\mathrm{reg}}$) the open subscheme of $\mathcal{C}$ (resp. $\widetilde{\mathcal{C}}$) defined by $y_i \neq y_j$ for $i \neq j$; since -2 is not a square in $\mathbb{F}_{29}$, we in fact have $\mathcal{C}(\mathbb{F}_{29}) = \mathcal{C}^{\mathrm{reg}}(\mathbb{F}_{29})$ and $\widetilde{\mathcal{C}}(\mathbb{F}_{29}) = \widetilde{\mathcal{C}}^{\mathrm{reg}}(\mathbb{F}_{29})$.

We have introduced the formalism above for the sake of the following statements:

- There is a canonical bijection between the set $\mathfrak{S}_{25} \backslash \widetilde{\mathrm{C}}^{\mathrm{reg}}_{\mathrm{A}_{24}}(\mathbb{F}_{29})$, quotient of the action of $\mathfrak{S}_{25}$ on $\widetilde{\mathrm{C}}^{\mathrm{reg}}_{\mathrm{A}_{24}}(\mathbb{F}_{29})$, and the set $\mathfrak{S}_4 \backslash \widetilde{\mathcal{C}}^{\mathrm{reg}}(\mathbb{F}_{29})$, quotient of the obvious action of $\mathfrak{S}_4$ on $\widetilde{\mathcal{C}}^{\mathrm{reg}}(\mathbb{F}_{29})$. Moreover, these two quotients are endowed with natural actions of $\mathbb{F}^\times_{29}$, and the bijection is equivariant.
- There is a canonical bijection

$$\kappa\colon \mathfrak{S}_4 \backslash \mathcal{C}^{\mathrm{reg}}(\mathbb{F}_{29}) \xrightarrow{\ \cong\ } \mathfrak{S}_{25} \backslash \mathrm{C}^{\mathrm{reg}}_{\mathrm{A}_{24}}(\mathbb{F}_{29})\ .$$

Moreover, for every $\mathfrak{S}_4$-orbit $\mathcal{O}$ of $\mathcal{C}^{\mathrm{reg}}(\mathbb{F}_{29})$, the stabilizers of $\mathcal{O}$ and $\kappa(\mathcal{O})$, which can both be identified with subgroups of $\mathbb{F}^\times_{29}$, are canonically isomorphic.

Finally, consider the action of $\mathfrak{S}_4$ on the set $\mathcal{C}(\mathbb{F}_{29}) = \mathcal{C}^{\mathrm{reg}}(\mathbb{F}_{29})$; it may be useful to note that the fact that -3 is not a square in $\mathbb{F}_{29}$ implies that every element (y_1, y_2, y_3, y_4) of $\widetilde{\mathcal{C}}(\mathbb{F}_{29})$ satisfies $y_i \neq 0$ for every i.

We observe that the action of the group $\mathfrak{S}_4$ on the set $\mathcal{C}(\mathbb{F}_{29})$ with 30 elements has exactly two orbits:

- the orbit $\mathcal{O}_1$ of the class of the point $(1, 12, -1, -12)$ of $\mathbb{F}^4_{29}$, whose stabilizer is isomorphic to $\mu_4(\mathbb{F}_{29})$ (note that $\{1, 12, -1, -12\} \subset \mathbb{F}^\times_{29}$ is the subgroup $\mu_4(\mathbb{F}_{29})$),
- the orbit $\mathcal{O}_2$ of the class of the point $(1, 4, 6, -11)$, which is free.

It follows that the action of the group $\mathfrak{S}_{25}$ on the set $\mathrm{C}^{\mathrm{reg}}_{\mathrm{A}_{24}}(\mathbb{F}_{29})$ has exactly two orbits, namely $\Omega_1 = \kappa(\mathcal{O}_1)$, whose stabilizer is isomorphic to $\mu_4(\mathbb{F}_{29})$, and $\Omega_2 = \kappa(\mathcal{O}_2)$, which is free. Note that this confirms Corollary-Definition 3.4.3.2 because we have $\mathrm{D}_{29}(\mathrm{A}_{24}^+) = 4$.

PARI tells us that the lattices $\mathrm{vois}_{29}(\mathrm{A}_{24}^+;\Omega_1)$ and $\mathrm{vois}_{29}(\mathrm{A}_{24}^+;\Omega_2)$ (the abuse of notation is venial) are both isomorphic to the Leech lattice. We finally obtain

$$\mathrm{N}_{29}(\mathrm{A}_{24}^+,\mathrm{Leech}) \;=\; \frac{5}{4}\,|\mathrm{W}(\mathbf{A}_{24})| \;=\; 19389012554163732480000000$$

(or, equivalently, $\mathrm{n}_{29}(\mathrm{A}_{24}^+) = 5$ in the notation introduced in Scholium-Definition 3.4.3.3).

We can, in fact, avoid turning to PARI by invoking the following ad hoc proposition.

Proposition 3.4.3.4. *Let L be a Niemeier lattice with roots, ρ a Weyl vector, and α a root of L. We denote the integer $2h+1-\rho.\alpha$ by d. Then:*

(1) *We have $d \geq h+2$.*
(2) *We have $\mathrm{q}(\rho - h\alpha) = hd$.*
(3) *There exists a β in B (that is, in the basis of $\mathrm{R}(L)$ determined by ρ) with $\alpha.\beta = 0$.*
(4) *The element $\rho - h\alpha$ of L is primitive.*
(5) *The lattice $\mathrm{vois}_d(L;\rho-h\alpha)$ (which is well defined by points* (2) *and* (4)) *is isomorphic to the lattice $\mathrm{vois}_h(L;\rho)$ (which is isomorphic to the Leech lattice by Theorem* 3.4.2.10 (c)).

Proof. Property (1) follows from the inequality $|\rho.\alpha| \leq h-1$; property (2) is immediate. Property (3) is obvious if the system of R is not irreducible; when R is irreducible, that is, $R = \mathbf{A}_{24}$ or $R = \mathbf{D}_{24}$, this trivially holds as well. Property (3) implies property (4): note that we have $(\rho - h\alpha).\beta = 1$. Finally, property (5) follows from Proposition 3.1.14 (b). □

To apply this proposition to the case we are interested in, namely $L = \mathrm{A}_{24}^+$ and $d = 29$, we must choose ρ and α with $\rho.\alpha = 22$. Following Bourbaki, we take $\rho = \sum_{i=1}^{25}(13-i)\varepsilon_i$; there are then three possible choices for α, namely $\alpha_i = \varepsilon_i - \varepsilon_{i+22}$ for $i = 1,2,3$. Let c_i be the class of $\rho - 25\alpha_i$ in $\mathrm{C}^{\mathrm{reg}}_{\mathrm{A}_{24}}(\mathbb{F}_{29})$ (c_i is necessarily 29-regular because we have $\mathrm{vois}_{29}(\mathrm{A}_{24}^+;\rho) \simeq \mathrm{Leech}$). We see that c_2 is in the orbit Ω_1 and c_1 and c_3 are in the orbit Ω_2 (let $w_0 \in \mathfrak{S}_{25}$ be the permutation $i \mapsto 26-i$; it fixes c_2 and interchanges c_1 and c_3).

(2) *The lattices* $\mathrm{vois}_{31}(\mathrm{A}_{24}^+;c)$ *for c in* $\mathrm{C}^{\mathrm{reg}}_{\mathrm{A}_{24}^+}(\mathbb{F}_{31})$

We can apply the same method as above to determine $\mathrm{N}_{31}(\mathrm{A}_{24}^+,\mathrm{Leech})$. This time, $\mathcal{C}$ is the projective quadric defined by the equations

$$y_1+y_2+y_3+y_4+y_5+y_6 = 0 \quad \text{and} \quad y_1^2+y_2^2+y_3^2+y_4^2+y_5^2+y_6^2 = 0\,.$$

The cardinalities of the sets $\mathcal{C}(\mathbb{F}_{31})$ and $\mathcal{C}^{\mathrm{reg}}(\mathbb{F}_{31})$ are 30784 and 18864, respectively, so that the volume of the computations is greater. We only give the result of these computations, and spare the reader the details.

The nontrivial stabilizers for the action of $\mathfrak{S}_6$ on $\mathcal{C}^{\mathrm{reg}}(\mathbb{F}_{31})$ are the subgroups $\mu_6(\mathbb{F}_{31})$, $\mu_5(\mathbb{F}_{31})$, $\mu_3(\mathbb{F}_{31})$, and $\mu_2(\mathbb{F}_{31})$ (this is consistent with the equality $\mathrm{D}_{31}(\mathrm{A}_{24}^{+}) = 30$).

- There exists a single orbit with stabilizer $\mu_6(\mathbb{F}_{31})$; the neighbor of A_{24}^{+} associated with it is Leech.
- There exists a single orbit with stabilizer $\mu_5(\mathbb{F}_{31})$; the neighbor associated with it is again Leech.
- There exist exactly four orbits with stabilizer $\mu_3(\mathbb{F}_{31})$; one leads to Leech, two to the Niemeier lattice with root system $24\mathbf{A}_1$, and one to the Niemeier lattice with root system $12\mathbf{A}_2$.
- There exists a single orbit with stabilizer $\mu_2(\mathbb{F}_{31})$; the associated neighbor is Leech.
- There exist exactly 24 free orbits; 8 lead to Leech, 15 to the Niemeier lattice with root system $24\mathbf{A}_1$, and a single one to the Niemeier lattice with root system $12\mathbf{A}_2$.

(Incidentally, the inventory we just made shows, in particular, that Scholium 3.4.1.4 (a) does not admit a converse.)

From the above, we deduce

$$\mathrm{N}_{31}(\mathrm{A}_{24}^{+}, \mathrm{Leech}) = \frac{46}{5}\,|\mathrm{W}(\mathbf{A}_{24})| = 142703132398645071052800000$$

(or, equivalently, $\mathrm{n}_{31}(\mathrm{A}_{24}^{+}) = 276$).

3.4.4 Complement: On the 2-Neighbors of a Niemeier Lattice with Roots, Associated with a Weyl Vector

This section follows [29]; in particular, the comments at the end are comparable to the arguments that Borcherds gives in this article to prove the a priori existence of an even unimodular lattice of dimension 24 with no roots.

Let L be a Niemeier lattice with roots and ρ one of its Weyl vectors. The equality $\mathrm{q}(\rho) = h(h+1)$ (Proposition 3.4.2.5) implies the congruence $\mathrm{q}(\rho) \equiv 0 \bmod 2$, so that we can consider the lattice $\mathrm{vois}_2(L;\rho)$ (ρ is primitive, hence a fortiori 2-primitive). We study this 2-neighbor of L below.

We extend the definition of the Coxeter number of a Niemeier lattice with roots (Proposition-Definition 2.3.3) to all Niemeier lattices, by agreeing that the Coxeter number of the Leech lattice is 0.

Proposition 3.4.4.1. *Let L be a Niemeier lattice with roots and ρ a Weyl vector of L. We have the inequality*

$$\mathrm{h}(\mathrm{vois}_2(L;\rho)) \;\leq\; \frac{\mathrm{h}(L)+1}{2}\,.$$

Proof. We set $h = \mathrm{h}(L)$ and $h' = \mathrm{h}(\mathrm{vois}_2(L;\rho))$.

We first suppose that h is even. Let $\widetilde{\rho}$ be an element of L with $\widetilde{\rho} \equiv \rho \bmod h$ and $\mathrm{q}(\widetilde{\rho}) \equiv 0 \bmod h^2$. While studying the neighbor algorithm, we saw that such an element exists; here, we can take $\widetilde{\rho} = \rho - h\alpha$, where α denotes a root of the basis of $\mathrm{R}(L)$ defined by ρ. Proposition 3.1.13 says that $\widetilde{\rho}/2$ is an $h/2$-primitive element of $\mathrm{vois}_2(L;\rho)$ and that in $\mathbb{Q} \otimes_{\mathbb{Z}} L$, we have

$$\mathrm{vois}_h(L;\rho) \;=\; \mathrm{vois}_{h/2}\Big(\mathrm{vois}_2(L;\rho);\frac{\widetilde{\rho}}{2}\Big)\,.$$

This equality shows that the lattice $\mathrm{vois}_2(L;\rho)$ admits an $h/2$-neighbor with no roots, because Theorem 3.4.2.10 (c) says that the lattice $\mathrm{vois}_h(L;\rho)$ has no roots. We deduce $h/2 \geq h'$ thanks to Proposition 3.4.1.1.

The case h odd is similar. We now consider the equality

$$\mathrm{vois}_{h+1}(L;\rho) \;=\; \mathrm{vois}_{(h+1)/2}\Big(\mathrm{vois}_2(L;\rho);\frac{\widetilde{\rho}}{2}\Big)$$

with, for example, $\widetilde{\rho} = \rho + (h+1)\alpha$. □

We have just seen that the proof of Proposition 3.4.4.1 relies on Proposition 3.4.1.1 and Theorem 3.4.2.10 (c) ("holy constructions"); let us now give a proof ab initio of the statement below that refines Proposition 3.4.4.1.

Proposition 3.4.4.2. *Let L be a Niemeier lattice with roots and ρ a Weyl vector of L. We have the equality*

$$\mathrm{h}(\mathrm{vois}_2(L;\rho)) \;=\; \frac{\mathrm{h}(L)}{2} - \frac{\iota(\mathrm{R}(L))}{8} + 2\,,$$

where $\iota(\mathrm{R}(L))$ denotes the number of odd exponents of the Weyl group of $\mathrm{R}(L)$ [39, Chap. V, Sect. 6, Définition 2].

Before explaining the proof of this proposition, let us give some information on the number $\iota(R)$ of odd exponents of the Weyl group of a root system R:

(1) The invariant ι is additive in the following sense: $\iota(R_1 \coprod R_2) = \iota(R_1) + \iota(R_2)$.
(2) The value of ι on the irreducible root systems of type ADE is as follows: $\iota(\mathbf{A}_l) = [(l+1)/2]$, $\iota(\mathbf{D}_l) = 2[l/2]$, $\iota(\mathbf{E}_6) = 4$, $\iota(\mathbf{E}_7) = 7$, $\iota(\mathbf{E}_8) = 8$.
(3) Let R be a root system of rank l and Coxeter number h. Since the set of exponents of R is stable under the involution $m \mapsto h-m$ and has cardinality l [39, Chap. V, Sect. 6, n°2], we have $l = 2\iota(R)$ if h is odd.

Let R be a root system of type ADE of rank l and Coxeter number h. Points (1) and (2) show that we have the inequality $l \geq 2\iota(R)$ and that we have equality if and only if all irreducible components of R are of type $\mathbf{A}_d$ with d even. This last condition is equivalent to h being odd, which is consistent with point (3).

The above shows that Proposition 3.4.4.2 indeed implies the inequality of Proposition 3.4.4.1 and that we have equality in the latter if and only if $\mathrm{R}(L)$ is isomorphic to the direct sum of $24/d$ copies of $\mathbf{A}_d$ with d an even divisor of 24 (recall that $\mathrm{R}(L)$ is equi-Coxeter). Incidentally, we observe that the following three conditions are equivalent:

- The Coxeter number $\mathrm{h}(L)$ is odd.
- The set of roots $\mathrm{R}(L)$ is isomorphic to the direct sum of $24/d$ copies of $\mathbf{A}_d$ with d an even divisor of 24.
- We have $\mathrm{h}(\mathrm{vois}_2(L;\rho)) = (\mathrm{h}(L)+1)/2$.

The proof of Proposition 3.4.4.2 is based on Propositions 3.4.4.3 and 3.4.4.4 below.

Proposition 3.4.4.3. *Let L be a Niemeier lattice with roots and ρ a Weyl vector of L. Then the number of roots α of L with $\rho.\alpha$ even is the difference* $12\,\mathrm{h}(L) - \iota(\mathrm{R}(L))$.

Proof. Let R be a root system, C a chamber of R, and $B \subset R$ the basis and $R_+ \subset R$ the subset of positive roots defined by C. The proof of Proposition 3.4.4.3 follows from the relation given by Bertram Kostant in [128] between the height function $\mathrm{R}_+ \to \mathbb{N} - 0$ and the exponents of the Weyl group of R. Let us recall that theory below.

The height function $\mathrm{H}\colon R_+ \to \mathbb{N} - \{0\}$ sends a positive root to the sum of its coordinates in the basis B. We denote the set of exponents of R by $\mathrm{Exp}(R)$.

Let A be an abelian group and $f\colon \mathbb{N} - \{0\} \to A$ a (set-theoretic) map. Let $F\colon \mathbb{N} - \{0\} \to A$ be the "primitive" of f, that is, the map defined by

$$F(m) \ = \ \sum_{k=1}^{m} f(k)\,.$$

Then, in A, we have the equality

$$\sum_{\alpha \in R_+} f(\mathrm{H}(\alpha)) \ = \ \sum_{m \in \mathrm{Exp}(R)} F(m)\,. \qquad \text{(Ko)}$$

This equality follows from Kostant's relation mentioned above. Let us explain why. Let i be an element of $\mathbb{N} - \{0\}$ and $\delta_{(i)}\colon \mathbb{N} - \{0\} \to \mathbb{Z}$ the corresponding "Dirac function." Equality (Ko) says that in the case $f = \delta_{(i)}$, the cardinality of $\mathrm{H}^{-1}(i)$ is equal to the cardinality of the subset of $\mathrm{Exp}(R)$ consisting of the m with $m \geq i$; this is Kostant's result. The general case follows by linearity.

Let us now return to the proof of Proposition 3.4.4.3. Let ν (resp. ν_+) be the number of roots (resp. positive roots, for the chamber associated with ρ) α of L with $\rho.\alpha$ even; it is clear that we have $\nu = 2\nu_+$. In the context of Proposition 3.4.4.3, we have $\mathrm{H}(\alpha) = \rho.\alpha$. By taking, in equality (Ko), f equal to the function $\mathbb{N} - \{0\} \to \mathbb{Z}$ defined by $k \mapsto (-1)^k$, we obtain

$$|\mathrm{R}_+(L;\rho)| - 2\,\nu_+ \;=\; \iota(\mathrm{R}(L))\ ,$$

where $\mathrm{R}_+(L;\rho)$ is the subset consisting of the positive roots for the chamber associated with ρ. This equality is equivalent to

$$\nu \;=\; \frac{|\mathrm{R}(L)|}{2} - \iota(\mathrm{R}(L))\ .$$

As we have $|\mathrm{R}(L)| = 24\,\mathrm{h}(L)$ by Proposition-Definition 2.3.3 (c), this concludes the proof. □

Let Λ be an integral lattice and $k \geq 0$ an integer. We denote by $\mathrm{r}_k(\Lambda)$ the number of representations of k by Λ, that is, the number of elements x of Λ with $x.x = k$.

Proposition 3.4.4.4 (Borcherds). *Let B be an odd unimodular lattice of dimension 24, and let L_1 and L_2 be the two even 2-neighbors of B. Then we have*

$$\mathrm{r}_2(L_1) + \mathrm{r}_2(L_2) \;=\; 3\,\mathrm{r}_2(B) \;-\; 24\,\mathrm{r}_1(B) \;+\; 48\ .$$

Proof (Sketch). Let $n > 0$ be an integer divisible by 8 and B a unimodular lattice of dimension n. We consider the theta series

$$\vartheta_B(\tau) \;=\; \sum_{x \in B} e^{\imath\pi\tau x.x}$$

(with τ in the upper half-plane). The function ϑ_B is a modular form of weight $n/2$ for the subgroup Γ' of $\Gamma := \mathrm{SL}_2(\mathbb{Z})/\{\pm\mathrm{I}\}$ generated by the transformations $\tau \mapsto \tau + 2$ and $\tau \mapsto -1/\tau$. It is, moreover, modular for Γ if B is even. We denote by $\mathrm{M}_{n/2}(\Gamma)$ and $\mathrm{M}_{n/2}(\Gamma')$, respectively, the $\mathbb{C}$-vector spaces consisting of the modular forms of weight $n/2$ for the groups Γ and Γ'. Since Γ' has finite index (namely 3) in Γ, there is a transfer homomorphism, which we denote by $\mathrm{tr}\colon \mathrm{M}_{n/2}(\Gamma') \to \mathrm{M}_{n/2}(\Gamma)$. The proof of the following statement is left to the reader.

Lemma 3.4.4.5. *Let $n > 0$ be an integer divisible by 8 and B an odd unimodular lattice of dimension n. Let L_1 and L_2 be the two even 2-neighbors of B. Then we have $\vartheta_{L_1} + \vartheta_{L_2} \;=\; \mathrm{tr}(\vartheta_B)$.*

We can verify that $\mathcal{B} := (\mathbb{E}_4^3\,, \mathbb{E}_4^2\,\vartheta_{\mathrm{I}_8}\,, \mathbb{E}_4\,\vartheta_{\mathrm{I}_{16}}\,, \Delta)$ is a basis of $\mathrm{M}_{12}(\Gamma')$ (recall that $\mathbb{E}_4$ is the normalized Eisenstein series that is modular of weight 4 for Γ, that we have $\vartheta_{\mathrm{E}_8} = \mathbb{E}_4$, that Δ is the unique normalized cusp form of weight 12 for Γ, and that $(\mathbb{E}_4^3\,, \Delta)$ is a basis of $\mathrm{M}_{12}(\Gamma)$).

Since $\mathbb{E}_4$ and Δ are modular for Γ, we have $\mathrm{tr}(\mathbb{E}_4^3) = 3\mathbb{E}_4^3$ and $\mathrm{tr}(\Delta) = 3\,\Delta$. On the other hand, Lemma 3.4.4.5 implies $\mathrm{tr}(\vartheta_{\mathrm{I}_8}) = 2\,\mathbb{E}_4$ and $\mathrm{tr}(\vartheta_{\mathrm{I}_{16}}) = 2\,\mathbb{E}_4^2$. Since the transfer is $\mathrm{M}(\Gamma)$-linear, where $\mathrm{M}(\Gamma)$ denotes the graded $\mathbb{C}$-algebra of modular forms for Γ, we conclude that the image of the basis $\mathcal{B}$ by the transfer homomorphism is $(3\,\mathbb{E}_4^3\,, 2\,\mathbb{E}_4^3\,, 2\,\mathbb{E}_4^3\,, 3\,\Delta)$.

Now, let B be an odd unimodular lattice of dimension 24. Let (c_0, c_1, c_2, c_3) be the coordinates of ϑ_B in the basis $\mathcal{B}$:

$$\vartheta_B = c_0\,\mathrm{E}_4^3 + c_1\,\mathrm{E}_4^2\,\vartheta_{\mathrm{I}_8} + c_2\,\mathrm{E}_4\,\vartheta_{\mathrm{I}_{16}} + c_3\,\Delta\,.$$

Since the constant term of the Fourier series expansion of ϑ_B is 1, we have $c_0 + c_1 + c_2 = 1$. Since B is odd, Lemma 3.4.4.5 implies, in particular, that the constant term of the Fourier series expansion of $\mathrm{tr}(\vartheta_B)$ is 2; we therefore have $3\,c_0 + 2\,c_1 + 2\,c_2 = 2$, and consequently $c_0 = 0$.

Let $\mathrm{M}_{12}^0(\Gamma')$ be the linear subspace of $\mathrm{M}_{12}(\Gamma')$ generated by $\mathrm{E}_4^2\,\vartheta_{\mathrm{I}_8}$, $\mathrm{E}_4\,\vartheta_{\mathrm{I}_{16}}$, and Δ. Let f be an element of $\mathrm{M}_{12}^0(\Gamma')$ and

$$f \;=\; \mathrm{r}_0(f) \;+\; \mathrm{r}_1(f)\,e^{\imath\pi\tau} \;+\; \mathrm{r}_2(f)\,e^{2\imath\pi\tau} \;+\; \ldots$$

the beginning of the Fourier series expansion. It is easy to check that the linear map $\mathrm{M}_{12}^0(\Gamma') \to \mathbb{C}^3$ defined by $f \mapsto (\mathrm{r}_0(f), \mathrm{r}_1(f), \mathrm{r}_2(f))$ is an isomorphism. It follows that the coefficient of $e^{2\imath\pi\tau}$ in the Fourier series expansion of $\mathrm{tr}(f)$ is a linear combination of $\mathrm{r}_0(f), \mathrm{r}_1(f), \mathrm{r}_2(f)$. By solving a linear system, we find that this coefficient is $48\,\mathrm{r}_0(f) - 24\,\mathrm{r}_1(f) + 3\,\mathrm{r}_2(f)$. In view of Lemma 3.4.4.5, we obtain Proposition 3.4.4.4 by taking $f = \vartheta_B$. □

Proof of Proposition 3.4.4.2 *Using Propositions* 3.4.4.3 *and* 3.4.4.4

Let B be the odd unimodular lattice of dimension 24 whose two even 2-neighbors are L and $\mathrm{vois}_2(L;\rho)$. As we have $\mathrm{r}_2(\Lambda) = 24\,\mathrm{h}(\Lambda)$ for every even unimodular lattice Λ of dimension 24 (part (c) of Proposition-Definition 2.3.3 for $\mathrm{r}_2(\Lambda) \neq 0$ and convention for $\mathrm{r}_2(\Lambda) = 0$), Proposition 3.4.4.4 gives

$$24\,\mathrm{h}(L) + 24\,\mathrm{h}(\mathrm{vois}_2(L;\rho)) \;=\; 3\,\mathrm{r}_2(B) \;-\; 24\,\mathrm{r}_1(B) \;+\; 48\,. \qquad (1)$$

By construction, the submodule of B consisting of the elements x with $x.x$ even is the lattice $\mathrm{M}_2(L;\rho)$; we therefore have $\mathrm{r}_2(B) = \mathrm{r}_2(\mathrm{M}_2(L;\rho))$. Again by construction, $\mathrm{r}_2(\mathrm{M}_2(L;\rho))$ is the number of roots α of L with $\rho.\alpha$ even; we therefore have

$$\mathrm{r}_2(\mathrm{M}_2(L;\rho)) \;=\; 12\,\mathrm{h}(L) - \iota(\mathrm{R}(L)) \qquad (2)$$

by Proposition 3.4.4.3. The equalities (1) and (2) imply

$$\mathrm{h}(\mathrm{vois}_2(L;\rho)) \;=\; \frac{\mathrm{h}(L)}{2} - \frac{\iota(\mathrm{R}(L))}{8} + 2 - \mathrm{r}_1(B)\,. \qquad (3)$$

It remains to show $\mathrm{r}_1(B) = 0$. We proceed by contradiction. If we have $\mathrm{r}_1(B) \neq 0$, then we have $\mathrm{r}_1(B) \geq 2$ and the equality (3) implies the inequality $\mathrm{h}(\mathrm{vois}_2(L;\rho)) < \mathrm{h}(L)$, which shows that the lattices L and $\mathrm{vois}_2(L;\rho)$ are not isomorphic. But Corollary 3.1.16 shows that if we have $\mathrm{r}_1(B) \neq 0$, then the lattices L and $\mathrm{vois}_2(L;\rho)$ are isomorphic. □

Comments

Let XR_{24} be the subset of X_{24} consisting of the isomorphism classes of even unimodular lattices of dimension 24 with roots. We denote by YR_{24} the set of isomorphism classes of equi-Coxeter root systems of rank 24 and by Y_{24} the disjoint union of YR_{24} and the singleton $\{\emptyset\}$. By parts (a) and (b) of Proposition-Definition 2.3.3, the map $L \mapsto \mathrm{R}(L)$ induces maps $\mathrm{XR}_{24} \to \mathrm{YR}_{24}$ and $\mathrm{X}_{24} \to \mathrm{Y}_{24}$, where the second extends the first, which we again denote by R. Below, we forget that we know that the map $\mathrm{R}\colon \mathrm{X}_{24} \to \mathrm{Y}_{24}$ is bijective.

Let L be an even unimodular lattice of dimension 24 and ρ a Weyl vector of L. Since the group $\mathrm{W}(\mathrm{R}(L))$ permutes the Weyl vectors of L transitively, the map $L \mapsto \mathrm{vois}_2(L;\rho)$ induces a map $\mathrm{XR}_{24} \to \mathrm{X}_{24}$; we denote the latter by φ.

Let R be an element of YR_{24}; we set

$$\mathrm{h}'(R) \;=\; \frac{\mathrm{h}(R)}{2} - \frac{\iota(R)}{8} + 2\,;$$

we easily verify that $\mathrm{h}'(R)$ belongs to $\mathbb{N}$. Proposition 3.4.4.2 tells us that we have the following equality:

$$\mathrm{h}(\mathrm{R}(\varphi([L]))) \;=\; \mathrm{h}'(\mathrm{R}([L]))\,. \tag{1}$$

This equality suffices to determine $\mathrm{R}(\varphi([L]))$ if $\mathrm{h}'(\mathrm{R}(L))$ is not 12, 10, or 6. Indeed, the fibers $\mathrm{h}^{-1}(k)$ of the map $\mathrm{h}\colon \mathrm{Y}_{24} \to \mathbb{N}$ have 0 or 1 element unless $k = 12, 10, 6$, in which cases we have $\mathrm{h}^{-1}(12) = \{\mathbf{A}_{11}\mathbf{D}_7\mathbf{E}_6, 4\mathbf{E}_6\}$, $\mathrm{h}^{-1}(10) = \{2\mathbf{A}_9\mathbf{D}_6, 4\mathbf{D}_6\}$, and $\mathrm{h}^{-1}(6) = \{4\mathbf{A}_5\mathbf{D}_4, 6\mathbf{D}_4\}$. We easily check that $\mathrm{h}'^{-1}(12)$ is empty and that we have $\mathrm{h}'^{-1}(10) = \{2\mathbf{D}_{12}\}$ and $\mathrm{h}'^{-1}(6) = \{3\mathbf{D}_8, \mathbf{A}_{11}\mathbf{D}_7\mathbf{E}_6, 4\mathbf{E}_6\}$. In the cases $R([L]) = 2\mathbf{D}_{12}, 3\mathbf{D}_8, \mathbf{A}_{11}\mathbf{D}_7\mathbf{E}_6, 4\mathbf{E}_6$, we can still determine $\mathrm{R}(\varphi([L]))$ using condition (2) below.

Let R be a root system of type ADE endowed with a chamber C or, equivalently, endowed with a Weyl vector ρ. We denote by $R/2$ the sub-root system of R consisting of the roots of even height for the height function defined by C. The root system $R/2$ is again of type ADE; it is canonically endowed with a chamber: the positive roots of this chamber are those that are positive for C. The isomorphism class of $R/2$ is clearly independent of the choice of C. At the level of isomorphism classes, the map $R \mapsto R/2$ is determined by the following properties:

- We have $(R_1 \coprod R_2)/2 = R_1/2 \coprod R_2/2$.
- For R irreducible, the root system $R/2$ is the following: $\mathbf{A}_{2m}/2 = \mathbf{A}_m \coprod \mathbf{A}_{m-1}$, $\mathbf{A}_{2m+1}/2 = \mathbf{A}_m \coprod \mathbf{A}_m$, $\mathbf{D}_{2m}/2 = \mathbf{D}_m \coprod \mathbf{D}_m$, $\mathbf{D}_{2m+1}/2 = \mathbf{D}_{m+1} \coprod \mathbf{D}_m$, $\mathbf{E}_6/2 = \mathbf{A}_5 \coprod \mathbf{A}_1$, $\mathbf{E}_7/2 = \mathbf{A}_7$, $\mathbf{E}_8/2 = \mathbf{D}_8$ (with the natural conventions $\mathbf{A}_0 = \emptyset$, $\mathbf{D}_2 = \mathbf{A}_1 \coprod \mathbf{A}_1$, and $\mathbf{D}_3 = \mathbf{A}_3$).

By definition, we have $\mathrm{R}(L)/2 = \mathrm{R}(\mathrm{M}_2(L;\rho))$ and therefore

$$\mathrm{R}([L])/2 \;\subset\; \mathrm{R}(\varphi([L])) \tag{2}$$

for every $[L]$ in XR_{24}. This inclusion allows us to determine $\mathrm{R}(\varphi([L]))$ for $\mathrm{h}'(\mathrm{R}([L])) \in \{10, 6\}$. Indeed, we have

- $(2\mathbf{D}_{12})/2 = 4\mathbf{D}_6 \not\subset 2\mathbf{A}_9\,\mathbf{D}_6$;
- $(3\mathbf{D}_8)/2 = 6\mathbf{D}_4 \not\subset 4\mathbf{A}_5\,\mathbf{D}_4$;
- $(\mathbf{A}_{11}\mathbf{D}_7\mathbf{E}_6)/2 = \mathbf{A}_1\,\mathbf{A}_3\,\mathbf{A}_4\,3\mathbf{A}_5 \not\subset 6\mathbf{D}_4$;
- $(4\mathbf{E}_6)/2 = 4\mathbf{A}_1\,4\mathbf{A}_5 \not\subset 6\mathbf{D}_4$.

The previous discussion leads to the following statements.

Proposition-Definition 3.4.5. *There exists a unique map*

$$\psi \colon \mathrm{YR}_{24} \to \mathrm{Y}_{24}$$

such that we have $\mathrm{h}(\psi(R)) = \mathrm{h}'(R)$ *and* $R/2 \subset \psi(R)$.

Remark. Let R be an element of YR_{24}. The following conditions are equivalent:

- We have $\psi(R) = R/2$.
- We have $\iota(R) = 24$.
- The irreducible components of R are of type $\mathbf{A}_1$, $\mathbf{D}_l$ with l even, $\mathbf{E}_7$, or $\mathbf{E}_8$.

We extend ψ to a map $\psi \colon \mathrm{Y}_{24} \to \mathrm{Y}_{24}$ by setting $\psi(\emptyset) = \emptyset$. Likewise, we extend φ to a map $\varphi \colon \mathrm{X}_{24} \to \mathrm{X}_{24}$ by setting $\varphi([L]) = [L]$ if L has no roots.

Proposition 3.4.5.1. *Let R be an element of Y_{24} and k a positive integer. We have* $\psi^k(R) = \emptyset$ *for* $\mathrm{h}(R) < 2^k + 1$.

(Note that we have $\mathrm{h}(\psi(R)) - 1 \leq (\mathrm{h}(R) - 1)/2$.)

Proposition 3.4.5.2. *The following diagram is commutative:*

$$\begin{array}{ccc} \mathrm{X}_{24} & \xrightarrow{\varphi} & \mathrm{X}_{24} \\ {\scriptstyle \mathrm{R}}\downarrow & & {\scriptstyle \mathrm{R}}\downarrow \\ \mathrm{Y}_{24} & \xrightarrow{\psi} & \mathrm{Y}_{24}. \end{array}$$

Scholium 3.4.5.3. *Let L be an even unimodular lattice of dimension 24 and k a positive integer. Then $\varphi^k([L])$ has no roots for* $\mathrm{h}(L) < 2^k + 1$.

The oriented graph in Fig. 3.1 gives the map ψ explicitly; its vertices are the elements of Y_{24} and its edges are the ordered pairs (x, y) in $\mathrm{Y}_{24} \times \mathrm{Y}_{24}$ – diagonal with $x \in \mathrm{YR}_{24}$ and $y = \psi(x)$.

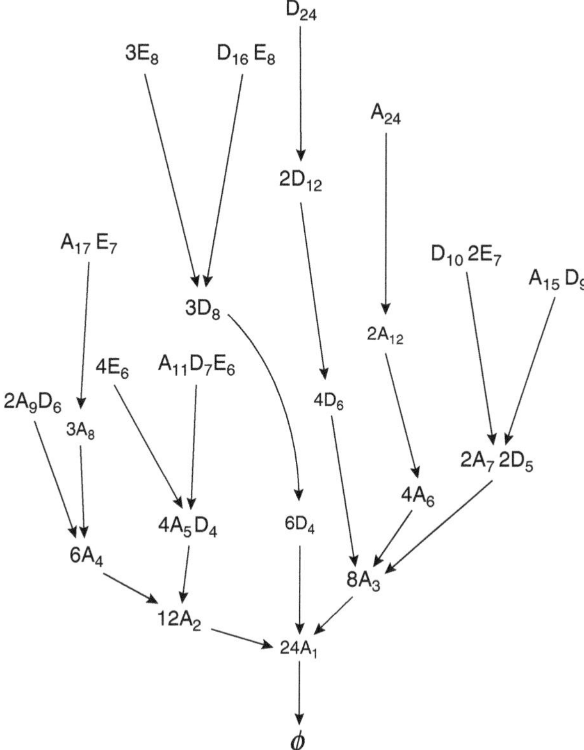

Fig. 3.1 Root system of the 2-neighbor associated with a Weyl vector of an even unimodular lattice of dimension 24, in terms of the root system of the latter

Chapter 4
Automorphic Forms and Hecke Operators

4.1 Lattices and Class Sets of $\mathbb{Z}$-groups

Let P be the set of prime numbers. Set $\widehat{\mathbb{Z}} = \prod_{p \in \mathrm{P}} \mathbb{Z}_p$, and let $\mathbb{A}_f = \mathbb{Q} \otimes \widehat{\mathbb{Z}}$ be the set of finite adeles of $\mathbb{Q}$. Fix a $\mathbb{Z}$-group G, that is, an affine group scheme of finite type over $\mathbb{Z}$. The group $G(\mathbb{A}_f)$ can be canonically identified with the subgroup of $\prod_{p \in \mathrm{P}} G(\mathbb{Q}_p)$ whose elements (g_p) satisfy $g_p \in G(\mathbb{Z}_p)$ for *almost all* p, in other words, for all $p \in \mathrm{P}$ except possibly a finite number. The groups $G(\mathbb{Q})$ and $G(\widehat{\mathbb{Z}})$ embed naturally into $G(\mathbb{A}_f)$ and satisfy $G(\widehat{\mathbb{Z}}) = \prod_{p \in \mathrm{P}} G(\mathbb{Z}_p)$ and $G(\mathbb{Z}) = G(\mathbb{Q}) \cap G(\widehat{\mathbb{Z}})$. The $G(\mathbb{A}_f)$-set

$$\mathcal{R}(G) = G(\mathbb{A}_f)/G(\widehat{\mathbb{Z}})$$

will play an important role in this chapter. We denote it by $\mathcal{R}$, for the French word for lattice, “réseau”, because it can, in general, be identified with the set of lattices of a certain type in a $\mathbb{Q}$-vector space.

A classical result of Borel [32, Sect. 5] asserts that the *class set of* G:

$$\mathrm{Cl}(G) = G(\mathbb{Q}) \backslash G(\mathbb{A}_f)/G(\widehat{\mathbb{Z}}) = G(\mathbb{Q}) \backslash \mathcal{R}(G)$$

is finite. Its cardinality $\mathrm{h}(G) = |\mathrm{Cl}(G)|$ is called the *class number* of G. In this section, we describe $\mathcal{R}(G)$ and $\mathrm{Cl}(G)$ in several standard cases we are interested in (see, for example, [32, Sect. 2]).

4.1.1 Linear Groups

Let us begin with the case of GL_n. If V is a vector space of finite dimension n over the field of fractions of a principal ideal domain A, we denote by $\mathcal{R}_A(V)$ the set of

G. Chenevier, J. Lannes, *Automorphic Forms and Even Unimodular Lattices*, Ergebnisse der Mathematik und ihrer Grenzgebiete. 3. Folge / A Series of Modern Surveys in Mathematics 69, https://doi.org/10.1007/978-3-319-95891-0_4

lattices in V with respect to A, that is, the set of free sub-A-modules of V of rank n (Sect. 2.1). It is endowed with a transitive action of $\mathrm{GL}(V)$; the stabilizer of L in $\mathcal{R}_A(V)$ is $\mathrm{GL}(L)$.

Let V be a $\mathbb{Q}$-vector space of dimension n. If p is prime and we set $V_p = V \otimes \mathbb{Q}_p$, then there is a natural map $\mathcal{R}_{\mathbb{Z}}(V) \to \mathcal{R}_{\mathbb{Z}_p}(V_p)$ defined by $M \mapsto M_p := M \otimes \mathbb{Z}_p$. We fix $L \in \mathcal{R}_{\mathbb{Z}}(V)$ and set $G = \mathrm{GL}_L$. We easily verify, following Eichler [78, Sect. 13], that the map

$$\mathcal{R}_{\mathbb{Z}}(V) \to \prod_{p \in \mathrm{P}} \mathcal{R}_{\mathbb{Z}_p}(V_p)\ , \quad M \mapsto (M_p)\ , \tag{4.1.1}$$

is an injection from $\mathcal{R}_{\mathbb{Z}}(V)$ to the subset $\prod'_{p \in \mathrm{P}} \mathcal{R}_{\mathbb{Z}_p}(V_p) \subset \prod_{p \in \mathrm{P}} \mathcal{R}_{\mathbb{Z}_p}(V_p)$ consisting of the families (M_p) such that $M_p = L_p$ for almost all p (this subset does not depend on the choice of L). The natural action of $G(\mathbb{A}_f)$ on $\prod_{p \in \mathrm{P}} \mathcal{R}_{\mathbb{Z}_p}(V_p)$ preserves $\prod'_{p \in \mathrm{P}} \mathcal{R}_{\mathbb{Z}_p}(V_p)$, and it is transitive on the latter. Therefore, if we identify $\mathcal{R}_{\mathbb{Z}}(V)$ with $\prod'_{p \in \mathrm{P}} \mathcal{R}_{\mathbb{Z}_p}(V_p)$ using the map (4.1.1), which we will do systematically from now on, then by transport of structure, we obtain a transitive action of $G(\mathbb{A}_f)$ on $\mathcal{R}_{\mathbb{Z}}(V)$ that extends the obvious action of $G(\mathbb{Q}) = \mathrm{GL}(V)$. Since the stabilizer of the lattice L is $G(\widehat{\mathbb{Z}})$, this leads to an isomorphism of $G(\mathbb{A}_f)$-sets

$$\mathcal{R}(G) \xrightarrow{\sim} \mathcal{R}_{\mathbb{Z}}(V).$$

Since $G(\mathbb{Q})$ also acts transitively on $\mathcal{R}_{\mathbb{Z}}(V)$, it follows, in particular, that we have

$$\mathrm{h}(\mathrm{GL}_n) = 1\ .$$

In the case $G = \mathrm{PGL}_L$ (resp. $G = \mathrm{SL}_L$), the set $\mathcal{R}(G)$ can also be viewed as the quotient of $\mathcal{R}_{\mathbb{Z}}(V)$ by $\mathbb{Q}^\times$ for the action by homotheties (resp. as the subset of $\mathcal{R}_{\mathbb{Z}}(V)$ consisting of the M that have a $\mathbb{Z}$-basis of determinant 1 with respect to a $\mathbb{Z}$-basis of L). We again have $\mathrm{h}(\mathrm{PGL}_n) = \mathrm{h}(\mathrm{SL}_n) = 1$.

4.1.2 Orthogonal and Symplectic Groups

We now assume that the $\mathbb{Q}$-vector space V is endowed with a nondegenerate bilinear form φ that is symmetric or alternating. Let $L \in \mathcal{R}_{\mathbb{Z}}(V)$. Recall that the *dual lattice* of L is the lattice $L^\sharp \in \mathcal{R}_{\mathbb{Z}}(V)$ defined by (Sect. 2.1)

$$L^\sharp = \{v \in V\ ;\ \varphi(v, x) \in \mathbb{Z}\ \forall x \in L\}\ .$$

We call L *homodual*, for "homothetic to its dual," if there exists a $\lambda \in \mathbb{Q}^\times$ such that we have $L^\sharp = \lambda L$; there then exists a unique strictly positive λ with this property; we denote it by λ_L. The lattice L is called *self-dual* if we have $L^\sharp = L$. If L is homodual and φ is symmetric (resp. alternating), then the bilinear form $\lambda_L \varphi$ gives L the structure of a b-module (resp. a-module) over $\mathbb{Z}$ in the sense of Sect. 2.1. We then say that L is *even* if $\lambda_L \varphi(x,x) \in 2\mathbb{Z}$ for every $x \in L$. This is automatic if φ is alternating, and if φ is symmetric, this allows us to view L as a q-module over $\mathbb{Z}$ by setting $\mathrm{q}(x) = \lambda_L\ \varphi(x,x)/2$ for $x \in L$. We denote by

$$\mathcal{R}^{\mathrm{a}}_{\mathbb{Z}}(V) \subset \mathcal{R}^{\mathrm{h}}_{\mathbb{Z}}(V)$$

the subsets of $\mathcal{R}_{\mathbb{Z}}(V)$ consisting of the even self-dual (resp. homodual) lattices.

Set $n = \dim V$, and fix $L \in \mathcal{R}^{\mathrm{a}}_{\mathbb{Z}}(V)$. By reduction modulo 2, the existence of such an L induces the congruence $n \equiv 0 \bmod 2$. Consider the sub-$\mathbb{Z}$-group $G \subset \mathrm{GL}_L$ defined by

$$G = \begin{cases} \mathrm{Sp}_L & \text{if } \varphi \text{ is alternating,} \\ \mathrm{O}_L & \text{else.} \end{cases}$$

We denote by $\widetilde{G}$ the corresponding similitude $\mathbb{Z}$-group, so that we have $G \subset \widetilde{G} \subset \mathrm{GL}_L$, and by $\mathrm{P}\widetilde{G}$ the *projective similitude* $\mathbb{Z}$-group, which is the quotient of $\widetilde{G}$ by its central sub-$\mathbb{Z}$-group isomorphic to $\mathbb{G}_m$ consisting of the homotheties (Sect. 2.1).

Lemma 4.1.3. *The restriction of the action of* $\mathrm{GL}_L(\mathbb{A}_f)$ *on* $\mathcal{R}_{\mathbb{Z}}(V)$ *to* $\widetilde{G}(\mathbb{A}_f)$ *(resp.* $G(\mathbb{A}_f)$*) preserves* $\mathcal{R}^{\mathrm{h}}_{\mathbb{Z}}(V)$ *(resp.* $\mathcal{R}^{\mathrm{a}}_{\mathbb{Z}}(V)$*).*

Before giving the proof, let us introduce the local analogs of the previous definitions. Let p be prime. For $M \in \mathcal{R}_{\mathbb{Z}_p}(V_p)$, the dual lattice $M^\sharp \in \mathcal{R}_{\mathbb{Z}_p}(V_p)$ (with respect to $\mathbb{Z}_p$; see Sect. 2.1) is well defined. We denote by $\mathcal{R}^{\mathrm{h}}_{\mathbb{Z}_p}(V_p) \subset \mathcal{R}_{\mathbb{Z}_p}(V_p)$ the subset of lattices M such that there exists $\lambda \in \mathbb{Q}_p^\times$ with $M^\sharp = \lambda M$ and $\lambda\varphi(x,x) \in 2\mathbb{Z}_p$ for every $x \in M$. Furthermore, we denote by $\mathcal{R}^{\mathrm{a}}_{\mathbb{Z}_p}(V_p) \subset \mathcal{R}^{\mathrm{h}}_{\mathbb{Z}_p}(V_p)$ the subset of lattices M such that we have $M^\sharp = M$. For $M \in \mathcal{R}^{\mathrm{h}}_{\mathbb{Z}_p}(V_p)$, there exists a unique $\lambda_M \in p^{\mathbb{Z}}$ with $M^\sharp = \lambda_M M$. If φ is symmetric (resp. alternating), the quadratic form $x \mapsto \lambda_M\ \varphi(x,x)/2$ (resp. the alternating form $\lambda_M \varphi$) then gives M the structure of a q-module (resp. a-module) over $\mathbb{Z}_p$.

Proof. Let $M \in \mathcal{R}_{\mathbb{Z}}(V)$. We begin by noting that M is in $\mathcal{R}^{\mathrm{h}}_{\mathbb{Z}}(V)$ if and only if M_p is in $\mathcal{R}^{\mathrm{h}}_{\mathbb{Z}_p}(V_p)$ for every prime p, in which case we, moreover, have $\lambda_M = \prod_p \lambda_{M_p}$ (of course, λ_{M_p} is 1 for almost all p). Indeed, this follows from the identity $\mathbb{A}_f^\times = \mathbb{Q}^\times \cdot \widehat{\mathbb{Z}}^\times$ (that is, $\mathrm{h}(\mathbb{G}_m) = 1$) and the immediate relation $(N^\sharp)_p = (N_p)^\sharp$, which holds for every prime p and every $N \in \mathcal{R}_{\mathbb{Z}}(V)$. In particular, we have $M \in \mathcal{R}^{\mathrm{a}}_{\mathbb{Z}}(V)$ if and only if we have $M_p \in \mathcal{R}^{\mathrm{a}}_{\mathbb{Z}_p}(V_p)$ for every p.

To conclude the proof, it suffices to note that if $g \in \widetilde{G}(\mathbb{Q}_p)$ has similitude factor $\nu(g)$ (Sect. 2.1) and we have $M \in \mathcal{R}_{\mathbb{Z}_p}(V_p)$, then we have the relation $g(M)^\sharp = \nu(g)^{-1} g(M^\sharp)$. □

Note that the action of the homotheties $\mathbb{Q}^\times$ on $\mathcal{R}_{\mathbb{Z}}(V)$ preserves $\mathcal{R}^{\mathrm{h}}_{\mathbb{Z}}(V)$. By Lemma 4.1.3, the quotient set

$$\underline{\mathcal{R}}^{\mathrm{h}}_{\mathbb{Z}}(V) := \mathbb{Q}^\times \backslash \mathcal{R}^{\mathrm{h}}_{\mathbb{Z}}(V)$$

is therefore endowed with an action of $\mathrm{P}\widetilde{G}(\mathbb{A}_f)$ that extends the obvious action of $\mathrm{P}\widetilde{G}(\mathbb{Q})$. We denote the homothety class of $M \in \mathcal{R}_{\mathbb{Z}}(V)$ by $\underline{M}$. In summary, we have the following commutative diagram:

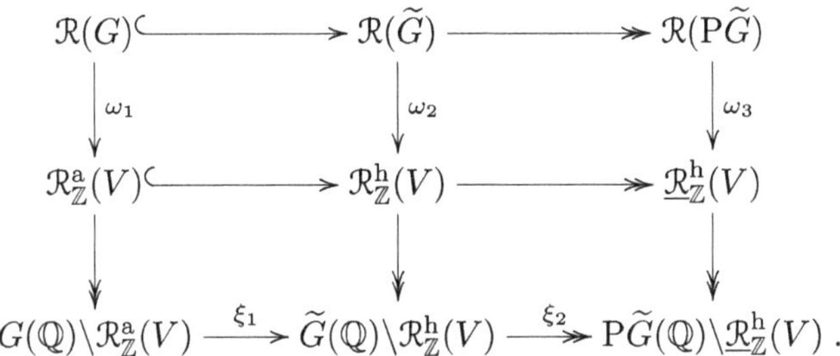

The ω_i, for $i = 1, 2, 3$, are, respectively, the "orbit" maps of L, L, and $\underline{L}$ under the actions of $G(\mathbb{A}_f)$, $\widetilde{G}(\mathbb{A}_f)$, and $\mathrm{P}\widetilde{G}(\mathbb{A}_f)$. All other arrows denote canonical maps.

Proposition 4.1.4. *The maps ω_i and ξ_j are bijective. In particular, the action of $G(\mathbb{A}_f)$ on $\mathcal{R}^{\mathrm{a}}_{\mathbb{Z}}(V)$ is transitive; the orbit of L defines an isomorphism of $G(\mathbb{A}_f)$-sets $\mathcal{R}(G) \overset{\sim}{\to} \mathcal{R}^{\mathrm{a}}_{\mathbb{Z}}(V)$.*

Proof. The injectivity of the ω_i is obvious. Let us begin by verifying the last assertion, which is nothing more than the surjectivity of ω_1. If φ is symmetric, Scholium 2.2.5 asserts that for every $M \in \mathcal{R}^{\mathrm{a}}_{\mathbb{Z}}(V)$, the q-module M_p over $\mathbb{Z}_p$ is hyperbolic. It is, in particular, isomorphic to L_p, which concludes the proof of the last assertion because every isometry $L_p \to M_p$ is necessarily induced by an element of $\mathrm{O}(V_p) = G(\mathbb{Q}_p)$. Let us therefore suppose that φ is alternating. It is well known that if A is a principal ideal domain, there exists, up to equivalence, a unique nondegenerate alternating bilinear form on the A-module A^n (n even). We conclude by considering the case $A = \mathbb{Z}_p$.

The surjectivity of ω_3 (resp. ω_2) follows from that of ω_2 (resp. from those of ω_1 and ξ_1). Let us show the surjectivity of ξ_1. For $M \in \mathcal{R}^{\mathrm{h}}_{\mathbb{Z}}(V)$ and $g \in \widetilde{G}(\mathbb{Q})$ with similitude factor $\nu(g)$, we have $\lambda_{g(M)} = \pm\,\nu(g)^{-1}\,\lambda_M$. It therefore suffices to see that $\nu(\widetilde{G}(\mathbb{Q}))$ contains the set $\mathbb{Q}_{>0}$ of strictly positive rational numbers. This is obvious in the alternating case and, more generally, when V is hyperbolic. In the symmetric case, we must show that for $\lambda \in \mathbb{Q}_{>0}$, the vector spaces V and $V \otimes \langle\lambda\rangle$ (obtained by multiplying the quadratic form on V by λ) are isomorphic as q-vector spaces over $\mathbb{Q}$. But they are so over $\mathbb{Q}_p$ for every prime p because the $V \otimes \mathbb{Q}_p$ are hyperbolic by Scholium 2.2.5, and they are so over $\mathbb{R}$ because we have $\lambda > 0$. We conclude using the Hasse–Minkowski theorem.

The map ξ_2 is bijective because of the equality $\mathrm{P}\widetilde{G}(\mathbb{Q}) = \widetilde{G}(\mathbb{Q})/\mathbb{Q}^\times$. Finally, let us verify the injectivity of ξ_1. We may assume that φ is symmetric because the argument given in the first paragraph shows that we have $h(G) = 1$ if φ is alternating. Let us therefore assume that there exist $M \in \mathcal{R}^{\mathrm{a}}_{\mathbb{Z}}(V)$ and $g \in \widetilde{G}(\mathbb{Q})$ such that $g(M) = L$. We then have $\nu(g) = \pm 1$. If $\nu(g) = 1$, then we have $g \in G(\mathbb{Q})$, and we are done. Otherwise, M is isometric to the q-module $L \otimes \langle -1\rangle$, which has underlying space L but opposite quadratic form. This implies that $V \otimes \mathbb{R}$ is hyperbolic, and thus that L and M are isomorphic by Theorem 2.2.7. □

Corollary 4.1.5. *We have* $\mathrm{h}(G) = \mathrm{h}(\widetilde{G}) = \mathrm{h}(\mathrm{P}\widetilde{G})$.

When φ is alternating, the classification given above of the nondegenerate alternating forms applied to the ring $\mathbb{Z}$ implies[1] $\mathrm{h}(G) = 1$, and therefore $\mathrm{h}(\mathrm{Sp}_{2g}) = \mathrm{h}(\mathrm{GSp}_{2g}) = \mathrm{h}(\mathrm{PGSp}_{2g}) = 1$ for every $g \geq 1$.

Let us assume that φ is symmetric. If the q-vector space $L \otimes \mathbb{R}$ is indefinite, then Theorem 2.2.7 implies $\mathrm{h}(\mathrm{O}_L) = 1$. The situation is quite different if $L \otimes \mathbb{R}$ is positive definite, which we will assume from now on. Recall that L can then be viewed as an even unimodular lattice in the Euclidean space $V \otimes \mathbb{R}$ of dimension n. In particular, we have $n \equiv 0 \bmod 8$. In this case, $\mathcal{R}^{\mathrm{a}}_{\mathbb{Z}}(V)$ is, by definition, the set of even unimodular lattices in $V \otimes \mathbb{R}$ that are contained in $L \otimes \mathbb{Q}$. Recall that X_n denotes the set of isometry classes of even unimodular lattices in the Euclidean space $V \otimes \mathbb{R}$. By Scholium 2.2.1, the natural inclusion $\mathrm{O}(V)\backslash\mathcal{R}^{\mathrm{a}}_{\mathbb{Z}}(V) \to \mathrm{X}_n$ is bijective and therefore induces an isomorphism $\mathrm{Cl}(\mathrm{O}_L) \xrightarrow{\sim} \mathrm{X}_n$. In particular, if O_n denotes the orthogonal $\mathbb{Z}$-group of the lattice $L = \mathrm{E}_n$ (Sect. 1.3), we obtain the equality

$$\mathrm{h}(\mathrm{O}_n) = |\mathrm{X}_n| \,,$$

which shows that $\mathrm{h}(\mathrm{O}_n)$ is a quite interesting number.

4.1.6 SO_L *Versus* O_L

We continue the analysis of the previous subsection by assuming that φ is symmetric, so that G, $\widetilde{G}$, and $\mathrm{P}\widetilde{G}$ are, respectively, O_L, GO_L, and PGO_L. We are interested in their respective sub-$\mathbb{Z}$-groups SO_L, GSO_L, and PGSO_L (Sect. 2.1). The groups $\mathrm{SO}_L(\mathbb{A}_f)$, $\mathrm{GSO}_L(\mathbb{A}_f)$, and $\mathrm{PGSO}_L(\mathbb{A}_f)$ act on, respectively, $\mathcal{R}^{\mathrm{a}}_{\mathbb{Z}}(V)$, $\mathcal{R}^{\mathrm{h}}_{\mathbb{Z}}(V)$, and $\underline{\mathcal{R}}^{\mathrm{h}}_{\mathbb{Z}}(V)$ (Proposition 4.1.3). Let us consider the following commutative diagram, which extends that of Sect. 4.1.2:

[1] The assertions $\mathrm{h}(\mathrm{SL}_n) = \mathrm{h}(\mathrm{Sp}_{2g}) = 1$ recalled above are also very particular cases of Kneser's strong approximation theorem (see [123], [162, Theorem 7.12]). It asserts that we have $\mathrm{h}(G) = 1$ whenever the $\mathbb{C}$-group $G_{\mathbb{C}}$ is semisimple and simply connected and the topological group $G(\mathbb{R})$ does not have a nontrivial connected, compact, normal subgroup.

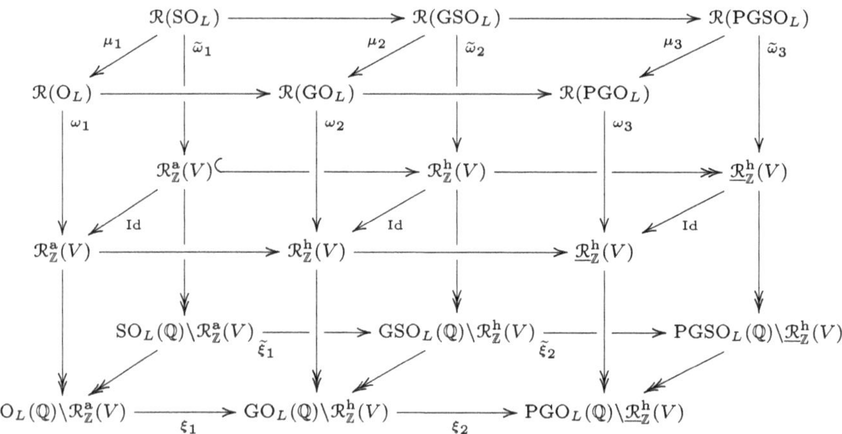

The vertical maps $\widetilde{\omega}_i$ are again the "orbit" maps of L (resp. L, resp. $\underline{L}$), and the other arrows are the canonical maps.

Proposition 4.1.7. *The maps $\widetilde{\omega}_i$, μ_i, and $\widetilde{\xi}_j$ are bijective. In particular, the action of $\mathrm{SO}_L(\mathbb{A}_f)$ on $\mathcal{R}^{\mathrm{a}}_{\mathbb{Z}}(V)$ is again transitive; the orbit of L defines an isomorphism $\mathcal{R}(\mathrm{SO}_L) \xrightarrow{\sim} \mathcal{R}^{\mathrm{a}}_{\mathbb{Z}}(V)$.*

Proof. We have already seen that the natural action of $\mathrm{O}_L(\mathbb{A}_f)$ on $\mathcal{R}^{\mathrm{a}}_{\mathbb{Z}}(V)$ is transitive (Proposition 4.1.4). The same holds for the restriction of this action to its subgroup $\mathrm{SO}_L(\mathbb{A}_f)$ because the orthogonal group of a nontrivial hyperbolic q-module over $\mathbb{Z}_p$ always has an element of determinant -1. The same reasoning shows that the μ_i are bijective because $\mathrm{O}_L(\mathbb{Z}_p)/\mathrm{SO}_L(\mathbb{Z}_p) \to \mathrm{GO}_L(\mathbb{Q}_p)/\mathrm{GSO}_L(\mathbb{Q}_p)$ is bijective for every prime p (Sect. 2.1). Since the ω_i are bijective, the bijectivity of the $\widetilde{\omega}_i$ follows.

The bijectivity of $\widetilde{\xi}_2$ is obvious. The surjectivity of $\widetilde{\xi}_1$ follows from that of ξ_1 and from the fact that we have $-1 \in \det(\mathrm{O}(V))$. Finally, the injectivity of $\widetilde{\xi}_1$ can be shown similarly to that of ξ_1 (Proposition 4.1.4), using that we have $-1 \in \det(\mathrm{O}(\mathrm{H}(\mathbb{Z}^{n/2})))$. □

Corollary 4.1.8. *If L is a q-module over $\mathbb{Z}$, we have $\mathrm{h}(\mathrm{SO}_L) = \mathrm{h}(\mathrm{GSO}_L) = \mathrm{h}(\mathrm{PGSO}_L)$. If, moreover, $L \otimes \mathbb{R}$ is indefinite, then these integers are equal to* 1.

Proof. The first assertion follows from the bijectivity of the maps ξ_i (Proposition 4.1.7). When $L \otimes \mathbb{R}$ is indefinite, we already explained the equality $\mathrm{h}(\mathrm{O}_L) = 1$ in Sect. 4.1.2. It remains to show that there exists an $s \in \mathrm{O}(L)$ with $\det s = -1$. The assumption on L and Theorem 2.2.7 show that there exists a q-module L' over $\mathbb{Z}$ such that $L \simeq L' \oplus \mathrm{H}(\mathbb{Z})$ (orthogonal sum). This concludes the proof because $\mathrm{H}(\mathbb{Z})$ contains an automorphism of determinant -1. □

Finally, let us assume that L is positive definite. As before, we then have a canonical bijection $\mathrm{Cl}(\mathrm{SO}_L) \xrightarrow{\sim} \widetilde{\mathrm{X}}_n$, where $\widetilde{\mathrm{X}}_n$ denotes the set of *direct* isometry classes of even unimodular lattices in $V \otimes \mathbb{R}$ (in other words, the set of orbits of the action of $\mathrm{SO}(V \otimes \mathbb{R})$ on the latter). The isometry class of an even unimodular lattice $M \subset V \otimes \mathbb{R}$ admits exactly one or two inverse images under the canonical projection

$$\widetilde{\mathrm{X}}_n \to \mathrm{X}_n ,$$

depending on whether $\mathrm{O}(M)$ has an element of determinant -1 or not. It has one if, for example, M has at least one root, that is, an $\alpha \in M$ such that $\alpha \cdot \alpha = 2$, because the associated orthogonal reflection is in $\mathrm{O}(M)$ (Sect. 2.3). On the other hand, if M is the Leech lattice, then we have $\mathrm{O}(M) = \mathrm{SO}(M)$ by Conway [65]. The results recalled in Sect. 2.3 imply the following corollary. For $n \equiv 0 \bmod 8$, we set $\mathrm{SO}_n = \mathrm{SO}_{\mathrm{E}_n}$.

Corollary 4.1.9. *We have* $\mathrm{h}(\mathrm{SO}_8) = 1$, $\mathrm{h}(\mathrm{SO}_{16}) = 2$, *and* $\mathrm{h}(\mathrm{SO}_{24}) = 25$.

4.1.10 Orthogonal Groups in Odd Dimensions

We return to the setting of Sect. 4.1.2, where we assume that φ is symmetric. We now consider the set

$$\mathcal{R}_{\mathbb{Z}}^{\mathrm{b}}(V) \subset \mathcal{R}_{\mathbb{Z}}(V)$$

consisting of the $L \in \mathcal{R}_{\mathbb{Z}}(V)$ with $\varphi(x,x) \in 2\mathbb{Z}$ for every $x \in L$ and $L^\sharp/L \simeq \mathbb{Z}/2\mathbb{Z}$. This last condition is equivalent to requiring that $\varphi_{|L\times L}$ have determinant ± 2. We refer to Appendix B for an analysis of these lattices.

We fix $L \in \mathcal{R}_{\mathbb{Z}}^{\mathrm{b}}(V)$, which requires the dimension n of V to be odd. Then $\mathrm{SO}_L(\mathbb{A}_f)$ acts transitively on $\mathcal{R}_{\mathbb{Z}}^{\mathrm{b}}(V)$ by Proposition B.2.5, and the stabilizer of L is $\mathrm{SO}_L(\widehat{\mathbb{Z}})$. If $L \otimes \mathbb{R}$ is indefinite, the number of classes of SO_L is 1; this is a classical result that would not be difficult to deduce from Proposition B.2.5 (iii) and Theorem 2.2.7. The situation is more interesting when $L \otimes \mathbb{R}$ is definite, say positive definite to fix the ideas; we will assume that this is the case from now on.

In this case, we have the congruence $n \equiv \pm 1 \bmod 8$ and $\mathrm{Cl}(\mathrm{SO}_L)$ can be identified with the set of isometry classes of even lattices of determinant 2 in $\mathbb{R}^n$ (Sect. B.2). Here, we do not need to distinguish between direct and indirect isometries because $x \mapsto -x$ is in $\mathrm{O}(M)$ and has determinant -1 for every $M \in \mathcal{R}_{\mathbb{Z}}^{\mathrm{b}}(V)$. If $n \equiv 1 \bmod 8$, we set $\mathrm{L}_n = \mathrm{E}_{n-1} \oplus \mathrm{A}_1$. If $n \equiv -1 \bmod 8$, we denote by L_n the orthogonal complement of an arbitrary root of E_{n+1}; since these roots are permuted transitively by the orthogonal group of E_{n+1}, the isometry class of such a lattice does not depend on any choice we make. If $n \equiv \pm 1 \bmod 8$, the lattice L_n is therefore even of determinant 2 (Sect. B.2), and we set $\mathrm{SO}_n = \mathrm{SO}_{\mathrm{L}_n}$ (Sect. B.1).

The known values of $\mathrm{h}(\mathrm{SO}_n)$ with n odd are gathered in the following corollary (see also [68]). The cases $n \leq 23$ are treated, for example, in Appendix B, Sect. B.2; the case $n = 25$ is due to Borcherds [29, Table -2].

Corollary 4.1.11. *We have* $\mathrm{h}(\mathrm{SO}_1) = \mathrm{h}(\mathrm{SO}_7) = \mathrm{h}(\mathrm{SO}_9) = 1$, $\mathrm{h}(\mathrm{SO}_{15}) = 2$, $\mathrm{h}(\mathrm{SO}_{17}) = 4$, $\mathrm{h}(\mathrm{SO}_{23}) = 32$, *and* $\mathrm{h}(\mathrm{SO}_{25}) = 121$.

4.2 Hecke Correspondences

4.2.1 General Formalism

Let Γ be an (abstract) group, and let X be a transitive Γ-set. The ring of *Hecke correspondences* (*or operators*) of X is the ring

$$\mathrm{H}(X) = \mathrm{End}_{\mathbb{Z}[\Gamma]}(\mathbb{Z}[X]) .$$

With each $T \in \mathrm{End}_{\mathbb{Z}}(\mathbb{Z}[X])$ is associated a matrix $(T_{x,y})_{(x,y)\in X\times X}$ that determines it uniquely; the matrix is defined by the formula

$$\forall y \in X , \quad T(y) = \sum_{x\in X} T_{x,y}\, x .$$

By definition, such an element T is in the ring $\mathrm{H}(X)$ if and only if the function $X \times X \to \mathbb{Z}$ given by $(x, y) \mapsto T_{x,y}$ is constant on the orbits of the group Γ acting diagonally on $X \times X$. The resulting function $\Gamma\backslash(X \times X) \to \mathbb{Z}$ then has finite support, by the finiteness of $\{x \in X\,;\, T_{x,y} \neq 0\}$ for $y \in X$ and by the transitivity of X. We therefore have an injective map

$$\mathrm{H}(X) \to \mathrm{Hom}_{\mathrm{fs}}(\Gamma\backslash(X \times X), \mathbb{Z}) , \quad T \mapsto ((x, y) \mapsto T_{x,y}) , \tag{4.2.1}$$

where $\mathrm{Hom}_{\mathrm{fs}}(Y, \mathbb{Z})$ denotes the abelian group of functions with finite support on the set Y and values in $\mathbb{Z}$.

For $x \in X$, we denote the stabilizer of x by $\Gamma_x \subset \Gamma$. We assume that the following properties hold:

(i) For every orbit Ω of Γ in $X\times X$ and every $x \in X$, the intersection $\Omega\cap(X\times\{x\})$ is finite.
(ii) For every $x \in X$, the orbits of Γ_x on X are finite. In other words, for every $x, y \in X$, the intersection $\Gamma_x \cap \Gamma_y$ has finite index in Γ_x.

These conditions ensure that the map (4.2.1) is bijective. In particular, $\mathrm{H}(X)$ is a free $\mathbb{Z}$-module with natural basis the characteristic functions of the orbits of Γ on $X \times X$.

Fix $x \in X$. The transitivity of X ensures that the map $\Gamma \to X \times X$ given by $\gamma \mapsto (\gamma(x), x)$ induces bijections

$$\Gamma_x\backslash\Gamma/\Gamma_x \xrightarrow{\sim} \Gamma_x\backslash(X \times \{x\}) \xrightarrow{\sim} \Gamma\backslash(X \times X) . \tag{4.2.2}$$

In particular, this identifies $\mathrm{H}(X)$ with $\mathrm{Hom}_{\mathrm{fs}}(\Gamma_x\backslash\Gamma/\Gamma_x, \mathbb{Z})$. By transport of structure, the latter inherits a ring structure from $\mathrm{H}(X)$: we recover the more standard presentations of the Hecke rings, such as those in [174], [188, Sect. 3], [48], [88, Kap. IV], or [97]. Note that depending on the reference, the ring structure considered on $\mathrm{Hom}_{\mathrm{fs}}(\Gamma_x\backslash\Gamma/\Gamma_x, \mathbb{Z})$ (defined, in general, by an explicit convolution product) may differ slightly from ours; this is, in particular, the case in the articles of Cartier and

Gross, to which we refer in Sect. 6.2, in which the ring $\mathrm{H}(X)$ is exactly the opposite of ours.

Since the second formulation of condition (ii) is symmetric in x, y, condition (i) is also equivalent to requiring that for every orbit Ω of Γ in $X \times X$ and every $x \in X$, the intersection $\Omega \cap (\{x\} \times X)$ be finite. Thus, if we have $T \in \mathrm{H}(X)$, there exists a unique $T^{\mathrm{t}} \in \mathrm{H}(X)$ such that $T^{\mathrm{t}}_{x,y} = T_{y,x}$ for every $x, y \in X$. The endomorphism

$$T \mapsto T^{\mathrm{t}}$$

of $\mathrm{H}(X)$ is an anti-involution, that is, satisfies $(ST)^{\mathrm{t}} = T^{\mathrm{t}} S^{\mathrm{t}}$ and $(T^{\mathrm{t}})^{\mathrm{t}} = T$ for every $S, T \in \mathrm{H}(X)$; this endomorphism simply corresponds to taking the transpose of the associated matrices. This anti-involution is the identity if and only if the Γ-orbits of $X \times X$ are invariant under $(x, y) \mapsto (y, x)$, in which case $\mathrm{H}(X)$ is commutative; this is a special case of Gelfand's criterion.

4.2.2 A Functor from Γ-Modules to $\mathrm{H}(X)^{\mathrm{opp}}$-Modules

Let X be a transitive Γ-set that satisfies conditions (i) and (ii) of Sect. 4.2.1. The ring $\mathrm{H}(X)$ appears as follows in the representation theory of Γ. If M is a $\mathbb{Z}[\Gamma]$-module, then the abelian group

$$M_X = \mathrm{Hom}_{\mathbb{Z}[\Gamma]}(\mathbb{Z}[X], M)$$

inherits a right action of $\mathrm{H}(X)$ by composition at the source. It is obvious that $M \mapsto M_X$ is a functor from Γ-modules (on the left) to $\mathrm{H}(X)$-modules on the right.

For a $\mathbb{Z}[\Gamma]$-module M and $x \in X$, the map $\varphi \mapsto \varphi(x)$ identifies M_X with the subgroup of invariants $M^{\Gamma_x} \subset M$, which also endows this subgroup with the structure of an $\mathrm{H}(X)$-module. Suppose that the matrix of $T \in \mathrm{H}(X)$ is the characteristic function of the double coset $\Gamma_x \gamma \Gamma_x$ through the identification $\Gamma_x \backslash \Gamma / \Gamma_x \xrightarrow{\sim} \Gamma \backslash (X \times X)$ chosen in Sect. 4.2.1. We have the classical formula

$$T(m) = \sum_i \gamma_i(m) \quad \forall m \in M^{\Gamma_x} \tag{4.2.3}$$

for every decomposition $\Gamma_x \gamma \Gamma_x = \coprod_i \gamma_i \Gamma_x$ (this is a finite union).

In this context, the anti-involution $T \mapsto T^{\mathrm{t}}$ defined in Sect. 4.2.1 takes on the following meaning. Let M and M' be two $\mathbb{Z}[\Gamma]$-modules, N an abelian group, and $(-|-)\colon M \times M' \to N$ a bilinear map with $(\gamma m | \gamma m') = (m | m')$ for every $\gamma \in \Gamma$ and every $(m, m') \in M \times M'$. For $(\varphi, \varphi') \in M_X \times M'_X$, $(\varphi(x) | \varphi'(x))$ does not depend on the choice of $x \in X$ hence

$$(\varphi | \varphi') := (\varphi(x) | \varphi'(x))$$

defines a bilinear form from $M_X \times M'_X$ to N. If we identify M_X with M^{Γ_x} as before, this bilinear map is nothing more than the restriction of $(-|-)$ to $M^{\Gamma_x} \times M'^{\Gamma_x}$.

We will say that X is *symmetric* if, in addition to verifying conditions (i) and (ii) of Sect. 4.2.1, it has the following equivalent properties[2]:

(iii) For every orbit Ω of Γ in $X \times X$ and every $x \in X$, we have the equality $|\Omega \cap (X \times \{x\})| = |\Omega \cap (\{x\} \times X)|$.
(iv) For every $x, y \in X$, the intersection $\Gamma_x \cap \Gamma_y$ has the same index in Γ_x and Γ_y.

Lemma 4.2.3. *Suppose that X is symmetric. For $T \in \mathrm{H}(X)$ and $(\varphi, \varphi') \in M_X \times M'_X$, we have $(T(\varphi)|\varphi') = (\varphi|T^{\mathrm{t}}(\varphi'))$.*

Proof. Let $\psi\colon X \times X \to N$ be a map that is constant on every Γ-orbit in $X \times X$ and zero outside a finite number of them. The symmetry of X implies, for every $x \in X$, the relation $\sum_{y \in X} \psi(y, x) = \sum_{y \in X} \psi(x, y)$. We apply this to the function $(x, y) \mapsto T_{x,y} \cdot (\varphi(x)|\varphi'(y))$. □

Remark 4.2.4. Suppose that V is a right $\mathrm{H}(X)$-module. The map $\mathrm{H}(X) \times V \to V$ given by $(T, v) \mapsto T^{\mathrm{t}}\, v$ defines the structure of a (left) $\mathrm{H}(X)$-module on V, which we denote by V^{t}.

4.2.5 The Hecke Ring of a $\mathbb{Z}$-group

Let G be a $\mathbb{Z}$-group. We will apply the definitions given above to $\Gamma = G(\mathbb{A}_f)$ and $X = \mathcal{R}(G)$. The *Hecke ring of* G is the ring

$$\mathrm{H}(G) := \mathrm{H}(\mathcal{R}(G)) .$$

Recall that for every prime p, the group $G(\mathbb{Q}_p)$ inherits from $\mathbb{Q}_p$ the structure of a locally compact topological group (that is, moreover, separated and the union of a countable number of compact groups). The subgroup $G(\mathbb{Z}_p)$ is both compact and open. The group $G(\mathbb{A}_f)$ is also a locally compact topological group for the topology whose base of open neighborhoods of the identity consists of the open sets of the form $\prod_{p \in \mathrm{P}} U_p$, where U_p for p prime is an open neighborhood of the identity in $G(\mathbb{Q}_p)$ and we have $U_p = G(\mathbb{Z}_p)$ for almost all p. In particular, $G(\widehat{\mathbb{Z}})$ is a compact open subgroup of $G(\mathbb{A}_f)$. Consequently, $\mathcal{R}(G)$ has property (ii) of Sect. 4.2.1, as do the $G(\mathbb{Q}_p)$-sets

$$\mathcal{R}_p(G) := G(\mathbb{Q}_p)/G(\mathbb{Z}_p) .$$

The $G(\mathbb{A}_f)$-set $\mathcal{R}(G)$ and the $\mathcal{R}_p(G)$ are symmetric in the sense of Sect. 4.2.2 if $G(\mathbb{A}_f)$ is unimodular, which is, in particular, the case if the neutral component of $G(\mathbb{C})$ is reductive [32, Sect. 5.5].

[2] This property is not automatic if X is infinite. Consider, for example, the group $\Gamma = \mathbb{Q} \rtimes \mathbb{Q}^\times$ of affine transformations of $\mathbb{Q}$ and the Γ-set X consisting of the subsets of $\mathbb{Q}$ of the form $a\mathbb{Z} + b$ with $a \in \mathbb{Q}^\times$ and $b \in \mathbb{Q}$.

For p prime, we also define $\mathrm{H}_p(G)$ as the Hecke ring of the $G(\mathbb{Q}_p)$-set $\mathcal{R}_p(G)$. The $G(\mathbb{A}_f)$-set $\mathcal{R}(G)$ can be canonically identified with the subset of $\prod_{p\in\mathrm{P}} \mathcal{R}_p(G)$ consisting of the (x_p) with $x_p = G(\mathbb{Z}_p)$ for almost all p. We have already seen a manifestation of this fact in the Eichler embedding (4.1.1). In particular, for every prime p, we have a canonical injective ring homomorphism

$$\mathrm{H}_p(G) \to \mathrm{H}(G)$$

that takes $T \in \mathrm{H}_p(G)$ to the endomorphism of $\mathbb{Z}[\mathcal{R}(G)]$ that sends $y = (y_\ell)_{\ell\in\mathrm{P}}$ to $\sum_x T_{x_p,y_p}\, x$, where the sum is taken over the elements x of $\mathcal{R}(G)$ with $x_\ell = y_\ell$ in $\mathcal{R}_\ell(G)$ for every $\ell \neq p$. We will simply write

$$\mathrm{H}_p(G) \subset \mathrm{H}(G)\ .$$

If $p \neq q$, then for $S \in \mathrm{H}_p(G)$ and $T \in \mathrm{H}_q(G)$, we have $TS = ST$.

If for every prime p, we take a $G(\mathbb{Q}_p)$-orbit $\Omega_p \subset \mathcal{R}_p(G) \times \mathcal{R}_p(G)$ and if, moreover, Ω_p is the orbit of $G(\mathbb{Z}_p) \times G(\mathbb{Z}_p)$ for almost all p, then the subset of elements (ω_p) of $\prod_p \Omega_p$ with $\omega_p = G(\mathbb{Z}_p) \times G(\mathbb{Z}_p)$ for almost all p can be naturally identified with a $G(\mathbb{A}_f)$-orbit in $\mathcal{R}(G) \times \mathcal{R}(G)$. Conversely, every $G(\mathbb{A}_f)$-orbit $\Omega \subset \mathcal{R}(G) \times \mathcal{R}(G)$ is of this form for a unique family (Ω_p), where the $G(\mathbb{Q}_p)$-orbit Ω_p is the image of Ω by the canonical projection $\mathcal{R}(G) \times \mathcal{R}(G) \to \mathcal{R}_p(G) \times \mathcal{R}_p(G)$. From these observations and the surjectivity of the map (4.2.1) it follows that $\mathrm{H}(G)$ is isomorphic to the tensor product of its subrings $\mathrm{H}_p(G)$:

$$\bigotimes_{p\in\mathrm{P}} \mathrm{H}_p(G) \xrightarrow{\sim} \mathrm{H}(G)\ .$$

Understanding $\mathrm{H}(G)$ therefore completely reduces to understanding the $\mathrm{H}_p(G)$.

The ring $\mathrm{H}_p(G)$ depends only on the $\mathbb{Z}_p$-group $G_{\mathbb{Z}_p} = G \times_{\mathbb{Z}} \mathbb{Z}_p$. When $G_{\mathbb{Z}_p}$ is reductive, general results of Satake and Bruhat–Tits imply that $\mathrm{H}_p(G)$ is commutative; we will come back to this in Sect. 6.2. As a consequence, the same holds for $\mathrm{H}(G)$ if G is reductive over $\mathbb{Z}$. However, this property is elementary in the most classical cases, which we recall below.

4.2.6 Some Classical Hecke Rings

First, suppose $G = \mathrm{PGL}_n$. We have seen that $\mathcal{R}(G)$ can be identified with

$$\underline{\mathcal{R}_{\mathbb{Z}}}(V) := \mathbb{Q}^\times \backslash \mathcal{R}_{\mathbb{Z}}(V)\ ,$$

where $V = \mathbb{Q}^n$. Recall that $\underline{M} \in \underline{\mathcal{R}_{\mathbb{Z}}}(V)$ denotes the homothety class of a lattice $M \in \mathcal{R}_{\mathbb{Z}}(V)$.

For $M, N \in \mathcal{R}_{\mathbb{Z}}(V)$, there exists a least integer $d \geq 1$ with $dN \subset M$. The isomorphism class of the abelian group M/dN depends only on the $G(\mathbb{A}_f)$-orbit of

$(\underline{N}, \underline{M})$ in $\underline{\mathcal{R}_{\mathbb{Z}}}(V) \times \underline{\mathcal{R}_{\mathbb{Z}}}(V)$. The theory of elementary divisors then shows that the resulting map

$$G(\mathbb{A}_f)\backslash(\underline{\mathcal{R}_{\mathbb{Z}}}(V) \times \underline{\mathcal{R}_{\mathbb{Z}}}(V)) \to \mathrm{AF} ,$$

where AF is the set of isomorphism classes of finite abelian groups, is an injection whose image consists of the groups generated by $n-1$ elements. If A is such a group, the associated Hecke operator $\mathrm{T}_A \in \mathrm{H}(G)$ satisfies, by definition,

$$\mathrm{T}_A(\underline{M}) = \sum_N \underline{N} ,$$

where the sum is taken over the subgroups N of M with $M/N \simeq A$. When A runs through the finite abelian groups generated by $n-1$ elements, these operators T_A therefore form a $\mathbb{Z}$-basis of $\mathrm{H}(G)$. It is clear that we have $\mathrm{T}_{A\times B} = \mathrm{T}_A\mathrm{T}_B$ if $|A|$ and $|B|$ are relatively prime and that we have $\mathrm{T}_A \in \mathrm{H}_p(G)$ if and only if A is a p-group.

If $n = 2$, we easily verify that $\mathrm{T}_A^{\mathrm{t}} = \mathrm{T}_A$ for every A; in particular, $\mathrm{H}(G)$ is commutative (the notation T^{t} is defined in Sect. 4.2.1). The first statement no longer holds for $n > 2$, but $\mathrm{H}(G)$ remains commutative. We can see this simply by endowing V with a nondegenerate symmetric bilinear form. The map $\underline{M} \mapsto \underline{M^\sharp}$ is an involution of $\underline{\mathcal{R}_{\mathbb{Z}}}(V)$. It induces a linear involution of $\mathbb{Z}[\,\underline{\mathcal{R}_{\mathbb{Z}}}(V)\,]$ and then, by conjugation, an involution ι of $\mathrm{H}(G)$, which is nothing more than $(T_{\underline{N},\underline{M}}) \mapsto (T_{\underline{N^\sharp},\underline{M^\sharp}})$ on the associated matrices. But for $N \subset M$, the quotient $N^\sharp/M^\sharp$ is in perfect duality with M/N and therefore ι coincides with the canonical anti-involution of $\mathrm{H}(G)$: $\iota(T) = T^{\mathrm{t}}$ for every $T \in \mathrm{H}(G)$ (see also [188, Sect. 3]).

Let us now discuss the case of orthogonal and symplectic $\mathbb{Z}$-groups, which is particularly important for this book [174, 88, 5]. We use the notation of Sect. 4.1.2; in particular, $V = L \otimes \mathbb{Q}$ has even dimension n, φ is a bilinear form on V that is symmetric (resp. alternating), for which L is self-dual and even, and $G \subset \mathrm{GL}_L$ is the group O_L (resp. Sp_L).

In this case, we have seen that $\mathcal{R}(G)$ can be identified with the $G(\mathbb{A}_f)$-set $\mathcal{R}_{\mathbb{Z}}^{\mathrm{a}}(V)$ of self-dual lattices in V (Proposition 4.1.4). For (N, M) in the product $\mathcal{R}_{\mathbb{Z}}^{\mathrm{a}}(V) \times \mathcal{R}_{\mathbb{Z}}^{\mathrm{a}}(V)$, the isomorphism class of the abelian group $M/(N \cap M)$ depends only on the $G(\mathbb{A}_f)$-orbit of (N, M). We have thus defined a natural map

$$G(\mathbb{A}_f)\backslash(\mathcal{R}_{\mathbb{Z}}^{\mathrm{a}}(V) \times \mathcal{R}_{\mathbb{Z}}^{\mathrm{a}}(V)) \to \mathrm{AF} , \quad (N, M) \mapsto M/(N \cap M) . \tag{4.2.4}$$

Proposition 4.2.7. *The map* (4.2.4) *is an injection whose image consists of the groups generated by* $n/2$ *elements.*

This proposition is well known; we will recall a proof at the end of this subsection for the sake of the reader. Let A be a finite abelian group generated by at most $n/2$ elements. To this group corresponds a Hecke operator

$$\mathrm{T}_A \in \mathrm{H}(G)$$

defined by $\mathrm{T}_A(M) = \sum_N N$, where the sum is taken over the N such that $M/(N \cap M) \simeq A$ or, equivalently, over the A-neighbors of M in the sense of Scholium-Definition 3.1.2 in the quadratic case. These operators T_A therefore form a $\mathbb{Z}$-basis of $\mathrm{H}(G)$. We, of course, still have $\mathrm{T}_{A\times B} = \mathrm{T}_A \mathrm{T}_B$ if $|A|$ and $|B|$ are relatively prime, and $\mathrm{T}_A \in \mathrm{H}_p(G)$ if and only if A is a p-group. From the point of view of Chap. 3, an operator that is particularly important for us is $\mathrm{T}_{\mathbb{Z}/d\mathbb{Z}}$ for $d \geq 1$, which we also denote simply by T_d.

Proposition 4.2.8. *Let A be a finite abelian group generated by $n/2$ elements. Then we have $\mathrm{T}_A^{\mathrm{t}} = \mathrm{T}_{A^\vee} = \mathrm{T}_A$. In particular, the ring $\mathrm{H}(G)$ is commutative.*

Proof. The first assertion follows from Scholium-Definition 3.1.2 when φ is symmetric, and from a similar argument in the alternating case. The second assertion follows from the first by the end of Sect. 4.2.1. See also [174, Chap. III], [88, Kap. IV], and Sect. 6.2.8. □

Finally, let us discuss the group of projective similitudes $\mathrm{P}\widetilde{G}$. Let p be a prime and $\mathcal{R}^{\mathrm{h}}_{\mathbb{Z}_p}(V_p)$ the set of even homodual lattices in V_p, introduced after Lemma 4.1.3. Recall that if φ is symmetric (resp. alternating), a lattice $M \in \mathcal{R}_{\mathbb{Z}_p}(V_p)$ is homodual if and only if there exists a $\lambda_M \in p^{\mathbb{Z}}$, necessarily unique, such that $x \mapsto \lambda_M\, \varphi(x,x)/2$ (resp. $\lambda_M \varphi$) endows M with the structure of a q-module (resp. a-module) over $\mathbb{Z}_p$. Since the q-vector space V_p is hyperbolic by Scholium 2.2.5, the same holds for $M \in \mathcal{R}^{\mathrm{h}}_{\mathbb{Z}_p}(V_p)$ as a q-module over $\mathbb{Z}_p$, by Proposition 2.1.2. This shows that the map $g \mapsto g(L)$ induces isomorphisms $\mathcal{R}_p(\widetilde{G}) \xrightarrow{\sim} \mathcal{R}^{\mathrm{h}}_{\mathbb{Z}_p}(V_p)$ and $\mathcal{R}_p(G) \xrightarrow{\sim} \mathcal{R}^{\mathrm{a}}_{\mathbb{Z}_p}(V_p)$. In particular, the set $\underline{\mathcal{R}}^{\mathrm{h}}_{\mathbb{Z}_p}(V_p) := \mathbb{Q}_p^\times \backslash \mathcal{R}^{\mathrm{h}}_{\mathbb{Z}_p}(V_p)$ can be naturally identified with $\mathcal{R}_p(\mathrm{P}\widetilde{G})$.

Consider $M \in \mathcal{R}^{\mathrm{h}}_{\mathbb{Z}_p}(V_p)$. We denote by $\mathrm{v}_M \in \mathbb{Z}$ the unique integer such that $\lambda_M = p^{-\mathrm{v}_M}$. For $g \in \widetilde{G}(\mathbb{Q}_p)$, we have $\mathrm{v}_{g(M)} = \mathrm{v}_M + v$, where v is the p-adic valuation of $\nu(g)$. Let $(\underline{N}, \underline{M})$ be an ordered pair of elements of $\underline{\mathcal{R}}^{\mathrm{h}}_{\mathbb{Z}_p}(V_p)$. After changing the representative N if necessary, we may assume $\mathrm{v}_M - \mathrm{v}_N \in \{0,1\}$. The pair $(M/N \cap M, \mathrm{v}_M - \mathrm{v}_N)$ then depends only on the $\mathrm{P}\widetilde{G}(\mathbb{Q}_p)$-orbit of $(\underline{N}, \underline{M})$, which defines a map

$$\mathrm{P}\widetilde{G}(\mathbb{Q}_p)\backslash(\underline{\mathcal{R}}^{\mathrm{h}}_{\mathbb{Z}_p}(V_p) \times \underline{\mathcal{R}}^{\mathrm{h}}_{\mathbb{Z}_p}(V_p)) \to \mathrm{AF} \times \{0,1\}\,. \tag{4.2.5}$$

Proposition 4.2.9. *The map (4.2.5) is an injection whose image is the set of pairs $(A,-)$ with A an abelian p-group generated by $n/2$ elements.*

We push back the proof of this proposition to Sect. 6.2.8. Consider $(A,i) \in \mathrm{AF} \times \{0,1\}$, where A is a p-group generated by at most $n/2$ elements. We say that $\underline{N} \in \underline{\mathcal{R}}^{\mathrm{h}}_{\mathbb{Z}_p}(V_p)$ is an A-neighbor of type i of $\underline{M} \in \underline{\mathcal{R}}^{\mathrm{h}}_{\mathbb{Z}_p}(V_p)$ if the image of $(\underline{N}, \underline{M})$ by the map (4.2.5) is (A,i). The corresponding Hecke operator is denoted by

$$\mathrm{T}_{(A,i)} \in \mathrm{H}_p(\mathrm{P}\widetilde{G})\;;$$

these operators form a $\mathbb{Z}$-basis of $\mathrm{H}_p(\mathrm{P}\widetilde{G})$. If we have $M^\sharp = M$, then $\underline{N}$ is an A-neighbor of type 0 of $\underline{M}$ if and only if $\underline{N}$ has a self-dual representative, which is then unique, and if the latter is an A-neighbor of M in the previous sense. The notion of an A-neighbor of type 1 of $\underline{M}$ is, on the other hand, "new." The following example will be particularly important in this book.

Consider $M, N \in \mathcal{R}^{\mathrm{h}}_{\mathbb{Z}}(V)$. Following Koch and Venkov in the quadratic case [127], we say that N is a *perestroika* of M with respect to p if we have

$$pM \subsetneq N \subsetneq M \ .$$

We easily verify that N is a perestroika of M with respect to p if and only if we have $\mathrm{v}_M - \mathrm{v}_{p^{-1}N} = 1$ and $\underline{N}$ is a 0-neighbor of $\underline{M}$ of type 1. Moreover, the following proposition is immediate.

Proposition 4.2.10. *Consider $M \in \mathcal{R}^{\mathrm{h}}_{\mathbb{Z}}(V)$, and let p be a prime number. The map $N \mapsto N/pM$ defines a bijection from the set of perestroikas of M with respect to p onto the set of Lagrangians of $M \otimes \mathbb{F}_p$.*

The perestroika operator with respect to p is the operator

$$\mathrm{K}_p := \mathrm{T}_{(0,1)} \in \mathrm{H}_p(\mathrm{P}\widetilde{G}) \ .$$

For $(N, M) \in \mathcal{R}^{\mathrm{h}}_{\mathbb{Z}}(V)$, N is a perestroika of M with respect to p if and only if pM is a perestroika of N with respect to p. In particular, we have $\mathrm{K}_p^{\mathrm{t}} = \mathrm{K}_p$. In fact, we have $T^{\mathrm{t}} = T$ for every $T \in \mathrm{H}(\mathrm{P}\widetilde{G})$, as we will see in Sect. 6.2.8.

Let us conclude this subsection, as announced, with a proof of Proposition 4.2.7.

Proof of Proposition 4.2.7. We place ourselves in the quadratic setting, that is, φ symmetric and $\mathrm{q}(x) = \varphi(x, x)/2$, in which case L is a q-module over $\mathbb{Z}$. The proof in the alternating setting is similar (and even simpler).

We must show that if U is a hyperbolic q-vector space over $\mathbb{Q}_p$ and (L_1, L_2) and (L'_1, L'_2) are two ordered pairs of self-dual lattices in U such that $L_1/(L_1 \cap L_2) \simeq L'_1/(L'_1 \cap L'_2)$, then there exists an $\alpha \in \mathrm{O}(U)$ with $\alpha(L_i) = L'_i$ for $i = 1, 2$. We use induction on $\dim(U)$.

The cases $U = 0$ and $L_1 = L_2$ are trivial. We assume $L_1 \neq L_2$; the annihilator of the quotient $L_1/(L_1 \cap L_2)$ is therefore of the form $p^\nu \mathbb{Z}_p$ with $\nu \geq 1$. Moreover, there exist an element e_1 of L_1 and an element e_2 of L_2 such that we have

$$\mathrm{q}(e_1) = 0 \ , \quad \mathrm{q}(e_2) = 0 \ , \quad e_1.e_2 = p^{-\nu} \ .$$

Indeed, it is first of all easy to see that there exist an element ϵ_1 of L_1 and an element ϵ_2 of L_2 with $\epsilon_1.\epsilon_2 = p^{-\nu}$. Hensel's lemma then shows that there exists a matrix

$$P = \begin{bmatrix} a_{1,1} & a_{1,2} \\ a_{2,1} & a_{2,2} \end{bmatrix}$$

with coefficients in $\mathbb{Z}_p$, with $P \equiv \mathrm{I} \bmod p^\nu$, such that we have

$$ {}^{\mathrm{t}}P \begin{bmatrix} 2\mathrm{q}(\epsilon_1) & p^{-\nu} \\ p^{-\nu} & 2\mathrm{q}(\epsilon_2) \end{bmatrix} P = \begin{bmatrix} 0 & p^{-\nu} \\ p^{-\nu} & 0 \end{bmatrix} . $$

We take $e_1 = a_{1,1}\epsilon_1 + a_{2,1}\epsilon_2$ and $e_2 = a_{1,2}\epsilon_1 + a_{2,2}\epsilon_2 \in L_2$ (the congruence $P \equiv \mathrm{I} \bmod p^\nu$ implies $e_1 \in L_1$ and $e_2 \in L_2$). This concludes the proof of the statement.

Let us now finish the induction. We denote by H, H_1, and H_2, respectively, the linear subspace of U generated by e_1 and e_2, the submodule of L_1 generated by e_1 and $p^\nu e_2$, and the submodule of L_2 generated by $p^\nu e_1$ and e_2. We endow H, H_1, and H_2 with the quadratic forms induced by those on U. By construction, we have $H \approx \mathrm{H}(\mathbb{Q}_p)$ and $H_i \approx \mathrm{H}(\mathbb{Z}_p)$ for $i = 1, 2$. We denote by W, M_1, and M_2, respectively, the orthogonal complement of H in U, the orthogonal complement of H_1 in L_1, and the orthogonal complement of H_2 in L_2. We have decompositions into orthogonal sums

$$ U = H \oplus W , \quad L_1 = H_1 \oplus M_1 , \quad L_2 = H_2 \oplus M_2 $$

and isomorphisms

$$ L_1/(L_1 \cap L_2) \cong H_1/(H_1 \cap H_2) \oplus M_1/(M_1 \cap M_2) , \quad H_1/(H_1 \cap H_2) \cong \mathbb{Z}_p/p^\nu \mathbb{Z}_p . $$

We replace the ordered pair (L_1, L_2) by the ordered pair (L_1', L_2') and introduce the q-vector spaces H' and W' and the q-modules H_1', H_2', M_1', and M_2' analogously. We obtain the desired automorphism $\alpha \colon U \to U$ as the orthogonal sum of suitable isomorphisms of q-vector spaces $H \to H'$ and $W \to W'$; the existence of the second is ensured by the induction hypothesis. □

4.2.11 $\mathrm{H}(\mathrm{SO}_L)$ *Versus* $\mathrm{H}(\mathrm{O}_L)$

Let L be a q-module over $\mathbb{Z}$. Let us briefly discuss the link between $\mathrm{H}(\mathrm{SO}_L)$ and $\mathrm{H}(\mathrm{O}_L)$. The cases PGSO_L and PGO_L can be treated similarly.

By Proposition 4.1.7, the inclusion $\mathrm{SO}_L \to \mathrm{O}_L$ induces an $\mathrm{SO}_L(\mathbb{A}_f)$-equivariant bijection $\mathcal{R}(\mathrm{SO}_L) \xrightarrow{\sim} \mathcal{R}(\mathrm{O}_L)$. It follows that $\mathrm{H}(\mathrm{O}_L)$ can be canonically identified with a subring of $\mathrm{H}(\mathrm{SO}_\mathrm{L})$: these are the subrings of $\mathrm{End}_{\mathbb{Z}}(\mathbb{Z}[\mathcal{R}^{\mathrm{a}}_{\mathbb{Z}}(V)])$ consisting of the $\mathrm{O}_L(\mathbb{A}_f)$-equivariant and $\mathrm{SO}_L(\mathbb{A}_f)$-equivariant endomorphisms, respectively. The quotient group

$$ \mathrm{O}_L(\mathbb{A}_f)/\mathrm{SO}_L(\mathbb{A}_f) \simeq (\mathbb{Z}/2\mathbb{Z})^{\mathrm{P}} $$

acts naturally by conjugation on $\mathrm{H}(\mathrm{SO}_L)$, with ring of invariants $\mathrm{H}(\mathrm{O}_L)$. This action respects the decomposition of $\mathrm{H}(G)$ as a tensor product of the $\mathrm{H}_p(G)$ over the $p \in \mathrm{P}$ and also identifies $\mathrm{H}_p(\mathrm{O}_L)$ with $\mathrm{H}_p(\mathrm{SO}_L)^{\mathbb{Z}/2\mathbb{Z}}$.

Let us give an example of an element of $\mathrm{H}_p(\mathrm{SO}_L)$ that is not in $\mathrm{H}_p(\mathrm{O}_L)$. Consider $A = (\mathbb{Z}/p\mathbb{Z})^{n/2}$, where n is the rank of L. Let Ω be the set of pairs (N, M) of elements of $\mathcal{R}^{\mathrm{a}}_{\mathbb{Z}}(V)$ such that N is an A-neighbor of M. Proposition 4.2.7 asserts that Ω is an $\mathrm{O}_L(\mathbb{Q}_p)$-orbit. However, it is the disjoint union of two orbits under the action of $\mathrm{SO}_L(\mathbb{Q}_p)$. To see this, we begin by verifying, using arguments similar to those in Sect. 3.1, that the map

$$N \mapsto (M \cap N)/pM$$

induces a surjection (that is not bijective in general) between the A-neighbors of M and the Lagrangians of the hyperbolic q-vector space $M \otimes \mathbb{F}_p$. But it is well known that for every field k and every hyperbolic q-vector space V over k, there are exactly two orbits of Lagrangians of V under the action of $\mathrm{SO}(V)$ (and only one under $\mathrm{O}(V)$, by Witt's theorem). By the smoothness of SO_M over $\mathbb{Z}_p$, each of these two orbits therefore defines an $\mathrm{SO}(M)$-orbit of A-neighbors of M and, consequently, two distinct Hecke operators $\mathrm{T}^{\pm}_A \in \mathrm{H}(\mathrm{SO}_L)$ with sum T_A, which are interchanged under the action of $\mathrm{O}_L(\mathbb{Q}_p)/\mathrm{SO}_L(\mathbb{Q}_p) = \mathbb{Z}/2\mathbb{Z}$.

4.2.12 Isogenies

We will now discuss the isogenies between transitive Γ-sets, by presenting a variant of the considerations in [174, Chap. II, Sect. 7].

Let X be a Γ-set and X' a Γ'-set. Recall that a morphism $X \to X'$ is a pair (f, g), where $g\colon X \to X'$ is a map and $f\colon \Gamma \to \Gamma'$ is a group morphism such that we have $g(\gamma x) = f(\gamma)g(x)$ for every $x \in X$ and every $\gamma \in \Gamma$. In what follows, we conveniently assume that a transitive set is nonempty.

Lemma 4.2.13. *Let X be a transitive Γ-set, X' a Γ'-set, and (f, g) a morphism $X \to X'$ such that $f(\Gamma)$ is normal in Γ'. Let S be the stabilizer of $g(X)$ in Γ', that is, $S = \{\gamma \in \Gamma' ; \gamma g(X) \subset g(X)\}$.*

(i) *For every $x \in g(X)$, we have $S = f(\Gamma)\Gamma'_x$.*
(ii) *We have $S = \{\gamma \in \Gamma' ; \gamma g(X) \cap g(X) \neq \emptyset\}$.*

Proof. Take $x \in g(X)$. Since the subgroup $f(\Gamma)$ is normal in Γ', the subset $E_x := f(\Gamma)\Gamma'_x \subset \Gamma'$ is a subgroup. The transitivity of X then shows that

- E_x does not depend on the choice of $x \in g(X)$;
- E_x is the set of $\gamma \in \Gamma'$ with $\gamma(x) \in g(X)$.

We consequently have the identities $S = \bigcap_{x \in g(X)} E_x = \bigcup_{x \in g(X)} E_x = \{\gamma \in \Gamma' ; \gamma g(X) \cap g(X) \neq \emptyset\}$. □

Let X be a transitive Γ-set, X' a Γ'-set, and (f, g) a morphism $X \to X'$. We assume, as in the lemma above, that $f(\Gamma)$ is normal in Γ' and, moreover that the map

g is injective.[3] Let S be the stabilizer of $g(X)$ in Γ'. The map $(s, x) \mapsto g^{-1}(s(g(x))$, which is well defined by the injectivity of g, defines an action of S on X whose restriction to $f: \Gamma \to S$ is the Γ-set X. It therefore induces an action of $S/f(\Gamma)$ on $\mathrm{H}(X)$ by ring automorphisms; we denote by $\mathrm{H}(X)^{\mathrm{inv}} \subset \mathrm{H}(X)$ the subring of invariants, which is also $\mathrm{End}_{\mathbb{Z}[S]}(\mathbb{Z}[X])$.

Proposition-Definition 4.2.14. *Let $u = (f, g): X \to X'$ be a morphism between the transitive Γ-set X and the transitive Γ'-set X'. We assume that $f(\Gamma)$ is normal in Γ' and that g is injective.*

(i) *For $T \in \mathrm{H}(X)^{\mathrm{inv}}$, there exists a unique $T' \in \mathrm{H}(X')$ that vanishes on $(X' - g(X)) \times g(X)$ and satisfies $T'_{g(x),g(y)} = T_{x,y}$ for every $x, y \in X$.*
(ii) *The resulting map $\mathrm{H}(u): \mathrm{H}(X)^{\mathrm{inv}} \to \mathrm{H}(X')$ defined by $T \mapsto T'$ is an injective ring homomorphism.*

Proof. The uniqueness assertion in part (i) follows from the injectivity of g and the transitivity of X'. Assertion (ii) immediately follows from part (i). We are therefore left with justifying the existence of T' in part (i). But part (ii) of Lemma 4.2.13 shows that the injection $g: X \to X'$ induces a bijection $\mathrm{Ind}_S^{\Gamma'} X \overset{\sim}{\to} X'$ and therefore an isomorphism $\mathbb{Z}[\Gamma'] \otimes_{\mathbb{Z}[S]} \mathbb{Z}[X] \overset{\sim}{\to} \mathbb{Z}[X']$. Thus, when composed with $g: \mathbb{Z}[X] \to \mathbb{Z}[X']$, every S-equivariant linear map $T: \mathbb{Z}[X] \to \mathbb{Z}[X]$ extends uniquely to a Γ'-equivariant map $T': \mathbb{Z}[X'] \to \mathbb{Z}[X']$; this has the desired properties. □

In all the examples we consider, it turns out that the group S preserves every Γ-orbit of $X \times X$, so that we have $\mathrm{H}(X)^{\mathrm{inv}} = \mathrm{H}(X)$. A particularly simple case is that where we have $\Gamma' = \Gamma$ and $X' = X$ and f and g are bijective. In this case, we have $S = f(\Gamma)$ and $\mathrm{H}(u)$ is, by definition, the automorphism of $\mathrm{H}(X)$ whose matrix is given by $(T_{x,y}) \mapsto (T_{g^{-1}x,g^{-1}y})$.

Let us assume that the hypotheses of Proposition-Definition 4.2.14 hold. For a Γ'-module M, we denote by $M_{|\Gamma}$ the Γ-module obtained by restricting M via $f: \Gamma \to \Gamma'$. We then have a canonical injective map

$$M_{X'} \to (M_{|\Gamma})_X\,, \quad \varphi \mapsto \varphi_{|X} := \varphi \circ g\,.$$

The following lemma is immediate.

Lemma 4.2.15. *Under the assumptions of Proposition-Definition 4.2.14, let M be a Γ'-module, and take $T \in \mathrm{H}(X)^{\mathrm{inv}}$ and $\varphi \in M_{X'}$. Then we have $T(\varphi_{|X}) = \mathrm{H}(u)(T)(\varphi)$.*

Example 4.2.16. By way of example, we return to the context of the similitude groups (Sect. 4.1.2) and consider the natural $\mathbb{Z}$-morphism $\mu: G \to \mathrm{P}\widetilde{G}$. The results of this

[3] We refer to the article of Satake for a variant without the injectivity assumption on g. The reader will not miss much in the current discussion by assuming $\Gamma \subset \Gamma'$ and $X \subset X'$, with f and g the corresponding inclusions.

section apply and define a ring morphism

$$\mathrm{H}(\mu)\colon \mathrm{H}(G) \to \mathrm{H}(\mathrm{P}\widetilde{G})$$

with $\mathrm{H}(\mu)(\mathrm{T}_A) = \mathrm{T}_{(A,0)}$ for every finite abelian group A generated by at most $n/2$ elements.

Indeed, consider $\Gamma = G(\mathbb{A}_f)$, $X = \mathcal{R}(G)$, $\Gamma' = \mathrm{P}\widetilde{G}(\mathbb{A}_f)$, and $X' = \mathcal{R}(\mathrm{P}\widetilde{G})$, and for f and g, take the natural maps deduced from μ. The group Γ is a normal subgroup of $\widetilde{G}(\mathbb{A}_f)$; likewise, $f(\Gamma)$ is a normal subgroup of Γ'. Moreover, g can be identified with the natural injection $\mathcal{R}_{\mathbb{Z}}^{\mathrm{a}}(V) \to \mathcal{R}_{\mathbb{Z}}^{\mathrm{h}}(V)$ defined by $M \mapsto \underline{M}$, by Proposition 4.1.4. The group S is the subgroup of elements $g \in \widetilde{G}(\mathbb{A}_f)$ such that $\nu(g)$ is of the form $a^2 b$ with $a \in \mathbb{A}_f^\times$ and $b \in \widehat{\mathbb{Z}}^\times$. It acts trivially on $\Gamma\backslash(\mathcal{R}_{\mathbb{Z}}^{\mathrm{a}}(V) \times \mathcal{R}_{\mathbb{Z}}^{\mathrm{a}}(V))$. Indeed, given $N, M \in \mathcal{R}_{\mathbb{Z}}^{\mathrm{a}}(V)$, $g \in \widetilde{G}(\mathbb{A}_f)$, and a prime p, the map g induces an isomorphism $M_p/(N_p \cap M_p) \simeq g(M)_p/(g(N)_p \cap g(M)_p)$, which allows us to conclude using Proposition 4.2.7. The assertion on T_A follows from the discussion following Proposition 4.2.9.

4.3 Automorpic Forms of a $\mathbb{Z}$-group

The ring of adeles of $\mathbb{Q}$ is the ring $\mathbb{A} = \mathbb{R} \times \mathbb{A}_f$. Let G be a $\mathbb{Z}$-group. The group $G(\mathbb{R})$ is naturally a Lie group, and the group

$$G(\mathbb{A}) = G(\mathbb{R}) \times G(\mathbb{A}_f)$$

is locally compact and separated for the product topology; we already recalled the topology on $G(\mathbb{A}_f)$ in Sect. 4.2.5. There is a natural diagonal embedding of the group $G(\mathbb{Q})$ in $G(\mathbb{A})$; the image is a discrete closed subgroup (see [92, Chap. II, Sect. 3] for the basics on these constructions).

4.3.1 Square-Integrable Automorphic Forms

Let us recall some classical results due to Borel and Harish-Chandra, for which we refer to [32, Sect. 5]. We assume that the neutral component of $G(\mathbb{C})$ is semisimple [103, 34]. The locally compact group $G(\mathbb{A})$ is then unimodular. By Weil, the homogeneous space

$$G(\mathbb{Q})\backslash G(\mathbb{A})$$

inherits a positive (nonzero) Radon measure μ invariant under the action of $G(\mathbb{A})$ by right translations [211, Chap. II], [172, Chap. 2]. It has finite measure.

The space of *square-integrable automorphic forms for G* is the subspace

$$\mathcal{A}^2(G) \subset \mathrm{L}^2(G(\mathbb{Q})\backslash G(\mathbb{A}), \mu)$$

of elements that are invariant under $G(\widehat{\mathbb{Z}})$ for right translations [92, Chap. 3], [36, Sect. 4]. It is a Hilbert space for the Hermitian inner product

$$\langle f, f' \rangle_{\mathrm{Pe}} := \int \overline{f} f' \, \mathrm{d}\mu ,$$

also called the *Petersson inner product*. Alternatively, $\mathcal{A}^2(G)$ can be viewed as the space of square-integrable functions on $G(\mathbb{Q})\backslash G(\mathbb{A})/G(\widehat{\mathbb{Z}})$ endowed with the Radon measure that is the image of μ by the canonical (proper) map $G(\mathbb{Q})\backslash G(\mathbb{A}) \to G(\mathbb{Q})\backslash G(\mathbb{A})/G(\widehat{\mathbb{Z}})$. The space $\mathcal{A}^2(G)$ is endowed with two important additional structures that we will now describe.

On the one hand, since the space $\mathcal{A}^2(G)$ is the space of $G(\widehat{\mathbb{Z}})$-invariants of the $G(\mathbb{A}_f)$-module $\mathrm{L}^2(G(\mathbb{Q})\backslash G(\mathbb{A}), \mu)$ for the right translations, it is endowed with a right action of the Hecke ring $\mathrm{H}(G)$ (Sects. 4.2.2, 4.2.5). This action is a $\star$*-action* for the Petersson inner product. By this, we mean that the adjoint of $T \in \mathrm{H}(G)$ is the operator T^{t} defined in Sect. 4.2.1: for $f, f' \in \mathcal{A}^2(G)$ and $T \in \mathrm{H}(G)$, we have

$$\langle T(f), f' \rangle_{\mathrm{Pe}} = \langle f, T^{\mathrm{t}}(f') \rangle_{\mathrm{Pe}} . \tag{4.3.1}$$

Indeed, this is a consequence of Lemma 4.2.3 and the unimodularity of $G(\mathbb{A}_f)$.

On the other hand, $\mathcal{A}^2(G)$ is stable under the action of $G(\mathbb{R})$ by right translations, and this action commutes with that of $\mathrm{H}(G)$. It turns $\mathcal{A}^2(G)$ into a unitary representation of the Lie group $G(\mathbb{R})$ (we refer to [119] as a general reference on unitary representations). A more classical description of this representation is obtained by writing

$$G(\mathbb{A}_f) = \coprod_{i=1}^{\mathrm{h}(G)} G(\mathbb{Q}) g_i G(\widehat{\mathbb{Z}}) \tag{4.3.2}$$

for certain elements $g_i \in G(\mathbb{A}_f)$, by the finiteness of the class set of G. For every i, the double coset $G(\mathbb{Q})g_i G(\widehat{\mathbb{Z}})$ is an open subset of $G(\mathbb{A}_f)$ and the *congruence subgroup*

$$\Gamma_i = G(\mathbb{Q}) \cap g_i G(\widehat{\mathbb{Z}}) g_i^{-1}$$

is a discrete subgroup of $G(\mathbb{R})$ that is commensurable with $G(\mathbb{Z})$. The map $f \mapsto (f_{|G(\mathbb{R}) \times g_i})_i$ induces a $G(\mathbb{R})$-equivariant isomorphism

$$\mathcal{A}^2(G) \xrightarrow{\sim} \prod_{i=1}^{\mathrm{h}(G)} \mathrm{L}^2(\Gamma_i \backslash G(\mathbb{R})) , \tag{4.3.3}$$

where each $\Gamma_i\backslash G(\mathbb{R})$ naturally inherits a strictly positive Radon measure that is right invariant for $G(\mathbb{R})$, has finite mass, and is uniquely determined by μ. This representation of $G(\mathbb{R})$ in general has a "discrete" part that is notoriously difficult to describe, as well as a "continuous" part whose study was reduced by Langlands to that of discrete subsets for auxiliary groups G' [138].

4.3.2 *The Set* $\Pi_{\mathrm{disc}}(G)$

Here, we are interested only in the discrete part of $\mathcal{A}^2(G)$, that is, in the subspace

$$\mathcal{A}_{\mathrm{disc}}(G) \subset \mathcal{A}^2(G)$$

defined as the closure of the sum of the closed and topologically irreducible sub-$G(\mathbb{R})$-representations of $\mathcal{A}^2(G)$. It is a representation of $G(\mathbb{R})$ that is, by construction, an orthogonal sum of irreducible representations,[4] where each component has a finite multiplicity by a fundamental result due to Harish-Chandra (see the introduction of [101], as well as Theorem 1 of Chap. 1 of the same reference; see also [36]). In other words, if U is a unitary irreducible representation of $G(\mathbb{R})$, then the space

$$\mathcal{A}_U(G) := \mathrm{Hom}_{G(\mathbb{R})}(U, \mathcal{A}_{\mathrm{disc}}(G)) = \mathrm{Hom}_{G(\mathbb{R})}(U, \mathcal{A}^2(G))$$

has finite dimension over $\mathbb{C}$. We have, of course, a canonical isomorphism

$$\widehat{\bigoplus_{U \in \mathrm{Irr}(G(\mathbb{R}))}} U \otimes \mathcal{A}_U(G) \xrightarrow{\sim} \mathcal{A}_{\mathrm{disc}}(G)\ , \tag{4.3.4}$$

where $\mathrm{Irr}(H)$ is the set of isomorphism classes of topologically irreducible unitary representations of the locally compact group H.

The right $\mathrm{H}(G)$-module structure on $\mathcal{A}^2(G)$ naturally induces the structure of a right $\mathrm{H}(G)$-module on $\mathcal{A}_U(G)$. The latter also inherits a Hermitian inner product for which the action of $\mathrm{H}(G)$ is again a $\star$-action. For example, for a fixed nonzero $e \in U$ and $\varphi, \varphi' \in \mathcal{A}_U(G)$, we can set $\langle \varphi, \varphi' \rangle = \langle \varphi(e), \varphi'(e) \rangle_{\mathrm{Pe}}$. But it is well known that a sub-$\mathbb{C}$-algebra of $\mathrm{M}_n(\mathbb{C})$ that is stable under $M \mapsto {}^{\mathrm{t}}\overline{M}$ is semisimple: if X is in its Jacobson radical, then the Hermitian matrix $X\,{}^{\mathrm{t}}\overline{X}$ is nilpotent, hence zero, which implies that X is zero. In particular, $\mathcal{A}_U(G)$ is semisimple when viewed as a representation of the $\mathbb{C}$-algebra $\mathrm{H}(G)^{\mathrm{opp}} \otimes \mathbb{C}$.

We define a *representation of* $(G(\mathbb{R}), \mathrm{H}(G))$ to be a Hilbert space endowed with a unitary representation of $G(\mathbb{R})$, together with the structure of a right $\mathrm{H}(G)$-module, such that the action of any element of $G(\mathbb{R})$ commutes with that of any element of $\mathrm{H}(G)$. These representations naturally form a $\mathbb{C}$-linear category: a morphism $E \to F$ is a continuous $\mathbb{C}$-linear map $E \to F$ that commutes with the actions of $G(\mathbb{R})$ and $\mathrm{H}(G)$. For a unitary representation U of $G(\mathbb{R})$ and a $\mathrm{H}(G)^{\mathrm{opp}} \otimes \mathbb{C}$-module V of finite dimension as a $\mathbb{C}$-vector space, $U \otimes V$ is naturally a representation of $(G(\mathbb{R}), \mathrm{H}(G))$ (where the tensor product is taken over $\mathbb{C}$). We denote by $\Pi(G)$ the set of isomorphism classes of representations of $(G(\mathbb{R}), \mathrm{H}(G))$ of this form

[4] At this point, it is useful to recall the following version of Schur's lemma. Let U and V be Hilbert spaces endowed with unitary representations of a group Γ. We assume that U is topologically irreducible and that $u\colon U \to V$ is a nonzero, Γ-equivariant, continuous linear map. Then the adjoint $u^*\colon V \to U$ (which is Γ-equivariant) satisfies $u^* \circ u = \lambda \mathrm{Id}_U$ for some $\lambda \in \mathbb{R}^\times$. Indeed, $u^* \circ u \in \mathrm{End}(U)$ is Hermitian and nonzero and commutes with Γ; by the spectral theorem, its spectrum is therefore reduced to a point $\{\lambda\}$. It follows that V is the orthogonal sum of $\mathrm{Im}(u)$ (which is closed) and $\mathrm{Ker}(u^*)$.

such that, moreover, U is topologically irreducible and V is simple. The restriction to $G(\mathbb{R})$ of such a unitary representation π is isomorphic to $U^{\dim V}$, so that the isomorphism class π_∞ of the unitary representation U is fully determined by the unitary representation of $G(\mathbb{R})$ underlying π. Likewise, the $\mathrm{H}(G)^{\mathrm{opp}} \otimes \mathbb{C}$-module underlying π is semisimple and V-isotypical, so that the isomorphism class π_f of the $\mathrm{H}(G)^{\mathrm{opp}} \otimes \mathbb{C}$-module V is uniquely determined by that of π. In particular, we have $\pi \simeq \pi_\infty \otimes \pi_f$ for every $\pi \in \Pi(G)$. Finally, Schur's lemma implies that every $\pi \in \Pi(G)$ is topologically irreducible as a representation of $(G(\mathbb{R}), \mathrm{H}(G))$.

By the discussion above, for $U \in \mathrm{Irr}(G(\mathbb{R}))$, the space $U \otimes \mathcal{A}_U(G)$ is naturally a representation of $(G(\mathbb{R}), \mathrm{H}(G))$, as is $\mathcal{A}_{\mathrm{disc}}(G)$, where the isomorphism (4.3.4) trivially commutes with the actions of $G(\mathbb{R})$ and $\mathrm{H}(G)$. It follows that we have a decomposition into a Hilbert sum of elements of $\Pi(G)$ that refines the decomposition (4.3.4):

$$\mathcal{A}_{\mathrm{disc}}(G) \simeq \widehat{\bigoplus_{\pi \in \Pi(G)}} m(\pi)\, \pi\ , \tag{4.3.5}$$

where $m(\pi) \geq 0$ is an integer that is called the *multiplicity of* π. By definition, if $\pi \in \Pi(G)$ and $U \simeq \pi_\infty$, then $\mathrm{m}(\pi)$ is the multiplicity of π_f in the $\mathrm{H}(G)^{\mathrm{opp}} \otimes \mathbb{C}$-module $\mathcal{A}_U(G)$, which is semisimple and of finite dimension. We denote by

$$\Pi_{\mathrm{disc}}(G) \subset \Pi(G)$$

the subsets consisting of the π with $m(\pi) \neq 0$.

The elements of $\Pi_{\mathrm{disc}}(G)$ are called the *discrete automorphic representations*[5] of G. The only truly obvious example of such a representation is the *trivial representation*, denoted 1_G, realized as the subspace (of dimension 1) of constant functions in $\mathcal{A}^2(G)$ (note that μ has finite mass). The action of $G(\mathbb{R})$ in 1_G is, of course, the trivial action, while that of $\mathrm{H}(G)$ is the multiplication by the "degree" (see Example 6.2.3). In general, the set $\Pi_{\mathrm{disc}}(G)$ is countably infinite, which is not the case for $\Pi(G)$. We will give a few concrete examples in the following chapters.

An element $F \in \mathcal{A}_U(G)$ is called an *eigenform* if it is nonzero and generates an irreducible $\mathrm{H}(G)^{\mathrm{opp}} \otimes \mathbb{C}$-module. When $\mathrm{H}(G)$ is commutative, this is equivalent to requiring that $F \neq 0$ be an eigenvector of all Hecke operators in $\mathrm{H}(G)$. If F is an eigenform and $V \subset \mathcal{A}_U(G)$ denotes the $\mathrm{H}(G)^{\mathrm{opp}} \otimes \mathbb{C}$-module generated by F, the image of $U \otimes V$ in $\mathcal{A}_{\mathrm{disc}}(G)$ by the canonical map (4.3.4) is a topologically irreducible subrepresentation of $(G(\mathbb{R}), \mathrm{H}(G))$, which we denote by π_F; it is the (*automorphic, discrete*) *representation generated by* F. We often also denote its isomorphism class by π_F; this is an element of $\Pi_{\mathrm{disc}}(G)$.

[5] The reader should be aware that the definition we use here depends not only on $G_{\mathbb{Q}}$ but also on G as a $\mathbb{Z}$-group. In the literature, our discrete automorphic representations are more commonly called "discrete automorphic representations of $G(\mathbb{A})$ that are spherical (or unramified) with respect to $G(\widehat{\mathbb{Z}})$." The apparent loss of generality in our presentation is, however, at this point illusory, because every open compact subgroup of $G(\mathbb{A}_f)$ is of the form $G'(\widehat{\mathbb{Z}})$ for a well-chosen $\mathbb{Z}$-group G' with $G'_{\mathbb{Q}} \simeq G_{\mathbb{Q}}$.

Finally, following Gelfand, Graev, and Piatetski-Shapiro in [92, Chap. 3, Sect. 7], we consider the subspace $\mathcal{A}_{\mathrm{cusp}}(G) \subset \mathcal{A}^2(G)$ consisting of the *cusp forms* (the definition of a cusp form is recalled below). This is a closed subspace that is stable under the actions of $G(\mathbb{R})$ and $\mathrm{H}(G)$. Gelfand, Graev, and Piatetski-Shapiro show the inclusion

$$\mathcal{A}_{\mathrm{cusp}}(G) \subset \mathcal{A}_{\mathrm{disc}}(G) \tag{4.3.6}$$

(see also [35, Theorem 16.2]). We denote by

$$\Pi_{\mathrm{cusp}}(G) \subset \Pi_{\mathrm{disc}}(G)$$

the set of $\pi \in \Pi(G)$ that occur in the subspace $\mathcal{A}_{\mathrm{cusp}}(G)$; these representations are called the *cuspidal automorphic representations* of G.

When $G_{\mathbb{Q}}$ does not admit a strict parabolic sub-$\mathbb{Q}$-group, which is equivalent to saying that $G(\mathbb{Q})$ does not have any nontrivial unipotent elements, we have the obvious equality $\mathcal{A}_{\mathrm{cusp}}(G) = \mathcal{A}^2(G)$. In this case,[6] the inclusion (4.3.6) implies $\mathcal{A}_{\mathrm{disc}}(G) = \mathcal{A}^2(G)$.

Let us recall the definition of a cusp form. Let $P \subset G_{\mathbb{Q}}$ be a *strict parabolic* sub-$\mathbb{Q}$-group, that is, such that $P(\mathbb{C})$ is connected, contains a Borel subgroup of the neutral component of $G(\mathbb{C})$, and is not equal to that component [103, 34]. If N denotes the unipotent radical of P, then the locally compact group $N(\mathbb{A})$ is unimodular and its subgroup $N(\mathbb{Q})$ is discrete and cocompact. We denote by $\mathrm{d}n$ a strictly positive $\mathrm{N}(\mathbb{A})$-invariant Radion measure on $N(\mathbb{Q})\backslash N(\mathbb{A})$. Let $f\colon G(\mathbb{Q})\backslash G(\mathbb{A}) \to \mathbb{C}$ be a Borel function that is square-integrable and take $g \in G(\mathbb{A})$. The function $n \mapsto f(ng)$, $N(\mathbb{Q})\backslash N(\mathbb{A}) \to \mathbb{C}$, is then a square-integrable Borel function for almost all $g \in G(\mathbb{A})$. We say that f is a cusp form if for every strict parabolic sub-$\mathbb{Q}$-group P of $G_{\mathbb{Q}}$, we have $\int_{N(\mathbb{Q})\backslash N(\mathbb{A})} f(ng)\, \mathrm{d}n = 0$ for almost all $g \in N(\mathbb{A})\backslash G(\mathbb{A})$. We can show that the subset of $\mathrm{L}^2(G(\mathbb{Q})\backslash G(\mathbb{A}), \mu)$ consisting of the classes of cusp forms is a closed linear subspace (see, for example, [35, Proposition 8.2]). It is trivially stable under right translations by the elements of $G(\mathbb{A})$.

4.4 Automorphic Forms for O_n

4.4.1 Automorphic Forms for the $\mathbb{Z}$-groups G with $G(\mathbb{R})$ Compact

We return to the setting of Sect. 4.3.1. Suppose that the $\mathbb{Z}$-group G has the property that $G(\mathbb{R})$ is compact. Then, the groups $\Gamma_i = G(\mathbb{Q}) \cap g_i G(\widehat{\mathbb{Z}}) g_i^{-1}$ of formula (4.3.2) are finite subgroups of $G(\mathbb{R})$ because they are discrete subgroups of a compact group. Moreover, the quotient $G(\mathbb{Q})\backslash G(\mathbb{A})$ is compact because it is homeomorphic to the disjoint union of the $\Gamma_i\backslash G(\mathbb{R})$. Formula (4.3.3) then implies $\mathcal{A}_{\mathrm{disc}}(G) = \mathcal{A}^2(G)$,

[6] In fact, a famous result of Godement shows that under this same hypothesis on G, the group $G(\mathbb{Q})$ is cocompact in $G(\mathbb{A})$, which implies the equality $\mathcal{A}_{\mathrm{disc}}(G) = \mathcal{A}^2(G)$ more directly in this specific case (see, for example, [35, Lemma 16.1]).

by the Peter–Weyl theorem. We will give another description of the $\mathrm{H}(G)$-modules $\mathcal{A}_U(G)$.

For a $\mathbb{Z}[G(\mathbb{Q})]$-module U, we denote by $\mathrm{M}_U(G)$ the space of functions

$$F\colon \mathcal{R}(G) \longrightarrow U$$

such that we have $F(\gamma x) = \gamma \cdot F(x)$ for all $\gamma \in G(\mathbb{Q})$ and $x \in \mathcal{R}(G)$. It can be canonically identified with $\mathrm{Hom}_{\mathbb{Z}[G(\mathbb{Q})]}(\mathbb{Z}[\mathcal{R}(G)], U)$, which endows it with a right action of the ring $\mathrm{H}(G)$. Even better, $U \mapsto \mathrm{M}_U(G)$ defines a functor from the $G(\mathbb{Q})$-modules to the $\mathrm{H}(G)^{\mathrm{opp}}$-modules. Its additive structure is very simple because $F \mapsto (F(g_i))$ induces an isomorphism

$$\mathrm{M}_U(G) \longrightarrow \prod_{i=1}^{\mathrm{h}(G)} U^{\Gamma_i} . \tag{4.4.1}$$

In particular, we have $\mathrm{M}_{U\oplus V}(G) \simeq \mathrm{M}_U(G) \oplus \mathrm{M}_V(G)$. Observe, incidentally, that the construction so far makes sense for an arbitrary $\mathbb{Z}$-group G.

Next, assume that U is a finite-dimensional, continuous, complex representation of $G(\mathbb{R})$, and denote its dual by U^*. For $F \in \mathrm{M}_U(G)$ and $u \in U^*$, we denote by $\varphi_F(u)$ the function $G(\mathbb{R}) \times \mathcal{R}(G) \to \mathbb{C}$ defined by $(h, x) \mapsto \langle u, h^{-1}F(x)\rangle$. This function is invariant under the diagonal action of $G(\mathbb{Q})$. This is a continuous function of its first variable; it is therefore in $\mathcal{A}^2(G)$ because $G(\mathbb{Q})\backslash(G(\mathbb{R}) \times \mathcal{R}(G))$ is compact by (4.3.3). The obvious relation $\varphi_F(gu) = g \cdot (\varphi_F(u))$, which holds for $u \in U^*$ and $g \in G(\mathbb{R})$, shows that the function φ_F defined by $u \mapsto \varphi_F(u)$ is an element of $\mathcal{A}_{U^*}(G)$. The proof of the following lemma is immediate and is left to the reader.

Lemma 4.4.2. *Let U be an irreducible representation of $G(\mathbb{R})$. Then $F \mapsto \varphi_F$ is an $\mathrm{H}(G)$-equivariant isomorphism $\mathrm{M}_U(G) \overset{\sim}{\to} \mathcal{A}_{U^*}(G)$.*

Since the article of Gross [98], the elements of $\mathrm{M}_U(G)$ are sometimes called *algebraic modular forms* of weight U for the $\mathbb{Z}$-group G; we will not use this terminology, which conflicts with the notion of algebraicity introduced in Sect. 8.2.6. For example, if $U = \mathbb{C}$ is the trivial representation, then the $\mathrm{H}(G)^{\mathrm{opp}}$-module $\mathrm{M}_{\mathbb{C}}(G)$ can be canonically identified with the space of functions $\mathrm{Cl}(G) \to \mathbb{C}$ or, equivalently, with the dual of the $\mathrm{H}(G)$-module $\mathbb{C}[\mathrm{Cl}(G)]$.

Let us conclude these basic results with an assertion of compatibility with certain morphisms of $\mathbb{Z}$-groups. Let $\mu\colon G \to G'$ be a morphism of $\mathbb{Z}$-groups. It induces, in an obvious way, a morphism (f_μ, g_μ) from the $G(\mathbb{A}_f)$-set $\mathcal{R}(G)$ to the $G'(\mathbb{A}_f)$-set $\mathcal{R}(G')$, in the sense of Sect. 4.2.12. We assume that $f_\mu(G(\mathbb{A}_f))$ is a normal subgroup of $G'(\mathbb{A}_f)$, that g_μ is injective, and, moreover, that the action of the group S defined loc. cit. on $\mathcal{R}(G)$ is trivial. This is, for example, trivially the case if μ is an isomorphism. We then have an injective ring homomorphism $\mathrm{H}(\mu)\colon \mathrm{H}(G) \to \mathrm{H}(G')$ defined loc. cit. Let U' be a $G'(\mathbb{Q})$-module, and let U be its restriction to $G(\mathbb{Q})$. The following lemma paraphrases Lemma 4.2.15.

Lemma 4.4.3. *The morphism* $\mu^* : \mathrm{M}_{U'}(G') \longrightarrow \mathrm{M}_U(G)$ *defined by* $\varphi \mapsto (x \mapsto \varphi(g_\mu(x)))$ *satisfies* $T \circ \mu^* = \mu^* \circ \mathrm{H}(\mu)(T)$ *for every* $T \in \mathrm{H}(G)$.

4.4.4 The Case of the Groups O_n *and* SO_n

Let us now specify this construction for the orthogonal $\mathbb{Z}$-group O_n of the even unimodular lattice $\mathrm{E}_n \subset \mathbb{R}^n$, for $n \equiv 0 \bmod 8$ (Sect. 2.3, choosing another lattice would lead to a theory equivalent to the one we now present).

In this case, we saw in Sect. 4.1.2 that the $\mathrm{O}_n(\mathbb{A}_f)$-set $\mathcal{R}(\mathrm{O}_n)$ can be canonically identified with the set of even unimodular lattices in $\mathbb{R}^n$ contained in $\mathrm{E}_n \otimes \mathbb{Q}$ and that we have $\mathrm{Cl}(\mathrm{O}_n) \xrightarrow{\sim} \mathrm{X}_n$. In particular, we have

$$\mathrm{M}_{\mathbb{C}}(\mathrm{O}_n) = \mathbb{C}[\mathrm{X}_n]^* \ .$$

The right action of $\mathrm{H}(\mathrm{O}_n)$ on $\mathrm{M}_{\mathbb{C}}(\mathrm{O}_n)$ defines by transposition a left action of $\mathrm{H}(\mathrm{O}_n)$ on $\mathbb{C}[\mathrm{X}_n]$. In particular, the operator $\mathrm{T}_{\mathbb{Z}/d\mathbb{Z}} \in \mathrm{H}(\mathrm{O}_n)$ defined in Sect. 4.2.6, viewed as an endomorphism of $\mathbb{C}[\mathrm{X}_n]$, is the operator T_d of Sect. 3.2. The description of the structure of the $\mathrm{H}(\mathrm{O}_n)^{\mathrm{opp}}$-module $\mathrm{M}_{\mathbb{C}}(\mathrm{O}_n)$ when $n \leq 24$ is therefore the main theme of this book.

The ring $\mathrm{H}(\mathrm{O}_n)$ is commutative by Proposition 4.2.8. Let us fix a (finite-dimensional, continuous, complex) representation U of $\mathrm{O}_n(\mathbb{R})$. By Lemma 4.4.2 and the general results recalled in Sect. 4.3.1, the action of $\mathrm{H}(\mathrm{O}_n)$ is therefore codiagonalizable on each $\mathrm{M}_U(\mathrm{O}_n)$. The eigenvalues of these operators have an important arithmetic meaning. In Corollary 8.2.20, we will see that they are linked, in an a priori rather surprising manner, to the representations of the absolute Galois group of $\mathbb{Q}$. The line of constant functions in $\mathrm{M}_{\mathbb{C}}(\mathrm{O}_n)$ is, for example, trivially stable under T_A for every A, where the eigenvalue of T_p on this line is, of course, $\mathrm{c}_n(p)$ (Proposition-Definition 3.2.1). We will give markedly more interesting examples in the next chapters.

Remark 4.4.5. Let $\mathcal{L}_n$ be the set of all even unimodular lattices in $\mathbb{R}^n$, which we already considered in the introduction (Chap. 1). It contains $\mathcal{R}(\mathrm{O}_n)$ and the natural action of $\mathrm{O}_n(\mathbb{R})$ on $\mathcal{L}_n$ extends the natural action of $\mathrm{O}_n(\mathbb{Q})$ on $\mathcal{R}(\mathrm{O}_n)$. The map $\mathrm{O}_n(\mathbb{R}) \times \mathrm{O}_n(\mathbb{A}_f) \to \mathcal{L}_n$ defined by $(g_\infty, g_f) \mapsto g_\infty^{-1}(g_f(\mathrm{E}_n))$ therefore factors through a map

$$\mathrm{O}_n(\mathbb{Q})\backslash\mathrm{O}_n(\mathbb{A})/\mathrm{O}_n(\widehat{\mathbb{Z}}) \to \mathcal{L}_n \ .$$

This is a bijection: the surjectivity follows from Scholium 2.2.1 and the injectivity is immediate.

Let us turn to the case of SO_n. By Proposition 4.1.7 and Sect. 4.2.11, the inclusion $\mathrm{SO}_n \to \mathrm{O}_n$ induces a bijection $\mathcal{R}(\mathrm{SO}_n) \xrightarrow{\sim} \mathcal{R}(\mathrm{O}_n)$ and $\mathrm{H}(\mathrm{O}_n)$ is naturally a subring of $\mathrm{H}(\mathrm{SO}_n)$. Let U be an $\mathrm{SO}_n(\mathbb{Q})$-module, and consider

$$U' = \mathrm{Ind}_{\mathrm{SO}_n(\mathbb{Q})}^{\mathrm{O}_n(\mathbb{Q})} U \ .$$

The universal property of induced modules provides a canonical isomorphism $\mathrm{ind}\colon \mathrm{Hom}_{\mathbb{Z}[\mathrm{SO}_n(\mathbb{Q})]}(\mathbb{Z}[\mathcal{R}(\mathrm{O}_n)]_{|\mathrm{SO}_n(\mathbb{Q})}, U) \overset{\sim}{\to} \mathrm{Hom}_{\mathbb{Z}[\mathrm{O}_n(\mathbb{Q})]}(\mathbb{Z}[\mathcal{R}(\mathrm{O}_n)], U')$, which can also be written as

$$\mathrm{ind}\colon \mathrm{M}_U(\mathrm{SO}_n) \overset{\sim}{\to} \mathrm{M}_{U'}(\mathrm{O}_n)\,.$$

This isomorphism is trivially $\mathrm{H}(\mathrm{O}_n)$-equivariant, so that studying the $\mathrm{H}(\mathrm{O}_n)$-modules $\mathrm{M}_U(\mathrm{SO}_n)$ reduces to studying $\mathrm{M}_W(\mathrm{O}_n)$, where W is an $\mathrm{O}_n(\mathbb{Q})$-module. Let us add that if U is the restriction to $\mathrm{SO}_n(\mathbb{Q})$ of an $\mathrm{SO}_n(\mathbb{R})$-module V and V' denotes the $\mathrm{O}_n(\mathbb{R})$-module induced by V, then we have $V'_{|\mathrm{O}_n(\mathbb{Q})} = \mathrm{Ind}_{\mathrm{SO}_n(\mathbb{Q})}^{\mathrm{O}_n(\mathbb{Q})} U$.

Finally, let W be an $\mathrm{O}_n(\mathbb{Q})$-module, and let W' denote its restriction to $\mathrm{SO}_n(\mathbb{Q})$. The group $\mathrm{O}_n(\mathbb{Q})$ has a natural action on $\mathrm{M}_{W'}(\mathrm{SO}_n)$, by $(\gamma, f) \mapsto (x \mapsto \gamma(f(\gamma^{-1}(x))))$, where the subgroup $\mathrm{SO}_n(\mathbb{Q})$ acts trivially. Let $s \in \mathrm{End}(\mathrm{M}_{W'}(\mathrm{SO}_n))$ be the operator induced by the nontrivial element of the quotient $\mathrm{O}_n(\mathbb{Q})/\mathrm{SO}_n(\mathbb{Q}) \simeq \mathbb{Z}/2\mathbb{Z}$. The restriction of the functions via the bijective map $\mathcal{R}(\mathrm{SO}_n) \to \mathcal{R}(\mathrm{O}_n)$ then defines an $\mathrm{H}(\mathrm{O}_n)$-equivariant injection

$$\mathrm{res}\colon \mathrm{M}_W(\mathrm{O}_n) \to \mathrm{M}_{W'}(\mathrm{SO}_n)$$

whose image is $\mathrm{M}_{W'}(\mathrm{SO}_n)^{s=\mathrm{id}}$.

Example 4.4.6. The isomorphism ind induces a canonical decomposition

$$\mathrm{M}_{\mathbb{C}}(\mathrm{SO}_n) \simeq \mathrm{M}_{\mathbb{C}}(\mathrm{O}_n) \oplus \mathrm{M}_{\det}(\mathrm{O}_n)\,,$$

where det is the representation of dimension 1 given by the determinant. If we, moreover, view $\mathbb{C}$ as the restriction to $\mathrm{SO}_n(\mathbb{R})$ of the trivial representation of $\mathrm{O}_n(\mathbb{R})$, this endows $\mathrm{M}_{\mathbb{C}}(\mathrm{SO}_n)$ with a symmetry s that preserves the decomposition given above, with fixed points $\mathrm{M}_{\mathbb{C}}(\mathrm{O}_n)$.

We refer to [55, Sect. 2] for a discussion of the spaces $\mathrm{M}_U(\mathrm{SO}_8)$, and in particular their dimension, in terms of the representation U; see also Sect. 7.4 for examples.

4.4.7 An Invariant Hermitian Inner Product

Let us consider the case of a general $\mathbb{Z}$-group G with $G(\mathbb{R})$ compact. Let U be a finite-dimensional, continuous, complex representation of $G(\mathbb{R})$. By transport of structure, the isomorphism $\mathrm{M}_U(G) \overset{\sim}{\to} \mathcal{A}_{U^*}(G)$ endows $\mathrm{M}_U(G)$ with a natural Hermitian inner product, for which the action of $\mathrm{H}(G)$ is a $\star$-action, by Sect. 4.3.1, which we now only need to make explicit. For this, fix a $G(\mathbb{R})$-invariant Hermitian inner product $\langle -, - \rangle_U$ on U. Also choose elements $g_i \in G(\mathbb{A}_f)$ satisfying equality (4.3.2); recall that $\Gamma_i = G(\mathbb{Q}) \cap g_i G(\widehat{\mathbb{Z}}) g_i^{-1}$ is a finite group.

Proposition 4.4.8. *For $F, F' \in \mathrm{M}_U(G)$, the formula*

$$(F|F') = \sum_{i=1}^{\mathrm{h}(G)} \frac{1}{|\Gamma_i|} \langle F(g_i), F'(g_i) \rangle_U$$

defines a Hermitian inner product on $\mathrm{M}_U(G)$ that does not depend on the choice of the g_i and for which the action of $\mathrm{H}(G)$ is a $\star$-action.

We include a proof because we could not find any adequate reference for this result.

Proof. Fix a nonzero $e \in U^*$. By the isomorphism (4.4.2) and Sect. 4.3,

$$(F|F') := \int_{G(\mathbb{Q})\backslash G(\mathbb{A})} \overline{\varphi_F(e)} \varphi_{F'}(e) \, \mathrm{d}m$$

is a Hermitian inner product on $\mathrm{M}_U(G)$ for which the action of $\mathrm{H}(G)$ is a $\star$-action. We will verify that it is proportional to the inner product of Proposition 4.4.8.

Let $\Omega_i \subset G(\mathbb{A})$ be the compact open set $g_i(G(\mathbb{R}) \times G(\widehat{\mathbb{Z}}))$, let $\pi \colon G(\mathbb{A}) \to G(\mathbb{Q})\backslash G(\mathbb{A})$ be the canonical projection, and set $\overline{\Omega_i} = \pi(\Omega_i)$. By definition, $G(\mathbb{Q})\backslash G(\mathbb{A})$ is the (finite) disjoint union of the $\overline{\Omega_i}$. Let us first verify that there exists a Haar measure m on $G(\mathbb{A})$ such that for every continuous function ψ on (the compact set) $G(\mathbb{Q})\backslash G(\mathbb{A})$, we have

$$\int_{G(\mathbb{Q})\backslash G(\mathbb{A})} \psi \, \mathrm{d}\mu = \sum_{i=1}^{\mathrm{h}(G)} \frac{1}{|\Gamma_i|} \int_{\Omega_i} \psi \circ \pi \, \mathrm{d}m \, . \tag{4.4.2}$$

Indeed, recall that if f is continuous with compact support on $G(\mathbb{A})$, then $\tilde{f}(g) := \sum_{\gamma \in G(\mathbb{Q})} f(\gamma g)$ is continuous with compact support on $G(\mathbb{Q})\backslash G(\mathbb{A})$. Moreover, by the characteristic property of the quotient measure μ, there exists a unique Haar measure m on $G(\mathbb{A})$ such that for every continuous function f on $G(\mathbb{A})$ with compact support, we have $\int_{G(\mathbb{A})} f \, \mathrm{d}m = \int_{G(\mathbb{Q})\backslash G(\mathbb{A})} \tilde{f} \, \mathrm{d}\mu$ (see [211, Chap. II]).

For $g \in G(\mathbb{A})$, set $n_i(g) = |G(\mathbb{Q})g \cap \Omega_i|$. We clearly have $n_i(\gamma g k) = n_i(g)$ for every $\gamma \in G(\mathbb{Q})$ and every $k \in 1 \times G(\widehat{\mathbb{Z}})$. By definition, we also have $n_i(g_j) = |\Gamma_i| \delta_{i,j}$, where $\delta_{i,j}$ is the Kronecker delta. Let ψ be a continuous function on $G(\mathbb{Q})\backslash G(\mathbb{A})$. The function $G(\mathbb{A}) \to \mathbb{C}$ defined by $f_i = 1_{\Omega_i} \times \psi \circ \pi$ is continuous with support in Ω_i and satisfies $\tilde{f}_i(g) = \psi(\pi(g))\, n_i(g)$ for every $g \in G(\mathbb{A})$ (we denote the characteristic function of the set A by 1_A). In other words, we have $\psi \times 1_{\overline{\Omega_i}} = (1/|\Gamma_i|) \tilde{f}_i$. This proves formula (4.4.2).

Let us apply this formula to the function $\psi = \overline{\varphi_F(e)}\, \varphi_{F'}(e)$. Note that if $U = \mathbb{C}$, so that ψ is constant, equal to $|e(1)|^2 \overline{F(g_i)} F'(g_i)$ on Ω_i, the proposition follows from the fact that $m(\Omega_i) = m(G(\mathbb{R}) \times G(\mathbb{Z}))$ is independent of i. In general, we introduce the Haar measure $\mathrm{d}g$ on $G(\mathbb{R})$ of total mass 1 and the Haar measure m_f on $G(\mathbb{A}_f)$ such that $\mathrm{d}m = \mathrm{d}g \times \mathrm{d}m_f$. The right invariance of ψ under $1 \times G(\widehat{\mathbb{Z}})$

and Fubini's theorem imply

$$\int_{\Omega_i} \psi \circ \pi \, \mathrm{d}m = m_f(G(\widehat{\mathbb{Z}})) \int_{G(\mathbb{R})} \overline{\langle e, g^{-1}F(g_i)\rangle} \langle e, g^{-1}F'(g_i)\rangle \, \mathrm{d}g \, .$$

Let $E \in U$ be such that we have $\langle E, x\rangle_U = \langle e, x\rangle$ for every $x \in U$. The orthogonality relations of the matrix entries for the irreducible representations of the compact groups imply that we have

$$\int_{G(\mathbb{R})} \overline{\langle e, g^{-1}F(g_i)\rangle} \, \langle e, g^{-1}F'(g_i)\rangle \, \mathrm{d}m_\infty = \frac{1}{\dim U} \langle E, E\rangle_U \langle F(g_i), F'(g_i)\rangle_U \ ,$$

which concludes the proof of the proposition. □

Assume, for example, that we have $G = \mathrm{O}_n$ and $U = \mathbb{C}$. If $L_i \in \mathcal{R}^{\mathrm{a}}_{\mathbb{Z}}(\mathrm{E}_n \otimes \mathbb{Q})$ denotes the lattice $g_i(L)$, we have $\Gamma_i = \mathrm{O}(L_i) \subset \mathrm{O}_n(\mathbb{Q})$. The relation $\mathrm{T}_A = \mathrm{T}_A^{\mathrm{t}}$ of Proposition 4.2.8 and Proposition 4.4.8 can then be written as

$$\mathrm{N}_A(L, M)|\mathrm{O}(M)| = \mathrm{N}_A(M, L)|\mathrm{O}(L)| \ ,$$

where $\mathrm{N}_A(L, M)$ denotes the number of A-neighbors of L isometric to M (with $L, M \in \mathcal{R}(\mathrm{O}_n)$). This is the generalization of Proposition 3.2.3 we announced earlier.

Corollary 4.4.9. *The bilinear form on* $\mathrm{M}_{U^*}(G) \times \mathrm{M}_U(G)$ *defined by*

$$(F|F') = \sum_i \frac{1}{|\Gamma_i|} \langle F(g_i), F'(g_i)\rangle$$

is independent of the choice of the g_i *and is nondegenerate. It satisfies the identity* $(T(F)|F') = (F|T^{\mathrm{t}}(F'))$ *for all* $T \in \mathrm{H}(G)$, $F \in \mathrm{M}_{U^*}(G)$, *and* $F' \in \mathrm{M}_U(G)$. *In particular, it defines a canonical isomorphism between the* $\mathrm{H}(G)$*-module* $\mathrm{M}_{U^*}(G)^*$ *and the* $\mathrm{H}(G)$*-module* $\mathrm{M}_U(G)^{\mathrm{t}}$ *(see Remark* 4.2.4).

Proof. For a $\mathbb{C}$-vector space V, we denote by $\overline{V}$ the conjugate $\mathbb{C}$-vector space (that is, the abelian group V endowed with the action $\mathbb{C} \times V \to V$ of $\mathbb{C}$ defined by $(\lambda, v) \mapsto \overline{\lambda} v$). For U as in the corollary, $\overline{U}$ is naturally a representation of $G(\mathbb{R})$ and the map $v \mapsto (u \mapsto \langle v, u\rangle_U)$ induces an isomorphism of representations $\overline{U} \xrightarrow{\sim} U^*$. We therefore have a natural isomorphism $\mathrm{M}_{U^*}(G) \xrightarrow{\sim} \mathrm{M}_{\overline{U}}(G) = \overline{\mathrm{M}_U(G)}$. Via this isomorphism, the bilinear form of the corollary coincides with the form $\overline{\mathrm{M}_U(G)} \times \mathrm{M}_U(G) \to \mathbb{C}$ defined by $(F, F') \mapsto \sum_i (1/|\Gamma_i|)\langle F(g_i), F'(g_i)\rangle_U$, which is none other than the Hermitian form on $\mathrm{M}_U(G)$ given by Proposition 4.4.8. The first two assertions follow; the last is obvious. □

Let us conclude with one last observation. For $L \in \mathcal{R}(G)$ and $u \in U$, the map $F \mapsto \langle F(L), u\rangle$ is a linear form on $\mathrm{M}_{U^*}(G)$, which we denote by $\mathrm{ev}_{L,u}$. We have a unique linear map

$$\mathbb{Z}[\mathcal{R}(G)] \otimes U \to \mathrm{M}_{U^*}(G)^*$$

that sends $[L] \otimes u$ to $\mathrm{ev}_{L,u}$ for every $L \in \mathcal{R}(G)$ and every $u \in U$. The $\mathbb{C}$-vector space $\mathbb{Z}[\mathcal{R}(G)] \otimes U$ is endowed with a diagonal action of $G(\mathbb{Q})$, and the map above is constant on the orbits of this action. It therefore factors through a linear map

$$(\mathbb{Z}[\mathcal{R}(G)] \otimes U)_{G(\mathbb{Q})} \to \mathrm{M}_{U^*}(G)^* , \tag{4.4.3}$$

where V_Γ denotes the coinvariants of the Γ-module V. This is an isomorphism: this follows simply from the finiteness of $G(\mathbb{Q})\backslash\mathcal{R}(G)$ and of the natural isomorphism $(U^*)^\Gamma \overset{\sim}{\to} (U_\Gamma)^*$, which holds for every finite subgroup Γ of $G(\mathbb{R})$. The isomorphism (4.4.3) trivially commutes with the natural (left) actions of $\mathrm{H}(G)$. If we compose it with the isomorphism $\mathrm{M}_{U^*}(G)^* \longrightarrow \mathrm{M}_U(G)^{\mathrm{t}}$ given by Corollary 4.4.9, we obtain a canonical isomorphism of $\mathrm{H}(G)$-modules

$$(\mathbb{Z}[\mathcal{R}(G)] \otimes U)_{G(\mathbb{Q})} \overset{\sim}{\to} \mathrm{M}_U(G)^{\mathrm{t}} . \tag{4.4.4}$$

It sends (the class of) the element $[L] \otimes u$ to an element of $\mathrm{M}_U(G)$ that we denote by $[L, u]$. Concretely, $[L, u]$ is the unique function $F \in \mathrm{M}_U(G)$ that is zero outside of $G(\mathbb{Q}) \cdot L$ that satisfies $F(L) = \sum_{\gamma\in\Gamma} \gamma(u)$, where $\Gamma = G(\mathbb{Q})_L$ is the stabilizer of L in $G(\mathbb{Q})$. The isomorphism (4.4.4) will play a (small) role in our discussion of the theta series in Sect. 5.4.1 and Chap. 7.

4.5 Siegel Modular Forms

Let us begin by recalling some results on Siegel modular forms (see [5, 45, 46, 88]). We will closely follow the exposition of Van der Geer [89], to which we refer, in particular, for a history of the subject.

4.5.1 The Classical Point of View

Let $g \geq 1$ be an integer. For a ring R, we denote by $\mathrm{Mat}_g(R)$ the set of $g \times g$ matrices with coefficients in R and by $\mathrm{Sym}_g(R) \subset \mathrm{Mat}_g(R)$ the subset of symmetric matrices. We denote by 1_g the identity matrix in $\mathrm{Mat}_g(R)$ and by $\mathrm{J}_{2g} \in \mathrm{Mat}_{2g}(R)$ the element

$$\mathrm{J}_{2g} = \begin{pmatrix} 0 & 1_g \\ -1_g & 0 \end{pmatrix} .$$

The Siegel half-space of genus g is the open subset

$$\mathbb{H}_g \subset \mathrm{Sym}_g(\mathbb{C})$$

of matrices with positive definite imaginary part. We view the $\mathbb{Z}$-group GSp_{2g} as the sub-group scheme of GL_{2g} consisting of the γ with $\gamma\, \mathrm{J}_{2g}\, {}^{\mathrm{t}}\gamma = \nu(\gamma) \mathrm{J}_{2g}$, where the morphism $\nu\colon \mathrm{GSp}_{2g} \to \mathbb{G}_m$ is the similitude factor. Its elements are of the form

$$\gamma = \begin{pmatrix} a_\gamma & b_\gamma \\ c_\gamma & d_\gamma \end{pmatrix}$$

with $a_\gamma, b_\gamma, c_\gamma, d_\gamma \in \mathrm{Mat}_g$ satisfying the relations $a_\gamma\, {}^{\mathrm{t}}b_\gamma = b_\gamma\, {}^{\mathrm{t}}a_\gamma, c_\gamma\, {}^{\mathrm{t}}d_\gamma = d_\gamma\, {}^{\mathrm{t}}c_\gamma$, and $a_\gamma\, {}^{\mathrm{t}}d_\gamma - b_\gamma\, {}^{\mathrm{t}}c_\gamma = \nu(\gamma)1_g$.

Let $\mathrm{GSp}_{2g}(\mathbb{R})^+$ be the subgroup of $\mathrm{GSp}_{2g}(\mathbb{R})$ consisting of the elements with strictly positive similitude factor. For $\gamma \in \mathrm{GSp}_{2g}(\mathbb{R})^+$ and $\tau \in \mathbb{H}_g$, we can show that the element $\mathrm{j}(\gamma, \tau) := c_\gamma\tau + d_\gamma$ is in $\mathrm{GL}_g(\mathbb{C})$ and that

$$(\gamma, \tau) \mapsto \gamma\tau = (a_\gamma\tau + b_\gamma)(c_\gamma\tau + d_\gamma)^{-1}$$

defines a transitive action of $\mathrm{GSp}_{2g}(\mathbb{R})^+$ on $\mathbb{H}_g$ by biholomorphic transformations. Moreover, we easily verify the 1-cocyle relation $\mathrm{j}(\gamma\gamma', \tau) = \mathrm{j}(\gamma, \gamma'\tau)\mathrm{j}(\gamma', \tau)$ for all $\gamma, \gamma' \in \mathrm{GSp}_{2g}(\mathbb{R})^+$ and every $\tau \in \mathbb{H}_g$.

Let W be a finite-dimensional $\mathbb{C}$-vector space endowed with a $\mathbb{C}$-representation $\rho\colon \mathrm{GL}_g \to \mathrm{GL}_W$. A *Siegel modular form of weight W and genus $g \geq 1$* is a holomorphic function $f\colon \mathbb{H}_g \to W$ with

$$f(\gamma\tau) = \rho(\,\mathrm{j}(\gamma, \tau)\,) \cdot f(\tau) \quad \forall \tau \in \mathbb{H}_g \,, \; \forall \gamma \in \mathrm{Sp}_{2g}(\mathbb{Z}) \,.$$

For $g = 1$, we add the assumption that f is bounded on $\{\tau \in \mathbb{H}_1 \,;\, \Im(\tau) > 1\}$. These functions form a $\mathbb{C}$-vector space that we denote by

$$\mathrm{M}_W(\mathrm{Sp}_{2g}(\mathbb{Z}))\,,$$

whose dimension is finite, as shown by Siegel.

When we have $(\rho, W) = (\det^k, \mathbb{C})$ for $k \in \mathbb{Z}$, we speak of *classical*, or scalar-valued *Siegel forms* of weight k; we speak of vector-valued forms otherwise. In the former case, we also denote the space $\mathrm{M}_W(\mathrm{Sp}_{2g}(\mathbb{Z}))$ by $\mathrm{M}_k(\mathrm{Sp}_{2g}(\mathbb{Z}))$. When $g = 1$, we recover, as a special case, the usual modular forms for the group $\mathrm{SL}_2(\mathbb{Z})$, which are, for example, treated in detail in Serre's book [177]. Finally, note that the presence of the element $-1_{2g} \in \mathrm{Sp}_{2g}(\mathbb{Z})$ and the relation $\mathrm{j}(-1_{2g}, \tau) = -1_g$ imply $\mathrm{M}_W(\mathrm{Sp}_{2g}(\mathbb{Z})) = 0$ if $\rho(-1_g) = -\mathrm{id}_W$.

Let us conclude this subsection with a reformulation of the notion of a Siegel modular form. Assume that the representation (ρ, W) is irreducible or, more generally, that there exists an element $\mathrm{m}_W \in \mathbb{Z}$, necessarily unique, such that $\rho(z1_g) = z^{\mathrm{m}_W}\,\mathrm{id}_W$ for every $z \in \mathbb{C}^\times$. For a map $f\colon \mathbb{H}_g \to W$, we set

$$f_{|_W}\gamma\colon \mathbb{H}_g \to W\,, \quad \tau \mapsto \nu(\gamma)^{\mathrm{m}_W/2}\; \rho(\mathrm{j}(\gamma, \tau))^{-1}\; f(\gamma\tau)\,.$$

The map $(\gamma, f) \mapsto f_{|_W}\gamma$ defines a right action of the group $\mathrm{GSp}_{2g}(\mathbb{R})^+$ on the space of holomorphic functions $\mathbb{H}_g \to W$; by construction, this action is trivial on the subgroup of homotheties with strictly positive factor in $\mathrm{GSp}_{2g}(\mathbb{R})^+$. A *Siegel modular form of weight W and genus $g \geq 2$* is, by definition, an $\mathrm{Sp}_{2g}(\mathbb{Z})$-invariant element for this action.

4.5.2 Fourier Series Expansions and Cusp Forms

For $n \in \mathrm{Sym}_g(\mathbb{C})$, we set

$$q^n = e^{2i\pi\,\mathrm{tr}(n\,\tau)} = \prod_{1\leq i,j\leq g} e^{2i\pi\, n_{i,j}\tau_{i,j}} \ ;$$

this is a holomorphic function on $\mathbb{H}_g$. If n is *semi-integral*, that is, if $n \in \frac{1}{2}\mathrm{Sym}_g(\mathbb{Z})$, and if $n_{i,i} \in \mathbb{Z}$ for every $i = 1, \ldots, g$, then q^n is invariant under translations by $\mathrm{Sym}_g(\mathbb{Z})$. It can be shown that every $f \in \mathrm{M}_W(\mathrm{Sp}_{2g}(\mathbb{Z}))$ admits a Fourier series expansion, which normally converges on every compact subset of $\mathbb{H}_g$, of the form

$$f = \sum_{n\geq 0} a_n q^n \ ,$$

where the sum is taken over the positive semi-integral elements $n \in \frac{1}{2}\mathrm{Sym}_g(\mathbb{Z})$ (in the sense of real symmetric matrices) and where the a_n are in W [89, Sect. 4]. For $g \geq 2$, the Siegel operator is the map

$$\Phi_g\colon \mathrm{M}_W(\mathrm{Sp}_{2g}(\mathbb{Z})) \longrightarrow \mathrm{M}_{W'}(\mathrm{Sp}_{2g-2}(\mathbb{Z}))$$

defined by $\Phi_g(\sum_n a_n q^n) = \sum_{n'} a_{n'} q^{n'}$, where we view $\mathrm{Sym}_{g-1}(-)$ as a subset of $\mathrm{Sym}_g(-)$ with last line and column consisting of zeros, and we have $W' = W_{|\mathrm{GL}_{g-1}\times 1}$ [89, Sect. 5]. The subspace of cusp forms is

$$\mathrm{S}_W(\mathrm{Sp}_{2g}(\mathbb{Z})) := \mathrm{Ker}(\Phi_g) \subset \mathrm{M}_W(\mathrm{Sp}_{2g}(\mathbb{Z})) \ .$$

A Siegel form is therefore cuspidal if its Fourier series expansion $\sum_n a_n q^n$ satisfies $a_n = 0$ for every n with $\det(n) = 0$. When we have $(W, \rho) = (\mathbb{C}, \det^k)$ for $k \in \mathbb{Z}$, we write $\mathrm{S}_k(\mathrm{Sp}_{2g}(\mathbb{Z}))$ for $\mathrm{S}_W(\mathrm{Sp}_{2g}(\mathbb{Z}))$.

4.5.3 The Relation Between $\mathrm{S}_W(\mathrm{Sp}_{2g}(\mathbb{Z}))$ and $\mathcal{A}^2(\mathrm{PGSp}_{2g})$

We will now recall the classical link between $\mathrm{S}_W(\mathrm{Sp}_{2g}(\mathbb{Z}))$ and the space $\mathcal{A}_{\mathrm{cusp}}(\mathrm{PGSp}_{2g})$. A nice recent reference on this subject is the article [14], to which we will refer as soon as we can formulate the statement (see also [195, Sect. 5]).

Set $G = \mathrm{PGSp}_{2g}$. The similitude factor $\nu\colon \mathrm{GSp}_{2g} \to \mathbb{G}_m$ induces a homomorphism $\nu_\infty\colon G(\mathbb{R}) \to \mathbb{R}^\times/\mathbb{R}^\times_{>0}$ whose kernel we denote by $G(\mathbb{R})^+$. The canonical morphism $\mathrm{Sp}_{2g}(\mathbb{R}) \to G(\mathbb{R})$ induces an isomorphism

$$\mathrm{Sp}_{2g}(\mathbb{R})/\{\pm 1\} \overset{\sim}{\to} G(\mathbb{R})^+ \ .$$

We also set $G(A)^+ = G(A) \cap G(\mathbb{R})^+$ when A is a subring of $\mathbb{R}$.

By Sect. 4.1.2, we have $\mathrm{h}(G) = 1$. Since we have $\nu_\infty(G(\mathbb{Z})) = \{\pm 1\}$, we obtain the equality

$$G(\mathbb{A}) = G(\mathbb{Q})(G(\mathbb{R})^+ \times G(\widehat{\mathbb{Z}})) \tag{4.5.1}$$

and, from (4.3.3), it follows that the restriction $f \mapsto f_{|G^+(\mathbb{R})\times 1}$ induces a $G(\mathbb{R})^+$-equivariant isomorphism

$$\mathcal{A}^2(\mathrm{PGSp}_{2g}) \stackrel{\sim}{\to} \mathrm{L}^2(G(\mathbb{Z})^+\backslash G(\mathbb{R})^+) \,. \tag{4.5.2}$$

The action of $\mathrm{GSp}_{2g}(\mathbb{R})^+$ on $\mathbb{H}_g$ recalled in Sect. 4.5.1 factors through an action of $G(\mathbb{R})^+$. The latter is faithful and transitive, and its stabilizers are the maximal compact subgroups of $G(\mathbb{R})^+$. If K denotes the stabilizer in $\mathrm{Sp}_{2g}(\mathbb{R})$ of the element $i1_g \in \mathbb{H}_g$ and K^+ denotes its image in $G(\mathbb{R})^+$, we therefore have a natural identification

$$G(\mathbb{R})^+/K^+ \stackrel{\sim}{\to} \mathbb{H}_g \,.$$

Let (ρ, W) be a $\mathbb{C}$-representation of GL_g as in Sect. 4.5.1, which we now assume to be irreducible and satisfy $\mathrm{m}_W \equiv 0 \bmod 2$. Fix $w \in W^*$ and $f \in \mathrm{S}_W(\mathrm{Sp}_{2g}(\mathbb{Z}))$; we will associate a function $\varphi_{w,f} \in \mathcal{A}^2(G)$ with w and f. Consider the function $\varphi\colon G(\mathbb{R})^+ \longrightarrow \mathbb{C}$ defined by

$$\varphi(\gamma) = \langle w, (f_{|_W}\gamma)(i1_g)\rangle \,.$$

By construction, φ is continuous and left invariant under $G(\mathbb{Z})^+$. By formula (4.5.1), it is therefore the restriction to $G(\mathbb{R})^+ \times 1$ of a unique continuous function $\varphi'\colon G(\mathbb{Q})\backslash G(\mathbb{A}) \to \mathbb{C}$ that is invariant under right translations by $G(\widehat{\mathbb{Z}})$. Set

$$\varphi_{w,f} := \varphi' \,.$$

By Asgari and Schmidt [14, Lemma 5], we have $\varphi_{w,f} \in \mathcal{A}_{\mathrm{cusp}}(G)$.

Before stating the final proposition, we still need to define the notion of a holomorphic element of $\mathcal{A}^2(G)$. Let $\mathfrak{g}$ and $\mathfrak{k}$ be the Lie algebras of $G(\mathbb{R})^+$ and K, respectively, and let $\mathfrak{g} = \mathfrak{k} \oplus \mathfrak{p}$ be the associated Cartan decomposition. Let $d\colon \mathfrak{g} \to \mathrm{T}_{i1_g}$ be the differential in the identity of the map $G(\mathbb{R})^+ \to \mathbb{H}_g$ defined by $h \mapsto h(i1_g)$. It induces an $\mathbb{R}$-linear isomorphism

$$\mathfrak{p} \stackrel{\sim}{\to} \mathrm{T}_{i1_g} = \mathrm{Sym}_g(\mathbb{C}) \,.$$

The $\mathbb{C}$-vector space structure of $\mathrm{Sym}_g(\mathbb{C})$ therefore endows $\mathfrak{p}$ with the structure of a $\mathbb{C}$-vector space that decomposes $\mathfrak{p} \otimes_{\mathbb{R}} \mathbb{C}$ into $\mathfrak{p}^+ \oplus \mathfrak{p}^-$, so that d induces a $\mathbb{C}$-linear isomorphism $\mathfrak{p}^+ \stackrel{\sim}{\to} \mathrm{T}_{i1_g}$. An element $f \in \mathcal{A}^2(G)$ is called *holomorphic* if it is continuous and if for every $g \in G(\mathbb{A})$, the function $G(\mathbb{R}) \to \mathbb{C}$ defined by $h \mapsto f(gh)$ is infinitely differentiable and annihilated by $\mathfrak{p}^-$.

Proposition 4.5.4. *The map $(w, f) \mapsto \varphi_{w,f}$ defines a $\mathbb{C}[K]$-linear injection*

$$W^* \otimes \mathrm{S}_W(\mathrm{Sp}_{2g}(\mathbb{Z})) \longrightarrow \mathcal{A}_{\mathrm{cusp}}(\mathrm{PGSp}_{2g})$$

whose image is the set of $f \in \mathcal{A}_{\mathrm{cusp}}(\mathrm{PGSp}_{2g})$ *that are holomorphic and* W^**-isotypical under the action of* K.

Let us make this statement more precise. The map $h \mapsto \mathrm{j}(h, i1_g)$ is a group morphism $K \to \mathrm{GL}_g(\mathbb{C})$ that realizes $\mathrm{GL}_g(\mathbb{C})$ as the complexification of the compact unitary group K. This, in particular, allows us to view W as a representation of K by restriction; it is irreducible because W is so as a representation of GL_g. We refer to [14, Sect. 4.5, Theorem 1] for a proof of this proposition, up to the assertion of surjectivity, which is verified in [195, Sect. 5.2].

4.5.5 The Action of Hecke Operators

It follows from Proposition 4.5.4 that the image of the map in that statement is stable under the action of $\mathrm{H}(\mathrm{PGSp}_{2g})$, so that the space $\mathrm{S}_W(\mathrm{Sp}_{2g}(\mathbb{Z}))$ inherits an action of $\mathrm{H}(\mathrm{PGSp}_{2g})$ from $\mathcal{A}^2(\mathrm{PGSp}_{2g})$. Up to normalization constants sometimes introduced by different authors for integrality reasons, this action coincides with the action traditionally defined on $\mathrm{S}_W(\mathrm{Sp}_{2g}(\mathbb{Z}))$, and even on $\mathrm{M}_W(\mathrm{Sp}_{2g}(\mathbb{Z}))$, which we recall below (see also [88, Kap. IV], [89, Sect. 16], and [14, Sect. 4.3]). Without going into details, let us mention that is it particularly natural when we view $\mathrm{Sp}_{2g}(\mathbb{Z})\backslash\mathbb{H}_g$ as the space of complex abelian varieties of dimension g endowed with a principal polarization[7] [89, Sect. 10].

Let (W, ρ) be an irreducible $\mathbb{C}$-representation of GL_g, p a prime, and G the $\mathbb{Z}$-group PGSp_{2g}. The natural map

$$a\colon G(\mathbb{Z}[\frac{1}{p}])^+/G(\mathbb{Z})^+ \to G(\mathbb{Q}_p)/G(\mathbb{Z}_p)$$

is bijective because we have $\mathrm{h}(G) = 1$ (Corollary 4.1.5) and $\nu_\infty(G(\mathbb{Z})) = \{\pm 1\}$ (Sect. 4.5.3). It therefore induces, in an obvious way, an injective homomorphism between the ring $\mathrm{H}_p(G)$ and the Hecke ring of the $G(\mathbb{Z}[1/p])^+$-set $G(\mathbb{Z}[1/p])^+/G(\mathbb{Z})^+$. This homomorphism is an isomorphism; this follows from the isomorphism (4.2.1) and the fact that a also induces a bijection

$$G(\mathbb{Z})^+\backslash G(\mathbb{Z}[1/p])^+/G(\mathbb{Z})^+ \to G(\mathbb{Z}_p)\backslash G(\mathbb{Q}_p)/G(\mathbb{Z}_p)\,, \qquad (4.5.3)$$

as shown by the theory of elementary divisors (Propositions 4.2.7 and 4.2.9, see also Sect. 6.2.5).

Suppose that the matrix of the element $T \in \mathrm{H}_p(G)$ is the characteristic function of the class set $G(\mathbb{Z}_p)\gamma G(\mathbb{Z}_p)$ with $\gamma \in G(\mathbb{Z}[1/p])^+$, in the sense of the identifica-

[7] A principal polarization on a lattice $L \subset \mathbb{C}^g$ consists of a nondegenerate alternating bilinear form $\eta\colon L \times L \to \mathbb{Z}$ whose extension of scalars $\eta_\mathbb{R}$ to $L \otimes \mathbb{R} = \mathbb{C}^g$ satisfies $\eta_\mathbb{R}(ix, iy) = \eta_\mathbb{R}(x, y)$ for every $x, y \in \mathbb{C}^g$ and whose associated Hermitian form $(x, y) \mapsto \eta_\mathbb{R}(ix, y) + i\eta_\mathbb{R}(x, y)$ on $\mathbb{C}^g$ is positive definite. Riemann's theory allows us to naturally identify $\mathrm{Sp}_{2g}(\mathbb{Z})\backslash\mathbb{H}_g$ with the set of $\mathrm{GL}_g(\mathbb{C})$-orbits of pairs (L, η), where $L \subset \mathbb{C}^g$ is a lattice and η is a principal polarization on L.

tion (4.2.2). If we write

$$G(\mathbb{Z})^+ \,\gamma\, G(\mathbb{Z})^+ = \coprod_i \gamma_i\, G(\mathbb{Z})^+ \ ,$$

we immediately see, using formula (4.2.3), that the following diagram is commutative:

$$\begin{array}{ccc} \mathrm{S}_W(\mathrm{Sp}_{2g}) & \longrightarrow & \mathrm{Hom}_K(W, \mathcal{A}_{\mathrm{cusp}}(\mathrm{PGSp}_{2g})) \\ {\scriptstyle f \mapsto \sum_i f_{|_W} \gamma_i^{-1}} \downarrow & & \downarrow {\scriptstyle T} \\ \mathrm{S}_W(\mathrm{Sp}_{2g}) & \longrightarrow & \mathrm{Hom}_K(W, \mathcal{A}_{\mathrm{cusp}}(\mathrm{PGSp}_{2g}))\ , \end{array} \qquad (4.5.4)$$

where the vertical maps are those defined by Proposition 4.5.4 (see [14, Lemma 9] for the details of the argument). Given the equality $T = T^{\mathrm{t}}$ for every $T \in \mathrm{H}(G)$, we will not need to remember the inversion of the γ_i in (4.5.4).

Formula (4.5.4) allows us to determine the link between the Hecke operators considered here and different definitions given in the literature. We will just give the translation of the definitions of Serre [177, Chap. VII, Sects. 2, 5] in the case $g = 1$. We will consider specific cases in genus $g = 2$ in Chap. 9.

Let $k \geq 0$ be an even integer. In [177, Chap. VII, Sect. 5.3], Serre defines, for every integer $n \geq 1$, an endomorphism of $\mathrm{M}_k(\mathrm{SL}_2(\mathbb{Z}))$ that he denotes by $\mathrm{T}(n)$ and whose effect on the q-expansions he determines. We also have another endomorphism, given by the action defined above of the operator $\mathrm{T}_A \in \mathrm{H}(\mathrm{PGL}_2)$ introduced in Sect. 4.2.6, where A is a cyclic group. The translation is then as follows:

$$n^{-(k-1)/2}\mathrm{T}(n) = n^{-1/2} \sum_{d^2|n} \mathrm{T}_{\mathbb{Z}/(n/d^2)\mathbb{Z}}\ . \qquad (4.5.5)$$

This comes, in particular, from the fact that in Serre's book, the correspondence $\mathrm{T}(n)$ sends a lattice to the set of its subgroups of index n rather than the set of those with quotient $\mathbb{Z}/n\mathbb{Z}$.

4.5.6 $\mathcal{A}_{\mathrm{disc}}(\mathrm{Sp}_{2g})$ *May Be Deduced from* $\mathcal{A}_{\mathrm{disc}}(\mathrm{PGSp}_{2g})$

By restriction of the functions, the morphism $\mathrm{Sp}_{2g}(\mathbb{A}) \to \mathrm{PGSp}_{2g}(\mathbb{A})$ induces an isomorphism

$$\mathrm{Res}\colon \mathcal{A}^2(\mathrm{PGSp}_{2g}) \xrightarrow{\sim} \mathcal{A}^2(\mathrm{Sp}_{2g})\ .$$

This follows from formula (4.3.3), taking into account that we have

$$\mathrm{h}(\mathrm{Sp}_{2g}) = \mathrm{h}(\mathrm{PGSp}_{2g}) = 1$$

and that the natural homomorphism $\mathrm{Sp}_{2g}(\mathbb{R}) \to \mathrm{PGSp}_{2g}(\mathbb{R})$ induces a homeomorphism $\mathrm{Sp}_{2g}(\mathbb{Z})\backslash\mathrm{Sp}_{2g}(\mathbb{R}) \xrightarrow{\sim} \mathrm{PGSp}_{2g}(\mathbb{Z})\backslash\mathrm{PGSp}_{2g}(\mathbb{R})$.

Recall that in Sect. 4.2.6, we defined an injective ring homomorphism $\mathrm{H}(\mathrm{Sp}_{2g}) \to \mathrm{H}(\mathrm{PGSp}_{2g})$, which we will from now on view as an inclusion, by a slight abuse of language. The source and target of the morphism Res are therefore both $\mathrm{H}(\mathrm{Sp}_{2g})$-modules.

Proposition 4.5.7. *The map* Res *commutes with the action of* $\mathrm{Sp}_{2g}(\mathbb{R})$ *and that of* $\mathrm{H}(\mathrm{Sp}_{2g})$. *It sends* $\mathcal{A}^2_{\mathrm{disc}}(\mathrm{PGSp}_{2g})$ *onto* $\mathcal{A}^2_{\mathrm{disc}}(\mathrm{Sp}_{2g})$.

Proof. The first assertion is obvious; the second follows from Lemma 4.2.15. The last is a consequence of the first and the fact that the image of $\mathrm{Sp}_{2g}(\mathbb{R})$ in $\mathrm{PGSp}_{2g}(\mathbb{R})$ has finite index (equal to 2). □

Chapter 5
Theta Series and Even Unimodular Lattices

5.1 Siegel Theta Series

Let $L \subset \mathbb{R}^n$ be an even unimodular lattice and $g \geq 1$ an integer. For a g-tuple v of elements of L, that is, $v = (v_i) \in L^g$, we denote the associated Gram matrix by $v.v := (v_i \cdot v_j)_{i,j} \in \mathrm{Sym}_g(\mathbb{Z})$; it is positive and $v.v/2$ is semi-integral in the sense of Sect. 4.5.2. The *Siegel theta series of genus g of L* is the holomorphic function on $\mathbb{H}_g$ defined by

$$\vartheta_g(L) = \sum_{v \in L^g} q^{v.v/2} .$$

It depends only on the isometry class of L. Its Fourier series expansion can be written as $\sum_{n \geq 0} a_n q^n$, where a_n is the number of g-tuples of elements of L with Gram matrix n. When $g = 1$, we, of course, recover the classical theta series, treated in [177], where a_n is simply the number of $x \in L$ such that $x \cdot x = 2n$. Siegel proved

$$\vartheta_g(L) \in \mathrm{M}_{n/2}(\mathrm{Sp}_{2g}(\mathbb{Z}))$$

[88, Kap. I, Sect. 0]. Let X_n be the set of isometry classes of even unimodular lattices in $\mathbb{R}^n$ already introduced in Chap. 1. It is worthwhile to linearize the construction above by considering the linear map

$$\vartheta_g \colon \mathbb{C}[\mathrm{X}_n] \longrightarrow \mathrm{M}_{n/2}(\mathrm{Sp}_{2g}(\mathbb{Z})) , \quad [L] \mapsto \vartheta_g(L) .$$

Furthermore, we denote by $\vartheta_0 \colon \mathbb{C}[\mathrm{X}_n] \to \mathbb{C}$ the linear map that sends $[L]$ to 1 for every L. Recall that the space $\mathbb{C}[\mathrm{X}_n]$ can be canonically identified with the dual of the space of algebraic modular forms $\mathrm{M}_{\mathbb{C}}(\mathrm{O}_n)$ (Sect. 4.4.4). It is therefore endowed with a (left) action of the ring $\mathrm{H}(\mathrm{O}_n)$, by transposition.

An important fact for understanding the $\mathrm{H}(\mathrm{O}_n)^{\mathrm{opp}}$-module $\mathrm{M}_{\mathbb{C}}(\mathrm{O}_n)$ is that the map ϑ_g intertwines the action of $\mathrm{H}(\mathrm{O}_n)$ on $\mathbb{C}[\mathrm{X}_n]$ and that of $\mathrm{H}(\mathrm{Sp}_{2g}) \subset \mathrm{H}(\mathrm{PGSp}_{2g})$ (Sect. 4.5.3), and this according to very precise recipes. These relations

G. Chenevier, J. Lannes, *Automorphic Forms and Even Unimodular Lattices*, Ergebnisse der Mathematik und ihrer Grenzgebiete. 3. Folge / A Series of Modern Surveys in Mathematics 69, https://doi.org/10.1007/978-3-319-95891-0_5

were discovered by Eichler in certain cases where $g = 1$ [78, Satz 21.3], [169] and have since become known as *the Eichler commutation relations*. They imply, in particular, that the kernel of ϑ_g is stable under the action of $\mathrm{H}(\mathrm{O}_n)$. Since Eichler, the case of arbitrary genus g has been studied, in various aspects, by many authors, including Rallis [170], Freitag [88, Chap. IV, Sect. 5], Yoshida [215], and Andrianov [5, Chap. V]. For now, we will restrict ourselves to stating the following special case, made explicit by Walling in [210]. We refer the reader to Sect. 5.5 for a proof in the case of genus $g = 1$, which is a quite simple exercise.

Recall that the Hecke operator $\mathrm{T}_p \in \mathrm{H}(\mathrm{O}_n)$ is the operator $\mathrm{T}_{\mathbb{Z}/p\mathbb{Z}}$ associated with the ordered pairs of p-neighbor lattices (Sects. 4.2.6 and 4.4.4). We define a Hecke operator

$$\mathrm{S}_p \in \mathrm{H}(\mathrm{Sp}_{2g})$$

by considering the ordered pairs $(M, N) \in \mathcal{R}_{\mathbb{Z}}^{\mathrm{a}}(\mathbb{Q}^{2g})$ such that either $M \cap N$ has index p in M and N, or $M = N$ (Sect. 4.2.6). In other words, this is the operator $\mathrm{T}_{\mathbb{Z}/p\mathbb{Z}} + 1$ in the notation loc. cit.

Proposition 5.1.1. *Let* $1 \leq g \leq n/2$. *For every prime p, the diagram*

$$\begin{array}{ccc} \mathbb{C}[\mathrm{X}_n] & \xrightarrow{\ \vartheta_g\ } & \mathrm{M}_{n/2}(\mathrm{Sp}_{2g}(\mathbb{Z})) \\ \downarrow{\scriptstyle \mathrm{T}_p} & & \downarrow{\scriptstyle p^{n/2-1-g}\,\mathrm{S}_p + p^g\,(p^{n-2g-1}-1)/(p-1)} \\ \mathbb{C}[\mathrm{X}_n] & \xrightarrow{\ \vartheta_g\ } & \mathrm{M}_{n/2}(\mathrm{Sp}_{2g}(\mathbb{Z})) \end{array}$$

is commutative. In particular, $\mathrm{Ker}\,\vartheta_g$ *is stable under* T_p.

Proof. This is a particular case of [210, Theorem 2.1] once we observe that our operator S_p coincides with the operator denoted by $\mathrm{T}_1(p^2) + 1$ by Walling. □

If we replace T_p by a more general Hecke operator $\mathrm{T}_A \in \mathrm{H}(\mathrm{O}_n)$, the analogous relations given, for example, by Andrianov and Freitag, take on a rather abstruse form. On the other hand, they are particularly transparent (especially those above) in the presentation of Rallis, which, however, requires the points of view of Satake and Langlands on Hecke operators. We will come back to this in Sect. 7.1.1.

Note that if Φ_g is the Siegel operator recalled in Sect. 4.5.2, we have the obvious relation $\Phi_g(\vartheta_g(L)) = \vartheta_{g-1}(L)$ for $g \geq 2$. This relation extends to $g = 1$ if we set $\Phi_1(\sum_{n \geq 0} a_n q^n) = a_0$. It follows that the sequence of subspaces

$$\mathrm{Ker}\,\vartheta_g \subset \mathbb{C}[\mathrm{X}_n]\ ,$$

for $g \geq 0$, is decreasing. Moreover, if $g \geq 1$, then ϑ_g induces a $\mathbb{C}$-linear injection

$$\mathrm{Ker}\,\vartheta_{g-1}/\mathrm{Ker}\,\vartheta_g \longrightarrow \mathrm{S}_{n/2}(\mathrm{Sp}_{2g}(\mathbb{Z}))\ .$$

It is clear that we have $\mathrm{Ker}\,\vartheta_n = \{0\}$ and that $\mathrm{Ker}\,\vartheta_0$ has codimension 1 in $\mathbb{C}[\mathrm{X}_n]$. Moreover, the vector $\sum_{L \in \mathrm{X}_n} (1/|\mathrm{O}(L)|)\,[L]$ is an eigenvector for the action of $\mathrm{H}(\mathrm{O}_n)$, with explicit eigenvalues (Proposition 3.2.4 and Sect. 4.4.7); the line it generates is a complement of $\mathrm{Ker}\,\vartheta_0$. On the other hand, the matter of determining the entire filtration $\mathrm{Ker}\,\vartheta_g$, as well as the structure of the $\mathrm{H}(\mathrm{O}_n)$-modules $\mathrm{Ker}\,\vartheta_{g-1}/\mathrm{Ker}\,\vartheta_g$ when $g \geq 1$, is completely nontrivial whenever $n > 8$. This is obviously a more delicate problem than that of understanding the $\mathrm{H}(\mathrm{O}_n)$-module $\mathrm{M}_{\mathbb{C}}(\mathrm{O}_n)$, the aim of this book... For $n = 16$ and $n = 24$, this filtration has been studied in detail by several authors, whose contributions we recall below. This will lead to a direct proof of Theorem A (the case $n = 16$), as well as a starting point for our proof of Theorem E (the case $n = 24$).

5.2 Theta Series of $E_8 \oplus E_8$ and E_{16}

Recall that we have $\mathrm{X}_{16} = \{E_8 \oplus E_8, E_{16}\}$ (Witt). Since the space $\mathrm{M}_8(\mathrm{SL}_2(\mathbb{Z}))$ has dimension 1, we have the well-known identity

$$\vartheta_1(E_8 \oplus E_8) = \vartheta_1(E_{16})\,. \tag{5.2.1}$$

In particular, the element $[E_8 \oplus E_8] - [E_{16}]$ generates $\mathrm{Ker}\,\vartheta_1 = \mathrm{Ker}\,\vartheta_0$ and is an eigenvector of the T_p. An absolutely remarkable fact, conjectured by Witt in [213], is that the identity (5.2.1) persists up to genus 3:

$$\vartheta_g(E_8 \oplus E_8) = \vartheta_g(E_{16}) \quad \text{if} \quad g = 1, 2, 3\,. \tag{5.2.2}$$

This was proved by Witt op. cit. for $g = 2$, and much later by Igusa and Kneser, independently, for $g = 3$ [106, 124]. Igusa proves $\mathrm{S}_8(\mathrm{Sp}_{2g}(\mathbb{Z})) = 0$ for $g \leq 3$. We refer to Appendix A for an exposition of Kneser's remarkable proof, which is very different. In summary, we have $\mathrm{Ker}\,\vartheta_3 = \mathrm{Ker}\,\vartheta_0$, and if

$$F := \vartheta_4(E_8 \oplus E_8) - \vartheta_4(E_{16})\,,$$

then we have $F \in \mathrm{S}_8(\mathrm{Sp}_8(\mathbb{Z}))$. It is well known that we have $F \neq 0$; let us be more precise about this nonvanishing.

Proposition 5.2.1. *Let c_Q be the Fourier coefficient of F corresponding to a Gram matrix of a $\mathbb{Z}$-basis of an even lattice Q of rank 4 (it does not depend on the choice of such a basis). We have*

$$\frac{c_{D_4}}{|\mathrm{O}(D_4)|} = 4480 \quad \textit{and} \quad \frac{c_{A_4}}{|\mathrm{O}(A_4)|} = -21504\,.$$

In particular, we have $c_{D_4} = -c_{A_4}$.

Proof. Indeed, an examination of the root systems $\mathbf{D}_n$ and $\mathbf{A}_m$ shows that the number of sublattices of D_n that are isometric to D_4 (resp. A_m) is $\binom{n}{4}$ (resp. $2^m\binom{n}{m+1}$). On the other hand, if $R = \mathbf{D}_4$ or $\mathbf{A}_4$, then under the action of $\mathrm{O}(\mathrm{E}_8)$, there exists exactly one orbit of sublattices of E_8 isometric to $\mathrm{Q}(R)$, with cardinality $|\mathrm{O}(\mathrm{E}_8)|/\big(|\mathrm{W}(R)|\cdot|\mathrm{A}(R)|\big)$. Let us briefly indicate how to justify this last statement, from which the proposition follows through a simple calculation. We treat the two cases $R = \mathbf{D}_4$ or $\mathbf{A}_4$ simultaneously. We set $Q = \mathrm{Q}(R)$, $E = \operatorname{res} Q$, and $\Gamma = \mathrm{A}(R)/\mathrm{W}(R)$ (the notation res was defined in Sect. 2.1).

We begin by observing that an even Euclidean lattice $L \subset \mathbb{R}^4$ such that $\operatorname{res} L \simeq E$ is necessarily isomorphic to Q. We leave it to the reader to verify that this can be deduced from the following well-known fact: every integral and unimodular lattice in $\mathbb{R}^d$ for $d \in \{4, 5\}$ is isometric to the square lattice I_d (Sect. 2.2). Proposition B.2.2 shows that if $L \subset \mathrm{E}_8$ is isometric to Q, its orthogonal complement $L^\perp$, which is even of rank 4, admits a residue isomorphic to $-E$. But in both cases, $E \simeq -E$, so $L^\perp$ is isometric to Q. Finally, we consider the Lagrangians of the residue $E \oplus E$ of $Q \oplus Q$; by Proposition 2.1.1, they all give rise to a q-module over $\mathbb{Z}$ of rank 8 containing $Q \oplus Q$ and necessarily isomorphic to E_8. Since the qe-module E is anisotropic, the Lagrangians of $E \oplus -E$ are the graphs of the automorphisms of E, and in both cases we see that they are simply permuted transitively by $1 \times \Gamma$, because the natural homomorphism $\Gamma \to \mathrm{Aut}(E)$ is bijective. It follows from these observations that the sublattices of E_8 isometric to Q form a single orbit under the action of $\mathrm{O}(\mathrm{E}_8)$, whose stabilizer is isomorphic to $\mathrm{A}(R) \times \mathrm{W}(R)$. □

Thus, we have $\operatorname{Ker} \vartheta_4 = 0$. This concludes the description of the filtration of $\mathbb{C}[\mathrm{X}_{16}]$, and we are reduced to understanding the action of $\mathrm{H}(\mathrm{Sp}_8)$ on the eigenform $F \in \mathrm{S}_8(\mathrm{Sp}_8(\mathbb{Z}))$. This form is particularly interesting. Indeed, Igusa proved in [107] that it is proportional to the famous Schottky form. In [167], Poor and Yen obtained another proof of this result, by verifying that we have

$$\dim \mathrm{S}_8(\mathrm{Sp}_8(\mathbb{Z})) = 1 \,. \tag{5.2.3}$$

We refer to [77] for a second proof of this equality, as well as to Theorem 9.5.9 for a third!

Let $\tau(n)$ be the Ramanujan function, defined by

$$\Delta = q \prod_{n\geq 1} (1-q^n)^{24} = \sum_{n\geq 1} \tau(n) q^n \,.$$

An elementary computation shows that Theorem A of the introduction is an immediate consequence of part (i) of the following theorem, where the terms 286 and 405 come from the relation $|\mathrm{O}(\mathrm{E}_{16})|/|\mathrm{O}(\mathrm{E}_8 \oplus \mathrm{E}_8)| = 286/405$. Recall that the Hecke operator $\mathrm{S}_p \in \mathrm{H}_p(\mathrm{Sp}_{2g})$ was introduced in Sect. 5.1.

Theorem 5.2.2. *Let p be a prime.*

(i) *The eigenvalue of* T_p *on* $[\mathrm{E}_8 \oplus \mathrm{E}_8] - [\mathrm{E}_{16}]$ *is*

$$p^4 \frac{p^7 - 1}{p - 1} + p^7 + \tau(p) \frac{p^4 - 1}{p - 1}.$$

(ii) *The eigenvalue of* S_p *on the line* $\mathrm{S}_8(\mathrm{Sp}_8(\mathbb{Z}))$ *is*

$$p^4 + \tau(p)\, p^{-3} \frac{p^4 - 1}{p - 1}.$$

Proof. Above, we saw that ϑ_4 induces an isomorphism $\mathrm{Ker}\,\vartheta_0 \xrightarrow{\sim} \mathrm{S}_8(\mathrm{Sp}_8(\mathbb{Z}))$. Assertions (i) and (ii) are then equivalent by Proposition 5.1.1 (the Eichler commutation relations). Assertion (ii) is an immediate consequence of the work of Ikeda [108] (proof of the Duke–Imamoğlu conjecture [40]). Indeed, if k and g are even integers such that $k \equiv g \bmod 4$, Ikeda constructs op. cit. an injective linear map

$$\mathrm{I}_g \colon \mathrm{S}_k(\mathrm{SL}_2(\mathbb{Z})) \to \mathrm{S}_{(k+g)/2}(\mathrm{Sp}_{2g}(\mathbb{Z}))$$

with the following compatibility property with the Hecke operators. If $f = q + \sum_{n \geq 2} a_n q^n \in \mathrm{S}_k(\mathrm{SL}_2(\mathbb{Z}))$ is an eigenfunction for $\mathrm{H}(\mathrm{PGL}_2)$, then for every prime p, the form $\mathrm{I}_g(f)$ is an eigenform for S_p, with eigenvalue

$$p^g \left(1 + a_p\, p^{(-(k+g)/2)+1} \frac{p^g - 1}{p - 1}\right). \tag{5.2.4}$$

We refer neophytes to Sect. 7.2 for an explanation of the passage from Ikeda's statement to the above. In the literature, the form $\mathrm{I}_g(f)$ is often called the *Ikeda lift* of f in genus g. If $k = 12$ and $g = 4$, and if f is the modular form $\Delta \in \mathrm{S}_{12}(\mathrm{SL}_2(\mathbb{Z}))$, we note that its Ikeda lift $\mathrm{I}_4(\Delta)$ is a (nonzero) element of $\mathrm{S}_8(\mathrm{Sp}_8(\mathbb{Z}))$, with eigenvalue that of assertion (ii), which completes the proof. □

Let us mention that Breulmann and Kuss had already observed [40, Sect. 3] that when $p = 2$, the eigenvalue of the Hecke operator S_p on the Schottky form is indeed given by the formula of Theorem 5.2.2 (ii), as predicted by the Duke–Imamoğlu conjecture. Their method consists in realizing F as a theta series with harmonic coefficients, constructed from E_8 (which we will, in fact, also do further on!). A similar verification, which seems more economical, is given by the computation of T_2 carried out in Sect. 3.3.1.

The occurrence of the form Δ in the argument above is very indirect, a consequence of the deep result of Ikeda. In Sect. 5.4, we will give another proof of Theorem 5.2.2 (ii), which we discovered and which is independent of the work of Ikeda. It is based, in particular, on the triality for the group $\mathrm{PGSO}_{\mathrm{E}_8}$ over $\mathbb{Z}$. Further on, we will also obtain a third proof of the same statement, which relies on Arthur's theory: it is the (particularly simple) particular case where $k = 8$ in the statement of Theorem 9.5.9; this proof is significantly more sophisticated than the first two. Finally, we will explain how assertion (i) of Theorem 5.2.2 trivially follows from a very general conjecture concerning Arthur's theory, which we will present in Chap. 8; see the examples in Sect. 8.5.7.

5.3 Theta Series of the Niemeier Lattices

Let us now consider the case $n = 24$. By Erokhin [80], the theta series of genus 12 of the 24 Niemeier lattices are linearly independent, that is, $\mathrm{Ker}\,\vartheta_{12} = 0$. This does not hold in genus 11. Indeed, as observed by Borcherds, Freitag, and Weissauer [31] using an ingenious construction, $\mathrm{Ker}\,\vartheta_{11}$ has dimension 1. This is a spectacular analog in dimension 24 of the discovery of Witt studied in Sect. 5.2.

A more detailed study of the filtered $\mathrm{H}(\mathrm{O}_{24})$-module $\mathbb{C}[\mathrm{X}_{24}]$ was initiated by Nebe and Venkov in their delightful article [156]. Their starting point is the computation of T_2 recalled in Sect. 3.3.3, which they deduce from Borcherds' Ph.D. thesis [29]. They note that T_2 has 24 distinct integral eigenvalues and have an explicit eigenvector for each of them, which is necessarily an eigenvector for the action of all of $\mathrm{H}(\mathrm{O}_n)$. For the sake of convenience, like these authors, we number these eigenvectors v_i for $i = 1, \ldots, 24$ in such a way that the associated eigenvalues λ_i of T_2 are in decreasing order (see Table 5.1). Determining the filtration of $\mathbb{C}[\mathrm{X}_{24}]$ is then equivalent to determining the *degree* of each v_i, that is, the least integer $\mathrm{g}_i \geq 0$ with $\vartheta_{\mathrm{g}_i}(\mathrm{v}_i) \neq 0$. Nebe and Venkov succeed for 22 of the 24 eigenvalues and propose a conjecture for the degree of the remaining ones, namely those with eigenvalues 17280 and −7920, which have been made bold in Tables 5.1 and 5.2 below.

Table 5.1 The filtration of $\mathbb{C}[\mathrm{X}_{24}]$ according to Nebe and Venkov

λ	8390655	4192830	2098332	1049832	533160	519120	268560	244800
degree	0	1	2	3	4	4	5	5
λ	145152	126000	99792	91152	89640	69552	51552	45792
degree	6	6	6	7	8	7	8	7
λ	35640	21600	17280	5040	−7920	−16128	−48528	−98280
degree	8	8	**9**	9	**10**	10	11	12

Let us say a few words about this table. The trivial eigenvalue

$$(2^{12} - 1)(2^{11} + 1) = 8390655 = \lambda_1$$

is, of course, associated with the unique eigenvector of degree 0. Moreover, it is easy to verify that ϑ_1 induces a surjection $\mathrm{Ker}\,\vartheta_0 \to \mathrm{S}_{12}(\mathrm{SL}_2(\mathbb{Z}))$, so that one, and only one, of the eigenvectors of T_2 has degree 1. By Theorem 5.5.1, it is the one with eigenvalue

$$\tau(2)^2 - 2^{11} + 2\,(2^{21} - 1) = 4192830 = \lambda_2 \; .$$

The other eigenvalues are significantly more subtle to understand. For example, at the end of their paper, Nebe and Venkov mention that for $i = 3, 5, 11, 13$, and 24 (in which case g_i is $2, 4, 6, 8$, and 12, respectively), $\vartheta_{\mathrm{g}_i}(\mathrm{v}_i)$ is proportional to the Ikeda lift

$$\mathrm{I}_{\mathrm{g}_i}(\Delta_{23-\mathrm{g}_i}) \; ,$$

where Δ_w denotes a generator of $\mathrm{S}_{w+1}(\mathrm{SL}_2(\mathbb{Z}))$ when $w \in \{11, 15, 17, 19, 21\}$ (see Corollary 7.3.4 for a justification). The action of $\mathrm{H}(\mathrm{O}_{24})$ on these v_i is thus known explicitly (modulo the Eichler relations) in terms of the Fourier coefficients of the five modular forms Δ_w mentioned above. For $i = 24$, this had already been noted by Borcherds, Freitag, and Weissauer immediately after the announcement of [108] (see also [40]). For example, we have

$$\tau(2)(2^{12} - 1) = -98280 = \lambda_{24} \; ;$$

more generally, the eigenvalue of T_p on v_{24} is $\tau(p)(p^{12} - 1)(p - 1)^{-1}$.

A spectacular additional step was then again obtained by Ikeda, in his paper [109], as a consequence of the results of Nebe–Venkov mentioned above and his partial resolution of a conjecture of Miyawaki. Ikeda succeeds in expressing the action of $\mathrm{H}(\mathrm{O}_{24})$, again in terms of the Δ_w above, on all but four of the v_i; the exceptions are listed here:

Table 5.2 The four mysterious eigenvectors

λ	126000	51552	17280	−7920
degree	6	8	**9**	**10**

We will return to the exact statement proved by Ikeda in Sect. 9.2, where we also explain the missing eigenvalues mentioned above.

Let us conclude this section with a discussion of $\mathrm{M}_{\det}(\mathrm{O}_{24})$. We already noted, in Sect. 4.1.6, that the natural map $\widetilde{\mathrm{X}}_n \to \mathrm{X}_n$ is bijective for $n < 24$, but that for $n = 24$, the inverse image of the class of the Leech lattice (and only of that class) has two elements, which we denote by $\mathrm{Leech}^{\pm}$. It follows that $\mathrm{M}_{\det}(\mathrm{O}_{24})$ has dimension 1; it consists of functions that are zero on the 23 Niemeier lattices with roots and take on opposite values on Leech^{+} and Leech^{-}.

Proposition 5.3.1. *The eigenvalue of* T_p *on the line* $\mathrm{M}_{\det}(\mathrm{O}_{24})$ *is*

$$\tau(p)\,\frac{p^{12} - 1}{p - 1} \; .$$

We will prove this result in Sect. 7.5; more precisely, we will prove that the ring $\mathrm{H}(\mathrm{O}_{24})$ acts similarly on the lines $\mathbb{C}\mathrm{v}_{24} \subset \mathrm{M}_{\mathbb{C}}(\mathrm{O}_{24})$ and $\mathrm{M}_{\det}(\mathrm{O}_{24})$. In particular, this will answer a question posed by Schulze-Pillot [175, Remark, Sect. 1].

This proposition admits a striking translation in terms of lattices. A p-neighbor M of an even unimodular lattice L in $\mathbb{R}^n$ is called *proper* if there exists a $g \in \mathrm{SO}(\mathbb{R}^n)$ such that $g(M) = L$. We denote by $\mathrm{N}_p^+(L, M)$ the number of proper p-neighbors of L isometric to M, and we set $\mathrm{N}_p^-(L, M) = \mathrm{N}_p(L, M) - \mathrm{N}_p^+(L, M)$. If L has an isometry of determinant -1, all its p-neighbors are proper. This is, of course, the case for all Niemeier lattices with roots. The case of the Leech lattice is, on the other hand, more interesting, as shown by the following immediate corollary of Proposition 5.3.1.

Corollary 5.3.2. *For every prime p, we have the relation*

$$\mathrm{N}_p^+(\mathrm{Leech}, \mathrm{Leech}) - \mathrm{N}_p^-(\mathrm{Leech}, \mathrm{Leech}) = \tau(p)\frac{p^{12}-1}{p-1}\,.$$

An amusing consequence of this corollary is that the famous Lehmer conjecture is equivalent to "$\mathrm{N}_p^+(\mathrm{Leech}, \mathrm{Leech}) \neq \mathrm{N}_p^-(\mathrm{Leech}, \mathrm{Leech})$ for every prime p" !

5.4 An Alternative Construction of $\mathrm{I}_4(\Delta)$ by Triality

As promised in Sect. 5.2, we will now give a second proof of Theorem 5.2.2 (and therefore of Theorem A from Chap. 1), which does not depend on Ikeda's theorem [108]. This will allow us, in passing, to give nontrivial examples of automorphic forms for O_8 and illustrate the techniques of Chap. 4.

5.4.1 Harmonic Theta Series

We place ourselves in the Euclidean space $V = \mathbb{R}^n$. Let $1 \leq g \leq n/2$ be an integer. The vector space $V^g = V \otimes \mathbb{R}^g$ is endowed with a natural $\mathbb{R}$-linear representation of $\mathrm{O}(V) \times \mathrm{GL}_g(\mathbb{R})$. For every integer $d \geq 0$, consider the space $\mathrm{H}_{d,g}(V)$ of polynomials $P \colon V^g \to \mathbb{C}$ such that

(i) P is harmonic with respect to the Euclidean Laplacian of V^g;
(ii) $P \circ h = \det(h)^d P$ for every $h \in \mathrm{GL}_g(\mathbb{R})$.

This space is stable under the action of $\mathrm{O}(V)$. We construct elements of $\mathrm{H}_{d,g}(V)$ as follows. Let $I \subset V \otimes \mathbb{C}$ be an isotropic subspace of dimension g, and let $e_1, \ldots, e_g$ be a $\mathbb{C}$-basis of I. One easily verifies that

$$(v_1, \ldots, v_g) \mapsto \det[e_i \cdot v_j]^d_{1 \leq i,j \leq g} \tag{5.4.1}$$

is an element of $\mathrm{H}_{d,g}(V)$. It generates a line that depends only on I. These lines are permuted transitively by $\mathrm{O}(V)$ and generate $\mathrm{H}_{d,g}(V)$, which is an irreducible representation of $\mathrm{O}(V)$ by Kashiwara and Vergne [116, (0.9) and (5.7)].

For an even unimodular lattice $L \subset V$ and $P \in \mathrm{H}_{d,g}(V)$, we set

$$\vartheta_g(L,P) = \sum_{v \in L^g} P(v) q^{v.v/2} .$$

The functional equation of the Jacobi ϑ-function allows us to prove $\vartheta_g(L,P) \in \mathrm{M}_{(n/2)+d}(\mathrm{Sp}_{2g}(\mathbb{Z}))$ [88, Kap. III, Sect. 3]. Note that

$$\vartheta_g(L,P) = \vartheta_g(\gamma(L), \gamma(P)) \quad \forall \gamma \in \mathrm{O}(V) . \tag{5.4.2}$$

In particular, $\vartheta_g(L,P) = 0$ for every P if $\mathrm{H}_{d,g}(V)^{\mathrm{O}(L)} = 0$. Let us begin by giving an example, undoubtedly well-known, where this space of invariants is nonzero for $n = 8$.

Lemma 5.4.2. *Let $R \subset V$ be a root system, $W \subset \mathrm{O}(V)$ its Weyl group, and $W^+ = W \cap \mathrm{SO}(V)$. Then*

$$H_W(t) := \sum_{d \geq 0} \left(\dim \mathrm{H}_{d,1}(V)^W\right) t^d = (1 - t^2) \prod_i \left(1 - t^{m_i+1}\right)^{-1} ,$$

where the m_i are the exponents of W, and

$$\sum_{d \geq 0} \left(\dim \mathrm{H}_{d,1}(V)^{W^+}\right) t^d = (1 + t^{|R|/2)}) H_W(t) .$$

Proof. Let A and B be the respective generating series of the sequences $\dim \mathrm{Pol}_d(V)^W$ and $\dim(\mathrm{Pol}_d(V) \otimes \det)^W$ (the "anti-invariants") for $d \geq 0$. By Bourbaki [39, Chap. V, Sect. 6], we have $B = t^{|R|/2} A$ and $A = \prod_i (1 - t^{m_i+1})^{-1}$. Let $\mathrm{Pol}_d(V)$ be the space of homogeneous polynomials $V \to \mathbb{C}$ of degree d. We denote by $\boldsymbol{\Delta} = \sum_{i=1}^n \partial^2/\partial x_i^2$ the standard Laplacian of $\mathbb{R}^n$. For every $d \in \mathbb{Z}$, we have an $\mathrm{O}(V)$-equivariant exact sequence

$$0 \longrightarrow \mathrm{H}_{d,1}(V) \longrightarrow \mathrm{Pol}_d(V) \xrightarrow{\boldsymbol{\Delta}} \mathrm{Pol}_{d-2}(V) \longrightarrow 0 ,$$

where the surjectivity of $\boldsymbol{\Delta}$ is a classical result (see, for example, [95, Sect. 5.2.3]). From this, we deduce statement (i) and then statement (ii). □

Consider, for example, the lattice $\mathrm{E}_8 \subset \mathbb{R}^8$, whose root system $\mathrm{R}(\mathrm{E}_8)$ is exactly of type $\mathbf{E}_8$. Its exponents are the eight integers $1 \leq m \leq 30$ relatively prime to 30. Since we have $\mathrm{W}(\mathbf{E}_8) = \mathrm{O}(\mathrm{E}_8)$, we find

$$\sum_{d \geq 0} \left(\dim \mathrm{H}_{d,1}(V)^{\mathrm{O}(\mathrm{E}_8)}\right) t^d = \\ 1 + t^8 + t^{12} + t^{14} + t^{16} + t^{18} + 2t^{20} + \cdots . \tag{5.4.3}$$

The smallest invariant is therefore the one for $d = 8$.

Proposition 5.4.3. *The polynomial* $A(x) = -30\,(x\cdot x)^4 + \sum_{\alpha\in \mathrm{R}(\mathrm{E}_8)}(\alpha\cdot x)^8$ *is in* $\mathrm{H}_{8,1}(V)^{\mathrm{O}(E_8)}$. *It satisfies* $A(\alpha) = 144$ *for every root* $\alpha \in \mathrm{R}(\mathrm{E}_8)$. *In particular, we have the equality* $\vartheta_1(\mathrm{E}_8, A) = 240 \cdot 144\ \Delta$.

Proof. The invariance of A under $\mathrm{O}(\mathrm{E}_8)$ is obvious. Let us verify that A is harmonic. In the notation of the proof of Lemma 5.4.2, the polynomial $\mathbf{\Delta} A \in \mathrm{Pol}_6(V)$ is invariant. For every $d \geq 2$, we have the decomposition

$$\mathrm{Pol}_d(V) = (x\cdot x)\,\mathrm{Pol}_{d-2}(V) \oplus \mathrm{H}_{d,1}(V)\ .$$

The Poincaré series (5.4.3) therefore shows that $\mathbf{\Delta} A$ is proportional to $(x\cdot x)^3$. It remains to see that $\mathbf{\Delta} A$ vanishes on the roots of E_8. This is an easy verification, left to the readers. It uses, on the one hand, the formulas

$$\mathbf{\Delta}(\alpha{\cdot}x)^k = k(k{-}1)(\alpha{\cdot}\alpha)(\alpha{\cdot}x)^{k-2}\ ,\ \mathbf{\Delta}(x{\cdot}x)^k = 2k(\dim(V){+}2k{-}2)(x{\cdot}x)^{k-1}\ ,$$

and, on the other hand, the following property of the root system $\mathbf{E}_8$ proved in [39, Chap. VI, Sect. 1.11, Proposition 32]: given a root α_0, there are exactly 114 roots in $\mathbf{E}_8$ that are nonorthogonal to α_0, and for 112 of those 114 roots, their inner product with α_0 is ± 1. This same property also implies $A(\alpha) = 144$ for every $\alpha \in \mathrm{R}(\mathrm{E}_8)$, and then the last assertion. □

It will be useful to express the identity above in terms of automorphic forms for $\mathrm{O}_8 = \mathrm{O}_{\mathrm{E}_8}$ (Sect. 4.4.4). For this, we will linearize the definition of harmonic theta series in the manner of Sect. 5.1. Recall that for every integer $n \equiv 0 \bmod 8$, the $\mathrm{O}_n(\mathbb{A}_f)$-set $\mathcal{R}(\mathrm{O}_n)$ can be naturally identified with that of the even unimodular lattices in $\mathbb{R}^n$ contained in $\mathrm{E}_n \otimes \mathbb{Q}$ (Sect. 4.1.2).

If $L \in \mathcal{R}(\mathrm{O}_n)$, the map $P \mapsto \vartheta_g(L, P)$ given by $\mathrm{H}_{d,g}(V) \to \mathrm{M}_{(n/2)+d}(\mathrm{O}_n)$ is $\mathbb{C}$-linear. Moreover, we have $\vartheta_g(\gamma L, \gamma P) = \vartheta_g(L, P)$ for every $\gamma \in \mathrm{O}_n(\mathbb{Q})$ (formula (5.4.2)). We therefore have a unique linear map

$$(\mathbb{Z}[\mathcal{R}(\mathrm{O}_n)] \otimes \mathrm{H}_{d,g}(V))_{\mathrm{O}_n(\mathbb{Q})} \to \mathrm{M}_{(n/2)+d}(\mathrm{Sp}_{2g}(\mathbb{Z}))$$

that sends the class of $[L] \otimes P$ to $\vartheta_g(L, P)$ for every $L \in \mathcal{R}(\mathrm{O}_n)$ and every $P \in \mathrm{H}_{d,g}(V)$. Through the isomorphism (4.4.3), the $\mathrm{H}(\mathrm{O}_n)$-module on the left can be canonically identified with $\mathrm{M}_{\mathrm{H}_{d,g}(V)^*}(\mathrm{O}_n)^*$. Since this double duality is not so fortunate, we prefer to remove it by using the canonical isomorphism between the $\mathrm{H}(\mathrm{O}_n)$-modules $\mathrm{M}_{\mathrm{H}_{d,g}(V)^*}(\mathrm{O}_n)^*$ and $\mathrm{M}_{\mathrm{H}_{d,g}}(V)(\mathrm{O}_n)^{\mathrm{t}}$ (Corollary 4.4.9). Recall that the t in the exponent of the latter means that the action of $\mathrm{H}(\mathrm{O}_n)$ is twisted by $T \mapsto T^{\mathrm{t}}$, in the sense of Remark 4.2.4. Since we have $T^{\mathrm{t}} = T$ for every $T \in \mathrm{H}(\mathrm{O}_n)$ (Proposition 4.2.8), this torsion has no effect here, and will cheerfully be ignored.

Finally, recall that we denote by $[L, P] \in \mathrm{M}_{\mathrm{H}_{d,g}(V)}(\mathrm{O}_n)$ the image of $[L] \otimes P$ by the canonical isomorphism $(\mathbb{Z}[\mathcal{R}(\mathrm{O}_n)] \otimes \mathrm{H}_{d,g}(V))_{\mathrm{O}_n(\mathbb{Q})} \xrightarrow{\sim} \mathrm{M}_{\mathrm{H}_{d,g}(V)}(\mathrm{O}_n)$ (see the end of Sect. 4.4.7).

Proposition-Definition 5.4.4. *There exists a unique linear map*

$$\vartheta_{d,g}\colon \mathrm{M}_{\mathrm{H}_{d,g}(V)}(\mathrm{O}_n) \to \mathrm{M}_{(n/2)+d}(\mathrm{Sp}_{2g}(\mathbb{Z}))$$

that sends $[L,P]$ *to* $\vartheta_g(L,P)$ *for every* $L \in \mathcal{R}(\mathrm{O}_n)$ *and every* $P \in \mathrm{H}_{d,g}(V)$. *If* $d > 0$, *then* $\mathrm{Im}(\vartheta_{d,g}) \subset \mathrm{S}_{(n/2)+d}(\mathrm{Sp}_{2g}(\mathbb{Z}))$.

Proof. The existence and uniqueness of such a map follow from the discussion above. Concretely, for $F \in \mathrm{M}_{\mathrm{H}_{d,g}(V)}(\mathrm{O}_n)$, we have

$$\vartheta_{d,g}(F) = \sum_i \frac{1}{|\mathrm{O}(L_i)|}\,\vartheta_g(L_i,P_i)\;,$$

where the L_i form a system of representatives of the $\mathrm{O}_n(\mathbb{Q})$-orbits of $\mathcal{R}(\mathrm{O}_n)$ and we have set $P_i = F(L_i)$. □

For $d = 0$, we have an isomorphism $\mathrm{H}_{d,g}(V) \simeq \mathbb{C}$ (trivial representation) and the map $\vartheta_{d,g}$ is just the composition of the isomorphism $\mathrm{M}_{\mathbb{C}}(\mathrm{O}_n) \xrightarrow{\sim} \mathbb{C}[\mathrm{X}_n]$ given by Corollary 4.4.9 and the map ϑ_g of Sect. 5.1.

Corollary 5.4.5. *The map* $\vartheta_{8,1}\colon \mathrm{M}_{\mathrm{H}_{8,1}(\mathbb{R}^8)}(\mathrm{O}_8) \to \mathrm{S}_{12}(\mathrm{SL}_2(\mathbb{Z}))$ *is an isomorphism between* 1*-dimensional spaces.*

Proof. It is well known that we have $\dim \mathrm{S}_{12}(\mathrm{SL}_2(\mathbb{Z})) = 1$. Recall that, by Mordell, we have $\mathrm{X}_8 = \{\mathrm{E}_8\}$, so that for every integer $d \geq 0$, we have $\dim \mathrm{M}_{\mathrm{H}_{d,1}(\mathbb{R}^8)}(\mathrm{O}_8) = \dim \mathrm{H}_{d,1}(\mathbb{R}^8)^{\mathrm{O}(\mathrm{E}_8)}$. An examination of the series (5.4.3) shows that we have $\dim \mathrm{M}_{\mathrm{H}_{8,1}(\mathbb{R}^8)}(\mathrm{O}_8) = 1$. It therefore suffices to see that $\vartheta_{8,1}$ is nonzero, but this follows from Proposition 5.4.3. □

We also have the Eichler commutation relations for the map $\vartheta_{d,g}$ from $\mathrm{M}_{\mathrm{H}_{d,g}(V)}(\mathrm{O}_n)$ to $\mathrm{M}_{(n/2)+d}(\mathrm{Sp}_{2g}(\mathbb{Z}))$ (see Sect. 7.1.1 for a justification). In particular, for every prime p, we have

$$\vartheta_{d,g} \circ \mathrm{T}_p = \left(p^{(n/2)-1-g}\,\mathrm{S}_p + p^g\,\frac{p^{n-2g-1}-1}{p-1}\right) \circ \vartheta_{d,g}\;. \tag{5.4.4}$$

We refer to Sect. 5.5 for a proof of this formula for $g = 1$. Combining this with Corollary 5.4.5 leads to the following result.

Corollary 5.4.6. *The eigenvalue of* T_p *on the line* $\mathrm{M}_{\mathrm{H}_{8,1}(\mathbb{R}^8)}(\mathrm{O}_8)$ *is*

$$p^{-8}\big(\tau(p)^2 - p^{11}\big) + p\,\frac{p^5-1}{p-1}\;.$$

5.4.7 Hecke Operators Corresponding to Perestroikas

Recall that in Example 4.2.16, we defined a natural injection

$$\mathrm{H}(\mu)\colon \mathrm{H}(\mathrm{O}_n) \to \mathrm{H}(\mathrm{PGO}_n)$$

associated with the canonical $\mathbb{Z}$-morphism $\mu\colon \mathrm{O}_n \to \mathrm{PGO}_n$. From now on, we will denote $\mathrm{H}(\mu)$ as an inclusion $\mathrm{H}(\mathrm{O}_n) \subset \mathrm{H}(\mathrm{PGO}_n)$ to avoid overloading the notation. Let W be a representation of $\mathrm{PGO}_n(\mathbb{R})$, and let W' be its restriction to $\mathrm{O}_n(\mathbb{R})$. By Proposition 4.4.3, the restriction of the functions via $\mathcal{R}(\mathrm{O}_n) \to \mathcal{R}(\mathrm{PGO}_n)$ defines an $\mathrm{H}(\mathrm{O}_n)$-equivariant map

$$\mu^*\colon \mathrm{M}_W(\mathrm{PGO}_n) \to \mathrm{M}_{W'}(\mathrm{O}_n)\ .$$

Lemma 5.4.8. *The map μ^* is an isomorphism.*

Proof. The map $\mathrm{O}_n(\mathbb{Q})\backslash\mathcal{R}(\mathrm{O}_n) \to \mathrm{PGO}_n(\mathbb{Q})\backslash\mathcal{R}(\mathrm{PGO}_n)$ is bijective by Proposition 4.1.4 (it is $\xi_2 \circ \xi_1$). We can now conclude because if $M \subset \mathbb{R}^n$ is a Euclidean lattice or, more generally, a positive definite $\widetilde{\mathrm{b}}$-module over $\mathbb{Z}$, then we have $\mathrm{O}(M) = \mathrm{GO}(M)$. □

One way to view this lemma is to say that the action of $\mathrm{H}(\mathrm{O}_n)$ on $\mathrm{M}_{W'}(\mathrm{O}_n)$ extends to an action of the larger ring $\mathrm{H}(\mathrm{PGO}_n)$. We will apply this to the $\mathrm{H}_{d,1}(\mathbb{R}^n)$. These spaces are endowed with a natural representation of $\mathrm{GO}(\mathbb{R}^n)$, on which the homothety of ratio λ acts by the scalar λ^{-d}. In particular, if d is even,

$$\mathrm{H}_{d,1}(\mathbb{R}^n) \otimes \nu^{d/2}$$

factors through a representation of $\mathrm{PGO}_n(\mathbb{R})$ whose restriction to $\mathrm{O}_n(\mathbb{R})$ is simply $\mathrm{H}_{d,1}(\mathbb{R}^n)$. Recall that the perestroika operator $\mathrm{K}_p \in \mathrm{H}_p(\mathrm{PGO}_n)$ with respect to p was defined in Sect. 4.2.6 (and $\mathrm{K}_p = \mathrm{K}_p^{\mathrm{t}}$).

Lemma 5.4.9. *Consider $d \equiv 0 \bmod 2$, $W = \mathrm{H}_{d,1}(\mathbb{R}^n) \otimes \nu^{d/2}$, a prime p, and $\ell_{2r}(p) = \prod_{i=0}^{r-1}(1+p^i)$ (that is, the number of Lagrangians of $\mathrm{H}(\mathbb{F}_p^r)$). The following diagram is commutative:*

$$\begin{array}{ccc}
\mathrm{M}_W(\mathrm{PGO}_n) & \xrightarrow{\ \vartheta_{d,1}\circ\mu^*\ } & \mathrm{M}_{(n/2)+d}(\mathrm{SL}_2(\mathbb{Z})) \\
\Big\downarrow{\scriptstyle \mathrm{K}_p} & & \Big\downarrow{\scriptstyle \ell_{n-2}(p)\,p^{-d/2}\,\mathrm{T}(p)} \\
\mathrm{M}_W(\mathrm{PGO}_n) & \xrightarrow{\ \vartheta_{d,1}\circ\mu^*\ } & \mathrm{M}_{(n/2)+d}(\mathrm{SL}_2(\mathbb{Z}))\ .
\end{array}$$

Proof. This is a harmonic variant of Eichler's result [78, Satz 21.3] (see also [169]). Let us recall the argument.

We fix a unimodular lattice M in $\mathrm{E}_n \otimes \mathbb{Q}$ (or, more generally, a homodual lattice in $\mathrm{E}_n \otimes \mathbb{Q}$), as well as a polynomial P in the space $\mathrm{H}_{d,1}(\mathbb{R}^n) \otimes \nu^{d/2}$. Let $\mathrm{Per}_p(M)$ be the set of perestroikas of M with respect to p.

For $N \in \mathrm{Per}_p(M)$, we have $\mu^*([\underline{N}, P]) = [\gamma(N), p^{-d/2}\gamma(P)]$, where $\gamma \in \mathrm{GO}(\mathbb{Q})$ is an arbitrary element with similitude factor p^{-1}. Since the lattice $p^{-1/2}N \subset \mathbb{R}^n$ is isometric to $\gamma(N)$, the relation (5.4.2) implies

$$\vartheta_{d,1} \circ \mu^*([N, P]) = \vartheta_{d,1}\left(\frac{N}{\sqrt{p}}, P\right).$$

For $m \geq 0$, the mth Fourier coefficient of $\vartheta_{d,1} \circ \mu^* \circ \mathrm{K}_p([M, P])$ is therefore

$$\sum_{(N,v)} P\left(\frac{v}{\sqrt{p}}\right), \tag{5.4.5}$$

where the sum is taken over all pairs (N, v) with $N \in \mathrm{Per}_p(M)$ and $v \in N$ such that $v \cdot v = 2mp$.

Suppose that m is relatively prime to p. Note that an element $v \in M$ such that $v \cdot v = 2mp$ is nonzero modulo pM and isotropic. It therefore belongs to exactly $\ell_{n-2}(p)$ perestroikas of M (that is, the number of Lagrangians of $M \otimes \mathbb{F}_p$ containing a given isotropic line). The sum (5.4.5) is therefore simply

$$p^{-d/2}\ell_{n-2}(p)a_{mp} ,$$

where a_r is the rth Fourier coefficient of $\vartheta_{d,1}(M, P)$. To conclude, we can treat the case where m is a multiple of p similarly or simply invoke Lemma 5.5.2. □

Corollary 5.4.10. *For every prime p, the eigenvalue of K_p on the line $\mathrm{M}_{\mathrm{H}_{8,1}(\mathbb{R}^8)\otimes\nu^4}(\mathrm{PGO}_8)$ is $2\,p^{-4}\,(p^4-1)(p-1)^{-1}\tau(p)$.*

Proof. This is a consequence of the identity $\ell_6(p) = 2(1+p)(1+p^2) = 2(p^4-1)(p-1)^{-1}$, Lemmas 5.4.8 and 5.4.9, and Corollary 5.4.5. □

5.4.11 Passage from PGO_n to PGSO_n

Since the map μ_3 from Proposition 4.1.7 is bijective and $\mathrm{PGSO}_n(\mathbb{A}_f)$-equivariant, the arguments given in Sects. 4.2.11 and 4.4.4 concerning the comparison between SO_n and O_n extend mutatis mutandis to the case of PGSO_n and PGO_n. In particular, we have an action of the group

$$\mathrm{PGO}_n(\mathbb{A}_f)/\mathrm{PGSO}_n(\mathbb{A}_f) \simeq (\mathbb{Z}/2\mathbb{Z})^{\mathrm{P}}$$

on $\mathrm{H}(\mathrm{PGSO}_n)$ that preserves every $\mathrm{H}_p(\mathrm{PGSO}_n)$ and whose invariants are exactly $\mathrm{H}(\mathrm{PGO}_n)$. Consequently, we have $\mathrm{H}(\mathrm{PGO}_n) \subset \mathrm{H}(\mathrm{PGSO}_n)$. If U is a representation of $\mathrm{PGSO}_n(\mathbb{R})$ and $\mathrm{Ind}\,U$ is the induced representation of $\mathrm{PGO}_n(\mathbb{R})$, we have a canonical $\mathrm{H}(\mathrm{PGO}_n)$-equivariant isomorphism

$$\mathrm{ind}\colon \mathrm{M}_U(\mathrm{PGSO}_n) \to \mathrm{M}_{\mathrm{Ind}\,U}(\mathrm{PGO}_n)\ .$$

Finally, if W is a representation of $\mathrm{PGO}_n(\mathbb{R})$ and W' denotes its restriction to $\mathrm{PGSO}_n(\mathbb{R})$, then the action of $\mathrm{PGO}_n(\mathbb{Q})/\mathrm{PGSO}_n(\mathbb{Q}) \simeq \mathbb{Z}/2\mathbb{Z}$ on $\mathrm{M}_{W'}(\mathrm{PGSO}_n)$ by conjugation endows the latter with a natural symmetry that we denote by s. The restriction of the functions via the bijection $\mathcal{R}(\mathrm{PGSO}_n) \to \mathcal{R}(\mathrm{PGO}_n)$ defines an $\mathrm{H}(\mathrm{PGO}_n)$-equivariant injection

$$\mathrm{res}\colon \mathrm{M}_W(\mathrm{PGO}_n) \to \mathrm{M}_{W'}(\mathrm{PGSO}_n)$$

with image $\mathrm{M}_{W'}(\mathrm{PGSO}_n)^{s=\mathrm{id}}$.

Let W be the representation $\mathrm{H}_{d,1}(\mathbb{R}^n)\otimes\nu^{d/2}$ of $\mathrm{PGO}_n(\mathbb{R})$ defined in Sect. 5.4.7 (for $d\equiv 0 \bmod 2$). If $n>2$, its restriction W' to $\mathrm{PGSO}_n(\mathbb{R})$ is irreducible [95, Sect. 5.2]. Finally, for general reasons, we have $\mathrm{Ind}\,W' \simeq W\oplus W\otimes\epsilon$, where ε is the character of order 2 of $\mathrm{PGO}_n(\mathbb{R})$ with kernel $\mathrm{PGSO}_n(\mathbb{R})$.

Lemma 5.4.12. *Let* $W = \mathrm{H}_{8,1}(\mathbb{R}^8)\otimes\nu^4$ *and* $W' = W_{|\mathrm{PGSO}_8(\mathbb{R})}$. *The restriction map* $\mathrm{res}\colon \mathrm{M}_W(\mathrm{PGO}_8)\to\mathrm{M}_{W'}(\mathrm{PGSO}_8)$ *is bijective.*

Proof. We must show that the space $\mathrm{M}_{W'}(\mathrm{PGSO}_8)^{s=-\mathrm{id}}$, which is naturally isomorphic to $\mathrm{M}_{W\otimes\varepsilon}(\mathrm{PGO}_8)$, is zero. By Lemma 5.4.8, this is equivalent to showing $\mathrm{M}_{\mathrm{H}_{8,1}(\mathbb{R}^8)\otimes\varepsilon}(\mathrm{O}_8) = 0$. This follows from the equalities $\mathrm{X}_8 = \{\mathrm{E}_8\}$ and $\mathrm{O}(\mathrm{E}_8) = \mathrm{W}(\mathbf{E}_8)$ and Lemma 5.4.2, which asserts that the $\mathrm{W}(\mathbf{E}_8)^+$-invariants of $\mathrm{H}_{8,1}(\mathbb{R}^8)$ are in fact $\mathrm{W}(\mathbf{E}_8)$-invariants (and of dimension 1). □

In Sect. 4.2.11, we saw that the element $\mathrm{T}_{(\mathbb{Z}/p\mathbb{Z})^{n/2}}$ of $\mathrm{H}_p(\mathrm{O}_n)$ decomposes naturally into the sum of two elements $\mathrm{T}^{\pm}_{(\mathbb{Z}/p\mathbb{Z})^{n/2}}$ of $\mathrm{H}_p(\mathrm{SO}_n)$. A similar phenomenon occurs with the perestroika operator $\mathrm{K}_p\in\mathrm{H}_p(\mathrm{PGO}_n)$, which may be refined when we view it in $\mathrm{H}_p(\mathrm{PGSO}_n)$.

Indeed, note that given $M\in\mathcal{R}^{\mathrm{a}}_{\mathbb{Z}}(\mathrm{E}_n\otimes\mathbb{Q})$, the set Ω of $\underline{N}\in\mathcal{R}^{\mathrm{h}}_{\mathbb{Z}}(\mathrm{E}_n\otimes\mathbb{Q})$ such that N is a perestroika of M with respect to p consists of exactly two orbits under the action of $\mathrm{GSO}(M)$ and a single one under $\mathrm{GO}(M)$. This follows from the smoothness of GO_M and GSO_M over $\mathbb{Z}$ and the fact that the action of $\mathrm{GO}(M\otimes\mathbb{F}_p)$ (resp. $\mathrm{GSO}(M\otimes\mathbb{F}_p)$) on the set of Lagrangians of $M\otimes\mathbb{F}_p\simeq\mathrm{H}(\mathbb{F}_p^{n/2})$ is transitive by Witt's theorem (resp. admits two orbits). These two orbits therefore lead to two Hecke operators $\mathrm{K}_p^{\pm}\in\mathrm{H}_p(\mathrm{PGSO}_n)$ with

$$\mathrm{K}_p = \mathrm{K}_p^+ + \mathrm{K}_p^-$$

that are interchanged under the action of $\mathrm{PGO}_n(\mathbb{Q}_p)/\mathrm{PGSO}_n(\mathbb{Q}_p)\simeq\mathbb{Z}/2\mathbb{Z}$.

Corollary 5.4.13. *Let* $W = (\mathrm{H}_{8,1}(\mathbb{R}^8)\otimes\nu^4)_{|\mathrm{PGSO}_8(\mathbb{R})}$. *For every prime p, the eigenvalue of* $\mathrm{K}_p^{\pm}$ *on the line* $\mathrm{M}_W(\mathrm{PGSO}_8)$ *is* $p^{-4}(p^4-1)(p-1)^{-1}\,\tau(p)$.

Proof. Let $s_0 \in \mathrm{PGO}(\mathrm{E}_8)$ be the image of a reflection with respect to a root. The conjugation by this element defines a $\mathbb{Z}$-automorphism of PGSO_8 and, in particular, induces an isomorphism $\mathrm{H}(s_0)$ of $\mathrm{H}(\mathrm{PGSO}_8)$. On the other hand, the proof of Lemma 5.4.12 shows that the symmetry s of $\mathrm{M}_W(\mathrm{PGSO}_8)$, also induced by s_0, is the identity. But Lemma 4.4.3 asserts that we have

$$T \circ s_0 = s_0 \circ \mathrm{H}(s_0)(T) \quad \forall\, T \in \mathrm{H}(\mathrm{PGSO}_8)\ .$$

But we have $\mathrm{H}(s_0)(\mathrm{K}_p^+) = \mathrm{K}_p^-$ because the image of s_0 in the quotient $\mathrm{PGO}_8(\mathbb{Q}_p)/\mathrm{PGSO}_8(\mathbb{Z}_p)$ is nontrivial for every prime p. It follows that K_p^+ and K_p^- have the same eigenvalue on the line $\mathrm{M}_W(\mathrm{PGSO}_8)$, namely half of that of K_p. We conclude using Corollary 5.4.10 and Lemma 5.4.12. □

5.4.14 *Triality for* PGSO_8

The next step relies on the triality for the $\mathbb{Z}$-group PGSO_8. The existence of the triality in this context is briefly discussed by Gross in [96, Sect. 4]. More precisely, he considers the case of the cover Spin_8 of PGSO_8 (we can then descend back to the group PGSO_8 by taking a quotient). Given its importance here, it does not seem excessive to give some more details on this construction.

We will follow the approach of [23] in the case of fields; this extends to any ring through the results of [126]. Recall that for a commutative ring A, an octonion A-algebra[1] C consists of a q-module over A of rank 8 endowed with the structure of an A-algebra with unit $(x, y) \mapsto x \star y$ such that $\mathrm{q}(x \star y) = \mathrm{q}(x)\mathrm{q}(y)$ for every $x, y \in C$. The starting point is to add the structure of an octonion $\mathbb{Z}$-algebra to the q-module E_8, on which the construction of a triality on PGSO_8 will depend. As observed by Van der Blij and Springer [22, (4.5)], there exists such a structure on E_8, and even exactly one modulo $\mathrm{O}(\mathrm{E}_8)$, namely the ring of Coxeter octonions [69].

Let C be an octonion A-algebra. Let us consider the following property concerning C and $\gamma \in \mathrm{GSO}(C)$, which we denote by $\mathcal{P}(C, \gamma)$ (*Cartan's triality principle*):

$$\exists\ \gamma', \gamma'' \in \mathrm{GSO}(C) \ \text{ s.t. } \gamma'(x \star y) = \gamma(x) \star \gamma''(y) \quad \forall x, y \in C\ . \tag{5.4.6}$$

Proposition 4.5 of [126] asserts the following:

(i) If $\mathcal{P}(C, \gamma)$ holds, then the ordered pair (γ', γ'') is unique modulo the diagonal action of $A^\times$.
(ii) If we have $\mathrm{Pic}(A) = 0$, then $\mathcal{P}(C, \gamma)$ holds for every $\gamma \in \mathrm{GSO}(C)$ (when A is a field, this is [23, Theorem 1]).
(iii) For $\gamma \in \mathrm{GSO}(C)$, there exists a partition of unity $1 = \sum_i f_i$ in A such that for every i, the property $\mathcal{P}(\gamma \otimes A_{f_i}, C \otimes A_{f_i})$ holds.

---

[1] In [126], the authors also call this a *Cayley algebra* or a *composition algebra of rank* 8; cf. pp. 51 and 56 op. cit. Let us point out that we do not require the associativity of $\star$, which in fact never holds.

Recall that, by definition, PGSO_C is the quotient of GSO_C by its central sub-A-group $\mathbb{G}_m$ consisting of the homotheties (Sect. 2.1). Properties (i) and (iii) therefore immediately justify the following definition.

Proposition-Definition 5.4.15 (Triality). *Let C be an octonion A-algebra, and let $\pi\colon \mathrm{GSO}_C \to \mathrm{PGSO}_C$ be the natural morphism. There exists a unique automorphism τ of the A-group PGSO_C with the following property: for every commutative A-algebra B and every $\gamma \in \mathrm{GSO}_C(B)$ such that $\mathcal{P}(C \otimes B, \gamma)$ holds, we have $\tau(\pi(\gamma)) = \pi(\gamma'')$.*

By Knus et al. [126, Proposition 4.6] and van der Blij and Springer [23, Sect. 1, Corollary 2], this automorphism τ satisfies $\tau^3 = 1$. We call it the *triality of the A-group* PGSO_C (which depends only on the q-module over A underlying C) *associated with the octonion structure C*. There exist many points of view on triality in the literature. A fascinating geometric property, discovered by E. Study and developed by E. Cartan, is the following.

Lemma 5.4.16. *Let C be an octonion algebra over the field k whose underlying* q-*module is hyperbolic. Let Q_1, Q_2, and Q_3 be the conjugacy classes of the subgroups of* $\mathrm{PGSO}(C)$ *that stabilize, respectively, an isotropic line of C and one of the two types of Lagrangians of C. The triality of* PGSO_C *permutes the three classes Q_i in* $\mathrm{PGSO}(C)$ *transitively.*

Proof. This is [23, Theorem 8]. Specifically, if $a \mapsto \overline{a} := -a + (a \cdot 1)$ denotes the canonical involution of C, then τ sends the stabilizer of the isotropic line $k\overline{a} \subset C$ to the stabilizer of the Lagrangian $a \star C$, and the latter to the stabilizer of the Lagrangian $C \star a$ (which has opposite type). □

Let $\tau \in \mathrm{Aut}(\mathrm{PGSO}_8)$ be the triality associated with a fixed octonion structure on E_8. As an automorphism of the $\mathbb{Z}$-group PGSO_8, it acts naturally on $\mathcal{R}(\mathrm{PGSO}_8)$, on the $\mathcal{R}_p(\mathrm{PGSO}_8)$, and on the ring $\mathrm{H}(\mathrm{PGSO}_8)$, preserving the $\mathrm{H}_p(\mathrm{PGSO}_8)$ (Lemma 4.4.3). The natural inclusions $\mathrm{H}(\mathrm{O}_8) \subset \mathrm{H}(\mathrm{PGO}_8) \subset \mathrm{H}(\mathrm{PGSO}_8)$ allow us to view the Hecke operator T_p as an element of $\mathrm{H}(\mathrm{PGSO}_8)$.

Corollary 5.4.17. *For every prime p, the map $T \mapsto \mathrm{H}(\tau)(T)$ induces a 3-cycle on the subset $\{\mathrm{T}_p, \mathrm{K}_p^+, \mathrm{K}_p^-\} \subset \mathrm{H}_p(\mathrm{PGSO}_8)$.*

Proof. Let $G = \mathrm{PGSO}_8$, $M = \mathrm{E}_8$, and $V_0 = \mathrm{E}_8 \otimes \mathbb{Q}$. The element $\underline{M} \in \underline{\mathcal{R}}_{\mathbb{Z}}^{\mathrm{h}}(V_0) = \mathcal{R}(G)$ of course has stabilizer $G(\widehat{\mathbb{Z}})$ under the action of $G(\mathbb{A}_f)$; it is therefore preserved by τ. Let Ω_1, Ω_2, and Ω_3 be the $G(\widehat{\mathbb{Z}})$-orbits in $\underline{\mathcal{R}}_{\mathbb{Z}}^{\mathrm{h}}(V_0)$ consisting of, respectively, the isometry classes of the p-neighbors of M and the two types of perestroikas of M with respect to p. These orbits clearly factor through $G(\mathbb{F}_p)$-orbits, and the corresponding conjugacy classes of the stabilizers of $G(\mathbb{F}_p)$ are exactly the classes of Q_i of Lemma 5.4.16 applied to $C = \mathrm{E}_8 \otimes \mathbb{F}_p$, which concludes the proof. □

If $d \geq 0$, we have already noted that the representation $\mathrm{H}_{d,r}(\mathbb{R}^{2r})$ is irreducible as a representation of $\mathrm{GO}(\mathbb{R}^{2r})$. However, it follows from [116, Theorem 6.13] that its restriction to $\mathrm{GSO}(\mathbb{R}^{2r})$ decomposes into a direct sum of two nonisomorphic irreducible representations that we will not try to distinguish:

$$\mathrm{H}_{d,r}(\mathbb{R}^{2r}) = \mathrm{H}_{d,r}(\mathbb{R}^{2r})^+ \oplus \mathrm{H}_{d,r}(\mathbb{R}^{2r})^- \ .$$

Concretely, if $e_1, \ldots, e_r$ is a basis of a Lagrangian $I \subset \mathbb{C}^{2r}$, then the function $(v_1, \ldots, v_r) \mapsto \det[e_i \cdot v_j]^d$ is in $\mathrm{H}_{d,r}(\mathbb{R}^{2r})^{\pm}$, where the sign $\pm$ is uniquely determined by the type of the Lagrangian I. Note that if $d = 0$, in which case we have $\mathrm{H}_{d,r}(\mathbb{R}^{2r}) \simeq \Lambda^r(\mathbb{R}^{2r})^*$, this phenomenon is well known!

Let Γ be a group, U a Γ-module, and let $\sigma \in \mathrm{Aut}(\Gamma)$. We denote by U^σ the Γ-module obtained by restricting U via $\sigma \colon \Gamma \to \Gamma$.

Corollary 5.4.18. *Let* $d \equiv 0 \bmod 2$. *The map* $U \mapsto U^\tau$ *induces a* 3*-cycle on the set consisting of the isomorphism classes of the three representations of* $\mathrm{PGSO}_8(\mathbb{R})$

$$\mathrm{H}_{d,1}(\mathbb{R}^8) \otimes \nu^{d/2} \quad \textit{and} \quad \mathrm{H}_{d/2,4}(\mathbb{R}^8)^{\pm} \otimes \nu^d \ .$$

Proof. Note that the spaces $\mathrm{H}_{d,g}(\mathbb{R}^8)$ are endowed with natural actions of $\mathrm{GO}_8(\mathbb{C})$ extending the actions of $\mathrm{GO}_8(\mathbb{R})$ considered before. In particular, the three representations of Corollary 5.4.18 factor through representations of $\mathrm{PGSO}_8(\mathbb{C})$ that are, of course, irreducible. Recall that we have $V = \mathbb{R}^8$.

Let $D \subset V \otimes \mathbb{C}$ be an isotropic line. The stabilizer $S_D \subset \mathrm{PGSO}_8(\mathbb{C})$ of D is a parabolic subgroup isomorphic to $\mathrm{GSO}(\mathbb{C}^6) \ltimes \mathbb{C}^6$. Its natural action on $V^{\otimes 2} \otimes \nu^{-1}$ preserves the line $D^{\otimes 2}$ on which it therefore acts by multiplication by a character that we denote by η_D. Let $\ell(D) \subset \mathrm{H}_{d,1}(\mathbb{R}^8) \otimes \nu^{d/2}$ be the line of harmonic polynomials associated with D (formula (5.4.1)). We see that, under the action of S_D, $\ell(D)$ is an eigenspace with character $\eta_D^{d/2}$.

Likewise, let $I \subset V \otimes \mathbb{C}$ be a Lagrangian. The stabilizer $S_I \subset \mathrm{PGSO}_8(\mathbb{C})$ of I is a parabolic subgroup isomorphic to $(\mathrm{GL}(\mathbb{C}^4)/\{\pm 1\}) \ltimes \mathrm{Sym}^2(\mathbb{C}^4)$. Its natural action on $\Lambda^4 V \otimes \nu^{-2}$ preserves the line $\Lambda^4 I$, on which it therefore acts via a character that we denote by η_I. The line $\ell(I) \subset \mathrm{H}^{\pm}_{d/2,4}(\mathbb{R}^8) \otimes \nu^d$ of harmonic polynomials associated with I, where the sign $\pm$ depends on the type of the Lagrangian I, is clearly an eigenspace under the action of S_I, with character $\eta_I^{d/2}$.

Recall that by the Cartan–Weyl theory of the highest weight, given a parabolic subgroup S of the semisimple group $G = \mathrm{PGSO}_8(\mathbb{C})$ and a polynomial character $\eta \colon S \to \mathbb{C}^\times$, there exists, up to isomorphism, at most one irreducible polynomial representation of G whose restriction to S contains the character η. Moreover, if such a representation exists for the pair (S, η) with $\eta \neq 1$, it does not exist for the pair (S, η^{-1}) (dominance property).

The observations above therefore uniquely characterize the three representations of the corollary. To conclude, we note that by Lemma 5.4.16, τ permutes the three types of parabolic subgroups considered above, each with their own character $S \to \mathbb{C}^\times$, denoted η_* above. This last property is automatic because if S is such a parabolic subgroup, then we see that the group of polynomial characters $S \to \mathbb{C}^\times$ is isomorphic to $\mathbb{Z}$, where the character η is the unique dominant generator by the case $d = 2$. □

Let d be an even integer. After interchanging the signs $\pm$ if necessary, we may assume $(\mathrm{H}_{d,1}(\mathbb{R}^8) \otimes \nu^{d/2})^\tau \simeq \mathrm{H}_{d/2,4}(\mathbb{R}^8)^+ \otimes \nu^d$ by Corollary 5.4.18. By Lemma 4.4.3, the automorphism τ induces an isomorphism

$$\tau^*\colon \mathrm{M}_{\mathrm{H}_{d/2,4}(\mathbb{R}^8)^+\otimes\nu^d}(\mathrm{PGSO}_8) \xrightarrow{\sim} \mathrm{M}_{\mathrm{H}_{d,1}(\mathbb{R}^8)\otimes\nu^{d/2}}(\mathrm{PGSO}_8)$$

such that $T \circ \tau^* = \tau^* \circ \mathrm{H}(\tau)(T)$ for every $T \in \mathrm{H}(\mathrm{PGSO}_8)$.

Corollary 5.4.19. *Let d be an even integer. We have a sequence of isomorphisms*

$$\begin{array}{ccccc} \mathrm{M}_{\mathrm{H}_{d/2,4}(\mathbb{R}^8)^+\otimes\nu^d}(\mathrm{PGSO}_8) & \xrightarrow[\sim]{\mathrm{ind}} & \mathrm{M}_{\mathrm{H}_{d/2,4}(\mathbb{R}^8)\otimes\nu^d}(\mathrm{PGO}_8) & \xrightarrow[\sim]{\mu^*} & \mathrm{M}_{\mathrm{H}_{d/2,4}(\mathbb{R}^8)}(\mathrm{O}_8) \\ \tau^* \downarrow \wr & & & & \\ \mathrm{M}_{\mathrm{H}_{d,1}(\mathbb{R}^8)\otimes\nu^{d/2}}(\mathrm{PGSO}_8) & \xrightarrow[\sim]{\mu^*} & \mathrm{M}_{\mathrm{H}_{d,1}(\mathbb{R}^8)}(\mathrm{SO}_8)\,. & & \end{array}$$

Proof. We have already described all of these isomorphisms, except for the one on the bottom line, denoted by μ^*. This is the morphism defined by the restriction of the functions via the bijection $\mathcal{R}(\mathrm{SO}_n) \to \mathcal{R}(\mathrm{PGSO}_n)$, which is an isomorphism for reasons identical to those invoked in the proof of Lemma 5.4.8. □

Corollary 5.4.20. *The eigenvalue of* T_p *on the line* $\mathrm{M}_{\mathrm{H}_{4,4}(\mathbb{R}^8)}(\mathrm{O}_8)$ *is*

$$p^{-4}\,\frac{p^4-1}{p-1}\,\tau(p)\,.$$

Proof. We apply Corollary 5.4.19 to $d = 8$. Corollary 5.4.13 and Lemma 5.4.17 show that the eigenvalue of the Hecke operator T_p on $\mathrm{M}_{\mathrm{H}_{4,4}(\mathbb{R}^8)^+\otimes\nu^8}(\mathrm{PGSO}_8)$ is that of the statement. This suffices to conclude because μ^* and ind are $\mathrm{H}(\mathrm{O}_8)$-equivariant (Lemma 5.4.8, Sect. 5.4.11). □

5.4.21 One Last Theta Series and the End of the Proof

To conclude, consider the map

$$\vartheta_{4,4}\colon \mathrm{M}_{\mathrm{H}_{4,4}(\mathbb{R}^8)}(\mathrm{O}_8) \longrightarrow \mathrm{S}_8(\mathrm{Sp}_8(\mathbb{Z}))\,.$$

Proposition 5.4.22. *The map* $\vartheta_{4,4}$ *is an isomorphism.*

Proof. Since the two spaces have dimension 1 (formula (5.2.3) and Corollary 5.4.20), it suffices to see that this map is nonzero, which has already been verified by Breulmann and Kuss in [40]. Let us briefly explain how to proceed.

Let $e = (e_1, \ldots, e_4)$ be a quadruple of elements of $E_8 \otimes \mathbb{C}$ generating a Lagrangian, and set $P_e(v_1, \ldots, v_4) = \det[e_i \cdot v_j]_{1 \le i,j \le 4}$; for every integer $d \ge 0$, we have $P_e^d \in H_{d,4}(\mathbb{R}^8)$. Let $Q \subset E_8$ be a sublattice of rank 4 and $v_1, \ldots, v_4$ a $\mathbb{Z}$-basis of Q. The relation $P_e(\gamma(v_1), \ldots, \gamma(v_4)) = \det(\gamma) P_e(v_1, \ldots, v_4)$ for every $\gamma \in \mathrm{GL}(Q)$ shows that $P_e(v_1, \ldots, v_4)^d$ does not depend on the choice of the v_i when d is even; hence, it makes sense to denote it by $P_e(Q)^d$. In particular, if d is even, the Fourier coefficient of the theta series $\vartheta_{d,4}(E_8, P_e^d)$ corresponding to the Gram matrix of a $\mathbb{Z}$-basis of Q is

$$c_Q(P_e^d) = |O(Q)| \sum_M P_e(M)^d ,$$

where the sum is taken over the sublattices $M \subset E_8$ isometric to Q. We give several numerical values in Table 5.3.

In this table, we have $Q \simeq Q(R)$, where R is a root system (of type ADE) of rank 4 (Sect. 1.3) and $(\varepsilon_j)_{1 \le j \le 8}$ denotes the canonical basis of $\mathbb{R}^8$. It is not difficult to enumerate the sublattices of E_8 isometric to Q using a computer. For example, if Φ denotes a positive system of $R(E_8)$ and $<$ denotes a fixed arbitrary total order on Φ, the sublattices of E_8 isometric to D_4 are in bijection with the quadruples (r_1, r_2, r_3, r_4) of elements of Φ such that we have $r_1 < r_2 < r_3$ and that the elements r_1, r_2 and r_3 are pairwise orthogonal and have inner product -1 with r_4. We refer to the source code [54] for an implementation of this algorithm in PARI [160] and for a justification of Table 5.3.

Table 5.3 Values of $c_Q(P_e^d)/|O(Q)|$, where $e = (\varepsilon_{2j-1} + i\varepsilon_{2j})_{1 \le j \le 4}$

$Q \backslash d$	0	2	4	6	8	10
D_4	3150	0	4800	−4800	43200	−81600
A_4	24192	0	−23040	−46080	−69120	−92160
$A_1 \oplus A_3$	151200	0	115200	1267200	6566400	7718400
${A_2}^2$	67200	0	115200	−1382400	4492800	−43084800
${A_1}^2 \oplus A_2$	302400	0	−691200	2073600	85017600	214963200
${A_1}^4$	122850	0	576000	−6796800	191808000	−343641600
$\dim M_{H_{d,4}(\mathbb{R}^8)}(O_8)$	1	0	1	1	1	2

The proposition follows from the fact that we have $c_{D_4}(P_e^4) \neq 0$. Let us also note, to reassure ourselves, that we indeed find the equality $c_{D_4}(P_e^4) = -c_{A_4}(P_e^4)$, which is consistent with Proposition 5.2.1. □

Remark 5.4.23. The last line of Table 5.3 follows from the isomorphism $M_{H_{d,4}(\mathbb{R}^8)}(O_8) \simeq M_{H_{2d,1}(\mathbb{R}^8)}(SO_8)$ ($\simeq H_{2d,1}(\mathbb{R}^8)^{W(E_8)^+}$) given by Corollary 5.4.20, together with Lemma 5.4.2. The vanishing of these spaces for $d = 2$

explains why the column $d = 2$ has entries 0. The table therefore shows that $\vartheta_{d,4}$ is injective when d is even and at most 8. By varying the Lagrangian basis e, we easily verify that $\vartheta_{10,4}$ is also injective.

This concludes the proof of part (ii) of Theorem 5.2.2, by virtue of Corollary 5.4.20 and the Eichler commutation relations (5.4.4). The following sequence of isomorphisms summarizes our proof quite well. We set $W = \mathrm{H}_{8,1}(\mathbb{R}^8)$, $U = \mathrm{H}_{4,4}(\mathbb{R}^8)$, and $U^+ = \mathrm{H}_{4,4}(\mathbb{R}^8)^+$.

$$\begin{array}{ccccccc}
\mathrm{M}_{W\otimes\nu^4}(\mathrm{PGSO}_8) & \xleftarrow[\sim]{\mathrm{res}} & \mathrm{M}_{W\otimes\nu^4}(\mathrm{PGO}_8) & \xrightarrow[\sim]{\mu^*} & \mathrm{M}_W(\mathrm{O}_8) & \xrightarrow[\sim]{\vartheta_{8,1}} & \mathrm{S}_{12}(\mathrm{SL}_2(\mathbb{Z})) \\
\uparrow {\scriptstyle \tau^*\,\wr} & & & & & & \\
\mathrm{M}_{U^+\otimes\nu^8}(\mathrm{PGSO}_8) & \xrightarrow[\mathrm{ind}]{\sim} & \mathrm{M}_{U\otimes\nu^8}(\mathrm{PGO}_8) & \xrightarrow[\mu^*]{\sim} & \mathrm{M}_U(\mathrm{O}_8) & \xrightarrow[\vartheta_{4,4}]{\sim} & \mathrm{S}_8(\mathrm{Sp}_8(\mathbb{Z}))\,.
\end{array}$$

5.5 Appendix: A Simple Example of the Eichler Relations

We will now prove formula (5.4.4) for $g = 1$. Let L be an even unimodular lattice of rank r and $P\colon L\otimes\mathbb{R}\to\mathbb{C}$ a homogeneous harmonic polynomial of degree d. Recall that the associated theta series $\vartheta(L,P) = \sum_{v\in L} P(v)q^{v\cdot v/2}$ is an element of $\mathrm{M}_{d+r/2}(\mathrm{SL}_2(\mathbb{Z}))$.

Theorem 5.5.1. *Let L be an even unimodular lattice of rank r, $P\colon L\otimes\mathbb{R}\to\mathbb{C}$ a homogeneous harmonic polynomial of degree d, and p a prime. We have the relation*

$$\sum_{L'}\vartheta(L',P) = \Big(p\,\frac{p^{r-3}-1}{p-1} + p^{-d}\,\mathrm{T}(p^2)\Big)\vartheta(L,P)\,,$$

where the sum is taken over the p-neighbors $L'\subset\frac{1}{p}L$ of L.

Proof. We denote by $\mathrm{a}_n(g)$ the nth Fourier coefficient of the modular form $g \in \mathrm{M}_k(\mathrm{SL}_2(\mathbb{Z}))$. Recall the relation [177, p. 164]

$$\mathrm{a}_n(\mathrm{T}(p^2)(g)) = \sum_{d\mid(p^2,n)} d^{k-1}\mathrm{a}_{np^2/d^2}(g)\,.$$

We set $f = \sum_{L'}\vartheta(L',P)$, where the sum is taken over the p-neighbors $L'\subset\frac{1}{p}L$ of L.

We fix an integer $n\geq 1$ that, for now, we take relatively prime to p, and set $\mathrm{q}(x) = x\cdot x/2$ for every $x\in L\otimes\mathbb{R}$. Consider the set X of pairs (L',w) where L' is a p-neighbor of L and w is an element of L' such that $\mathrm{q}(w) = n$. Let $Y = \{v\in L\,;\,\mathrm{q}(v) = np^2\}$. We have an obvious map

$$\pi\colon X\to Y\,,\quad (L',w)\mapsto pw\,.$$

We will see that π is surjective and examine its fibers. Given $v \in Y$, there are two cases:

(1) *v is in pL.* In this case, $w = v/p$ is in L and there are as many p-neighbors L' of L containing w as there are sublattices of L of index p that contain w and are the orthogonal complement of an isotropic line modulo p (the "M" of Sect. 3.1). Since we have $(\mathrm{q}(w), p) = 1$, the element w is nonisotropic modulo pL: we find $|\pi^{-1}(\{v\})| = (p^{r-2} - 1)(p-1)^{-1}$.

(2) *v is not in pL.* In this case, v generates an isotropic line in L/pL. If M is its orthogonal complement modulo p, that is, the set of $x \in L$ such that $x.v \in p\mathbb{Z}$, then $L' = M + \mathbb{Z}(v/p)$ is a p-neighbor of L because p^2 divides $\mathrm{q}(v)$. Even better, it is the unique p-neighbor of L that contains $w = v/p$. Indeed, if K is such a p-neighbor and $N = L \cap K$, then v is in N and for every x in K, we have $x.v/p \in \mathbb{Z}$. In particular, we have $N \subset M$, and therefore $N = M$ and $K = L'$. Consequently, we have $|\pi^{-1}(\{v\})| = 1$.

Let us first suppose $P = 1$ (and therefore $d = 0$). Our analysis shows

$$\mathrm{a}_n(f) = |X| = \frac{p^{r-2} - 1}{p-1} \mathrm{a}_n(\vartheta(L,1)) + \left(\mathrm{a}_{np^2}(\vartheta(L,1)) - \mathrm{a}_n(\vartheta(L,1))\right),$$

which is the desired formula, at least for the coefficients with index relatively prime to p. For P arbitrary, $\mathrm{a}_n(f)$ is the sum of the $p^{-d}P(v)$ for $(L', v/p)$ running through X, and we therefore have

$$\mathrm{a}_n(f) = \frac{p^{r-2} - 1}{p-1} \mathrm{a}_n(\vartheta(L,P)) + \left(p^{-d}\mathrm{a}_{np^2}(\vartheta(L,P)) - \mathrm{a}_n(\vartheta(L,P))\right).$$

We conclude using the following lemma.

Lemma 5.5.2. *Let $g \in \mathrm{M}_k(\mathrm{SL}_2(\mathbb{Z}))$ with $k > 0$, and let p be prime. We suppose $\mathrm{a}_n(g) = 0$ for every n relatively prime to p. Then we have $g = 0$.*

Proof. In this case, the holomorphic function $g(\tau)$ is invariant under the map $\tau \mapsto \tau + 1/p$, which suffices because the subgroup of $\mathrm{SL}_2(\mathbb{R})$ generated by this translation and $\mathrm{SL}_2(\mathbb{Z})$ is not discrete. □

We can also verify the formula for all coefficients, by again introducing $\pi\colon X \to Y$ as before. We see that the count is not changed in case (2), but is in case (1).

First subcase: $n/p \in \mathbb{Z} - p\mathbb{Z}$. Let v be an element of Y of the form pw with $w \in L$. Note that we have $w \notin pL$ because p^2 does not divide $\mathrm{q}(w) = n$; on the other hand, w is isotropic in L/pL. But if $x \in L/pL$ is an isotropic vector, there exist

$$1 + p\,\mathrm{c}_{r-2}(p) = \frac{p^{r-2} - 1}{p-1} + p^{r/2-1}$$

isotropic lines in L/pL orthogonal to x; hence this also equals $|\pi^{-1}(\{v\})|$. Here, $\mathrm{c}_i(p) = (p^{i-1} - 1)(p-1)^{-1} + p^{i/2-1}$ is the cardinality of the hyperbolic quadric of rank i over $\mathbb{Z}/p\mathbb{Z}$. It follows that we have

$$\mathrm{a}_n(f) = \left(\frac{p^{r-2}-1}{p-1} + p^{r/2-1}\right)\mathrm{a}_n(\vartheta(L,P)) + \left(p^{-d}\mathrm{a}_{np^2}(\vartheta(L,P)) - \mathrm{a}_n(\vartheta(L,P))\right) ,$$

which concludes the proof.

Second subcase: p^2 divides n. Let v be an element of Y of the form pw with w in L. Then w is isotropic in L/pL. If it is zero, it is in all p-neighbors of L; we then have $|\pi^{-1}(\{v\})| = \mathrm{c}_r(p)$. Else, we have, as above, $|\pi^{-1}(\{v\})| = (p^{r-2}-1)(p-1)^{-1} + p^{r/2-1}$. But we have

$$\mathrm{c}_r(p) - \frac{p^{r-2}-1}{p-1} - p^{r/2-1} = p^{r-2} ;$$

from this, we deduce the identity

$$\begin{aligned}\mathrm{a}_n(f) = {} & p^{d+r-2}\mathrm{a}_{n/p^2}(\vartheta(L,P)) + \left(\frac{p^{r-2}-1}{p-1} + p^{r/2-1}\right)\mathrm{a}_n(\vartheta(L,P)) \\ & + \left(p^{-d}\mathrm{a}_{np^2}(\vartheta(L,P)) - \mathrm{a}_n(\vartheta(L,P))\right) .\end{aligned}$$

This concludes the proof. □

Chapter 6
Langlands Parametrization

6.1 Basic Facts on Reductive k-Groups

Let k be an algebraically closed field. We refer to the treatises of Springer [191] and Borel [34] for the theory of reductive k-groups. Our convention is that such a k-group is connected. Recall that if k has characteristic zero, a connected k-group G is called reductive if the category of its finite-dimensional k-representations is semisimple.

For an arbitrary commutative ring A, an A-group[1] G is called reductive if it is smooth over A and if for every homomorphism from A to an algebraically closed field k, the group $G \times_A k$ is reductive; see [73] and [63]. The classical A-groups studied in Sect. 2.1 are therefore reductive [34, Sect. 23], [63, App. C], except for the orthogonal group in even dimension, which is not connected, as well as the associated similitude and projective similitude groups, for the same reason. An A-group is called semisimple if it is reductive and its center is finite over A. We denote the (scheme-theoretic) center of a reductive A-group G by $\mathrm{Z}(G)$. A *central isogeny* $G \to G'$ between two reductive A-groups is a finite, flat morphism of A-groups that is surjective and whose (scheme-theoretic) kernel is contained in $\mathrm{Z}(G)$. More generally, a morphism of A-groups $G \to G'$ is said to be *central* if the induced morphism $G \times \mathrm{Z}(G') \to G'$ is flat and surjective and has kernel contained in $\mathrm{Z}(G) \times \mathrm{Z}(G')$.

6.1.1 The Based Root Datum of a Reductive k-Group

Let k be an algebraically closed field. The theory of root systems of reductive k-groups, when suitably formulated, produces a canonical equivalence of categories

$$\Psi : \mathcal{C}_k \overset{\sim}{\to} \mathcal{D}$$

[1] Recall that an A-group is a group scheme over A which is affine and of finite type.

G. Chenevier, J. Lannes, *Automorphic Forms and Even Unimodular Lattices*, Ergebnisse der Mathematik und ihrer Grenzgebiete. 3. Folge / A Series of Modern Surveys in Mathematics 69, https://doi.org/10.1007/978-3-319-95891-0_6

between the category $\mathcal{C}_k$ of reductive k-groups "up to inner automorphisms" and the category $\mathcal{D}$ of *based root data*. This classification is due to Chevalley in the case of semisimple groups and to the Demazure–Grothendieck seminar [73, Exp. XXI] in general, where it is even studied over an arbitrary ring k. We restrict ourselves to stating the result below and refer to [191] for a detailed treatment, also summarized in [192] and [114, II, Chap 1], as well as to Kottwitz [129, Sect. 1] for the intrinsic formulation adopted here.

For a reductive k-group G, we denote by $\mathrm{Inn}(G)$ the group of *inner* automorphisms of G, that is, of the form $\mathrm{inn}_g \colon x \mapsto gxg^{-1}$ for $g \in G(k)$. For two reductive k-groups G and G', we denote the set of central morphisms from G to G' by $\mathrm{Hom}_c(G, G')$. It is endowed with an obvious action of $\mathrm{Inn}(G')$. Note that the category $\mathcal{C}_k$ whose objects are the reductive k-groups and whose morphisms $G \to G'$ are given by the quotient set $\mathrm{Hom}_c(G, G')/\mathrm{Inn}(G')$, where the composition of morphisms follows from that of central morphisms by passing to the quotient, is well defined.

A *based root datum* consists of

- two free abelian groups of finite rank X and $X^\vee$ endowed with a perfect pairing $\langle -, - \rangle \colon X \times X^\vee \to \mathbb{Z}$,
- finite subsets $\Phi \subset X$ and $\Phi^\vee \subset X^\vee$ endowed with a bijection $\Phi \to \Phi^\vee$ denoted by $\alpha \mapsto \alpha^\vee$,
- subsets $\Delta \subset \Phi$ and $\Delta^\vee \subset \Phi^\vee$ such that we have $\Delta^\vee = \{\alpha^\vee, \alpha \in \Delta\}$

that satisfy the following conditions:

- for every $\alpha \in \Phi$, we have $\langle \alpha, \alpha^\vee \rangle = 2$;
- if $\mathrm{s}_\alpha \in \mathrm{End}(X)$ denotes the reflection $x \mapsto x - \langle x, \alpha^\vee \rangle \alpha$ and if $\mathrm{s}_{\alpha^\vee} \in \mathrm{End}(X^\vee)$ is defined analogously after interchanging α and $\alpha^\vee$, then for every $\alpha \in \Phi$, we have $\mathrm{s}_\alpha(\Phi) = \Phi$ and $\mathrm{s}_{\alpha^\vee}(\Phi^\vee) = \Phi^\vee$.

It follows from these axioms that the abelian group $\mathrm{Q}(\Phi) \subset X$ generated by the elements of Φ is a root system in $\mathrm{Q}(\Phi) \otimes \mathbb{Q}$ in the sense of Bourbaki [39, Chap VI]. Finally, we assume that

- Φ is reduced[2] and Δ is a basis of Φ.

A morphism $\psi_1 \to \psi_2$ between two based root data $\psi_i = (X_i, \Phi_i, \Delta_i, X_i^\vee, \Phi_i^\vee, \Delta_i^\vee)$ consists of a linear map $X_2 \to X_1$ that induces a bijection $\Phi_2 \to \Phi_1$ and that sends Δ_2 to Δ_1, whose transpose $X_1^\vee \to X_2^\vee$ also induces a bijection $\Phi_1^\vee \to \Phi_2^\vee$, which sends $\Delta_1^\vee$ to $\Delta_2^\vee$. This defines the category $\mathcal{D}$. An *isogeny* $\psi_1 \to \psi_2$ is a morphism as above that induces an isomorphism $X_2 \otimes \mathbb{Q} \to X_1 \otimes \mathbb{Q}$. It remains to recall the definition of the functor Ψ.

[2] Let us emphasize that this assumption is not part of the axioms in the references listed above; it will help us avoid certain difficulties.

For a reductive k-group G, the associated based root datum $\Psi(G)$ is obtained as follows. We choose a maximal torus T of G and a Borel subgroup containing T. We denote by

$$\mathrm{X}^*(T) = \mathrm{Hom}(T, \mathbb{G}_m) \quad \text{and} \quad \mathrm{X}_*(T) = \mathrm{Hom}(\mathbb{G}_m, T)$$

the free abelian groups of finite rank consisting of the characters and cocharacters of the torus T, respectively. They are endowed with an obvious perfect pairing $\langle -, - \rangle \colon \mathrm{X}^*(T) \times \mathrm{X}_*(T) \longrightarrow \mathrm{Hom}(\mathbb{G}_m, \mathbb{G}_m) = \mathbb{Z}$. We then set

$$\Psi(G, T, B) = (X^*(T), \Phi(G, T), \Delta(G, T, B), X_*(T), \Phi^\vee(G, T), \Delta^\vee(G, T, B)) \, ,$$

where $\Phi(G, T)$ (resp. $\Phi^\vee(G, T)$) is the set of roots (resp. coroots[3]) of G with respect to T and $\Delta(G, T, B)$ is the basis of $\Phi(G, T)$ associated with the positive system of $\Phi(G, T)$ appearing in $\mathrm{Lie}(B)$. This is a based root datum.

If we change the pair (T, B) to (T', B'), there exists an element $g \in G(k)$, unique modulo $T(k)$, such that $gTg^{-1} = T'$ and $gBg^{-1} = B'$. The inner automorphism inn_g induces an isomorphism $\Psi(G, T, B) \xrightarrow{\sim} \Psi(G, T', B')$ in $\mathcal{D}$ that is independent of the choice of g.

Following Kottwitz [129, Sect. 1], we define $\Psi(G)$ as the direct (or inverse!) limit of the $\Psi(G, T, B)$, indexed by the pairs (T, B), with transition morphisms the isomorphisms induced by elements of $\mathrm{Inn}(G)$. The construction $G \mapsto \Psi(G)$ is functorial in the central morphisms and sends a central isogeny to an isogeny of root data. In particular the group $\mathrm{Aut}(G)$ of automorphisms of the k-group G acts on $\mathrm{Aut}_{\mathcal{D}}(\Psi(G))$, with the subgroup $\mathrm{Inn}(G)$ acting trivially.

Up to now, k was an algebraically closed field. However, the definition of the functor Ψ extends verbatim to the case of an arbitrary ring k if we restrict ourselves to the subcategory of $\mathcal{C}_k$ consisting of the *split* reductive k-groups [73, Chap. XXII, Proposition 1.14], [114, II, Chap. 1]. If k is an integral domain with $\mathrm{Pic}(k) = 0$ (for example a field or a principal ideal domain), these are the reductive k-groups that have a split maximal torus, that is, one isomorphic to a power of $\mathbb{G}_m$. In particular, the based root datum $\Psi(G)$ of such a k-group is well defined. In $\mathcal{D}$, it can be canonically identified with that of $G \times_k K$ for every homomorphism from k to an algebraically closed field K.

Vocabulary

When a reductive k-group G has a well-defined based root datum, we will speak freely of the root system of G, of simple or positive roots of G, of the Weyl group of G, etc... to indicate the analogous objects deduced from $\Psi(G)$. For example, the

[3] For $\alpha \in \Phi(G, T)$, let $T_\alpha \subset T$ be the neutral component of the kernel of $\alpha \colon T \to \mathbb{G}_m$, and let Z_α be the derived subgroup of the centralizer of T_α in G. It is a k-group isomorphic to SL_2 or PGL_2. Recall that the coroot $\alpha^\vee \in \mathrm{X}_*(T)$ is the unique cocharacter with image in Z_α such that $\langle \alpha, \alpha^\vee \rangle = 2$.

Weyl group of G is the subgroup $W \subset \mathrm{Aut}(X)$ generated by the set of s_α with $\alpha \in \Delta$, where $\Psi(G) = (X, \Phi, \Delta, X^\vee, \Phi^\vee, \Delta^\vee)$. An element of X is called a *weight* of G, and the abelian group X the *weight lattice* of G; likewise, $X^\vee$ is the *coweight lattice* of G.

6.1.2 Langlands Dual

If $\psi = (X, \Phi, \Delta, X^\vee, \Phi^\vee, \Delta^\vee)$ is a based root datum, then

$$\psi^\vee = (X^\vee, \Phi^\vee, \Delta^\vee, X, \Phi, \Delta)$$

is also one, in an obvious way; it is called the *dual datum* of ψ. The correspondence $\psi \mapsto \psi^\vee$ defines an involutive contravariant endofunctor of $\mathfrak{D}$. When k is algebraically closed, it induces an involution of $\mathfrak{C}_k$ via the equivalence of categories Ψ; this is the starting point of the notion of Langlands dual, up to the fact that we involve the field of complex numbers.

Specifically, if G is a split reductive k-group, then a *dual group* of G in the sense of Langlands consists of a reductive $\mathbb{C}$-group $\widehat{G}$ and an isomorphism $\Psi(\widehat{G}) \xrightarrow{\sim} \Psi(G)^\vee$ in $\mathfrak{D}$. The $\mathbb{C}$-group $\widehat{G}$ is then uniquely determined by G, up to inner isomorphisms. By abuse of language, we call it the *Langlands dual of* G.

6.1.3 Examples

We leave it to the reader to verify $\widehat{\mathrm{GL}_n}(\mathbb{C}) \simeq \mathrm{GL}_n(\mathbb{C})$ and $\widehat{\mathrm{PGL}_n}(\mathbb{C}) \simeq \mathrm{SL}_n(\mathbb{C})$. On the other hand, the details in the (very classical!) cases of the orthogonal and symplectic groups will be useful to us further on, so we give them below.

We will use the following construction several times. Consider a based root datum $\psi = (X, \Phi, \Delta, X^\vee, \Phi^\vee, \Delta^\vee)$, and let $Y \subset X \otimes \mathbb{Q}$ be a subgroup of finite type that contains Φ. We suppose

$$\Phi^\vee \subset Y^\sharp := \{x \in X^\vee \otimes \mathbb{Q};\ \langle y, x\rangle \subset \mathbb{Z} \quad \forall y \in Y\}\,.$$

The orthogonal complement $Y^\perp$ of Y in $X^\vee \otimes \mathbb{Q}$ then has intersection zero with $\mathrm{Q}(\Phi^\vee)$, and if $\pi\colon \mathrm{Q}(\Phi^\vee) \to Y^\sharp/Y^\perp$ is the canonical map, then

$$\psi' = (Y, \Phi, \Delta, Y^\sharp/Y^\perp, \pi(\Phi^\vee), \pi(\Delta^\vee))$$

is a based root datum, in an obvious way. An inclusion $Y \subset X$ (resp. $X \subset Y$) induces a morphism $\psi \to \psi'$ (resp. an isogeny $\psi' \to \psi$). Moreover, when applied to $\psi^\vee$, this construction provides a similar construction in which the characters and cocharacters are interchanged.

From now on, k is an arbitrary ring.

The Even Special Orthogonal Group and Its Variants

Let $r \geq 2$ be an integer, $U = k^r$, and $V = \mathrm{H}(U) = U \oplus U^*$ the hyperbolic q-module over U (Sect. 2.1). The k-group $\widetilde{G} = \mathrm{GSO_V}$ is reductive and split.

If $(e_i)_{i=1}^r$ is a k-basis of U and $e_i^* \in U^*$ is the dual basis, then the sub-k-group $\widetilde{T}$ of $\widetilde{G}$ that preserves each of the lines ke_i and ke_j^* is a split maximal torus of $\widetilde{G}$. The k-subgroup of $\widetilde{G}$ that preserves the full flag of U associated with $\{e_1\}, \{e_1, e_2\}, \ldots$ is a Borel subgroup that contains $\widetilde{T}$.

Let $\varepsilon_i \in \mathrm{X}^*(\widetilde{T})$ be the character of $\widetilde{T}$ acting on ke_i, let $\nu\colon \widetilde{G} \to \mathbb{G}_m$ be the similitude factor, and let ε_0 be the restriction of ν to $\widetilde{T}$. Then we see that $\widetilde{T}$ acts on ke_j^* by multiplication by the character $-\varepsilon_j + \varepsilon_0$. The ε_i for $i = 0, \ldots, r$ form a $\mathbb{Z}$-basis of $\mathrm{X}^*(\widetilde{T})$.

The set $\Phi(\widetilde{G}, \widetilde{T})$ consists of the $\pm(\varepsilon_i - \varepsilon_j)$ and $\pm(\varepsilon_i + \varepsilon_j - \varepsilon_0)$ for $1 \leq i < j \leq r$. Moreover, $\Delta(\widetilde{G}, \widetilde{T}, \widetilde{B})$ is the union of the $\varepsilon_i - \varepsilon_{i+1}$ for $i = 1, \ldots, r-1$ and $\varepsilon_{r-1} + \varepsilon_r - \varepsilon_0$. Let $\varepsilon_i^* \in \mathrm{X}_*(\widetilde{T})$ be the dual $\mathbb{Z}$-basis of the basis of $\mathrm{X}^*(\widetilde{T})$ consisting of the ε_i for $i = 0, \ldots, r$. For $1 \leq i < j$, we have $(\varepsilon_i - \varepsilon_j)^\vee = \varepsilon_i^* - \varepsilon_j^*$ and $(\varepsilon_i + \varepsilon_j - \varepsilon_0)^\vee = \varepsilon_i^* + \varepsilon_j^*$.

Let $s \in \mathrm{O}_V(k)$ be the element that fixes e_i and e_i^* for $i < r$ and interchanges e_r and e_r^*. The conjugation by s induces an automorphism of $\widetilde{G}$ that preserves $\widetilde{T}$ and $\widetilde{B}$. Let $\Psi(s)$ be the induced automorphism of $\Psi(\widetilde{G})$: it fixes ε_i for $i = 0, \cdots, r-1$ and sends ε_r to $\varepsilon_0 - \varepsilon_r$. If $r \neq 4$, it is the unique nontrivial involution of the "Dynkin diagram" of $\Delta(\widetilde{G}, \widetilde{T}, \widetilde{B})$.

Let us now consider the k-group $G = \mathrm{SO}_V$. Its based root datum associated with $T := \widetilde{T} \cap G$ and $B = \widetilde{B} \cap G$ can be deduced from that of $\widetilde{G}$ through the method recalled above, by considering the subgroup of cocharacters $\mathrm{X}_*(T) = \varepsilon_0^\perp = \oplus_{i=1}^r \mathbb{Z}\varepsilon_i^* \subset \mathrm{X}_*(\widetilde{T})$ and the group of characters $\mathrm{X}^*(T) = \mathrm{X}^*(\widetilde{T})/\mathbb{Z}\varepsilon_0$. In other words, "we impose $\varepsilon_0 = 0$ in $\Psi(\widetilde{G}, \widetilde{T}, \widetilde{B})$."

Let $\underline{\varepsilon_i}$ be the image of ε_i in $\mathrm{X}^*(T)$, so that we have $\mathrm{X}^*(T) = \oplus_{i=1}^r \mathbb{Z}\underline{\varepsilon_i}$. The linear map $\mathrm{X}_*(T) \to \mathrm{X}^*(T)$ that sends ε_i^* to $\underline{\varepsilon_i}$ induces an isomorphism $\Psi(\mathrm{SO}_V) \xrightarrow{\sim} \Psi(\mathrm{SO}_V)^\vee$; in particular, we have

$$\widehat{\mathrm{SO}}_V(\mathbb{C}) \simeq \mathrm{SO}_{2\mathrm{r}}(\mathbb{C}) \ ,$$

where SO_{2r} is the special orthogonal $\mathbb{C}$-group of the standard q-vector space $\mathbb{C}^{2r}$.

Likewise, the root datum of $\mathrm{P}\widetilde{G} = \mathrm{PGSO}_V$ associated with the respective images $\mathrm{P}\widetilde{T}$ and $\mathrm{P}\widetilde{B}$ of $\widetilde{T}$ and $\widetilde{B}$ in $\mathrm{P}\widetilde{G}$ is obtained by considering the subgroup of characters $\mathrm{X}^*(\mathrm{P}\widetilde{T}) = \zeta^\perp \subset \mathrm{X}^*(\widetilde{T})$, where ζ denotes the central cocharacter $\varepsilon_0^* + \sum_{i=1}^r \varepsilon_i^*$, and the group of cocharacters $\mathrm{X}_*(\mathrm{P}\widetilde{T}) = \mathrm{X}_*(\widetilde{T})/\mathbb{Z}\zeta$. In other words, we impose $-2\varepsilon_0^* = \sum_{i=1}^r \varepsilon_i^*$ in the datum of $\widetilde{G}$. The group $\widehat{\mathrm{PGSO}_V}(\mathbb{C})$ is isomorphic to the spin group $\mathrm{Spin}_{2r}(\mathbb{C})$ of the standard q-vector space $\mathbb{C}^{2r}$.

THE ODD SPECIAL ORTHOGONAL GROUP

Let $r \geq 1$ be an integer, $U = k^r$, and let V be the k-module $\mathrm{H}(U) \oplus k$ endowed with the quadratic form that is the orthogonal sum of the q-module $\mathrm{H}(U)$ and $x \mapsto x^2$. The k-group $G = \mathrm{SO}_\mathrm{V}$ is then semisimple and split (Sect. B.1).

We define a split maximal torus T from a k-basis (e_i) of U, a Borel subgroup B containing T, and a $\mathbb{Z}$-basis ε_i of $\mathrm{X}^*(T)$ as before. This time, $\Phi(G, T)$ is the union of the $\pm\varepsilon_i \pm \varepsilon_j$ for $1 \leq i < j \leq r$ and the $\pm\varepsilon_i$ for $i = 1, \ldots, r$. Moreover, $\Delta(G, T, B)$ is the union of the $\varepsilon_i - \varepsilon_{i+1}$ for $i < r$ and ε_r.

The similitude and projective similitude groups associated with V differ little from G in this setting; we will not consider them. On the other hand, the spin group of V will play a role. Following Chevalley, we define it using the Clifford algebra of V. Over an algebraically closed field, it suffices to describe its based root datum: it is the datum associated with the subgroup $Y = \mathrm{X}^*(T) + \mathbb{Z}\,\frac{1}{2}(\sum_{i=1}^{r} \varepsilon_i) \subset \mathrm{X}^*(T) \otimes \mathbb{Q}$.

THE SYMPLECTIC GROUP AND ITS VARIANTS

Finally, the k-groups of the symplectic series are also split and reductive. Let us first consider the k-group $\widetilde{G} = \mathrm{GSp}_{2g}$ of symplectic similitudes of the hyperbolic alternating form on $U = k^g$.

We define $\widetilde{T}$, $\widetilde{B}$, and the ε_i and ε_i^* for $i = 0, \cdots, g$ as in the even orthogonal case. This time, the set $\Phi(\widetilde{G}, \widetilde{T})$ consists of the $\pm(\varepsilon_i - \varepsilon_j)$ for $1 \leq i < j \leq g$ and the $\pm(\varepsilon_i + \varepsilon_j - \varepsilon_0)$ for $1 \leq i \leq j \leq g$. Moreover, $\Delta(\widetilde{G}, \widetilde{T}, \widetilde{B})$ is the union of the $\varepsilon_i - \varepsilon_{i+1}$ for $1 \leq i < g$ and $2\varepsilon_g - \varepsilon_0$. Finally, we have $(\varepsilon_i - \varepsilon_j)^\vee = \varepsilon_i^* - \varepsilon_j^*$ and $(\varepsilon_i + \varepsilon_j - \varepsilon_0)^\vee = \varepsilon_i^* + \varepsilon_j^*$ for $i < j$ and $(2\varepsilon_i - \varepsilon_0)^\vee = \varepsilon_i^*$.

The root data of the k-groups $G = \mathrm{Sp}_{2g}$ and $\mathrm{P}\widetilde{G} = \mathrm{PGSp}_{2g}$ can be deduced verbatim from those of $\widetilde{G}$ as in the even orthogonal case. Finally, we note that $\widehat{\mathrm{Sp}_{2g}}(\mathbb{C}) \simeq \mathrm{SO}_{2g+1}(\mathbb{C})$ and $\widehat{\mathrm{PGSp}_{2g}}(\mathbb{C}) \simeq \mathrm{Spin}_{2g+1}(\mathbb{C})$.

6.1.4 Representations of Split Reductive Groups in Characteristic Zero

Let k be an algebraically closed field of characteristic zero, let G be a reductive k-group, and let $\Psi(G) = (X, \Phi, \Delta, X^\vee, \Phi^\vee, \Delta^\vee)$ be its based root datum. Let

$$X_+ = \{\lambda \in X\,;\, \langle \lambda, \alpha^\vee \rangle \geq 0 \;\; \forall \alpha \in \Delta\}$$

be the additive submonoid of X consisting of the *dominant weights* of G. It is a fundamental domain for the action of the Weyl group W of G on the set X.

We endow X with a partial order for the so-called *dominance* relation: $\lambda \leq \mu \Leftrightarrow \mu - \lambda$ is a finite sum of elements of Δ [193]. A remarkable property of this relation is that if $\lambda, \mu \in X_+$ satisfy $\lambda < \mu$, there exists a root $\alpha \in \Phi$ which is positive with respect to Δ, such that $\mu - \alpha \in X_+$ and $\lambda \leq \mu - \alpha$ [193, Corollary 2.7]. For example, an element $\lambda \in X_+$ is minimal if and only if we have $\lambda - \alpha \notin X_+$ for every positive root $\alpha \in \Phi$.

A k-representation of G consists of a finite-dimensional k-vector space V and a morphism of k-groups $G \to \mathrm{GL}_V$. These form an abelian category in an obvious way; the category is semisimple because G is reductive. The tensor product of representations defines the structure of a commutative ring $\mathrm{Rep}(G)$ on the Grothendieck group of this category. The map $G \mapsto \mathrm{Rep}(G)$ defines, in a natural way, a functor from the category $\mathcal{C}_k$ to the commutative rings (Sect. 6.1.1).

For $\lambda \in X_+$, the Cartan–Weyl theory of the highest weight shows that there exists an irreducible k-representation V_λ of G, unique up to isomorphism, with highest weight λ. Moreover, every irreducible k-representation can be obtained this way. Let us briefly recall what is the highest weight of an irreducible k-representation. Let T be a maximal torus of G, and let B be a Borel subgroup containing T, so that $\Psi(G)$ can be canonically identified with $\Psi(G, T, B)$. The action of T on any k-representation V of G is diagonalizable, and we denote by $\mathrm{Weights}(V) \subset X$ the set of characters of T in V. It is stable under the action of W. If V is irreducible, one can prove that the space of invariants $V^{B(k)}$ is of dimension 1 and that the action T in this space is by an element of X_+: it is the *highest weight* of V. The highest weight λ of V then has the following property: for every $\mu \in \mathrm{Weights}(V)$, we have $\mu \leq \lambda$. Moreover, we have

$$\mathrm{Weights}(V) \cap X_+ = \{\mu \in X_+ \, ; \, \mu \leq \lambda\}$$

(see, for example, [102, Sects. 13.2 and 21.3]).

6.2 Satake Parametrization

6.2.1 The Satake Isomorphism

Let G be a $\mathbb{Z}_p$-group. As in Sect. 4.2.5, we denote by $\mathcal{R}_p(G)$ the $G(\mathbb{Q}_p)$-set $G(\mathbb{Q}_p)/G(\mathbb{Z}_p)$ and by $\mathrm{H}_p(G)$ the Hecke ring of $\mathcal{R}_p(G)$ (Sect. 4.2).

We assume that G is split reductive (Sect. 6.1.1). As observed by Gross [96, Proposition 1.1], this last assumption is satisfied if G arises from a reductive $\mathbb{Z}$-group by extension of scalars to $\mathbb{Z}_p$; this will always be the case in our applications. Let $\widehat{G}$ be the Langlands dual of G, that is, a reductive $\mathbb{C}$-group $\widehat{G}$ endowed with an isomorphism $\Psi(\widehat{G}) \xrightarrow{\sim} \Psi(G)^\vee$ (Sect. 6.1.2). Its Grothendieck ring $\mathrm{Rep}(\widehat{G})$ is then canonically defined (Sect. 6.1.4). The Satake isomorphism [174], revisited by

Langlands [136, Sect. 2], is a canonical ring isomorphism[4]

$$\mathrm{Sat}\colon \mathrm{H}_p(G) \otimes \mathbb{Z}[\,p^{-1/2}\,] \xrightarrow{\sim} \mathrm{Rep}(\widehat{G}) \otimes \mathbb{Z}[\,p^{-1/2}\,]\,.$$

We refer to the article of Satake [174], as well as the *survey* articles of Cartier [48, Sect. IV] and Gross [97], for the details of the definition and general properties of this isomorphism, which we only discuss briefly below. The original construction of Satake assumes certain axiomatic properties of the pair of groups $(G(\mathbb{Z}_p), G(\mathbb{Q}_p))$, which he verifies for the classical groups, and which were proved by Tits in general [198]. The point of view used here, in which the focus is on the "integral structure" $\mathrm{Rep}(\widehat{G})$ rather than on the central functions on $\widehat{G}$, has been borrowed from the article of Gross mentioned above. As observed by Gross, we may replace $\mathbb{Z}[\,p^{-1/2}\,]$ by $\mathbb{Z}[\,p^{-1}\,]$ in the Satake isomorphism when the half-sum of the positive roots of G is a weight of G.

Definition of the Satake Homomorphism. Let T be a split maximal $\mathbb{Z}_p$-torus of G, B a Borel sub-$\mathbb{Z}_p$-group of G containing T, and N the unipotent radical of B. If V is a $G(\mathbb{Q}_p)$-module, then the abelian group $V_{N(\mathbb{Q}_p)}$ of the coinvariants of V under the action of $N(\mathbb{Q}_p)$ is endowed with the structure of a $T(\mathbb{Q}_p)$-module because $T(\mathbb{Q}_p)$ normalizes $N(\mathbb{Q}_p)$; this defines a functor from the $G(\mathbb{Q}_p)$-modules to the $T(\mathbb{Q}_p)$-modules, called the *Jacquet functor*. The set-theoretic decompositions $G(\mathbb{Q}_p) = B(\mathbb{Q}_p)G(\mathbb{Z}_p)$ and $B(\mathbb{Q}_p) = T(\mathbb{Q}_p) \times N(\mathbb{Q}_p)$ ensure that the obvious inclusion $\mathcal{R}_p(T) \to \mathcal{R}_p(G)$ induces a bijection (a "horocyclic projection")

$$\mathcal{R}_p(T) \xrightarrow{\sim} N(\mathbb{Q}_p)\backslash\mathcal{R}_p(G)\,.$$

It follows that if we take $V = \mathbb{Z}[\mathcal{R}_p(G)]$, then $V_{N(\mathbb{Q}_p)}$ can be canonically identified with $\mathbb{Z}[\mathcal{R}_p(T)]$, which leads to a ring homomorphism

$$\mathrm{s}_1\colon \mathrm{H}_p(G) \to \mathrm{H}_p(T)\,.$$

Let $\mathrm{X}_*(T) = \mathrm{Hom}(\mathbb{G}_m, T)$ be the group of cocharacters of T. The natural map $\mathrm{X}_*(T) \to T(\mathbb{Q}_p)$ defined by $\lambda \mapsto \lambda(p)$ induces a bijection $\mathrm{X}_*(T) \xrightarrow{\sim} \mathcal{R}_p(T)$, and consequently a ring isomorphism $\eta\colon \mathbb{Z}[\mathrm{X}_*(T)] \xrightarrow{\sim} \mathrm{H}_p(T)$ by the commutativity of T. We then consider the homomorphism

$$s_2\colon \mathrm{H}_p(G) \to \mathbb{Z}[\mathrm{X}_*(T)][\,p^{-1/2}\,]$$

defined as the composition of $(\eta^{-1} \circ s_1) \otimes \mathbb{Z}[\,p^{-1/2}]$ and the automorphism of $\mathbb{Z}[\mathrm{X}_*(T)][\,p^{-1/2}\,]$ that sends a cocharacter λ to $p^{-\langle \lambda, \rho \rangle}\,\lambda$, where ρ is the half-sum of the positive roots of G with respect to (T, B). Satake shows that s_2 is an isomorphism on the ring of invariants $\mathbb{Z}[\mathrm{X}_*(T)]^W[\,p^{-1/2}]$, where W is the Weyl group of G. However, by the definition of the dual group $\widehat{G}$, the ring $\mathbb{Z}[\mathrm{X}_*(T)]^W$

[4] Strictly speaking, we should replace $\mathrm{H}_p(G)$ by the opposite ring $\mathrm{H}_p(G)^{\mathrm{opp}}$ in this isomorphism. Since the existence of the latter implies the commutativity of $\mathrm{H}_p(G)$, we will leave out this decoration.

can be canonically identified with $\mathbb{Z}[X^*(\widehat{T})]^W$, or with $\mathrm{Rep}(\widehat{G})$ by Chevalley; this defines the isomorphism Sat.

Let us denote by $\widehat{G}(\mathbb{C})_{\mathrm{ss}}$ the (well-defined!) set of conjugacy classes of semisimple elements of $\widehat{G}(\mathbb{C})$. Let $c \in \widehat{G}(\mathbb{C})_{\mathrm{ss}}$. The map $V \mapsto \mathrm{trace}(c \mid V)$ that sends a finite-dimensional $\mathbb{C}$-representation V of $\widehat{G}$ to the trace of c in V extends to a ring homomorphism $\mathrm{tr}(c)\colon \mathrm{Rep}(\widehat{G}) \to \mathbb{C}$. By a classical result due to Chevalley, the resulting map

$$\mathrm{tr}\colon \widehat{G}(\mathbb{C})_{\mathrm{ss}} \to \mathrm{Hom}_{\mathrm{ring}}(\mathrm{Rep}(\widehat{G}), \mathbb{C})$$

is a bijection. The following scholium, one of the starting points of the work of Langlands, immediately follows.

Scholium 6.2.2. *The map* $c \mapsto \mathrm{tr}(c) \circ \mathrm{Sat}$ *defines a bijection*

$$\widehat{G}(\mathbb{C})_{\mathrm{ss}} \xrightarrow{\sim} \mathrm{Hom}_{\mathrm{ring}}(\mathrm{H}_p(G), \mathbb{C}) \ .$$

Finally, let us mention that by the Satake homomorphism, the involution $T \mapsto T^{\mathrm{t}}$ of $\mathrm{H}_p(G)$ corresponds to the involution of $\mathrm{Rep}(\widehat{G})$ induced by the duality on the representations, and also to the inversion on $\widehat{G}(\mathbb{C})_{\mathrm{ss}}$.

Example 6.2.3. Let us first return to the general setting of Sects. 4.2.1 and 4.2.2, where X denotes an arbitrary transitive Γ-set. We view $\mathbb{Z}$ as a Γ-module for the trivial action. The $\mathrm{H}(X)^{\mathrm{opp}}$-module $\mathbb{Z}_X$ is free of rank 1 over $\mathbb{Z}$ and therefore defines a ring morphism

$$\deg\colon \mathrm{H}(X) \to \mathbb{Z} \ ,$$

called the *degree*, which is none other than $\deg(h) = \sum_{x \in X} h_{x,y}$, where $y \in X$ is an arbitrary element. For $X = \mathcal{R}_p(G)$, we can ask ourselves which element $s \in \widehat{G}(\mathbb{C})_{\mathrm{ss}}$ corresponds to the homomorphism deg by Scholium 6.2.2. Since the Jacquet functor of the trivial $G(\mathbb{Q}_p)$-module $\mathbb{Z}$ is the trivial $T(\mathbb{Q}_p)$-module $\mathbb{Z}$, it follows[5] from the definition of the Satake homomorphism recalled above that s is the conjugacy class of $\rho(p) = (2\rho)(p^{1/2})$, where 2ρ is viewed as a cocharacter of $\widehat{G}$.

Isogenies

Let G and G' be two split reductive $\mathbb{Z}_p$-groups, and let $f\colon G \to G'$ be a central morphism. On the one hand, this morphism induces a ring homomorphism $\mathrm{Rep}(f)\colon \mathrm{Rep}(\widehat{G}) \to \mathrm{Rep}(\widehat{G'})$ via the equivalence Ψ and the duality on the root data. On the other hand, in [174, Sect. 7], Satake defines a canonical ring homomorphism $\mathrm{H}(f)\colon \mathrm{H}_p(G) \longrightarrow \mathrm{H}_p(G')$.

[5] Let V be a $G(\mathbb{Q}_p)$-module and $\pi\colon V^{G(\mathbb{Z}_p)} \to V_{N(\mathbb{Q}_p)}$ the canonical projection. The ring $\mathrm{H}(G)$ acts on $V^{G(\mathbb{Z}_p)}$ (Sect. 4.2.2). By the construction of s_2, we have $\pi \circ T = s_2(T) \circ \pi$ for every $T \in \mathrm{H}_p(G)$. The assertion follows by considering $V = \mathbb{Z}$ and recalling the shift by ρ in the definition of the Satake homomorphism.

When f is a central isogeny, $\mathrm{H}(f)$ coincides with the homomorphism $\mathrm{H}_p(G) \to \mathrm{H}_p(G')$ that Proposition-Definition 4.2.14 associates with the obvious morphism $\mathcal{R}_p(G) \to \mathcal{R}_p(G')$ defined by f. Indeed, let us verify that the latter satisfies the assumptions of Sect. 4.2.12. On the one hand, a direct Galois-theoretic argument ensures that $f(G(\mathbb{Q}_p))$ contains the derived subgroup of $G'(\mathbb{Q}_p)$. Moreover, the Cartan–Tits decomposition (Sect. 6.2.5) shows that $G(\mathbb{Z}_p)$ is a maximal compact subgroup of G, equal to $f^{-1}(G'(\mathbb{Z}_p))$, giving the injectivity of $\mathcal{R}_p(G) \to \mathcal{R}_p(G')$. Even better, this decomposition implies the injectivity of $G(\mathbb{Z}_p)\backslash G(\mathbb{Q}_p)/G(\mathbb{Z}_p) \to G'(\mathbb{Z}_p)\backslash G'(\mathbb{Q}_p)/G'(\mathbb{Z}_p)$, and therefore that the action of $G'(\mathbb{Z}_p)$, and consequently of the group S defined loc. cit., on $\mathrm{H}_p(G)$, is trivial.

The second theorem of Satake [174, Sect. 7, Theorem 4] states the commutativity of the diagram

$$\begin{array}{ccc} \mathrm{H}_p(G) & \xrightarrow{\mathrm{H}(f)} & \mathrm{H}_p(G') \\ \downarrow{\scriptstyle \mathrm{Sat}} & & \downarrow{\scriptstyle \mathrm{Sat}} \\ \mathrm{Rep}(\widehat{G}) \otimes \mathbb{Z}[p^{-1/2}] & \xrightarrow[\mathrm{Rep}(f)]{} & \mathrm{Rep}(\widehat{G}') \otimes \mathbb{Z}[p^{-1/2}]\,, \end{array} \tag{6.2.1}$$

to which we will refer as the "compatibility of the Satake isomorphism with isogenies."

Example: The Even Special Orthogonal Group

Let us give an example of an application of the previous discussion in the case of an automorphism of G. Let $r \geq 1$ be an integer, let V be the hyperbolic q-module over $\mathbb{Z}_p^r$, and let $G = \mathrm{SO}_V$, so that $\widehat{G}$ is the $\mathbb{C}$-group SO_{2r} (Sect. 6.1.3). The group $\mathrm{O}(V)$ acts by $\mathbb{Z}_p$-automorphisms on G (by conjugation), and therefore likewise on $\Psi(G)$. The induced homomorphism $\mathrm{O}(V)/\mathrm{SO}(V) \to \mathrm{Aut}_{\mathcal{D}}(\Psi(G))$ is bijective, and the nontrivial element is induced by the element $\Psi(s)$ defined loc. cit. This group also acts on $\mathrm{Rep}(\widehat{G})$ by functoriality, and this action coincides with the natural action of $\mathrm{O}_{2r}(\mathbb{C})/\mathrm{SO}_{2r}(\mathbb{C})$, for the same reason. Denote by $\mathrm{H}_p(\mathrm{O}_V)$ and $\mathrm{H}_p(\mathrm{SO}_V)$ the Hecke rings of the $\mathbb{Z}_p$-groups O_V and SO_V, respectively. In Sect. 4.2.11, we defined a canonical homomorphism $\mathrm{H}_p(\mathrm{O}_V) \to \mathrm{H}_p(\mathrm{SO}_V)$ that identifies $\mathrm{H}_p(\mathrm{O}_V)$ with the ring of invariants $\mathrm{H}_p(\mathrm{SO}_V)^{\mathrm{O}(V)}$. By composition with the Satake isomorphism of SO_V, we deduce a canonical isomorphism

$$\mathrm{H}_p(\mathrm{O}_V) \otimes \mathbb{Z}[p^{-1/2}] \xrightarrow{\sim} (\mathrm{Rep}(\mathrm{SO}_{2r}(\mathbb{C})) \otimes \mathbb{Z}[p^{-1/2}])^{\mathrm{O}_{2r}(\mathbb{C})}\,. \tag{6.2.2}$$

Scholium 6.2.4. *The Satake isomorphism of* SO_V *induces a bijection between* $\mathrm{Hom}_{\mathrm{ring}}(\mathrm{H}_p(\mathrm{O}_V), \mathbb{C})$ *and the set of* $\mathrm{O}_{2r}(\mathbb{C})$*-conjugacy classes of semisimple elements of* $\mathrm{SO}_{2r}(\mathbb{C})$.

6.2.5 The Two Natural Bases of the Hecke Ring of G

Let G be a split reductive $\mathbb{Z}_p$-group, with Langlands dual $\widehat{G}$. Write $\Psi(\widehat{G}) = (X, \Phi, \Delta, X^\vee, \Phi^\vee, \Delta^\vee)$, and denote by $X_+ \subset X$ the ordered set of dominant weights of $\widehat{G}$, as in Sect. 6.1.4. Following Gross [97], we recall the two natural $\mathbb{Z}$-bases of $\mathrm{H}_p(G)$ and $\mathrm{Rep}(\widehat{G})$ indexed by X_+ and indicate several links between these bases, which we will need further on.

A consequence of the reductivity of G over $\mathbb{Z}_p$ is the existence of a *Cartan decomposition*, due to Tits in this generality but classical in our examples (the theory of "elementary divisors"). Let T be a split maximal torus of G and B a Borel subgroup of G containing T, which canonically identifies $\Psi(\widehat{G})$ with $\Psi(G, T, B)^\vee$ and, in particular, X with the group of cocharacters $\mathrm{X}_*(T) = \mathrm{Hom}(\mathbb{G}_m, T)$. The decomposition in question can be written[6]

$$G(\mathbb{Q}_p) = \coprod_{\lambda \in X_+} G(\mathbb{Z}_p)\, \lambda(p)\, G(\mathbb{Z}_p)\,.$$

For $\lambda \in X$, we denote by $\mathrm{c}_\lambda \in \mathrm{H}_p(G)$ the characteristic function of the double coset $G(\mathbb{Z}_p)\lambda(p)G(\mathbb{Z}_p)$ or, depending on the point of view, of the ordered pairs (x, y) in $G(\mathbb{Q}_p)/G(\mathbb{Z}_p)$ such that we have $y^{-1}x \in G(\mathbb{Z}_p)\lambda(p)G(\mathbb{Z}_p)$ (Sect. 4.2.1). The element $\mathrm{c}_\lambda \in \mathrm{H}_p(G)$ does not depend on the choice of (T, B). It is clear that we have $\mathrm{c}_\lambda^{\mathrm{t}} = \mathrm{c}_{-\lambda}$ (Sect. 4.2.1) and $\mathrm{c}_{w(\lambda)} = \mathrm{c}_\lambda$ for every $\lambda \in X$ and $w \in W$, where W is the Weyl group of G. By the Cartan–Tits decomposition, the c_λ with $\lambda \in X_+$ form a $\mathbb{Z}$-basis of $\mathrm{H}_p(G)$. For $\lambda, \mu \in X_+$, we have

$$\mathrm{c}_\lambda \cdot \mathrm{c}_\mu = \mathrm{c}_{\lambda+\mu} + \sum_{\nu < \lambda+\mu} n_{\lambda,\mu,\nu} \mathrm{c}_\nu \tag{6.2.3}$$

for certain integers $n_{\lambda,\mu,\nu}$ [97, (2.9)]. The ring $\mathrm{H}_p(G)$ therefore admits an obvious filtration indexed by the ordered monoid X_+, with associated graded ring $\mathbb{Z}[X_+]$. In particular, if we denote by $\Omega \subset X_+$ a generating family of X_+, then the ring homomorphism $\mathbb{Z}[\{x_\omega\}_{\omega\in\Omega}] \to \mathrm{H}_p(G)$ that sends x_ω to c_ω is surjective. If $X^\vee = \mathrm{Q}(\Phi^\vee)$, in which case $X_+ \simeq \mathbb{N}^r$, and if Ω is the basis of X_+ (fundamental coweights), this homomorphism is an isomorphism.

Likewise, the classes $[V_\lambda] \in \mathrm{Rep}(\widehat{G})$ of the irreducible representations V_λ for $\lambda \in X_+$ provide a $\mathbb{Z}$-basis of $\mathrm{Rep}(\widehat{G})$, by the theory of the highest weight recalled in Sect. 6.1.4. Although they are both indexed by X_+, the link between the $\mathrm{Sat}(\mathrm{c}_\lambda)$ and the $[V_\lambda]$ is nontrivial. We refer to the article [97] of Gross for a detailed discussion of this matter, in which the work of Lusztig [142] plays an essential role.

Let $\widehat{T} \subset \widehat{G}$ be a maximal torus and $\widehat{B} \subset \widehat{G}$ a Borel subgroup containing $\widehat{T}$, so that $\Psi(\widehat{G})$ can be identified with $\Psi(\widehat{G}, \widehat{T}, \widehat{B})$. For a $\mathbb{C}$-representation V of $\widehat{G}$ and $\mu \in X$, we denote by $V(\mu) \subset V$ the eigenspace for the character μ under the action of $\widehat{T}$. As explained by Gross [97, Sect. 3], for every $\lambda \in X_+$, we have an identity of

[6] We use $\lambda(p)$ to denote the image of p by the morphism $\mathbb{Q}_p^\times \to T(\mathbb{Q}_p)$ induced by λ.

the form

$$p^{\langle \lambda, \rho \rangle}[V_\lambda] = \mathrm{Sat}(\mathrm{c}_\lambda) + \sum_{\{\mu \in X_+ \,;\, \mu < \lambda\}} d_\lambda(\mu)\, \mathrm{Sat}(\mathrm{c}_\mu) \tag{6.2.4}$$

for certain integers $d_\lambda(\mu)$ depending on p.

Proposition 6.2.6 (Gross). *Let G be a split semisimple $\mathbb{Z}_p$-group and X_+ the ordered set of dominant weights of $\widehat{G}$. Let $\lambda \in X_+$.*

(i) *If λ is a minimal element, then we have $p^{\langle \lambda, \rho \rangle}[V_\lambda] = \mathrm{Sat}(\mathrm{c}_\lambda)$, where 2ρ is the sum of the positive roots of G.*
(ii) *For $\mu \in X_+$ with $\dim(V_\lambda(\mu)) = 1$, we have $d_\lambda(\mu) = 1$.*
(iii) (Lusztig) *If $V_\lambda = \mathrm{Lie}(G)$ is the adjoint representation, then we have $d_\lambda(0) = \sum_i p^{m_i - 1}$, where the m_i are the exponents of the Weyl group of G.*

Proof. Part (i) follows immediately from formula (6.2.4). Following Lusztig and S. Kato, Gross also gives an explicit (though difficult to use in practice) formula for $d_\lambda(\mu)$ under the assumption that G is adjoint, that is, has trivial center. From this, he deduces parts (ii) and (iii) in the adjoint case [97, Sect. 4, Formulas (4.5) and (4.6)]. To conclude the proof of the proposition, it remains to explain how to reduce to this case for a general semisimple G. The following lemma is an immediate consequence of the definitions (see [174, (7.4)]).

Lemma 6.2.7. *Let $f : G \to G'$ be a central homomorphism between split reductive $\mathbb{Z}_p$-groups, and let X and X' be the weight lattices of $\widehat{G}$ and $\widehat{G'}$, respectively. For every dominant weight $\mu \in X$, we have $\mathrm{H}(f)(\mathrm{c}_\mu) = \mathrm{c}_{\mu'}$ and $\mathrm{Rep}(f)([V_\mu]) = [V_{\mu'}]$, where μ' is the image of μ by the map $\Psi(f)^\vee : X \to X'$.*

In the notation of this lemma and if f is, moreover, a central isogeny, so that the map $\mathrm{H}(f)$ is injective, the linear independence of the $\mathrm{c}_{\mu'}$ in $\mathrm{H}_p(G')$ therefore implies $d_\lambda(\mu) = d_{\lambda'}(\mu')$ for every $\lambda, \mu \in X_+$. We conclude the proof of the proposition by considering the canonical isogeny $G \to G/\mathrm{Z}(G)$. This argument also shows that the formula of Kato and Lusztig mentioned above holds for every semisimple $\mathbb{Z}_p$-group G, by reduction to the adjoint case. □

6.2.8 The Classical Groups: A Collection of Formulas

THE EVEN ORTHOGONAL GROUP AND ITS VARIANTS

Let $r \geq 2$ be an integer and L the hyperbolic q-module over $\mathbb{Z}_p^r$. We have a commutative square of natural injections (Sect. 4.2.11, Example 4.2.16, and Sect. 6.2.1)

$$\begin{array}{ccc} \mathrm{H}_p(\mathrm{SO}_L) & \hookrightarrow & \mathrm{H}_p(\mathrm{PGSO}_L) \\ \uparrow & & \uparrow \\ \mathrm{H}_p(\mathrm{O}_L) & \hookrightarrow & \mathrm{H}_p(\mathrm{PGO}_L)\,. \end{array}$$

The top injection commutes with the natural actions of the group with two elements $\mathrm{O}_L(\mathbb{Z}_p)/\mathrm{SO}_L(\mathbb{Z}_p)$, the bottom one is then the injection deduced from it on the invariants. To alleviate the notation, we will view these injections as inclusions. We begin by describing $\mathrm{H}_p(\mathrm{PGSO}_L)$, which will be useful further on.

We take the notation of Sect. 6.1.3 with respect to the $\mathbb{Z}_p$-groups GSO_L, SO_L, and PGSO_L. Let $\lambda \in \mathrm{X}_*(\mathrm{P}\widetilde{T})$. It admits a unique inverse image under the canonical map $\mathrm{X}_*(\widetilde{T}) \to \mathrm{X}_*(\mathrm{P}\widetilde{T})$, which we denote by

$$\widetilde{\lambda} = \sum_{i=0}^{r} m_i \varepsilon_i^* ,$$

such that we have $m_0 = \langle \varepsilon_0, \widetilde{\lambda} \rangle \in \{0,1\}$. The $p^{m_i} e_i$ and $p^{m_0 - m_i} e_i^*$ for $1 \leq i \leq r$ form a $\mathbb{Z}_p$-basis of the homodual lattice $M = \widetilde{\lambda}(p)L$. The latter therefore satisfies

$$M/M \cap L \simeq \prod_{i=1}^{r} (\mathbb{Z}/p^{d_i}\mathbb{Z}) ,$$

where we have $d_i = \max(m_i - m_0, -m_i) = |m_i - m_0/2| - m_0/2$ for $i \in \{1, \cdots, r\}$ and $M^\sharp = p^{-m_0} M$. We denote the isomorphism class of the abelian group above by A_λ and set $v_\lambda = \langle \varepsilon_0, \widetilde{\lambda} \rangle = m_0$. The map

$$\eta \colon \lambda \mapsto (A_\lambda, v_\lambda)$$

obviously induces a surjection from $\mathrm{X}_*(\mathrm{P}\widetilde{T})$ to the set of pairs (A, v) with A the isomorphism class of a finite abelian p-group generated by r elements and $v \in \{0,1\}$. It is not difficult to verify that η is constant on the orbits of the subgroup of $\mathrm{Aut}(\mathrm{X}_*(\mathrm{P}\widetilde{T}))$ generated by W and the automorphism $\tau := \Psi(s)$ introduced in Sect. 6.1.3. Moreover, the coweight λ is dominant if and only if we have $m_1 \geq m_2 \geq \cdots \geq m_{r-1} \geq d_r$. Two dominant coweights λ, λ' therefore have the same image by η if and only if we have $\lambda' \in \{\lambda, \tau(\lambda)\}$. If we compare this discussion with that of Sect. 4.2.6, we deduce the following.

Scholium 6.2.9. *If λ is a coweight of PGSO_L, then we have the equality $\mathrm{T}_{(A_\lambda, v_\lambda)} = \sum_{\mu \in \{\lambda, \tau(\lambda)\}} \mathrm{c}_\mu$. Moreover, we have $T^{\mathrm{t}} = T$ for every $T \in \mathrm{H}_p(\mathrm{PGO}_L)$.*

The natural injection $\mathrm{X}_*(T) \to \mathrm{X}_*(\mathrm{P}\widetilde{T})$ identifies the coweights of SO_L with those of PGSO_L that satisfy $v_\lambda = 0$. If λ is a coweight of SO_L, we deduce from this the equality $\mathrm{T}_{A_\lambda} = \sum_{\mu \in \{\lambda, \tau(\lambda)\}} \mathrm{c}_\mu$ in $\mathrm{H}_p(\mathrm{O}_L)$.

Given that $\mathrm{H}_p(\mathrm{PGO}_L)$ can be identified with the invariants of $\mathrm{H}_p(\mathrm{PGSO}_L)$ under the action of $\mathrm{O}_L(\mathbb{Q}_p)/\mathrm{SO}_L(\mathbb{Q}_p)$ by conjugation, Proposition 4.2.9 follows from the fact that the c_λ with λ dominant form a $\mathbb{Z}$-basis of $\mathrm{H}_p(\mathrm{PGSO}_L)$. For $1 \leq i \leq r$, we denote by $\lambda_i \in \mathrm{X}_*(\mathrm{P}\widetilde{T})$ the image of $\varepsilon_1^* + \cdots + \varepsilon_i^* \in \mathrm{X}_*(\widetilde{T})$, and we denote by $\lambda_{r+1} \in \mathrm{X}_*(\mathrm{P}\widetilde{T})$ the image of $-\varepsilon_0^*$. In particular, we have $2\lambda_{r+1} = \lambda_r$. By Scholium 6.2.9, we have the relations $\mathrm{c}_{\lambda_i} = \mathrm{T}_{(\mathbb{Z}/p\mathbb{Z})^i}$ in $\mathrm{H}_p(\mathrm{O}_L)$ for $i < r$ and the relations $\mathrm{c}_{\lambda_r} + \mathrm{c}_{\tau(\lambda_r)} = \mathrm{T}_{(\mathbb{Z}/p\mathbb{Z})^r}$ and $\mathrm{c}_{\lambda_{r+1}} + \mathrm{c}_{\tau(\lambda_{r+1})} = \mathrm{K}_p$ in $\mathrm{H}_p(\mathrm{PGO}_L)$. The following statement is well known [174], [169, Sect. 4].

Corollary 6.2.10. (i) *The homomorphism* $\mathbb{Z}[X_1, \cdots, X_r] \to \mathrm{H}_p(\mathrm{PGO}_L)$ *that sends* X_i *to* $\mathrm{T}_{(\mathbb{Z}/p\mathbb{Z})^i}$ *for* $1 \leq i \leq r-1$ *and* X_r *to* K_p *is a ring isomorphism.*
(ii) *The homomorphism* $\mathbb{Z}[Y_1, \cdots, Y_r] \to \mathrm{H}_p(\mathrm{O}_L)$ *that sends* Y_i *to* $\mathrm{T}_{(\mathbb{Z}/p\mathbb{Z})^i}$ *for* $1 \leq i \leq r$ *is a ring isomorphism.*

Proof. Since the group PGSO_L is adjoint, the discussion in Sect. 6.2.5 implies that $\mathrm{H}_p(\mathrm{PGSO}_L)$ is the polynomial ring in the c_ω, where ω runs through the fundamental coweights of PGSO_L. These are the elements λ_{r+1} and $\tau(\lambda_{r+1})$ and the λ_i for $i = 1, \cdots, r-2$, by Bourbaki [39, Planche IV]. The last $r-2$ are invariant under τ, and the first two are interchanged. Recall that if A is a commutative ring, the subring of $A[U, V]$ consisting of the $P(U, V)$ such that $P(U, V) = P(V, U)$ is $A[UV, U+V]$. Thus, $\mathrm{H}_p(\mathrm{PGO}_L)$ is the polynomial ring in the c_{λ_i} for $1 \leq i < r-1$, K_p, and $\mathrm{c}_{\lambda_{r+1}}\mathrm{c}_{\tau(\lambda_{r+1})}$. But the only dominant coweights of PGSO_L that are strictly less than $\lambda_{r+1} + \tau(\lambda_{r+1}) = \lambda_{r-1}$ are the λ_i with $0 \leq i < r-1$ and $i \equiv r-1 \bmod 2$, with the convention $\lambda_0 = 0$. Hence, there exist integers $a_j \in \mathbb{Z}$ such that we have an identity of the form

$$\mathrm{c}_{\lambda_{r+1}}\mathrm{c}_{\tau(\lambda_{r+1})} = \mathrm{c}_{\lambda_{r-1}} + \sum_{0 \leq j < r-1} a_j \mathrm{c}_{\lambda_j} .$$

This proves part (i). Part (ii) is proved using similar arguments. The monoid of dominant coweights of SO_L is generated by the λ_i for $i \leq r-1$, λ_r, and $\tau(\lambda_r)$. The subring $\mathrm{H}_p(\mathrm{SO}_L) \subset \mathrm{H}_p(\mathrm{PGSO}_L)$ is therefore generated by the c_λ, where λ runs through this list. But if $S = \{\lambda_{r+1}, \tau(\lambda_{r+1})\}$ and $s, t \in S$, then, as above, $\mathrm{c}_s\mathrm{c}_t - \mathrm{c}_{s+t}$ is an integral linear combination of the c_{λ_i} for $0 \leq i \leq r-2$. This implies that $\mathrm{H}_p(\mathrm{SO}_L)$ is also generated by the ring $\mathbb{Z}[\mathrm{c}_{\lambda_1}, \cdots, \mathrm{c}_{\lambda_{r-2}}]$ and the three elements $\mathrm{c}_s\mathrm{c}_t$, where $s, t \in S$. We conclude by noting that if A is a commutative ring, the subring of $A[U^2, V^2, UV] \subset A[U, V]$ consisting of the symmetric polynomials in U and V is $A[UV, (U+V)^2]$. □

By Scholium 6.2.9, for $m \geq 0$, the operator $\mathrm{T}_{p^m} \in \mathrm{H}_p(\mathrm{O}_L)$ of p^m-neighbors coincides with $\mathrm{c}_{m\lambda_1}$. The operator $\mathrm{T}_{(\mathbb{Z}/p\mathbb{Z})^2}$ appears several times in what follows; we also denote it by $\mathrm{T}_{p,p}$.

Example 6.2.11. In $\mathrm{H}_p(\mathrm{O}_L)$, we have the relation

$$(\mathrm{T}_p)^2 = \mathrm{T}_{p^2} + (p+1)\ \mathrm{T}_{p,p} + \frac{(p^r-1)(p^{r-1}+1)}{p-1} .$$

Proof. The dominant coweights of SO_L strictly less than $2\lambda_1$ are λ_2 and 0, which implies the existence of $a, b \in \mathbb{Z}$ such that $(\mathrm{T}_p^2) = \mathrm{T}_{p^2} + a\,\mathrm{T}_{p,p} + b$ (formula (6.2.3)). Since L is a p-neighbor of each of its p-neighbors, the integer b is simply the number $b = \mathrm{c}_{2r}(p)$ of p-neighbors of a self-dual lattice (Scholium-Definition 3.1.2). Let us compute a using the *degree* homomorphism $\deg \colon \mathrm{H}(\mathrm{SO}_L) \to \mathbb{Z}$ introduced in

Example 6.2.3. The degree of T_A is the number of A-neighbors of L; in particular, we have $\deg(\mathrm{T}_p) = \mathrm{c}_{2r}(p)$ and $\deg(\mathrm{T}_{p^2}) = p^{2r-2}\mathrm{c}_{2r}(p)$ by Proposition 3.1.4. In the same spirit as in Sect. 3.1, we easily verify that for $1 \leq i \leq r$, the number of $(\mathbb{Z}/p\mathbb{Z})^i$-neighbors of L is the product of the number of isotropic subspaces of rank i of $L \otimes \mathbb{F}_p$ and the number of Lagrangians of the hyperbolic q-vector space over $(\mathbb{Z}/p\mathbb{Z})^i$ that are transverse to $(\mathbb{Z}/p\mathbb{Z})^i$ (that is, $p^{i(i-1)/2}$ by Proposition-Definition 2.1.3 (b)). For $i = 2$, we therefore obtain $\mathrm{c}_{2r}(p)\mathrm{c}_{2r-2}(p)(p+1)^{-1} \cdot p$. A short calculation leads to $a = p + 1$. $\square$

Remark 6.2.12. It would be interesting to know whether the T_{p^i} for $i = 1, \cdots, r$ generate the $\mathbb{Q}$-algebra $\mathrm{H}_p(\mathrm{O}_L) \otimes \mathbb{Q}$.

Let us conclude this collection of formulas for the even orthogonal groups with certain properties of the Satake isomorphism. The half-sum of the positive roots of GSO_L is $\rho = (r-1)\varepsilon_1 + (r-2)\varepsilon_2 + \cdots + \varepsilon_{r-1} - r(r+1)/4\varepsilon_0$. The only minimal dominant coweight of SO_L is λ_1, and PGSO_L admits two other ones, namely λ_{r+1} and $\tau(\lambda_{r+1})$. The first is the dominant weight of the standard representation V_{St} (of dimension $2r$) of $\widehat{\mathrm{SO}_L}(\mathbb{C}) = \mathrm{SO}_{2r}(\mathbb{C})$, and the other two are the dominant weights of the two spin representations $V_{\mathrm{Spin}}^{\pm}$ of $\widehat{\mathrm{PGSO}_L}(\mathbb{C}) = \mathrm{Spin}_{2r}(\mathbb{C})$. Proposition 6.2.6 (i) and Scholium 6.2.9 imply the identities

$$p^{r-1}[V_{\mathrm{St}}] = \mathrm{Sat}(\mathrm{T}_p) \quad \text{and} \quad p^{r(r-1)/4}([V_{\mathrm{Spin}}^+] + [V_{\mathrm{Spin}}^-]) = \mathrm{Sat}(\mathrm{K}_p)\,. \quad (6.2.5)$$

Let us now consider the representation $\Lambda^2 V_{\mathrm{St}}$, which is nothing more than the adjoint representation of $\mathrm{SO}_{2r}(\mathbb{C})$. Its highest weight is λ_2; the unique dominant weight strictly less than λ_2 is the weight 0. Points (i) and (iii) of Proposition 6.2.6 therefore imply

$$p^{2r-3}[\Lambda^2 V_{\mathrm{St}}] = \mathrm{Sat}(\mathrm{T}_{p,p}) + p^{r-2} + \sum_{i=0}^{r-2} p^{2i}\,. \quad (6.2.6)$$

(We could also invoke Example 6.2.3 instead of part (iii) of Proposition 6.2.6.)

The Symplectic Group and Its Variants

Let $g \geq 1$ be an integer and L the hyperbolic a-module over $\mathbb{Z}_p^g$. We use the notation of Sect. 6.1.3 with respect to the $\mathbb{Z}_p$-groups GSp_{2g}, Sp_{2g}, and PGSp_{2g}. As above, if A is a finite abelian p-group generated by g elements, say $A \simeq \prod_{i=1}^g \mathbb{Z}/p^{m_i}\mathbb{Z}$ with $m_1 \geq \cdots \geq m_g \geq 0$, then we have

$$\mathrm{T}_A = \mathrm{c}_{\sum_{i=1}^g m_i \varepsilon_i^*} \quad \text{and} \quad \mathrm{T}_{(A,1)} = \mathrm{c}_{\varepsilon_0^* + \sum_{i=1}^g (m_i+1)\varepsilon_i^*}\,. \quad (6.2.7)$$

By Shimura [187], the ring $\mathrm{H}_p(\mathrm{PGSp}_{2g})$ is the polynomial ring in K_p and the $\mathrm{T}_{(\mathbb{Z}/p\mathbb{Z})^i}$ for $i < g$; the subring $\mathrm{H}_p(\mathrm{Sp}_{2g})$ is generated by the $\mathrm{T}_{(\mathbb{Z}/p\mathbb{Z})^i}$ for $i \leq g$ (the situation is in fact simpler than that of Corollary 6.2.10, as the monoids of dominants weights of Sp_{2g} and PGSp_{2g} are free).

The half-sum of the positive roots ρ of GSp_{2g} is $-(g(g+1)/4)\,\varepsilon_0^* + g\,\varepsilon_1^* + (g-1)\,\varepsilon_2^* + \cdots + \varepsilon_g^*$. Let V_{St} be the standard representation (of dimension $2g+1$) of $\widehat{\mathrm{Sp}_{2g}}(\mathbb{C}) = \mathrm{SO}_{2g+1}(\mathbb{C})$, and let V_{Spin} be the spin representation of $\mathrm{Spin}_{2g+1}(\mathbb{C})$. The highest weight of V_{St} is ε_1^*, whose only strictly smaller dominant weight is 0. The highest weight of V_{Spin} is $-\varepsilon_0^*$, which is minimal. Proposition 6.2.6 therefore implies

$$p^g[V_{\mathrm{St}}] = \mathrm{Sat}(\mathrm{T}_p) + 1 \quad \text{and} \quad p^{g(g+1)/4}[V_{\mathrm{Spin}}] = \mathrm{Sat}(\mathrm{K}_p)\,. \tag{6.2.8}$$

THE ODD SPECIAL ORTHOGONAL GROUP

We will only study the differences with the other cases, which are minor. Let $r \geq 1$ be an integer, let L be the $\mathbb{Z}_p$-module $\mathbb{Z}_p^r \oplus (\mathbb{Z}_p^r)^* \oplus \mathbb{Z}_p$ endowed with the quadratic form obtained by taking the orthogonal sum of the hyperbolic q-module over $\mathbb{Z}_p^r$ and $x \mapsto x^2$, let $V = L \otimes \mathbb{Q}_p$, and let G be the $\mathbb{Z}_p$-group SO_L (Sect. B.1). We leave it as an exercise to verify that the map $g \mapsto g(L)$ identifies $G(\mathbb{Q}_p)/G(\mathbb{Z}_p)$ with the subset $\mathcal{R}^{\mathrm{b}}_{\mathbb{Z}_p}(V) \subset \mathcal{R}_{\mathbb{Z}_p}(V)$ of lattices $M \subset V$ such that $M^\sharp = M$ if $p > 2$, or such that M is a sublattice of index 2 of $M^\sharp$ if $p = 2$ (Sect. 4.2.6).

Let V_{St} be the standard representation of $\widehat{\mathrm{SO}_L}(\mathbb{C}) = \mathrm{Sp}_{2r}(\mathbb{C})$ (of dimension $2r$); its highest weight is ε_1^*, which is minimal (we use the notation of Sect. 6.1.3 with respect to the $\mathbb{Z}_p$-group G). The Hecke operator $\mathrm{c}_{\varepsilon_1^*}$ is associated with the ordered pairs $(N, M) \in \mathcal{R}^{\mathrm{b}}_{\mathbb{Z}_p}(V)^2$ such that $M \cap N$ has index p in M: this is the operator T_p of p-neighbors in the sense of Sect. B.3. Since the half-sum of the positive roots of SO_L is $\frac{1}{2}(2r-1)\,\varepsilon_1 + \frac{1}{2}(2r-3)\,\varepsilon_2 + \cdots + \frac{1}{2}\,\varepsilon_r$, we therefore have

$$p^{(2r-1)/2}[V_{\mathrm{St}}] = \mathrm{T}_p\,.$$

6.3 The Harish-Chandra Isomorphism

6.3.1 The Center of the Universal Enveloping Algebra of a Reductive $\mathbb{C}$-group

Let G be a reductive $\mathbb{C}$-group, $\mathfrak{g}$ the Lie $\mathbb{C}$-algebra of G, $\mathrm{U}(\mathfrak{g})$ its universal enveloping algebra, and $\mathrm{Z}(\mathrm{U}(\mathfrak{g}))$ the center of $\mathrm{U}(\mathfrak{g})$ [76, Chap. 2]. Let V be a $\mathrm{U}(\mathfrak{g})$-module. We say that V admits a central character if there exists a homomorphism of $\mathbb{C}$-algebras

$$\mathrm{c}_V : \mathrm{Z}(\mathrm{U}(\mathfrak{g})) \to \mathbb{C}$$

such that we have $z \cdot v = c_V(z)v$ for every $v \in V$ and every $z \in \mathrm{Z}(\mathrm{U}(\mathfrak{g}))$; we then call c_V the *central character* of V. By Dixmier [76, Proposition 2.6.8], every simple $\mathrm{U}(\mathfrak{g})$-module admits a central character. In this subsection, following Harish-Chandra and Langlands, we recall how to view these central characters as semisimple conjugacy classes in the Lie $\mathbb{C}$-algebra $\widehat{\mathfrak{g}}$ of the dual reductive $\mathbb{C}$-group $\widehat{G}$ of G.

Let $\mathrm{Pol}(\widehat{\mathfrak{g}}) = \mathrm{Sym}(\widehat{\mathfrak{g}}^*)$ be the $\mathbb{C}$-algebra of the polynomial functions over $\widehat{\mathfrak{g}}$. It is endowed with a natural action of $\widehat{G}(\mathbb{C})$ arising from the adjoint action on $\widehat{\mathfrak{g}}$, whose algebra of invariants we denote by $\mathrm{Pol}(\widehat{\mathfrak{g}})^{\widehat{G}}$. The Harish-Chandra isomorphism is a canonical isomorphism

$$\mathrm{HC}\colon \mathrm{Z}(\mathrm{U}(\mathfrak{g})) \xrightarrow{\sim} \mathrm{Pol}(\widehat{\mathfrak{g}})^{\widehat{G}}$$

[76, Theorems 7.4.5 and 7.3.5], [136, Sect. 2]. Let $\widehat{\mathfrak{g}}_{\mathrm{ss}}$ be the set of conjugacy classes of semisimple elements of $\widehat{\mathfrak{g}}$. Each such class $X \in \widehat{\mathfrak{g}}_{\mathrm{ss}}$ defines, by evaluation, a homomorphism of $\mathbb{C}$-algebras $\mathrm{Pol}(\widehat{\mathfrak{g}})^{\widehat{G}} \to \mathbb{C}$, $P \mapsto P(X)$. A classical result of Chevalley asserts that the resulting map $\widehat{\mathfrak{g}}_{\mathrm{ss}} \to \mathrm{Hom}_{\mathbb{C}-\mathrm{alg}}(\mathrm{Pol}(\widehat{\mathfrak{g}})^{\widehat{G}}, \mathbb{C})$ is bijective.

Scholium 6.3.2. *The Harish-Chandra isomorphism induces a canonical bijection* $\mathrm{Hom}_{\mathbb{C}-\mathrm{alg}}(\mathrm{Z}(\mathrm{U}(\mathfrak{g})), \mathbb{C}) \xrightarrow{\sim} \widehat{\mathfrak{g}}_{\mathrm{ss}}$.

If X is the weight lattice of G, then the elements of $X \otimes \mathbb{C}$ can be viewed as elements of $\widehat{\mathfrak{g}}_{\mathrm{ss}}$. Indeed, let $\widehat{T}$ be a maximal torus of $\widehat{G}$ and $\widehat{B} \subset \widehat{G}$ a Borel subgroup containing $\widehat{T}$. The datum $\Psi(G)^\vee$ can be identified with $\Psi(\widehat{G}, \widehat{T}, \widehat{B})$; in particular, X can be identified with $\mathrm{X}_*(\widehat{T})$. The exponential map defines a natural map between $\mathrm{X} \otimes \mathbb{C}$ and the complex Lie algebra $\widehat{\mathfrak{t}}$ of $\widehat{T}$. If W is the Weyl group of G, we deduce from this a canonical bijection

$$(X \otimes \mathbb{C})/W \xrightarrow{\sim} \widehat{\mathfrak{g}}_{\mathrm{ss}}\,. \tag{6.3.1}$$

Example 6.3.3. Let $\lambda \in X_+$ be a dominant weight of G and V_λ the irreducible $\mathbb{C}$-representation of G with highest weight λ (Sect. 6.1.4). This representation endows V_λ with the structure of a $\mathrm{U}(\mathfrak{g})$-module. This module is simple, and its central character corresponds to the conjugacy class of $\lambda + \rho$, where ρ is the half-sum of the positive roots of G.

More generally, fix a pair $T \subset B$ in G that identifies $\Psi(G)$ with $\Psi(G, T, B)$. Let $\mathfrak{t} \subset \mathfrak{b}$ be their respective Lie $\mathbb{C}$-algebras and V a $\mathrm{U}(\mathfrak{g})$-module generated by an element $e \in V$ such that $\mathfrak{b}\, e \subset \mathbb{C}e$ (*module of highest weight*). Let $\lambda \in \mathfrak{t}^*$ be the linear form defined by $he = \lambda(h)e$ for every $h \in \mathfrak{t}$ (we can also view it, dually, as an element of $\widehat{\mathfrak{t}}$). Then, V admits a central character by Dixmier [76, Proposition 7.1.8], and it follows rather directly from the definition of the Harish-Chandra homomorphism that the corresponding conjugacy class is that of $\lambda + \rho$ [76, Sect. 7.4.6].

6.3.4 The Infinitesimal Character of a Unitary Representation

Let G be a reductive $\mathbb{R}$-group. We apply the considerations and notation of the previous subsection to the $\mathbb{C}$-group $G_{\mathbb{C}} := G \times_{\mathbb{R}} \mathbb{C}$. We refer to [119] and [209] for a general expository treatment of the theory of unitary representations of reductive Lie groups.

Let V be a Hilbert space endowed with a unitary representation of the Lie group of $G(\mathbb{R})$. Let $V^{\infty} \subset V$ be the subspace of $\mathcal{C}^{\infty}$-vectors, that is, of the $v \in V$ such that the map $g \mapsto gv, G(\mathbb{R}) \to V$ is of class $\mathcal{C}^{\infty}$; it is dense in V (Gårding) and stable under $G(\mathbb{R})$. If the unitary representation V is irreducible, then the $\mathrm{U}(\mathfrak{g})$-module V^{∞} admits a central character [209, Sect. 1.6.5], called the *infinitesimal character* of V; we denote it by $\inf_V$. As proved by Harish-Chandra, this is a rather fine invariant of the representation V: up to isomorphism, there are only a finite number (possibly zero) of irreducible unitary representations of G with a given infinitesimal character (this is a difficult result; see [119, Corollary 10.37]). The Harish-Chandra isomorphism allows us to view $\inf_V$ as a semisimple conjugacy class in $\widehat{\mathfrak{g}}$. We will give two examples.

Let us first suppose that $G(\mathbb{R})$ is a compact group, in which case it is necessarily connected by Chevalley [34, Chap. V, Sect. 24.6 (c) (ii)]. Every $\mathbb{C}$-representation V of $G_{\mathbb{C}}$ defines, by restriction, a representation $V_{|G(\mathbb{R})}$ of $G(\mathbb{R})$. The functor $V \mapsto V_{|G(\mathbb{R})}$ is an equivalence of categories between $\mathbb{C}$-representations of $G_{\mathbb{C}}$ and finite-dimensional, continuous, complex representations of $G(\mathbb{R})$. In particular, every irreducible representation of $G(\mathbb{R})$ is isomorphic to $(V_{\lambda})_{|G(\mathbb{R})}$ for a unique dominant weight λ of $G_{\mathbb{C}}$; we will, in general, denote it by V_{λ} to alleviate the notation. By Example 6.3.3, its infinitesimal character is the conjugacy class of $\lambda + \rho$ in $\widehat{\mathfrak{g}}$. In particular, this character determines V_{λ} uniquely.

Let us now suppose that G is the $\mathbb{R}$-group Sp_{2g}. We use the notation of Sect. 4.5.3, except that $\mathfrak{g}$ now denotes the *complexified* Lie algebra of $G(\mathbb{R})$. For the maximal compact subgroup, we choose the stabilizer $K \subset G(\mathbb{R})$ of $i1_g$ in $\mathbb{H}_g$, which has Lie algebra $\mathfrak{k}$. This is a unitary group with g variables: the homomorphism $K \to \mathrm{GL}_g(\mathbb{C})$ given by $k \mapsto \mathrm{j}(k, i1_g)$ identifies $\mathrm{GL}_g(\mathbb{C})$ with the complexification of K, and then (by differentiation) $\mathfrak{k}_{\mathbb{C}}$ with the Lie algebra $\mathfrak{gl}_g(\mathbb{C})$. The complexified Cartan decomposition can be written as

$$\mathfrak{g} = \mathfrak{k}_{\mathbb{C}} \oplus \mathfrak{p}^{+} \oplus \mathfrak{p}^{-} ,$$

where $\mathfrak{p}^{\pm}$ are *abelian* Lie subalgebras stable under $\mathrm{ad}(K)$. The key point is that the adjoint action on $\mathfrak{p}$ of the element $(1/\sqrt{2})\left(\begin{smallmatrix} & 1 \\ -1 & 1 \end{smallmatrix}\right)$ of the center of K (which is isomorphic to $\mathrm{U}(1)$) induces the natural complex structure of the $\mathbb{R}$-vector space $\mathfrak{p} \simeq \mathrm{Sym}_g(\mathbb{C})$.

Let T be a maximal torus of GL_g and B a Borel subgroup containing T, with respective complex Lie algebras $\mathfrak{t} \subset \mathfrak{b} \subset \mathfrak{gl}_g(\mathbb{C}) = \mathfrak{k}_{\mathbb{C}}$. The properties of $\mathfrak{p}^{-}$ mentioned above ensure that $\mathfrak{t}$ is a Cartan subalgebra of $\mathfrak{g}$ and that $\mathfrak{b} \oplus \mathfrak{p}^{-}$ is a Borel subalgebra of $\mathfrak{g}$.

Proposition 6.3.5 (Harish-Chandra). *Let V be a unitary representation of $\mathrm{Sp}_{2g}(\mathbb{R})$, e an element of V^∞, and U an irreducible $\mathbb{C}$-representation of GL_g such that*

(i) $\mathfrak{p}^- e = 0$;
(ii) *the representation of K generated by e is isomorphic to $U_{|K}$.*

Then:

(a) *The $\mathrm{U}(\mathfrak{g})$-module $\mathrm{U}(\mathfrak{g})e \subset V^\infty$ admits a central character. Its associated semisimple conjugacy class is $\lambda + \rho$, where $\lambda \in \mathfrak{t}^*$ is the highest weight of U with respect to B and ρ is the half-sum of the roots of $\mathfrak{t}$ in $\mathfrak{b} \oplus \mathfrak{p}^-$.*
(b) *The closed subrepresentation $V' \subset V$ generated by e under the action of $\mathrm{Sp}_{2g}(\mathbb{R})$ is irreducible. Moreover, if $f \in (V')^\infty$ has properties* (i) *and* (ii), *then we have $f \in \mathbb{C}[K].e$.*

Up to isomorphism, there exists at most one irreducible unitary representation of the group $\mathrm{Sp}_{2g}(\mathbb{R})$ admitting a vector e that is in $\mathcal{C}^\infty$ and has properties (i) *and* (ii).

This result is well known to specialists in the theory of unitary representations of Lie groups; we provide a proof for the sake of the reader.

Proof. By property (ii), the space $E = \mathbb{C}[K].e \subset V^\infty$ is a representation of K isomorphic to $U_{|K}$. It is stable under $\mathfrak{k}_\mathbb{C}$ and annihilated by $\mathfrak{p}^-$ because of the inclusion $\mathrm{ad}(K)\mathfrak{p}^- \subset \mathfrak{p}^-$. It is therefore also stable under the parabolic subalgebra $\mathfrak{q} = \mathfrak{k}_\mathbb{C} \oplus \mathfrak{p}^-$ of $\mathfrak{g}$. Let Y be the induced $\mathrm{U}(\mathfrak{g})$-module $\mathrm{U}(\mathfrak{g}) \otimes_{\mathrm{U}(\mathfrak{q})} E$. The inclusion of E in V^∞ therefore induces a $\mathrm{U}(\mathfrak{g})$-equivariant morphism

$$u\colon Y \longrightarrow V^\infty ,$$

whose image we denote by $X = u(Y)$. Since K is connected, E is irreducible as an $\mathrm{U}(\mathfrak{k}_\mathbb{C})$-module. After replacing e with a suitable element of E, if necessary, we may assume $\mathfrak{b}e \subset \mathbb{C}e$ and $he = \lambda(h)e$ for every $h \in \mathfrak{t}$ (Cartan–Weyl theory of the highest weight). The element $1 \otimes e$ therefore generates Y and satisfies $(\mathfrak{b} \oplus \mathfrak{p}^-)(1 \otimes e) \subset \mathbb{C}(1 \otimes e)$ by condition (i). It follows that Y, and therefore $X = \mathrm{U}(\mathfrak{g})e$, admits an infinitesimal character satisfying assertion (a), by the second paragraph of Example 6.3.3.

Also note that X is stable under K. Moreover, the adjoint action of K on $\mathrm{U}(\mathfrak{g})$, as well as its natural action on E, defines the structure of a K-module on Y such that u is K-equivariant. These structures turn Y and X into $(\mathfrak{g}, K)$-modules, which we denote by Y' and X', and turn u into a morphism of $(\mathfrak{g}, K)$-modules (we refer to [209, Sect. 3.3] for these notions). We have already seen that Y is a $\mathrm{U}(\mathfrak{g})$-module of highest weight. By Dixmier [76, Proposition 7.1.8], this implies, on the one hand, that Y' admits a unique simple quotient and, on the other hand, that Y' is admissible (this means that every irreducible representation of K occurs with finite multiplicity; this property follows from the fact that every weight of Y has finite multiplicity; see loc. cit.). But X' admits an invariant Hermitian product (it is unitary in the sense of [209, Sect. 9.3.3]) because the representation V is unitary by assumption. Since X' is admissible as a quotient of Y', it is irreducible and therefore the unique irreducible quotient of Y'. Since X' admits a central character, a result of Harish-Chandra ensures that all its vectors are in fact analytic [209, Sects. 1.6 and 3.4.9], [119, Chap. VIII, Sect. 8.7] and therefore that the closure $\overline{X}$ of X in V is stable under $G(\mathbb{R})$ [209, Sect. 1.6.6]. It admits X' as a $(\mathfrak{g}, K)$-module: it is therefore the unique unitary irreducible representation of $G(\mathbb{R})$ with $(\mathfrak{g}, K)$-module X' [209, Sect. 3.4.11]. This proves the first part of statement (b). The two remaining assertions follow from the already proved fact that X' is the unique irreducible quotient of Y'. □

Let $(X, \Phi, \Delta, X^\vee, \Phi^\vee, \Delta^\vee)$ be the based root datum associated with the triple (GL_g, T, B). As usual, we write $X = \oplus_{i=1}^g \mathbb{Z}\varepsilon_i$, $\Phi = \{\pm(\varepsilon_i - \varepsilon_j)\,;\, 1 \leq i < j \leq g\}$, $\Delta = \{\varepsilon_i - \varepsilon_{i+1}\,;\, 1 \leq i < g\}$, $X^\vee = \oplus_{i=1}^g \mathbb{Z}\varepsilon_i^*$, and $(\varepsilon_i - \varepsilon_j)^\vee = \varepsilon_i^* - \varepsilon_j^*$ for $i < j$. The dominant weights of GL_g are therefore the $\lambda \in X$ such that we have $\lambda = \sum_{i=1}^g m_i \varepsilon_i$ with $m_1 \geq m_2 \geq \cdots \geq m_g$.

Corollary 6.3.6. *Let W be the irreducible $\mathbb{C}$-representation of GL_g of highest weight $\sum_{i=1}^g m_i \varepsilon_i$. Suppose that there exists an irreducible unitary representation π'_W of $\mathrm{Sp}_{2g}(\mathbb{R})$ satisfying the conditions of Proposition 6.3.5 for $U = W^*$. The eigenvalues of the semisimple conjugacy class of $\mathfrak{so}_{2g+1}(\mathbb{C})$ that corresponds to $\inf_{\pi'_W}$ are the $2g+1$ integers*

$$\pm(m_i - i) \text{ for } i = 1, \cdots, g \text{ and } 0\,.$$

Proof. A simple computation shows that the adjoint representation of K on $\mathfrak{p}^-$ is isomorphic to the restriction, via the homomorphism j, of the representation $(g, X) \mapsto g\,X\,{}^{\mathrm{t}}g$ of $\mathrm{GL}_g(\mathbb{C})$ on $\mathrm{Sym}_g(\mathbb{C})$ (this is the symmetric square of the standard representation). Its set of weights is therefore

$$\{\varepsilon_i + \varepsilon_j\,;\, 1 \leq i \leq j \leq g\}\,.$$

This description shows that the basis of the root system of $G_{\mathbb{C}}$ associated with T corresponding to $\mathfrak{b} \oplus \mathfrak{p}^-$ is none other than the standard basis introduced in Sect. 6.1.3. In particular, the element ρ of Proposition 6.3.5 (a) is none other than $g\varepsilon_1 + (g-1)\varepsilon_2 + \cdots + \varepsilon_g$. The dominant weight λ of W^* with respect to B is $\sum_{i=1}^g -m_{g+1-i}\varepsilon_i$, and we therefore have

$$\lambda + \rho = \sum_{i=1}^{g} (i - m_i)\varepsilon_{g+1-i}\,.$$

Since by Sect. 6.1.3, the weights of $\widehat{\mathrm{Sp}_{2g}}(\mathbb{C}) = \mathrm{SO}_{2g+1}(\mathbb{C})$ in its standard representation on $\mathbb{C}^{2g+1}$ are 0 and the $\pm\varepsilon_i^*$, we are done. □

Let W be an irreducible $\mathbb{C}$-representation of GL_g in which -1_g acts trivially. Let $f \in \mathrm{S}_W(\mathrm{Sp}_{2g}(\mathbb{Z}))$ be a nonzero Siegel cusp form of weight W. We apply Proposition 6.3.5 below to $U = \mathcal{A}_{\mathrm{cusp}}(\mathrm{PGSp}_{2g})$, every element e in the image of $W^* \otimes f$ (Proposition 4.5.4), and $U = W^*$. It shows that if $w \in W^*$ is nonzero, then under the action of $\mathrm{Sp}_{2g}(\mathbb{R})$, the function $\varphi_{w,f} \in \mathcal{A}_{\mathrm{cusp}}(\mathrm{PGSp}_{2g})$ defined in Sect. 4.5.3 generates, topologically, an irreducible subrepresentation of $\mathcal{A}_{\mathrm{cusp}}(\mathrm{PGSp}_{2g})$ that is necessarily isomorphic to the representation π'_W of Corollary 6.3.6. This proves the existence of π'_W when $\mathrm{S}_W(\mathrm{Sp}_{2g}(\mathbb{Z})) \neq 0$. In fact, Harish-Chandra has proved the existence of π'_W for every W whose highest weight satisfies $m_g > g$ (this is the *holomorphic discrete series*; see [119, Chap. VI, Sect. 4, Theorem 6.6]). If this assumption on W is satisfied, we say that W is *positive*; it is the only case that interests us in this book. Note that if W is positive, the $2g+1$ integers of Corollary 6.3.6 are distinct.

Assume that W is positive and that -1_g acts trivially in W (that is, we have $\sum_i m_i \equiv 0 \bmod 2$), so that π'_W factors through $\mathrm{Sp}_{2g}(\mathbb{R})/\{\pm 1_{2g}\}$. It is not difficult to verify that π'_W is not isomorphic to its outer conjugate[7] by an element of $\mathrm{PGSp}_{2g}(\mathbb{R})\backslash\mathrm{Sp}_{2g}(\mathbb{R})$. In other words, the unitary representation of $\mathrm{PSGp}_{2g}(\mathbb{R})$

$$\pi_W = \mathrm{Ind}_{\mathrm{Sp}_{2g}(\mathbb{R})}^{\mathrm{PGSp}_{2g}(\mathbb{R})} \pi'_W$$

induced by a subgroup of index 2 is irreducible. Of course, π_W and π'_W have the same infinitesimal character because $\mathrm{Sp}_{2g}(\mathbb{R})$ and $\mathrm{PGSp}_{2g}(\mathbb{R})$ have the same Lie algebra.

We fix a nonzero $\mathcal{C}^\infty$-vector $v_W \in \pi_W$ that is annihilated by $\mathfrak{p}^-$, generates W^* under the action of K, and is an eigenvector for the action of $\mathfrak{b}$. Such a vector is unique up to multiplication by an element of $\mathbb{C}^*$, by Proposition 6.3.5. Likewise, we fix a nonzero $e_W \in W^*$ of highest weight with respect to B.

Corollary 6.3.7. *Assume that W is positive. For $F \in \mathrm{S}_W(\mathrm{Sp}_{2g}(\mathbb{Z}))$, there exists a unique $u_F \in \mathcal{A}_{\pi_W}(\mathrm{PGSp}_{2g})$ such that $u_F(v_W) = \varphi_{e_W,F}$. The map $F \mapsto u_F$ defines an $\mathrm{H}(\mathrm{PGSp}_{2g})^{\mathrm{opp}}$-equivariant isomorphism*

$$\mathrm{S}_W(\mathrm{Sp}_{2g}(\mathbb{Z})) \xrightarrow{\sim} \mathcal{A}_{\pi_W}(\mathrm{PGSp}_{2g}) .$$

Proof. Propositions 4.5.4 and 6.3.5 show that the map of the corollary induces an $\mathrm{H}^{\mathrm{opp}}(\mathrm{PGSp}_{2g})$-equivariant isomorphism between $\mathrm{S}_W(\mathrm{Sp}_{2g}(\mathbb{Z}))$ and the subspace $\mathrm{Hom}_{G(\mathbb{R})}(\pi_W, \mathcal{A}_{\mathrm{cusp}}(\mathrm{PGSp}_{2g})) \subset \mathcal{A}_{\pi_W}(\mathrm{PGSp}_{2g})$. We conclude using the following general fact: if G is a $\mathbb{Z}$-group such that $G_{\mathbb{Q}}$ is semisimple and if U is a discrete series in $G(\mathbb{R})$, then the inclusion $\mathrm{Hom}_{G(\mathbb{R})}(U, \mathcal{A}_{\mathrm{cusp}}(G)) \subset \mathcal{A}_U(G)$ is an equality [208, Theorem 4.3]. □

Thus, if $F \in \mathrm{S}_W(\mathrm{Sp}_{2g}(\mathbb{Z}))$ is nonzero and an eigenvector of all Hecke operators in $\mathrm{H}(\mathrm{PGSp}_{2g})$, the representation $\pi_F \in \Pi_{\mathrm{disc}}(\mathrm{PGSp}_{2g})$ generated by F following the general definition of Sect. 4.3.2 is well defined. It satisfies $(\pi_F)_\infty = \pi_W$.

---

[7] This is because this outer conjugate admits a vector $\mathcal{C}^\infty$ that is annihilated by $\mathfrak{p}^+$ and generates W under the action of K (*lowest weight*). Its $(\mathfrak{g}, K)$-module can be studied in a manner completely analogous to that of π'_W: it can be isomorphic to that of π'_W only if it is finite-dimensional, that is, if π'_W (and therefore W) is trivial. This does not occur because the trivial representation of $\mathrm{Sp}_{2g}(\mathbb{R})$ does not occur in $\mathcal{A}_{\mathrm{cusp}}(\mathrm{PGSp}_{2g})$.

The discourse above can also be held for Sp_{2g} instead of PGSp_{2g} and shows the existence of an $\mathrm{H}^{\mathrm{opp}}(\mathrm{Sp}_{2g})$-equivariant isomorphism between the spaces $\mathrm{S}_W(\mathrm{Sp}_{2g}(\mathbb{Z}))$ and $\mathcal{A}_{\pi'_W}(\mathrm{Sp}_{2g})$; its contents are only somewhat coarser, by Proposition 4.5.7.

Exceptional Isomorphisms in Genus 1 and 2

In the following, we assume that W is positive, of highest weight $\sum_i m_i \varepsilon_i$, and that -1_g acts trivially in W.

Suppose $g = 1$. In this case, W is the representation $\det^k$ with $k = m_1 > 1$ and $k \equiv 0 \bmod 2$. The isomorphism $\mathfrak{sl}_2(\mathbb{C}) \simeq \mathfrak{so}_3(\mathbb{C})$ (symmetric square) allows us to view the infinitesimal character of π_W as the semisimple conjugacy class in $\mathfrak{sl}_2(\mathbb{C})$ with eigenvalues $\pm(k-1)/2$. In fact, the well-known classification of the unitary dual of $\mathrm{SL}_2(\mathbb{R})$ (Bargmann [18]) shows that, up to isomorphism, the unique irreducible unitary representation of $\mathrm{PGL}_2(\mathbb{R})$ that has an infinitesimal character with eigenvalues $\pm(k-1)/2$ with $k > 3$ an even integer, is the representation $\pi_{\det^k}$. When $k = 2$, we must add the two representations of dimension 1.

Suppose $g = 2$. In this case, in the notation of [89], W is the representation $\mathrm{Sym}^j(\mathbb{C}^2) \otimes \det^k$ with $j = m_1 - m_2$ and $k = m_2$; moreover, we have $k > 2$ and $j \equiv 0 \bmod 2$. The exceptional isomorphism $\mathfrak{sp}_4(\mathbb{C}) \simeq \mathfrak{so}_5(\mathbb{C})$ allows us to view the infinitesimal character of π_W as the semisimple conjugacy class in $\mathfrak{sp}_4(\mathbb{C})$ with the following eigenvalues for its action on $\mathbb{C}^4$:

$$\pm\frac{w_1}{2}, \quad \pm\frac{w_2}{2},$$

where $w_1 = m_1 + m_2 - 3 = 2k + j - 3$ and $w_2 = m_1 - m_2 + 1 = j + 1$.

6.4 The Arthur–Langlands Conjecture

6.4.1 Langlands Parametrization of $\Pi(G)$ for G Semisimple over $\mathbb{Z}$

Let H be a $\mathbb{C}$-group, with neutral component H^0 and complex Lie algebra $\mathfrak{h}$. We denote by $H(\mathbb{C})_{\mathrm{ss}}$ (resp. $\mathfrak{h}_{\mathrm{ss}}$) the set of $H(\mathbb{C})$-conjugacy classes of semisimple elements of $H^0(\mathbb{C})$ (resp. $\mathfrak{h}$) and consider the set

$$\mathcal{X}(H)$$

of families $(c_v)_{v \in \mathrm{P} \cup \{\infty\}}$, where $c_\infty \in \mathfrak{h}_{\mathrm{ss}}$ and $c_p \in H(\mathbb{C})_{\mathrm{ss}}$ for every $p \in \mathrm{P}$. In the discussion that follows, H will be connected (and even semisimple), but we will later encounter nonconnected examples associated with the even orthogonal groups.

Every morphism of $\mathbb{C}$-groups $r\colon H \to H'$ defines a map that we also denote by $r\colon \mathcal{X}(H) \to \mathcal{X}(H')$, which sends (c_v) to $(r(c_v))$.

Let G be a semisimple $\mathbb{Z}$-group. As we have already mentioned, for every prime p, the $\mathbb{Z}_p$-group $G_{\mathbb{Z}_p}$ is split and reductive [96, Proposition 1.1]; it therefore admits a based root datum $\Psi(G_{\mathbb{Z}_p})$. Moreover, if $\overline{\mathbb{Q}}$ (resp. $\overline{\mathbb{Q}_p}$) is an algebraic closure of $\mathbb{Q}$ (resp. $\mathbb{Q}_p$), and if $\overline{\mathbb{Q}} \to \overline{\mathbb{Q}_p}$ and $\overline{\mathbb{Q}} \to \mathbb{C}$ are two embeddings, then the associated isomorphisms of based root data

$$\Psi(G_{\mathbb{Z}_p}) \xrightarrow{\sim} \Psi(G_{\overline{\mathbb{Q}_p}}) \xleftarrow{\sim} \Psi(G_{\overline{\mathbb{Q}}}) \xrightarrow{\sim} \Psi(G_{\mathbb{C}})$$

do not depend on any of the choices of embeddings.[8] The Langlands dual of $G_{\overline{\mathbb{Q}}}$ is therefore canonically the Langlands dual of the $G_{\mathbb{Z}_p}$ for every p and of $G_{\mathbb{C}}$; we denote it by $\widehat{G}$.

Following Langlands [136], we have a canonical map

$$\mathrm{c}\colon \Pi(G) \to \mathcal{X}(\widehat{G})\ , \quad \pi \mapsto (\mathrm{c}_v(\pi))\ ,$$

defined as follows. Set $\pi = \pi_\infty \otimes \pi_f \in \Pi(G)$. Let $\mathrm{c}_\infty(\pi)$ be the infinitesimal character of π_∞ (Sect. 6.3.4). The Satake isomorphism implies that $\mathrm{H}(G) = \bigotimes_p \mathrm{H}_p(G)$ is commutative, so that π_f has dimension 1 and can be viewed as a ring homomorphism from $\mathrm{H}(G)^{\mathrm{opp}} = \mathrm{H}(G)$ to $\mathbb{C}$ or, equivalently, as a collection of ring morphisms

$$\pi_p\colon \mathrm{H}_p(G) \to \mathbb{C}\ ,$$

where π_p is the restriction of π_f to $\mathrm{H}_p(G)$ in the sense of Sect. 4.2.5. Consequently, by Scholium 6.2.2, to each π_p, there corresponds a unique element $\mathrm{c}_p(\pi) \in \widehat{G}(\mathbb{C})_{\mathrm{ss}}$. By definition, $\mathrm{c}(\pi)$ determines π_f and the infinitesimal character of π_∞; the map c therefore has finite fibers (Harish-Chandra, Sect. 6.3.4).

Example 6.4.2 (Trivial Representation). Let $\pi = 1_G \in \Pi_{\mathrm{disc}}(G)$ be the trivial representation of G (Sect. 4.3.2). By Example 6.3.3, $2\mathrm{c}_\infty(\pi)$ is the conjugacy class of the coweight 2ρ of $\widehat{G}$. Likewise, by Example 6.2.3, the conjugacy class $\mathrm{c}_p(\pi)$ is that of $\rho(p) = (2\rho)(p^{1/2})$.

6.4.3 A Few Formulas

We first consider the $\mathbb{Z}$-group PGL_2, with dual group SL_2. Let $k > 0$ be an even integer and $F = \sum_{n \geq 1} a_n q^n \in \mathrm{S}_k(\mathrm{SL}_2(\mathbb{Z}))$ a modular eigenform for all

[8] Gross' argument is the following. It is a general fact that the natural action of $\mathrm{Gal}(\overline{\mathbb{Q}}/\mathbb{Q})$ on $\Psi(G_{\overline{\mathbb{Q}}})$ factors into a faithful action of the Galois group of a number field K that is Galois over $\mathbb{Q}$. The reductivity of G over $\mathbb{Z}_p$ implies that K is unramified at p, and therefore $K = \mathbb{Q}$ by a famous result of Minkowski. This, in turn, implies that G is split over $\mathbb{Z}_p$ and the rest of the assertions above.

the Hecke operators of $\mathrm{H}(\mathrm{PGL}_2)$ with $a_1 = 1$ (these form a basis of $\mathrm{S}_k(\mathrm{SL}_2(\mathbb{Z}))$ [177, Chap. VII, Sect. 5.4]). Let $\pi \in \Pi_{\mathrm{cusp}}(\mathrm{PGL}_2)$ be the representation generated by F (see Sects. 4.3.2 and 6.3.4). We already determined $\mathrm{c}_\infty(\pi)$, that is, $\inf_{\pi_W}$, in terms of k loc. cit. A $\mathbb{Z}$-isomorphism $\mathrm{PGL}_2 \simeq \mathrm{PGSp}_2$ induces an isomorphism $\mathrm{H}_p(\mathrm{PGSp}_2) \overset{\sim}{\to} \mathrm{H}_p(\mathrm{PGL}_2)$ that sends K_p to $\mathrm{T}_{\mathbb{Z}/p\mathbb{Z}}$. The relations (6.2.8) and (4.5.5), as well as [177, Chap. VII, Theorem 7], therefore show that for every prime p,

$$p^{(k-1)/2}\,\mathrm{Trace}(\mathrm{c}_p(\pi)) = a_p\ .$$

Let us now consider the $\mathbb{Z}$-group PGSp_4, with dual group the $\mathbb{C}$-group Sp_4 (which is also Spin_5). Let W be the representation $\mathrm{Sym}^j(\mathbb{C}^2)\otimes\det^k$ of $\mathrm{GL}_2(\mathbb{C})$ with $k \geq 3$, let $F \in \mathrm{S}_W(\mathrm{Sp}_4(\mathbb{Z}))$ be a (nonzero) eigenform, and let $\pi \in \Pi_{\mathrm{cusp}}(\mathrm{PGSp}_4)$ be the representation generated by F. We already determined $\mathrm{c}_\infty(\pi)$ (that is, $\inf_{\pi_W}$) in terms of j and k in Sect. 6.3.4. For a prime p, the element $\mathrm{c}_p(\pi) \in \mathrm{Sp}_4(\mathbb{C})_{\mathrm{ss}}$ is uniquely characterized by its trace and that of the second exterior power of the tautological representation $V_{\mathrm{Spin}} \simeq \mathbb{C}^4$ of $\mathrm{Sp}_4(\mathbb{C})$. If $\mathrm{K}_p(F) = a_pF$ and $\mathrm{T}_p(F) = b_pF$, the relations (6.2.8) show

$$p^{3/2}\,\mathrm{Trace}(\mathrm{c}_p(\pi)\,|\,V_{\mathrm{Spin}}) = a_p \quad \text{and} \quad p^2\,\mathrm{Trace}(\mathrm{c}_p(\pi)\,|\,\Lambda^2 V_{\mathrm{Spin}}) = b_p + p^2 + 1\ .$$

For a general $g \geq 1$, we consider an eigenform $F \in \mathrm{S}_W(\mathrm{Sp}_{2g}(\mathbb{Z}))$ such that $\mathrm{T}_p(F) = b_pF$. If $\pi \in \Pi_{\mathrm{cusp}}(\mathrm{Sp}_{2g})$ denotes the representation generated by F, then $\mathrm{c}_\infty(\pi) \subset \mathfrak{so}_{2g+1}(\mathbb{C})_{\mathrm{ss}}$ is given in terms of W by Proposition 6.3.6. If $V_{\mathrm{St}} \simeq \mathbb{C}^{2g+1}$ denotes the tautological representation of $\mathrm{SO}_{2g+1}(\mathbb{C})$, then for every prime p,

$$p^g\,\mathrm{Trace}(\mathrm{c}_p(\pi)\,|\,V_{\mathrm{st}}) = b_p + 1\ .$$

Now, take $n \equiv 0 \bmod 8$ and $G = \mathrm{SO}_n$, the special orthogonal group of E_n (Sect. 4.4.4), so that we have $\widehat{G}(\mathbb{C}) = \mathrm{SO}_n(\mathbb{C})$. Let W be the irreducible representation of highest weight $\sum_{i=1}^{n/2} m_i\varepsilon_i$, with $m_1 \geq \cdots m_{n/2-1} \geq |m_{n/2}|$ in the notation of Sect. 6.1.3. Let $F \in \mathrm{M}_W(\mathrm{SO}_n)$ be an eigenform and $\pi \in \Pi_{\mathrm{disc}}(\mathrm{SO}_n)$ the representation it generates. By definition, we have $\pi_\infty \simeq W^*$, but $W^* \simeq W$ holds because $n \equiv 0 \bmod 4$, so that the n eigenvalues of $\mathrm{c}_\infty(\pi) \in \mathfrak{so}_n(\mathbb{C})_{\mathrm{ss}}$ are

$$\pm\left(m_i + \frac{n}{2} - i\right) \ \text{for}\ i = 1, \cdots, \frac{n}{2}$$

by Sect. 6.3.4. Let p be a prime. Suppose $\mathrm{T}_p(F) = \lambda_pF$, $\mathrm{T}_{p^2}(F) = \lambda_{p^2}F$, and $\mathrm{T}_{p,p}(F) = \lambda_{p,p}F$ (Sect. 6.2.8). The relations (6.2.5) and (6.2.6), and that of Example 6.2.11, can then be written as follows:

$$p^{n/2-1}\,\mathrm{Trace}(\mathrm{c}_p(\pi)\,|\,V_{\mathrm{St}}) = \lambda_p\ ,$$

$$p^{n-3}\,\mathrm{Trace}(\mathrm{c}_p(\pi)\,|\,\Lambda^2 V_{\mathrm{St}}) = \lambda_{p,p} + p^{(n/2)-2} + \frac{p^{n-2}-1}{p^2-1}\ ,$$

$$(p+1)\lambda_{p,p} = \lambda_p^2 - \lambda_{p^2} - \frac{(p^{n/2}-1)(p^{(n/2)-1}+1)}{p-1}\ .$$

6.4.4 The Arthur–Langlands Conjecture

Let G be a semisimple $\mathbb{Z}$-group and $r\colon \widehat{G} \to \mathrm{SL}_n$ a $\mathbb{C}$-representation. This representation induces a map $\mathcal{X}(\widehat{G}) \to \mathcal{X}(\mathrm{SL}_n)$ defined by $(c_v) \mapsto (r(c_v))$, which we also denote by r. With any $\pi \in \Pi(G)$, we associate the element

$$\psi(\pi, r) := r(\mathrm{c}(\pi)) \in \mathcal{X}(\mathrm{SL}_n)\ .$$

This element is called the *Langlands parameter of the pair* (π, r). For $\pi \in \Pi_{\mathrm{disc}}(G)$, the conjectures of Langlands [135], made more precise by Arthur [9], state that $\psi(\pi, r)$ can be expressed in terms of the $\Pi_{\mathrm{cusp}}(\mathrm{PGL}_m)$ for $m \geq 1$. Before recalling how, we need to introduce some notation.

- We denote by St_m the tautological $\mathbb{C}$-representation of SL_m over $\mathbb{C}^m$. For integers a and b, the direct sum and the tensor product of the representations St_a and St_b define $\mathbb{C}$-representations of $\mathrm{SL}_a \times \mathrm{SL}_b$ of respective dimensions $a+b$ and ab, hence also natural maps

  $$\mathcal{X}(\mathrm{SL}_a) \times \mathcal{X}(\mathrm{SL}_b) \to \mathcal{X}(\mathrm{SL}_{a+b}) \quad \text{and} \quad \mathcal{X}(\mathrm{SL}_a) \times \mathcal{X}(\mathrm{SL}_b) \to \mathcal{X}(\mathrm{SL}_{ab})\ .$$

  We denote these maps by $(c, c') \mapsto c \oplus c'$ and $(c, c') \mapsto c \otimes c'$, respectively. These operations are commutative, associative, and distributive in the obvious sense.
- Following Arthur [9], we consider the element $e \in \mathcal{X}(\mathrm{SL}_2)$ defined by

  $$e_p = \begin{bmatrix} p^{-1/2} & 0 \\ 0 & p^{1/2} \end{bmatrix} \quad \forall p \in \mathrm{P} \quad \text{and} \quad e_\infty = \begin{bmatrix} -1/2 & 0 \\ 0 & 1/2 \end{bmatrix}.$$

  For every integer $d \geq 1$, the element e gives rise to the element $\mathrm{Sym}^{d-1}(e) \in \mathcal{X}(\mathrm{SL}_d)$, where Sym^{d-1} denotes the representation $\mathrm{Sym}^{d-1}\mathrm{St}_2$ of SL_2. We denote this new element by $[d]$; for example, we have $[2] = e$. These elements will later play a particularly important role. Let us already note that we have $[d] = \mathrm{c}(1_{\mathrm{PGL}_d})$, by Example 6.4.2. More generally, for integers $m, d \geq 1$ and $c \in \mathcal{X}(\mathrm{SL}_m)$, we set

  $$c[d] := c \otimes [d]\ .$$
- For $\pi \in \Pi_{\mathrm{cusp}}(\mathrm{PGL}_m)$, the element $\mathrm{c}(\pi) \in \mathcal{X}(\mathrm{SL}_m)$ will simply be denoted by π. This abuse of notation will, in general, be innocent because $\mathrm{c}(\pi)$ determines π by the *strong multiplicity* 1 *theorem* of Piatetski-Shapiro, Jacquet, and Shalika [112]. (Note that the injectivity of the parametrization map c is very specific to the $\mathbb{Z}$-groups PGL_m.)

Thus, if $n_1, \ldots, n_k$ and $d_1, \ldots, d_k$ are integers that are at least 1, if we have $\pi_i \in \Pi_{\mathrm{cusp}}(\mathrm{PGL}_{n_i})$ for every $i = 1, \ldots, k$, and if we set $n = \sum_{i=1}^k n_i d_i$, then we have a well-defined element

$$\pi_1[d_1] \oplus \pi_2[d_2] \oplus \cdots \oplus \pi_k[d_k] \ \in\ \mathcal{X}(\mathrm{SL}_n)\ .$$

It depends only on the multi-set $\{(\pi_i, d_i) \,;\, i = 1, \ldots, k\}$. We denote by

$$\mathcal{X}_{\mathrm{AL}}(\mathrm{SL}_n)$$

the subset of $\mathcal{X}(\mathrm{SL}_n)$ consisting of the elements of this form, for an arbitrary quadruple $(k, (n_i), (d_i), (\pi_i))$ with $n = \sum_{i=1}^{k} n_i d_i$. We have the following remarkable uniqueness result, due to Jacquet and Shalika [113] (see also [139]).

Proposition 6.4.5. *Let $k, l \geq 1$ be integers. For $1 \leq i \leq k$ (resp. $1 \leq j \leq l$), consider integers $n_i, d_i \geq 1$ (resp. $n'_j, d'_j \geq 1$) and a representation π_i (resp. π'_j) in $\Pi_{\mathrm{cusp}}(\mathrm{PGL}_{n_i})$ (resp. $\Pi_{\mathrm{cusp}}(\mathrm{PGL}_{n'_j})$). Suppose that we have $n := \sum_i n_i d_i = \sum_j n'_j d'_j$ and*

$$\oplus_{i=1}^{k} \pi_i[d_i] = \oplus_{j=1}^{l} \pi'_j[d'_j]$$

in $\mathcal{X}(\mathrm{SL}_n)$. Then $k = l$ and there exists a $\sigma \in \mathfrak{S}_k$ such that for every $1 \leq i \leq k$, we have $(n'_i, \pi'_i, d'_i) = (n_{\sigma(i)}, \pi_{\sigma(i)}, d_{\sigma(i)})$.

The particular case of the conjectures of Arthur and Langlands that we wish to highlight is the following.

Conjecture 6.4.6 (Langlands [135], Arthur [9]). Let G be a semisimple $\mathbb{Z}$-group and $r \colon \widehat{G} \to \mathrm{SL}_n$ a $\mathbb{C}$-representation. For $\pi \in \Pi_{\mathrm{disc}}(G)$, we have $\psi(\pi, r) \in \mathcal{X}_{\mathrm{AL}}(\mathrm{SL}_n)$.

In other words, for every $\pi \in \Pi_{\mathrm{disc}}(G)$, there exist an integer $k \geq 1$, integers $n_1, \ldots, n_k$, $d_1, \ldots, d_k$, and representations $\pi_i \in \Pi_{\mathrm{cusp}}(\mathrm{PGL}_{n_i})$ for every $1 \leq i \leq k$, such that we have $\psi(\pi, r) = \oplus_{i=1}^{k} \pi_i[d_i]$ (and this decomposition is unique up to permutation of the factors, by Proposition 6.4.5). Concretely, this says that for every $v \in \mathrm{P} \cup \{\infty\}$, the n eigenvalues of the semisimple conjugacy class $\rho(\mathrm{c}_v(\pi))$ are the $\lambda\, p^{\mu}$ for $v \in \mathrm{P}$ (resp. $\lambda + \mu$ for $v = \infty$), where

- λ runs through the eigenvalues of $\mathrm{St}_{r_i}(\mathrm{c}_v(\pi_i))$, counted with multiplicities;
- μ takes on the values $(1 - d_i)/2$, $(3 - d_i)/2$, $\ldots$, $(d_i - 3)/2$, $(d_i - 1)/2$; and
- i runs through $\{1, \ldots, k\}$.

We refer to the preface for another point of view on this conjecture, where it is motivated by the existence of a conjectural group with wondrous properties (the *Langlands group* of $\mathbb{Z}$). Let us add that given the group G and r, the philosophy of Langlands and Arthur also suggests a description of the image of $\pi \mapsto \psi(\pi, r)$, which is much more difficult to formulate in general; one ingredient is the *Arthur–Langlands multiplicity formula* already encountered in the preface. In the examples that follow, we will discuss only much simpler cases where G is either PGL_n or a classical group, and where r is its "tautological" representation.

6.4.7 A Few Examples

THE CASE OF PGL_n

For $\pi \in \Pi_{\mathrm{cusp}}(\mathrm{PGL}_n)$, we tautologically have $\mathrm{c}(\pi) = \psi(\pi, \mathrm{St}_n) = \pi$. For a divisor d of n and $\varpi \in \Pi_{\mathrm{cusp}}(\mathrm{PGL}_{n/d})$, using residues of Eisenstein series, Speh constructed a $\pi \in \Pi_{\mathrm{disc}}(\mathrm{PGL}_n)$ such that we have $\psi(\pi, \mathrm{St}_n) = \varpi[d]$. The conjecture of Jacquet, proved by Moeglin and Waldspurger [151], asserts that every $\pi \in \Pi_{\mathrm{disc}}(\mathrm{PGL}_n)$ is of this form. This proves Conjecture 6.4.6 for $G = \mathrm{PGL}_n$ and $r = \mathrm{St}_n$.

Another famous and well-known case of Conjecture 6.4.6 concerns the symmetric square representation $\mathrm{Sym}^2 \colon \widehat{\mathrm{PGL}_2} = \mathrm{SL}_2 \to \mathrm{SL}_3$. More precisely, Gelbart and Jacquet have proved that for $\pi \in \Pi_{\mathrm{cusp}}(\mathrm{PGL}_2)$, there exists a unique $\pi' \in \Pi_{\mathrm{cusp}}(\mathrm{PGL}_3)$ such that we have $\psi(\pi, \mathrm{Sym}^2) = \pi'$ [90]. By abuse of notation, we write $\pi' = \mathrm{Sym}^2 \pi$. For example, if π is generated by an eigenform $F \in \mathrm{S}_k(\mathrm{SL}_2(\mathbb{Z}))$ as in Sect. 6.4.1, then $\mathrm{c}_\infty(\mathrm{Sym}^2 \pi)$ has eigenvalues 0 and $\pm(k-1)$, and $\mathrm{c}_p(\mathrm{Sym}^2 \pi) \in \mathrm{SL}_3(\mathbb{C})_{\mathrm{ss}}$ has characteristic polynomial

$$X^3 - (p^{1-k} a_p^2 - 1)(X^2 - X) + 1 .$$

CLASSICAL GROUPS

The *Arthur classification* [13], to which we will come back next chapter, proves Conjecture 6.4.6 when G is a *classical group* (in the slightly restrictive sense defined below) and r is the *standard representation* of $\widehat{G}$; Arthur also describes the image of $\pi \mapsto \psi(\pi, r)$. For the moment, we restrict ourselves to explaining the terminology written in italics.

We denote by $\mathrm{Class}_{\mathbb{C}}$ the set consisting of the $\mathbb{C}$-groups Sp_{2g} for integers $g \geq 1$ and the $\mathbb{C}$-groups SO_m for integers $m \neq 2$. For example, PGSp_{2g} is not isomorphic to a group in $\mathrm{Class}_{\mathbb{C}}$ if $g > 2$. The reader will note that the $\mathbb{C}$-groups in the two families mentioned above are pairwise nonisomorphic. Every $\mathbb{C}$-group in $\mathrm{Class}_{\mathbb{C}}$ has a tautological distinguished $\mathbb{C}$-representation, over $\mathbb{C}^{2g}$ for Sp_{2g} and over $\mathbb{C}^m$ for SO_m, called the *standard* representation and denoted by St. It is irreducible and faithful, and of minimal dimension for these properties.

The semisimple $\mathbb{Z}$-groups G to which the work of Arthur mentioned above applies are those such that $G_{\mathbb{C}}$ is isomorphic to an element of $\mathrm{Class}_{\mathbb{C}}$, in which case the same holds for $\widehat{G}$ (Sect. 6.1.3). These include the $\mathbb{Z}$-group Sp_{2g} and the $\mathbb{Z}$-group SO_L, where L is a q-module over $\mathbb{Z}$ of dimension greater than 2 (Sect. 2.1). They also include the special orthogonal $\mathbb{Z}$-groups of odd rank studied in Appendix B (see Proposition B.1.7). We can prove that, up to isomorphism, there are no other such groups [96].

Definition 6.4.8. Let G be a semisimple $\mathbb{Z}$-group such that $G_{\mathbb{C}}$ is isomorphic to an element of $\mathrm{Class}_{\mathbb{C}}$, and let $\mathrm{St} : \widehat{G} \to \mathrm{SL}_n$ be the standard representation of $\widehat{G}$. For any π in $\Pi(G)$, the *standard parameter* of π is the element $\psi(\pi, \mathrm{St})$ of $\mathcal{X}(\mathrm{SL}_n)$.

The Case of the $\mathbb{Z}$-groups O_n and PGO_n

Consider the orthogonal $\mathbb{Z}$-group $G' = \mathrm{O}_n$ of the q-module E_n (Sect. 4.4.4). Since it is not connected, we cannot apply the arguments of Sect. 6.4.1 to it verbatim. However, we already observed in Corollary 6.2.4 that for every prime p, the Satake isomorphism of $G = \mathrm{SO}_n$ induces a bijection

$$\mathrm{Hom}_{\mathrm{ring}}(\mathrm{H}_p(\mathrm{O}_n), \mathbb{C}) \xrightarrow{\sim} \mathrm{O}_n(\mathbb{C})_{\mathrm{ss}} \ ;$$

let us stress that the right-hand side denotes the set of $\mathrm{O}_n(\mathbb{C})$-conjugacy classes of semisimple elements of $\mathrm{SO}_n(\mathbb{C})$ (Sect. 6.4.1). Likewise, we easily verify that if V' is an irreducible representation of $G'(\mathbb{R})$, its restriction V to $G(\mathbb{R})$ is either irreducible or the sum of two nonisomorphic representations that are outer conjugates under the action of $\mathrm{O}_n(\mathbb{R})$. The $\mathrm{O}_n(\mathbb{C})$-orbit of the element of $(\mathfrak{so}_n)_{\mathrm{ss}}$ associated with the infinitesimal character of each of the components of V is therefore independent of the chosen component; by abuse of language, we will call it the infinitesimal character of V'. We have thus defined a parametrization map

$$\mathrm{c} \colon \Pi(\mathrm{O}_n) \longrightarrow \mathcal{X}(\mathrm{O}_n(\mathbb{C})) \ ,$$

given by $\pi \mapsto (\mathrm{c}_v(\pi))$. It is therefore natural to set

$$\mathcal{X}(\widehat{\mathrm{O}_n}) := \mathcal{X}(\mathrm{O}_n(\mathbb{C})) \ .$$

By Sect. 6.4.1, every $\mathbb{C}$-representation $r \colon \mathrm{O}_n(\mathbb{C}) \to \mathrm{GL}_n(\mathbb{C})$ induces a map

$$r \colon \mathcal{X}(\widehat{\mathrm{O}_n}) \to \mathcal{X}(\mathrm{GL}_n) \ ,$$

so that the element $\psi(\pi, r) := r(\mathrm{c}(\pi))$ of $\mathcal{X}(\mathrm{GL}_n)$ is well defined. This element is actually in the subset $\mathcal{X}(\mathrm{SL}_n)$ of $\mathcal{X}(\mathrm{GL}_n)$. This construction applies, in particular, to the *standard* (tautological) representation $\mathrm{St} \colon \mathrm{O}_n(\mathbb{C}) \to \mathrm{GL}_n(\mathbb{C})$, and we have the following definition, which is parallel to Definition 6.4.8.

Definition 6.4.9. For any π in $\Pi(\mathrm{O}_n)$, the *standard parameter* of π is the element $\psi(\pi, \mathrm{St})$ of $\mathcal{X}(\mathrm{SL}_n)$.

For O_n, we have a conjecture analogous to Conjecture 6.4.6; it actually follows from the latter applied to the $\mathbb{Z}$-group SO_n; let us explain how. Consider $\pi' \in \Pi_{\mathrm{disc}}(\mathrm{O}_n)$, $V' = (\pi')^*_\infty$, and $V = V'_{|\mathrm{SO}_n(\mathbb{R})}$. By Sect. 4.4.4, there is a natural $\mathrm{H}(\mathrm{O}_n)$-equivariant injection

$$\mathrm{res} \colon \mathrm{M}_{V'}(\mathrm{O}_n) \longrightarrow \mathrm{M}_V(\mathrm{SO}_n) \ .$$

Let $\pi \in \Pi_{\rm disc}(\mathrm{SO}_n)$ be the representation generated by an arbitrary eigenform belonging to the $\mathrm{H}(\mathrm{SO}_n)$-module generated by $\mathrm{res}(F)$, where $F \in \mathrm{M}_{V'}(\mathrm{O}_n)$ is an arbitrary eigenform generating π' (Sect. 4.3.2). The resulting representations $\pi \in \Pi_{\rm disc}(\mathrm{SO}_n)$ will be called *the components of the restriction of* π' *to* SO_n; they form a nonempty finite set. If π is such a component, then by definition, $\mathrm{c}(\pi')$ is the image of $\mathrm{c}(\pi)$ by the canonical homomorphism $\mathcal{X}(\widehat{\mathrm{SO}_n}) \to \mathcal{X}(\widehat{\mathrm{O}_n})$. The following proposition is therefore obvious.

Proposition 6.4.10. *Let* $\pi' \in \Pi_{\rm disc}(\mathrm{O}_n)$*, let* $\pi \in \Pi_{\rm disc}(\mathrm{SO}_n)$ *be a component of the restriction of* π' *to* SO_n*, let* $r' \colon \mathrm{O}_n(\mathbb{C}) \to \mathrm{GL}_m(\mathbb{C})$ *be a* $\mathbb{C}$*-representation, and let* r *be the restriction of* r' *to* $\mathrm{SO}_n(\mathbb{C})$*. Then* $\psi(\pi', r') = \psi(\pi, r)$*. In particular, the Arthur–Langlands conjecture is true for* (π, r) *if and only if it is for* (π', r)*.*

This proposition is especially useful in the case of the tautological representation St of $\mathrm{O}_n(\mathbb{C})$ on $\mathbb{C}^n$, whose restriction to $\mathrm{SO}_n(\mathbb{C})$ is the latter's standard representation.

Finally, the discussion above admits a natural analog for the $\mathbb{Z}$-group PGO_n, for which we have a parametrization map

$$\mathrm{c} \colon \Pi(\mathrm{PGO}_n) \to \mathcal{X}(\widehat{\mathrm{PGO}_n}) := \mathcal{X}(\mathrm{Pin}_n) \ ,$$

where Pin_n is the $\mathbb{C}$-group Pin of the standard q-vector space V_n of dimension n over $\mathbb{C}$. Following [15], we recall that this is the subgroup of elements x of the Clifford algebra $\mathrm{C}(V_n)$ of V_n such that $xx^t = 1$ and $\alpha(x)V_n x^{-1} \subset V_n$, where $x \mapsto \alpha(x)$ and $x \mapsto x^t$ denote the canonical involution and anti-involution of $\mathrm{C}(V_n)$, respectively. Its neutral component, defined by $\alpha = \mathrm{id}$, is the $\mathbb{C}$-group Spin_n; it has index 2. Every element $e \in V_n$ such that $\mathrm{q}(e) = 1$ (that is, $e^2 = 1$) belongs to $\mathrm{Pin}(V_n) = \mathrm{Pin}_n(\mathbb{C})$ and defines a section of the canonical morphism $\mathrm{Pin}_n \to \mathbb{Z}/2\mathbb{Z}$. Finally, we have a natural surjective morphism $\mathrm{Pin}_n \to \mathrm{O}_{V_n}$ given by $x \mapsto (v \mapsto \alpha(x)vx^{-1})$. Its kernel is ± 1, and the image of any element $e \in V_n$ such that $\mathrm{q}(e) = 1$ is the orthogonal reflection with respect to e.

The Case of the Trivial Representation

Let us now suppose that G is an arbitrary semisimple $\mathbb{Z}$-group and consider the trivial representation $1 \in \Pi_{\rm disc}(G)$. Let $\mu \colon \mathrm{SL}_2 \to \widehat{G}$ be a principal $\mathbb{C}$-morphism in the sense of Kostant. An equivalent formulation of the description of $\mathrm{c}(1)$ given in Example 6.4.2 is

$$\mu(e) = \mathrm{c}(1)$$

(this observation is given explicitly in [97, Sect. 7], but it is undoubtedly older). In particular, if $r \colon \widehat{G} \to \mathrm{SL}_n$ is an arbitrary $\mathbb{C}$-morphism and we decompose the representation $r \circ \mu$ of SL_2 as $\oplus_{i=1}^k \mathrm{Sym}^{d_i - 1}\, \mathbb{C}^2$ with $d_1, \ldots, d_k \geq 1$ integers, we therefore have

$$\psi(1, r) = \oplus_{i=1}^k [d_i] \ .$$

This agrees with the Arthur–Langlands conjecture (as it happens, with the conjectures of Arthur [9]), which is therefore true for the pair $(1, r)$ for any representation r.

Consider, for example, $G = \mathrm{SO}_n$ with $n \equiv 0 \bmod 8$. We then have $\mathrm{St} \circ \mu = \mathrm{Sym}^{n-2}\, \mathbb{C}^2 \oplus 1$, in other words, $\psi(1, \mathrm{St}) = [n-1] \oplus [1]$. This relation also holds if $G = \mathrm{O}_n$ because the trivial representation of SO_n is clearly the restriction of the trivial representation of O_n to SO_n; they therefore have the same standard parameter by Proposition 6.4.10.

6.4.11 Relations with L*-Functions*

Let G be a semisimple $\mathbb{Z}$-group, $r \colon \widehat{G} \to \mathrm{GL}_n$ a $\mathbb{C}$-representation, and $\pi \in \Pi_{\mathrm{disc}}(G)$. By Langlands [136, Sect. 3], the Euler product

$$\mathrm{L}(s, \pi, r) = \prod_{p \in \mathrm{P}} \det\bigl(1 - p^{-s} r(\mathrm{c}_p(\pi))\bigr)^{-1}$$

is absolutely convergent for $\Re(s)$ sufficiently large; see also [178, Sect. 2.5].

For $\pi \in \Pi_{\mathrm{cusp}}(\mathrm{PGL}_n)$, we set $\mathrm{L}(s, \pi) = \mathrm{L}(s, \pi, \mathrm{St}_n)$ (recall that St_n is the tautological representation of SL_n, Sect. 6.4.4). From the work of Godement and Jacquet, we know that $\mathrm{L}(s, \pi)$ admits a holomorphic extension to all of $\mathbb{C}$, unless $n = 1$, in which case $\pi = 1_{\mathrm{PGL}_1}$ and this function $\mathrm{L}(s, \pi)$ is none other than the Riemann $\zeta(s)$-function. By Jacquet and Shalika [112], the Euler product $\mathrm{L}(s, \pi)$ is even absolutely convergent for $\Re(s) > 1$. If the Arthur–Langlands conjecture holds for π and r, then we may write $\psi(\pi, r) = \oplus_{i=1}^{k} \pi_i[d_i]$ with $\pi_i \in \Pi_{\mathrm{cusp}}(\mathrm{PGL}_{n_i})$ for each i, and it follows from the definitions that we have

$$\mathrm{L}(s, \pi, r) = \prod_{i=1}^{k} \prod_{j=0}^{d_i - 1} \mathrm{L}\Bigl(s + j + \frac{1 - d_i}{2}, \pi_i\Bigr)\,,$$

and therefore $\mathrm{L}(s, \pi, r)$ also admits a meromorphic extension to all of $\mathbb{C}$, whose poles are explained by the appearance of the trivial representation in $\psi(\pi, r)$.

The reader should be aware that the normalizations used here make $s = \frac{1}{2}$ into the natural center of the functional equations that are involved. Suppose, for example, that $\pi \in \Pi_{\mathrm{cusp}}(\mathrm{PGL}_2)$ denotes the representation associated with a normalized eigenform $F = \sum_{n \geq 0} a_n q^n$ of weight k, as in Sect. 6.4.3. We then have $\mathrm{L}(s - (k-1)/2, \pi) = \sum_{n \geq 1} a_n / n^s$ for $\Re(s) > 1$, as well as the Hecke relation

$$(2\pi)^{-s} \Gamma(s) \mathrm{L}\Bigl(s - \frac{k-1}{2}, \pi\Bigr) = \int_0^{\infty} F(it) t^s \frac{dt}{t} \quad \forall s \in \mathbb{C}\,. \tag{6.4.1}$$

6.4.12 The Generalized Ramanujan Conjecture

Let G be a semisimple $\mathbb{Z}$-group, and let $\pi \in \Pi(G)$. We say that π satisfies the Ramanujan conjecture, or that it is *tempered*, if[9] for every $p \in \mathrm{P}$, the eigenvalues of $\mathrm{c}_p(\pi)$ in some (and therefore every) faithful $\mathbb{C}$-representation of $\widehat{G}$ all have absolute value 1.

The *generalized Ramanujan conjecture* asserts that if $\pi \in \Pi_{\mathrm{cusp}}(\mathrm{PGL}_n)$, then π is tempered. The typical example of a representation that is not tempered is the trivial representation $1 \in \Pi_{\mathrm{disc}}(\mathrm{PGL}_2)$; the eigenvalues of $\mathrm{c}_p(1) = e_p$ are $p^{\pm 1/2}$. More generally, the trivial representation 1_G of G is not tempered if $G \neq 1$ (Example 6.4.2). Conjecture 6.4.6 therefore expresses, in particular, the defect of the Ramanujan conjecture in general.

The generalized Ramanujan conjecture is still open, even for $G = \mathrm{PGL}_2$. Thanks to the work of many authors, it is, however, known for the important class of representations $\pi \in \Pi_{\mathrm{cusp}}(\mathrm{PGL}_n)$ called *polarized regular algebraic* (see Sect. 8.2.16). In general, we do have the Jacquet–Shalika estimate [112]: for every $\pi \in \Pi_{\mathrm{cusp}}(\mathrm{PGL}_n)$, every $p \in \mathrm{P}$, and every eigenvalue λ of $\mathrm{c}_p(\pi)$, we have $p^{-1/2} < |\lambda| < p^{1/2}$.

[9] In general, a condition that is conjecturally automatic is added on π_∞; we omit it here.

Chapter 7
A Few Cases of the Arthur–Langlands Conjecture

7.1 The Eichler Relations Revisited

In this entire chapter, g and n are fixed integers greater than or equal to 1, with $n \equiv 0 \bmod 8$. We consider the $\mathbb{Z}$-groups Sp_{2g} and O_n (recall that the latter is the orthogonal $\mathbb{Z}$-group of the lattice E_n).

7.1.1 The Point of View of Rallis

The Jacobi theta series allows us to construct a natural $\mathbb{C}$-linear map

$$\vartheta \colon \mathrm{M}_U(\mathrm{O}_n) \longrightarrow \mathrm{M}_V(\mathrm{Sp}_{2g})$$

for certain pairs (U, V), where U is an irreducible $\mathbb{C}$-representation of $\mathrm{O}_n(\mathbb{C})$ and V is an irreducible $\mathbb{C}$-representation of $\mathrm{GL}_g(\mathbb{C})$ [116], [86]. The admissible pairs (U, V) are called *compatible*; we describe them further on. Two particular cases of this construction have already played a role in this book: the pair $(1, \det^{n/2})$ in Sect. 5.1 and the pair $(\mathrm{H}_{d,g}, \det^{n/2+d})$ in Sect. 5.4 for $2g \leq n$. An important property of the map ϑ, already discussed in various cases loc. cit., is that it intertwines certain Hecke operators of O_n and Sp_{2g} ("Eichler commutation relations"). The aim of this subsection is to recall the point of view of Rallis [170] on these formulas.

Set $\mathfrak{z}_{\mathrm{Sp}_{2g}} = \mathrm{Z}(\mathrm{U}(\mathfrak{sp}_{2g}(\mathbb{C})))$ and $\mathfrak{z}_{\mathrm{O}_n} = \mathrm{Z}(\mathrm{U}(\mathfrak{so}_n(\mathbb{C})))^{\mathrm{O}_n(\mathbb{C})}$. In [170], Rallis constructs a surjective morphism of $\mathbb{C}$-algebras

$$\mathrm{Ral}\colon \begin{cases} \mathrm{H}(\mathrm{O}_n) \otimes \mathfrak{z}_{\mathrm{O}_n} \to \mathrm{H}(\mathrm{Sp}_{2g}) \otimes \mathfrak{z}_{\mathrm{Sp}_{2g}} & \text{if } n > 2g\,, \\ \mathrm{H}(\mathrm{Sp}_{2g}) \otimes \mathfrak{z}_{\mathrm{Sp}_{2g}} \to \mathrm{H}(\mathrm{O}_n) \otimes \mathfrak{z}_{\mathrm{O}_n} & \text{if } n \leq 2g \end{cases}$$

G. Chenevier, J. Lannes, *Automorphic Forms and Even Unimodular Lattices*, Ergebnisse der Mathematik und ihrer Grenzgebiete. 3. Folge / A Series of Modern Surveys in Mathematics 69, https://doi.org/10.1007/978-3-319-95891-0_7

that respects the subrings $\mathrm{H}_p(*) \otimes \mathbb{C}$ (for p prime) and $\mathfrak{z}_*$ on either side. It has the following properties:

(i) If (U, V) is compatible, then $\mathrm{Inf}_V \circ \mathrm{Ral}_{|\mathfrak{z}_{\mathrm{O}_n}} = \mathrm{Inf}_U$ if $n > 2g$ and $\mathrm{Inf}_U \circ \mathrm{Ral}_{|\mathfrak{z}_{\mathrm{Sp}_{2g}}} = \mathrm{Inf}_V$ otherwise.
(ii) (Eichler–Rallis commutation relations) We have $\vartheta \circ T = \mathrm{Ral}(T) \circ \vartheta$ for every $T \in \mathrm{H}(\mathrm{O}_n)$ if $n > 2g$, and $\vartheta \circ \mathrm{Ral}(T) = T \circ \vartheta$ for every $T \in \mathrm{H}(\mathrm{Sp}_{2g})$ otherwise.

Let us add that if we assume $n > 2g$, and if the necessary condition for the admissibility of a pair (U, V) given in part (i) holds, then exactly one of the pairs (U, V) and $(U \otimes \det, V)$ is compatible; see [116] §6 for the precise condition in general.

Finally, Rallis gives an interpretation of the morphism Ral in terms of the Satake and Harish-Chandra isomorphisms of O_n and Sp_{2g}, which we will also recall. For $a \geq 1$, we denote by O_a (resp. SO_a) the standard orthogonal (resp. special orthogonal) $\mathbb{C}$-group in a variables. This notation conflicts, a priori, with that of the $\mathbb{Z}$-groups O_n and SO_n, defined only for $n \equiv 0 \bmod 8$, but this is irrelevant because when the symbols coincide, they denote the same object over $\mathbb{C}$. The group $\mathrm{O}_a(\mathbb{C})$ acts by conjugation on $\mathcal{X}(\mathrm{SO}_a)$, and this action is nontrivial if a is even. If $a < b$ are integers greater than or equal to 1 with $a \not\equiv b \bmod 2$, there exists a $\mathbb{C}$-morphism

$$\rho_{a,b} \colon \mathrm{O}_a \times \mathrm{SL}_2 \longrightarrow \mathrm{O}_b \ ,$$

uniquely determined modulo conjugation by $\mathrm{O}_b(\mathbb{C})$ at the target, such that the representation $\mathrm{St} \circ \rho_{a,b}$ is isomorphic to the direct sum of the standard representation of the factor O_a and the representation $\mathrm{Sym}^{b-a-1}\mathrm{St}_2$ of SL_2. The respective Langlands duals of SO_n and Sp_{2g} are the $\mathbb{C}$-groups SO_n and SO_{2g+1}. The discussion in [170, Sect. 6] translates into the following statement (recall that $e \in \mathcal{X}(\mathrm{SL}_2)$ denotes the Arthur element defined in Sect. 6.4.4). Observe that for $G = \mathrm{O}_n$ or $G = \mathrm{Sp}_{2g}$, the Satake isomorphism and the Harish-Chandra isomorphism (Sects. 6.2 and 6.3) allow the identification of $\mathrm{Hom}_{\mathbb{C}-\mathrm{alg}}(\mathrm{H}(G) \otimes \mathfrak{z}_G, \mathbb{C})$ with $\mathcal{X}(\widehat{G})$ (Sect. 6.4.7).

Proposition 7.1.2. (i) *For $n > 2g$, the map $\mathcal{X}(\widehat{\mathrm{Sp}_{2g}}) \to \mathcal{X}(\widehat{\mathrm{O}_n})$ induced by* Ral *is given by $x \mapsto \rho_{2g+1,n}(x, e)$.*
(ii) *For $n \leq 2g$, the map $\mathcal{X}(\widehat{\mathrm{O}_n}) \to \mathcal{X}(\widehat{\mathrm{Sp}_{2g}})$ induced by* Ral *is given by $x \mapsto \rho_{n,2g+1}(x, e)$.*

This result, combined with the diagonalizability of $\mathrm{H}(\mathrm{O}_n)$ over the spaces $\mathrm{M}_U(\mathrm{O}_n)$, reduces the study of the Eichler commutation relations to that of properties of the Satake isomorphisms of SO_n and Sp_{2g}. For example, Proposition 5.1.1 and the relation (5.4.4) immediately follow from the formulas (6.2.5) and (6.2.8).

Likewise, the archimedean part of Proposition 7.1.2 substantially clarifies the significance of the numbers associated with the infinitesimal characters of the compatible pairs (U, V). Let us illustrate this on the pair $(\mathrm{H}_{d,g}, \det^{n/2+d})$. For $g < n/2$, the highest weight of $\mathrm{H}_{d,g}(\mathbb{R}^n)$ is clearly $d\sum_{i=1}^{g} \varepsilon_i$, and that of $\mathrm{H}_{d,n/2}(\mathbb{R}^n)^{\pm}$

is $d(\pm\varepsilon_{n/2} + \sum_{i=1}^{n/2-1} \varepsilon_i)$ (see Sects. 5.4.14 and 6.4.3), which gives an infinitesimal character in $\mathfrak{so}_n(\mathbb{C})$ with eigenvalues $\pm(d + n/2 - i)$ for $i = 1, \ldots, g$ and $\pm(n/2 - i)$ for $i = g+1, \ldots, n/2$ if $g < n/2$. On the other hand, the representation $\pi'_{\det^k}$ (Sect. 6.3.4) has an infinitesimal character in $\mathfrak{so}_{2g+1}(\mathbb{C})$ with eigenvalues 0 and $\pm(k-i)$ for $i = 1, \ldots, g$, which is indeed compatible with Proposition 7.1.2.

We will say that the representations $\pi_{\mathrm{O}} \in \Pi_{\mathrm{disc}}(\mathrm{O}_n)$ and $\pi_{\mathrm{Sp}} \in \Pi_{\mathrm{disc}}(\mathrm{Sp}_{2g})$ are ϑ-correspondent if there exist a compatible pair (U, V) and eigenforms $F \in \mathrm{M}_U(\mathrm{O}_n)$ and $G \in \mathrm{S}_V(\mathrm{Sp}_{2g})$ respectively generating π_{O} and π_{Sp}, such that $\vartheta(F) = G$.

Corollary 7.1.3. *Suppose that $\pi_{\mathrm{O}} \in \Pi_{\mathrm{disc}}(\mathrm{O}_n)$ and $\pi_{\mathrm{Sp}} \in \Pi_{\mathrm{disc}}(\mathrm{Sp}_{2g})$ are ϑ-correspondent. Then we have $\psi(\pi_{\mathrm{O}}, \mathrm{St}) = \psi(\pi_{\mathrm{Sp}}, \mathrm{St}) \oplus [n - 2g - 1]$ if $n > 2g$ and $\psi(\pi_{\mathrm{Sp}}, \mathrm{St}) = \psi(\pi_{\mathrm{O}}, \mathrm{St}) \oplus [2g + 1 - n]$ otherwise.*

7.1.4 A Refinement: Passage to the Spin *Groups*

In this subsection, we discuss a refinement of Proposition 7.1.2, probably well known to the specialists, for which we have not found a reference in the literature (it is, however, implicit in [169] in the case $g = 1$). Assume that the pair (U, V) is compatible. Note that if $\mathrm{M}_U(\mathrm{O}_n) \neq 0$, then $-1 \in \mathrm{O}_n(\mathbb{Z})$ acts trivially in U, which we assume from now on. This implies that U factors into a representation U' of $\mathrm{PGO}_n(\mathbb{C})$, and Lemma 5.4.8 provides a natural $\mathrm{H}(\mathrm{O}_n)$-equivariant isomorphism

$$\mathrm{M}_{U'}(\mathrm{PGO}_n) \xrightarrow{\sim} \mathrm{M}_U(\mathrm{O}_n) .$$

In other words, $\mathrm{M}_U(\mathrm{O}_n)$ is naturally endowed with an action of the largest Hecke ring $\mathrm{H}(\mathrm{PGO}_n)$. Likewise, $\mathrm{M}_V(\mathrm{Sp}_{2g}(\mathbb{Z}))$ is endowed with an action of $\mathrm{H}(\mathrm{PGSp}_{2g})$, and we can ask ourselves how Rallis' statement extends to these operators.

More precisely, let $F \in \mathrm{M}_U(\mathrm{PGO}_n)$ be an eigenform for $\mathrm{H}(\mathrm{PGO}_n)$ with $\vartheta(F) \neq 0$. Rallis' relations ensure that $\vartheta(F)$ is an eigenform for $\mathrm{H}(\mathrm{Sp}_{2g})$. By Sect. 6.2.8, for every prime p, the ring $\mathrm{H}_p(\mathrm{PGSp}_{2g})$ (resp. $\mathrm{H}_p(\mathrm{PGO}_n)$) is generated by $\mathrm{H}(\mathrm{Sp}_{2g})$ (resp. $\mathrm{H}(\mathrm{O}_n)$) and the corresponding perestroika operator K_p. But we have an additional Eichler relation, in fact the simplest one of all [87, Theorem 4.5], which takes on the following form when $n > 2g$:

$$\vartheta \circ \mathrm{K}_p = p^{\frac{g(n/2-g-1)}{2}} \left[\prod_{i=0}^{n/2-g+1} (p^i + 1) \right] \mathrm{K}_p \circ \vartheta . \tag{7.1.1}$$

The case of this formula we will need in the application to Theorem 7.2.1 is that of Lemma 5.4.9 (compare with formula (4.5.5)). This shows that $\vartheta(F)$ is an eigenform for $\mathrm{H}(\mathrm{PGSp}_{2g})$. We will say that $\pi_{\mathrm{PGO}} \in \Pi_{\mathrm{disc}}(\mathrm{PGO}_n)$ and $\pi_{\mathrm{PGSp}} \in \Pi_{\mathrm{disc}}(\mathrm{PGSp}_{2g})$ are ϑ-correspondent if there exist a compatible pair (U, V)

and eigenforms $F \in \mathrm{M}_U(\mathrm{PGO}_n)$ and $G \in \mathrm{S}_V(\mathrm{Sp}_{2g}(\mathbb{Z}))$, generating π_{PGO} and π_{PGSp}, respectively, such that $\vartheta(F) = G$.

If $n > 2g$, then the $\mathbb{C}$-morphism $\rho_{2g+1,n} \colon \mathrm{SO}_{2g+1} \times \mathrm{SL}_2 \to \mathrm{SO}_n$ lifts to a $\mathbb{C}$-morphism $\widetilde{\rho}_{2g+1,n} \colon \mathrm{Spin}_{2g+1} \times \mathrm{SL}_2 \to \mathrm{Spin}_n$. Likewise, if $n \leq 2g$, then the $\mathbb{C}$-morphism $\rho_{n,2g+1}$ lifts to a morphism $\widetilde{\rho}_{n,2g+1} \colon \mathrm{Pin}_n \times \mathrm{SL}_2 \to \mathrm{Pin}_{2g+1}$ (in Sect. 6.4.7, we recall some results concerning the group Pin).

Proposition 7.1.5. *Let* $\pi_{\mathrm{PGO}} \in \Pi_{\mathrm{disc}}(\mathrm{PGO}_n)$ *and* $\pi_{\mathrm{PGSp}} \in \Pi_{\mathrm{disc}}(\mathrm{PGSp}_{2g})$ *be* ϑ*-correspondent representations.*

(i) *If* $n > 2g$, *then* $\mathrm{c}(\pi_{\mathrm{PGO}})$ *is the image of* $\widetilde{\rho}_{2g+1,n}(\mathrm{c}(\pi_{\mathrm{PGSp}}), e)$ *by the natural map* $\mathcal{X}(\mathrm{Spin}_n) \to \mathcal{X}(\mathrm{Pin}_n)$.
(ii) *If* $n \leq 2g$, *then* $\mathrm{c}(\pi_{\mathrm{PGSp}}) = \widetilde{\rho}_{n,2g+1}(\mathrm{c}(\pi_{\mathrm{PGO}}), e)$.

Proof. Suppose $n > 2g$. The equality we want to show holds after projection into $\mathcal{X}(\widehat{\mathrm{O}_n})$ by Rallis, Proposition 7.1.2 (i). By an observation made above, it remains to verify the equality after applying the Hecke operator $\mathrm{K}_p \in \mathrm{H}(\mathrm{PGO}_n)$, viewed via the Satake isomorphism as a function on $\mathrm{Spin}_n(\mathbb{C})_{\mathrm{ss}}$, invariant under the action of $\mathrm{Pin}_n(\mathbb{C})$. By formula (6.2.5), we have

$$\mathrm{Sat}(\mathrm{K}_p) = p^{\frac{n/2(n/2-1)}{4}}\left([V_{\mathrm{Spin}}^+] + [V_{\mathrm{Spin}}^-]\right),$$

where $V_{\mathrm{Spin}^{\pm}}$ are the two spin representations of $\mathrm{Spin}_n(\mathbb{C})$ (conjugate to each other under $\mathrm{Pin}_n(\mathbb{C})$). But it is well known that the restriction of each of these to $\mathrm{Spin}_{2g+1}(\mathbb{C}) \times \mathrm{Spin}_{n-2g-1}(\mathbb{C})$ is isomorphic to the tensor product of the spin representations of each of the two factors. Let $r_a \colon \mathrm{SL}_2 \to \mathrm{SL}_{2^a}$ be the $\mathbb{C}$-representation obtained by lifting the irreducible representation of odd dimension $2a+1$ to Spin_{2a+1} and then composing with the spin representation of the latter. We leave it as an exercise to verify

$$\mathrm{trace}(\, r_a(e_p)\,) \;=\; \prod_{i=1}^{a} (p^{-i/2} + p^{i/2})\,.$$

We conclude with a calculation that is immediate from the formulas (6.2.5), (6.2.8), and (7.1.1). The case $n \leq 2g$ is similar. The Eichler relation shown by Freitag [87, Theorem 4.5] is

$$\mathrm{K}_p \circ \vartheta = p^{-\frac{g(n/2-g-1)}{2}} \left[\prod_{i=1}^{g-n/2} (p^{-i} + 1)\right] \vartheta \circ \mathrm{K}_p\,. \tag{7.1.2}$$

We conclude as before, by using that the restriction of the spin representation of $\mathrm{Spin}_{2g+1}(\mathbb{C})$ to $\mathrm{Spin}_n(\mathbb{C}) \times \mathrm{Spin}_{2g+1-n}(\mathbb{C})$ is the tensor product of the representation $V_{\mathrm{Spin}}^+ \oplus V_{\mathrm{Spin}}^-$ of $\mathrm{Spin}_n(\mathbb{C})$ and the spin representation of $\mathrm{Spin}_{2g+1-n}(\mathbb{C})$.

Remark 7.1.6. To finish this subsection, we note that if $n > 2g$, then an element (c_v) in the image of the composition $\mathcal{X}(\mathrm{Spin}_{2g+1}) \times \mathcal{X}(\mathrm{SL}_2) \to \mathcal{X}(\mathrm{Spin}_n) \to \mathcal{X}(\mathrm{Pin}_n)$ has the property that for every v, the $\mathrm{Pin}_n(\mathbb{C})$-conjugacy class of c_v is in fact a simple $\mathrm{Spin}_n(\mathbb{C})$-conjugacy class. One way to see this, for example for v prime, is to note that if the image γ' of $\gamma \in \mathrm{Spin}(V_n)$ in $\mathrm{SO}(V_n)$ admits the eigenvalue 1, then there exists an $e \in V_n$ with $\mathrm{q}(e) = 1$ such that we have $e\gamma = \gamma e$ (see Sect. 6.4.7 for the notation). Indeed, it suffices to choose an arbitrary e in the space $V_n^{\gamma'=1}$ (which is nondegenerate and nonzero) with $\mathrm{q}(e) = 1$. Note that we have $\alpha(\gamma)e\gamma^{-1} = \gamma'(e) = e$ and $\alpha(\gamma) = \gamma$, and therefore $\gamma e = e\gamma$.

7.2 $\Pi_{\rm disc}(\mathrm{O}_8)$ and Triality

The first part of the following result is due to Waldspurger [205]; Proposition 5.4.3 is an elementary verification of it in the particular case $k = 12$. The second part of the theorem is a form of the main idea of Sect. 5.4 that is both more precise and more conceptual. Recall that we introduced the irreducible representation $\mathrm{H}_{d,g}(\mathbb{R}^n)$ of $\mathrm{O}_n(\mathbb{R})$ in Sect. 5.4.1.

Theorem 7.2.1. *Let $\pi \in \Pi_{\rm cusp}(\mathrm{PGL}_2)$ be the representation generated by an eigenform of $\mathrm{S}_k(\mathrm{SL}_2(\mathbb{Z}))$, where $k \geq 12$ is an even integer.*

(i) *There exists a $\pi' \in \Pi_{\rm disc}(\mathrm{O}_8)$ such that $\pi'_\infty \simeq \mathrm{H}_{k-4,1}(\mathbb{R}^8)$ and*

$$\psi(\pi', \mathrm{St}) = \mathrm{Sym}^2\pi \oplus [5]\ .$$

(ii) *There exists a $\pi'' \in \Pi_{\rm disc}(\mathrm{O}_8)$ such that $\pi''_\infty \simeq \mathrm{H}_{k/2-2,4}(\mathbb{R}^8)$ and*

$$\psi(\pi'', \mathrm{St}) = \pi[4]\ .$$

Proof. Let $U = \mathrm{H}_{k-4,1}(\mathbb{R}^8)$. By Waldspurger [205, Theorem 1], we have

$$\vartheta_{k-4,1}(\mathrm{M}_U(\mathrm{O}_8)) = \mathrm{M}_k(\mathrm{SL}_2(\mathbb{Z}))\ .$$

By the Eichler commutation relations, we can therefore find an eigenform $F \in \mathrm{M}_U(\mathrm{O}_8)$ whose image $G = \vartheta_{k-4,1}(F)$ generates π. Let $\pi' \in \Pi_{\rm cusp}(\mathrm{SL}_2)$ be the representation generated by G. Consider the isogeny $i\colon \mathrm{SL}_2(\mathbb{C}) = \widehat{\mathrm{PGL}_2}(\mathbb{C}) \to \widehat{\mathrm{SL}_2}(\mathbb{C}) = \mathrm{SO}_3(\mathbb{C})$. Proposition 4.5.7 and the compatibility of the Satake isomorphism with isogenies ensure that we have $\mathrm{c}(\pi') = i(\mathrm{c}(\pi))$. But $\mathrm{St} \circ i$ is none other than the representation $\mathrm{Sym}^2\mathrm{St}_2$ of $\mathrm{SL}_2(\mathbb{C})$, so that we have $\mathrm{St}(\mathrm{c}(\pi')) = \mathrm{Sym}^2\mathrm{c}(\pi)$. Part (i) then follows from Corollary 7.1.3.

Let us verify part (ii). After modifying F if necessary, we may assume that $F_0 = F$ is an eigenform for $\mathrm{H}(\mathrm{PGO}_8)$, as in Sect. 7.1.4. Let $\pi_0 \in \Pi_{\rm disc}(\mathrm{PGO}_8)$ be the representation it generates, and let $U' = U \otimes \nu^{k/2-2}$. Let F_1 be the image of F_0 by the natural map $\mathrm{res}\colon \mathrm{M}_{U'}(\mathrm{PGO}_8) \to \mathrm{M}_{U'}(\mathrm{PGSO}_8)$ (Sect. 5.4.11). Since this map is injective and $\mathrm{H}(\mathrm{PGSO}_8)$-equivariant, by loc. cit., the form F_1 is nonzero, and if $\pi_1 \in \Pi_{\rm disc}(\mathrm{PGSO}_8)$ denotes the representation it generates, then $\mathrm{c}(\pi_0) \in \mathcal{X}(\mathrm{Pin}_8)$ is the image of $\mathrm{c}(\pi_1) \in \mathcal{X}(\mathrm{Spin}_8)$ by the natural homomorphism. This last property uniquely determines $\mathrm{c}(\pi_1)$, by Remark 7.1.6, so that Proposition 7.1.5 can be written as

$$\mathrm{c}(\pi_1) = \widetilde{\rho}_{3,8}(\mathrm{c}(\pi), e) . \tag{7.2.1}$$

Following Sect. 5.4.14, let us now consider the triality automorphism τ of PGSO_8 defined using a structure of Coxeter octonions on E_8. In particular, $(U')^\tau$ is isomorphic to the representation $V = \mathrm{H}^{\pm}_{k/2-2,4} \otimes \nu^{k-4}$ by Corollary 5.4.18. Let $\mathrm{tri}\colon \mathrm{M}_{U'}(\mathrm{PGSO}_8) \overset{\sim}{\to} \mathrm{M}_V(\mathrm{PGSO}_8)$ be the isomorphism denoted by $(\tau^*)^{-1}$ loc. cit. Finally, we set

$$F_2 = \mathrm{tri}(F_1) \ , \ F_3 = \mathrm{ind}\,(F_2) \in \mathrm{M}_{\mathrm{Ind}(V)}(\mathrm{PGO}_8) \ , \ F_4 = \mu^*(F_3) \in \mathrm{M}_{\mathrm{Ind V}}(\mathrm{O}_8)$$

(Sects. 4.4.4 and 5.4.11). These functions are nonzero eigenforms and therefore generate automorphic representations π_2, π_3, and π_4 of the $\mathbb{Z}$-groups PGSO_8, PGO_8, and O_8, respectively.

The compatibility of the Satake isomorphism with isogenies ensures that $\mathrm{c}(\pi_2)$ is the image of $\mathrm{c}(\pi_1)$ by $\tau^{\pm 1}$, and consequently that $\mathrm{c}(\pi_4)$ is the image of $\mathrm{c}(\pi_2)$ by the natural homomorphism $\eta\colon \mathrm{Spin}_8(\mathbb{C}) \to \mathrm{SO}_8(\mathbb{C})$. But it is well known that the representation $\mathrm{St} \circ \eta \circ \tau^{\pm 1}$ is none other than the representation $V_{\mathrm{Spin}^{\pm}}$ of Spin_8. We already mentioned that the restriction of $V_{\mathrm{Spin}^{\pm}}$ to $\mathrm{Spin}_3 \times \mathrm{Spin}_5$ is the tensor product of the Spin representations of $\mathrm{Spin}_3 \simeq \mathrm{SL}_2$ (of dimension 2) and $\mathrm{Spin}_5 \simeq \mathrm{Sp}_4$ (of dimension 4); in particular, it does not depend on the sign $\pm$. But the representation $\mathrm{Sym}^4\,\mathrm{St}_2$ of SL_2, viewed in SO_5 and then lifted to $\mathrm{Spin}_5 \simeq \mathrm{Sp}_4$ and composed with the standard representation of Sp_4, is the representation $\mathrm{Sym}^3\,\mathrm{St}_2$. We have thus proved the sequence of equalities

$$\pi[4] = \psi(\pi_1, V^{\pm}_{\mathrm{Spin}}) = \psi(\pi_2, \mathrm{St} \circ \eta) = \psi(\pi_4, \mathrm{St}) \ ,$$

and the representation $\pi'' = \pi_4$ satisfies the conditions of part (ii) of the theorem. □

Note that the Gelbart–Jacquet theorem (see the examples concerning PGL_n in Sect. 6.4.7) implies that the pair (π', St) satisfies the Arthur–Langlands conjecture. It is, moreover, clear that the pair (π'', St) also satisfies this conjecture.

Let us give a second formulation of the previous result. Recall that the homomorphism $\mathrm{SO}_8 \to \mathrm{PGSO}_8$ determines, by Langlands duality, a $\mathbb{C}$-morphism $\eta\colon \widehat{\mathrm{PGSO}}_8 \to \widehat{\mathrm{SO}}_8$; the three irreducible representations of dimension 8 of $\widehat{\mathrm{PGSO}}_8(\mathbb{C}) \simeq \mathrm{Spin}_8(\mathbb{C})$ are therefore $\mathrm{St} \circ \eta$, V^{+}_{Spin}, and V^{-}_{Spin}.

Theorem 7.2.2. *Let $\pi \in \Pi_{\mathrm{cusp}}(\mathrm{PGL}_2)$ be the representation generated by an eigenform of $\mathrm{S}_k(\mathrm{SL}_2(\mathbb{Z}))$, where $k \geq 12$ is an even integer. There exists a $\pi' \in \Pi_{\mathrm{disc}}(\mathrm{PGSO}_8)$ such that*

$$\psi(\pi', \mathrm{St} \circ \eta) = \pi[4] \ , \ \psi(\pi', V^{+}_{\mathrm{Spin}}) = \pi[4] \ , \ \textit{and } \psi(\pi', V^{-}_{\mathrm{Spin}}) = \mathrm{Sym}^2\pi \oplus [5] \ .$$

Proof. The representation π_2 from the proof of Theorem 7.2.1 satisfies $\psi(\pi_2, \mathrm{St} \circ \eta) = \pi[4]$ and

$$\{\psi(\pi_2, V_{\mathrm{Spin}}^+), \psi(\pi_2, V_{\mathrm{Spin}}^-)\} = \{\pi[4], \mathrm{Sym}^2\pi \oplus [5]\} .$$

Let F be an eigenform for PGSO_8 that generates π_2; we define π_2' as the discrete automorphic representation of PGSO_8 generated by $\mathrm{s}(F)$. Then we can take one of the two representations π_2 and π_2' for π'. □

The principle of the proof of Theorem 7.2.1 has a greater reach and can, in particular, be applied to theta series of higher genus. It allows us to produce representations of O_8 with interesting standard Langlands parameters, which are functions of those of the elements of $\Pi_{\mathrm{cusp}}(\mathrm{PGSp}_{2g})$ for $1 \leq g \leq 3$.

Theorem 7.2.3. *Suppose that* $\pi \in \Pi_{\mathrm{cusp}}(\mathrm{PGSp}_{2g})$ *admits a* ϑ*-correspondent in* $\Pi_{\mathrm{disc}}(\mathrm{PGO}_8)$.

(i) *Suppose* $g = 2$*. Let* V_4 *and* V_5 *be the irreducible representations of* $\mathrm{Sp}_4(\mathbb{C}) = \widehat{\mathrm{PGSp}_4}(\mathbb{C})$ *of respective dimensions* 4 *and* 5*; that is,* V_4 *is the standard representation and* $\Lambda^2 V_4 \simeq V_5 \oplus 1$*. Then there exist*

- $\pi' \in \Pi_{\mathrm{disc}}(\mathrm{SO}_8)$ *such that* $\psi(\pi', \mathrm{St}) = \psi(\pi, V_5) \oplus [3]$,
- $\pi'' \in \Pi_{\mathrm{disc}}(\mathrm{SO}_8)$ *such that* $\psi(\pi'', \mathrm{St}) = \psi(\pi, V_4)[2]$.

(ii) *Suppose* $g = 3$*. Let* V_{Spin} *be the spin representation of* $\mathrm{Spin}_7(\mathbb{C}) = \widehat{\mathrm{PGSp}_6}(\mathbb{C})$ *and* V_7 *its natural representation of dimension* 7*. Then there exist*

- $\pi' \in \Pi_{\mathrm{disc}}(\mathrm{SO}_8)$ *such that* $\psi(\pi', \mathrm{St}) = \psi(\pi, V_7) \oplus [1]$,
- $\pi'' \in \Pi_{\mathrm{disc}}(\mathrm{SO}_8)$ *such that* $\psi(\pi'', \mathrm{St}) = \psi(\pi, V_{\mathrm{Spin}})$.

Proof. The existence of π' in the two cases is classical and follows from Corollary 7.1.3. As far as the existence of π'' is concerned, its proof is very similar to that of the existence of the representation of the same name in Theorem 7.2.1; it is therefore left as an exercise for the reader. For example, in case (i), we first show the existence of $\pi_0 \in \Pi_{\mathrm{disc}}(\mathrm{PGSO}_8)$ such that $\psi(\pi_0, V_{\mathrm{Spin}}^{\pm}) = \psi(\pi, V_4)[2]$ and $\psi(\pi_0, \mathrm{St} \circ \eta) = \psi(\pi, V_5) \oplus [3]$; the application of the triality to π_0 then leads to the representation π''. □

Remark 7.2.4 (Work of Böcherer). Let us say a few words on the assumption of the theorem and the associated question of the surjectivity of the map ϑ in general, which is a classical problem going back to Eichler (the *Eichler basis problem*). We have the following remarkable result due to Böcherer [25, 27], which generalizes the work of Waldspurger for $g = 1$ mentioned above: for $d > 0$, the map

$$\vartheta_{d,g} \colon \mathrm{M}_{\mathrm{H}_{d,g}(\mathbb{R}^n)}(\mathrm{O}_n) \to \mathrm{S}_{n/2+d}(\mathrm{Sp}_{2g}(\mathbb{Z}))$$

is surjective provided $n > 4g$ (see also [24] for the case $d = 0$, as well as Remark 8.6.3). More precisely, Böcherer gives a necessary and sufficient condition for an eigenform $F \in \mathrm{S}_{n/2+d}(\mathrm{Sp}_{2g}(\mathbb{Z}))$ to be in the image of $\vartheta_{d,g}$, when $n \geq 2g$. It concerns the function $\mathrm{L}(s,\pi,\mathrm{St})$, where $\pi \in \Pi_{\mathrm{cusp}}(\mathrm{Sp}_{2g})$ is generated by F, of which we know that it admits a meromorphic continuation to all of $\mathbb{C}$ (see Sect. 8.7). If $n > 2g$ (resp. $n = 2g$), he shows that F is in the image of $\vartheta_{d,g}$ if and only if $\mathrm{L}(s,\pi,\mathrm{St})$ is nonzero at $s = n/2 - g$ (resp. if and only if $\mathrm{L}(s,\pi,\mathrm{St})$ admits a simple pole at $s = 1$); see [27, Theorems 4_1 and 5]. This condition is automatically satisfied if $n > 4g$. Böcherer has also studied the question of the injectivity of $\vartheta_g = \vartheta_{0,g}$, for which he obtains criteria of the same type [28].

Let us return to the statement of Theorem 7.2.2.

Corollary 7.2.5. *Suppose that $\pi \in \Pi_{\mathrm{cusp}}(\mathrm{PGL}_2)$ and $\pi' \in \Pi_{\mathrm{disc}}(\mathrm{PGSO}_8)$ satisfy the hypotheses and conclusions of Theorem 7.2.2. Suppose, moreover, that π' has a ϑ-correspondent, that is, that there exists an element $F \in \mathrm{M}_{\mathrm{H}^{\pm}_{k/2-2,4}(\mathbb{R}^8)}(\mathrm{PGSO}_8) \simeq \mathrm{M}_{\mathrm{H}_{k/2-2,4}(\mathbb{R}^8)}(\mathrm{PGO}_8)$ that generates π' and has the property that $\vartheta(F)$ is a nonzero element of $\mathrm{S}_{k/2+2}(\mathrm{Sp}_8(\mathbb{Z}))$. Denote by π'' this ϑ-correspondent, generated by $\vartheta(F)$. Then we have*

$$\psi(\pi'', V_{\mathrm{St}}) = \pi[4] \oplus [1] \quad \textit{and} \quad \psi(\pi'', \mathrm{V}_{\mathrm{Spin}}) = \pi[4] \oplus \mathrm{Sym}^2\pi \oplus [5]\ .$$

Proof. This immediately follows from Theorem 7.2.2 and the refined Eichler–Rallis relations (Proposition 7.1.5 in the case $g = 4 = n/2$), because the restriction of the spin representation of $\mathrm{Spin}_9(\mathbb{C})$ to $\mathrm{Spin}_8 \to \mathrm{Spin}_9$ is the representation $V^+_{\mathrm{Spin}} \oplus V^-_{\mathrm{Spin}}$ of $\mathrm{Spin}_8(\mathbb{C})$. □

When π is generated by $\Delta \in \mathrm{S}_{12}(\mathrm{SL}_2(\mathbb{Z}))$, we verified in Proposition 5.4.22 that the assumption on π' is satisfied (this could also have been deduced from a harmonic variant of [28]; see [27, Sect. XI]). Recall that $\mathrm{S}_8(\mathrm{Sp}_8(\mathbb{Z}))$ is of dimension 1, generated by the Schottky form J (Sect. 5.2). We denote the representation generated by the modular form Δ by $\Delta_{11} \in \Pi_{\mathrm{cusp}}(\mathrm{PGL}_2)$.

Corollary 7.2.6. (i) *If $\pi_J \in \Pi_{\mathrm{cusp}}(\mathrm{PGSp}_8)$ denotes the representation generated by the Schottky form, then $\psi(\pi_J, V_{\mathrm{St}}) = \Delta_{11}[4] \oplus [1]$ and*

$$\psi(\pi_J, V_{\mathrm{Spin}}) = \Delta_{11}[4] \oplus \mathrm{Sym}^2\Delta_{11} \oplus [5]\ .$$

(ii) *Let $\pi \in \Pi_{\mathrm{disc}}(\mathrm{PGSO}_{16})$ be the unique nontrivial representation such that $\pi_\infty = \mathbb{C}$. Then $\psi(\pi, V_{\mathrm{St}}) = \Delta_{11}[4] \oplus [7] \oplus [1]$ and*

$$\psi(\pi, V^{\pm}_{\mathrm{Spin}}) = \psi(\pi_J, V_{\mathrm{Spin}}) \oplus \psi(\pi_J, V_{\mathrm{Spin}})[7]\ .$$

Proof. Assertion (i) follows from Corollary 7.2.5 and the discussion preceding it. The second assertion follows from the first, given the relation $J = \vartheta_4(\mathrm{E}_8 \oplus \mathrm{E}_8) - \vartheta_4(\mathrm{E}_{16})$ (Sect. 5.2) and Proposition 7.1.5 (ii). Note that if the $\mathbb{C}$-morphism $g\colon \mathrm{SL}_2 \to \mathrm{SO}_7$ satisfies $\mathrm{St} \circ g \simeq \mathrm{Sym}^6\mathrm{St}_2$ and if $f\colon \mathrm{SL}_2 \to \mathrm{Spin}_7$ is a lift of g, then the restriction of the spin representation of Spin_7 to f is isomorphic to $\mathrm{Sym}^6\mathrm{St}_2 \oplus 1$ (see, for example, [99, Sect. 7]).

Corollary 7.2.7. *For every prime p, the number of perestroikas of $\mathrm{E}_8 \oplus \mathrm{E}_8$ with respect to p that are isomorphic to E_{16} is*

$$\frac{405}{691}\left(\prod_{i=0}^{3}(p^i+1)\right)\left(p^{11}+p^7+p^6+p^5+p^4+1+\tau(p)\right)\left(p^{11}+1-\tau(p)\right).$$

7.3 A Few Consequences of the Work of Ikeda and Böcherer

As already explained in Sect. 5.2, the first assertion of Corollary 7.2.6 is also a consequence of the following theorem due to Ikeda [108] (proof of the *Duke–Imamoğlu conjecture*). It extends a result of Andrianov, Maass, and Zagier in the case of genus $g = 2$ (proof of the *Saito–Kurokawa conjecture*; see [132], [216], [79]).

Theorem 7.3.1 ([108]). *Let $\pi \in \Pi_{\mathrm{cusp}}(\mathrm{PGL}_2)$ be the representation generated by an eigenform of weight k for $\mathrm{SL}_2(\mathbb{Z})$, and let $g \geq 1$ be an integer such that $k \equiv g \bmod 4$; then there exists a representation $\pi' \in \Pi_{\mathrm{cusp}}(\mathrm{Sp}_{2g})$, generated by a scalar-valued Siegel modular form of weight $(k+g)/2$ for $\mathrm{Sp}_{2g}(\mathbb{Z})$, such that $\psi(\pi', \mathrm{St}) = \pi[g] \oplus [1]$.*

Suppose π and π' as in Theorem 7.3.1. Given that we know the function $\mathrm{L}(s, \pi', \mathrm{St})$, the results of Böcherer mentioned above (Remark 7.2.4, [27]) give a necessary and sufficient condition for π' to admit a ϑ-correspondent.

Theorem 7.3.2. *Let k, g, and n be nonzero even integers such that $k \equiv g \bmod 4$, $n \equiv 0 \bmod 8$, and $2g \leq n \leq k+g$. Let $\pi \in \Pi_{\mathrm{cusp}}(\mathrm{PGL}_2)$ be the representation generated by an eigenform of weight k for $\mathrm{SL}_2(\mathbb{Z})$ and $\pi' \in \Pi_{\mathrm{cusp}}(\mathrm{Sp}_{2g})$ a representation satisfying the conclusions of Ikeda's theorem with respect to π.*

(i) *The representation π' admits a ϑ-correspondent $\pi'' \in \Pi_{\mathrm{disc}}(\mathrm{O}_n)$ such that $\pi''_\infty \simeq \mathrm{H}_{(k+g-n)/2,g}(\mathbb{R}^n)$ if and only if $n > 3g$ or $\mathrm{L}(1/2, \pi) \neq 0$.*
(ii) *Suppose $n > 3g$ or $\mathrm{L}(1/2, \pi) \neq 0$ (in which case $k \equiv g \equiv 0 \bmod 4$). If $n > 2g$ (resp. $n = 2g$), then there exists a $\pi'' \in \Pi_{\mathrm{cusp}}(\mathrm{O}_n)$ such that $\psi(\pi'', \mathrm{St}) = \pi[g] \oplus [n-2g-1] \oplus [1]$ (resp. $\psi(\pi'', \mathrm{St}) = \pi[g]$) and $\pi''_\infty \simeq \mathrm{H}_{(k+g-n)/2,g}(\mathbb{R}^n)$.*

Proof. Let $\mathcal{D} = \frac{1}{2}\mathbb{Z} - \mathbb{Z}$ (the set of half-integers). Since we have $\psi(\pi', \mathrm{St}) = \pi[g] \oplus [1]$, the function $\mathrm{L}(s, \pi', \mathrm{St})$ is the product of the Riemann $\zeta(s)$-function and the functions $\mathrm{L}(s+j, \pi)$, where j runs through the elements of $\mathcal{D}$ such that $|j| \leq (g-1)/2$ (note that $g \equiv 0 \bmod 2$). We consider the compatible pair $(\mathrm{H}_{(k+g-n)/2,g}(\mathbb{R}^n), \det^{(k+g)/2})$, which is well defined because $2g \leq n \leq k+g$. If $n > 2g$ (resp. $n = 2g$), Böcherer shows that π' admits a ϑ-correspondent as in the statement above if and only if $\mathrm{L}(n/2 - g, \pi', \mathrm{St}) \neq 0$ (resp. $\mathrm{L}(s, \pi', \mathrm{St})$ admits a simple pole at $s = 1$). Since $\zeta(s)$ admits a simple pole at $s = 1$ and is nonzero if $\Re(s) > 1$, it is equivalent to requiring that $\mathrm{L}(s, \pi) \neq 0$ for every $s \in \mathcal{D}$ such that $|s - \delta - n/2 + g| \leq (g-1)/2$, where $\delta = 1$ if $n = 2g$ and $\delta = 0$ otherwise.

If $s \in \mathcal{D}$ and $s \neq 1/2$, then $\mathrm{L}(s,\pi) \neq 0$. Indeed, for $\Re(s) > 1$, this follows from the absolute convergence of the Euler product defining $\mathrm{L}(s,\pi)$ (for example, by Deligne or Rankin–Selberg). In general, recall that by Hecke, the function $\xi(s,\pi) = (2\pi)^{-s-(k-1)/2}\Gamma(s+(k-1)/2)\mathrm{L}(s,\pi)$ is an entire function of s that satisfies (see Sect. 6.4.11)

$$\xi(1-s,\pi) = i^k \xi(s,\pi)\,. \tag{7.3.1}$$

This allows us to conclude because the function $\Gamma(s)$ does not have any poles at elements of $\mathcal{D}$. To conclude for assertion (i), we note that $|1/2 - \delta - n/2 + g| \leq (g-1)/2$ if and only if $n \leq 3g$.

Finally, let us verify assertion (ii). Under the assumptions of the statement, we have a ϑ-correspondent $\pi'' \in \Pi_{\mathrm{disc}}(\mathrm{O}_n)$ of the representation π', by part (i). Let us apply Corollary 7.1.3. Under the assumption $n > 2g$, it implies $\psi(\pi'',\mathrm{St}) = \psi(\pi',\mathrm{St}) \oplus [n-2g-1]$, which we wanted. In the case of the equality $n = 2g$, it can be written as

$$\psi(\pi'',\mathrm{St}) \oplus [1] = \psi(\pi',\mathrm{St}) = \pi[g] \oplus [1]\,,$$

which is clearly equivalent to $\psi(\pi'',\mathrm{St}) = \pi[g]$. Finally, we note that the assumption $\mathrm{L}(1/2,\pi) \neq 0$ implies $k \equiv 0 \bmod 4$ by the functional equation (7.3.1). □

Remark 7.3.3. Suppose that $\pi \in \Pi_{\mathrm{cusp}}(\mathrm{PGL}_2)$ is generated by an eigenform F of weight $k \equiv 0 \bmod 4$ for $\mathrm{SL}_2(\mathbb{Z})$. If $12 \leq k \leq 20$, then $F = \Delta\,\vartheta_1(\mathrm{E}_8)^{(k-12)/4}$; since the latter takes on only strictly positive values on the imaginary axis, we have $\Gamma(s+(k-1)/2)\mathrm{L}(s,\pi) > 0$ for every $s \in \mathbb{R}$ (formula (6.4.1)). In particular, $\mathrm{L}(1/2,\pi) \neq 0$. It seems that there is no known example where $\mathrm{L}(1/2,\pi) = 0$; see [64], in which the authors verify this for every $k \leq 500$ (this is related to Theorem 7.3.2, which they, however, do not give explicitly).

It is interesting to confront Theorem 7.3.2 with the results above. First of all, if we apply it to $n = 16$, $g = 4 < n/3$, and $k = 12$ (so that $d = 0$), we easily deduce from part (i) that $\vartheta_4(\mathrm{E}_8 \oplus \mathrm{E}_8) - \vartheta_4(\mathrm{E}_{16}) \in \mathrm{M}_8(\mathrm{Sp}_8(\mathbb{Z}))$ is a nonzero cusp form with standard parameter $\Delta_{11}[4] \oplus [1]$. This "sledgehammer" argument therefore re-proves both the Witt conjecture (Eq. (5.2.1)) and the assertion concerning $\psi(\pi, V_{\mathrm{St}})$ in part (ii) of Corollary 7.2.6 (and therefore Theorem 5.2.2!). Likewise, Theorem 7.3.2 can be applied for $n = 24 = k + g$, which produces the five ordered pairs

$$(k,g) \in \{(12,12),(16,8),(18,6),(20,4),(22,2)\}$$

by taking for π the representation generated by the unique normalized eigenform of weight k for $\mathrm{SL}_2(\mathbb{Z})$ when $k \leq 22$. We denote this representation by $\Delta_{k-1} \in \Pi_{\mathrm{cusp}}(\mathrm{PGL}_2)$. The assumption of the theorem is satisfied in the last three cases because $g < 24/3 = 8$, and also in the first two cases because $\mathrm{L}(1/2,\Delta_{k-1}) \neq 0$ when $k = 12$ or 16.

Corollary 7.3.4. *For every $k \in \{12,16,18,20,22\}$, there exists a representation $\pi \in \Pi_{\mathrm{disc}}(\mathrm{O}_{24})$ such that we have $\pi_\infty = \mathbb{C}$ and $\psi(\pi,\mathrm{St}) = \Delta_{k-1}[24-k] \oplus [2k-25] \oplus [1]$ if $k > 12$, $\psi(\pi,\mathrm{St}) = \Delta_{11}[12]$ if $k = 12$.*

On the other hand, we see that the case $n = 2g = 8$ leads to a weakened, and paradoxically more costly, version of Theorem 7.2.1 (ii), because it proves it only under the additional condition $\mathrm{L}(1/2, \pi) \neq 0$ (and, in particular, $k \equiv 0 \bmod 4$). The case $k \equiv 2 \bmod 4$ of Theorem 7.2.1 (ii) therefore seems particularly interesting from this point of view. This slightly troubling phenomenon, as well as the somewhat particular numbers in the statement of Theorem 7.3.2, will be greatly clarified when we explain the results of Arthur in Chap. 8 (see, in particular, Sects. 8.5.7 and 8.6).

Let us conclude this section with a last example showing that, in general, the inclusion $\vartheta_g(\mathbb{C}[\mathrm{X}_n]) \subset \mathrm{M}_{n/2}(\mathrm{Sp}_{2g}(\mathbb{Z}))$ is strict.

Corollary 7.3.5. *The map* $\vartheta_{14} : \mathbb{C}[\mathrm{X}_{32}] \to \mathrm{M}_{16}(\mathrm{Sp}_{28}(\mathbb{Z}))$ *is not surjective.*

Proof. Consider the case $n = 32$, $k = 18$, and $g = 14$ (and therefore again $d = 0$). Ikeda's theorem ensures the existence of an eigenform F in $\mathrm{S}_{16}(\mathrm{Sp}_{28}(\mathbb{Z}))$ that generates a representation with standard parameter $\Delta_{17}[14] \oplus [1]$. We have $n < 3g$ and $\mathrm{L}(1/2, \Delta_{17}) = 0$ because $18 \equiv 2 \bmod 4$, so that Theorem 7.3.2 (i) ensures $F \notin \mathrm{Im}(\vartheta_{14})$. □

This example seems to have remained unnoticed by Nebe and Venkov [156, Sect. 2.2], who present the question of equality between $\vartheta_g(\mathbb{C}[\mathrm{X}_n])$ and $\mathrm{M}_{n/2}(\mathrm{Sp}_{2g}(\mathbb{Z}))$, for every $n \equiv 0 \bmod 8$ and every $g \geq 1$, as an open problem. As we will see in Sect. 8.5.7, Arthur's theory in fact suggests that there does not exist a $\pi \in \Pi_{\mathrm{disc}}(\mathrm{O}_{32})$ such that $\psi(\pi, \mathrm{St}) = \Delta_{17}[14] \oplus [3] \oplus [1]$.

7.4 A Table of the First Elements of $\Pi_{\mathrm{disc}}(\mathrm{SO}_8)$

By Chenevier and Renard [55, Chap 2], we have a formula for the dimension of $\mathrm{M}_{U_\lambda}(\mathrm{SO}_8)$ in terms of the highest weight $\lambda = \sum_{i=1}^4 m_i \varepsilon_i$ of the representation U_λ (Sect. 6.4.3). For small values of λ, for example as long as $m_1(\lambda) := m_1$ is at most 9, we see that these dimensions are at most 1, and even almost always zero [55, App. C, Table 2]. When this dimension is 1, there consequently exists a unique representation $\pi \in \Pi_{\mathrm{disc}}(\mathrm{SO}_8)$ such that $\pi_\infty \simeq U_\lambda$; we denote it by $\pi(\lambda)$.

The considerations of this chapter allow us to prove the existence of a certain number of elements of $\Pi_{\mathrm{disc}}(\mathrm{O}_8)$ or $\Pi_{\mathrm{disc}}(\mathrm{SO}_8)$. We can ask ourselves whether these elements suffice to explain all $\pi(\lambda)$ above. The answer to this question is given by Table 7.1, which gives the list of all $\psi(\pi(\lambda), \mathrm{St})$ for $m_1(\lambda) \leq 8$. For numerical reasons, it is more meaningful to include in this table the element $\lambda + \rho$, the infinitesimal character of U_λ (Sect. 6.4.3), which we encode by the quadruple $z(\lambda) = (2m_1 + 6, 2m_2 + 4, 2m_3 + 2, 2|m_4|)$ if $\lambda = \sum_{i=1}^4 m_i \varepsilon_i$.

Let us say a few words about this table. The representation with parameter $[7] \oplus [1]$ is, of course, the trivial representation. Recall that the notation $\Delta_w \in \Pi_{\mathrm{cusp}}(\mathrm{PGL}_2)$ was introduced in Sect. 7.3. The four elements $\psi_{j,k} \in \mathcal{X}(\mathrm{SL}_4)$ will be explained in Sect. 9.1.17.

(i) The existence of the representations with parameter $\Delta_w[4]$ follows from Theorem 7.2.1. In the case of the quadruple $(14, 12, 10, 8)$, it is the representation used in the proof of Theorem 5.2.

Table 7.1 Standard parameters of the $\pi(\lambda)$ when $m_1(\lambda) \leq 8$

$z(\lambda)$	$\psi(\pi(\lambda), \mathrm{St})$	$z(\lambda)$	$\psi(\pi(\lambda), \mathrm{St})$
$(6, 4, 2, 0)$	$[7] \oplus [1]$	$(22, 16, 14, 0)$	$\mathbf{Sym}^2\Delta_{11} \oplus \Delta_{15}[2] \oplus [1]$
$(14, 12, 10, 8)$	$\Delta_{11}[4]$	$(22, 18, 16, 0)$	$\mathbf{Sym}^2\Delta_{11} \oplus \Delta_{17}[2] \oplus [1]$
$(18, 16, 2, 0)$	$\Delta_{17}[2] \oplus [3] \oplus [1]$	$(22, 20, 2, 0)$	$\Delta_{21}[2] \oplus [3] \oplus [1]$
$(18, 16, 14, 12)$	$\Delta_{15}[4]$	$(22, 20, 6, 4)$	$\psi_{4,10}[2]$
$(20, 18, 8, 6)$	$\psi_{6,8}[2]$	$(22, 20, 10, 8)$	$\psi_{8,8}[2]$
$(20, 18, 16, 14)$	$\Delta_{17}[4]$	$(22, 20, 14, 12)$	$\psi_{12,6}[2]$
$(22, 4, 2, 0)$	$\mathrm{Sym}^2\Delta_{11} \oplus [5]$	$(22, 20, 18, 0)$	$\mathrm{Sym}^2\Delta_{11} \oplus \Delta_{19}[2] \oplus [1]$
$(22, 12, 10, 0)$	$\mathbf{Sym}^2\Delta_{11} \oplus \Delta_{11}[2] \oplus [1]$	$(22, 20, 18, 16)$	$\Delta_{19}[4]$

(ii) The existence of a representation of $\Pi_{\mathrm{disc}}(\mathrm{O}_8)$ with parameter $\Delta_w[2] \oplus [3] \oplus [1]$ for $w \equiv 1 \bmod 4$ follows from Theorem 7.3.2: it is the case $n = 8$, $k = w + 1$, and $g = 2$, which satisfies the necessary conditions because we have $k \equiv g \bmod 4$ and $n > 3g$. In these particular cases, an important role is played by the eigenforms in $\mathrm{S}_{(w+3)/2}(\mathrm{Sp}_4(\mathbb{Z}))$ with standard parameter $\Delta_w[2] \oplus [1]$ (the two "first" forms of Saito–Kurokawa, the case $g = 2$ of Theorem 7.3.1). When $w = 17, 21$, the surjectivity of $\vartheta_{(w-5)/2,2}\colon \mathrm{M}_{\mathrm{H}_{(w-5)/2,2}(\mathbb{R}^8)}(\mathrm{O}_8) \to \mathrm{S}_{(w+3)/2}(\mathrm{Sp}_4(\mathbb{Z}))$ can be verified through a simple calculation of the coefficient of the theta series, given that, since Igusa [105], we know that $\mathrm{S}_{(w+3)/2}(\mathrm{Sp}_4(\mathbb{Z}))$ has dimension 1. This calculation will be justified in Proposition 9.1.2.

(iii) The case $z(\lambda) = (22, 20, 18, 0)$ also has an interesting history, because it was studied by Miyawaki in [149]. He showed that $\pi(\lambda)$ admits a ϑ-correspondent in $\mathrm{S}_{12}(\mathrm{Sp}_6(\mathbb{Z}))$ that he conjectured to have standard parameter $\mathrm{Sym}^2\Delta_{11} \oplus \Delta_{19}[2]$, which was later proved by Ikeda [109]. Although it looks similar, the case of the other parameters of the form $\mathrm{Sym}^2\Delta_{11} \oplus \Delta_w[2] \oplus [1]$ with $w \in \{11, 15, 17\}$ is more subtle. When $w = 11, 15$, we could justify it as for $w = 19$ if we had an analog of Ikeda's construction for nonscalar forms, because for the two pertinent values of λ, we could certainly verify that $\pi(\lambda)$ admits a ϑ-correspondent for Sp_8 (see Sect. 8.6). These parameters (bold in the table) are predicted by Arthur's theory, as we will see in Chap. 8.

We refer to Sect. 8.5.7 for a direct, but conditional, confirmation of all of Table 7.1 using Arthur's theory, and to [55, Chaps. 2 and 7] for much more extensive tables. Although it is undoubtedly possible to do so using constructions of theta series and the methods of Sect. 9.5, we will not give an unconditional justification of the three bold parameters in Table 7.1 (in these cases, $z(\lambda)$ is $(22, 12, 10, 0)$, $(22, 16, 14, 0)$ or $(22, 18, 16, 0)$).

7.5 The Space $\mathrm{M}_{\det}(\mathrm{O}_{24})$

In this section, we prove Proposition 5.3.1.

Proposition 7.5.1. *The standard parameter of the unique representation* π *in* $\Pi_{\mathrm{disc}}(\mathrm{O}_{24})$ *such that* $\pi_\infty \simeq \det$ *is* $\Delta_{11}[12]$.

Proof. By Corollary 7.3.4, there exists a $\pi \in \Pi_{\mathrm{disc}}(\mathrm{O}_{24})$ such that $\pi_\infty = \mathbb{C}$ and $\psi(\pi, \mathrm{St}) = \Delta_{11}[12]$. We have

$$\tau(2)(2^{12} - 1) = 2^{11}\,\mathrm{trace}(\mathrm{c}_2(\pi), V_{\mathrm{St}})\ ,$$

which is none other than the eigenvalue λ_{24} in the notation of Sect. 5.3. Let $f_{24} \in \mathrm{M}_{\mathbb{C}}(\mathrm{O}_{24})$ be the function that is the image of the element v_{24} by the isomorphism of $\mathrm{H}(\mathrm{O}_{24})$-modules $\mathrm{M}_{\mathbb{C}}(\mathrm{O}_{24}) \simeq \mathbb{C}[\mathrm{X}_{24}]$ defined in Corollary 4.4.9.

Recall that $\mathrm{M}_{\mathbb{C}}(\mathrm{SO}_{24})$ is endowed with an action of $\mathrm{H}(\mathrm{SO}_{24})$, as well as an involution s (the "change of orientation," see the end of Sect. 4.4.4) whose decomposition in eigenspaces can be written as

$$\mathrm{M}_{\mathbb{C}}(\mathrm{SO}_{24}) = \mathrm{M}_{\mathbb{C}}(\mathrm{O}_{24}) \oplus \mathrm{M}_{\det}(\mathrm{O}_{24})\ .$$

The two summands of this decomposition, however, are not necessarily stable under the action of $\mathrm{H}(\mathrm{SO}_{24})$, but only by $\mathrm{H}(\mathrm{O}_{24})$, since the relation we have is $T \circ s = s \circ \mathrm{H}(s)(T)$ for every $T \in \mathrm{H}(\mathrm{SO}_{24})$.

Let $V \subset \mathrm{M}_{\mathbb{C}}(\mathrm{SO}_{24})$ be the $\mathrm{H}(\mathrm{SO}_{24})[s]$-module generated by f_{24}, and let $f'_{24} \in V$ be an eigenform for $\mathrm{H}(\mathrm{SO}_{24})$. It naturally generates a representation $\pi' \in \Pi_{\mathrm{disc}}(\mathrm{SO}_{24})$ with standard parameter $\Delta_{11}[12]$. In particular, the conjugacy class $\mathrm{c}_2(\pi') \subset \mathrm{SO}_{24}(\mathbb{C})$ does not have eigenvalue ± 1, because the eigenvalues of $\mathrm{c}_2(\Delta_{11})$ are not real (they are the roots of $x^2 + (24/2^{11/2})x + 1$). This conjugacy class under $\mathrm{SO}_{24}(\mathbb{C})$ is therefore not stable under the action of $\mathrm{O}_{24}(\mathbb{C})$ by conjugation. The compatibility of the Satake isomorphism with isomorphisms therefore shows that if $f''_{24} = s(f'_{24}) \in V$ and if $\pi'' \in \Pi_{\mathrm{disc}}(\mathrm{SO}_{24})$ is generated by f''_{24}, then we have

$$\mathrm{c}_2(\pi') \neq \mathrm{c}_2(\pi'')\ .$$

In particular, f'_{24} and f''_{24} are not proportional and have the same eigenvalues as f_{24} under the action of $\mathrm{H}(\mathrm{O}_{24})$. This allows us to conclude because the nonzero element $f'_{24} - f''_{24} = (1 - s)f'_{24}$ generates the line $\mathrm{M}_{\det}(\mathrm{O}_{24})$. □

Chapter 8
Arthur's Classification for the Classical $\mathbb{Z}$-groups

8.1 Standard Parameters for the Classical Groups

The aim of this chapter is to explain the description of $\Pi_{\rm disc}(G)$ stemming from the work of Arthur [13] when G is a *classical* $\mathbb{Z}$-group. *By this we mean, from now on, that G is of the form* Sp_{2g} *for* $g \geq 1$ *or* SO_L*, where L is either a* q*-module over* $\mathbb{Z}$ *with* $\dim L \neq 2$ *or a* q-i*-module over* $\mathbb{Z}$ *in the sense of Appendix B* (*Sect.* 6.4.7). For $r \geq 1$ an integer, we set

$$\mathrm{SO}_{r,r} = \mathrm{SO}_{\mathrm{H}(\mathbb{Z}^r)} \quad \text{and} \quad \mathrm{SO}_{r+1,r} = \mathrm{SO}_{\mathrm{H}(\mathbb{Z}^r)\oplus \mathrm{A}_1} \,.$$

The classical $\mathbb{Z}$-groups that are Chevalley groups[1] are therefore Sp_{2g} for $g \geq 1$, $\mathrm{SO}_{r,r}$ for $r \geq 2$, and $\mathrm{SO}_{r+1,r}$ for $r \geq 1$. It will be convenient to view the trivial $\mathbb{Z}$-group as a classical Chevalley group, which we also denote by $\mathrm{SO}_{1,0}$. Moreover, an important role will be played by the $\mathbb{Z}$-groups SO_n, defined as follows for every integer $n \geq 1$ such that $n \equiv -1, 0, 1 \bmod 8$:

$$\mathrm{SO}_n = \mathrm{SO}_{\mathrm{L}_n} \,,$$

where $\mathrm{L}_n = \mathrm{E}_n$ if $n \equiv 0 \bmod 8$, $\mathrm{L}_n = \mathrm{E}_{n-1} \oplus \mathrm{A}_1$ if $n \equiv 1 \bmod 8$, and where L_n denotes the orthogonal complement of a root[2] of E_{n+1} if $n \equiv -1 \bmod 8$ (Sects. 4.1.2 and 4.1.10).

[1] In this chapter, we use the term *Chevalley group* as a synonym for *split semisimple* $\mathbb{Z}$*-group*.

[2] The choice of this root is fixed once and for all and will not play any role in the rest of this book. For example, since all these roots are permuted transitively by $\mathrm{W}(\mathbf{D}_{n+1}) \subset \mathrm{O}(\mathrm{E}_{n+1})$, the isomorphism class of the $\mathbb{Z}$-group SO_n depends only on n.

G. Chenevier, J. Lannes, *Automorphic Forms and Even Unimodular Lattices*, Ergebnisse der Mathematik und ihrer Grenzgebiete. 3. Folge / A Series of Modern Surveys in Mathematics 69, https://doi.org/10.1007/978-3-319-95891-0_8

If G is a classical $\mathbb{Z}$-group, then G is semisimple over $\mathbb{Z}$ and the $\mathbb{C}$-group $\widehat{G}$ is a complex classical group and has a distinguished irreducible representation that is its standard representation (Sects. 6.4.1 and 6.4.7)

$$\mathrm{St}\colon \widehat{G} \to \mathrm{SL}_n(\mathbb{C}) .$$

Theorem⋆ 8.1.1 (Arthur). *If G is a classical Chevalley $\mathbb{Z}$-group and π is in $\Pi_{\mathrm{disc}}(G)$, then $\psi(\pi, \mathrm{St})$ is an element of $\mathcal{X}_{\mathrm{AL}}(\mathrm{SL}_n)$.*

This result, which is a particular case of the general Conjecture 6.4.6, is also a very specific case of [13, Theorem 1.5.2] (the case of representations that are "unramified at all primes"). It relies on a rather formidable collection of difficult results, in particular multiple variants of the Arthur–Selberg trace formula (Arthur), the spectral decomposition of the spaces of automorphic forms (Langlands), the theory of endoscopy (Langlands, Shelstad, Kottwitz), and the proof of the famous *fundamental lemma* (Waldspurger [206], Ngô [157], Laumon and Chaudouard [49, 50]). As explained by Arthur in his book, the results of [13] depend on a variant "with torsion" of his work on the "stabilization" of the trace formula. The required formula was recently established by Moeglin and Waldspurger, in a long series of articles [207, 152] (see also [134]). Arthur also mentions another hypothesis concerning an extension of work of Shelstad on the twisted endoscopy for real Lie groups, which has since been the object of work of Shelstad [184, 185, 186] and Mezo [145, 146]. The statements in this book that rely on these recent works, through the statements of [13], will be indicated with a star ⋆.

The work of Arthur [13] concerns, as mentioned, the classical Chevalley $\mathbb{Z}$-groups. However, his previous work allows the deduction from this of a classification of $\Pi_{\mathrm{disc}}(G)$ when G is an arbitrary classical $\mathbb{Z}$-group. This classification is announced in Chap. 9 of [13] but has not yet been redacted completely, which is why we announce it in the form of a conjecture.[3] As we will see, these results will not be needed to establish our main results. On the other hand, they substantially clarify the questions that preoccupy us, so that it would be a shame not to mention them. For example, as we will see, they lead to a direct and very precise description of $\Pi_{\mathrm{disc}}(\mathrm{SO}_n)$, the main theme of this book.

Conjecture 8.1.2 ([13, Chap. 9]). The statement of Theorem 8.1.1 still holds if G is an arbitrary classical $\mathbb{Z}$-group.

Notation. As in Sect. 6.4.7, we denote by $\mathrm{Class}_{\mathbb{C}}$ the set consisting of the $\mathbb{C}$-groups Sp_{2g} ($g \geq 1$) and SO_m ($m \geq 0$, $m \neq 2$). A group $H \in \mathrm{Class}_{\mathbb{C}}$ is uniquely determined by the associated pair $(\mathrm{n}_H, \mathrm{w}_H) \in \mathbb{N} \times \{0, 1\}$ defined as follows:

- n_H is the dimension of the standard representation of H;
- we have $\mathrm{w}_H = 0$ if and only if there is an isomorphism $H \simeq \mathrm{SO}_m$ for an integer $m \geq 0$.

[3] Let us mention that none of the difficulties stated by Arthur in his Chap. 9 seem to apply to the situation we are interested in, which concerns only "pure inner forms" of Chevalley groups [115].

8.2 Self-Dual Representations of PGL_n

Arthur's classification and, more generally, the Arthur–Langlands conjecture involve representations in $\Pi_{\mathrm{cusp}}(\mathrm{PGL}_m)$ for every $m \geq 1$ that, up to now, we have encountered essentially only in the case $m = 2$ and rather from the point of view of the identity $\mathrm{GSp}_2 = \mathrm{GL}_2$. Let us return to these.

8.2.1 Duality in $\Pi_{\mathrm{disc}}(\mathrm{PGL}_n)$

Let $n \geq 1$ be an integer and $\mathcal{R}_n$ the set of discrete subgroups of $\mathbb{R}^n$ of rank n. This set is endowed with a natural transitive action of $\mathrm{GL}_n(\mathbb{R})$, and the orbit of the lattice $\mathbb{Z}^n$ can be identified with $\mathrm{GL}_n(\mathbb{R})/\mathrm{GL}_n(\mathbb{Z}) \xrightarrow{\sim} \mathcal{R}_n$. The subgroup $\mathrm{GL}_n(\mathbb{Q})$ preserves the subspace $\mathcal{R}_{\mathbb{Z}}(\mathbb{Q}^n) \subset \mathcal{R}_n$ of lattices in $\mathbb{Q}^n$; this action is transitive and extends naturally to an action of $\mathrm{GL}_n(\mathbb{A}_f)$ (Sect. 4.1.1). The map $\mathrm{GL}_n(\mathbb{R})\times\mathrm{GL}_n(\mathbb{A}_f) \to \mathcal{R}_n$ given by $(g_\infty, g_f) \mapsto g_\infty^{-1}(g_f(\mathbb{Z}^n))$ is therefore well defined; it induces a bijection $\mathrm{GL}_n(\mathbb{Q})\backslash\mathrm{GL}_n(\mathbb{A})/\mathrm{GL}_n(\widehat{\mathbb{Z}}) \xrightarrow{\sim} \mathcal{R}_n$ and, consequently, a natural isomorphism

$$\mathcal{A}^2(\mathrm{PGL}_n) \xrightarrow{\sim} \mathrm{L}^2(\underline{\mathcal{R}_n}) ,$$

where $\underline{\mathcal{R}_n}$ is the quotient of $\mathcal{R}_n$ by the group $\mathbb{R}^\times$ of homotheties, endowed with a nonzero $\mathrm{GL}_n(\mathbb{R})$-invariant measure (Sect. 4.3.1). The natural actions of $\mathrm{PGL}_n(\mathbb{R})$ and $\mathrm{H}(\mathrm{PGL}_n)$ on $\mathrm{L}^2(\underline{\mathcal{R}_n})$ deduced by transport of structure are then the obvious actions. In particular, if $f \in \mathrm{L}^2(\underline{\mathcal{R}_n})$ is continuous, $\mathrm{T}_A \in \mathrm{H}(\mathrm{PGL}_n)$ is the operator defined in Sect. 4.2.6, and $L \in \mathcal{R}_n$, then we have $\mathrm{T}_A(f)(\underline{L}) = \sum_M f(\underline{M})$, where the sum is taken over the subgroups M of L with $L/M \simeq A$ and $\underline{N}$ denotes the homothety class of $N \in \mathcal{R}_n$.

The $\mathbb{Z}$-group GL_n has automorphism $g \mapsto {}^t g^{-1}$, which therefore also acts on $\mathcal{A}^2(\mathrm{PGL}_n)$ by an involution that we denote by θ, which preserves the subspaces $\mathcal{A}_{\mathrm{disc}}(\mathrm{PGL}_n)$ and $\mathcal{A}_{\mathrm{cusp}}(\mathrm{PGL}_n)$. Concretely, we endow the $\mathbb{Z}$-module $\mathbb{Z}^n$ with the standard nondegenerate symmetric bilinear form and denote the dual lattice of $L \in \mathcal{R}_n$ for this form by $L^\sharp$. The involution of $\mathrm{L}^2(\underline{\mathcal{R}_n})$ in question, which we also denote by θ, is simply defined by $\theta(f)(\underline{L}) = f(\underline{L^\sharp})$. It therefore satisfies $\theta(T(f)) = \iota(T)(\theta(f))$ for every $T \in \mathrm{H}(\mathrm{PGL}_n)$ and every $f \in \mathrm{L}^2(\underline{\mathcal{R}_n})$, where ι is the involutive automorphism of $\mathrm{H}(\mathrm{PGL}_n)$ defined in Sect. 4.2.6.

For $\pi = \pi_\infty \otimes \pi_f$ in $\Pi(\mathrm{PGL}_n)$, we denote by $\pi^\vee \in \Pi(\mathrm{PGL}_n)$ the element defined as follows. On the one hand, $(\pi^\vee)_\infty$ is the representation with the same space as π_∞ but composed with the automorphism $g \mapsto {}^t g^{-1}$ of $\mathrm{PGL}_n(\mathbb{R})$. On the other hand, if we view $(\pi^\vee)_f$ and π_f as homomorphisms $\mathrm{H}(\mathrm{PGL}_n) \to \mathbb{C}$, then we have $(\pi^\vee)_f = \pi_f \circ \iota$.

If c is the conjugacy class of a semisimple element g of $\mathrm{SL}_n(\mathbb{C})$, we denote the conjugacy class of g^{-1} (resp. of the complex conjugate of g) by c^{-1} (resp. $\overline{c}$). If c is the conjugacy class of a semisimple element X of $\mathfrak{sl}_n(\mathbb{C})$, we denote the conjugacy class of $-X$ by $-c$. The following proposition is well known.

Proposition 8.2.2. *Let $\pi \in \Pi(\mathrm{PGL}_n)$.*

(i) *If π is an element of $\Pi_{\mathrm{cusp}}(\mathrm{PGL}_n)$ (resp. $\Pi_{\mathrm{disc}}(\mathrm{PGL}_n)$), then the same holds for $\pi^\vee$.*
(ii) *For every prime p, we have the equality $\mathrm{c}_p(\pi^\vee) = \mathrm{c}_p(\pi)^{-1}$. If we, moreover, have $\pi \in \Pi_{\mathrm{disc}}(\mathrm{PGL}_n)$, then we also have $\mathrm{c}_p(\pi^\vee) = \overline{\mathrm{c}_p(\pi)}$.*
(iii) *The representation $(\pi^\vee)_\infty$ is the dual of the unitary representation π_∞, and we have the equality $\mathrm{c}_\infty(\pi^\vee) = -\mathrm{c}_\infty(\pi)$.*

Proof. The paragraphs before the proposition justify assertion (i). Let p be a prime. For every element $T \in \mathrm{H}(\mathrm{PGL}_n)$, we have the relation $\iota(T) = T^{\mathrm{t}}$, by Sect. 4.2.6. The discussion following Scholium 6.2.2 therefore shows the equality $\mathrm{c}_p(\pi^\vee) = \mathrm{c}_p(\pi)^{-1}$. On the other hand, since the action of $\mathrm{H}(\mathrm{PGL}_n)$ on $\mathcal{A}_{\mathrm{cusp}}(\mathrm{PGL}_n)$ is a $\star$-action for the Petersson product on the latter, by Sect. 4.3.1, it also follows that the elements π_p and $\pi_p^\vee$, viewed as ring morphisms $\mathrm{H}_p(\mathrm{PGL}_n) \to \mathbb{C}$, are each other's complex conjugates. From the point of view of the Satake isomorphism, this can be written as $\mathrm{c}_p(\pi^\vee) = \overline{\mathrm{c}_p(\pi)}$. Part (iii) follows from the Harish-Chandra theory of characters and the fact that every element of $\mathrm{GL}_n(\mathbb{R})$ is conjugate to its transpose. Using the definition of the Harish-Chandra isomorphism, it is not difficult to verify that we have $\mathrm{c}_\infty(\pi^\vee) = -\mathrm{c}_\infty(\pi)$. □

Definition 8.2.3. Let $\pi \in \Pi(\mathrm{PGL}_n)$. The representation $\pi^\vee$ is called the *dual representation* of π. We say that π is *self-dual* if $\pi^\vee \simeq \pi$. We denote by $\Pi_{\mathrm{cusp}}^\perp(\mathrm{PGL}_n) \subset \Pi_{\mathrm{cusp}}(\mathrm{PGL}_n)$ the subset of self-dual representations.

Note that by the multiplicity 1 theorem of Jacquet–Shalika, for $\pi \in \Pi_{\mathrm{cusp}}(\mathrm{PGL}_n)$ to be self-dual, it suffices to have $\mathrm{c}_p(\pi) = \mathrm{c}_p(\pi)^{-1}$ for every prime p (or even for every prime p except finitely many).

The trivial representation of PGL_1 is, of course, self-dual. Moreover, since $g \mapsto {}^t g^{-1}$ is an inner automorphism when $n = 2$, the inclusion $\Pi_{\mathrm{cusp}}^\perp(\mathrm{PGL}_2) \subset \Pi_{\mathrm{cusp}}(\mathrm{PGL}_2)$ is an equality. This is no longer true for $n > 2$. The main interest of self-dual representations for our concerns comes from the following theorem, which refines the statement of Theorem 8.1.1.

Theorem* 8.2.4 ([13, Theorem 1.5.2]). *Let G be a classical Chevalley $\mathbb{Z}$-group and $\pi \in \Pi_{\mathrm{disc}}(G)$. Then $\psi(\pi, \mathrm{St})$ is of the form $\oplus_{i=1}^k \pi_i[d_i]$ with $\pi_i \in \Pi_{\mathrm{cusp}}^\perp(\mathrm{PGL}_{n_i})$ for every $i = 1, \ldots, k$ and $\sum_{i=1}^k n_i d_i = \mathrm{n}_{\widehat{G}}$. Moreover, this decomposition is unique and the pairs (π_i, d_i) for $i = 1, \ldots, k$ are pairwise distinct.*

A similar refinement of Conjecture 8.1.2 is also expected. Note that all Langlands parameters encountered in Chaps. 6 and 7 satisfy the conclusions of Theorem 8.2.4. In particular, the Gelbart–Jacquet representation $\mathrm{Sym}^2\pi \in \Pi_{\mathrm{cusp}}(\mathrm{PGL}_3)$, where $\pi \in \Pi_{\mathrm{cusp}}(\mathrm{PGL}_2)$, is self-dual. A partial justification of these self-duality properties is given by the following elementary proposition.

Proposition 8.2.5. *Let G be a classical $\mathbb{Z}$-group and $\pi \in \Pi(G)$. Suppose $\psi(\pi, \mathrm{St}) = \oplus_{i=1}^k \pi_i[d_i]$, where $\pi_i \in \Pi_{\mathrm{cusp}}(\mathrm{PGL}_{n_i})$ for every $i = 1, \ldots, k$. Then for every i, there exists a j such that $\pi_j = \pi_i^\vee$ and $d_j = d_i$.*

Proof. Since the representation St of $\widehat{G}$ is self-dual, we have the equality $\mathrm{St}(\mathrm{c}_p(\pi)) = \mathrm{St}(\mathrm{c}_p(\pi))^{-1}$ for every prime p and $\mathrm{St}(\mathrm{c}_\infty(\pi)) = -\mathrm{St}(\mathrm{c}_\infty(\pi))$, and therefore also $\psi(\pi, \mathrm{St}) = \oplus_{i=1}^k \pi_i^\vee[d_i]$. We conclude using Proposition 6.4.5 (the Jacquet–Shalika theorem). □

8.2.6 Regular Algebraic Representations

Let $\pi \in \Pi_{\mathrm{cusp}}(\mathrm{PGL}_n)$. The *weights* of π are the eigenvalues of the semisimple conjugacy class $\mathrm{c}_\infty(\pi) \subset \mathrm{M}_n(\mathbb{C})$. We denote this set of weights by $\mathrm{Weights}(\pi) \subset \mathbb{C}$.

Definition 8.2.7. Let $\pi \in \Pi_{\mathrm{cusp}}(\mathrm{PGL}_n)$. We say that π is *algebraic*[4] if $\mathrm{Weights}(\pi) \subset \frac{1}{2}\mathbb{Z}$ and if for all $w, w' \in \mathrm{Weights}(\pi)$, we have $w - w' \in \mathbb{Z}$.

If $\pi \in \Pi_{\mathrm{cusp}}(\mathrm{PGL}_n)$ is algebraic, its *motivic weight* is the greatest $w \in \mathbb{Z}$ such that $-w/2 \in \mathrm{Weights}(\pi)$; we denote it by $\mathrm{w}(\pi)$. In particular, we have $\mathrm{Weights}(\pi) \subset \mathrm{w}(\pi)/2 + \mathbb{Z}$.

Although the algebraic representations form a tiny part of $\Pi_{\mathrm{cusp}}(\mathrm{PGL}_n)$, they will be the only ones to play a role in this work. An indication of this is given by the following proposition.

Proposition 8.2.8. *Let G be a semisimple $\mathbb{Z}$-group, $\pi \in \Pi(G)$, and let $r : \widehat{G} \to \mathrm{SL}_n$ be a $\mathbb{C}$-representation. We suppose that*

(i) *π_∞ has the same infinitesimal character as a finite-dimensional irreducible $\mathbb{C}$-representation of $G_\mathbb{C}$;*
(ii) *we have $\psi(\pi, r) = \oplus_{i=1}^k \pi_i[d_i]$ with $\pi_i \in \Pi_{\mathrm{cusp}}(\mathrm{PGL}_{n_i})$ for $i = 1, \ldots, k$ (Sect. 6.4.4).*

Then π_i is algebraic for every i. Moreover, the class of $\mathrm{w}(\pi_i) + d_i - 1$ in $\mathbb{Z}/2\mathbb{Z}$ depends only on r (and not on the integer i or even on π).

Proof. Let μ be the highest weight of r (a coweight of $G_\mathbb{C}$). The infinitesimal character of π_∞ is of the form $\lambda + \rho$, where λ is a dominant weight of $G_\mathbb{C}$ and ρ is the half-sum of the positive roots (Sect. 6.3.4). The eigenvalues of $r(\mathrm{c}_\infty(\pi))$ are, by definition, of the form $\langle \lambda + \rho, \mu' \rangle$, where μ' is a weight of r. But 2ρ is a weight of $G_\mathbb{C}$ and $\langle \mu - \mu', \rho \rangle \in \mathbb{Z}$ if $\mu' \leq \mu$; these eigenvalues are therefore all in $\frac{1}{2}\mathbb{Z}$, and pairwise they differ by an element of $\mathbb{Z}$. This property is inherited by the weights of the π_i. □

This proposition applies, in particular, for every $\pi \in \Pi(G)$ if $G(\mathbb{R})$ is compact (Sect. 6.3.4). It also applies if $G = \mathrm{Sp}_{2g}$ and π is generated by an eigenform in

[4] The reader should be aware that there are several notions of algebraic automorphic representations in the literature. Definition 8.2.7, which is essentially the one considered, for example, in [33, Sect. 18.2], but which is not the one used by Clozel in [59], is reminiscent of the notion of Hecke character of type A_0 in the sense of Weil (see [43] for a clarification of the various notions).

$\mathrm{S}_W(\mathrm{Sp}_{2g}(\mathbb{Z}))$ with W positive, by Corollary 6.3.6 (as well as the discussion that follows it) and Sect. 6.1.3.

Let G be a classical $\mathbb{Z}$-group and $\mathrm{St}\colon \widehat{G} \to \mathrm{SL}_n$ the standard representation of $\widehat{G}$. We will need to specify the analysis above in this context. Let $\mathrm{Irr}(G_{\mathbb{C}})$ be the set of isomorphism classes of finite-dimensional irreducible $\mathbb{C}$-representations of $G_{\mathbb{C}}$. For each $V \in \mathrm{Irr}(G_{\mathbb{C}})$, we consider the semisimple conjugacy class $\mathrm{St}(\mathrm{Inf}_V) \subset \mathrm{M}_n(\mathbb{C})$ (recall that $\mathrm{Inf}_V \in \widehat{\mathfrak{g}}_{\mathrm{ss}}$ denotes the infinitesimal character of V). Through a careful examination of the root data (Sect. 6.1.3), we deduce that there are three clearly distinct cases:

I. If $\widehat{G} = \mathrm{SO}_n(\mathbb{C})$ with $n = 2g+1$ odd, then $V \mapsto \mathrm{St}(\mathrm{inf}_V)$ induces a bijection between $\mathrm{Irr}(\widehat{G})$ and the set of semisimple conjugacy classes $X \subset \mathrm{M}_n(\mathbb{C})$ such that $-X = X$ and that the eigenvalues of X are all distinct and in $\mathbb{Z}$.
II. If $\widehat{G} = \mathrm{Sp}_n(\mathbb{C})$ (and therefore n is even), then $V \mapsto \mathrm{St}(\mathrm{Inf}_V)$ induces a bijection between $\mathrm{Irr}(\widehat{G})$ and the set of semisimple conjugacy classes $X \subset \mathrm{M}_n(\mathbb{C})$ such that $-X = X$ and that the eigenvalues of X are all distinct and in $\frac{1}{2}\mathbb{Z} - \mathbb{Z}$.
III. If $\widehat{G} = \mathrm{SO}_n(\mathbb{C})$ with n even, then $V \mapsto \mathrm{St}(\mathrm{Inf}_V)$ induces a surjection from $\mathrm{Irr}(\widehat{G})$ to the set of semisimple conjugacy classes $X \subset \mathrm{M}_n(\mathbb{C})$ such that $-X = X$ and that the eigenvalues of X are all in $\mathbb{Z}$ and distinct, with the possible exception of the eigenvalue 0, whose multiplicity is at most 2. Moreover, $\mathrm{St}(\mathrm{Inf}_V) = \mathrm{St}(\mathrm{Inf}_{V'})$ if and only if V and V' are each other's conjugates under the outer action of $\mathrm{O}_n(\mathbb{C})$ (which implies $V = V'$ if and only if 0 is an eigenvalue of $\mathrm{Inf}_V = \mathrm{Inf}_{V'}$).

Thus, in all cases, the eigenvalues of $\mathrm{St}(\mathrm{Inf}_V)$ are in $\mathrm{w}_{\widehat{G}}/2 + \mathbb{Z}$.

Definition 8.2.9. A representation $\pi \in \Pi_{\mathrm{cusp}}(\mathrm{PGL}_n)$ is called *regular* if we have $|\mathrm{Weights}(\pi)| = n$.

Proposition 8.2.10. *Let G be a classical $\mathbb{Z}$-group and* $\mathrm{St}\colon \widehat{G} \to \mathrm{SL}_n$ *the standard representation of $\widehat{G}$. Suppose* $\psi = \oplus_{i=1}^{r} \pi_i[d_i] \in \mathcal{X}_{\mathrm{AL}}(\mathrm{SL}_n)$*, where* $\pi_i \in \Pi_{\mathrm{cusp}}(\mathrm{PGL}_{n_i})$ *for every* $i = 1, \ldots, k$*, and* $\psi_\infty = \mathrm{St}(\mathrm{Inf}_V)$*, where* $V \in \mathrm{Irr}(G_{\mathbb{C}})$.

Let $i \in \{1, \ldots, k\}$*. Then* π_i *is algebraic, with motivic weight* $\mathrm{w}(\pi_i) \equiv d_i - 1 + \mathrm{w}_{\widehat{G}} \bmod 2$*, and self-dual. Moreover,* π_i *is regular unless we are in the following exceptional case:*

(a) $\widehat{G}(\mathbb{C}) \simeq \mathrm{SO}_{2r}(\mathbb{C})$ *and* ψ_∞ *admits* 0 *as a double eigenvalue;*
(b) $d_i = 1$, $n_i \equiv 0 \bmod 2$, *and* $|\mathrm{Weights}(\pi_i)| = n_i - 1$*; and*
(c) *for every* $j \neq i$*, the representation* π_j *is regular and* $n_j \equiv 0 \bmod 2$.

Proof. With the exception of the assertion on the self-duality of π_i, the proposition immediately follows from the analysis of the cases I, II, and III above and Proposition 8.2.8.

Let us verify the self-duality of π_i. By Proposition 8.2.5, there exists a j such that $\pi_j = \pi_i^\vee$ and $d_j = d_i$. By Proposition 8.2.11 below, $\pi_j = \pi_i^\vee$ implies $\mathrm{c}_\infty(\pi_j) = \mathrm{c}_\infty(\pi_i)$. In view of the assumption on ψ_∞ and the analysis of the cases I, II, and III, this implies $j = i$ or $n_i = n_j = 1$. In the latter case, we necessarily have $\pi_j = \pi_i = 1$, and therefore we indeed have $\pi_i^\vee = \pi_i$ in all cases. □

In the proof above, we have called upon the following proposition (which is obvious if π is supposed to be self-dual).

Proposition 8.2.11. *If* $\pi \in \Pi_{\mathrm{cusp}}(\mathrm{PGL}_n)$ *is algebraic, then* π_∞ *is isomorphic to its dual. In particular, we have* $\mathrm{c}_\infty(\pi) = \mathrm{c}_\infty(\pi^\vee) = -\mathrm{c}_\infty(\pi)$ *and*

(i) $w \mapsto -w$ *is a bijection of* $\mathrm{Weights}(\pi)$*;*
(ii) *if we have* $n \equiv 1 \bmod 2$*, then* 0 *is in* $\mathrm{Weights}(\pi)$ *and we have* $\mathrm{w}(\pi) \equiv 0 \bmod 2$.

To explain this, we will need to study the Archimedean components of the elements of $\Pi_{\mathrm{cusp}}(\mathrm{PGL}_n)$ in more detail. Another motivation for this is that the self-dual regular algebraic representations, as well as those that intervene in the exceptional case of the Proposition 8.2.10, satisfy certain hidden additional properties, which we will need to specify. This analysis will also be necessary to apply Arthur's statements. Indeed, the latter will involve *ε-factors of pairs of algebraic representations*, which can also be read (as Γ-factors) on their Archimedean components. This work has already been carried out in [55, Sect. 3.11], from which we recall several results in the next subsections.

8.2.12 Representations of $\mathrm{GL}_n(\mathbb{R})$

Let $\mathrm{W}_\mathbb{R}$ be the Weil group of the field $\mathbb{R}$ [196, Sect. 1]. This is a topological group, a nontrivial extension of $\mathrm{Gal}(\mathbb{C}/\mathbb{R}) = \mathbb{Z}/2\mathbb{Z}$ by $\mathbb{C}^\times$ for the natural action by conjugation. It is generated by its open subgroup $\mathbb{C}^\times$ together with an element j, with relations $j^2 = -1$ and $jzj^{-1} = \overline{z}$ for every $z \in \mathbb{C}^\times$.

Following Langlands [137], the continuous and semisimple representations $\mathrm{W}_\mathbb{R} \to \mathrm{GL}_n(\mathbb{C})$ will play an important role. Let us recall the form of the irreducible representations, which are of dimension 1 or 2. Let

$$\eta\colon \mathrm{W}_\mathbb{R} \to \mathbb{R}^\times$$

be the unique group morphism such that $\eta(j) = -1$ and $\eta(z) = z\overline{z}$ for every $z \in \mathbb{C}^\times$; it induces an isomorphism $\mathrm{W}_\mathbb{R}^{\mathrm{ab}} \xrightarrow{\sim} \mathbb{R}^\times$. The continuous morphisms $\mathrm{W}_\mathbb{R} \to \mathbb{C}^\times$ are therefore the $|\eta|^s$ and $\epsilon_{\mathbb{C}/\mathbb{R}}|\eta|^s$, where $s \in \mathbb{C}$ and $\epsilon_{\mathbb{C}/\mathbb{R}} = \eta/|\eta|$. For an integer $w \geq 0$, consider the induced representation

$$\mathrm{I}_w = \mathrm{Ind}_{\mathbb{C}^\times}^{\mathrm{W}_\mathbb{R}} \left(z \mapsto \left(\frac{z}{|z|}\right)^w \right).$$

It is irreducible if and only if $w \neq 0$, and moreover $\mathrm{I}_0 \simeq 1 \oplus \epsilon_{\mathbb{C}/\mathbb{R}}$. The irreducible representations of dimension 2 of $\mathrm{W}_\mathbb{R}$ are the $\mathrm{I}_w \otimes |\eta|^s$ with $w \neq 0$ and $s \in \mathbb{C}$.

We denote by $\Phi(\mathrm{GL}_n(\mathbb{R}))$ the set of isomorphism classes of semisimple continuous representations $\mathrm{W}_\mathbb{R} \to \mathrm{GL}_n(\mathbb{C})$. The *Langlands parametrization* associates with each irreducible unitary representation U of $\mathrm{GL}_n(\mathbb{R})$ (and, more generally, with every irreducible $(\mathfrak{g}, K)$-module of $\mathrm{GL}_n(\mathbb{R})$) an element $\mathrm{L}(U) \in \Phi(\mathrm{GL}_n(\mathbb{R}))$

that determines U up to isomorphism [137, 120]. Although the map $U \mapsto \mathrm{L}(U)$ can be made completely explicit [120], this is not really relevant; the following compatibility statements, which hold for every irreducible unitary representation U of $\mathrm{GL}_n(\mathbb{R})$, suffice:

(i) (Duality) $\mathrm{L}(U^*) \simeq \mathrm{L}(U)^*$.
(ii) (Central character) $\det \mathrm{L}(U) = \chi_U \circ \eta$, where $\chi_U : \mathbb{R}^\times \to \mathbb{C}^\times$ is the central character of U.
(iii) (Infinitesimal character) Let us write $\mathrm{L}(U)_{|\mathbb{C}^\times} \simeq \oplus_{i=1}^n \chi_i$, where the χ_i for $1 \leq i \leq n$ are characters $\mathbb{C}^\times \to \mathbb{C}^\times$. For every i, there consequently exists a unique ordered pair $(\lambda_i, \mu_i) \in \mathbb{C}^2$ with $\lambda_i - \mu_i \in \mathbb{Z}$, such that[5] $\chi_i(z) = (z/|z|)^{\lambda_i - \mu_i} |z|^{\lambda_i + \mu_i}$. Then Inf_U is the semisimple conjugacy class of $\mathrm{M}_n(\mathbb{C})$ whose eigenvalues are the λ_i for $i = 1, \ldots, n$.

The parametrization above applies, in particular, to irreducible unitary representations of $\mathrm{PGL}_n(\mathbb{R})$, viewed as representations of $\mathrm{GL}_n(\mathbb{R})$ with trivial central character.

Assertion (i) of the following proposition is the so-called *purity lemma* of Clozel [59, Lemma 4.9]; it implies Proposition 8.2.11.

Proposition 8.2.13. *Let $\pi \in \Pi_{\mathrm{cusp}}(\mathrm{PGL}_n)$ be algebraic with weights the $\omega_i/2$ (counted with multiplicity), where $\omega_1 \geq \cdots \geq \omega_n$. Let E and F be the subsets of $\{1, \ldots, n\}$ defined by $E = \{i\,;\, \omega_i > 0\}$ and $F = \{i\,;\, \omega_i = 0\}$.*

(i) *There exists a unique $(m_j) \in \{0,1\}^F$ such that*

$$\mathrm{L}(\pi_\infty) \simeq \bigoplus_{i \in E} \mathrm{I}_{\omega_i} \oplus \bigoplus_{j \in F} \epsilon_{\mathbb{C}/\mathbb{R}}^{m_j}\,;$$

in particular, $|F| \equiv n \bmod 2$ and π_∞ is isomorphic to its dual.
(ii) *If $\mathrm{w}(\pi) \equiv 0 \bmod 2$, then $\sum_{j \in F} m_j \equiv |E| \bmod 2$.*
(iii) *If π is regular and $\mathrm{w}(\pi) \equiv n \equiv 0 \bmod 4$, then $n \equiv 0 \bmod 4$.*
(iv) *If $|\mathrm{Weights}(\pi)| = n-1$, then $n \equiv 0 \bmod 2$ and $F = \{n/2, n/2+1\}$. Moreover, $n \equiv 0 \bmod 4$ if and only if $\mathrm{L}(\pi_\infty) \simeq \oplus_{i=1}^{n/2} \mathrm{I}_{\omega_i}$.*

Proof (See [55, Sect. 3.11]). Let us recall the argument of Clozel's purity lemma. Suppose that $\mathrm{I}_w \otimes |\eta|^{s/2}$ (resp. $|\eta|^{s/2}$ or $\varepsilon_{\mathbb{C}/\mathbb{R}} |\eta|^{s/2}$) is a subrepresentation of $\mathrm{L}(\pi_\infty)$, with $w \in \mathbb{Z}$ and $s \in \mathbb{C}$. In particular, $(s \pm w)/2$ (resp. $s/2$) is a weight of π, by the compatibility of the Langlands parametrization with the infinitesimal character. The assumption $\mathrm{Weights}(\pi) \subset \frac{1}{2}\mathbb{Z}$ therefore implies $s \in \mathbb{Z}$. But by Jacquet and Shalika, the assumption $\pi \in \Pi_{\mathrm{cusp}}(\mathrm{PGL}_n)$ implies $|\Re\, s| < 1/2$, which shows $s = 0$. Thus, $\mathrm{L}(\pi_\infty)$ is a direct sum of representations of the form I_w, 1, and $\epsilon_{\mathbb{C}/\mathbb{R}}$. The first assertion of part (i) is then a consequence of the compatibility of the Langlands parametrization with the infinitesimal character. The congruence $|F| \equiv n \bmod 2$

[5] Following Langlands, it is suggestive to write $z^\lambda \bar{z}^\mu$ for the element $(z/|z|)^{\lambda-\mu}|z|^{\lambda+\mu}$ when $z \in \mathbb{C}^\times$ and $\lambda, \mu \in \mathbb{C}$ are such that $\lambda - \mu \in \mathbb{Z}$.

follows. The self-duality of π_∞ follows from that of $\mathrm{L}(\pi_\infty)$ (as a representation of $\mathrm{W}_\mathbb{R}$) and from the compatibility of the Langlands parametrization with the dual.

For part (ii), note that $\det(\mathrm{L}(\pi_\infty)) = 1$ (compatibility with the central character), which suffices because $\det(\mathrm{I}_w) = \epsilon_{\mathbb{C}/\mathbb{R}}^{w+1}$ and $\omega_i \equiv \mathrm{w}(\pi) \bmod 2$ for every i. Part (iii) follows from part (ii) because if π is regular and $n \equiv 0 \bmod 2$, then part (i) implies $F = \emptyset$ and $|E| = n/2$. If $|\mathrm{Weights}(\pi)| = n - 1$, then part (i) shows that 0 is the unique double weight and $n \equiv 0 \bmod 2$; hence $\mathrm{w}(\pi) \equiv 0 \bmod 2$, $|F| = 2$, and $|E| = n/2 - 1$. Finally, part (ii) implies part (iv). □

Remark 8.2.14. Let $\pi \in \Pi_{\mathrm{cusp}}(\mathrm{PGL}_n)$ be such that we have $\lambda - \mu \in \mathbb{Z}$ for all $\lambda, \mu \in \mathrm{Weights}(\pi)$. Then a modification of the argument for part (i) shows that π is algebraic, that is, $\mathrm{Weights}(\pi) \subset \frac{1}{2}\mathbb{Z}$. Indeed, $\mathrm{L}(\pi_\infty)$ is a direct sum of representations of the form $r_i \otimes |\eta|^{s_i/2}$ with $|\Re\, s_i| < 1/2$ and r_i isomorphic to I_w, 1, or $\varepsilon_{\mathbb{C}/\mathbb{R}}$. The weights of π corresponding to the factor $r_i \otimes |\eta|^{s_i/2}$ are of the form $(s_i + m_i)/2$, with $m_i = \pm w$ or 0 according to whether or not we have $r_i \simeq \mathrm{I}_w$. It therefore suffices to see that we have $s_i = s_j$ for every i, j since this implies $s_i = 0$ for every i because of the relation $\det(\mathrm{L}(\pi_\infty)) = 1$. But for every i, j, we have $(s_i + m_i)/2 - (s_j + m_j)/2 \in \mathbb{Z}$ by assumption. Since we have $m_i \in \mathbb{Z}$ and $|\Re\, s_i| < 1/2$ for every i, we deduce $s_i - s_j \in \mathbb{Z}$ and $|\Re\,(s_i - s_j)| < 1$, and consequently $s_i = s_j$.

This statement and Proposition 8.2.10 show that if G is a classical group and if $\pi \in \Pi_{\mathrm{disc}}(G)$ is such that $\mathrm{Inf}_{\pi_\infty} = \mathrm{Inf}_V$ for some $V \in \mathrm{Irr}(\widehat{G})$, then $\psi(\pi, \mathrm{St})$ satisfies certain combinatorial constraints, which we summarize in the following statement [55, Lemma 3.23].

Corollary 8.2.15. *Let G be a classical $\mathbb{Z}$-group and $\mathrm{St} : \widehat{G} \to \mathrm{SL}_n$ the standard representation of $\widehat{G}$. Suppose $\psi = \oplus_{i=1}^r \pi_i[d_i] \in \mathcal{X}_{\mathrm{AL}}(\mathrm{SL}_n)$, where $\pi_i \in \Pi_{\mathrm{cusp}}(\mathrm{PGL}_{n_i})$ for every $i = 1, \ldots, k$, and $\psi_\infty = \mathrm{St}(\mathrm{Inf}_V)$, where $V \in \mathrm{Irr}(G_\mathbb{C})$.*

(i) *If $\widehat{G}(\mathbb{C}) \simeq \mathrm{SO}_{2g+1}(\mathbb{C})$, then there exists a unique $1 \le i_0 \le k$ such that $n_{i_0} d_{i_0} \equiv 1 \bmod 2$. Moreover, $n_i d_i \equiv 0 \bmod 4$ for every $i \ne i_0$.*
(ii) *If $\widehat{G}(\mathbb{C}) \simeq \mathrm{Sp}_{2g}(\mathbb{C})$, then $n_i d_i \equiv 0 \bmod 2$ for every i.*
(iii) *If $\widehat{G}(\mathbb{C}) \simeq \mathrm{SO}_n(\mathbb{C})$ with $n \equiv 0 \bmod 4$, then $n_i d_i \equiv 0 \bmod 4$ for every i unless we are in the following exceptional case: 0 is a double eigenvalue of $\mathrm{St}(\mathrm{c}_\infty(\pi))$ and there exist exactly two integers i, say i_1 and i_2, such that $n_i d_i \not\equiv 0 \bmod 4$. These integers satisfy $n_{i_1} d_{i_1} n_{i_2} d_{i_2} \equiv 3 \bmod 4$.*

Proof. Suppose $\widehat{G}(\mathbb{C}) \simeq \mathrm{SO}_{2g+1}(\mathbb{C})$. Since we have $\sum_{i=1}^k n_i d_i = 2g + 1$, there exists at least one integer i_0 such that $n_{i_0} d_{i_0} \equiv 1 \bmod 2$. For such an integer, $\mathrm{c}_\infty(\pi_{i_0})$ admits the eigenvalue 0 because $\pi_{i_0} = \pi_{i_0}^\vee$ and n_{i_0} is odd. Since d_{i_0} is also odd, $[d_{i_0}]_\infty$ also admits the eigenvalue 0, and therefore so does $(\pi_{i_0}[d_{i_0}])_\infty$. The first part of assertion (i) follows because 0 is a simple eigenvalue of ψ_∞ by case I of the analysis of Sect. 8.2.6. For the second part, we observe that for every $i = 1, \ldots, k$, we have $\mathrm{w}(\pi_i) \equiv d_i - 1 \bmod 2$, which suffices when $\mathrm{w}(\pi_i)$ is odd; the remaining case follows from part (iii) of Proposition 8.2.13.

If $\widehat{G}(\mathbb{C}) \simeq \mathrm{Sp}_{2g}(\mathbb{C})$, the relation $\mathrm{w}(\pi_i)+d_i-1 \equiv 1 \bmod 2$ for every $i = 1, \ldots, k$ shows that if d_i is odd, then so is $\mathrm{w}(\pi_i)$, and therefore n_i is even (Proposition 8.2.11). This proves part (ii).

Finally, suppose $\widehat{G}(\mathbb{C}) \simeq \mathrm{SO}_n(\mathbb{C})$ with $n \equiv 0 \bmod 4$. In particular, $\mathrm{w}(\pi_i) \equiv d_i - 1 \bmod 2$ for every $i = 1, \cdots, k$. If $\mathrm{w}(\pi_i) \equiv 1 \bmod 2$ (and therefore $n_i \equiv 0 \bmod 2$), then d_i is even and therefore $n_i d_i \equiv 0 \bmod 4$. If $\mathrm{w}(\pi_i) \equiv 0 \bmod 2$ and $n_i \equiv 0 \bmod 2$, Proposition 8.2.13 (iii) asserts that we have $n_i \equiv 0 \bmod 4$, except perhaps if π_i is not regular. If this happens, it does so for a unique i_0, by Proposition 8.2.10, and in this case $n_i \equiv 0 \bmod 2$ for every $i \neq i_0$, so that in the end, $n_i d_i \equiv 0 \bmod 4$ for every $i \neq i_0$. The result follows from the identity $n = \sum_{i=1}^{k} n_i d_i \equiv n_{i_0} d_{i_0} \bmod 4$ and the assumption $n \equiv 0 \bmod 4$. We can therefore rule out this exception and suppose that π_i is regular for every i; in particular, $n_i d_i \equiv 0 \bmod 4$ for every i such that n_i is even. Let $J \subset \{1, \ldots, k\}$ be the set of i such that n_i is odd (in which case d_i is odd and $\mathrm{w}(\pi_i)$ is even); we may assume J nonempty. By the argument given for part (i), this implies that 0 is a double eigenvalue of ψ_∞ and that we have $|J| \leq 2$. This concludes the proof because $n \equiv \sum_{j \in J} n_j d_j \bmod 4$. □

8.2.16 The Ramanujan Conjecture and Galois Representations

A particular case of the Langlands conjectures, in the spirit of the famous Shimura–Taniyama–Weil conjecture, is that the set of L-functions of the form $\mathrm{L}(s + \mathrm{w}(\pi)/2 + m, \pi)$, where $m \in \mathbb{Z}$ and π runs through the algebraic representations of $\Pi_{\mathrm{cusp}}(\mathrm{PGL}_n)$ for $n \geq 1$, should coincide exactly with the set of L-functions of the motives over $\mathbb{Q}$ with good reduction everywhere (and, say, "with coefficients in $\overline{\mathbb{Q}}$" and "simple") [139, 153]. The Ramanujan conjecture for an algebraic π (Sect. 6.4.12) would then be a consequence of the existence of the associated motive $\mathrm{M}(\pi)$ and the Weil conjectures, proved by Deligne. Owing to the work of numerous mathematicians (including Eichler–Shimura, Deligne, Langlands, Kottwitz, Clozel, Harris–Taylor, Waldspurger, Ngô, Laumon, Clozel–Harris–Labesse, Shin, Chenevier–Harris), we nowadays dispose of a weakened construction of $\mathrm{M}(\pi)$ for the regular and self-dual algebraic π that, nevertheless, suffices to prove the following theorem. If $\pi \in \Pi_{\mathrm{cusp}}(\mathrm{PGL}_n)$ is a regular, self-dual algebraic representation, it is known that the characteristic polynomial[6]

$$\mathrm{P}_p(\pi) = \det\big(t - \mathrm{c}_p(\pi)\, p^{w(\pi)/2}\big) \in \mathbb{C}[t]$$

has coefficients in the subfield $\overline{\mathbb{Q}} \subset \mathbb{C}$ of algebraic numbers. The following theorem is proved in [61, 189, 52, 60] (see also [44]).

[6] In this definition of $\mathrm{P}_p(\pi)$, it is understood that $\mathrm{c}_p(\pi)\, p^{w(\pi)/2}$ denotes the semisimple conjugacy class of $\mathrm{GL}_n(\mathbb{C})$ obtained by taking the product of the class $\mathrm{c}_p(\pi)$, viewed in $\mathrm{GL}_n(\mathbb{C}) \supset \mathrm{SL}_n(\mathbb{C})$, and the scalar $p^{w(\pi)/2} \in \mathbb{C}^* \subset \mathrm{GL}_n(\mathbb{C})$.

Theorem 8.2.17. *Let $\pi \in \Pi_{\mathrm{cusp}}^{\perp}(\mathrm{PGL}_n)$ be algebraic and regular.*

(i) *The representation π satisfies the Ramanujan conjecture.*
(ii) *Let ℓ be a prime, and let $\overline{\mathbb{Q}}_\ell$ be an algebraic closure of $\mathbb{Q}_\ell$ and $\iota\colon \overline{\mathbb{Q}} \to \overline{\mathbb{Q}}_\ell$ an embedding. There exists a continuous representation $\rho_{\pi,\iota}\colon \mathrm{Gal}(\overline{\mathbb{Q}}/\mathbb{Q}) \to \mathrm{GL}_n(\overline{\mathbb{Q}}_\ell)$, unique up to isomorphism, that is semisimple and unramified outside ℓ and satisfies*
$$\det(t - \rho_{\pi,\iota}(\mathrm{Frob}_p)) = \iota(\mathrm{P}_p(\pi))$$
for every prime $p \neq \ell$.

In this statement, Frob_p denotes a conjugacy class of arithmetic Frobenius elements at p. We, moreover, know by proc. cit. that the restriction of the representation $\rho_{\pi,\iota}$ to $\mathrm{Gal}(\overline{\mathbb{Q}}_\ell/\mathbb{Q}_\ell)$ is crystalline in the sense of Fontaine, with Hodge–Tate weights the $\lambda + \mathrm{w}(\pi)/2$, where $\lambda \in \mathrm{Weights}(\pi)$. It has been conjectured that $\rho_{\pi,\lambda}$ is irreducible, but this is known only for $n \leq 3$ (Ribet, Blasius-Rogawski). Note that the self-duality of π and Chebotarev's density theorem imply the isomorphism

$$\rho_{\pi,\iota}^* \simeq \rho_{\pi,\iota} \otimes \omega_\ell^{-\mathrm{w}(\pi)}\ , \tag{8.2.1}$$

where $\omega_\ell\colon \mathrm{Gal}(\overline{\mathbb{Q}}/\mathbb{Q}) \to \mathbb{Z}_\ell^\times$ denotes the ℓ-adic cyclotomic character. It can be proved that if $\mathrm{w}(\pi) \equiv 1 \bmod 2$ (resp. $\mathrm{w}(\pi) \equiv 0 \bmod 2$), then there exists a non-degenerate $\mathrm{Gal}(\overline{\mathbb{Q}}/\mathbb{Q})$-equivariant pairing $\rho_{\pi,\iota} \otimes \rho_{\pi,\iota} \to \omega_\ell^{\mathrm{w}(\pi)}$ that is alternating (resp. symmetric) [20].

Remark 8.2.18. Part (ii) of the theorem is expected to hold without assuming that π is regular or self-dual. Recent works of Harris–Lan–Taylor–Thorne and of Sholze show that the self-duality assumption can be removed (but these authors do not prove part (i) for these π). Finally, let us mention that if $\pi \in \Pi_{\mathrm{cusp}}^{\perp}(\mathrm{PGL}_n)$ is algebraic and satisfies $|\mathrm{Weights}(\pi)| = n-1$ and $n \equiv 0 \bmod 4$, it is also known how to prove part (ii) [94], but not part (i).

Corollary 8.2.19. *Let G be a classical $\mathbb{Z}$-group and $\pi \in \Pi_{\mathrm{disc}}(G)$ such that π_∞ has the same infinitesimal character as a finite-dimensional irreducible representation of $G(\mathbb{C})$. We suppose that the Arthur–Langlands conjecture is true for (π, St). If $\widehat{G}(\mathbb{C}) \simeq \mathrm{SO}_m(\mathbb{C})$, we moreover suppose $m \not\equiv 2 \bmod 4$.*

Let ℓ be a prime, and let $\overline{\mathbb{Q}}_\ell$ be an algebraic closure of $\mathbb{Q}_\ell$ and $\iota\colon \overline{\mathbb{Q}} \to \overline{\mathbb{Q}}_\ell$ an embedding. There exists a unique (up to isomorphism) continuous semisimple representation, unramified outside ℓ,

$$\rho_{\pi,\iota}\colon \mathrm{Gal}(\overline{\mathbb{Q}}/\mathbb{Q}) \longrightarrow \mathrm{GL}_{\mathrm{n}_{\widehat{G}}}(\overline{\mathbb{Q}}_\ell)$$

such that for all $p \in \mathrm{P} - \{\ell\}$, we have

$$\det(t - \rho_{\pi,\iota}(\mathrm{Frob}_p)) = \iota\big(\det\big(t - \ \mathrm{St}(\mathrm{c}_p(\pi))\, p^{\mathrm{w}_{\widehat{G}}/2}\big)\big)\ .$$

The fact that $\det\big(t - \ \mathrm{St}(\mathrm{c}_p(\pi))\, p^{\mathrm{w}_{\widehat{G}}/2}\big)$ is an element of $\overline{\mathbb{Q}}[t]$ is part of the assertion (and can easily be verified directly in the cases that interest us).

Proof. Let us write $\psi(\pi, \mathrm{St}) = \oplus_{i=1}^{k} \pi_i[d_i]$. Theorem 8.2.17 and Remark 8.2.18 apply to the automorphic representations π_i, by Proposition 8.2.10 and Corollary 8.2.15 (iii). It then suffices to set

$$\rho_{\pi,\iota} = \bigoplus_{i=1}^{k} \rho_{\pi_i,\iota} \otimes \left(\oplus_{j=0}^{d_i-1} \omega_\ell^j\right) \otimes \omega_\ell^{(\mathrm{w}_{\widehat{G}}-\mathrm{w}(\pi_i)+1-d_i)/2} .$$

The uniqueness follows from the Chebotarev density theorem. □

Let us specify this result in the case of O_n, using formula (6.2.5).

Corollary 8.2.20. *Let* $n \equiv 0 \bmod 8$, *and let* $F \in \mathrm{M}_U(\mathrm{O}_n)$ *be an eigenform for* $\mathrm{H}(\mathrm{O}_n)$, *say such that* $\mathrm{T}_p(F) = \lambda_p F$ *for every prime* p. *We suppose that the Arthur–Langlands conjecture is true for the pair* (π, St), *where* $\pi \in \Pi_{\mathrm{disc}}(\mathrm{O}_n)$ *is the representation generated by* F *(Sect.* 6.4.7).

Let ℓ *be a prime, and let* $\overline{\mathbb{Q}}_\ell$ *be an algebraic closure of* $\mathbb{Q}_\ell$ *and* $\iota\colon \overline{\mathbb{Q}} \to \overline{\mathbb{Q}}_\ell$ *an embedding. There exists a unique (up to isomorphism) representation* $\rho_{F,\iota}\colon \mathrm{Gal}(\overline{\mathbb{Q}}/\mathbb{Q}) \longrightarrow \mathrm{GL}_n(\overline{\mathbb{Q}}_\ell)$ *that is continuous, semisimple, and unramified outside* ℓ, *such that for every prime* $p \neq \ell$, *we have* $\mathrm{trace}\, \rho_{F,\iota}(\mathrm{Frob}_p) = \iota(\lambda_p)$.

8.2.21 L*-Functions of Pairs of Algebraic Representations*

Let $\pi \in \Pi_{\mathrm{cusp}}(\mathrm{PGL}_n)$ and $\pi' \in \Pi_{\mathrm{cusp}}(\mathrm{PGL}_{n'})$; the L-function of the pair $\{\pi, \pi'\}$ is defined by the Euler product

$$\mathrm{L}(s, \pi \times \pi') = \prod_{p \in \mathrm{P}} \det(\mathrm{I}_{nn'} - p^{-s}\, \mathrm{c}_p(\pi) \otimes \mathrm{c}_p(\pi'))^{-1} .$$

This is a particular case of Langlands' construction, recalled in Sect. 6.4.11, where $\mathrm{G} = \mathrm{PGL}_n \times \mathrm{PGL}_{n'}$ and where r is the tensor product of the standard representations of SL_n and $\mathrm{SL}_{n'}$. We recover $\mathrm{L}(s, \pi)$ when $\pi' = 1$ is the trivial representation of PGL_1. This L-function has been studied by Rankin and Selberg when $n = n' = 2$ and by Jacquet, Piatetski-Shapiro, and Shalika for all n, n'. These authors prove that the Euler product above is absolutely convergent when $\Re\, s > 1$ and that it admits a meromorphic continuation to all of $\mathbb{C}$. Moreover, if $\mathrm{L}_\infty(s, \pi \times \pi')$ is a suitable product of Γ-*factors* and if $\xi(s, \pi \times \pi') = \mathrm{L}_\infty(s, \pi \times \pi')\, \mathrm{L}(s, \pi \times \pi')$, then we have a functional equation of the form

$$\xi(s, \pi \times \pi') = \varepsilon(\pi \times \pi')\, \xi(1 - s, \pi^\vee \times (\pi')^\vee) ,$$

where $\varepsilon(\pi \times \pi') \in \mathbb{C}^\times$. We refer to the lectures of Cogdell [62, Sect. 9] for an overview of these results. Let us add that if π and π' are self-dual, the relation $\xi(s, \pi \times \pi') = \varepsilon(\pi \times \pi')\, \xi(1 - s, \pi \times \pi')$ implies that $\varepsilon(\pi \times \pi') = \pm 1$ is just a sign. Let us recall the exact method for obtaining $\varepsilon(\pi \times \pi')$ and $\mathrm{L}_\infty(s, \pi \times \pi')$, for later use. They both depend only on the Archimedean components of π and π'. To simplify, we restrict this discussion to the case where π and π' are algebraic representations, the only case we will need.

Let $\mathrm{Rep}_{\mathrm{alg}}(\mathrm{W}_{\mathbb{R}})$ be the set of isomorphism classes of continuous and semisimple representations of $\mathrm{W}_{\mathbb{R}}$ on finite-dimensional $\mathbb{C}$-vector spaces that are trivial on the subgroup $\mathbb{R}_{>0} \subset \mathbb{C}^\times$ of $\mathrm{W}_{\mathbb{R}}$. The elements of $\mathrm{Rep}_{\mathrm{alg}}(\mathrm{W}_{\mathbb{R}})$ are exactly the direct sums of representations of the form 1, $\epsilon_{\mathbb{C}/\mathbb{R}}$, or I_w for $w > 0$ (Sect. 8.2.12). According to Weil, there is a unique way to associate with every $\rho \in \mathrm{Rep}_{\mathrm{alg}}(\mathrm{W}_{\mathbb{R}})$ a fourth root of unity $\varepsilon(\rho) \in \{1, i, -1, -i\}$ and a meromorphic function $\Gamma(s, \rho)$ in the complex variable s such that for every $\rho, \rho' \in \mathrm{Rep}_{\mathrm{alg}}(\mathrm{W}_{\mathbb{R}})$, we have

$$\varepsilon(\rho \oplus \rho') = \varepsilon(\rho)\varepsilon(\rho')\,, \quad \Gamma(s, \rho \oplus \rho') = \Gamma(s, \rho)\Gamma(s, \rho')\,,$$

as well as

(i) $\varepsilon(\mathrm{I}_w) = i^{w+1}$ and $\Gamma(s, \mathrm{I}_w) = \Gamma_{\mathbb{C}}(s + w/2)$ for every $w \geq 0$,
(ii) $\varepsilon(1) = 1$ and $\Gamma(s, 1) = \Gamma_{\mathbb{R}}(s)$.

Recall that $\Gamma(s) = \int_0^\infty e^{-t} t^s \, \mathrm{d}t/t$ if $\Re s > 0$ and that it is customary to set

$$\Gamma_{\mathbb{R}}(s) = \pi^{-s/2}\Gamma(s/2) \quad \text{and} \quad \Gamma_{\mathbb{C}}(s) = 2(2\pi)^{-s}\Gamma(s)\,,$$

so that $\Gamma_{\mathbb{C}}(s) = \Gamma_{\mathbb{R}}(s)\Gamma_{\mathbb{R}}(s+1)$ (duplication formula). Note that from the case $w = 0$, we deduce $\varepsilon(\epsilon_{\mathbb{C}/\mathbb{R}}) = i$ and $\Gamma(s, \epsilon_{\mathbb{C}/\mathbb{R}}) = \Gamma_{\mathbb{R}}(s+1)$.

Proposition 8.2.22. *Let $\pi \in \Pi_{\mathrm{cusp}}(\mathrm{PGL}_n)$ and $\pi' \in \Pi_{\mathrm{cusp}}(\mathrm{PGL}_{n'})$ be algebraic. Set $\rho = \mathrm{L}(\pi_\infty) \otimes \mathrm{L}(\pi'_\infty)$. We have*

$$\varepsilon(\pi \times \pi') = \varepsilon(\rho) \quad \textit{and} \quad \mathrm{L}_\infty(s, \pi \times \pi') = \Gamma(s, \rho)\,.$$

Proof. The statement concerning $\mathrm{L}_\infty(s, \pi \times \pi')$ has a meaning because $\mathrm{L}(\pi_\infty)$ and $\mathrm{L}(\pi'_\infty)$ are in $\mathrm{Rep}_{\mathrm{alg}}(\mathrm{W}_{\mathbb{R}})$ by Proposition 8.2.13 (i). The assertion $\mathrm{L}_\infty(s, \pi \times \pi') = \Gamma(s, \rho)$ holds by definition [62, Chap. 9]. A close examination of the formulas in [196, Sect. 3] shows that for every $\rho \in \mathrm{Rep}_{\mathrm{alg}}(\mathrm{W}_{\mathbb{R}})$, the number $\varepsilon(\rho)$ defined above is exactly the one denoted by $\varepsilon(\rho, \psi, \mathrm{d}x)$ loc. cit., where $\mathrm{d}x$ is the Lebesgue measure on $\mathbb{R}$ and $\psi : \mathbb{R} \to \mathbb{C}^\times$ is the character $x \mapsto e^{2i\pi x}$. Since π and π' are "unramified at all finite places" in the usual terminology, this factor $\varepsilon(\rho)$ therefore coincides with $\varepsilon(\pi \times \pi')$ [62, Chap. 9]. □

It follows from these formulas that $\varepsilon(\pi \times \pi')$ is an explicit function of the weights of π and π'. It is useful, at this point, to note that we have $\mathrm{I}_w \otimes \epsilon_{\mathbb{C}/\mathbb{R}} \simeq \mathrm{I}_w$ and

$$\mathrm{I}_w \otimes \mathrm{I}_{w'} \simeq \mathrm{I}_{w+w'} \oplus \mathrm{I}_{|w-w'|}$$

for all integers $w, w' \geq 0$. In particular, $\varepsilon(\mathrm{I}_w \otimes \mathrm{I}_{w'}) = (-1)^{1+\max(w,w')}$.

8.3 Arthur's Multiplicity Formula

8.3.1 Arthur's Symplectic-Orthogonal Alternative

Recall that if H is a classical $\mathbb{C}$-group, we denote by n_H the dimension of its standard representation.

Theorem⋆ 8.3.2 (Arthur). *Let $\pi \in \Pi_{\mathrm{cusp}}^{\perp}(\mathrm{PGL}_n)$. There exists a classical Chevalley $\mathbb{Z}$-group G^{π}, unique up to isomorphism, with the following properties:*

(i) *We have $\mathrm{n}_{\widehat{G^{\pi}}} = n$.*
(ii) *There exists a $\pi' \in \Pi_{\mathrm{disc}}(G^{\pi})$ such that $\mathrm{c}(\pi) = \psi(\pi', \mathrm{St})$.*

This is a particular case of [13, Theorems 1.4.1 and 1.5.2] (see also the *descent method* of Ginzburg, Rallis, and Soudry [93] for a weakened statement). By definition, the group G^{π} satisfies $\mathrm{n}_{\widehat{G^{\pi}}} = n$. When n is odd, the only possibility is therefore $G^{\pi} \simeq \mathrm{Sp}_{n-1}$, but when n is even, G^{π} is isomorphic to $\mathrm{SO}_{n/2,n/2}$ or $\mathrm{SO}_{n/2+1,n/2}$ (exclusively). If $n = 2$, then $G^{\pi} \simeq \mathrm{SO}_{2,1} \simeq \mathrm{PGL}_2$ because $\mathrm{SO}_{1,1} \simeq \mathbb{G}_m$ is not semisimple. Finally, when $n = 1$, so that π is the trivial representation of PGL_1, we have $G^{\pi} = \mathrm{SO}_{1,0}$ (the trivial $\mathbb{Z}$-group).

The representation $\pi \in \Pi_{\mathrm{cusp}}^{\perp}(\mathrm{PGL}_n)$ is called *orthogonal* if $\widehat{G^{\pi}}(\mathbb{C}) \simeq \mathrm{SO}_n(\mathbb{C})$ (or, equivalently, if $\mathrm{w}_{\widehat{G^{\pi}}} = 0$ in the notation of Sect. 8) and *symplectic* otherwise.

Proposition⋆ 8.3.3. *Let $\pi \in \Pi_{\mathrm{cusp}}^{\perp}(\mathrm{PGL}_n)$ be algebraic. Suppose that π has at least one weight that is a simple eigenvalue of $\mathrm{c}_{\infty}(\pi)$. Then π is symplectic if and only if $\mathrm{w}(\pi) \equiv 1 \bmod 2$.*

Proof. This is a variant of [55, Corollary 3.8]. Following Arthur [13, Theorem 11.4.2], the representation $\mathrm{L}(\pi_{\infty})$ of $\mathrm{W}_{\mathbb{R}}$ on $\mathbb{C}^n$ preserves a nondegenerate bilinear form b that is alternating if π is symplectic and symmetric otherwise. The assumption on π implies that at least one of the representations 1, $\varepsilon_{\mathbb{C}/\mathbb{R}}$, and I_w with $w > 0$ occurs in $\mathrm{L}(\pi_{\infty})$ with multiplicity 1 (Proposition 8.2.13 (i)); we denote the corresponding subspace by $E \subset \mathbb{C}^n$. Since each of these representations is irreducible and self-dual, the restriction of b to E is nondegenerate. Given the relation $\det \mathrm{I}_w = \varepsilon_{\mathbb{C}/\mathbb{R}}^{w+1}$, we see that b is alternating if and only if we have $E \simeq \mathrm{I}_w$ with $w \equiv 1 \bmod 2$. □

Arthur's results also have consequences for the L-functions of pairs of self-dual representations; see [13, Theorem 1.5.3]. In particular, if $\pi \in \Pi_{\mathrm{cusp}}^{\perp}(\mathrm{PGL}_n)$ and $\pi' \in \Pi_{\mathrm{cusp}}^{\perp}(\mathrm{GL}_m)$ are either both symplectic or both orthogonal, then $\varepsilon(\pi \times \pi') = 1$ (this is a ⋆-theorem). In the case where $\pi' = 1$, we deduce from this that

$$\varepsilon(\pi) := \varepsilon(\pi \times 1)$$

equals 1 if π is orthogonal. When π is algebraic, self-dual, and orthogonal, this gives a nontrivial relation on its weights; see [55, Proposition 1.8].

8.3.4 The Multiplicity Formula: General Assumptions

Let G be a classical $\mathbb{Z}$-group and $n = \mathrm{n}_{\widehat{G}}$ the dimension of the standard representation St of $\widehat{G}$. Fix an integer $k \geq 1$, as well as a pair (π_i, d_i) for every $i = 1, \ldots, k$, where $d_i \geq 1$ is an integer and $\pi_i \in \Pi_{\mathrm{cusp}}^{\perp}(\mathrm{PGL}_{n_i})$. We suppose $n = \sum_{i=1}^{k} n_i d_i$ and consider the element

$$\psi = \oplus_{i=1}^{k} \pi_i[d_i]$$

of $\mathcal{X}(\mathrm{SL}_n)$.

Let U be an irreducible unitary representation of $G(\mathbb{R})$. Arthur's multiplicity formula, conjectured in full generality in [9] and proved in [13] when G is a classical Chevalley group, gives a necessary and sufficient condition for the existence of $\pi \in \Pi_{\mathrm{disc}}(G)$ such that $\pi_\infty \simeq U$ and $\psi(\pi, \mathrm{St}) = \psi$. It can be expressed as the equality of two characters on an elementary finite abelian 2-group C_ψ, which we give explicitly in the next subsections. The first of these characters, denoted by ε_ψ and described in Sect. 8.3.5, is independent of U. It is introduced in great generality by Arthur in [9] and takes into account the signs $\varepsilon(\pi_i \times \pi_j)$ (Sect. 8.2.21) according to very precise combinatorics. The origin of the second of these characters, which is the most delicate of the two, goes back to the work of Shelstad [180] (see also [2, 3, 9, 13, 121, 133, 140, 181, 186]). It essentially depends only on U and on a certain morphism $\mathrm{SL}_2(\mathbb{C}) \times \mathrm{W}_{\mathbb{R}} \to \widehat{G}$ associated with ψ; we describe it in Sects. 8.3.8 and 8.4.14.

The work of Arthur [13] is very general, and we will apply it only in very particular cases, for which the statements are substantially simplified. We assume that the following conditions hold:

(H1) If $\widehat{G} \simeq \mathrm{SO}_n$, then $n \not\equiv 2 \bmod 4$.
(H2) $\psi_\infty = \mathrm{St}(\mathrm{Inf}_V)$, where $V \in \mathrm{Irr}(G_{\mathbb{C}})$ (Sect. 8.2.6).

The first assumption is only a constraint if G is an even special orthogonal group. In this case, $G(\mathbb{R})$ has signature (p, q) with $p \equiv q \bmod 8$ by Scholium 2.2.2 (b), so that the assumption can also be written $p \equiv q \equiv 0 \bmod 2$ (it is, of course, satisfied if $G = \mathrm{SO}_n$ with $n \equiv 0 \bmod 8$).

The second assumption, on ψ_∞, has been made explicit in Sect. 8.2.6 (case I, II, or III), where we have given several combinatorial consequences concerning the π_i. In particular,

(a) for every $i = 1, \ldots, k$, the representation π_i is self-dual and algebraic (and even regular in all but one exceptional case);
(b) for every $i = 1, \ldots, k$, we have $\mathrm{w}(\pi_i) + d_i - 1 \equiv \mathrm{w}_{\widehat{G}} \bmod 2$;
(c) for all $i \neq j$, if $(n_i, d_i) = (n_j, d_j)$, then $\pi_j \not\simeq \pi_i$.

Parts (a) and (b) follow from Proposition 8.2.10. Part (c), which is nontrivial under condition (H2) only if $n_i = n_j = 1$ and $d_i = d_j = 1$, follows from condition (H1) and Corollary 8.2.15 (iii).

8.3.5 The Group C_ψ and the Character ε_ψ

We keep the assumptions and notation of the previous subsection. By part (b) above and Proposition 8.3.3, there exists a $\mathbb{C}$-morphism

$$\nu\colon \mathrm{SL}_2 \times \prod_{i=1}^{k} \widehat{G^{\pi_i}} \longrightarrow \widehat{G}$$

such that the $\mathbb{C}$-representation $\mathrm{St} \circ \nu$, with underlying space $V \simeq \mathbb{C}^n$, decomposes into a direct sum

$$V = \oplus_{i=1}^{k} V_i \ ,$$

where V_i is isomorphic to the tensor product of the representation $\mathrm{Sym}^{d_i-1}\mathrm{St}_2$ of SL_2 and the standard representation of $\widehat{G^{\pi_i}}$ (the action of the other factors $\widehat{G^{\pi_j}}$ for $j \neq i$ is trivial). Such a morphism ν is not unique: it is only unique modulo composition at the target with an automorphism of the $\mathbb{C}$-group $\widehat{G}$. We fix it once and for all; when the time comes, we will discuss how the final formula depends on this choice.

Let C_ν be the centralizer of the image of ν in $\widehat{G}(\mathbb{C})$. The representation St identifies it with the subgroup of $\mathrm{SL}(V)$ consisting of the elements g that preserve each V_i and satisfy $g_{|V_i} = \epsilon_i \,\mathrm{Id}_{V_i}$, where $(\epsilon_i) \in \{\pm 1\}^k$. Since we have $\dim(V_i) = n_i d_i$, the group C_ν is therefore in a natural exact sequence

$$1 \longrightarrow \mathrm{C}_\nu \xrightarrow{\mathrm{St}} \{\pm 1\}^k \xrightarrow{\delta} \{\pm 1\} \ ,$$

where $\delta(\epsilon_i) = \prod_{i=1}^{k} \epsilon_i^{n_i d_i}$. This abstract description of C_ν is clearly independent of the choice of ν, which is why we denote it simply by C_ψ.

The center $\mathrm{Z}_{\widehat{G}}$ of $\widehat{G}(\mathbb{C})$ is a subgroup of C_ψ. We denote by $I \subset \{1, \ldots, k\}$ the subset consisting of the integers i such that $n_i d_i \equiv 0 \bmod 2$, and for every $i \in I$, we denote by

$$s_i \in \mathrm{C}_\psi$$

the element that acts by -1 on V_i and by 1 on V_j for $j \neq i$. By assumptions (H1) and (H2), we can apply Corollary 8.2.15. It implies $|I| \geq k-1$, as well as the following lemma.

Lemma 8.3.6. *The group C_ψ is generated by $\mathrm{Z}_{\widehat{G}}$ and the s_i for $i \in I$.*

Next, Arthur defines [13, p. 47] a homomorphism $\varepsilon_\psi\colon \mathrm{C}_\psi \to \{\pm 1\}$ that is trivial on $\mathrm{Z}_{\widehat{G}}$. To describe it, it suffices to give its value on the elements s_i for $i \in I$. For this, Arthur considers the restriction to ν of the adjoint representation of $\widehat{G}$ on $\mathrm{Lie}\,\widehat{G}$; this is a representation of the product $\mathrm{C}_\nu \times \mathrm{SL}_2 \times (\prod_{i=1}^{k} \widehat{G^{\pi_i}})$. If we fix the integer $i \in I$, it is an exercise to verify that the subspace of $\mathrm{Lie}\,\widehat{G}$ on which s_i acts by -1 is isomorphic to $\bigoplus_{j \neq i} V_j \otimes V_i$ as a representation of $\mathrm{SL}_2 \times (\prod_{i=1}^{k} \widehat{G^{\pi_i}})$. But if for $d \geq 1$, we denote by r_d the representation $\mathrm{Sym}^{d-1}\mathrm{St}_2$ of SL_2 (where St_2 denotes the standard representation, Sect. 6.4.4) and if $a \geq b \geq 1$ are integers, then it is well known that

$$r_a \otimes r_b \simeq \oplus_{i=1}^{b} r_{a-b+2i-1} \ ;$$

in particular, $r_a \otimes r_b$ has $\min(a, b)$ irreducible factors for every $a, b \geq 1$. The method for obtaining ε_ψ described by Arthur loc. cit. therefore takes on the following form, where we have incorporated the *-result of Arthur asserting that we have $\varepsilon(\pi \times \pi') = 1$ if π and π' are either both symplectic or both orthogonal.

Proposition-Definition 8.3.7. *There exists a unique homomorphism* $\varepsilon_\psi : \mathrm{C}_\psi \to \{\pm 1\}$ *that is trivial on* $\mathrm{Z}_{\widehat{G}}$ *and satisfies for all* $i \in I$ *the equality*

$$\varepsilon_\psi(s_i) = \prod_{j \neq i} \varepsilon(\pi_i \times \pi_j)^{\min(d_i, d_j)}.$$

The product above is taken over all $j = 1, \ldots, k$ distinct from i. By the *-result of Arthur mentioned above, we can even restrict ourselves to the integers $j = 1, \ldots, k$ such that $\mathrm{w}(\pi_j) \not\equiv \mathrm{w}(\pi_i) \bmod 2$. To justify the existence of ε_ψ directly, the reader should note that the s_i for $i \in I$ are linearly independent over $\mathbb{F}_2$ in C_ψ and that if they generate a subgroup that meets $\mathrm{Z}_{\widehat{G}}$ nontrivially, then $|I| = k$ and $\mathrm{Z}_{\widehat{G}}$ is generated by $\prod_{i=1}^{k} s_i$.

8.3.8 The Case of the Chevalley Groups

We keep the notation and assumptions of Sects. 8.3.4 and 8.3.5. Arthur considers a group morphism

$$\nu_\infty \colon \mathrm{SL}_2(\mathbb{C}) \times \mathrm{W}_\mathbb{R} \to \widehat{G}(\mathbb{C})$$

defined as follows.

For every $i = 1, \ldots, k$, an argument similar to that given in the proof of Proposition 8.3.3, based on Proposition 8.2.13, ensures that there exists a group morphism $\mu_i \colon \mathrm{W}_\mathbb{R} \to \widehat{G^{\pi_i}}(\mathbb{C})$ whose composition with the standard representation of $\widehat{G^{\pi_i}}$ is isomorphic to $\mathrm{L}((\pi_i)_\infty)$. This property determines μ_i uniquely modulo composition with $\mathrm{Aut}(\widehat{G^{\pi_i}})$ at the target, but it will be useful to arbitrarily fix such a μ_i. The morphism ν_∞ is, by definition, the composition of the diagonal morphism $(g, w) \mapsto (g, \prod_{i=1}^{k} \mu_i(w))$ and the morphism ν. The $\mathrm{Aut}(\widehat{G})$-orbit of ν_∞ in the set $\mathrm{Hom}(\mathrm{SL}_2(\mathbb{C}) \times \mathrm{W}_\mathbb{R}, \widehat{G}(\mathbb{C}))$ will be denoted by $\psi_\mathbb{R}$; it should not be confused with the semisimple conjugacy class ψ_∞, which contains considerably coarser information. The orbit $\psi_\mathbb{R}$ depends only on ψ (and not on the choice of ν or of the μ_i) and, even better, only on the set of k pairs $((\pi_i)_\infty, d_i)$ for $i = 1, \ldots k$.

Let C_{ν_∞} be the centralizer of the image of ν_∞ in $\widehat{G}$. We clearly have

$$\mathrm{C}_\nu \subset \mathrm{C}_{\nu_\infty} \ .$$

It is easy to describe C_{ν_∞} in the same way as we described C_ν earlier.

Lemma 8.3.9. *The representation* $\mathrm{St} \circ \nu_\infty$ *of* $\mathrm{SL}_2(\mathbb{C}) \times \mathrm{W}_\mathbb{R}$ *is semisimple, without multiplicities, and all its irreducible components are self-dual. In particular,* C_{ν_∞} *is an elementary, finite, abelian* 2*-group.*

Proof. This is [55, Lemma 3.15]. The second assertion follows from the first. The only point of the latter that does not follow directly from the definitions and condition (H2) is the assertion of multiplicity 1, which is nontrivial when ψ_∞ has 0 as a double eigenvalue (and therefore $\mathrm{w}_{\widehat{G}} = 0$). In this case, $\mathrm{St} \circ \nu_\infty$ could contain either 1 or $\epsilon_{\mathbb{C}/\mathbb{R}}$ with multiplicity 2 (with trivial action of the factor $\mathrm{SL}_2(\mathbb{C})$ in both cases). By condition (H1), Corollary 8.2.15 (iii), and Proposition 8.2.13 (iii), these characters cannot occur with multiplicity 2 in one and the same $\mathrm{L}((\pi_i)_\infty)$. They would therefore occur (necessarily with multiplicity 1) in $\mathrm{L}((\pi_i)_\infty)$ and $\mathrm{L}((\pi_j)_\infty)$ with $i \neq j$; moreover, in this case, $d_i = d_j = 1$ and n_i and n_j are odd and not congruent modulo 4; see loc. cit. However, since $\mathrm{w}(\pi_i) = \mathrm{w}(\pi_j) = 0$, this contradicts Proposition 8.2.13 (ii). □

In order to continue our analysis of Arthur's formula, we first assume $G = \mathrm{Sp}_{2g}$ or $\mathrm{SO}_{r+1,r}$. We denote by $\Pi_{\mathrm{unit}}(H)$ the set of isomorphism classes of irreducible unitary representations of the real Lie group H. Arthur [13, Theorem 1.5.1] associates a finite set[7] $\Pi(\nu_\infty)$ with ν_∞, usually called an *Arthur packet*, endowed with two maps

$$\Pi_{\mathrm{unit}}(G(\mathbb{R})) \xleftarrow{\ \iota\ } \Pi(\nu_\infty) \xrightarrow{u \mapsto \chi_u} \mathrm{Hom}_{\mathrm{groups}}(\mathrm{C}_{\nu_\infty}, \mathbb{C}^\times)\,. \tag{8.3.1}$$

The set $\Pi(\nu_\infty)$ and ι, as well as, in fact, χ once we have clarified the dependence of C_{ν_∞} on the choice of ν_∞, depend only on the $\mathrm{Aut}(\widehat{G})$-conjugacy class $\psi_\mathbb{R}$ of ν_∞, which is why we also write $\Pi(\psi_\mathbb{R})$ for $\Pi(\nu_\infty)$. Arthur proves a property that fully characterizes the triple $(\Pi(\psi_\mathbb{R}), \iota, \chi)$ [13, Theorem 2.2.1], without, however, describing this triple concretely. We will return to this point in the next subsection. As explained by Arthur [13, p. 42], we expect ι to be injective, so that $\Pi(\psi_\mathbb{R})$ would be defined as a subset of $\Pi_{\mathrm{unit}}(G(\mathbb{R}))$ and ι would simply be ignored.

Theorem⋆ 8.3.10 (Arthur's Multiplicity Formula [13, Theorem 1.5.2]). *Suppose* $G = \mathrm{Sp}_{2g}$ *or* $\mathrm{SO}_{r+1,r}$. *Fix* $\psi \in \mathcal{X}_{\mathrm{AL}}(\mathrm{SL}_{\mathrm{n}_{\widehat{G}}})$ *satisfying condition* (H2)*, as well as* $U \in \Pi_{\mathrm{unit}}(G(\mathbb{R}))$.

Let $\pi \in \Pi(G)$ *be the unique representation such that* $\pi_\infty \simeq U$ *and* $\mathrm{St}(\mathrm{c}_p(\pi)) = \psi_p$ *for every prime* p. *Then* $\pi \in \Pi_{\mathrm{disc}}(G)$ *if and only if there exists a* $u \in \Pi(\psi_\mathbb{R})$ *such that*

$$U = \iota(u) \quad \textit{and} \quad \chi_{u|\mathrm{C}_\nu} = \varepsilon_\psi\,.$$

[7] What we denote here by $\psi_\mathbb{R}$, C_{ν_∞}, and $\Pi(\psi_\mathbb{R})$ is denoted by ψ, S_ψ, and $\widetilde{\Pi}_\psi$, respectively, in Arthur's statement; moreover, Arthur does not give a name to ι and for $u \in \Pi(\nu_\infty)$, he writes the character χ_u as $x \mapsto \langle x, u\rangle$. The image of ι, a finite subset of $\Pi_{\mathrm{unit}}(G(\mathbb{R}))$, is commonly called the *Arthur packet associated with* $\psi_\mathbb{R}$.

More precisely, the multiplicity $\mathrm{m}(\pi)$ *of* π *in* $\mathcal{A}_{\mathrm{disc}}(G)$ (Sect. 4.3.2) *is exactly the number of elements* $u \in \Pi(\psi_{\mathbb{R}})$ *with the property above.*

Let us decipher the statement. First, note that the existence and uniqueness of π come from the fact that if $G = \mathrm{Sp}_{2g}$ or $\mathrm{SO}_{r+1,r}$, the map $\mathrm{St}\colon \widehat{G}(\mathbb{C})_{\mathrm{ss}} \to \mathrm{SL}_{\mathrm{n}_{\widehat{G}}}(\mathbb{C})_{\mathrm{ss}}$ is injective and its image is the set of classes equal to their inverse. In particular, the representation π of the statement is the unique possible candidate such that $\pi_\infty \simeq U$ and $\psi(\pi, \mathrm{St}) = \psi$. The theorem first asserts that if U is not in $\iota(\Pi(\psi_{\mathbb{R}}))$, then $\mathrm{m}(\pi) = 0$. Let us therefore suppose $U \in \iota(\Pi(\psi_{\mathbb{R}}))$ and also suppose, to simplify, that we know that $\iota^{-1}(U)$ is a singleton $\{u\}$. The theorem then asserts that $\mathrm{m}(\pi) \neq 0$ if and only if $\chi_{u|\mathrm{C}_\psi} = \varepsilon_\psi$, in which case $m(\pi) = 1$.

For the sake of completeness, let us now describe the remaining case of the group $G = \mathrm{SO}_{r,r}$, where $r \equiv 0 \mod 2$ by assumption (H1). In this case, $\widehat{G}(\mathbb{C}) = \mathrm{SO}_{2r}(\mathbb{C})$. The image of the map $\mathrm{St}\colon \widehat{G}(\mathbb{C})_{\mathrm{ss}} \to \mathrm{SL}_{2r}(\mathbb{C})_{\mathrm{ss}}$ is still the set of classes equal to their inverse, but the map is no longer injective: two semisimple elements of $\mathrm{SO}_{2r}(\mathbb{C})$ that are conjugate in $\mathrm{SL}_{2r}(\mathbb{C})$ are not always so in $\mathrm{SO}_{2r}(\mathbb{C})$, and may only be conjugate in $\mathrm{O}_{2r}(\mathbb{C})$. The nonempty fibers of the map above are therefore exactly the orbits of the natural action of

$$\mathrm{O}_{2r}(\mathbb{C})/\mathrm{SO}_{2r}(\mathbb{C}) = \mathrm{Out}(\widehat{G}) = \mathbb{Z}/2\mathbb{Z}$$

on $\widehat{G}(\mathbb{C})_{\mathrm{ss}}$. An analogous phenomenon occurs for $\mathrm{St}\colon \widehat{\mathfrak{g}}_{\mathrm{ss}} \to (\mathfrak{sl}_{2r})_{\mathrm{ss}}$. The action of $\mathrm{O}_{r,r}(\mathbb{R})$ on $G(\mathbb{R})$ by conjugation also defines an action of $\mathbb{Z}/2\mathbb{Z} = \mathrm{O}_{r,r}(\mathbb{R})/\mathrm{SO}_{r,r}(\mathbb{R})$ on $\Pi_{\mathrm{unit}}(G(\mathbb{R}))$, whose set of orbits Arthur denotes by $\widetilde{\Pi}_{\mathrm{unit}}(G(\mathbb{R}))$. His theorem [13, Theorem 1.5.1] then associates with ν_∞ a triple $(\Pi(\psi_{\mathbb{R}}), \iota, \chi)$ as above, with the only difference that ι is now a map $\Pi(\psi_{\mathbb{R}}) \to \widetilde{\Pi}_{\mathrm{unit}}(G(\mathbb{R}))$.

Theorem* 8.3.11 (Arthur's Multiplicity Formula for the Special Orthogonal $\mathbb{Z}$-group $\mathrm{SO}_{r,r}$ [13, Theorem 1.5.2]). *Suppose* $G = \mathrm{SO}_{r,r}$ *with* $r \equiv 0 \bmod 2$. *Fix* $\psi \in \mathcal{X}_{\mathrm{AL}}(\mathrm{SL}_{2r})$ *satisfying condition* (H2), *as well as* $U \in \widetilde{\Pi}_{\mathrm{unit}}(G(\mathbb{R}))$. *Let* $E \subset \Pi(G)$ *be the set of* $\pi \in \Pi(G)$ *such that* $\psi(\pi, \mathrm{St}) = \psi$ *and* $\pi_\infty \in U$. *Let* $F \subset \Pi(\psi_{\mathbb{R}})$ *be the set of elements* u *such that* $U \in \iota(u)$ *and* $\chi_{u|\mathrm{C}_\nu} = \varepsilon_\psi$. *Then we have*

$$\sum_{\pi \in E} \mathrm{m}(\pi) = m_\psi |F| ,$$

where $m_\psi = 1$ *unless* $\psi = \oplus_{i=1}^k \pi_i[d_i]$ *with* $d_i \equiv 0 \bmod 2$ *for every* i, *in which case* $m_\psi = 2$.

(Arthur still expects the injectivity of ι, and therefore $|F| \in \{0, 1\}$.) In order to apply Theorem 8.3.10 or 8.3.11, it is obviously crucial to know more about the triple $(\Pi(\psi_{\mathbb{R}}), \iota, \chi)$. The object of the subsections that follow is to recall what is known on this subject.

8.4 Discrete Series

8.4.1 Discrete Series, Following Harish-Chandra

The results of this subsection are due to Harish-Chandra. It will be convenient to refer to Knapp's book [119] for the proofs.

Let H be a semisimple $\mathbb{R}$-group. Recall that $\pi \in \Pi_{\mathrm{unit}}(H(\mathbb{R}))$ is a *discrete series* if it occurs as a subrepresentation of the regular representation $\mathrm{L}^2(H(\mathbb{R}))$ [119, Chap. IX, Sect. 3]. Harish-Chandra proves that $H(\mathbb{R})$ admits discrete series if and only if H has an *anisotropic maximal torus*, that is, a sub-$\mathbb{R}$-group $T \subset H$ such that $T_{\mathbb{C}} \subset H_{\mathbb{C}}$ is a maximal torus and $T(\mathbb{R})$ is compact [119, Theorem 12.20]. This is the case for Sp_{2g}, as well as for the real special orthogonal group with signature (p,q) if and only if $pq \equiv 0 \bmod 2$ or $p+q \equiv 1 \bmod 2$, so that condition (H1) of Sect. 8.3.4 is in fact equivalent to requiring that $G(\mathbb{R})$ admits discrete series.

From now on, we assume that $H(\mathbb{R})$ admits discrete series. The anisotropic maximal tori of H form a single orbit under the action of $H(\mathbb{R})$ by conjugation; we fix one of them, which we denote by T. Let us introduce several objects associated with the pair (H,T). We will simply write $\mathrm{X}_*(T)$ (resp. $\mathrm{X}^*(T)$) for $\mathrm{X}_*(T_{\mathbb{C}})$ (resp. $\mathrm{X}^*(T_{\mathbb{C}})$). Let

$$\Phi = \Phi(H_{\mathbb{C}}, T_{\mathbb{C}}) \subset \mathrm{X}^*(T) \quad \text{and} \quad W = \mathrm{W}(H_{\mathbb{C}}, T_{\mathbb{C}}) \; ;$$

the latter is the Weyl group of Φ. We also set

$$W_{\mathrm{r}} = \mathrm{W}(H,T) \stackrel{\mathrm{def}}{=} \mathrm{N}_{H(\mathbb{R})}(T(\mathbb{R}))/T(\mathbb{R}) \; ,$$

where $\mathrm{N}_{H(\mathbb{R})}(T(\mathbb{R}))$ is the normalizer of $T(\mathbb{R})$ in $H(\mathbb{R})$ (the *real Weyl group* of (H,T)). Finally, we have a unique maximal compact subgroup K of $H(\mathbb{R})$ containing $T(\mathbb{R})$ (and therefore $\mathrm{N}_{H(\mathbb{R})}(T(\mathbb{R}))$), and we denote by $\Phi_{\mathrm{c}} \subset \Phi$ the system of (so-called *compact*) roots of K with respect to $T(\mathbb{R})$. We have natural inclusions $\mathrm{W}(\Phi_{\mathrm{c}}) \subset W_{\mathrm{r}} \subset W$, which are in general strict.

For $V \in \mathrm{Irr}(H_{\mathbb{C}})$, we denote by $\Pi_V \subset \Pi_{\mathrm{unit}}(H(\mathbb{R}))$ the set of discrete series with the same infinitesimal character as V. Harish-Chandra proves that every discrete series of $H(\mathbb{R})$ belongs to Π_V for a unique V. For a basis Δ of the root system Φ, he defines a representation $\pi_{\Delta,V} \in \Pi_V$, uniquely determined by the values taken on by its character $\Theta_{\Delta,V}$ on the set $T(\mathbb{R})_{\mathrm{reg}} \subset T(\mathbb{R})$ of elements $t \in T(\mathbb{R})$ such that $\alpha(t) \neq 1$ for every $\alpha \in \Phi$. Specifically, if we write t^μ for $\mu(t)$ if $t \in T(\mathbb{R})$ and $\mu \in \mathrm{X}^*(T)$, we have

$$\Theta_{\Delta,V}(t) = (-1)^{\frac{1}{2}\dim H(\mathbb{R})/K} \frac{\sum_{w \in W_{\mathrm{r}}} \varepsilon(w) t^{w(\lambda+\rho)-\rho}}{\prod_{\alpha \in \Phi^+}(1-t^{-\alpha})} \quad \forall t \in T(\mathbb{R})_{\mathrm{reg}} \; ,$$

where $\lambda \in \mathrm{X}^*(T)$ denotes the highest weight of V with respect to the Borel subgroup $T_{\mathbb{C}} \subset B \subset H_{\mathbb{C}}$ defined by Δ, $\Phi^+ \subset \Phi$ is the positive system defined by Δ, and 2ρ is the sum of the elements of Φ^+ [119, Theorems 9.20 and 12.7]. Harish-Chandra proves that every element of Π_V is of the form $\pi_{\Delta,V}$ and that $\pi_{\Delta,V} \simeq \pi_{\Delta',V}$ if and only if $\Delta' = w(\Delta)$ with $w \in W_{\mathrm{r}}$ [119, Theorem 12.21]. In particular, $|\Pi_V| = |W_{\mathrm{r}} \backslash W|$ is independent of V.

Suppose that there exists a subgroup $S \subset K$ isomorphic to $\mathbb{S}^1$ such that the set of fixed points of S for the adjoint action on Lie $H(\mathbb{R})$ is exactly $\mathrm{Lie}(K)$ (in particular, S is in the center of the neutral component of K). This is the case, for example, when H is Sp_{2g} or the special orthogonal group with signature $(p, 2)$ with $p \geq 1$. In this case, $H(\mathbb{R})$ has *holomorphic* discrete series [119, Chap. VI]; we already encountered examples in Sect. 6.3.4. In the Harish-Chandra classification, these are exactly the representations $\pi_{\Delta,V}$ obtained when Δ is the basis of a positive system of the form $\{\alpha \in \Phi\,;\ \varphi(\alpha) > 0\}$, where $\varphi\colon \mathrm{X}^*(T) \to \mathbb{R}$ is a linear form such that

$$\forall\, \alpha \in \Phi_c\,,\ \forall\, \beta \in \Phi - \Phi_c\,,\quad 0 < |\varphi(\alpha)| < |\varphi(\beta)|$$

(the compact roots are "smaller" than the noncompact roots), by the remark following [119, Theorem 9.20].

8.4.2 Shelstad's Canonical Parametrization, the Case of Split Groups

We need to recall a second parametrization of the elements of Π_V, where $V \in \mathrm{Irr}(H_{\mathbb{C}})$ is fixed, that comes up in the statement of Arthur's multiplicity formula; as we already mentioned in Sect. 8.3.4, this parametrization is due to Shelstad. It comes from the existing identities between the characters of the discrete series of H and those of the discrete series of a collection of associated $\mathbb{R}$-groups, called *endoscopic* by Langlands [180]. A detailed exposition of these identities would go well beyond the scope of this book (and the authors' competence), and we will not venture into it. We follow the overview of Shelstad [183], who, in particular, describes the precise normalizations used by Arthur, while sometimes borrowing the illuminating point of view of Adams [1].

Let us, from now on, make the additional assumption that H is split over $\mathbb{R}$ (and semisimple and such that $H(\mathbb{R})$ has discrete series). Let T be an anisotropic maximal torus of H; we use the notation of Sect. 8.4.1 for the associated objects. Shelstad's parametrization is completely canonical only if H is adjoint. In general, it will depend on the choice of a W_{r}-orbit of the set $\mathcal{B}(T)$ of bases Δ of the root system Φ such that $\Delta \cap \Phi_{\mathrm{c}} = \emptyset$. This set is nonempty and endowed with a natural simply transitive action of the real Weyl group $W_{\mathrm{r}}^{\mathrm{ad}}$ of $(H/\mathrm{Z}(H))(\mathbb{R})$ with respect to the image of $T(\mathbb{R})$, which satisfies $W_{\mathrm{r}} \subset W_{\mathrm{r}}^{\mathrm{ad}} \subset W$. The $\mathrm{W}(H,T)$-set $\mathcal{B}(T)$ obviously depends on the choice of T, but the fact that the set of these T forms a single $H(\mathbb{R})$-conjugacy class ensures that the quotient map $\mathrm{W}(H,T)\backslash\mathcal{B}(T)$ depends only on H; we denote it by $\mathcal{B}(H)$. Results of Kostant and Vogan show that the choice of a W_{r}-orbit $O \subset \mathcal{B}(T)$, which we also denote by $O \in \mathcal{B}(H)$, is equivalent to that of an *equivalence class of Whittaker data* D for $H(\mathbb{R})$, a notion we will not introduce here, but which is exactly the reference datum for Arthur: the unique discrete series in Π_V that is *generic* for D is $\pi_{\Delta,V}$ for $\Delta \in O$.

Definition 8.4.3. Let T be an anisotropic maximal torus of H, $\Delta \in \mathcal{B}(T)$, and $\rho^\vee$ the half-sum of the coroots of T positive with respect to Δ. Shelstad's parametrization of Π_V with respect to (T, Δ) is the map

$$\kappa_\Delta \colon \Pi_V \longrightarrow \mathrm{X}_*(T) \otimes \mathbb{Z}/2\mathbb{Z}\,, \quad \pi \mapsto \kappa_\Delta(\pi)$$

defined by $\kappa_\Delta(\pi_{w^{-1}\Delta,V}) \equiv w\rho^\vee - \rho^\vee \bmod 2\mathrm{X}_*(T)$ for every $w \in W$.

Let us emphasize that $\rho^\vee \in \frac{1}{2}\mathrm{X}_*(T)$ is not, in general, in $\mathrm{X}_*(T)$. On the other hand, the term $w\rho^\vee - \rho^\vee$ is indeed in $\mathrm{X}_*(T)$: it even belongs to the subgroup generated by the coroots. We will see further on that $\pi \mapsto \kappa_\Delta(\pi)$ is injective.

Let us explain why this definition coincides with the one given by Shelstad [183, Sect. 8]. Following Langlands and Shelstad, we consider the first cohomology group $H^1(\mathbb{R}, T)$ of $T(\mathbb{C})$, the latter being viewed as a $\mathbb{Z}[\mathrm{Gal}(\mathbb{C}/\mathbb{R})]$-module. Since we have $\mathrm{Gal}(\mathbb{C}/\mathbb{R}) = \mathbb{Z}/2\mathbb{Z} = \langle 1, \sigma\rangle$, $H^1(\mathbb{R}, T)$ is simply the quotient of the abelian group

$$Z^1(\mathbb{R}, T) = \{t \in T(\mathbb{C})\,;\, t\sigma(t) = 1\}$$

by the subgroup of elements of the form $t\sigma(t)^{-1}$ with $t \in T(\mathbb{C})$. Let us first recall that $H^1(\mathbb{R}, T)$ can be canonically identified with the target of the map κ_Δ (Tate–Nakayama duality). Indeed, let $T_2 = \{t \in T(\mathbb{R})\,;\, t^2 = 1\} = T(\mathbb{R}) \cap Z^1(\mathbb{R}, T)$. The composition of the natural maps

$$T_2 \hookrightarrow Z^1(\mathbb{R}, T) \to H^1(\mathbb{R}, T) \tag{8.4.1}$$

is clearly an isomorphism because the torus T is $\mathbb{R}$-isomorphic to a finite product of copies of $\mathbb{S}^1$. For $\mu \in \mathrm{X}_*(T) \otimes \mathbb{C}$, we denote by e^μ the unique element $z \in T(\mathbb{C})$ such that $\lambda(z) = e^{\langle\lambda,\mu\rangle}$ for every $\lambda \in \mathrm{X}^*(T)$. The map $\mu \mapsto e^{i\pi\mu}$ therefore induces a natural isomorphism $\mathrm{X}_*(T) \otimes \mathbb{Z}/2\mathbb{Z} \xrightarrow{\sim} T_2$. By putting the isomorphisms above end-to-end, we obtain the natural isomorphism

$$H^1(\mathbb{R}, T) \xrightarrow{\sim} \mathrm{X}_*(T) \otimes \mathbb{Z}/2\mathbb{Z} \tag{8.4.2}$$

announced earlier.

We also have a natural action of W on $H^1(\mathbb{R}, T)$ induced by the inclusion $T \subset H$, defined as follows [37]. If N is the normalizer of $T(\mathbb{C})$ in $H(\mathbb{C})$, then $n\sigma(n)^{-1} \in T(\mathbb{C})$ for every $n \in N$, and therefore $(n, t) \mapsto n \star t := nt\sigma(n)^{-1}$ defines an action of N on $Z^1(\mathbb{R}, T)$. By passing to the quotient, it induces an action of $W = N/T(\mathbb{C})$ on $H^1(\mathbb{R}, T)$, which we also denote by $(w, x) \mapsto w \star x$. By definition, for $w \in W$, the element $\kappa_\Delta(\pi_{w^{-1}\Delta,V}) \in \mathrm{X}_*(T) \otimes \mathbb{Z}/2\mathbb{Z}$ considered by Shelstad [183, p. 15] is the image of $w \star 1 \in H^1(\mathbb{R}, T)$, where 1 denotes the identity element of the group $H^1(\mathbb{R}, T)$, by the isomorphism (8.4.2).

One should be aware that the isomorphism $\gamma \colon T_2 \to H^1(\mathbb{R}, T)$ defined by formula (8.4.1) does not intertwine the obvious action of W on T_2 by conjugation and that on $H^1(\mathbb{R}, T)$ defined above. However, the identity $gt\sigma(g)^{-1} = gtg^{-1}g\sigma(g)^{-1}$ shows that this does hold for the action of the subgroup W_{r}. To avoid confusion, we denote by $(w, t) \mapsto w(t)$ the usual action of W on $T(\mathbb{C})$ and by $(w, t) \mapsto w \cdot t = \gamma^{-1}(w \star \gamma(t))$ the "twisted" action on T_2. The exact relation between the two is given by part (ii) of the following lemma, essentially due to Langlands [179, Sect. 3], [130, Lemma 5.1], [1, Lemma 7.8]. Fix $\Delta \in \mathcal{B}(T)$, and let $\rho^\vee \in \frac{1}{2}\mathrm{X}_*(T)$ be the half-sum of the coroots of $(H_\mathbb{C}, T_\mathbb{C})$ positive with respect to Δ. Following [1], we set

$$t_\mathrm{b} = e^{i\pi\rho^\vee} \in T(\mathbb{C}) .$$

We have $t_\mathrm{b}^4 = 1$; hence $t_\mathrm{b} \in T(\mathbb{R})$.

Lemma 8.4.4. (i) *The centralizer of t_b in $H(\mathbb{R})$ is K, and we have $t_\mathrm{b}^2 \in \mathrm{Z}(H)$.*
(ii) *For every $t \in T_2$ and $w \in W$, we have $w \cdot t = w(tt_\mathrm{b})t_\mathrm{b}^{-1}$.*

The relation $w \cdot 1 = e^{i\pi(w\rho^\vee - \rho^\vee)}$ follows and concludes the proof of our claim that Definition 8.4.3 coincides with Shelstad's definition. Note that part (i) implies that the element t_b depends only on the W_r-orbit of Δ in $\mathcal{B}(T)$.

Proof. If $\alpha \in \Phi$, then $\alpha(t_\mathrm{b}) = (-1)^{\langle \rho^\vee, \alpha \rangle}$. Let $s \colon \Phi \to \mathbb{Z}/2\mathbb{Z}$ be the map such that $s^{-1}(0) = \Phi_c$. The Cartan decomposition of Lie $H(\mathbb{R})$ with respect to K shows that we have $s(\alpha + \beta) = s(\alpha) + s(\beta)$ whenever α, β, and $\alpha + \beta$ are in Φ. By induction on the "height" $|\langle \rho^\vee, \alpha \rangle|$, we have $s(\alpha) \equiv \langle \rho^\vee, \alpha \rangle \bmod 2$ for every $\alpha \in \Phi$. This shows that the inner automorphism of H defined by t_b is the Cartan involution of $H(\mathbb{R})$ with respect to K, and therefore part (i).

Since we have $t_\mathrm{b}^2 \in \mathrm{Z}(H)$, the function $f(w) = w(t_\mathrm{b})t_\mathrm{b}^{-1}$ defines a 1-cocycle of W with values in T_2. Likewise, $g(w) = w \cdot 1$ is also a 1-cocycle with values in T_2. Since $w \cdot t = w(t)\, w \cdot 1$, it suffices to see that f and g coincide on the s_α with $\alpha \in \Delta$ or, equivalently, that $\mathrm{s}_\alpha \cdot 1 = f(s_\alpha) = e^{i\pi\alpha^\vee}$. This is exactly the computation of Langlands [179, Proposition 2.1]. This proves part (ii). □

Finally, note that if $\Delta \in \mathcal{B}(T)$, $w \in W_\mathrm{r}$, and $\pi \in \Pi_V$, then we have

$$\kappa_{w\Delta}(\pi) = w\kappa_\Delta(\pi) , \tag{8.4.3}$$

because of the immediate identity $(ww') \cdot 1 = w \cdot (w' \cdot 1)$ for every $w' \in W$. In particular, the W_r-orbit of $\kappa_\Delta(\pi)$ depends only on that of Δ in $\mathcal{B}(T)$.

8.4.5 Dual Interpretation and Link with Arthur Packets

We keep the notation and assumptions of Sects. 8.4.1 and 8.4.2. Before giving an example, let us give the useful dual interpretation of Shelstad's parametrization. Let $\widehat{H}$ be the dual $\mathbb{C}$-group of $H_\mathbb{C}$ (recall that H is split over $\mathbb{R}$). Following Langlands [137], there exists a natural bijection between $\mathrm{Irr}(H_\mathbb{C})$ and the set of $\widehat{H}$-conjugacy classes of *discrete parameters* $\varphi\colon \mathrm{W}_{\mathbb{R}} \to \widehat{H}$. Recall that, by definition, such an object is a continuous group morphism such that the subgroup $\varphi(\mathrm{W}_{\mathbb{R}}) \subset \widehat{H}$ consists of semisimple elements and has a finite centralizer in $\widehat{H}$, which we denote by C_φ. This bijection is characterized by the fact that the infinitesimal character of such a φ coincides with that of the corresponding representation in $\mathrm{Irr}(H_{\mathbb{C}})$.

Let us explain this assertion. The subgroup $\varphi(\mathbb{C}^\times) \subset \widehat{H}$ is commutative and connected, and consists of semisimple elements. Its Zariski closure in $\widehat{H}$ is therefore a torus, and its centralizer in $\widehat{H}$, which we denote by S, is a Levi subgroup (of a parabolic subgroup) of $\widehat{H}$. By definition, $\mathrm{C}_\varphi \subset S$ is the subgroup fixed by conjugation by $\varphi(j)$. Since it is finite, this forces S to be a maximal torus and $\varphi(j)$ to act by the inversion $s \mapsto s^{-1}$ (in particular, -1 is an element of the Weyl group of $\widehat{H}$, which is indeed the case under our assumption on H). Let $\lambda_\phi, \mu_\phi \in \mathrm{X}_*(S) \otimes \mathbb{C}$ be the unique elements such that $\lambda_\phi - \mu_\phi \in \mathrm{X}_*(S)$ and $\xi(\varphi(z)) = z^{\langle \xi, \lambda_\phi \rangle} \overline{z}^{\langle \xi, \mu_\phi \rangle}$ for every $z \in \mathbb{C}^\times$ and $\xi \in \mathrm{X}^*(S)$ (see the footnote in Definition 8.2.7, p. 195). By definition, the infinitesimal character of φ is the $\widehat{H}$-conjugacy class of λ_ϕ, viewed in $\mathrm{Lie}\,\widehat{H}$ (formula (6.3.1)). Moreover, we have $\mu_\phi = -\lambda_\phi$.

Let us now fix a discrete parameter $\varphi\colon \mathrm{W}_{\mathbb{R}} \to \widehat{H}$. Note that if φ' is in the $\widehat{H}$-conjugacy class φ and if $h \in \widehat{H}$ satisfies $\varphi' = h\varphi h^{-1}$, the isomorphism $\mathrm{C}_\varphi \to \mathrm{C}_{\varphi'}$ induced by conjugation by h is independent of the choice of h, so that C_φ is a canonical abelian group. Let S be the maximal torus of $\widehat{H}$ containing $\varphi(\mathbb{C}^\times)$ and $B \subset \widehat{H}$ the unique Borel subgroup containing S such that the element $\lambda_\phi \in \frac{1}{2}\mathrm{X}_*(S)$ defined above is dominant with respect to B. Together, $\Delta \in \mathcal{B}(T)$, which defines a unique Borel subgroup $T_{\mathbb{C}} \subset Q \subset H_{\mathbb{C}}$, and $\widehat{H}$ define a unique isomorphism $\Psi(H_{\mathbb{C}}, T, Q)^\vee \simeq \Psi(\widehat{H}, S, B)$ and, in particular, a distinguished isomorphism $\widehat{T} \to S$ or, equivalently, an isomorphism

$$i_\Delta\colon \mathrm{X}_*(T) \xrightarrow{\sim} \mathrm{X}^*(S)\,.$$

That being said, $\mathrm{C}_\varphi = S_2 = \{s \in S\,;\, s^2 = 1\}$ and the natural map $\beta\colon \mathrm{X}_*(S) \otimes \mathbb{Z}/2\mathbb{Z} \to \mathrm{Hom}(S_2, \mathbb{C}^\times)$ is a group isomorphism. To conclude, Shelstad's map κ_Δ (Definition 8.4.3) induces a natural map

$$\chi_O\colon \Pi_V \longrightarrow \mathrm{Hom}_{\mathrm{groups}}(\mathrm{C}_\varphi, \mathbb{C}^\times)\,, \tag{8.4.4}$$

defined by $\chi_O := \beta \circ (i_\Delta \otimes \mathbb{Z}/2\mathbb{Z}) \circ \kappa_\Delta$, where $O \in \mathcal{B}(T)$ denotes the W_{r}-orbit of Δ. In fact, for $w \in W$, we have $i_{w\Delta} = i_\Delta \circ w^{-1}$, so that the map χ_O indeed depends only on O, by the relation (8.4.3). Every homomorphism in the image of χ_O is trivial on $\mathrm{Z}(\widehat{H})$ because $w\rho^\vee - \rho^\vee$ is a sum of roots of $(\widehat{H}, S)$ for every $w \in W$. The map χ_O is injective but not surjective in general.

The link with Sect. 8.3.8 is that if the homomorphism ν_∞ defined loc. cit. is trivial on the factor $\mathrm{SL}_2(\mathbb{C})$, which is equivalent to requiring $d_i = 1$ for every $i = 1, \ldots, k$, then ν_∞ is a discrete parameter $\mathrm{W}_{\mathbb{R}} \to \widehat{G}$ (Lemma 8.3.9) with the same infinitesimal character as the representation $V \in \mathrm{Irr}(G_{\mathbb{C}})$ fixed by condition (H2) in Sect. 8.3.4. To proceed, we need to treat the case $\mathrm{SO}_{r,r}$ separately.

(a) Suppose $G = \mathrm{Sp}_{2g}$ or $\mathrm{SO}_{r+1,r}$. Then by [183, 184] and [146], the set $\Pi(\psi_{\mathbb{R}})$ considered by Arthur is Π_V, the map ι is the obvious inclusion $\Pi_V \subset \Pi_{\mathrm{unit}}(G(\mathbb{R}))$, and the map χ defined in Diagram (8.3.1) is the map χ_O defined above. When $G = \mathrm{SO}_{r+1,r}$ (adjoint), there is only one possible choice of $O \in \mathcal{B}(G_{\mathbb{R}})$, so that everything is canonically defined. When $G = \mathrm{Sp}_{2g}$, there are exactly two choices (see below), and we must, of course, make the same choice as Arthur [13, p. 55]. Since the natural map $\mathrm{PGSp}_{2g}(\mathbb{Z}) \to \pi_0(\mathrm{PGSp}_{2g}(\mathbb{R})) = \mathbb{Z}/2\mathbb{Z}$ is surjective, this choice will not play any role in our applications.

(b) Finally, suppose $G = \mathrm{SO}_{r,r}$ with $r \equiv 0 \bmod 2$. The group with two elements $\mathrm{O}_{r,r}(\mathbb{R})/\mathrm{SO}_{r,r}(\mathbb{R}) = \{1, \theta\}$ has a natural action on $\Pi_{\mathrm{unit}}(G(\mathbb{R}))$ and $\mathrm{Irr}(G_{\mathbb{C}})$. Note that θ induces a bijection $\Pi_V \xrightarrow{\sim} \Pi_{\theta(V)}$.

Lemma 8.4.6. *Let $H = \mathrm{SO}_{r,r}$ with $r \equiv 0 \bmod 2$, let T be an anisotropic maximal torus of H, $\Phi = \Phi(H_{\mathbb{C}}, T_{\mathbb{C}})$, Δ a basis of Φ, and $V \in \mathrm{Irr}(H_{\mathbb{C}})$. Then we have $\theta(\pi_{\Delta,V}) \simeq \pi_{\Delta,\theta(V)}$.*

Proof. It is equivalent to fix T or a decomposition of $\mathrm{H}(\mathbb{R}^r)$ as an orthogonal sum $\oplus_{i\in I} P_i$ of planes P_i supposed (positive or negative) definite. Having done this, it is equivalent to fix Δ, a Borel subgroup of $H_{\mathbb{C}}$ containing $T_{\mathbb{C}}$, or a total order on the set I with, for each $i \in I$, a choice of one of the two isotropic lines in $P_i \otimes \mathbb{C}$ (Sect. 6.1.3). Let i_0 be the greatest element of I. Let $s \in \mathrm{O}(\mathrm{H}(\mathbb{R}^r))$ be the unique element that acts as the identity on P_i for $i < i_0$ and interchanges the two isotropic lines in P_{i_0}. This is a representative of θ that preserves T as well as the basis $\Delta \subset \Phi$. This allows us to conclude the proof because the characteristic property of $\pi_{\Delta,V}$ shows $\theta(\pi_{\Delta,V}) = \pi_{s(\Delta),\theta(V)}$ and we have $s(\Delta) = \Delta$. □

In particular, if $\theta(V) \simeq V$, then θ acts by the identity on Π_V: every $\pi \in \Pi_V$ extends to $\mathrm{O}(\mathrm{H}(\mathbb{R}^r))$. Let us return to the $\mathbb{Z}$-group $G = \mathrm{SO}_{r,r}$. By the results of Shelstad and Mezo mentioned above, the set $\Pi(\psi_{\mathbb{R}})$ considered by Arthur is the image $\widetilde{\Pi}_V$ of Π_V in $\widetilde{\Pi}_{\mathrm{unit}}(G(\mathbb{R}))$, and the map ι is the obvious inclusion. For $\pi \in \Pi_V$, Lemma 8.4.6 asserts that π and $\theta(\pi)$ have the same Shelstad character, which, by passing to the quotient, provides a well-defined map $\widetilde{\Pi}_V \to \mathrm{Hom}_{\mathrm{groups}}(\mathrm{C}_{\psi_\infty}, \mathbb{C}^\times)$: this is the map considered by definition by Arthur. To be completely exact, as in the case $G = \mathrm{PGSp}_{2g}$, the set $\mathcal{B}(G_{\mathbb{R}})$ contains two elements, and we must choose the one that corresponds to the Whittaker datum fixed by Arthur, but, again, this choice will not play a role in our applications.

8.4.7 Example: The Holomorphic Discrete Series of $\mathrm{Sp}_{2g}(\mathbb{R})$

Consider the group $H = \mathrm{Sp}_{2g}$, with dual $\mathbb{C}$-group $\widehat{H} = \mathrm{SO}_{2g+1}$. Set $E = \mathrm{H}(\mathbb{R}^g)$ endowed with its hyperbolic alternating form a, so that $H = \mathrm{Sp}_E$. Recall that if $I \in \mathrm{Sp}(E)$ is such that $I^2 = -\mathrm{id}_E$, it endows E with a complex structure, as well as a nondegenerate Hermitian form h for this structure, with associated bilinear form

$$(e,f) \mapsto \mathrm{a}(Ie,f) + i\mathrm{a}(e,f) .$$

The centralizer of I in Sp_E is then the unitary $\mathbb{R}$-group U_h. Choose the element I such that h is positive definite, in which case $\mathrm{U}_\mathrm{h}(\mathbb{R})$ is a maximal compact subgroup of Sp_E. There exists a unique conjugacy class of such elements under $\mathrm{Sp}(E)$. For example, we can take $I = \mathrm{J}_{2g}$ in the notation of Sect. 4.5.1; $\mathrm{U}_\mathrm{h}(\mathbb{R})$ is then the group K of Sect. 4.5.3 (with complexification $\mathrm{j}(-, i1_g)\colon K \to \mathrm{GL}_g(\mathbb{C})$ defined loc. cit.). The choice of a decomposition of the Hermitian space (E, h) into an orthogonal sum of lines determines an anisotropic maximal torus T such that $T(\mathbb{R}) \subset K$. In the notation of Sect. 8.4.1 (and in accordance with Sect. 6.3.4), we have $\mathrm{X}^*(T) = \oplus_{i=1}^{g} \mathbb{Z}\varepsilon_i$,

$$\Phi = \{\pm 2\varepsilon_i\,;\, i = 1,\ldots,g\} \cup \{\pm\varepsilon_i \pm \varepsilon_j\,;\, 1 \leq i < j \leq g\} ,$$

and $\Phi_\mathrm{c} = \{\pm(\varepsilon_i - \varepsilon_j)\,;\, 1 \leq i < j \leq g\}$. In particular, $W_\mathrm{r} = \mathrm{W}(\Phi_c)$ is none other than the symmetric group $\mathfrak{S}_g$ acting on $\mathrm{X}^*(T) \simeq \mathbb{Z}^g$ in the usual way, and $W_\mathrm{r}^\mathrm{ad} = W_\mathrm{r} \times \{\pm \mathrm{id}\}$. The set of bases $\mathcal{B}(T)$ consists of two W_r-orbits, interchanged by $x \mapsto -x$; for example, one of them is

$$\Delta = \{2\varepsilon_g, -\varepsilon_g - \varepsilon_{g-1}, \varepsilon_{g-1} + \varepsilon_{g-2}, -\varepsilon_{g-2} - \varepsilon_{g-3}, \ldots, (-1)^{g-1}(\varepsilon_2 + \varepsilon_1)\} .$$

Note that if $\varepsilon_i^* \in \mathrm{X}_*(T)$ is the dual basis of (ε_i), then Δ is the basis of the positive system $\{\alpha \in \Phi\,;\, f(\alpha) > 0\}$, where $f = \varepsilon_g^* - 2\,\varepsilon_{g-1}^* + 3\,\varepsilon_{g-2}^* + \cdots + (-1)^{g-1}\,(g-1)\,\varepsilon_1^*$. A short calculation then shows that we have $\rho^\vee = \sum_{i=1}^{g} (-1)^{i-g}\,\frac{1}{2}(2g - 2i + 1)\,\varepsilon_i^*$, and therefore

$$\rho^\vee \equiv \frac{1}{2}\sum_{i=1}^{g} \varepsilon_i^* \bmod 2\mathrm{X}_*(T) .$$

Note that this element is indeed invariant under W_r. Let $A \subset W$ be the subgroup consisting of the elements $a \in W$ such that $a(\varepsilon_i) = \pm\varepsilon_i$; it is clearly isomorphic to $\{\pm 1\}^g$. We see that every element $\mathrm{X}_*(T)$ is congruent modulo 2 to an element of the form $a\rho^\vee - \rho^\vee$ for a unique $a \in A$: the action of A on $H^1(\mathbb{R}, T)$ is therefore simply transitive, and κ_Δ (resp. χ_O) is bijective. Finally, note that if we replace O by the W_r-orbit $-O$, which is equivalent to changing w to $-w$, we have

$$\kappa_\Delta \equiv \kappa_{-\Delta} + 2\rho^\vee \bmod 2\mathrm{X}_*(T) .$$

Now, consider the holomorphic discrete series in Π_V; see Sects. 6.3.4 and 8.4.1. One easily verifies that there exist exactly two W_r-orbits of bases of Φ whose associated positive system is such that the compact roots are smaller than the noncompact roots, namely the orbits of the basis $\pm\Delta'$, where

$$\Delta' = \{2\varepsilon_g\} \cup \{\varepsilon_{i+1} - \varepsilon_i\,;\, i = 1,\ldots,g-1\}$$

(consider, for example, the linear form $\sum_{i=1}^{g}(g+i)\,\varepsilon_i^*$). By the chosen conventions, which we do not need to specify, one of these bases leads to the representations denoted by π'_W in Sect. 6.3.4, while its opposite leads to the outer conjugates under $\mathrm{PGSp}_{2g}(\mathbb{R})$ (in the literature, we sometimes encounter the names *holomorphic* and *antiholomorphic* discrete series to distinguish between these two types). We then have $\Delta' = w^{-1}\Delta$, where $w \in W$ is the element that sends ε_i to $(-1)^{g-i}\varepsilon_i$. Hence

$$\kappa_\Delta(\pi_{\Delta',V}) \equiv w\rho^\vee - \rho^\vee \equiv \sum_{i \not\equiv g \bmod 2} \varepsilon_i^* \bmod 2\ , \tag{8.4.5}$$

which agrees with the computation carried out in [55, Lemma 9.1]. Likewise, we obtain $\kappa_\Delta(\pi_{-\Delta',V}) \equiv \kappa_\Delta(\pi_{\Delta',V}) + 2\rho^\vee \bmod 2\mathrm{X}_*(T)$.

8.4.8 Pure Forms of the Split Groups

In this subsection, we recall how the parametrization of Sect. 8.4.2 extends to all pure forms of the split $\mathbb{R}$-group H, following Vogan, Kottwitz, Arthur, Shelstad, and Adams [2], [10, Sect. 1]. Our exposition is largely inspired by the pleasant presentation of Adams [1], as well as the notes of Shelstad [182, 183]. We refer to [176, Chap. III, Sect. 1], [2, Sect. 2], and [37] for general results on the forms of real groups.

Let us first consider an arbitrary $\mathbb{R}$-group G. The set $G(\mathbb{C})$ is endowed with an action of $\mathrm{Gal}(\mathbb{C}/\mathbb{R}) = \{1, \sigma\}$. Let

$$Z^1(\mathbb{R}, G) = \{x \in G(\mathbb{C})\,;\, x\sigma(x) = 1\}\ .$$

The group $G(\mathbb{C})$ acts on $Z^1(\mathbb{R}, G)$ by $(g, x) \mapsto gx\sigma(g)^{-1}$; the quotient set is the usual cohomology set $H^1(\mathbb{R}, G)$. To every element $x \in Z^1(\mathbb{R}, G)$ corresponds an involution

$$\sigma_x = \mathrm{inn}_x \circ \sigma$$

of $G(\mathbb{C})$. This is the Galois involution of a unique real structure on the $\mathbb{C}$-group $G_\mathbb{C}$, whose associated real group we denote by G_x. In particular,

$$G_x(\mathbb{R}) = \{g \in G(\mathbb{C})\,;\, \sigma(g) = x^{-1}gx\}\ .$$

Such a real form of G is called *pure*. The following lemma is obvious.

Lemma 8.4.9. *The stabilizer in $G(\mathbb{C})$ of $x \in Z^1(\mathbb{R}, G)$ is $G_x(\mathbb{R})$.*

If x and $x' \in Z^1(\mathbb{R}, G)$ have the same class in $H^1(\mathbb{R}, G)$ and if $h \in G(\mathbb{C})$ is such that $hx\sigma(h)^{-1} = x'$, then $\sigma_{x'} \circ \mathrm{inn}_h = \mathrm{inn}_h \circ \sigma_x$, so that

$$\mathrm{inn}_h\colon G_x \to G_{x'}$$

is an isomorphism defined over $\mathbb{R}$. An important point is that it does not depend on the h chosen above, at least *modulo the* inn_g *with* $g \in G_x(\mathbb{R})$, by the lemma above. In other words, if x and x' are equivalent in $H^1(\mathbb{R}, G)$, the $\mathbb{R}$-groups G_x and $G_{x'}$ are naturally isomorphic, and even canonically so modulo "inner automorphisms." Hence, while some caution is necessary, it makes sense to speak of the $\mathbb{R}$-group G_c defined "up to inner automorphisms" as being the $\mathbb{R}$-group G_x for any $x \in Z^1(\mathbb{R}, G)$ in the class $c \in H^1(\mathbb{R}, G)$. In particular, there is an obvious definition of $\Pi_{\mathrm{unit}}(G_c(\mathbb{R}))$ for $c \in H^1(\mathbb{R}, G)$.

Example 8.4.10. Let us give some classical examples [176, Chap. III, Sect. 1.2]. If E is a finite-dimensional $\mathbb{R}$-vector space, then $\mathrm{Z}^1(\mathbb{R}, \mathrm{GL}_E)$ is the set of semilinear involutions of $E \otimes_{\mathbb{R}} \mathbb{C}$. By Hilbert's theorem 90, $x \mapsto E_x = (E \otimes_{\mathbb{R}} \mathbb{C})^{x=\mathrm{Id}}$ identifies $\mathrm{Z}^1(\mathbb{R}, \mathrm{GL}_E)$ with the set of real structures on the $\mathbb{C}$-vector space $E \otimes_{\mathbb{R}} \mathbb{C}$ and we have $(\mathrm{GL}_E)_x = \mathrm{GL}_{E_x}$ for every $x \in \mathrm{Z}^1(\mathbb{R}, \mathrm{GL}_E)$. One immediately sees that if E is a q-vector space over $\mathbb{R}$, then $\mathrm{Z}^1(\mathbb{R}, \mathrm{O}_E) \subset \mathrm{Z}^1(\mathbb{R}, \mathrm{GL}_E)$ can be identified with the real structures $E_x \subset E \otimes_{\mathbb{R}} \mathbb{C}$ such that $\mathrm{q}(E_x) \subset \mathbb{R}$. For $x \in \mathrm{Z}^1(\mathbb{R}, \mathrm{O}_E)$, we therefore have a q-vector space E_x for the form $\mathrm{q}_{|E_x}$, and we have $(\mathrm{O}_E)_x = \mathrm{O}_{E_x}$. The map $x \mapsto E_x$ then induces a bijection between $H^1(\mathbb{R}, \mathrm{O}_E)$ and the set of isomorphism classes of q-vector spaces over $\mathbb{R}$ of rank $\dim\, E$ (the signature), which has $\dim(E) + 1$ elements. Finally, we see that $x \in \mathrm{Z}^1(\mathbb{R}, \mathrm{SO}_E)$ if and only if E and E_x have the same discriminant, in which case $(\mathrm{SO}_E)_x = \mathrm{SO}_{E_x}$, and then that $x \mapsto E_x$ induces a bijection between $H^1(\mathbb{R}, \mathrm{SO}_E)$ and the set of isomorphism classes of q-vector spaces over $\mathbb{R}$ with same rank and discriminant as E. The analog holds in the alternating case, which is just simpler, because there is only one nondegenerate alternating form of each even dimension ($H^1(\mathbb{R}, \mathrm{Sp}_{2g}) = 1$).

Let us now return to our split $\mathbb{R}$-group H, endowed with an anisotropic maximal torus T. For $c \in H^1(\mathbb{R}, H)$, we recall the Shelstad parametrization of the set Π_V^c of discrete series of $H_c(\mathbb{R})$ with the same infinitesimal character as $V \in \mathrm{Irr}(H_{\mathbb{C}}) = \mathrm{Irr}((H_c)_{\mathbb{C}})$. The inclusion $T \subset H$ induces a natural injection $Z^1(\mathbb{R}, T) \to Z^1(\mathbb{R}, H)$, as well as a map

$$W \backslash H^1(\mathbb{R}, T) \longrightarrow H^1(\mathbb{R}, H)\ , \tag{8.4.6}$$

where the action of W on $H^1(\mathbb{R}, T)$ is that recalled in Sect. 8.4.2. This map is bijective by Shelstad (see also [37]). In particular, every pure real form of H is isomorphic, as an $\mathbb{R}$-group, to H_t for $t \in T_2$.

The forms H_t with $t \in T_2$, and more generally $t \in Z^1(\mathbb{R}, T)$, have the nice property that they all share T as anisotropic maximal torus, because σ_t coincides with σ on T. Since, moreover, $H_{\mathbb{C}} = (H_t)_{\mathbb{C}}$, the root system Φ of $(H_{\mathbb{C}}, T_{\mathbb{C}})$ is canonically that of $((H_t)_{\mathbb{C}}, T_{\mathbb{C}})$, and likewise for its Weyl group W (on the other hand, the real Weyl subgroup $\mathrm{W}(H_t, T) \subset W$ does, of course, depend on t). If Δ is a basis of Φ and $t \in T_2$, it consequently makes sense to consider the discrete series $\pi_{\Delta, V, t}$ of $H_t(\mathbb{R})$ associated by Harish-Chandra with the basis Δ of Φ, in the notation of Sect. 8.4.1. Recall that the set $\mathcal{B}(T)$ defined in Sect. 8.4.2 is defined with respect to the pair (H, T).

Definition 8.4.11. Let $c \in H^1(\mathbb{R}, H)$ and $V \in \mathrm{Irr}(H_{\mathbb{C}})$. Let T be an anisotropic maximal torus of H, $\Delta \in \mathcal{B}(T)$, and $\rho^\vee$ the half-sum of the coroots of T positive with respect to Δ. The Shelstad parametrization map with respect to Δ is the unique map

$$\kappa_\Delta^c : \Pi_V^c \longrightarrow \mathrm{X}_*(T) \otimes \mathbb{Z}/2\mathbb{Z} \,, \quad \pi \mapsto \kappa_\Delta^c(\pi)$$

such that for every $t \in T_2$ in the class c and every $w \in W$, we have $\kappa_\Delta^c(\pi_{w^{-1}\Delta,V,t}) \equiv w(\mu + \rho^\vee) - \rho^\vee \mod 2\mathrm{X}_*(T)$, where $\mu \in \mathrm{X}_*(T)$ is such that $t = e^{i\pi\mu}$.

Hidden behind this definition is the following fact: let $w \in W$ and $t \in T_2$, and let $n \in H(\mathbb{C})$ be a representative of w such that $n \star t = w \cdot t$, so that $\mathrm{inn}_n : H_t \to H_{w\cdot t}$ defines an $\mathbb{R}$-isomorphism; then the restriction of the representation $\pi_{w\Delta,V,w\cdot t}$ of $H_{w\cdot t}(\mathbb{R})$ to inn_n is isomorphic to the representation $\pi_{\Delta,V,t}$ of $H_t(\mathbb{R})$. When $c = 1$, we have $\kappa_\Delta^1 = \kappa_\Delta$, and we, of course, recover Definition 8.4.3. We easily verify, as in Sect. 8.4.2, that $\kappa_{w\Delta}^c = w \circ \kappa_\Delta^c$ for every $w \in \mathrm{W}(H,T)$. Part (i) of Lemma 8.4.12 below moreover asserts that κ_Δ^c is injective, so that the map

$$\coprod_{c\in H^1(\mathbb{R},G)} \kappa_\Delta^c : \coprod_{c\in H^1(\mathbb{R},H)} \Pi_V^c \longrightarrow \mathrm{X}_*(T) \otimes \mathbb{Z}/2\mathbb{Z}$$

is bijective.

The agreement of this presentation with the definition given by Shelstad [182, 183], which sends $\pi_{w^{-1}\Delta,V,t}$ to the class of $w \cdot t$ in $H^1(\mathbb{R}, T)$, immediately follows from Lemma 8.4.4. This is essentially the point of view given by Adams in [1], up to the fact that his starting point is an $\mathbb{R}$-group with real points that is compact, rather than split (and that he considers general inner forms). The definition above admits a dual interpretation identical to that mentioned in Sect. 8.4.5, in terms of discrete parameters of H, and leads to a canonical map

$$\chi_O^c = \beta \circ (i_\Delta \otimes \mathbb{Z}/2\mathbb{Z}) \circ \kappa_\Delta^c \,, \quad \Pi_V^c \to \mathrm{Hom}_{\mathrm{groups}}(\mathrm{C}_\varphi, \mathbb{C}^\times) \,,$$

which depends only on the W_{r}-orbit O of Δ in $\mathcal{B}(T)$ (let us emphasize again that this choice is with respect to H and not H_c). Let us note that in her exposition, following Arthur and Kottwitz, Shelstad limits herself to the $c \in H^1(\mathbb{R}, H)$ that are in the image of the natural map $H^1(\mathbb{R}, H_{\mathrm{sc}}) \to H^1(\mathbb{R}, H)$, where $H_{\mathrm{sc}} \to H$ denotes the simply connected cover of H. The disjoint union of the H_c indexed by such classes c then forms a K-*group* in the sense of Arthur. This corresponds to restricting oneself to elements $t \in T_2$ of the form $e^{i\pi\mu}$, where $\mu \in \mathrm{X}_*(T)$ is a sum of coroots of $(H_{\mathbb{C}}, T_{\mathbb{C}})$ or, equivalently, to the elements of $\mathrm{X}^*(\widehat{T}) \otimes \mathbb{Z}/2\mathbb{Z}$ that are trivial on the center of $\widehat{H}$.

Lemma 8.4.12. *Let T be an anisotropic maximal torus of H and $t \in T_2$.*

(i) $\mathrm{W}(H_t, T) = \{w \in W\,;\, w \cdot t = t\}$.
(ii) $\mathrm{inn}_{tt_{\mathrm{b}}}$ *is a Cartan involution of H_t, where $t_{\mathrm{b}} \in T_2$ is the element defined in Sect.* 8.4.2 *and associated with a W_{r}-orbit in $\mathcal{B}(T)$.*

Proof. We have $w \cdot t = t$ if and only if there exists a representative $n \in H(\mathbb{C})$ of w that fixes $t \in Z^1(\mathbb{R}, H)$, that is, is an element of $H_t(\mathbb{R})$ by Lemma 8.4.9. Part (ii) follows from the case $t = 1$, which is Lemma 8.4.4 (i). □

If $c \in H^1(\mathbb{R}, H)$ and $V \in \mathrm{Irr}(H_{\mathbb{C}})$, then V can be viewed, by restriction, as a finite-dimensional irreducible representation of $H_c(\mathbb{R})$ that is well defined up to isomorphism. When $H_c(\mathbb{R})$ is compact, it is the unique element of the singleton Π_V^c (Sect. 8.4.1). Part (ii) of Lemma 8.4.12 shows that this occurs if and only if $t_{\mathrm{b}} \in T_2$ and c is the class of a $t \in t_b \mathrm{Z}(H)$. Recall that $\mathrm{Z}(H) \subset T_2$ because $-1 \in W$ (this follows from the fact that H is split over $\mathbb{R}$ and has a compact maximal torus).

Corollary 8.4.13. *Let $c \in H^1(\mathbb{R}, H)$ be such that $H_c(\mathbb{R})$ is compact, let $V \in \mathrm{Irr}(H_{\mathbb{C}})$ and $\Delta \in \mathcal{B}(T)$, and let $\rho^\vee \in \frac{1}{2}\mathrm{X}_*(T)$ be associated with Δ. Then $\rho^\vee \in \mathrm{X}_*(T)$ and $\kappa_\Delta^c(V) \equiv \rho^\vee + \nu \bmod 2\mathrm{X}_*(T)$, where $e^{i\pi\nu} \in \mathrm{Z}(H)$.*

Let us consider the interesting example of the holomorphic discrete series of the special orthogonal group with signature $(m, 2)$ for $m \geq 1$ odd (it is split only if $m \leq 3$). We begin with the split $\mathbb{R}$-group $H = \mathrm{SO}_{r+1,r}$. We write $\mathrm{H}(\mathbb{R}^r) \oplus \mathbb{R}$ as an orthogonal sum of a line D and planes P_i for $i = 1, \ldots r$, each definite and with a sign that we will give further on. This decomposition defines a unique anisotropic torus T of H that preserves D and each of the P_i. For $i = 1, \ldots, r$, we choose, arbitrarily, one of the isotropic lines ℓ_i of $\mathrm{P}_i \otimes \mathbb{C}$; we denote the character of T on ℓ_i by $\varepsilon_i \in \mathrm{X}^*(T_{\mathbb{C}})$. As in Sect. 6.1.3, the sequence $\ell_1, \ldots, \ell_r$ defines a unique Borel subgroup of $H_{\mathbb{C}}$ containing $T_{\mathbb{C}}$, which corresponds to the standard basis

$$\Delta = \{\varepsilon_r\} \cup \{\varepsilon_i - \varepsilon_{i+1}\,;\, 1 \leq i < r\}$$

of $\Phi = \Phi(H_{\mathbb{C}}, T_{\mathbb{C}})$. Now, suppose that we have chosen the P_i with sign $(-1)^{i-1}$ and D with sign $(-1)^r$, which is allowed. We see that no element of Δ is compact, so that $\Delta \in \mathcal{B}(T)$ (it is not difficult to see that $\mathcal{B}(T)$ forms a single W_r-orbit). Moreover, Sect. 6.1.3 shows that the half-sum $\rho^\vee$ of the coroots of $(H_{\mathbb{C}}, T_{\mathbb{C}})$ positive with respect to Δ is

$$\rho^\vee = \sum_{i=1}^{r} (r - i + 1)\varepsilon_i^* \in \mathrm{X}_*(T_{\mathbb{C}})\,,$$

where $\varepsilon_i^* \in \mathrm{X}_*(T_{\mathbb{C}})$ denotes the dual basis of (ε_i), so that $\varepsilon_i(t_{\mathrm{b}}) = (-1)^{r-i+1}$.

Having said this, let us turn to the real forms of H. By Example 8.4.10, there exists a unique class $c \in H^1(\mathbb{R}, H)$ such that H_c is isomorphic to the special orthogonal group with signature $(2, 2r - 1)$. Concretely, if $t \in T_2$ acts by multiplication by $s_j = \pm 1$ on the plane P_j and by 1 on D, then the real form of the q-vector space $\mathrm{H}(\mathbb{R}^r) \oplus \mathbb{R}$ associated with t is the direct sum of D, the P_j such that $s_j = 1$, and the iP_j such that $s_j = -1$. In particular, H_t is the special orthogonal group with signature $(2a, b)$, where a is the number of $1 \leq j \leq r$ such that $s_j(-1)^{j-1} \neq (-1)^r$ and $2a + b = 2r + 1$. In other words, $H_t \simeq H'$ if and only if there exists an integer $1 \leq s \leq r$ such that $t = t_{\mathrm{b}} e^{i\pi\varepsilon_s^*}$. Since these elements of T_2 form only one W-orbit for the twisted action, they indeed belong to the same class $c \in \mathrm{H}^1(\mathbb{R}, H)$. Let us, for example, identify H' with $H_{t'}$, where

$$t' = t_{\mathrm{b}} e^{i\pi\varepsilon_1^*}$$

(since every $\mathbb{R}$-isomorphism of H' is of the form inn_h with $h \in H'(\mathbb{R})$, the choice of the identification does not matter). The set $\Phi'_{\mathrm{c}} \subset \Phi$ of compact roots of $(H_{t'}, T)$ is

$$\Phi'_{\mathrm{c}} = \{\pm\varepsilon_i\,;\, i = 2, \ldots, r\} \cup \{\pm\varepsilon_i \pm \varepsilon_j\,;\, 2 \leq i < j \leq r\}\,,$$

by Lemma 8.4.12. By considering the linear form $(2r-2)\,\varepsilon_1^* + \sum_{i=2}^{r}(r-i+1)\,\varepsilon_i^*$, we see that the positive system of the basis Δ above is such that every element of Φ'_{c} is smaller than every element of $\Phi - \Phi'_{\mathrm{c}}$ (Sect. 8.4.1). It is even the unique basis with this property modulo the action of the real Weyl group of $(H_{t'}, T)$. Hence, there exists a unique holomorphic discrete series $\pi_{\mathrm{hol},V}$ of H' with the same infinitesimal character as $V \in \mathrm{Irr}(H_{\mathbb{C}})$, and we have

$$\kappa_{\Delta}^{\mathrm{c}}(\pi_{\mathrm{hol,V}}) \equiv (r-1)(\varepsilon_1^* + \varepsilon_2^*) + \sum_{i=3}^{r}(r-i+1)\varepsilon_i^* \bmod 2\mathrm{X}_*(T_{\mathbb{C}})\,. \qquad (8.4.7)$$

Indeed, this is Definition 8.4.11 applied to $w = 1$ and $t = t' = e^{i\pi(\rho^\vee + \varepsilon_1^*)}$.

8.4.14 Adams–Johnson Packets

Let H be a split semisimple $\mathbb{R}$-group admitting discrete series and T an anisotropic maximal torus of H. We again denote its roots system by $\Phi = \Phi(H_{\mathbb{C}}, T_{\mathbb{C}}) \subset \mathrm{X}^*(T)$ and the Weyl group of Φ by W.

In this subsection, we briefly recall certain sets, or *packets*, of unitary irreducible representations of $H(\mathbb{R})$ defined by Adams and Johnson in [3], which play an important role in Arthur's theory [9, Sect. 5] (see also [2] for a very general context). The starting point consists of an *Adams–Johnson parameter*

$$\varphi\colon \mathrm{SL}_2(\mathbb{C}) \times \mathrm{W}_{\mathbb{R}} \to \widehat{H}(\mathbb{C})\,,$$

which is a group morphism with certain properties that we first discuss informally and then specify further on. Adams and Johnson associate with it a finite subset

$$\Pi_{\mathrm{AJ}}(\varphi) \subset \Pi_{\mathrm{unit}}(H(\mathbb{R}))$$

that depends only on the $\widehat{H}(\mathbb{C})$-conjugacy class of φ.

The parameter φ first determines a representation $V_\varphi \in \mathrm{Irr}(H_{\mathbb{C}})$. For example, the Adams–Johnson parameters that are trivial on the factor $\mathrm{SL}_2(\mathbb{C})$ can be identified exactly with the discrete Langlands parameters recalled in Sect. 8.4.5, and for such a φ, by definition, $\Pi_{\mathrm{AJ}}(\varphi) = \Pi_{V_\varphi}$. In general, $\Pi_{\mathrm{AJ}}(\varphi)$ consists of representations with the same infinitesimal character as V_φ; even better, they have $(\mathfrak{h}, K)$-cohomology with coefficients in V_φ^* [204]. Concretely, with each basis $\Delta \subset \Phi$, the parameter φ associates a parabolic subgroup $P_{\Delta,\varphi} \subset H_\mathbb{C}$ containing $T_\mathbb{C}$. Let $L_{\Delta,\varphi} \subset H$ be the Levi subgroup of $P_{\Delta,\varphi}$ containing T; it is necessarily defined over $\mathbb{R}$ because $T(\mathbb{R})$ is compact (of course, none of the proper parabolic subgroups of $H_\mathbb{C}$ containing T are defined over $\mathbb{R}$). Finally, φ determines a representation ρ of $L_{\Delta,\varphi}(\mathbb{R})$ of dimension 1. Its exact description by Adams and Johnson is rather delicate, at least when $L_{\Delta,\varphi}(\mathbb{R})$ is not connected,[8] but for the most part, we will not need to understand it for our discussion. The data of $P_{\Delta,\varphi}$ and ρ then allows us to define a representation

$$\pi_{\Delta,\varphi} \in \Pi_{\mathrm{unit}}(H(\mathbb{R}))$$

by cohomological induction in the relevant degree [202, 203]. The set of these representations, when Δ runs through the bases of Φ, is by definition the packet $\Pi_{\mathrm{AJ}}(\varphi)$. Let $W_\mathrm{r} = \mathrm{W}(H,T) \subset W$ (Sect. 8.4.1). If we fix a basis $\Delta \subset \Phi$ and $L = L_{\Delta,\varphi}$, then the map $W \to \Pi_{\mathrm{AJ}}(\varphi)$ defined by $w \mapsto \pi_{w\Delta,\varphi}$ induces a bijection (that depends on Δ)

$$W_\mathrm{r}\backslash W/\mathrm{W}(L_\mathbb{C}, T_\mathbb{C}) \xrightarrow{\sim} \Pi_{\mathrm{AJ}}(\varphi) .$$

Let us now state the axioms (AJ1) and (AJ2) that defines the Adams–Johnson parameters, following [3], [130, p. 195], [55, App. A]. Let $\varphi\colon \mathrm{SL}_2(\mathbb{C}) \times \mathrm{W}_\mathbb{R} \to \widehat{H}(\mathbb{C})$ be a group morphism assumed to be continuous, algebraic on the factor $\mathrm{SL}_2(\mathbb{C})$, and such that $\varphi(1 \times \mathrm{W}_\mathbb{R})$ consists of semisimple elements. Consider the homomorphism $\widetilde{\varphi}\colon \mathrm{W}_\mathbb{R} \to \widehat{H}(\mathbb{C})$ obtained by composing φ with the following morphism introduced by Arthur:

$$\mathrm{W}_\mathbb{R} \to \mathrm{SL}_2(\mathbb{C}) \times \mathrm{W}_\mathbb{R} , \quad g \mapsto \begin{bmatrix} |\eta(g)|^{1/2} & 0 \\ 0 & |\eta(g)|^{-1/2} \end{bmatrix} \times g ,$$

where $\eta\colon \mathrm{W}_\mathbb{R} \to \mathbb{R}^\times$ is the character recalled in Sect. 8.2.12. The subgroup $\widetilde{\varphi}(\mathbb{C}^\times) \subset \widehat{H}(\mathbb{C})$ is connected and consists of semisimple elements; we can therefore embed it in a maximal torus

$$S \subset \widehat{H} .$$

There then exist unique $\lambda, \mu \in \mathrm{X}_*(S) \otimes \mathbb{C}$ such that $\lambda - \mu \in \mathrm{X}_*(S)$ and $\xi(\widetilde{\varphi}(z)) = z^{\langle \xi,\lambda\rangle}\overline{z}^{\langle \xi,\mu\rangle}$ for every $\xi \in \mathrm{X}^*(S)$ and every $z \in \mathbb{C}^\times$ (see the footnote at the beginning of Sect. 8.2.12, p. 198).

The $\widehat{H}(\mathbb{C})$*-conjugacy class of* λ*, viewed in* Lie $\widehat{H}$*, is the infinitesimal character of a finite-dimensional representation* $V_\varphi \in \mathrm{Irr}(H_\mathbb{C})$. (AJ1)

[8] When $L_{\Delta,\varphi}(\mathbb{R})$ is connected, the character ρ is determined by its differential at the identity, itself characterized by the property that the representation $\pi_{\Delta,\varphi}$ defined in the text must have the same infinitesimal character as V.

This implies, in particular, that S is the unique maximal torus of $\widehat{H}$ containing $\widetilde{\varphi}(\mathbb{C}^\times)$. This also endows $\widehat{H}$ with a unique Borel subgroup B containing S such that λ is dominant with respect to B. Next, we consider the centralizer $M \subset \widehat{H}(\mathbb{C})$ of the commutative connected subgroup consisting of the semisimple elements $\varphi(1 \times \mathbb{C}^\times)$; this is therefore a Levi subgroup (of a parabolic subgroup) of $\widehat{H}$. It contains S.

Choosing a basis Δ of Φ allows us to identify the based root datum $(\mathrm{X}^*(T), \Phi, \Delta, \cdots)^\vee$ with $\Psi(\widehat{H}, S, B)$ and, in particular, provides a privileged isomorphism

$$i_\Delta \colon \mathrm{X}_*(T) \xrightarrow{\sim} \mathrm{X}^*(S)$$

that sends $\Delta^\vee$ onto the basis of $\Phi(\widehat{H}, S)$ associated with B. Let $L_{\Delta,\varphi} \subset H_\mathbb{C}$ be the unique Levi subgroup (of parabolic subgroups) containing T such that $i_\Delta(\Phi^\vee(L_{\Delta,\varphi}, T_\mathbb{C})) = \Phi(M, S)$ (in particular, $M \simeq \widehat{L_{\Delta,\varphi}}$). Let $P_{\Delta,\Phi} \subset H_\mathbb{C}$ be the unique parabolic subgroup with Levi subgroup $L_{\Delta,\varphi}$ containing the Borel subgroup of $H_\mathbb{C}$ containing T and associated with Δ; it is the subgroup mentioned in the informal description above. The remaining axiom serves to define the character χ (see [3] and the work of Taïbi [195, Sect. 4.2.2] for more details on this subject).

The homomorphism $\mathrm{SL}_2(\mathbb{C}) \to M$ *induced by* φ *is principal, that is, induces an* $\mathfrak{sl}_2$*-triple of* $\mathrm{Lie}\, M$ *that is* principal *in the sense of Kostant* [128]. *Moreover, the centralizer* C_φ *of* $\mathrm{Im}\,\varphi$ *in* $\widehat{H}(\mathbb{C})$ *is finite.* (AJ2)

The first assumption implies that the centralizer of $\varphi(\mathrm{SL}_2(\mathbb{C}) \times \mathbb{C}^\times)$ in $\widehat{H}(\mathbb{C})$ is the center $\mathrm{Z}(M)$ of M. The group C_φ is therefore the subgroup of $\mathrm{Z}(M)$ fixed by conjugation by $\varphi(1 \times j)$. The second assumption asserts that

$$\mathrm{C}_\varphi = \mathrm{Z}(M)_2 \stackrel{\mathrm{def}}{=} \{z \in \mathrm{Z}(M)\,;\, z^2 = 1\}$$

(see, for example, [55, Lemma A.1]). As noted by Taïbi [195, Sect. 4.2.2], under the first assumption of (AJ2), the second one is also equivalent to requiring that φ be trivial on $1 \times \mathbb{R}_{>0}$ (this is obviously necessary, because $\mathbb{R}^\times$ is the center of $\mathrm{W}_\mathbb{R}$, but it is also sufficient). When H is a classical group, it is easy to determine all its Adams–Johnson parameters; see Example 8.4.15 and Lemma 8.4.16.

The Case of Pure Real Forms

To conclude this subsection, let us consider the general case of the pure real forms of H. Let $c \in H^1(\mathbb{R}, H)$. The construction of Adams and Johnson in [3], which is not specific to split groups, also associates with every Adams–Johnson parameter φ of H a set of representations

$$\Pi^c_{\mathrm{AJ}}(\varphi) \subset \Pi_{\mathrm{unit}}(H_c(\mathbb{R}))\ .$$

If $t \in T_2$ is in the class c and Δ is a basis of Φ, we again have the parabolic subgroup $T_\mathbb{C} \subset P_{\Delta,\varphi} \subset (H_t)_\mathbb{C} = H_\mathbb{C}$. Its Levi subgroup $T \subset L_{\Delta,\varphi,t} \subset H_t$, defined over $\mathbb{R}$, is the pure real form of $L_{\Delta,\varphi}$ associated with σ_t (Sect. 8.4.8). Adams and Johnson associate a character $\rho\colon L_{\Delta,\varphi,t}(\mathbb{R}) \to \mathbb{C}^*$ with this group and from ρ define a representation

$$\pi_{\Delta,\varphi,t} \in \Pi_{\mathrm{unit}}(H_t(\mathbb{R})) \xrightarrow{\sim} \Pi_{\mathrm{unit}}(H_c(\mathbb{R}))$$

by cohomological induction to $H_t(\mathbb{R})$ with respect to $P_{\Delta,\varphi}$, in a suitable degree. They set $\Pi^c_{\mathrm{AJ}}(\varphi) = \{\pi_{\Delta,\varphi,t}\}$, where Δ runs through the bases of Φ and $t \in T_2$ through the representatives of c. As in Sect. 8.4.8, the variables Δ and t are redundant and linked by the relation $\pi_{w\Delta,\varphi,w\cdot t} \simeq \pi_{\Delta,\varphi,t}$ that holds for every basis $\Delta \subset \Phi$, every $t \in T_2$, and every $w \in W$. If we fix $t \in T_2$ and a basis $\Delta \subset \Phi$, and if $L = L_{\Delta,\varphi,t} \subset H_t$ is the associated Levi subgroup, then, this time, the map $W \to \Pi_{\mathrm{unit}}(H(\mathbb{R}))$ defined by $w \mapsto \pi_{w\Delta,\varphi,t}$ induces a bijection

$$\mathrm{W}(H_t, T)\backslash W/\mathrm{W}(L_{\mathbb{C}}, T_{\mathbb{C}}) \xrightarrow{\sim} \Pi^c_{\mathrm{AJ}}(\varphi)\ .$$

The particular case $c = 1$ gives another point of view on $\Pi^1_{\mathrm{AJ}}(\varphi) = \Pi_{\mathrm{AJ}}(\varphi)$.

8.4.15 Example: Adams–Johnson Parameters of Sp_{2g}

Consider the $\mathbb{R}$-group $H = \mathrm{Sp}_{2g}$. Let $V = \mathbb{C}^{2g+1}$ be the underlying space of the standard representation of $\widehat{H} = \mathrm{SO}_{2g+1}$, endowed with the standard quadratic form q. Let $\varphi\colon \mathrm{SL}_2(\mathbb{C}) \times \mathrm{W}_{\mathbb{R}} \to \mathrm{SO}(V)$ be a continuous homomorphism that is algebraic on the factor $\mathrm{SL}_2(\mathbb{C})$ and trivial on $1 \times \mathbb{R}_{>0}$. Since $\mathrm{W}_{\mathbb{R}}/\mathbb{R}_{>0}$ is compact, the group $\mathrm{SL}_2(\mathbb{C}) \times \mathrm{W}_{\mathbb{R}}$ acts in a semisimple manner on V, which therefore decomposes into an orthogonal sum

$$V = \bigoplus_{j \in J} V_j$$

of irreducible subspaces V_j. As a representation of $\mathrm{SL}_2(\mathbb{C}) \times \mathrm{W}_{\mathbb{R}}$ that is trivial on $1 \times \mathbb{R}_{>0}$, V_j is necessarily self-dual. Also note that if $\mathrm{q}_{|V_j}$ is degenerate, then $\mathrm{q}(V_j) = 0$ and there exists $j' \neq j$ such that $V_{j'} \simeq V_j^* \simeq V_j$.

Suppose that φ satisfies (AJ1); we will analyze φ and, in particular, see that (AJ2) is automatically satisfied. By case I of Sect. 8.2.6, the V_j are pairwise nonisomorphic (and therefore nondegenerate), and only one of them has odd dimension; we may assume that it is V_0. It is then obvious that C_φ is the finite subgroup of $\mathrm{SL}(V)$ consisting of the elements g such that $g(V_j) \subset V_j$ for every j and $g_{|V_j} = \pm\mathrm{id}_{V_j}$. Moreover, V_0 is an irreducible representation of $\mathrm{SL}_2(\mathbb{C})$, with $\mathrm{W}_{\mathbb{R}}$ acting by multiplication by a character χ_0, and for $j \neq 0$, we can write

$$V_j \simeq Q_j \otimes R_j\ , \quad Q_j \simeq \mathrm{Sym}^{d_j-1}\mathrm{St}_2\ , \quad R_j \simeq \mathrm{I}_{r_j}\ ,$$

with $r_j > 0$ and $d_j + r_j \equiv 1 \bmod 2$. We can endow Q_j and R_j with nondegenerate bilinear forms that are preserved by $\mathrm{SL}_2(\mathbb{C})$ and $\mathrm{W}_{\mathbb{R}}$, respectively, and whose tensor product is the bilinear form on V_j associated with q. The restriction of R_j to

$\mathbb{C}^\times \subset \mathrm{W}_\mathbb{R}$ is the direct sum of the two stable and isotropic lines, say ℓ_j and ℓ'_j; the element $z \in \mathbb{C}^\times$ acts on ℓ_j by multiplication by $(z/|z|)^{r_j}$. Thus, $M \subset \mathrm{SO}_V$ is the subgroup

$$\mathrm{SO}_{V_0} \times \prod_{j \neq 0} M_j \ ,$$

where $M_j \subset \mathrm{SO}_{V_j}$ is the stabilizer of the transverse Lagrangians $Q_j \otimes \ell_j$ and $Q_j \otimes \ell'_j$. In particular, $M_j \simeq \mathrm{GL}_{d_j}$, M is indeed the Levi subgroup of a parabolic subgroup of SO_V, and $\mathrm{C}_\varphi = \mathrm{Z}(M)_2$. Recall that a $\mathbb{C}$-morphism $\mathrm{SL}_2 \to L$ with L classical (resp. GL_d) is principal if and only if the representation of SL_2 composed of f and the standard (resp. tautological) representation of L is irreducible or the sum of the trivial representation and a nontrivial irreducible representation if $L(\mathbb{C}) \simeq \mathrm{SO}_{2r}(\mathbb{C})$. This shows that (AJ2) is satisfied. More generally, an analysis similar to the one just carried out shows the following lemma.

Lemma 8.4.16. *Suppose $H_\mathbb{C} \in \mathrm{Class}_\mathbb{C}$, and let $\mathrm{St}\colon \widehat{H}(\mathbb{C}) \to \mathrm{SL}_n(\mathbb{C})$ be the standard representation. Let $\varphi\colon \mathrm{SL}_2(\mathbb{C}) \times \mathrm{W}_\mathbb{R} \to \widehat{H}(\mathbb{C})$ be a continuous morphism that is algebraic on the factor $\mathrm{SL}_2(\mathbb{C})$ and trivial on $1 \times \mathbb{R}_{>0}$. Then φ is an Adams–Johnson parameter if and only if it satisfies* (AJ1) *and if the representation* $\mathrm{St} \circ \varphi\colon \mathrm{W}_\mathbb{R} \to \mathrm{SL}_n(\mathbb{C})$ *has no multiplicities.*

Let us continue studying the previous example: we describe the torus S and the Borel subgroup B of SO_V associated with φ. We set $Q_0 := V_0$. Note that the restriction of Q_j to the diagonal torus of $\mathrm{SL}_2(\mathbb{C})$ is the direct sum of d_j canonical stable lines $\ell_{j,n}$ for $n = (d_j - 1)/2, \cdots, (1 - d_j)/2$, where the element $\begin{bmatrix} |z|^{1/2} & 0 \\ 0 & |z|^{-1/2} \end{bmatrix}$ acts on $\ell_{j,n}$ by multiplication by $|z|^n$. On the one hand, we have an orthogonal decomposition

$$V = V^+ \oplus V^- \oplus \ell_{0,0} \ ,$$

where V^+ (resp. V^-) is the Lagrangian that is the direct sum of the $Q_j \otimes \ell_j$ (resp. $Q_j \otimes \ell'_j$) and the $\ell_{0,n}$ with $n > 0$ (resp. $n < 0$). On the other hand, the space V^+ is itself the direct sum of g lines $\ell_{0,n}$ (with $n > 0$) and $\ell_{j,m} \otimes \ell_j$ ($j \neq 0$, m arbitrary). Condition (AJ1) determines a unique way to order these g isotropic lines, say $\mathbb{C}e_1, \ldots, \mathbb{C}e_g \subset V^+$, so that for every $z \in \mathbb{C}^\times$, the element $\widetilde{\varphi}(z)$ acts by multiplication by $z^{w_i}\overline{z}^{w'_i}$ on e_i, where (w_i, w'_i) is an ordered pair of integers such that the w_i satisfy

$$w_1 > w_2 > \cdots > w_g > 0$$

(these can, of course, be expressed simply in terms of the r_j and d_j, but it will not be necessary to specify how). The torus S is the stabilizer in SO_V of the lines $\mathbb{C}e_i$ for $i = 1, \ldots, g$ and of V^-. The Borel subgroup $S \subset B \subset \mathrm{SO}_{2g+1}$ is the stabilizer of the flag associated with the e_i as in Sect. 6.1.3; the element λ_φ is dominant with respect to S by the decreasing order and positivity of the w_i.

Finally, let us describe the $L_{\Delta,\varphi}$. Consider the hyperbolic a-vector space $E = \mathrm{H}(\mathbb{R}^g)$, an element $I \in \mathrm{Sp}(E)$ with square $-\mathrm{id}_E$, and the associated positive definite

Hermitian form h on E as in Sect. 8.4.7. Choose a decomposition of the Hermitian space (E, h) as an orthogonal sum of I-stable $\mathbb{R}$-planes, and let $T \subset \mathrm{U}_{\mathrm{h}}$ be the associated torus, consisting of the elements of Sp_E that stabilize each of these planes; it is an anisotropic maximal torus of H. If $P \subset E$ is an I-stable $\mathbb{R}$-plane, we have a canonical decomposition

$$P \otimes_{\mathbb{R}} \mathbb{C} = P^+ \oplus P^-$$

as a sum of 1-dimensional eigenspaces of I for the respective eigenvalues $+i$ and $-i$. The choice of a basis Δ of $\Phi(H_{\mathbb{C}}, T_{\mathbb{C}})$ is equivalent to that of a numbering $P_1, \ldots, P_g$ of the $\mathbb{R}$-planes defining T, as well as a sign $s_i \in \{+,-\}$ for every $i = 1, \ldots, g$: the Borel subgroup containing T associated with such data is the stabilizer of the flag formed on the spaces

$$P_1^{s_1}, P_2^{s_2}, \ldots, P_g^{s_g}$$

(see Sect. 6.1.3). Having made such a choice, let us describe the associated $\mathbb{R}$-group $L_{\Delta,\varphi}$. Let $\eta_i \in \mathrm{X}^*(T_{\mathbb{C}})$ be the character of $T_{\mathbb{C}}$ on $P_i^{s_i}$, and let (η_i^*) be the dual basis of the (η_i) in $\mathrm{X}_*(T_{\mathbb{C}})$. By Sect. 6.1.3, the isomorphism $i_\Delta \colon \mathrm{X}_*(T_{\mathbb{C}}) \simeq \mathrm{X}^*(S)$ sends η_i^* onto the character of S on the line $\mathbb{C}e_i$, and by definition, $i_\Delta(\Phi^\vee(L_{\Delta,\varphi}, T_{\mathbb{C}})) = \Phi(M, S)$. The $\mathbb{C}$-group $L_{\Delta,\varphi}$ immediately follows. Concretely, there exists a unique decomposition $E = \oplus_{j \in J} E_j$, where E_j is the direct sum of the $\mathbb{R}$-planes P_i for the indices i such that $\mathbb{C}e_i \subset V_j$. In particular, $\dim_{\mathbb{R}} E_j = \dim_{\mathbb{C}} V_j$ if $j \neq 0$ and $\dim_{\mathbb{R}} E_0 = \dim_{\mathbb{C}} V_0 - 1$. Moreover, if $j \neq 0$, there exists a decomposition into a sum of transverse Lagrangians

$$E_j \otimes \mathbb{C} = E_j^+ \oplus E_j^- ,$$

where E_j^+ (resp. E_j^-) is the direct sum of the $P_i^{s_i}$ (resp. $P_i^{-s_i}$) belonging to $E_j \otimes \mathbb{C}$. By definition, the $\mathbb{C}$-group $L_{\Delta,\varphi}$ is the subgroup

$$\mathrm{Sp}_{E_0 \otimes \mathbb{C}} \times \prod_{j \neq 0} L_j$$

of $\mathrm{Sp}_{E \otimes \mathbb{C}}$, where $L_j \simeq \mathrm{GL}_{E_i^+}$ is the stabilizer in $\mathrm{Sp}_{E_i \otimes \mathbb{C}}$ of the subspaces E_j^+ and E_j^-. It remains to give the real structure. Let $I' \in T(\mathbb{R}) \subset \mathrm{Sp}(E)$ be the element with square $-\mathrm{id}_E$ coinciding with $s_i I$ on P_i. Let h' be the Hermitian form on E associated with I', defined by $\mathrm{h}'(u) = \mathrm{a}(I'u, u)$, and let $\mathrm{h}'_j \colon E_j \to \mathbb{R}$ be the restriction of h' to E_j; it has signature (p_j, q_j), where q_j is the number of indices i such that $P_i^- \subset E_j^+$. The element I' induces a central element of L_j for every $j \neq 0$, and therefore $L_j = \mathrm{U}_{\mathrm{h}'_j}$. To conclude, we have the isomorphism of $\mathbb{R}$-groups

$$L_{\Delta,\varphi} \simeq \mathrm{Sp}_{E_0} \times \prod_{j \neq 0} \mathrm{U}_{\mathrm{h}'_j} .$$

In particular, $L_{\Delta,\varphi}(\mathbb{R})$ is connected.

8.4.17 Dual Parametrization of $\Pi^c_{\mathrm{AJ}}(\varphi)$

Let φ be an Adams–Johnson parameter of the split $\mathbb{R}$-group H and $c \in H^1(\mathbb{R}, H)$ a pure real form of H. Following [3, Sect. 3] and [9, Sect. 5], the packet $\Pi^c_{\mathrm{AJ}}(\varphi)$ is again endowed with a natural parametrization map

$$\Pi^c_{\mathrm{AJ}}(\varphi) \longrightarrow \mathrm{Hom}_{\mathrm{groups}}(\mathrm{C}_\varphi, \mathbb{C}^\times)\,, \quad \pi \mapsto \chi^c_{O,\varphi}(\pi)\,,$$

induced by that of Shelstad, which we now recall (see also [130, Sect. 8], [55, App. A], [195, Sect. 4.2.2]). As in Sect. 8.4.2, it depends only on the choice of an $O \in \mathcal{B}(H)$, which we assume fixed from now on.

Let $M \supset S \subset B \subset \widehat{H}$ be associated with φ as in Sect. 8.4.14. In particular, $\mathrm{C}_\varphi = \mathrm{Z}(M)_2$. Fix $\Delta \in O$, which defines $\rho^\vee \in \frac{1}{2}\mathrm{X}_*(T)$ (Sect. 8.4.2), a privileged isomorphism $i_\Delta \colon \mathrm{X}_*(T) \xrightarrow{\sim} \mathrm{X}^*(S)$ (Sect. 8.4.5), as well as a Levi subgroup $L = L_{\Delta,\varphi} \subset H$ (Sect. 8.4.14). The inclusion $T \to L$ induces a bijection

$$\mathrm{W}(L_{\mathbb{C}}, T_{\mathbb{C}})\backslash H^1(\mathbb{R}, T) \xrightarrow{\sim} H^1(\mathbb{R}, L)\,,$$

following Shelstad (see also [37]). This bijection and Lemma 8.4.12 (i) show that the map

$$f_\Delta \colon \Pi^c_{\mathrm{AJ}}(\varphi) \to H^1(\mathbb{R}, L)$$

that sends $\pi_{w^{-1}\Delta,\varphi,t}$, for $w \in W$ and $t \in T_2$ in the class c, onto the image of the element $w \star t \in H^1(\mathbb{R}, T)$ in $H^1(\mathbb{R}, L)$, is well defined. Moreover, it is injective, with image equal to the fiber of the natural map $H^1(\mathbb{R}, L) \to H^1(\mathbb{R}, H)$ above the class c. We also have a commutative diagram

$$\begin{array}{ccc} H^1(\mathbb{R}, L) & \xrightarrow{\;g_\Delta\;} & \mathrm{Hom}_{\mathrm{groups}}(\mathrm{Z}(M)_2, \mathbb{C}^\times) \\ \uparrow{\scriptstyle \mathrm{can}} & & \uparrow{\scriptstyle \mathrm{can}} \\ H^1(\mathbb{R}, T) & \xrightarrow{\;h_\Delta\;} & \mathrm{X}^*(S) \otimes \mathbb{Z}/2\mathbb{Z}\,. \end{array}$$

The vertical maps are the obvious ones (in particular, the one on the right is induced by the inclusion $\mathrm{Z}(M)_2 \subset S$), the map h_Δ is the composition of $i_\Delta \otimes \mathbb{Z}/2\mathbb{Z}$ and the canonical isomorphism $H^1(\mathbb{R}, T) \xrightarrow{\sim} \mathrm{X}_*(T) \otimes \mathbb{Z}/2\mathbb{Z}$ recalled in Sect. 8.4.5, and the map g (a special case of the general constructions of Kottwitz [129]) is the unique map that makes the diagram commute. Concretely, for $x \in H^1(\mathbb{R}, L)$ and $t \in T_2$ in the class x, written as $t = e^{i\pi\mu}$ with $\mu \in \mathrm{X}_*(T)$, the image $g(x)$ is the restriction to $\mathrm{Z}(M)_2$ of the character $i_\Delta(\mu) \in \mathrm{X}^*(S)$ (it depends neither on the choice of t in the class x nor, of course, on that of μ). Finally, we set

$$\chi^c_{O,\varphi}(\pi) \stackrel{\mathrm{def}}{=} g_\Delta \circ f_\Delta(\pi)$$

(it depends only on O and not on $\Delta \in O$). Concretely, for every $w \in W$ and $t = e^{i\pi\mu} \in T_2$ in the class c, the character $\chi^c_{O,\varphi}(\pi_{w^{-1}\Delta,\varphi,t})$ is the restriction to $\mathrm{Z}(M)_2$ of $i_\Delta(w(\mu + \rho^\vee) - \rho^\vee)$, where $\rho^\vee$ is with respect to Δ.

We should note that as remarked by Adams and Johnson, $\pi \mapsto \chi^{\mathrm{c}}_{O,\varphi}(\pi)$ is not, in general, injective. Let us conclude with a few simple but important observations on the behavior of discrete series, following [130, Sect. 8] and [55, App. A]. First of all, we have already said that when φ is trivial on the factor SL_2, the construction of Adams and Johnson recovers the equality $\Pi^{\mathrm{c}}_{\mathrm{AJ}}(\varphi) = \Pi_{V_\varphi}$. More precisely, in this case we have $M = S$ and $L_{\Delta,\varphi} = T$ for every basis Δ of Φ, and the representation $\pi_{\Delta,\varphi,t}$ coincides with π_{Δ,V_φ}. It is then clear that the two parametrizations $\chi^{\mathrm{c}}_{O,\varphi}$ and χ^{c}_{O} coincide. When correctly formulated, this property extends to the discrete series of $H_c(\mathbb{R})$ that appear in $\Pi^{\mathrm{c}}_{\mathrm{AJ}}(\varphi)$ for every φ, as observed by Kottwitz [130, p. 196]; let us recall how. For this, we fix an Adams–Johnson parameter φ of H. With it is associated a series of inclusions

$$\mathrm{C}_\varphi = \mathrm{Z}(M)_2 \subset S \subset B \subset \widehat{H} \tag{8.4.8}$$

defined in Sect. 8.4.14. Following Kottwitz, note that there exists a discrete Langlands parameter $\varphi_{\mathrm{disc}}\colon \mathrm{W}_{\mathbb{R}} \to \widehat{H}(\mathbb{C})$, unique up to conjugation by $S(\mathbb{C})$, with the same infinitesimal character as V_φ, such that $\varphi_{\mathrm{disc}}(\mathbb{C}^\times) \subset S(\mathbb{C})$, and with infinitesimal character dominant with respect to B (Sect. 8.4.5). In particular, we have a canonical inclusion

$$\mathrm{C}_\varphi = \mathrm{Z}(M)_2 \to \mathrm{C}_{\varphi_{\mathrm{disc}}} = S_2\ . \tag{8.4.9}$$

Proposition 8.4.18. *Let φ be an Adams–Johnson parameter of H and $c \in H^1(\mathbb{R}, H)$. Let $T \subset H$ be an anisotropic maximal torus, O a W_{r}-orbit in $\mathcal{B}(T)$, $t_{\mathrm{b}} \in T_2$ the element associated with O, $t \in T_2$ in the class c, and Δ a basis of $\Phi(H_{\mathbb{C}}, T_{\mathbb{C}})$.*

(i) *The discrete series $\pi_{\Delta,V_\varphi,t}$ belongs to $\Pi^{\mathrm{c}}_{\mathrm{AJ}}(\varphi)$ if and only if $tt_{\mathrm{b}} \in \mathrm{Z}(L_{\Delta,\varphi})$, in which case $\pi_{\Delta,V_\varphi,t} \simeq \pi_{\Delta,\varphi,t}$.*
(ii) *If $\pi \in \Pi^{\mathrm{c}}_{\mathrm{AJ}}(\varphi)$ is a discrete series, then $\chi^{\mathrm{c}}_{O,\varphi}(\pi)$ is the restriction of $\chi^{\mathrm{c}}_{O}(\pi)$ to C_φ via the canonical homomorphism* (8.4.9).

Proof. This follows from [55, Lemmas A.3 and A.5] and [130, p. 196]. □

Example 8.4.19. Let $c \in H^1(\mathbb{R}, H)$ be such that $H_c(\mathbb{R})$ is compact. Then $\Pi^{\mathrm{c}}_{\mathrm{AJ}}(\varphi) = \{V_\varphi\}$ for every Adams–Johnson parameter φ. Moreover, by Corollary 8.4.13 and Proposition 8.4.18, the character $\chi^{\mathrm{c}}_{O,\varphi}(V_\varphi)$ is the restriction to $\mathrm{C}_\varphi = S_2 \to S$ of $\rho^\vee + \nu \in \mathrm{X}^*(S)$. Here, S denotes the maximal torus of $\widehat{H}$ associated with φ, $\rho^\vee$ denotes the half-sum of the positive roots of $(\widehat{H}, S)$ with respect to the unique Borel subgroup with respect to which the infinitesimal character of φ is dominant (Sect. 8.4.14), and the image of ν in $\mathrm{X}^*(S) \otimes \mathbb{Z}/2\mathbb{Z}$ is defined in Corollary 8.4.13. Alternatively, the element ν belongs to the kernel of the natural map $\mathrm{X}^*(S) \otimes \mathbb{Z}/2\mathbb{Z} \to \mathrm{X}^*(S_{\mathrm{sc}}) \otimes \mathbb{Z}/2\mathbb{Z}$, where the torus S_{sc} denotes the inverse image of S in the universal cover of $\widehat{H}$ [55, Lemma A.6].

Example 8.4.20. Let $H = \mathrm{Sp}_{2g}(\mathbb{R})$, $V \in \mathrm{Irr}(H_{\mathbb{C}})$, and let π be the holomorphic or antiholomorphic discrete series of $H(\mathbb{R})$ with the same infinitesimal character as V (Sect. 8.4.7). Let $T \subset H$ be an anisotropic maximal torus and $K \subset H(\mathbb{R})$ the maximal compact subgroup containing $T(\mathbb{R})$, chosen, for example, as in Sect. 8.4.7. We already saw in that subsection that $\pi \simeq \pi_{\pm\Delta',V_\varphi,1}$, where Δ' is the basis given there. Let $\varphi\colon \mathrm{SL}_2(\mathbb{C}) \times \mathrm{W}_\mathbb{R} \to \widehat{H}(\mathbb{C})$ be an Adams–Johnson parameter such that $V_\varphi \simeq V$. Proposition 8.4.18 therefore shows that $\pi \in \Pi_{\mathrm{AJ}}(\varphi)$ if and only if $t_\mathrm{b} \in \mathrm{Z}(L_{\pm\Delta',\varphi})$. By Lemma 8.4.4 (i), this is also equivalent to requiring $L_{\pm\Delta',\varphi}(\mathbb{R}) \subset K$. But $L_{\pm\Delta',\varphi}$ is obtained as described in Sect. 8.4.15; knowing that, by construction, the basis Δ' has the property that all s_i have the same sign, we see that $L_{\pm\Delta',\varphi}(\mathbb{R}) \subset K$ if and only if $\dim E_0 = d_0 - 1 = 0$ in the notation of that subsection. To conclude, $\pi \in \Pi_{\mathrm{AJ}}(\varphi)$ if and only if the only component of odd dimension of the semisimple representation $\mathrm{St} \circ \varphi$ is of dimension 1 (we have given a new proof of [55, Lemma 9.4]).

8.4.21 Adams–Johnson Packets and Arthur Packets

The following conjecture is part of folklore [9, Sect. 5], [13, p. 43].

Conjecture 8.4.22. Let H be the split $\mathbb{R}$-group Sp_{2g} or $\mathrm{SO}_{r+1,r}$, and let φ be a $\widehat{H}(\mathbb{C})$-conjugacy class of Adams–Johnson parameters of H. If $(\Pi(\varphi), \iota, \chi)$ denotes the triple associated with it by Arthur [13, Theorem 1.5.1], in the notation of Sect. 8.3.8, then

(a) the map $\iota\colon \Pi(\varphi) \to \Pi_{\mathrm{unit}}(H(\mathbb{R}))$ is an injection with image $\Pi_{\mathrm{AJ}}(\varphi)$;
(b) the character $\chi \circ \iota^{-1}\colon \Pi_{\mathrm{AJ}}(\varphi) \to \mathrm{Hom}_{\mathrm{groups}}(\mathrm{C}_\varphi, \mathbb{C}^\times)$, which is well defined by part (a), coincides with the character $\pi \mapsto \chi^1_{O,\varphi}(\pi)$ defined in Sect. 8.4.17.

In part (b) above, $O \in \mathcal{B}(H)$ corresponds with the Whittaker datum chosen by Arthur [13, p. 55].

Remark 8.4.23. As already mentioned in Sect. 8.4.5, this conjecture is known if φ is trivial on the factor $\mathrm{SL}_2(\mathbb{C})$, by the work of Shelstad and Mezo. In the general setting, progress concerning this conjecture has recently been made by Colette Moeglin and Nicolas Arancibia (Ph.D. thesis [8]). More precisely, Arancibia announces the proof of Conjecture 8.4.22 in the particular case where each irreducible component of the representation $\mathrm{St} \circ \varphi$, which we can write as $U \otimes V$ with U (resp. V) an irreducible representation of $\mathrm{SL}_2(\mathbb{C})$ (resp. $\mathrm{W}_\mathbb{R}$), satisfies $\dim V = 1$ or $\dim U \leq 4$.

Remark 8.4.24. Conjecture 8.4.22 (and the theorem of Arancibia) admit a variant for the $\mathbb{R}$-group $H = \mathrm{SO}_{r,r}$ with $r \equiv 0 \bmod 2$, in which $\Pi_{\mathrm{unit}}(H(\mathbb{R}))$ is replaced by $\widetilde{\Pi}_{\mathrm{unit}}(H(\mathbb{R}))$ (Sect. 8.3.8) and $\Pi_{\mathrm{AJ}}(\varphi)$ by its image in $\widetilde{\Pi}_{\mathrm{unit}}(H(\mathbb{R}))$ (we have an analog of Lemma 8.4.6 for $\Pi_{\mathrm{AJ}}(\varphi)$).

As announced by Arthur [13, Chap. 9], his description of $\Pi_{\mathrm{disc}}(G)$ and the multiplicity formula stated in Sect. 8.3.8 for the Chevalley $\mathbb{Z}$-groups G admit an analog for other classical $\mathbb{Z}$-groups, namely the special orthogonal $\mathbb{Z}$-groups SO_L introduced in Sect. 8.1. Let us therefore fix such an L and set $G = \mathrm{SO}_L$; we, moreover, suppose that $G(\mathbb{R})$ has discrete series, that is, that $L \otimes \mathbb{R}$ has an even signature if it has even rank (the important case for further on is $G = \mathrm{SO}_n$). We denote by $G^\star$ the (special orthogonal) Chevalley group such that $G_\mathbb{C} \simeq G^\star_\mathbb{C}$.

In the generality considered by Arthur, the statement itself of his formula depends on an additional datum, namely a "realization" of G as an inner form of $G^\star$. The special case of the $\mathbb{Z}$-group $G = \mathrm{SO}_L$ is particularly nice because G can be constructed as a pure inner form of $G^\star$ "over $\mathbb{Z}$"; let us simply say that this follows from the fact that $\langle \pm 1\rangle \otimes L$ is locally isomorphic to $\mathrm{H}(\mathbb{Z}^r)$ for the étale topology on $\mathrm{Spec}(\mathbb{Z})$ if L has even rank, and to $\mathrm{H}(\mathbb{Z}^r) \oplus \mathrm{A}_1$ otherwise (Sect. 2.2, Appendix B). This property seems to significantly simplify the situation for questions of normalization of the transfer factors invoked by Arthur loc. cit. and studied by Kaletha [115].

More concretely, a *real realization* of G is defined to be a pair $\xi = (c, f)$ with $c \in \mathrm{H}^1(\mathbb{R}, G^\star_\mathbb{R})$ and $f\colon G_\mathbb{R} \overset{\sim}{\to} (G^\star_\mathbb{R})_c$ an isomorphism. There always exist real realizations of G: we saw this in the example treated at the end of Sect. 8.4.8 when L has odd rank; the case of even rank is analogous (it uses that the signature of $L \otimes \mathbb{R}$ is even). Given such a realization, we can define, for every Adams–Johnson parameter φ of $G_\mathbb{R}$ and $O \in \mathcal{B}(G^\star_\mathbb{R})$, a pair

$$(\Pi^\xi_{\mathrm{AJ}}(\varphi), \chi^\xi_{O,\varphi})$$

with $\Pi^\xi_{\mathrm{AJ}}(\varphi) \subset \Pi_{\mathrm{unit}}(G(\mathbb{R}))$ and $\chi^\xi_{O,\varphi}\colon \Pi_{\mathrm{AJ}}(\varphi) \to \mathrm{Hom}_{\mathrm{groups}}(\mathrm{C}_\varphi, \mathbb{C}^\times)$, by simply transporting the pair $(\Pi^c_{\mathrm{AJ}}(\varphi), \chi^c_{O,\varphi})$ by the bijection $\Pi_{\mathrm{unit}}(G(\mathbb{R})) \overset{\sim}{\to} \Pi_{\mathrm{unit}}((G^\star_\mathbb{R})_c(\mathbb{R}))$ induced by f.

A second interesting phenomenon is that the projective similitude group $\mathrm{P}\widetilde{G}(\mathbb{Z})$ meets all connected components of $\mathrm{P}\widetilde{G}(\mathbb{R})$ (to see this, use, for example, Theorem 2.2.7). As already observed in [55], this suggests that Arthur's final formula must be completely canonical. As in Sect. 8.3.8, it involves certain sets of unitary representations of $G(\mathbb{R})$ (depending on the choice of a real realization of G) whose existence, as well as a characterization, was announced by Arthur. We expect that those that contain discrete series of $G(\mathbb{R})$ are exactly the Adams–Johnson packets. To avoid multiplying the statements and because we have already treated the case of Chevalley groups in detail, we will only state the expected final conjecture, it being understood that it is a concatenation of two statements. For $U, U' \in \Pi_{\mathrm{unit}}(G(\mathbb{R}))$, we write $U \sim U'$ if $U \simeq U'$ or if $G(\mathbb{R})$ is an even special orthogonal group and U and U' are outer conjugates of each other by the corresponding even real orthogonal group. Finally, for $X \subset \Pi_{\mathrm{unit}}(G(\mathbb{R}))$, we write $\widetilde{X} = \{U \in \Pi_{\mathrm{unit}}(G(\mathbb{R}))\,;\, \exists U' \in X, U \sim U'\}$.

Conjecture 8.4.25. Let G be a classical $\mathbb{Z}$-group, $\mathrm{St}\colon \widehat{G} \to \mathrm{SL}_n$ the standard representation, $\psi \in \mathcal{X}_{\mathrm{AL}}(\mathrm{SL}_n)$, and $U \in \Pi_{\mathrm{unit}}(G(\mathbb{R}))$ a discrete series such that $\mathrm{St}(\mathrm{Inf}_U) = \psi_\infty$.

Let ξ be a real realization of G, $E \subset \Pi(G)$ the set of representations π such that $\pi_\infty \sim U$, and $\psi(\pi, \mathrm{St}) = \psi$ (this is a singleton if G is not an even special orthogonal group), and let $\nu_\infty\colon \mathrm{SL}_2(\mathbb{C}) \times \mathrm{W}_\mathbb{R} \longrightarrow \widehat{G}$ be an Adams–Johnson parameter associated with ψ as described in Sect. 8.3.8. Then $E \cap \Pi_{\mathrm{disc}}(G) = \emptyset$ unless

$U \in \Pi^{\xi}_{\mathrm{AJ}}(\nu_\infty)$ and

$$(\chi^{\xi}_{O,\nu_\infty}(U))_{|\mathrm{C}_\psi} = \varepsilon_\psi\ , \tag{8.4.10}$$

in which case $\sum_{\pi\in E} \mathrm{m}(\pi) = m_\psi$, where the integer m_ψ is 1 unless G is an even special orthogonal group and $\psi = \oplus_i \pi_i[d_i]$ with $d_i \equiv 0 \bmod 2$ for every i, in which case $m_\psi = 2$.

Specifically, the canonicity mentioned above means that even though the character $\chi^{\xi}_{O,\nu_\infty}$ depends on the choices of O and ξ, its restriction to $\mathrm{C}_\psi \subset \mathrm{C}_{\nu_\infty}$ does not depend on them. Indeed, it is clear if G is an odd special orthogonal group because every $\mathbb{R}$-automorphism of $G_{\mathbb{R}}$ is an inner automorphism and $|\mathcal{B}(G_{\mathbb{R}})| = 1$. If $G \simeq G^* \simeq \mathrm{Sp}_{2g}$, the group $\mathrm{Aut}_{\mathbb{R}}(G_{\mathbb{R}})/\mathrm{Int}(G(\mathbb{R})) = \mathbb{Z}/2\mathbb{Z}$ acts simply transitively on $\mathcal{B}(G_{\mathbb{R}})$ and $|\mathcal{B}(G_{\mathbb{R}})| = 2$, so that there are exactly two choices to consider; as already observed in [55, Lemmas 9.5 and 9.6], criterion (8.4.10) is in fact the same in the two cases (this will be clear in the proof of Theorem 8.5.2). The situation is similar if G is an even special orthogonal group, for which we must also take into account the outer automorphism derived from the corresponding orthogonal group (it is, in fact, defined over $\mathbb{Z}$ whenever L admits a root, that is, an element α such that $\mathrm{q}(\alpha) = 1$).

8.5 Explicit Multiplicity Formulas

If we confront Arthur's general theorem, Theorem 8.3.10 (or Conjecture 8.4.25) with the considerations and examples in Sect. 8.4, we obtain explicit forms of Arthur's multiplicity formula. In this section, we propose to describe them, in the manner of [55, Sect. 3.29], in the cases that are particularly important for this book, namely where $G = \mathrm{Sp}_{2g}$ and π_∞ is an Archimedean component in the holomorphic discrete series, or where $G = \mathrm{SO}_n$ for $n \equiv -1, 0, 1 \bmod 8$.

8.5.1 Explicit Formula for Sp_{2g}

In this subsection, $g \geq 1$ denotes an integer. Let

$$\psi = \oplus_{i=1}^{k} \pi_i[d_i] \in \mathcal{X}_{\mathrm{AL}}(\mathrm{SL}_{2g+1})\ ,$$

where $k \geq 1$ is an integer and we have $\pi_i \in \Pi^{\perp}_{\mathrm{cusp}}(\mathrm{PGL}_{n_i})$ and $d_i \geq 1$ for every $i = 1, \ldots, k$. We suppose that ψ_∞ satisfies condition (H2) with respect to Sp_{2g} (Sect. 8.3.4), which means that the eigenvalues of ψ_∞ are $2g+1$ distinct integers

$$w_1 > \cdots > w_g > 0 > -w_g > \cdots > -w_1$$

(Sect. 8.2.6, case I). By Lemma 8.2.15 (i), there exists a unique integer $i_0 \in \{1, \ldots, k\}$ such that $n_{i_0} d_{i_0}$ is odd. After reindexing the π_i if necessary, we may assume $i_0 = k$ without loss of generality. By the same lemma, we also have $n_i \equiv 0 \bmod 2$ and $n_i d_i \equiv 0 \bmod 4$ for every $i \neq k$, and $n_k d_k \equiv 2g+1 \bmod 4$.

Consider the homomorphism of multiplicative groups $\chi\colon \{\pm 1\}^{k-1} \to \{\pm 1\}$ defined as follows. Fix $1 \leq i \leq k-1$, and let $s_i \in \{\pm 1\}^{k-1}$ be the element defined by $(s_i)_j = -1$ if and only if $j = i$. There are two cases:

(i) If $d_i \equiv 0 \bmod 2$, we set $\chi(s_i) = (-1)^{n_i d_i/4}$.
(ii) If $d_i \equiv 1 \bmod 2$, we set $\chi(s_i) = (-1)^{|K_i|}$, where K_i is the set of odd indices $1 \leq j \leq g$ such that $w_j \in \mathrm{Weights}(\pi_i)$.

Theorem⋆ 8.5.2. *Let $\psi = \oplus_{i=1}^k \pi_i[d_i] \in \mathcal{X}_{\mathrm{AL}}(\mathrm{SL}_{2g+1})$ and χ be as above. Let $\pi \in \Pi(\mathrm{Sp}_{2g})$ be the unique representation such that $\psi(\pi, \mathrm{St}) = \psi$ and π_∞ is a holomorphic discrete series. Suppose that Conjecture 8.4.22 is true for Sp_{2g} and the morphism ν_∞ associated with ψ defined in Sect. 8.3.8 (this holds, for example, if we have $d_i = 1$ for every $i = 1, \ldots, k$).*

Then $\pi \in \Pi_{\mathrm{disc}}(\mathrm{Sp}_{2g})$ if and only if the following two conditions are satisfied:

(a) $d_k = 1$.
(b) *For every $i = 1, \ldots, k-1$, we have*

$$\chi(s_i) = \prod_{1 \leq j \leq k, j \neq i} \varepsilon(\pi_i \times \pi_j)^{\min(d_i, d_j)} .$$

Finally, if these conditions are satisfied, then $\mathrm{m}(\pi) = 1$.

Proof. Let us apply a few constructions from Sect. 8.3.8. In particular, we choose $\nu\colon \mathrm{SL}_2(\mathbb{C}) \times \prod_{i=1}^k \widehat{G^{\pi_i}}(\mathbb{C}) \to \mathrm{SO}_{2g+1}(\mathbb{C})$ associated with ψ as in that subsection, and for each $i = 1, \ldots, k$, we choose a homomorphism $\mu_i\colon \mathrm{W}_{\mathbb{R}} \to \widehat{G^{\pi_i}}(\mathbb{C})$ such that $\mathrm{St} \circ \mu_i \simeq \mathrm{L}((\pi_i)_\infty)$. The group C_ν defined in Sect. 8.3.5 can be naturally identified with $\{\pm 1\}^{k-1}$, with their respective distinguished elements $s_1, \ldots, s_{k-1}$. We, moreover, have a homomorphism

$$\nu_\infty\colon \mathrm{SL}_2(\mathbb{C}) \times \mathrm{W}_{\mathbb{R}} \longrightarrow \mathrm{SO}_{2g+1}(\mathbb{C})$$

deduced from ν and the μ_i. Condition (H2) and Lemmas 8.4.16 and 8.3.9 imply that this is an Adams–Johnson parameter of Sp_{2g} (and even a discrete Langlands parameter if $d_i = 1$ for every i). As a consequence, Conjecture 8.4.22 applies to ν_∞.

By Example 8.4.20, the holomorphic discrete series π_{hol} of $\mathrm{Sp}_{2g}(\mathbb{R})$, with infinitesimal character z such that $\mathrm{St}(z) = \psi_\infty$, is in $\Pi(\nu_\infty)$ if and only if $\mathrm{St} \circ \nu_\infty$ does not contain a representation of the form $\mathrm{Sym}^{d-1}\mathrm{St}_2 \otimes \chi$ with $d > 1$ and $\chi \in \{1, \epsilon_{\mathbb{C}/\mathbb{R}}\}$. This is equivalent to requiring $d_k = 1$. Indeed, for $i = 1, \ldots, k$, we see that $\mathrm{c}_\infty(\pi_i)$ admits the eigenvalue 0 if and only if n_i is odd, that is, $i = k$ (Sect. 8.2.6). By Theorem 8.3.10 and Conjecture 8.4.22 for the pair $(\mathrm{Sp}_{2g}, \nu_\infty)$, it remains to prove

$$\chi^1_{O,\nu_\infty}(\pi_{\mathrm{hol}})_{|\mathrm{C}_\nu} = \chi \,.$$

For this, we will specify the constructions of Sect. 8.4.14 with respect to ν_∞, in the manner of the analysis carried out in Example 8.4.15. We consider the homomorphism $\widetilde{\nu_\infty}\colon \mathrm{W}_\mathbb{R} \to \mathrm{SO}_{2g+1}(\mathbb{C})$ deduced by composing ν_∞ and Arthur's morphism as in Sect. 8.4.14. The analysis of Sect. 8.4.15 shows that there exists a unique ordered pair (V^+, V^-) of transverse Lagrangians of the q-vector space $V = \mathbb{C}^{2g+1}$ such that

- V^+ and V^- are stable under $\widetilde{\nu_\infty}(\mathbb{C}^\times)$;
- V^+ has a $\mathbb{C}$-basis $e_1, \ldots, e_g$ such that for every $j = 1, \ldots, g$ and every $z \in \mathbb{C}^\times$, the relation $\widetilde{\nu_\infty}(z)(e_j) = z^{w_j}\overline{z}^{w'_j}e_j$ holds.

In this relation, $w_1, \ldots, w_g$ are the integers deduced from ψ_∞ defined before Theorem 8.5.2, and the w'_i are also uniquely determined integers (that we will not need to specify). As in Sect. 8.4.15, giving this Lagrangian $\mathbb{C}$-basis $(e_j)_{1\leq j\leq g}$ determines a unique maximal torus $S \subset \mathrm{SO}_{2g+1}$, as well as a unique Borel subgroup of B containing S. As in Sect. 6.1.3, we write

$$\mathrm{X}^*(S) = \oplus_{j=1}^{g}\mathbb{Z}\varepsilon_j \ ,$$

where ε_j is the character of S over $\mathbb{C}e_j$. Let $\varepsilon_j^* \in \mathrm{X}_*(S)$ be the dual basis of (ε_j). Let $\lambda \in \frac{1}{2}\mathrm{X}_*(S)$ be the element associated with $\widetilde{\nu_\infty}$ appearing in axiom (AJ1) of Sect. 8.4.14. We clearly have

$$\lambda = \sum_{j=1}^{g} w_j\varepsilon_j^* \ ,$$

so that λ is dominant with respect to B, by the inequalities $w_1 > w_2 > \cdots > w_g > 0$. Up to here, we have made explicit the sequence of inclusions

$$\mathrm{C}_\nu \subset \mathrm{C}_{\nu_\infty} \subset S \subset B \subset \mathrm{SO}_{2g+1}$$

associated with ν and ν_∞. Proposition 8.4.18 (ii), as well as the example of Sect. 8.4.7, implies that the character $\chi^1_{O,\nu_\infty}(\pi_{\mathrm{hol}})\colon \mathrm{C}_{\nu_\infty} \to \mathbb{C}^\times$ is the restriction to C_{ν_∞} of one of the following two elements of $\mathrm{X}^*(S)$:

$$\chi_0 = \sum_{j\equiv 0 \bmod 2} \varepsilon_j \quad \text{or} \quad \chi_1 = \sum_{j\equiv 1 \bmod 2} \varepsilon_j \ ,$$

where the sums are taken over the $j \in \{1, \ldots, g\}$ with the given parity.

Finally, let us verify that the restriction to C_ν of either of these two characters χ_u for $u \in \{0, 1\}$ coincides with the character χ. Fix $1 \leq i \leq k-1$. Let J_i be the subset of $\{1, \ldots, g\}$ consisting of the integers j such that w_j is an eigenvalue of $(\pi_i[d_i])_\infty$. Let us first note that, by construction, the image of the element $s_i \in \mathrm{C}_\nu = \{\pm 1\}^{k-1}$ by the natural inclusion $\mathrm{C}_\nu \subset S$ is determined by the following relation, satisfied for every $j \in \{1, \ldots, g\}$:

$$\varepsilon_j(s_i) = -1 \ \Leftrightarrow \ j \in J_i \ .$$

Let us write $J_i = J_i^0 \coprod J_i^1$, where $J_i^u = \{j \in J_i \,;\, j \equiv u \bmod 2\}$. By definition,

$$\chi_u(s_i) = (-1)^{|J_i^u|} .$$

If d_i is even, in which case π_i is symplectic and n_i is also even, J_i is the disjoint union of $|J_i|/2 = n_i d_i/4$ pairs of consecutive integers, so that

$$\chi_0(s_i) = \chi_1(s_i) = (-1)^{n_i d_i/4} = \chi(s_i) .$$

Finally, suppose that d_i is odd, so that π_i is orthogonal and $n_i \equiv 0 \bmod 4$ by Corollary 8.2.15. Let $P_i \subset J_i$ be the subset of $j \in J_i$ such that $w_j \in \mathrm{Weights}(\pi_i)$. Then J_i is the disjoint union of P_i and its translates by

$$\pm 1, \pm 2, \ldots, \pm \frac{d_i - 1}{2} .$$

But for $1 \leq d \leq (d_i - 1)/2$, we necessarily have $w_i + d = w_{i-d}$ and $w_i - d = w_{i+d}$. Since the indices $i - d$ and $i + d$ are congruent modulo 2, it follows that

$$\chi_u(s_i) = (-1)^{|P_i^u|} ,$$

where $P_i^u = \{j \in P_i \,;\, j \equiv u \bmod 2\}$. We therefore have $\chi_1(s_i) = \chi(s_i)$ and also $\chi_0(s_i) = \chi_1(s_i)$ because $|P_i| = n_i/2 \equiv 0 \bmod 2$. □

Example 8.5.3. By way of example, let us compare the statements of Theorem 8.5.2 and Ikeda's Theorem 7.3.1. Let $k' > 0$ be an even integer, and let $\pi \in \Pi_{\mathrm{cusp}}(\mathrm{PGL}_2)$ be the representation generated by an eigenform in $\mathrm{S}_{k'}(\mathrm{SL}_2(\mathbb{Z}))$. Consider the parameter $\psi = \pi[g] \oplus [1]$. Since $\mathrm{Weights}(\pi) = \{\pm(k'-1)/2\}$, we see that ψ satisfies condition (H2) if and only if $g \equiv 0 \bmod 2$ and $k' > g$, in which case the eigenvalues of ψ_∞ are

$$\frac{k'-1}{2}, \frac{k'-3}{2}, \ldots, \frac{k'-g}{2}, 0, -\frac{k'-g}{2}, \ldots, -\frac{k'-3}{2}, -\frac{k'-1}{2} .$$

We, of course, have $k = 2$ and $d_2 = 1$, and $\mathrm{C}_\psi = \{\pm 1\}$ is generated by the element s_1. We also have

$$\varepsilon_\psi(s_1) = \varepsilon(\pi \times 1) = \varepsilon(\pi) = i^{k'} = (-1)^{k'/2} .$$

Moreover, since g is even, we have $\chi(s_1) = (-1)^{g/2}$. Under Conjecture 8.4.22, the necessary and sufficient condition for the existence of $\pi' \in \Pi_{\mathrm{cusp}}(\mathrm{Sp}_{2g})$ such that π'_∞ is a holomorphic discrete series and $\psi(\pi', \mathrm{St}) = \psi$ can therefore be written as

$$(-1)^{k'/2} = (-1)^{g/2}$$

or, equivalently, $k' \equiv g \bmod 4$. This is indeed the condition in Ikeda's statement. Ikeda's result is in fact stronger, first because it is unconditional, but also because we do not need to assume $k' > g$ (and it would be interesting to also study the Arthur packets corresponding to this more general case). Let us also mention that in his supplement [110] to [109], Ikeda proves that if $k' \equiv g \bmod 4$ (resp. $k' \not\equiv g \bmod 4$), then $\mathrm{m}(\pi') = 1$ (resp. $\mathrm{m}(\pi') = 0$); see [110, Theorem 7.1, Sect. 15].

Let us conclude this subsection with a translation of the multiplicity 1 assertion in Theorem 8.5.2 to the classical language.

Corollary* 8.5.4. *Let W be the $\mathbb{C}$-representation of GL_g of highest weight $\sum_{i=1}^{g} m_i\varepsilon_i$ with $m_1 > m_2 > \cdots > m_g > g+1$ (Sect. 6.3.4). If $F, G \in \mathrm{S}_W(\mathrm{Sp}_{2g}(\mathbb{Z}))$ are two eigenforms for $\mathrm{H}(\mathrm{Sp}_{2g})$ and if every element of $\mathrm{H}(\mathrm{Sp}_{2g})$ admits the same eigenvalue on F and G, then F and G are proportional. When $g = 2$, the same assertion holds by supposing only $m_1 > m_2 > 2$.*

Proof. Recall (Corollary 6.3.7 and the remark that follows it) that we have an isomorphism of $\mathrm{H}^{\mathrm{opp}}(\mathrm{Sp}_{2g})$-modules $\mathrm{S}_W(\mathrm{Sp}_{2g}(\mathbb{Z})) \xrightarrow{\sim} \mathcal{A}_{\pi'_W}(\mathrm{Sp}_{2g})$. It therefore suffices to show that if π is the representation generated by an eigenform of $\mathrm{S}_W(\mathrm{Sp}_{2g}(\mathbb{Z}))$ under the action of $\mathrm{H}(\mathrm{Sp}_{2g})$, then $\mathrm{m}(\pi) = 1$. Recall that $\pi_\infty \simeq \pi'_W$ and $\mathrm{St}(\mathrm{Inf}_{\pi'_W})$ has eigenvalues 0 and the $\pm(m_r - r)$ for $r = 1, \ldots, g$ (Corollary 6.3.6): the latter are therefore pairwise distinct and nonconsecutive by assumption. If we write $\psi(\pi, \mathrm{St}) = \oplus_i \pi_i[d_i]$, which is allowed by Theorem 8.1.1, we therefore have $d_i = 1$ for every i, and we conclude using Theorem 8.5.2. When $g = 2$ and $m_1 > m_2 = 2$, we can also conclude because the only other possibility $\psi(\pi, \mathrm{St}) = \pi_1 \oplus [3]$ with $\pi_1 \in \Pi_{\mathrm{cusp}}(\mathrm{PGL}_2)$ does not occur because such a π_1 is necessarily symplectic (Sect. 8.3.1 or Proposition 8.2.13 (i)). □

We expect this corollary to hold for every W: this follows from Conjecture 8.4.22 whenever $m_g > g$.

8.5.5 Explicit Formula for SO_n *with* $n \equiv \pm 1 \bmod 8$

Now, suppose that n is an integer congruent to $\pm 1 \bmod 8$, and consider the $\mathbb{Z}$-group SO_n. Let

$$\psi = \oplus_{i=1}^{k} \pi_i[d_i] \in \mathcal{X}_{\mathrm{AL}}(\mathrm{SL}_{n-1}),$$

where $k \geq 1$ is an integer and we have $\pi_i \in \Pi_{\mathrm{cusp}}^{\perp}(\mathrm{PGL}_{n_i})$ and $d_i \geq 1$ for every $i = 1, \ldots, k$. We suppose that ψ_∞ satisfies condition (H2) with respect to SO_n (Sect. 8.3.4), which means that the eigenvalues of ψ_∞ are $n-1$ distinct (nonintegral) half-integers

$$w_1 > \cdots > w_{(n-1)/2} > -w_{(n-1)/2} > \cdots > -w_1$$

(Sect. 8.2.6, case II). By Lemma 8.2.15 (ii), for every $i = 1, \ldots, k$, we have $n_i d_i \equiv 0 \bmod 2$.

Consider the homomorphism of multiplicative groups $\chi\colon \{\pm 1\}^k \to \{\pm 1\}$ defined as follows. Fix $1 \le i \le k$, and let $s_i \in \{\pm 1\}^k$ be the element defined by $(s_i)_j = -1$ if and only if $j = i$. There are two cases:

(i) *The integer d_i is even.* If n_i is even, we set $\chi(s_i) = (-1)^{(n_i d_i)/4}$. If n_i is odd, we set $\chi(s_i) = \epsilon_i \cdot (-1)^{(n_i-1)d_i/4}$, where $\epsilon_i = (-1)^{[3d_i/4]}$ is -1 if $d_i/2 \equiv 1, 2 \bmod 4$ and 1 otherwise. (In all cases, $\chi(s_i) = (-1)^{[3\,n_i d_i/4]}$.)
(ii) *The integer d_i is odd.* We set $\chi(s_i) = (-1)^{|K_i|}$, where K_i is the set of indices $1 \le j \le (n-1)/2$ congruent to $(n-1)/2 \bmod 2$ such that $w_j \in \mathrm{Weights}(\pi_i)$.

Theorem 8.5.6 (Case $n \equiv \pm 1 \bmod 8$). *Let $\psi = \oplus_{i=1}^k \pi_i[d_i] \in \mathcal{X}_{\mathrm{AL}}(\mathrm{SL}_{n-1})$ and χ be as above. Let $\pi \in \Pi(\mathrm{SO}_n)$ be the unique representation such that $\psi(\pi, \mathrm{St}) = \psi$. Suppose that Conjecture* 8.4.25 *is true for* SO_n *and the morphism ν_∞ associated with ψ defined in Sect.* 8.3.8. *Then $\pi \in \Pi_{\mathrm{disc}}(\mathrm{SO}_n)$ if and only if for every $i = 1, \ldots, k$, we have*

$$\chi(s_i) = \prod_{1 \le j \le k, j \ne i} \varepsilon(\pi_i \times \pi_j)^{\min(d_i, d_j)} .$$

Finally, if these conditions hold, then $\mathrm{m}(\pi) = 1$.

Proof (See [55, Sect. 3.30.1]). The proof is similar to that of Theorem 8.5.2, which is why we will only mention the differences with the latter. An analysis of ν_∞ similar to that of Sect. 8.4.15, keeping in mind that this time, the dual group is Sp_{n-1}, leads to a specification of the sequence of canonical inclusions

$$\mathrm{C}_\nu = \{\pm 1\}^k \subset \mathrm{C}_{\nu_\infty} \subset S \subset B \subset \mathrm{Sp}_{n-1} \,.$$

We then invoke Corollary 8.4.13 instead of Example 8.4.7 (and Conjecture 8.4.25 instead of Conjecture 8.4.22 and Theorem 8.3.10). Since the center of SO_n is trivial, this corollary asserts that the character $\chi^{\xi}_{O,\nu_\infty}(\pi_\infty)$ is the restriction to C_{ν_∞} of the half-sum $\rho^\vee$ of the positive roots of T with respect to B, which it therefore suffices to specify. Set $n = 2r+1$ and consider the standard based root datum of Sp_{2r} recalled in Sect. 6.1.3. In the notation loc. cit., we see that

$$\rho^\vee = \sum_{i=1}^{r} (r - i + 1)\varepsilon_i \equiv \varepsilon_r + \varepsilon_{r-2} + \cdots \bmod 2 \,.$$

We conclude as in the proof of Theorem 8.5.2 that the restriction of this character to C_ψ is the character χ of the theorem. We should take extra care with the case where the integer $i \in \{1, \ldots, k\}$ satisfies $d_i \equiv 0 \bmod 2$ and $n_i \equiv 1 \bmod 2$, because 0 is then a weight of π_i. In this case, the set J_i defined in that proof is the disjoint union of $(n_i - 1)d_i/4$ pairs of consecutive integers and the set $\{r, r-1, \ldots, r+1-d_i/2\}$, whence the need to modify the definition of χ in case (i). $\square$

8.5.7 Explicit Formula for SO_n with $n \equiv 0 \bmod 8$

Finally, let us consider the $\mathbb{Z}$-group SO_n for $n \equiv 0 \bmod 8$. Let

$$\psi = \oplus_{i=1}^k \pi_i[d_i] \in \mathcal{X}_{\mathrm{AL}}(\mathrm{SL}_n) ,$$

where $k \geq 1$ is an integer and we have $\pi_i \in \Pi_{\mathrm{cusp}}^{\perp}(\mathrm{PGL}_{n_i})$ and $d_i \geq 1$ for every $i = 1, \ldots, k$. We suppose that ψ_∞ satisfies condition (H2) with respect to SO_n (Sect. 8.3.4), which means that the eigenvalues of ψ_∞ are n integers

$$w_1 > \cdots > w_{n/2} \geq -w_{n/2} > \cdots > -w_1$$

(Sect. 8.2.6, case III). Let $I_1 \subset \{1, \ldots, k\}$ be the subset of indices i such that $n_i d_i \equiv 1 \bmod 2$, and set $I_0 = \{1, \ldots, k\} - I_1$. By Lemma 8.2.15 (ii), we have either $I_1 = \emptyset$ or $|I_1| = 2$ (the latter can only happen if $w_{n/2} = 0$). Moreover, $n_i d_i \equiv 0 \bmod 4$ if $i \in I_0$.

Consider the homomorphism of multiplicative groups $\chi\colon \{\pm 1\}^{I_0} \to \{\pm 1\}$ defined as follows. Fix $i \in I_0$, and let $s_i \in \{\pm 1\}^{I_0}$ be the element defined by $(s_i)_j = -1$ if and only if $j = i$.

(i) If $d_i \equiv 0 \bmod 2$, we set $\chi(s_i) = (-1)^{n_i d_i/4}$.
(ii) If $d_i \equiv 1 \bmod 2$, we set $\chi(s_i) = (-1)^{|K_i|}$, where K_i is the set of odd indices $1 \leq j \leq n/2$ such that $w_j \in \mathrm{Weights}(\pi_i)$.

Theorem 8.5.8 (Case $n \equiv 0 \bmod 8$). *Let $\psi = \oplus_{i=1}^k \pi_i[d_i] \in \mathcal{X}_{\mathrm{AL}}(\mathrm{SL}_n)$, the partition $\{1, \ldots, k\} = I_0 \coprod I_1$, and χ be as above. Let $\Pi \subset \Pi(\mathrm{SO}_n)$ be the subset of representations π such that $\psi(\pi, \mathrm{St}) = \psi$; this is a singleton if $I_1 \neq \emptyset$. Suppose that Conjecture 8.4.25 is true for SO_n and the morphism ν_∞ associated with ψ defined in Sect. 8.3.8. Then $\Pi \cap \Pi_{\mathrm{disc}}(\mathrm{SO}_n) \neq \emptyset$ if and only if we have*

$$\chi(s_i) = \prod_{1 \leq j \leq k, j \neq i} \varepsilon(\pi_i \times \pi_j)^{\min(d_i, d_j)} \quad \forall i \in I_0 . \tag{8.5.1}$$

Finally, if this condition holds, then we have $\sum_{\pi \in \Pi} \mathrm{m}(\pi) = 1$ if I_1 is not empty, and $\sum_{\pi \in \Pi} \mathrm{m}(\pi) = 2$ otherwise.

Proof (See [55, Sect. 3.30.2]). The proof is similar to that of Theorem 8.5.6. As in Sect. 8.4.15, we specify the sequence of canonical inclusions

$$\mathrm{C}_\nu \subset \mathrm{C}_{\nu_\infty} \subset S \subset B \subset \mathrm{SO}_n .$$

The homomorphism $\{\pm 1\}^{I_0} \to \mathrm{C}_\nu$ that sends the element s_i for $i \in I_0$ defined above to the element of the same name defined in Sect. 8.3.5 induces a surjection $\{\pm 1\}^{I_0} \to \mathrm{C}_\nu / \mathrm{Z}(\mathrm{SO}_n)$. In the standard based root datum of (SO_n, S, B) recalled in Sect. 6.1.3, we now see that the half-sum of the positive roots equals

$$\rho^\vee = \sum_{i=1}^r (r - i)\varepsilon_i \equiv \varepsilon_{r-1} + \varepsilon_{r-3} + \cdots \bmod 2 ,$$

where $r = n/2$. Since the center of SO_n is nontrivial, Corollary 8.4.13 only asserts that $\chi^{\xi}_{O,\nu_\infty}(\pi_\infty)$ is the restriction to $\mathrm{C}_{\nu_\infty} \subset S$ of either $\rho^\vee$ or $\rho^\vee + \nu$, where

$$\nu \equiv \sum_{i=1}^{r} \varepsilon_i \bmod 2 .$$

But $\nu_{|\mathrm{C}_\nu} = 1$ because $n_i d_i \equiv 0 \bmod 4$ for every $i \in I_0$, and the rest follows. □

Remark 8.5.9 (On Multiplicity 2). Let $\alpha \in \mathrm{E}_n$ be such that $\alpha \cdot \alpha = 2$, and let $\mathrm{s}_\alpha \in \mathrm{O}(\mathrm{E}_n)$ be the orthogonal symmetry associated with this root. Outer conjugation by s_α induces an involution of $\Pi(\mathrm{SO}_n)$, which we denote by S, of which we have already studied certain aspects in Sect. 4.4.4, as well as in the examples of Sects. 6.2.1 and 6.4.7. Those subsections show that if $\pi \in \Pi(\mathrm{SO}_n)$, then $\mathrm{m}(\pi) = \mathrm{m}(S(\pi))$, and for every $v \in \mathrm{P} \cup \{\infty\}$, the conjugacy classes $\mathrm{c}_v(\pi)$ and $\mathrm{c}_v(S(\pi))$ are each other's images under the action of the nontrivial element of $\mathrm{O}_n(\mathbb{C})/\mathrm{SO}_n(\mathbb{C})$. In particular, S preserves $\Pi_{\mathrm{disc}}(\mathrm{SO}_n)$ and $\psi(S(\pi), \mathrm{St}) = \psi(\pi, \mathrm{St})$. Moreover, this implies that $S(\pi)$ is isomorphic to π if and only if ± 1 (resp. 0) is an eigenvalue of $\mathrm{c}_p(\pi)$ for every prime p (resp. of $\mathrm{c}_\infty(\pi)$). Let us now assume the hypotheses of Theorem 8.5.8 and also assume that (8.5.1) holds. The previous observations show that the set Π of the theorem is stable under S. In particular, if S has no fixed point in Π, the theorem asserts that $\Pi \cap \Pi_{\mathrm{disc}}(\mathrm{SO}_n)$ consists of two representations interchanged by S, each of multiplicity 1. Here are two particular cases where this applies:

I. 0 *is not an eigenvalue of* ψ_∞: $S(\pi)_\infty \neq \pi_\infty$ for every $\pi \in \Pi$.
II. π_i *is symplectic for* $i = 1, \ldots k$: $S(\pi)_p \neq \pi_p$ for every $\pi \in \Pi$ and p prime. Indeed, the eigenvalues λ of $\mathrm{c}_p(\pi_i)$ satisfy $p^{-1/2} < |\lambda| < p^{1/2}$ by Jacquet–Shalika (and even $|\lambda| = 1$ by Ramanujan's conjecture, Sect. 6.4.12), while the eigenvalues μ of $[d_i]_p$ satisfy $|\mu| \geq p^{1/2}$ or $|\mu| \leq p^{-1/2}$ (because $d_i \equiv 0 \bmod 2$), so that $|\lambda\mu| \neq 1$ and ± 1 is not an eigenvalue of ψ_p.

In the general case, a combination of these ideas shows that for there to exist a representation of multiplicity greater than 1 in a $\Pi_{\mathrm{disc}}(\mathrm{SO}_n)$, it is necessary and sufficient that there exist a self-dual, algebraic, orthogonal $\pi \in \Pi_{\mathrm{cusp}}(\mathrm{PGL}_{4m})$ such that $|\mathrm{Weights}(\pi)| = 4m - 1$ and $\mathrm{c}_p(\pi)$ admits the eigenvalue ± 1 for every p. We can show that no such element exists when $m = 1$.

Two Criteria

Let us give two criteria for relation (8.5.1) to hold.

Criterion 8.5.10. Suppose that ψ is of the form $\left(\bigoplus_{i=1}^{k-2} \pi_i[d_i]\right) \oplus [d_{k-1}] \oplus [1]$ with $d_i \equiv 0 \bmod 2$ for every $i = 1, \ldots, k-2$ and $d_{k-1} > 1$ odd. Then relation (8.5.1) holds if and only if for every $i = 1, \ldots, k-2$, we have

$$(-1)^{n_i d_i/4} = \varepsilon(\pi_i)^{1+\min(d_i, d_{k-1})} .$$

Proof. This is an immediate application of the formulas and the fact that $\varepsilon(\pi_i \times \pi_j) = 1$ if $i, j \in I_0 = \{1, \ldots, k-2\}$. The latter can either be viewed as a particular case of Arthur's general result, because the representations π_i and π_j are symplectic (Sect. 8.3.1), or be proved directly because $\varepsilon(\mathrm{I}_a \otimes \mathrm{I}_b) = (-1)^{1+\max(a,b)} = 1$ if a and b are odd (Sect. 8.2.21). □

Criterion 8.5.11. Suppose that ψ is of the form $\left(\bigoplus_{i=1}^{k-2} \pi_i[d_i]\right) \oplus \pi_{k-1} \oplus [d_k]$ with

(i) $d_i \equiv 0 \bmod 2$ for every $i = 1, \ldots, k-2$ and
(ii) $\pi_{k-1} \in \Pi_{\mathrm{cusp}}(\mathrm{PGL}_3)$ such that $\mathrm{w}(\pi_{k-1}) > \max_{1 \le i \le k-2} \mathrm{w}(\pi_i)$.

Relation (8.5.1) holds if and only if for every $i = 1, \ldots, k-2$, we have

$$(-1)^{(n_i/2)((d_i/2)-1)} = \varepsilon(\pi_i)^{1+\min(d_i,d_k)} .$$

Proof. Recall that $\varepsilon(\mathrm{I}_w \otimes \mathrm{I}_{w'}) = (-1)^{1+\max(w,w')}$ and $\mathrm{I}_w \otimes \epsilon_{\mathbb{C}/\mathbb{R}} \simeq \mathrm{I}_w$. Since $\mathrm{w}(\pi_i)$ is odd if $i < k-1$ and even if $i = k-1$, the assumption on $\mathrm{w}(\pi_{k-1})$ ensures that for every $i \le k-2$,

$$\varepsilon(\pi_i \times \pi_{k-1}) = \varepsilon(\mathrm{L}(\pi_i)_\infty \otimes \mathrm{I}_{\mathrm{w}(\pi_{k-1})})\varepsilon(\mathrm{L}(\pi_i)_\infty \otimes \epsilon_{\mathbb{C}/\mathbb{R}}) = (-1)^{n_i/2}\varepsilon(\pi_i) .$$

This allows us to conclude because $\varepsilon(\pi_i \times \pi_j) = 1$ if $i, j \le k-2$. □

Examples

If we admit Conjecture 8.4.25, Theorem 8.5.8 becomes a powerful tool for verifying (and perhaps, in the near future, re-proving) the results obtained previously in this book. Its most immediate application concerns the case where ψ is such that either $I_0 = \emptyset$ or $k = 1$, because condition (8.5.1) then holds trivially. We recover, for example, the conclusion of Theorem 7.2.1 (ii), as well as that of part (i) if we simply require that the representation π' be in $\Pi_{\mathrm{disc}}(\mathrm{SO}_8)$ rather than $\Pi_{\mathrm{disc}}(\mathrm{O}_8)$. We also recover the assertion concerning $\Delta_{11}[12]$ in Corollary 7.3.4 (see Sect. 9.2.10).

For another example, consider the case where

$$\psi = \pi_1[d_1] \oplus [d_2] \oplus [1]$$

with $\pi_1 \in \Pi_{\mathrm{cusp}}(\mathrm{PGL}_2)$ of weight in $\{\pm(k-1)/2\}$, for k a strictly positive integer. We have $\varepsilon(\pi_1) = (-1)^{k/2}$. By Criterion 8.5.10, the relation (8.5.1) holds if and only if either

(I) $d_1 < d_2$ and $d_1 \equiv k \bmod 4$, or
(II) $d_1 > d_2$ and $d_1 \equiv 0 \bmod 4$.

For example, if $\psi = \Delta_{11}[4] \oplus [7] \oplus [1]$, in which case $n = 16$ and $\psi_\infty = \mathrm{St}(\mathrm{Inf}_1)$, we are in case (I) above. This immediately re-proves the assertion concerning $\psi(\pi, V_{\mathrm{St}})$ in Corollary 7.2.6, as well as part (i) of Theorem 5.2.2. We also recover the assertions of Corollary 7.3.4 for $k > 12$ (see Sect. 9.2.10): we are in case (II) for $k = 16$ because $(d_1, d_2) = (8, 7)$ and in case (I) for $k > 16$ because $24 - k \equiv k \bmod 4$ and

$2k - 25 > 24 - k$. As far as $\psi = \Delta_{17}[14] \oplus [3] \oplus [1]$ is concerned, we are neither in case (I) nor in case (II), which, as promised, corroborates Corollary 7.3.5.

Finally, let us explain Table 7.1 of Sect. 7.4. In view of the analysis above, it only remains to understand the bold cases of this table. But if $\psi = \Delta_w[2] \oplus \mathrm{Sym}^2\Delta_{11} \oplus [1]$ with $w < 22 = \mathrm{w}(\mathrm{Sym}^2\Delta_{11})$, then Criterion 8.5.11 shows that relation (8.5.1) holds, which suffices to conclude.

8.6 Compatibility with the Theta Correspondence

Let $n \equiv 0 \bmod 8$ and $g \geq 1$ be integers such that $n > 2g$. Let

$$\psi_S = \oplus_{i=1}^{k} \pi_i[d_i] \in \mathcal{X}_{\mathrm{AL}}(\mathrm{SL}_{2g+1}) \ ,$$

where $k \geq 1$ and $\pi_i \in \Pi_{\mathrm{cusp}}^{\perp}(\mathrm{PGL}_{n_i})$ for every $i = 1, \ldots, k$. Suppose that

$$\psi_O := \psi_S \oplus [n - 2g - 1] \in \mathcal{X}_{\mathrm{AL}}(\mathrm{SL}_n)$$

satisfies condition (H2) with respect to SO_n. This is equivalent to requiring that the eigenvalues of $(\psi_S)_\infty$ be $2g + 1$ integers

$$w_1 > \cdots > w_g > 0 > -w_g > \cdots > -w_1$$

with, moreover, $w_g \geq n/2 - g$. We choose the indexation of the π_i such that $n_k d_k \equiv 1 \bmod 2$ (Sect. 8.5.1). Finally, we assume $d_k = 1$, which is automatic if $n \neq 2g + 2$.

Let $\pi_O \in \Pi(\mathrm{SO}_n)$ (resp. $\pi_S \in \Pi(\mathrm{Sp}_{2g})$) be the unique representation such that $\psi(\pi_O, \mathrm{St}) = \psi_O$ (resp. $\psi(\pi_S, \mathrm{St}) = \psi_S$). Let $\mathrm{m}(\pi_O)$ and $\mathrm{m}(\pi_S)$ be the respective multiplicities of π_O and π_S in $\Pi_{\mathrm{disc}}(\mathrm{SO_n})$ and $\pi_{\mathrm{disc}}(\mathrm{Sp}_{2g})$). They are each 0 or 1 by Arthur's multiplicity formula, under Conjecture 8.4.25.

Proposition 8.6.1. *Assume Conjecture* 8.4.25. *We have* $\mathrm{m}(\pi_O) = \mathrm{m}(\pi_S)$ *if and only if for every* $i = 1, \ldots, k-1$ *such that* $d_i \equiv 0 \bmod 2$ *and* $d_i \geq n - 2g$, *we have* $\varepsilon(\pi_i) = 1$.

Proof. This immediately follows from the explicit formulas given in Sect. 8.5.1 and Sect. 8.5.7. Indeed, the natural injection

$$\mathrm{C}_{\psi_S} \to \mathrm{C}_{\psi_O}$$

induces an isomorphism $\mathrm{C}_{\psi_S} \overset{\sim}{\to} \mathrm{C}_{\psi_O}/\mathrm{Z}(\mathrm{SO}_n)$. The group C_{ψ_S} can be naturally identified with $\{\pm 1\}^{k-1}$ (see the proof of Theorem 8.5.2), and the injection above identifies it (with its distinguished elements s_i) with the subgroup $\{\pm 1\}^{I_0}$ defined in Sect. 8.5.7 (in particular, $\mathrm{I}_0 = \{1, \ldots, k-1\}$). Via this identification, we see that the character χ of $\{\pm 1\}^{k-1}$ defined in Sect. 8.5.1 coincides with the character χ of $\{\pm 1\}^{I_0}$ defined in Sect. 8.5.7. By Theorems 8.5.2 and 8.5.8, the condition for

$\mathrm{m}(\pi_O) = \mathrm{m}(\pi_S)$ is therefore equivalent to

$$\varepsilon(\pi_i)^{\min(d_i, n-2g-1)} = 1$$

for every $i = 1, \ldots, k-1$. This automatically holds when d_i is odd because π_i is orthogonal. $\square$

When $\mathrm{m}(\pi_S) = 1$, Böcherer's criterion (Remark 7.2.4) gives a necessary and sufficient condition, at least if the integers w_i are consecutive, for the eigenform of $\mathrm{S}_{w_1+1}(\mathrm{Sp}_{2g})$ (that is well defined up to a scalar) generating π_S to admit a ϑ-correspondent π' in $\Pi_{\mathrm{disc}}(\mathrm{O}_n)$: it is necessary and sufficient that $\mathrm{L}(s, \pi_S, \mathrm{St})$ be nonzero at $s = n/2 - g$. By restriction to SO_n (Sect. 4.4.4), the existence of π' implies that $\mathrm{m}(\pi_O)$ is nonzero (Corollary 7.1.3). It is therefore important to verify that Böcherer's criterion is compatible with Proposition 8.6.1. The following proposition shows that this is indeed the case.

Proposition 8.6.2. *The L-function* $\mathrm{L}(s, \pi_{\mathrm{S}}, \mathrm{St})$ *is nonzero at* $s = n/2 - g$ *if and only if for every* $i = 1, \ldots, k-1$ *such that* $d_i \geq n - 2g$ *and* $d_i \equiv 0 \bmod 2$, *we have* $\mathrm{L}(1/2, \pi_i) \neq 0$ *(and therefore* $\varepsilon(\pi_i) = 1$*). In particular,* $\mathrm{L}(n/2 - g, \pi_{\mathrm{S}}, \mathrm{St}) \neq 0$ *whenever* $n > 3g$.

Proof. The function $\mathrm{L}(s, \pi_{\mathrm{S}}, \mathrm{St})$ is the product of the $\mathrm{L}(s + j, \pi_i)$ for $i = 1, \ldots, k$ and $j \in (d_i - 1)/2 + \mathbb{Z}$ such that $|j| \leq (d_i - 1)/2$. Recall that if $\pi_i \neq 1$, the Euler product of $\mathrm{L}(s, \pi_i)$ is absolutely convergent for $\Re s > 1$ and that $\xi(s, \pi_i) = \Gamma(s, \mathrm{L}((\pi_i)_\infty))\mathrm{L}(s, \pi_i)$ admits a holomorphic extension to $\mathbb{C}$ such that $\xi(1-s, \pi_i) = \varepsilon(\pi_i)\xi(s, \pi_i)$ (Sect. 6.4.11). Moreover, we have $\mathrm{L}(1, \pi_i) \neq 0$ [111]. By assumption, if $\pi_i = 1$, then $i = k$ and $d_k = 1$, and $\mathrm{L}(s, \pi_i)$ is the Riemann ζ-function. Note that if this happens, then $n - 2g - 1 \equiv 3 \bmod 4$; in particular, if we set

$$s_0 = \frac{n}{2} - g\,,$$

then $s_0 \geq 2$ is neither a zero nor a pole of ζ. Since $d_k = 1$ and $s_0 \geq 1$, we have $\mathrm{L}(s_0, \pi_{\mathrm{S}}, \mathrm{St}) = 0$ if and only if there exist $1 \leq i \leq k-1$ and $j \in (d_i - 1)/2 + \mathbb{Z}$ with $|j| \leq (d_i - 1)/2$ such that $\mathrm{L}(s_0 + j, \pi_i) = 0$.

Fix $i < k$. Since the representation π_k is the only one of the π_s to have a weight equal to 0, the representation $\mathrm{L}((\pi_i)_\infty)$ is the direct sum of the I_w, where $w/2$ runs through the strictly positive weights of π_i. The function $\Gamma(s)$ is nonzero on the real axis and its only poles are the nonpositive integers. The description of $\Gamma(s, \mathrm{L}((\pi_i)_\infty))$ (Sect. 8.2.21) and the properties of $\xi(s, \pi_i)$ recalled above therefore show that if $\mathrm{L}(s, \pi_i) = 0$, say for $s \in \mathbb{R}$, then either $0 < s < 1$ or $s \leq -w$, where w is the lowest positive weight of π_i. But $w - (d_i - 1)/2 \geq w_g \geq n/2 - g = s_0$, so that for $j \geq (1 - d_i)/2$, we have

$$s_0 + j > -s_0 + \frac{1 - d_i}{2} \geq -w\,.$$

Finally, $0 < s_0 + j < 1$ is equivalent to $s_0 + j = 1/2$ if $j \in \frac{1}{2}\mathbb{Z}$. The first assertion follows because $1/2$ is of the form $n/2 - g + j$ with $(1 - d_i)/2 \leq j \leq (1 - d_i)/2$ and $j \in (d_i - 1)/2 + \mathbb{Z}$ if and only if $d_i \equiv 0 \bmod 2$ and $n - 2g \leq d_i$. The second assertion comes from the obvious relation $d_i \leq g$. □

Remark 8.6.3. Note that the sufficient condition $n > 3g$ obtained above, which is justified by Arthur's Theorem* 8.1.1, turns out to be better than Böcherer's general condition $n > 4g$.

In theory, there could exist parameters ψ_{S} such that $\mathrm{m}(\pi_O) = \mathrm{m}(\pi_S) = 1$, but such that π_O and π_S are not ϑ-correspondent. To produce such an example, one would need to find a symplectic, self-dual, regular algebraic representation ϖ such that $\mathrm{L}(1/2, \varpi) = 0$ but $\varepsilon(\varpi) = 1$. The authors do not know of such an example (compare with Remark 7.3.3). This is a fact that we use to our advantage several times in this book!

8.7 Compatibility with Böcherer's L-function

Let $g \geq 1$ be an integer, $k \in \mathbb{Z}$, $F \in \mathrm{S}_k(\mathrm{Sp}_{2g}(\mathbb{Z}))$ an eigenform, and $\pi_F \in \Pi_{\mathrm{cusp}}(\mathrm{Sp}_{2g})$ the representation generated by F. Böcherer proved, in [26], that the Euler product $\mathrm{L}(s, \pi_F, \mathrm{St})$ (defined in Sect. 6.4.11) is absolutely convergent if $\Re\, s > g + 1$ and that the function

$$\xi_{\mathrm{B}}(s, \pi_F, \mathrm{St}) := \Big(\Gamma(s, \epsilon_{\mathbb{C}/\mathbb{R}}^g) \prod_{i=1}^{g} \Gamma_{\mathbb{C}}(s + k - i)\Big) \mathrm{L}(s, \pi_F, \mathrm{St})$$

admits a meromorphic continuation to $\mathbb{C}$ as well as a functional equation

$$\xi_{\mathrm{B}}(s, \pi_F, \mathrm{St}) = \xi_{\mathrm{B}}(1 - s, \pi_F, \mathrm{St})$$

(see also [138, 6, 161]). Recall that we have $\Gamma(s, 1) = \Gamma_{\mathbb{R}}(s)$ and $\Gamma(s, \epsilon_{\mathbb{C}/\mathbb{R}}) = \Gamma_{\mathbb{R}}(s + 1)$ in the notation of Sect. 8.2.21. The poles of $\xi_{\mathrm{B}}(s, \pi_F, \mathrm{St})$ have been studied by Mizumoto [150, Corollary to Theorem 1]. For $k \geq g$ he proves that $\xi_{\mathrm{B}}(s, \pi_F, \mathrm{St})$ admits at most simple poles at $s = 0$ and $s = 1$, and it is holomorphic elsewhere.

Now, suppose that we have the relation

$$\psi(\pi_F, \mathrm{St}) = \oplus_{i=1}^{r} \pi_i[d_i] \ ,$$

where $\pi_i \in \Pi_{\mathrm{cusp}}^{\perp}(\mathrm{PGL}_{n_i})$ and $d_i \geq 1$ for every $1 \leq i \leq r$, which we may always do according to Arthur (Theorem 8.2.4). The theory of the standard L-functions of the elements of $\Pi_{\mathrm{cusp}}(\mathrm{PGL}_m)$, by Godement and Jacquet, shows that the function defined by $\xi(s, \pi_i) = \Gamma(s, \mathrm{L}((\pi_i)_\infty))\mathrm{L}(s, \pi_i)$ (Sect. 8.2.21) has a meromorphic continuation to $\mathbb{C}$ and a functional equation $s \mapsto 1-s$. This therefore provides a second natural way to complete $\mathrm{L}(s, \pi_F, \mathrm{St})$, by simply setting

$$\xi_{\mathrm{A}}(s, \pi_F, \mathrm{St}) \ := \ \prod_{i=1}^{r} \prod_{j=0}^{d_i-1} \xi\Big(s+j-\frac{d_i-1}{2}, \pi_i\Big) \,.$$

This function is also meromorphic on $\mathbb{C}$ and invariant under $s \mapsto 1-s$ (a priori up to a sign, but that sign is in fact equal to 1 because following Arthur [13, Theorem 1.5.3 (b)], we have $\varepsilon(\pi)=1$ for every self-dual orthogonal representation $\pi \in \Pi_{\mathrm{cusp}}(\mathrm{PGL}_m)$).

Thus, $\xi_{\mathrm{B}}(s, \pi_F, \mathrm{St})/\xi_{\mathrm{A}}(s, \pi_F, \mathrm{St})$ is an "explicit" quotient of products of Γ-factors. When $k > g+1$, it is easy to deduce from the respective descriptions of these factors that this quotient is 1, that is, $\xi_{\mathrm{B}}(s, \pi_F, \mathrm{St}) = \xi_{\mathrm{A}}(s, \pi_F, \mathrm{St})$. The situation turns out to be more interesting when $k \leq g+1$, in which case the comparison of these factors, combined with the properties of the poles of $\xi_{\mathrm{B}}(s, \pi_F, \mathrm{St})$ and $\xi_{\mathrm{A}}(s, \pi_F, \mathrm{St})$ recalled above, has nontrivial consequences for $\psi(\pi_F, \mathrm{St})$. Proposition 8.7.1 below is suggested by part (a) of Theorem 8.5.2 (and therefore by Conjecture 8.4.22) when $k = g+1$. To state it, we need to introduce several preliminary quantities.

For every integer $a \geq 1$, we set

$$\delta(\pi_F, a) \ = \ \mathrm{ord}_{s=a} \prod_{\{i\,;\,\pi_i \neq 1\}} \ \prod_{j=0}^{d_i-1} \xi\Big(s+j-\frac{d_i-1}{2}, \pi_i\Big) \,.$$

Recall that if $\pi \in \Pi_{\mathrm{cusp}}(\mathrm{PGL}_m)$ is such that $\pi \neq 1$, then $\xi(s,\pi)$ is an entire function of s. Moreover, if $s \in \frac{1}{2}\mathbb{Z}$ satisfies $\xi(s,\pi)=0$, then by Jacquet–Shalika, we have $s=1/2$ and $\mathrm{ord}_{s=1/2}\Gamma(s, \mathrm{L}(\pi_\infty)) = 0$. We therefore have the following equality for every integer $a \geq 1$:

$$\delta(\pi_F, a) \ := \sum_{\{i\,;\,d_i \equiv 0 \bmod 2,\, d_i \geq 2a\}} \mathrm{ord}_{s=1/2}\mathrm{L}(s, \pi_i) \,. \tag{8.7.1}$$

In particular, we have the inequalities $0 \leq \delta(\pi_F, b) \leq \delta(\pi_F, a)$ for $b \geq a \geq 1$. For use further on, we define, for every integer $n \geq 0$,

$$p_n(s) \ = \ \frac{\Gamma_{\mathbb{C}}(s+n)}{\Gamma_{\mathbb{C}}(s-n)} \quad \text{and} \quad \gamma_n(s) \ = \ \frac{\Gamma(s, \varepsilon_{\mathbb{C}/\mathbb{R}}^n)}{\Gamma_{\mathbb{R}}(s)} \prod_{i=1}^{n} \frac{\Gamma_{\mathbb{C}}(s+i)}{\Gamma_{\mathbb{R}}(s-i)\Gamma_{\mathbb{R}}(s+i)} \,.$$

Using the formulas $\Gamma_{\mathbb{C}}(s+1) = (s/2\pi)\,\Gamma_{\mathbb{C}}(s)$ and $\Gamma_{\mathbb{C}}(s) = \Gamma_{\mathbb{R}}(s)\Gamma_{\mathbb{R}}(s+1)$, we verify that the following equalities hold for every integer $n \geq 0$:

$$p_n(s) \ = \ (2\pi)^{-2n} \prod_{1-n \leq m \leq n} (s-m) \quad \text{and} \quad \gamma_n(s) = \prod_{0 \leq 2m \leq n} p_{n-2m}(s) \,.$$

In particular, p_n and γ_n are polynomials in s invariant under $s \mapsto 1-s$.

Proposition⋆ 8.7.1. *Let $F \in \mathrm{S}_k(\mathrm{Sp}_{2g}(\mathbb{Z}))$ be an eigenform with $k = g$ or $k = g+1$. Suppose $\psi(\pi_F, \mathrm{St}) = \oplus_{i=1}^r \pi_i[d_i]$ with $\pi_r = 1$ and $d_r > 1$. Then we have the inequality $\delta(\pi_F, (d_r+1)/2) > 0$.*

Proof. Let us first treat the case $k = g+1$. Since $\psi(\pi_F, \mathrm{St})_\infty$ admits 0 as simple eigenvalue (the eigenvalues are the $2g+1$ integers n such that $|n| \leq g$ by Sect. 6.3.6), we have $\pi_i \neq 1$ if $i < r$, and therefore the equality

$$\mathrm{ord}_{s=(d_r+1)/2}\, \xi_{\mathrm{A}}(s, \pi_F, \mathrm{St}) \;=\; \delta\Big(\pi, \frac{d_r+1}{2}\Big) - 1 \qquad (8.7.2)$$

by the above. We also see that if $i < r$, then π_i is regular. We therefore have, by definition, the equalities

$$\frac{\xi_{\mathrm{B}}(s, \pi_F, \mathrm{St})}{\xi_{\mathrm{A}}(s, \pi_F, \mathrm{St})} = \frac{\Gamma\big(s, \epsilon_{\mathbb{C}/\mathbb{R}}^g\big)}{\Gamma_{\mathbb{R}}(s)} \prod_{i=1}^{(d_r-1)/2} \frac{\Gamma_{\mathbb{C}}(s+i)}{\Gamma_{\mathbb{R}}(s+i)\Gamma_{\mathbb{R}}(s-i)}$$
$$= \frac{\Gamma\big(s, \epsilon_{\mathbb{C}/\mathbb{R}}^g\big)}{\Gamma\Big(s, \epsilon_{\mathbb{C}/\mathbb{R}}^{(d_r-1)/2}\Big)}\, \gamma_{(d_r-1)/2}(s)\,.$$

This term is nonzero and finite at $s = (d_r+1)/2$. Suppose $\delta(\pi, (d_r+1)/2) = 0$; we then have $\mathrm{ord}_{s=(d_r+1)/2}\xi_{\mathrm{B}}(s, \pi_F, \mathrm{St}) = -1$. But by Mizumoto, the only possible poles of $\xi_{\mathrm{B}}(s, \pi_F, \mathrm{St})$ are at $s = 0$ or $s = 1$, which implies $d_r = 1$.

The case $k = g$ is similar. In this case, $\psi(\pi_F, \mathrm{St})_\infty$ admits 0 as triple eigenvalue and the integers $\pm 1, \ldots, \pm(g-1)$ as simple eigenvalues. In particular, π_i may have a weight equal to 0 for $i < r$, but in that case we have $d_i = 1$. Equality (8.7.2) therefore still holds. Moreover, the same argument shows $\xi_{\mathrm{B}}(s, \pi_F, \mathrm{St})/\xi_{\mathrm{A}}(s, \pi_F, \mathrm{St}) = \mu(s)\, \gamma_{(d_r-1)/2}(s)$ with

$$\mu(s) = \frac{\Gamma_{\mathbb{C}}(s)\Gamma\big(s, \epsilon_{\mathbb{C}/\mathbb{R}}^g\big)}{\Gamma\Big(s, \epsilon_{\mathbb{C}/\mathbb{R}}^{(d_r-1)/2}\Big)\Gamma_{\mathbb{R}}(s+e_1)\Gamma_{\mathbb{R}}(s+e_2)}$$

for certain elements $e_1, e_2 \in \{0, 1\}$ that we will not need to specify (see Remark 8.7.2). The function $\mu(s)$ is finite and nonzero at $s = (d_r+1)/2$, and we conclude as in the case $k = g+1$ by using Mizumoto's result. □

Remark 8.7.2. Let $a, b \in \mathbb{Z}$. If the meromorphic function $\Gamma_{\mathbb{R}}(s)^a\Gamma_{\mathbb{R}}(s+1)^b$ is invariant under $s \mapsto 1-s$, then $a = b = 0$. Indeed, the vanishing order of this function at $s = 2, 1, 0$, and -1 is, respectively, $0, 0, -a$, and $-b$. The invariance under $s \mapsto 1-s$ of the functions $\xi_{\mathrm{A}}(s, \pi_F, \mathrm{St})$ and $\xi_{\mathrm{B}}(s, \pi_F, \mathrm{St})$, as well as $\gamma_n(s)$ for every $n \geq 0$, therefore allows us to complete the analysis made during the proof of Proposition 8.7.1. First of all, we deduce the congruence $(d_r-1)/2 \equiv g \bmod 2$ for $k = g+1$, already obtained another way in Sect. 8.5.1. In the case $k = g$, we also deduce the equality of the images of the sets $\{e_1, e_2\}$ and $\{(d_r+1)/2, g\}$ in $\mathbb{Z}/2\mathbb{Z}$.

Chapter 9
Proofs of the Main Theorems

9.1 Tsushima's Modular Forms of Genus 2

For integers $j \geq 0$ and k, we denote by $\mathrm{S}_{j,k}$ the space $\mathrm{S}_W(\mathrm{Sp}_4(\mathbb{Z}))$, where W is the representation $\mathrm{Sym}^j\, \mathbb{C}^2 \otimes \det^k$ of $\mathrm{GL}_2(\mathbb{C})$ (Sects. 4.5 and 6.3.4). It is zero if $j \equiv 1 \bmod 2$, because -1_2 then acts by $-\mathrm{id}$ on W, or if $k \leq 0$ (Freitag [89, Proposition 4.6]), which is why we will always assume $j \equiv 0 \bmod 2$ and $k > 0$.

9.1.1 *Tsushima's Dimension Formula*

An explicit formula for $\dim \mathrm{S}_{j,k}$ was determined by R. Tsushima for $k \geq 5$ [199], extending a result of Igusa concerning the scalar-valued forms (case $j = 0$, $k \in \mathbb{Z}$ arbitrary [105]). When $j + 2k - 3 \leq 21$, which will turn out to be the case that interests us in this book, Tsushima's formula shows $\mathrm{S}_{j,k} = 0$ for all except six values (j, k) given in the following table, for which $\dim \mathrm{S}_{j,k} = 1$. As we will see further on, $\dim \mathrm{S}_{j,k}$ is also zero when $k \leq 4$ and $j + 2k - 3 \leq 21$; see Remark 9.3.41. The line (w, v) in Table 9.1 will be explained in Sect. 9.1.3.

Table 9.1 The pairs (j, k) such that $\dim \mathrm{S}_{j,k} \neq 0$, for $j + 2k - 3 \leq 21$ and $k \geq 5$, according to Tsushima

(j,k)	$(0,10)$	$(6,8)$	$(0,12)$	$(4,10)$	$(8,8)$	$(12,6)$
(w,v)	$(17,1)$	$(19,7)$	$(21,1)$	$(21,5)$	$(21,9)$	$(21,13)$

G. Chenevier, J. Lannes, *Automorphic Forms and Even Unimodular Lattices*, Ergebnisse der Mathematik und ihrer Grenzgebiete. 3. Folge / A Series of Modern Surveys in Mathematics 69, https://doi.org/10.1007/978-3-319-95891-0_9

For each of the six pairs (j,k) above, let $F_{j,k}$ be a generator of $\mathrm{S}_{j,k}(\mathrm{Sp}_4(\mathbb{Z}))$. Given the important role played by these Siegel forms further on, let us explain how to show their existence directly, through a construction of theta series based on the lattice E_8.

Fix $j \geq 0$ even and $k \geq 4$, for now arbitrary integers. There exists a unique isomorphism of $\mathbb{C}$-algebras $\mathbb{C}[X,Y] \xrightarrow{\sim} \mathrm{Sym}\,\mathbb{C}^2$ that sends X and Y, respectively, onto the elements $(1,0)$ and $(0,1)$ of $\mathbb{C}^2$. By transport of structure, this isomorphism endows $\mathbb{C}[X,Y]$ with a representation of $\mathrm{GL}_2(\mathbb{C})$. The subspace $\mathbb{C}[X,Y]_j \subset \mathbb{C}[X,Y]$ of homogeneous polynomials of degree j is a subrepresentation isomorphic to $\mathrm{Sym}^j\,\mathbb{C}^2$.

Let $I \subset \mathrm{E}_8 \otimes \mathbb{C}$ be an isotropic subspace of dimension 2, and let u, v, and w be three elements of I. Consider the map $\mathrm{E}_8^2 \to \mathbb{C}[X,Y]_j$ defined by

$$\mathrm{P}_{j,k,u,v,w}(x,y) = \det{}^{k-4}\begin{bmatrix} x \cdot u & x \cdot v \\ y \cdot u & y \cdot v \end{bmatrix} \left((x \cdot w)X + (y \cdot w)Y\right)^j .$$

The functional equation of the Jacobi ϑ-function (in two variables) allows one to prove that the function

$$\vartheta_2(\mathrm{E}_8, \mathrm{P}_{j,k,u,v,w}) = \sum_{(x,y)\in \mathrm{E}_8 \times \mathrm{E}_8} \mathrm{P}_{j,k,u,v,w}(x,y)\; q^{\frac{1}{2}\left[\begin{smallmatrix} x\cdot x & x\cdot y \\ y\cdot x & y\cdot y\end{smallmatrix}\right]}$$

is a Siegel modular form for $\mathrm{Sp}_4(\mathbb{Z})$ with coefficients in the representation $\mathrm{Sym}^j \otimes \det^k$ [86, Sect. 2]. It is clearly cuspidal if $k > 4$, and if $k = 4$, the coefficient of X^j in its image by the Siegel operator Φ_1 is the theta series of the harmonic polynomial $x \mapsto (x \cdot w)^j$ on E_8 (Sect. 5.4.1), an element of $\mathrm{M}_{j+4}(\mathrm{SL}_2(\mathbb{Z}))$ that is cuspidal if $j > 0$.

Numerical Application. Using a computer, it is easy to determine the Fourier coefficients of $f_{j,k,u,v,w} = \vartheta_2(\mathrm{E}_8, \mathrm{P}_{j,k,u,v,w})$ in Gram matrices of small discriminant; we refer to the code [54], and its output therein, to justify the affirmations that follow. Let us describe the result of these computations in discriminant less than or equal to 12, obtained by taking $u = (2,i,i,i,i,0,0,0)$, $v = (0,0,0,i,-i,i,i,2)$, $w = u+v$, and simply listing all ordered pairs $(x,y) \in \mathrm{E}_8^2$ whose Gram matrix is one of the seven matrices in Table C.2. We see that for each of the six pairs (j,k) in question, all computed coefficients are nonzero, except for one when $(j,k) = (6,8)$. Table C.2 gives exactly the Fourier coefficients of $(1/\lambda_{j,k})f_{j,k,u,v,u+v}$, where $\lambda_{j,k} \in \mathbb{Z} - \{0\}$ is a constant that does not have any particular meaning and that we will not give explicitly. By way of verification, let us mention that for $(j,k) = (0,10)$, our computations are compatible with Table IV of [171]. Since we have $\mathrm{S}_{14}(\mathrm{SL}_2(\mathbb{Z})) = 0$, this re-proves $\mathrm{S}_{j,k} \neq 0$ in all cases.

Having fixed the pair (j,k), we can also verify that if we vary the parameters u, v, and w in the computation above (or even use a formal computation), the quadruple of computed coefficients (polynomials!) is modified by only a scalar, as it should be because $\dim \mathrm{S}_{j,k} = 1$. This can also be proved in another way, as follows.

Set $W_{j,k} = \mathrm{Sym}^j\mathbb{C}^2 \otimes \det^k$ and denote by $U_{j,k}$ the natural representation of $\mathrm{O}_8(\mathbb{C})$ on the space of polynomials $\mathrm{E}_8 \otimes \mathbb{C}^2 \to W_{j,k-4}$ that are $\mathrm{GL}_2(\mathbb{C})$-equivariant and *pluriharmonic* [116, 86]; the function $\mathrm{P}_{j,k,u,v,w}$ is a typical element of $U_{j,k}$. These references assert that if $k \geq 4$, the pair $(U_{j,k}, W_{j,k})$ is compatible in the sense of Sect. 7.1.1. Specifically, we have a linear map

$$\vartheta \colon \mathrm{M}_{U_{j,k}}(\mathrm{O}_8) \longrightarrow \mathrm{M}_{W_{j,k}}(\mathrm{Sp}_4(\mathbb{Z})) \tag{9.1.1}$$

that sends the element $[\mathrm{E}_8, \mathrm{P}_{j,k,u,v,w}]$, defined at the end of Sect. 4.4.7, to the theta series $\vartheta_2(\mathrm{E}_8, \mathrm{P}_{j,k,u,v,w})$, for every triple of elements u, v, w belonging to the same isotropic subspace of rank 2 of $\mathrm{E}_8 \otimes \mathbb{C}$.

One easily verifies that the representation $U_{j,k}$, which is irreducible when restricted to $\mathrm{SO}_8(\mathbb{C})$ by Kashiwara and Vergne [116], admits a highest weight of the form $(j+k-4)\varepsilon_1 + (k-4)\varepsilon_2$ in the notation of Sect. 6.4.3. But the tables[1] of [55] show that for the six pairs (j,k) that interest us, we have $\dim \mathrm{M}_{U_{j,k}}(\mathrm{O}_8) = 1$. It follows, as promised, that the space $\vartheta(\mathrm{M}_{U_{j,k}}(\mathrm{O}_8))$ is of dimension 1 for these pairs. Since we also have $\dim \mathrm{S}_{j,k} = 1$, we obtain the following proposition.

Proposition 9.1.2. *If* (j,k) *is one of the six pairs of Table* 9.1, *then the map* (9.1.1) *induces an isomorphism* $\mathrm{M}_{U_{j,k}}(\mathrm{O}_8) \xrightarrow{\sim} \mathrm{S}_{j,k}$ *between spaces of dimension* 1.

9.1.3 Standard Parameters of the First Six Forms of Genus 2

Let $F \in \mathrm{S}_{j,k}$ be an eigenform for the action of $\mathrm{H}(\mathrm{PGSp}_4)$. Denote by $\pi_F \in \Pi_{\mathrm{cusp}}(\mathrm{PGSp}_4)$ the representation generated by F (Corollary 6.3.7). Note that the Chevalley group PGSp_4 is a classical $\mathbb{Z}$-group, since it is isomorphic to the $\mathbb{Z}$-group $\mathrm{SO}_{3,2}$. Its Langlands dual is the $\mathbb{C}$-group Sp_4; it is endowed with its standard representation of dimension 4. By (the end of) Sect. 6.3.4, the semisimple conjugacy class $\mathrm{St}(\mathrm{c}_\infty(\pi_F)) \subset \mathrm{M}_4(\mathbb{C})$ has eigenvalues $\pm w/2$ and $\pm v/2$, where

$$(w,v) = (j+2k-3, j+1) \ ,$$

which explains the second line of Table 9.1. Note that the map $(j,k) \mapsto (w,v)$ is a bijection between the set of ordered pairs (j,k) with $j \geq 0$ even and $k \geq 3$, and the set of ordered pairs (w,v) with w, v odd such that $w > v > 0$.

For each of the six pairs (j,k) of Table 9.1, the action of $\mathrm{H}(\mathrm{PGSp}_4)$ on $\mathrm{S}_{j,k}$ is then trivially scalar, that is, $F_{j,k}$ is an eigenform; we will study the parameter

$$\psi_{j,k} = \psi(\pi_{F_{j,k}}, \mathrm{St}) \in \mathcal{X}(\mathrm{SL}_4) \ .$$

The case of the scalar-valued form $F_{0,10}$ has a famous history because it is the first Saito–Kurokawa form, associated with the modular form of weight 18 for $\mathrm{SL}_2(\mathbb{Z})$ ([132], Sect. 7.3). Because we view $\pi_{F_{0,10}}$ as a representation of PGSp_4, rather than

[1] See http://gaetan.chenevier.perso.math.cnrs.fr/table/dim_SO8_dom.txt.

Sp_4, we have the relation (see [79, 216])

$$\psi_{0,10} = \Delta_{17} \oplus [2] \,,$$

which is clearly compatible with the equality $(w, v) = (17, 1)$ (the notation Δ_w was introduced in Sect. 7.3). The case of the form $F_{0,12}$ is similar, and we have $\psi_{0,12} = \Delta_{21} \oplus [2]$ by Andrianov, Maass, and Zagier. As also guessed by Kurokawa and explained by Arthur [12], the situation is quite different for the four other representations.

Proposition* **9.1.4.** *Suppose $j > 0$ even and $k \geq 3$.*

(i) (Multiplicity 1) *If $F, G \in \mathrm{S}_{j,k}$ are two eigenforms for the action of* $\mathrm{H}(\mathrm{Sp}_4)$ *and if every element of* $\mathrm{H}(\mathrm{Sp}_4)$ *has the same eigenvalue on F and G, then F and G are proportional.*
(ii) *If $F \in \mathrm{S}_{j,k}$ is an eigenform for the action of* $\mathrm{H}(\mathrm{PGSp}_4)$*, then $\psi(\pi_F, \mathrm{St}) = \pi$ with $\pi \in \Pi_{\mathrm{cusp}}^{\perp}(\mathrm{PGL}_4)$.*
(iii) *The map $F \mapsto \psi(\pi_F, \mathrm{St})$ induces a bijection between the set of 1-dimensional eigenspaces of $\mathrm{S}_{j,k}(\mathrm{Sp}_4(\mathbb{Z}))$ under the action of $\mathrm{H}(\mathrm{PGSp}_4)$ and the set of $\pi \in \Pi_{\mathrm{cusp}}^{\perp}(\mathrm{PGL}_4)$ such that we have $\mathrm{Weights}(\pi) = \{\pm(j + 2k - 3)/2, \pm(j + 1)/2\}$.*

Proof. Part (i) is the particular case $g = 2$ of Corollary 8.5.4: in the notation loc. cit., we have $(m_1, m_2) = (j + k, k)$, so that $m_1 > m_2$.

Set $(w, v) = (j + 2k - 3, j + 1)$. Let $F \in \mathrm{S}_{j,k}(\mathrm{Sp}_4(\mathbb{Z}))$ be an eigenform for $\mathrm{H}(\mathrm{PGSp}_4)$. Apply Theorem 8.1.1 to the classical $\mathbb{Z}$-group $\mathrm{PGSp}_4 \simeq \mathrm{SO}_{3,2}$ and its representation $\pi_F \in \Pi_{\mathrm{disc}}(\mathrm{PGSp}_4)$. Since a representation $\pi \in \Pi_{\mathrm{cusp}}^{\perp}(\mathrm{PGL}_2)$ is symplectic (Proposition 9.1.5) and $\pm 1/2$ is not an eigenvalue of $\mathrm{St}(\mathrm{c}_\infty(\pi_F))$ because $j > 0$, there are only two possibilities for $\psi(\pi_F, \mathrm{St})$ (Corollary 8.2.15 (ii)): either

(a) $\psi(\pi_F, \mathrm{St}) = \pi_1 \in \Pi_{\mathrm{cusp}}^{\perp}(\mathrm{PGL}_4)$, or
(b) $\psi(\pi_F, \mathrm{St}) = \pi_1 \oplus \pi_2$ with $\pi_1, \pi_2 \in \Pi_{\mathrm{cusp}}(\mathrm{PGL}_2)$ such that $\mathrm{w}(\pi_1) = w$ and $\mathrm{w}(\pi_2) = v$.

To prove part (ii), we must therefore show that case (b) does not occur. Note that for the four pairs (j, k) of Table 9.1, we have $v \in \{5, 7, 9, 13\}$, so that this follows directly from Proposition 9.1.5 and the fact that we have $\dim \mathrm{S}_{v+1}(\mathrm{SL}_2(\mathbb{Z})) = 0$ for these values of v. For a general pair (j, k), this is, instead, a consequence of Arthur's multiplicity formula for $\mathrm{SO}_{3,2}$ (Theorem 8.3.10). Indeed, suppose that we have $\psi = \pi_1 \oplus \pi_2 \in \mathcal{X}_{\mathrm{AL}}(\mathrm{SL}_4)$, where π_1 and π_2 are as in case (b) above. Consider homomorphisms ν and ν_∞ associated with ψ as in Sect. 8.3.8. By definition, $\nu \colon \mathrm{SL}_2 \times (\mathrm{SL}_2 \times \mathrm{SL}_2) \to \mathrm{Sp}_4$ is trivial on the first factor SL_2 and $\mathrm{St} \circ \nu$ is the direct sum of the tautological representations of the two other factors SL_2, so that the inclusion $\{\pm 1\}^2 = \mathrm{C}_\nu \hookrightarrow \mathrm{C}_{\nu_\infty}$ is an equality. It follows that we have $\varepsilon_\psi = 1$ (because "$d_i = 1$ for every i") and that $\Pi(\nu_\infty)$ is the set consisting of the two discrete series in $\mathrm{SO}_{3,2}(\mathbb{R})$ with infinitesimal character ψ_∞, by Sect. 8.4.5. One is holomorphic, say π_{hol}, and the other is generic (here, the notion is canonical because $\mathrm{SO}_{3,2}$ is adjoint), so that the Shelstad character $\chi_{\pi_{\rm hol}}$ is the nontrivial character of C_{ν_∞} that is trivial on the center of $\widehat{\mathrm{SO}_{3,2}} = \mathrm{Sp}_4$, namely the diagonal subgroup $\{\pm 1\}$ in C_{ν_∞}. Another way to determine $\chi_{\pi_{\rm hol}}$ is to simply apply formula (8.4.7) in the particular case $r = 2$. It follows that the restriction of $\chi_{\pi_{\rm hol}}$ to $\mathrm{C}_\nu = \mathrm{C}_{\nu_\infty}$ is nontrivial, and Arthur's multiplicity formula asserts that the unique $\pi \in \Pi(\mathrm{PGSp}_4)$ such that we have $\pi \simeq \pi_{\rm hol}$ and $\psi(\pi, \mathrm{St}) = \psi$ is of multiplicity zero (Theorem 8.3.10). Since $\mathrm{m}(\pi_F) > 0$, we are in case (a) above, which proves part (ii) of the proposition. The same multiplicity formula of Arthur then asserts that we have $\mathrm{m}(\pi_F) = 1$ and, more generally, that for every $\pi \in \Pi_{\rm cusp}(\mathrm{PGL}_4)$ with weights $\{\pm w/2, \pm v/2\}$, there exists (a unique) $\pi \in \Pi_{\rm disc}(\mathrm{PGSp}_4)$ such that $\psi(\pi, \mathrm{St}) = \pi$ and $\pi_\infty \simeq \pi_{\rm hol}$, and that this π satisfies $\mathrm{m}(\pi) = 1$ (this is the case where nothing needs to be verified because $\mathrm{C}_\psi = \mathrm{Z}(\widehat{G})$). Assertion (iii) then follows from Corollary 6.3.7. □

We have used the following very classical result.

Proposition 9.1.5. *Let $k \geq 2$ be an even integer and $\mathcal{F}_k \subset \mathrm{S}_k(\mathrm{SL}_2(\mathbb{Z})$ the set of modular forms that are eigenforms for* $\mathrm{H}(\mathrm{PGL}_2)$ *and normalized (that is, the first Fourier coefficient equals* 1*). The map that sends $F \in \mathcal{F}_k$ to the representation $\pi_F \in \Pi_{\rm cusp}(\mathrm{PGL}_2)$ it generates induces a bijection between $\mathcal{F}_k$ and the set of $\pi \in \Pi_{\rm cusp}(\mathrm{PGL}_2)$ such that* $\mathrm{Weights}(\pi) = \{\pm(k-1)/2\}$.

Proof. Recall that the first Fourier coefficient of an eigenform F in $\mathrm{S}_k(\mathrm{SL}_2(\mathbb{Z}))$ is always nonzero and that if the eigenform is normalized, it is uniquely determined by its eigenvalues under $\mathrm{H}(\mathrm{PGL}_2)$ [177, Chap. VII, Theorem 7]; in particular, $\mathcal{F}_k$ is a basis of the vector space $\mathrm{S}_k(\mathrm{SL}_2(\mathbb{Z}))$. Let U_k be the discrete series of $\mathrm{PGL}_2(\mathbb{R})$ such that $\mathrm{Inf}_{U_k} \subset \mathrm{M}_2(\mathbb{C})$ has eigenvalues $\pm(k-1)/2$. A well-known special case of Proposition 6.3.7 is that we have an $\mathrm{H}(\mathrm{PGL}_2)$-equivariant isomorphism between $\mathrm{S}_k(\mathrm{SL}_2(\mathbb{Z}))$ and $\mathcal{A}_{U_k}(\mathrm{PGL}_2) = \mathrm{Hom}_{\mathrm{PGL}_2(\mathbb{R})}(U_k, \mathcal{A}_{\rm cusp}(\mathrm{PGL}_2))$ [92, Chap. I, Sect. 4]. This shows that the map in the proposition is well defined and injective.

A sophisticated justification of the surjectivity consists in invoking Proposition 8.2.13 (i). We can also use Bargmann's classification [18] of the unitary dual of $\mathrm{SL}_2(\mathbb{R})$. It shows that if U is an irreducible unitary representation of $\mathrm{PGL}_2(\mathbb{R})$ such that $\mathrm{Inf}_U \subset \mathrm{M}_2(\mathbb{C})$ has eigenvalues $\pm(k-1)/2$, then either $U \simeq U_k$, or $\dim U = 1$ and $k = 2$. Indeed, the representations of the principal series have an infinitesimal character whose eigenvalues are of the form $\pm is$ with $s \in \mathbb{R}$ (the "tempered" case) or of the form $\pm s$ with $s \in]-1/2, 1/2[$ (the "complementary series"); moreover, the infinitesimal character of the "limit of discrete series" is 0. To eliminate the case $\dim U = 1$, note that formula (4.5.1) implies that the only elements of $\mathcal{A}^2(\mathrm{PGL}_2)$ invariant under $\mathrm{PGL}_2(\mathbb{R})^+$ are the constant functions, which are not cuspidal, concluding the proof. □

Proposition 9.1.4 and Tsushima's table justify the following Definition-Proposition.

Proposition-Definition* 9.1.6. *Let (w, v) be one of $(19, 7)$, $(21, 5)$, $(21, 9)$, and $(21, 13)$; then there exists a unique representation in $\Pi_{\rm cusp}^{\perp}(\mathrm{PGL}_4)$ that is algebraic with weights $\{\pm w/2, \pm v/2\}$. We denote it by $\Delta_{w,v}$.*

Thus, if (j,k) is one of $(6,8)$, $(4,10)$, $(8,8)$, and $(12,6)$, we have the relation $\psi_{j,k} = \Delta_{w,v}$ with $(w,v) = (j+2k-3, j+1)$, and $\psi_{j,k}$ cannot be expressed using forms of genus 1.

9.1.7 A Few Eigenvalues of Hecke Operators

Let $(j,k) \in \{(6,8),(4,10),(8,8),(12,6)\}$ and $(w,v) = (j+2k-3, j+1)$, and let p be a prime and $n \geq 1$ an integer. Set

$$\tau_{j,k}(p^n) = p^{nw/2}\,\mathrm{trace}\,\mathrm{St}(\mathrm{c}_p(\pi_{F_{j,k}})^n) = p^{nw/2}\,\mathrm{trace}\,(\mathrm{c}_p(\Delta_{w,v})^n)\,. \quad (9.1.2)$$

The conjugacy class $\mathrm{St}(\mathrm{c}_p(\pi_{F_{j,k}})) \subset \mathrm{SL}_4(\mathbb{C})$ is equal to its inverse; the characteristic polynomial of $p^{w/2}\mathrm{c}_p(\pi_{F_{j,k}})$ is therefore

$$t^4 - \tau_{j,k}(p)\,t^3 + \frac{\tau_{j,k}(p)^2 - \tau_{j,k}(p^2)}{2}\,t^2 - \tau_{j,k}(p)p^{j+2k-3}\,t + p^{2j+4k-6}\,. \quad (9.1.3)$$

In particular, the complex number $\tau_{j,k}(p^n)$ is a polynomial with integer coefficients in $\tau_{j,k}(p)$ and $\frac{1}{2}(\tau_{j,k}(p)^2 - \tau_{j,k}(p^2))$.

The following proposition was known to Shimura [187]. As Gross already explained in [97, Sect. 6], it is also an immediate consequence of the relation (6.2.8) (see also Sect. 6.4.3).

Proposition 9.1.8. *Let j,k be as above and p a prime.*

(a) *The complex number $\tau_{j,k}(p)$ is the eigenvalue of the Hecke operator $p^{(j+2k-6)/2}\mathrm{K}_p$ acting on the line $\mathrm{S}_{j,k}$.*
(b) *The complex number $\frac{1}{2}(\tau_{j,k}(p)^2 - \tau_{j,k}(p^2))$ is the eigenvalue of the Hecke operator $p^{j+2k-5}(\mathrm{T}_p+1) + p^{j+2k-3}$ acting on the line $\mathrm{S}_{j,k}$.*

By Relation (4.5.4), the operator $p^{(j+2k-6)/2}\,\mathrm{K}_p$ coincides with the one denoted by $\mathrm{T}(p)$ by Van der Geer [89, Sect. 16], at least when the latter is defined by including the normalization between parentheses in Definition 16.5 loc. cit. The operator we denote by T_p is sometimes denoted by $\mathrm{T}_1(p^2)$ in the literature, up to a factor of a power of p depending on the authors.

The problem of determining the eigenvalues of the Hecke operators acting on the spaces $\mathrm{S}_{j,k}$ is significantly more difficult in practice than its analog in genus 1. One reason is the difficulty in determining the Fourier coefficients of the forms of genus 2, in particular those indexed by Gram matrices of large determinant. Moreover, the relation between the Fourier coefficients and eigenvalues, studied in this context by Andrianov [4] for the scalar-valued forms and extended to the vector-valued forms by Arakawa [7], is more subtle than in genus 1. In what follows, we recall this relation in the case of the Hecke operators K_p and T_p. From this, we will deduce both the following proposition and the computation of several values of the integers $\tau_{j,k}(p)$.

Proposition 9.1.9. *Let $(j,k) \in \{(6,8),(4,10),(8,8),(12,6)\}$, and let p be a prime and $n \geq 1$ an integer. We have $\tau_{j,k}(p^n) \in \mathbb{Z}$, as well as the congruence*

$$\tau_{j,k}(p^2) \;\equiv\; \tau_{j,k}(p)^2 \mod 2\,p^{k-2}\,.$$

Set $\Gamma = \mathrm{Sp}_4(\mathbb{Z})$, and consider the following elements of $\mathrm{GSp}_4(\mathbb{Z}[1/p])^+$:

$$\gamma := \begin{bmatrix} 1&0&0&0\\ 0&1&0&0\\ 0&0&p&0\\ 0&0&0&p \end{bmatrix} \quad \text{and} \quad \gamma' := \begin{bmatrix} 1&0&0&0\\ 0&p&0&0\\ 0&0&p^2&0\\ 0&0&0&p \end{bmatrix}.$$

Lemma 9.1.10. (a) *The Hecke operator $\mathrm{K}_p \in \mathrm{H}_p(\mathrm{PGSp}_4)$ is of degree $(1+p)(1+p^2)$. Its matrix is the characteristic function of the image of $\Gamma\gamma^{-1}\Gamma$ in $\mathrm{PGSp}_4(\mathbb{Z}[1/p])^+$, in the sense of the identifications* (4.2.2) *and* (4.5.3). *The double coset $\Gamma\gamma\Gamma$ is the disjoint union of the right-cosets $\Gamma\gamma_i$, where γ_i runs through the list of elements of the following form:*

$$\begin{bmatrix} p&0&0&0\\ 0&p&0&0\\ 0&0&1&0\\ 0&0&0&1 \end{bmatrix}, \quad \begin{bmatrix} 1&0&a&b\\ 0&1&b&c\\ 0&0&p&0\\ 0&0&0&p \end{bmatrix}, \quad \begin{bmatrix} p&0&0&0\\ -d&1&0&e\\ 0&0&1&d\\ 0&0&0&p \end{bmatrix}, \quad \text{or} \quad \begin{bmatrix} 1&0&f&0\\ 0&p&0&0\\ 0&0&p&0\\ 0&0&0&1 \end{bmatrix},$$

with a,b,c,d,e,f integers lying between 0 and $p-1$;

(b) *Likewise, $\mathrm{T}_p \in \mathrm{H}_p(\mathrm{PGSp}_4)$ is of degree $p\,(p^4-1)(p-1)^{-1}$, and its matrix is the characteristic function of the image of $\Gamma{\gamma'}^{-1}\Gamma$ in the group $\mathrm{PGSp}_4(\mathbb{Z}[1/p])^+$. The double coset $\Gamma\gamma'\Gamma$ is the disjoint union of the right-cosets $\Gamma\gamma'_i$, where γ'_i runs through the list of elements of the following form:*

$$\begin{bmatrix} p&0&0&0\\ 0&p^2&0&0\\ 0&0&p&0\\ 0&0&0&1 \end{bmatrix}, \quad \begin{bmatrix} p^2&0&0&0\\ -ap&p&0&0\\ 0&0&1&a\\ 0&0&0&p \end{bmatrix}, \quad \begin{bmatrix} p&0&b&c\\ 0&p&c&d\\ 0&0&p&0\\ 0&0&0&p \end{bmatrix},$$

$$\begin{bmatrix} p&0&0&pe\\ -f&1&e&ef+g\\ 0&0&p&pf\\ 0&0&0&p^2 \end{bmatrix}, \quad \text{or} \quad \begin{bmatrix} 1&0&h&i\\ 0&p&pi&0\\ 0&0&p^2&0\\ 0&0&0&p \end{bmatrix},$$

with a,b,c,d,e,f,i integers lying between 0 and $p-1$, such that $c^2 \equiv bd \bmod p$ and $(b,c,d) \neq (0,0,0)$, and with g and h integers lying between 0 and p^2-1.

Proof. The fact that the matrices of K_p and T_p are the characteristic functions of the images of the double cosets $\Gamma\gamma^{-1}\Gamma$ and $\Gamma{\gamma'}^{-1}\Gamma$, respectively, is formula (6.2.7).

The degree of $\mathrm{K}_p \in \mathrm{H}_p(\mathrm{PGSp}_{2g})$ is the number of Lagrangians of the hyperbolic a-vector space $\mathrm{H}(\mathbb{F}_p^g)$, namely $\prod_{i=1}^{g}(1+p^i)$. Likewise, the degree of T_p is the number of (isotropic) lines in $\mathrm{H}(\mathbb{F}_p^g)$, multiplied by the number of (isotropic) lines in $\mathrm{H}(\mathbb{F}_p)$ transverse to a given line, which gives $p(p^g-1)(p-1)^{-1}$.

The assertions concerning the decompositions of the double cosets are due to Andrianov [4, 5]. Let us justify them briefly, following the notes of Buzzard [42]. An element of GL_4, which we assume given by blocks of size 2×2 and of the form

$$\begin{bmatrix} a & b \\ 0 & d \end{bmatrix},$$

is in GSp_4 for the similitude factor ν if and only if $a\,{}^{\mathrm{t}}b = b\,{}^{\mathrm{t}}a$ and $a\,{}^{\mathrm{t}}d = \nu 1_g$ (Sect. 4.5.1). This shows that each of the elements of the statement is in $\mathrm{GSp}_4(\mathbb{Z}[1/p])$, with similitude factor p in case (a) and p^2 in case (b).

Let $h \in \mathrm{GSp}_4(\mathbb{Z}[1/p]) \cap \mathrm{M}_4(\mathbb{Z})$, and let $\overline{h} \in \mathrm{M}_4(\mathbb{Z}/p)$ be the reduction modulo p of h. The theory of "symplectic elementary divisors" shows that h is in $\Gamma\gamma'\Gamma$ (resp. $\Gamma\gamma\Gamma$) if and only if $\nu(h) = p^2$ and the rank of $\overline{h}$ is 1 (resp. $\nu(h) = p$). This shows $\gamma_i \in \Gamma\gamma\Gamma$ and $\gamma'_i \in \Gamma\gamma'\Gamma$ for every i.

Finally, we verify that $\gamma_i\gamma_j^{-1} \in \Gamma$ (resp. $\gamma'_i{\gamma'}_j^{-1} \in \Gamma$) implies $i = j$. To do this, it is useful to note that all elements above are in a same Borel subgroup of GSp_4 because the "projection onto the diagonal" is a homomorphism. This suffices to conclude because in both cases, the cardinality of the list is the degree of the Hecke operator. □

Let $j \geq 0$ be an integer. Denote by ρ_j the natural representation of $\mathrm{GL}_2(\mathbb{C})$ on the space $W_j := \mathrm{Sym}^j\mathbb{C}^2$. Recall that for $w \in W_j$ and $n \in \mathrm{M}_2(\mathbb{C})$, the notation $w\,\mathrm{q}^n$ is used for the function $\mathbb{H}_2 \to W_j$ defined by $\tau \mapsto e^{2i\pi\,\mathrm{tr}(n\tau)}\,w$ (Sect. 4.5.2).

Let $k \in \mathbb{Z}$, and let F be a Siegel modular form for $\mathrm{Sp}_4(\mathbb{Z})$ with coefficients in the representation $W_{j,k} := W_j \otimes \det^k$ of $\mathrm{GL}_2(\mathbb{C})$. By definition, this representation has underlying space W_j, and $\mathrm{GL}_2(\mathbb{C})$ acts on it by $g \mapsto \rho_j(g)\det(g)^k$. Recall that the form F admits a Fourier expansion, which we write here as

$$F = \sum_{n \in \mathcal{N}} \mathrm{a}(n;F)\,\mathrm{q}^n\,,$$

where $\mathcal{N} \subset \frac{1}{2}\mathrm{M}_2(\mathbb{Z})$ is the subset consisting of the matrices that are symmetric, positive, and with diagonal coefficients in $\mathbb{Z}$, and where we have $\mathrm{a}(n;F) \in W_j$ for every $n \in \mathcal{N}$ (Sect. 4.5.2). It will be convenient to set $\mathrm{a}(n;F) = 0$ if $n \in \mathrm{M}_2(\mathbb{Q}) - \mathcal{N}$.

Recall that in Sect. 4.5.1, we defined a right action of $\mathrm{GSp}_4(\mathbb{R})^+$ on the space of functions $\mathbb{H}_2 \to W_{j,k}$, which we denote by $(f,\gamma) \mapsto f_{|W_{j,k}}\gamma$. Let $w \in W_j$ and $n \in \mathrm{M}_2(\mathbb{C})$, and let $\gamma = \left[\begin{smallmatrix} a & b \\ 0 & d \end{smallmatrix}\right]$ in $\mathrm{GSp}_4(\mathbb{R})^+$ have similitude factor ν; we have

$$w\,\mathrm{q}^n{}_{|W_{j,k}}\gamma \;=\; \nu^{-(j+2k)/2} \cdot \det(a)^k \cdot e^{(2i\pi/\nu)\mathrm{tr}({}^{\mathrm{t}}anb)} \cdot \rho_j({}^{\mathrm{t}}a)w\;\mathrm{q}^{{}^{\mathrm{t}}ana/\nu} \tag{9.1.4}$$

(recall the relation $d^{-1} = \nu^{-1}\,{}^{\mathrm{t}}a$). It will be convenient to let the group $\mathrm{GL}_2(\mathbb{C})$ act on $\mathrm{M}_2(\mathbb{C})$ by $(g,s) \mapsto g \cdot s := g\,s\,{}^{\mathrm{t}}g$.

Corollary 9.1.11. *Let $j \geq 0$ and k be integers. Let $F \in S_{j,k}$, and let p be a prime and $n \in \mathcal{N}$. We have*

$$\begin{aligned} p^{(j+2k)/2}\mathrm{a}(n;\mathrm{K}_p F) &= p^{j+2k}\ \mathrm{a}\left(\frac{1}{p}n;F\right) + p^3\ \mathrm{a}(pn;F) \\ &+ p^{k+1}\sum_{d=0}^{p-1}\rho_j\left(\begin{bmatrix} p & -d \\ 0 & 1\end{bmatrix}\right)\mathrm{a}\left(\begin{bmatrix} 1 & d \\ 0 & p\end{bmatrix}\cdot\frac{1}{p}n;F\right) \\ &+ p^{k+1}\ \rho_j\left(\begin{bmatrix} 1 & 0 \\ 0 & p\end{bmatrix}\right)\mathrm{a}\left(\begin{bmatrix} p & 0 \\ 0 & 1\end{bmatrix}\cdot\frac{1}{p}n;F\right)\end{aligned}$$

and

$$\begin{aligned} p^{j+2k}\mathrm{a}(n;\mathrm{T}_p F) &= p^{j+3k}\ \rho_j\left(\begin{bmatrix} 1 & 0 \\ 0 & p\end{bmatrix}\right)\mathrm{a}\left(\begin{bmatrix} p & 0 \\ 0 & 1\end{bmatrix}\cdot\frac{1}{p^2}n;F\right) \\ &+ p^{j+3k}\sum_{a=0}^{p-1}\rho_j\left(\begin{bmatrix} p & -a \\ 0 & 1\end{bmatrix}\right)\mathrm{a}\left(\begin{bmatrix} 1 & a \\ 0 & p\end{bmatrix}\cdot\frac{1}{p^2}n;F\right) \\ &+ \delta(n,p)\ p^{j+2k}\ \mathrm{a}(n;F) \\ &+ p^{k+3}\sum_{f=0}^{p-1}\rho_j\left(\begin{bmatrix} p & -f \\ 0 & 1\end{bmatrix}\right)\mathrm{a}\left(\begin{bmatrix} 1 & f \\ 0 & p\end{bmatrix}\cdot n;F\right) \\ &+ p^{k+3}\ \rho_j\left(\begin{bmatrix} 1 & 0 \\ 0 & p\end{bmatrix}\right)\mathrm{a}\left(\begin{bmatrix} p & 0 \\ 0 & 1\end{bmatrix}\cdot n;F\right),\end{aligned}$$

where $\delta(n,p) \in \mathbb{Z}$ is defined by formula (9.1.5) *below; $\delta(n,p) \equiv -1 \bmod p$.*

Proof. By Lemma 9.1.10 and the diagram (4.5.4), we have

$$\mathrm{K}_p F = \sum_i F_{|W_{j,k}}\gamma_i \quad\text{and}\quad \mathrm{T}_p F = \sum_i F_{|W_{j,k}}\gamma_i'\,.$$

In view of the uniform convergence of the Fourier expansion of F on every compact subset of $\mathbb{H}_2$, the corollary is a direct application of formula (9.1.4). Note that if $a, d \in \mathrm{GL}_2(\mathbb{C})$, $b \in \mathrm{M}_2(\mathbb{C})$, $\nu \in \mathbb{C}^*$, and $m, n \in \mathrm{M}_2(\mathbb{C})$ are such that $n = \nu^{-1}\,{}^{\mathrm{t}}ama$ and $d^{-1} = \nu^{-1}\,{}^{\mathrm{t}}a$, then $m = d \cdot \nu^{-1}\,n$ and $\mathrm{tr}({}^{\mathrm{t}}amb) = \mathrm{tr}(bn\,{}^{\mathrm{t}}d)$.

By way of example, let us determine the contribution of the $p^2 - 1$ elements γ_i' of the form

$$\begin{bmatrix} p & 0 & b & c \\ 0 & p & c & d \\ 0 & 0 & p & 0 \\ 0 & 0 & 0 & p\end{bmatrix}$$

with b, c, d as in Lemma 9.1.10 (ii) to the sum defining $p^{j+2k}\ \mathrm{a}(n; \mathrm{T}_p F)$. The latter can be written $p^{j+2k}\ \delta(n,p)\ \mathrm{a}(n;F)$, where

$$\delta(n,p) := \sum_{v \in V-\{0\},\ \det(v)=0} e^{(2i\pi/p)\mathrm{tr}(vn)} \tag{9.1.5}$$

and $V \subset \mathrm{M}_2(\mathbb{Z}/p\mathbb{Z})$ is the subspace of symmetric matrices. The quadratic form $\det \colon V \to \mathbb{Z}/p\mathbb{Z}$ admits $p+1$ isotropic lines. If a is the number of these lines that are in the kernel of the linear form $v \mapsto \mathrm{tr}(vn)$, so that $a \in \{0, 1, 2, p+1\}$, then $\delta(n,p) = (p-1) \cdot a - (p+1-a) = p(a-1) - 1$. □

Since the monoid $\rho_j(\mathrm{M}_2(\mathbb{Z}) \cap \mathrm{GL}_2(\mathbb{C}))$ preserves the lattice $\mathrm{Sym}^j \mathbb{Z}^2 \subset W_j$, this leads to the following corollary.

Corollary 9.1.12. *Let j, k be integers with $j \geq 0$ and $k \geq 2$. For every prime p, the Hecke operators $p^{(j+2k-6)/2}\ \mathrm{K}_p$ and $p^{j+k-3}\ (\mathrm{T}_p + 1)$ preserve the subgroup of $\mathrm{S}_{j,k}$ consisting of the elements whose Fourier coefficients all have their values in the subgroup $\mathrm{Sym}^j \mathbb{Z}^2 \subset W_j$.*

Proof of Proposition 9.1.9. Let $\mathrm{S}_{j,k}^{\mathrm{int}} \subset \mathrm{S}_{j,k}$ be the subgroup defined in the statement. Since the $\mathbb{C}$-vector space $\mathrm{S}_{j,k}$ is finite dimensional, there exists a finite subset $N \subset \mathcal{N}$ such that the linear map

$$F \mapsto (\mathrm{a}(n;F))_{n \in N}\ , \quad \mathrm{S}_{j,k} \to W_j^N$$

is injective. It sends the $\mathbb{Z}$-module $\mathrm{S}_{j,k}^{\mathrm{int}}$ into $(\mathrm{Sym}^j \mathbb{Z}^2)^N$. This shows, on the one hand, that the $\mathbb{Z}$-module $\mathrm{S}_{j,k}^{\mathrm{int}}$ is free of finite rank and, on the other hand, that the natural map $\eta \colon \mathrm{S}_{j,k}^{\mathrm{int}} \otimes_{\mathbb{Z}} \mathbb{C} \to \mathrm{S}_{j,k}$ is injective, because this is the case for the natural map $\mathrm{Sym}^j \mathbb{Z}^2 \otimes \mathbb{C} \to W_j$.

The map η may be bijective in full generality, but we have not found a reference for this. Let us prove this when (j,k) is in the list given in Proposition 9.1.9. In this case, $\mathrm{S}_{j,k}$ is of dimension 1, so that it suffices to verify that $\mathrm{S}_{j,k}^{\mathrm{int}}$ is nonzero. Consider the modular form $f_{j,k,u,v,w} = \vartheta_2(\mathrm{E}_8, \mathrm{P}_{j,k,u,v,w})$ in $\mathrm{S}_{j,k}$, constructed in Sect. 9.1.1. Its coefficients are in $\mathbb{Z}[i][X,Y]$, where $\mathbb{Z}[i]$ is the ring of Gaussian integers, because the harmonic polynomial $\mathrm{P}_{j,k,u,v,w}$ sends E_8^2 into $\mathbb{Z}[i][X,Y]$; the few nonzero Fourier coefficients of $f_{j,k,u,v,w}$ that we have determined are even in $\mathbb{Z}$ (Table C.2). Note that for every $n \in \mathcal{N}$, we have $\overline{\mathrm{a}(n; f_{j,k,u,v,w})} = \mathrm{a}(n; f_{j,k,\overline{u},\overline{v},\overline{w}})$, where $z \mapsto \overline{z}$ denotes complex conjugation on $\mathbb{C}[X,Y]$ and $\mathrm{E}_8 \otimes \mathbb{C}$, respectively. Hence, $f_{j,k,u,v,w} + f_{j,k,\overline{u},\overline{v},\overline{w}}$ is a nonzero element of $\mathrm{S}_{j,k}^{\mathrm{int}}$.

To conclude this proof of Proposition 9.1.9, it suffices to apply Proposition 9.1.8 and Corollary 9.1.12. □

Let $n = \left[\begin{smallmatrix} n_{11} & n_{12}/2 \\ n_{12}/2 & n_{22} \end{smallmatrix}\right] \in \mathcal{N}$, and let p be a prime. We have the identities

$$\begin{bmatrix} p & 0 \\ 0 & 1 \end{bmatrix} \cdot \frac{1}{p} n = \begin{bmatrix} pn_{11} & n_{12}/2 \\ n_{12}/2 & n_{22}/p \end{bmatrix},$$

$$\begin{bmatrix} 1 & d \\ 0 & p \end{bmatrix} \cdot \frac{1}{p} n = \begin{bmatrix} (n_{11} + dn_{12} + d^2 n_{22})/p & (n_{12}/2) + dn_{22} \\ (n_{12}/2) + dn_{22} & pn_{22} \end{bmatrix}.$$

Hence, if the quadratic form on $\mathbb{Z}^2$ defined by n is anisotropic modulo the prime p, then neither n/p nor one of the two matrices above is in $\mathcal{N}$. Proposition 9.1.11 therefore has the following corollary.

Scholium 9.1.13. *Suppose that $F \in \mathrm{S}_{j,k}$ is an eigenvector of the operator $p^{(j+2k-6)/2}\,\mathrm{K}_p$, with eigenvalue λ. If $n \in \mathcal{N}$ and $2n$ is a Gram matrix of a quadratic form on $\mathbb{Z}^2$ that is anisotropic modulo the prime p, then we have the relation $\lambda\,\mathrm{a}(n;F) = \mathrm{a}(pn;F)$. In particular, this relation determines λ if $\mathrm{a}(n;F) \neq 0$.*

This scholium applies, for example, for

$$2\,n = \begin{bmatrix} 2 & -1 \\ -1 & 2 \end{bmatrix},$$

which is none other than the standard Gram matrix of the lattice A_2, when $\mathrm{A}_2 \otimes \mathbb{F}_p$ does not represent 0, that is, $p \equiv -1 \bmod 3$. We therefore deduce the following corollary from Table C.2.

Corollary 9.1.14. *The integers $\tau_{6,8}(2)$, $\tau_{4,10}(2)$, $\tau_{8,8}(2)$, and $\tau_{12,6}(2)$ equal, respectively, 0, -1680, 1344, and -240.*

Remark 9.1.15. Suppose that $F \in \mathrm{S}_{j,k}$ is an eigenvector of the operator $2^{j+2k-5}(\mathrm{T}_2 + 1) + 2^{j+2k-3}$, with eigenvalue λ, and set

$$n = \frac{1}{2}\begin{bmatrix} 2 & -1 \\ -1 & 2 \end{bmatrix} \quad \text{and} \quad m = \frac{1}{2}\begin{bmatrix} 2 & 0 \\ 0 & 6 \end{bmatrix}.$$

We leave it to the reader to deduce the following relation from Corollary 9.1.11:

$$(\lambda - 2^{j+2\,k-4})\ \mathrm{a}(n;F) = \\ 2^{k-2}\left(\rho_j\left(\begin{bmatrix} 2 & 0 \\ -1 & 1 \end{bmatrix}\right) + \rho_j\left(\begin{bmatrix} -1 & 1 \\ 2 & 0 \end{bmatrix}\right) + \rho_j\left(\begin{bmatrix} 1 & -1 \\ 1 & 1 \end{bmatrix}\right)\right)\ \mathrm{a}(m;F)\,.$$

(Verify, in particular, that we have $\delta(n,2) = -3$.) Using the values of Table C.2, this formula allows one to prove that the integers $\tau_{6,8}(4)$, $\tau_{4,10}(4)$, $\tau_{8,8}(4)$, and $\tau_{12,6}(4)$ are, respectively, 409600, -700160, 348160, and 4276480.

It turns out that the eigenvalues of the $F_{j,k}$ were studied by Faber and Van der Geer [83], [89, Sect. 24] in a completely different way, by counting curves of genus 2 over finite fields in the manner of Deligne. If q denotes a power of a prime, they were able to determine $\tau_{j,k}(q)$ for every $q \leq 37$ (by, however, admitting an expected property of the cohomology of certain sheaves on the Siegel space of genus 2 [89, Sect. 24]); loc. cit., they give several values, including the value $\tau_{j,k}(2)$ above. Further on, we will present a very different method to determine $\tau_{j,k}(q)$, which will lead to a proof of the following theorem.

Theorem* 9.1.16. *Let p be a prime and (j, k) one of the pairs $(6, 8)$, $(4, 10)$, $(8, 8)$, and $(12, 6)$.*

(i) *If $p \leq 113$, the integer $\tau_{j,k}(p)$ is given by Table C.3.*
(ii) *If $p \leq 29$, the integer $\tau_{j,k}(p^2)$ is given by Table C.4.*

Finally, let us mention that in a recent work [58, Sect. 8], Clery and Van der Geer have recovered the values $\tau_{6,8}(q)$ for $q \leq 49$ using yet another method.

9.1.17 Where We Explain the Occurrence of the $\psi_{j,k}$ in Table 7.1

Fix one of the four pairs (j, k) of Table 9.1 with $j > 0$. Let $U'_{j,k}$ be the irreducible representation of $\mathrm{SO}_8(\mathbb{C})$ of highest weight $(\frac{1}{2}j + k - 4)(\varepsilon_1 + \varepsilon_2) + \frac{1}{2}j(\varepsilon_3 + \varepsilon_4)$ (Sect. 6.4.3); it factors through $\mathrm{PGSO}_8(\mathbb{C})$. We then have natural isomorphisms

$$\mathrm{M}_{U'_{j,k}}(\mathrm{SO}_8) \xleftarrow{\sim} \mathrm{M}_{U'_{j,k}}(\mathrm{PGSO}_8) \xrightarrow{\sim} \mathrm{M}_{U_{j,k}}(\mathrm{PGSO}_8)$$
$$\xrightarrow{\sim} \mathrm{M}_{U_{j,k}}(\mathrm{SO}_8) \xleftarrow{\sim} \mathrm{M}_{U_{j,k}}(\mathrm{O}_8) \xrightarrow{\sim} \mathrm{S}_{j,k}\ .$$

Indeed, the first and third are general (a variant of Lemma 5.4.8 based on Proposition 4.1.4). The last isomorphism is that of Proposition 9.1.2. The before-last morphism is injective for general reasons (Sect. 4.4.4) and bijective because $\dim \mathrm{M}_{U_{j,k}}(\mathrm{SO}_8) = 1$ by Chenevier and Renard [55, Table 2]. The one in the middle is induced by the triality. Indeed, by a computation left to the reader, based on the well-known action of the triality on the Dynkin diagram of type $\mathbf{D}_4$, we see that if an irreducible $\mathbb{C}$-representation of $\mathrm{PGSO}_8(\mathbb{C})$ has highest weight $\sum_{i=1}^4 n_i\varepsilon_i$, then the two representations deduced from it by applying the triality automorphisms have highest weight $\sum_{i=1}^4 m_i\varepsilon_i$, where

$$m_1 = \frac{n_1 + n_2 + n_3 + n_4}{2}, \qquad m_2 = \frac{n_1 + n_2 - n_3 - n_4}{2},$$
$$m_3 = \frac{n_1 - n_2 + n_3 - n_4}{2}, \qquad \pm m_4 = \left|\frac{n_1 - n_2 - n_3 + n_4}{2}\right| .$$

The occurrence of the $\psi_{j,k}$ in Table 7.1 is therefore a consequence the series of isomorphisms above and Theorem 7.2.3 (i).

Finally, let us mention that this subsection suggests an alternative method for determining the $\tau_{j,k}(q)$, based on a computation of eigenvalues of Hecke operators for O_8. We refer to work by Mégarbané [143] on this subject.

9.2 $\Pi_{\rm disc}(\mathrm{SO}_{24})$ and the Nebe–Venkov Conjecture

9.2.1 A Characterization of Table 1.2

Consider the following subset of $\coprod_{n\geq 1}\Pi_{\rm alg}(\mathrm{PGL}_n)$:

$$\Pi = \{1, \Delta_{11}, \Delta_{15}, \Delta_{17}, \Delta_{19}, \Delta_{21}, \mathrm{Sym}^2\Delta_{11}, \Delta_{19,7}, \Delta_{21,5}, \Delta_{21,9}, \Delta_{21,13}\} .$$

The proposition below gives a direct "definition" of Table 1.2.

Proposition* **9.2.2.** *The set of* $\psi \in \mathcal{X}(\mathrm{SL}_{24})$ *such that*

(i) *the eigenvalues of* ψ_∞ *are the* 22 *integers* $\pm 11, \pm 10, \ldots, \pm 1$, *as well as the integer* 0 *with multiplicity* 2;
(ii) ψ *is of the form* $\oplus_{i=1}^k \pi_i[d_i]$ *with* $\pi_i \in \Pi$ *for every* i

is exactly the set given by Table 1.2. *It has* 24 *elements.*

Proof. This is a simple exercise in combinatorics that can be treated as follows. Consider, more generally, for every integer $n \geq 1$, the set Ψ_n of elements $\psi \in \mathcal{X}(\mathrm{SL}_n)$ satisfying assertion (ii) and such that

- the eigenvalues of ψ_∞ are the n integers $\pm(n-1)/2, \pm(n-3)/2, \ldots, \pm 1$, and 0 if n is odd;
- the eigenvalues of ψ_∞ are the $n-2$ integers $\pm(n-2)/2, \pm(n-4)/2, \ldots, \pm 1$, as well as the integer 0 with multiplicity 2 if n is even.

The problem is determining Ψ_{24}. We will, more generally, specify Ψ_n for every $1 \leq n \leq 24$, by induction on n. For $c \in \mathcal{X}(\mathrm{SL}_a)$ and $\Psi \subset \mathcal{X}(\mathrm{SL}_b)$, it will be convenient to denote by $c \oplus \Psi$ the set of elements of $\mathcal{X}(\mathrm{SL}_{a+b})$ of the form $c \oplus \psi$ with $\psi \in \Psi$.

Let $1 \leq n \leq 24$ be an even integer and $\psi \in \Psi_n$. Write $\psi = \oplus_{i=1}^k \pi_i[d_i]$, as in assertion (ii). The inequality $n \leq 24$ implies that for every i, the eigenvalues of $(\pi_i[d_i])_\infty$ are at most 11. Fix an integer i such that $(\pi_i[d_i])_\infty$ has eigenvalue 0. Lemma 9.2.3 shows that we are in one of the following cases:

- $\pi_i = 1$,
- $\pi_i = \Delta_{11}$, $d_i = 12$, $n = 24$, and therefore $\psi = \Delta_{11}[12]$,
- $\pi_i = \mathrm{Sym}^2\Delta_{11}$, in which case $d_i = 1$ and $n = 24$.

In particular, we see that under the assumption $n \leq 22$, there exist two integers i such that $\pi_i = 1$, of which exactly one moreover satisfies $d_i = 1$. We therefore have the equality $\Psi_n = [1] \oplus \Psi_{n-1}$ for n even and at most 22. Moreover, this analysis shows

$$\Psi_{24} = \{\ \Delta_{11}[12]\ \} \ \cup\ [1] \oplus \Psi_{23} \ \cup\ \mathrm{Sym}^2\Delta_{11} \oplus \Psi_{21} .$$

It remains to describe Ψ_n for n odd and at most 23. Since the nontrivial elements of Π have motivic weight at least 11, we obviously have $\Psi_n = \{\ [n]\ \}$ for every odd integer $1 \leq n \leq 11$. Since the only representation in Π of motivic weight less than 15 is Δ_{11}, we moreover have

$$\Psi_{13} = \{ [13],\ \Delta_{11}[2] \oplus [9] \} \quad \text{and} \quad \Psi_{15} = \{ [15],\ \Delta_{11}[4] \oplus [7] \}.$$

Likewise, we deduce the following assertions:

- Ψ_{17} is the union of $\{ [17],\ \Delta_{11}[6] \oplus [5] \}$ and $\Delta_{15}[2] \oplus \Psi_{13}$;
- Ψ_{19} is the union of $\{ [19],\ \Delta_{11}[8] \oplus [3],\ \Delta_{15}[4] \oplus [11] \}$ and $\Delta_{17}[2] \oplus \Psi_{15}$;
- Ψ_{21} is the union of the sets $\Delta_{17}[4] \oplus \Psi_{13}$, $\Delta_{19}[2] \oplus \Psi_{17}$, and

$$\{ [21],\ \Delta_{11}[10] \oplus [1],\ \Delta_{15}[6] \oplus [9],\ \Delta_{19,7}[2] \oplus \Delta_{15}[2] \oplus \Delta_{11}[2] \oplus [5] \};$$

- Ψ_{23} is union of $\{ [23],\ \Delta_{15}[8] \oplus [7],\ \Delta_{17}[6] \oplus [11],\ \mathrm{Sym}^2\Delta_{11} \oplus \Delta_{11}[10] \}$, $\Delta_{19}[4] \oplus \Psi_{15}$, $\Delta_{21}[2] \oplus \Psi_{19}$, and the set

$$\{ \Delta_{21,5}[2] \oplus \Delta_{17}[2] \oplus \Delta_{11}[4] \oplus [3],\ \Delta_{21,9}[2] \oplus \Delta_{15}[4] \oplus [7],\ \Delta_{21,13} \oplus \Delta_{17}[2] \oplus [11] \}.$$

To conclude, this analysis shows that the set Ψ_{24} is the set consisting of the 24 elements of Table 1.2. One way to show that these 24 elements are distinct is to invoke Proposition 6.4.5 (the Jacquet–Shalika theorem). Note that the intersection of $[1] \oplus \Psi_{23}$ and $\mathrm{Sym}^2\Delta_{11} \oplus \Psi_{21}$ is the singleton $\{ \mathrm{Sym}^2\Delta_{11} \oplus \Delta_{11}[10] \oplus [1]\}$. More generally, Table 9.2 gives the cardinality of Ψ_n in terms of $n \leq 24$.

Table 9.2 The cardinality of the subset $\Psi_n \subset \mathcal{X}(\mathrm{SL}_n)$ introduced in the proof of Proposition 9.2.2

n	≤ 12	13	14	15	16	17	18	19	20	21	22	23	24
$\|\Psi_n\|$	1	2	2	2	2	4	4	5	5	10	10	14	24

(Another way to show that the 24 elements of Table 1.2 are distinct, to which we will come back in Sect. 9.2.4, would consist in observing that the components of these elements at the prime 2 have distinct traces.) □

Lemma 9.2.3. *Let $\pi \in \Pi - \{1\}$, let $d \geq 1$ be an integer, $\psi = \pi[d]$, and $\Lambda \subset \mathbb{C}$ the set of eigenvalues of ψ_∞. Suppose $\Lambda \subset \mathbb{Z}$ and $|\lambda| \leq 11$ for every $\lambda \in \Lambda$. Then we have*

$$|\lambda| \geq \frac{d-1}{2}$$

for every $\lambda \in \mathrm{Weights}(\pi)$. Moreover, the following assertions hold:

(i) *If $0 \in \Lambda$, then we have either $\pi = \mathrm{Sym}^2\Delta_{11}$ and $d = 1$, or $\pi = \Delta_{11}$ and $d = 12$.*
(ii) *If $1 \in \Lambda$, then we have $\pi = \Delta_{11}$ and $d \in \{10, 12\}$.*

Proof. This immediately follows by studying the list Π. □

9.2.4 Statements and an Overview of the Proofs

Let ψ be one of the 24 elements listed in Table 1.2. By Proposition 9.2.2 and the examples of Sect. 6.4.3, we have $\psi_\infty = \mathrm{St}(\inf_V)$, where V is the trivial representation of $\mathrm{SO}_{24}(\mathbb{R})$. Since the set Π consists of self-dual representations, the following statement (which is also Theorem E) is not absurd!

Theorem* 9.2.5. *The standard parameters $\psi(\pi, \mathrm{St})$ of the 24 representations π in $\Pi_{\mathrm{disc}}(\mathrm{O}_{24})$ with $\pi_\infty = 1$ are the 24 elements of Table 1.2.*

Let us emphasize that in his work [109] (which, in particular, depends on [31, 108], and [156]), Ikeda succeeded in identifying 20 of these 24 parameters, namely those in the list that "do not contain" one of the four representations $\Delta_{w,v}$.

In order to say a bit more about this, let us recall some notation introduced in Sect. 5.3. For $i = 1, \ldots, 24$, we denote by λ_i the 24 distinct eigenvalues of the operator T_2 acting on $\mathbb{C}[\mathrm{X}_{24}]$, in decreasing order, as done by Nebe and Venkov (Table 5.1). Fix an eigenvector $\mathrm{v}_i \in \mathbb{Z}[\mathrm{X}_{24}]$ of T_2, and therefore of $\mathrm{H}(\mathrm{O}_{24})$, associated with λ_i. Denote by $\pi_i \in \Pi_{\mathrm{disc}}(\mathrm{O}_{24})$ the representation generated by v_i, and denote its standard parameter by

$$\psi_i = \psi(\pi_i, \mathrm{St}) \ .$$

Theorem 9.2.5 asserts that these 24 parameters ψ_i are those of Table 1.2. Given that the 24 eigenvalues of T_2 are distinct, the relation

$$\lambda_i \;=\; 2^{11}\,\mathrm{trace}\;(\psi_i)_2 \;=\; 2^{11}\,\mathrm{trace}\,\mathrm{St}\,\mathrm{c}_2(\pi_i)$$

uniquely characterizes the map $i \mapsto \psi_i$. Moreover, this provides a first verification of Theorem 9.2.5 because we can show that the four values $\tau_{j,k}(2)$ (Corollary 9.1.14, [89, Sects. 24 and 27]), as well as the coefficients $\tau_k(2)$ of q^2 in the normalized modular forms for $\mathrm{SL}_2(\mathbb{Z})$ of weight $k \leq 22$, are compatible with the computation of the λ_i by Nebe and Venkov.

By way of example, consider the parameter

$$\psi = \Delta_{21,5}[2] \oplus \Delta_{17}[2] \oplus \Delta_{11}[4] \oplus [1] \oplus [3] \ .$$

We have $\tau_{4,10}(2) = -1680$, $\tau_{12}(2) = -24$, and $\tau_{18}(2) = -528$, so that

$$\begin{aligned} 2^{11}\,\mathrm{trace}(\mathrm{St}(\psi_2)) &= (1+2)\cdot(-1680) + 2^2\cdot(1+2)\cdot(-528) \\ &\quad + 2^4\cdot(1+2+2^2+2^3)\cdot(-24) + 2^{11} \\ &\quad + 2^{10}\cdot(1+2+2^2) \\ &= -\,7920\ . \end{aligned}$$

We recover the eigenvalue λ_{21} of Nebe-Venkov; that is, $\psi = \psi_{21}$.

Finally, we denote by g_i the degree of v_i, defined in Sect. 5.3: since the eigenvalue λ_i has multiplicity 1, it is the least integer g such that π_i admits a ϑ-correspondent in $\Pi_{\mathrm{cusp}}(\mathrm{Sp}_{2g})$ in the sense of Sect. 7.1.1. By convention, we have $g_1 = 0$. As already explained in Sect. 5.3, the g_i were determined by Nebe and Venkov for $i \neq 19, 21$ in [156]; they, moreover, conjectured the values $g_{19} = 9$ and $g_{21} = 10$.

Theorem* **9.2.6.** (i) *The* g_i *are given by Table* C.5. *In particular, the Nebe–Venkov conjecture* [156, Sect. 3.3] *is true.*
(ii) *For every* $i \leq 23$, *the degree* g_i *is the least integer* $m \geq 0$ *such that* ψ_i *is of the form* $[23-2m] \oplus \psi_i'$ *with* $\psi_i' \in \mathcal{X}_{\mathrm{AL}}(\mathrm{SL}_{2m+1})$. *Finally, we have* $g_{24} = 12$.

One should be aware that in our Table C.5, the degrees g_i are in increasing order, but this does not quite hold for the indices i. We will first prove, in Sect. 9.2.8, that Theorem 9.2.5 implies Theorem 9.2.6. Then, we will give three proofs of Theorem 9.2.5, the first two of which are conditional:

- The first, and undoubtedly most natural, is a direct application of Arthur's multiplicity formula for SO_{24}. Its obvious disadvantage is that it depends on establishing Arthur's multiplicity formula for the $\mathbb{Z}$-groups SO_n, as well as the analog of Conjecture 8.4.22 for these groups; we combined the two into Conjecture 8.4.25 in Chap. 8. This conditional proof is explained in Sect. 9.2.10.
- Next, in Sect. 9.2.11, we give a second conditional proof, which this time only uses Arthur's theory for Chevalley groups, Conjecture 8.4.22, as well as "ϑ-correspondence" arguments. In this second approach, we in fact simultaneously prove the assertions (i) and (ii) of Theorem* 9.2.6 above.
- Finally, in Sect. 9.4.3, we give a last proof of Theorem 9.2.5, this time unconditional. This proof, rather different in spirit and already discussed in the introduction, does not depend on Arthur's multiplicity formula at all. It will give a deeper justification for the statement of Theorem 9.2.5.

Remark 9.2.7. (i) Curiously, the Hecke operator T_3 on $\mathbb{C}[\mathrm{X}_{24}]$ has eigenvalue 1827360 with multiplicity 2. This is a translation of the slightly miraculous equality of the traces of the components at $p = 3$ of the parameters $\psi_{19} = \mathrm{Sym}^2\Delta_{11} \oplus \Delta_{19,7}[2] \oplus \Delta_{15}[2] \oplus \Delta_{11}[2] \oplus [5]$ and $\psi_{21} = \Delta_{21,5}[2] \oplus \Delta_{17}[2] \oplus \Delta_{11}[4] \oplus [1] \oplus [3]$, as one can verify using Table C.3.

(ii) The Hecke operator T_2 acting on $\mathbb{C}[\mathrm{X}_{32}]$ has noninteger eigenvalues. Indeed, let Δ_{23} be one of the two normalized eigenforms of weight 24 for $\mathrm{SL}_2(\mathbb{Z})$. It is well known that its Fourier coefficients are in $\mathbb{Q}(\sqrt{144169})$; for example, the second is $540 \pm 12\sqrt{144169}$. This suffices to conclude because by Ikeda and Böcherer (Sect. 7.3), there exists a $\pi \in \Pi_{\mathrm{disc}}(\mathrm{O}_{32})$ such that we have $\psi(\pi, \mathrm{St}) = \Delta_{23}[8] \oplus [15] \oplus [1]$.

9.2.8 Theorem 9.2.5 Implies Theorem 9.2.6

We now explain how to deduce the values of the g_i from the list of the ψ_i in Table C.5.

Lemma 9.2.9. *We have the inequalities $g_{23} \leq 11$ and $g_i \leq 10$ for $i \leq 22$, as well as the equality $g_{24} = 12$.*

Proof. As recalled in Sect. 5.3, Erokhin proved $g_i \leq 12$ for every i in [80]. This result was recovered by Borcherds, Freitag, and Weissauer in [31], where they, moreover, verify the inequality $g_i \leq 11$ for all i except exactly one (this is the assertion that $\mathrm{Ker}\,\vartheta_{11}$ is of dimension 1). To do this, they compute, explicitly, the coefficients of the theta series of the Niemeier lattices corresponding to Gram matrices of the lattices of the form $\mathrm{Q}(R)$, where R is an irreducible root system of type ADE and rank at most 12 [31, Table p. 146]. As observed by Nebe and Venkov [156, Sect. 3.1, Lemma 3.3], these computations show, more precisely, the inequalities $g_{23} \leq 11$, $g_i \leq 10$ for every $i \leq 22$, and $g_{24} = 12$ (at least one of the g_i must equal 12 by the result of [31] mentioned above). □

Proof that Theorem 9.2.5 Implies Theorem 9.2.6. Let $1 \leq i \leq 23$, and let $\psi_i' \in \mathcal{X}(\mathrm{SL}_{2g_i+1})$ be the standard parameter of the ϑ-correspondent of π_i in $\Pi_{\mathrm{cusp}}(\mathrm{Sp}_{2g_i})$. By Lemma 9.2.9, we have $g_i \leq 11$, so that by Rallis (Corollary 7.1.3), we have the identity

$$\psi_i = \psi_i' \oplus [23 - 2g_i]\ .$$

By Arthur's Theorem 8.1.1, we have $\psi_i' \in \mathcal{X}_{\mathrm{AL}}(\mathrm{SL}_{2g_i+1})$. The uniqueness of the Arthur–Langlands parameters (Jacquet–Shalika, Proposition 6.4.5) therefore shows that g_i has the property that $[23 - 2g_i]$ is a "component" of ψ_i in Table C.5.

For $i \leq 22$, Lemma 9.2.9 implies $g_i \leq 10$, that is, $23 - 2g_i \geq 3$. In this case, $23 - 2g_i$ is the unique integer $m_i > 1$ such that $[m_i]$ is a component of ψ_i; the other possible component of ψ_i of the form $[d]$ is $[1]$. This trivially determines g_i for $i \leq 22$ by a direct examination of Table C.5. Likewise, the inequality $g_{23} \leq 11$ and the identity $\psi_{23} = \mathrm{Sym}^2\Delta_{11} \oplus \Delta_{11}[10] \oplus [1]$ show that we necessarily have $g_{23} = 11$. This concludes the proof. □

Note that the approach used here does not depend on the refined computations of Sects. 3.3 and 3.4 of [156], but "only" on the table [31, p. 146] of Borcherds–Freitag–Weissauer and the computation of T_2 by Nebe and Venkov.

9.2.10 First, Conditional, Proof of Theorem 9.2.5

Let us admit Conjecture 8.4.25 and apply Theorem 8.5.8 to each parameter ψ of Table 1.2 (the assumptions hold with $\psi_\infty = \mathrm{St}(\mathrm{Inf}_1)$). We assert that Equality (8.5.1) still holds, which is, in itself, a rather miraculous phenomenon. It is, of course, something one can simply verify in each of the 24 cases. We can also make the following remarks.

(a) If ψ does not contain $\mathrm{Sym}^2\Delta_{11}$ and $\psi \neq \Delta_{11}[12]$, then ψ satisfies the assumption of Criterion 8.5.10, that is, it is of the form

$$(\oplus_{i=1}^{k-2}\pi_i[d_i]) \oplus [d_{k-1}] \oplus [1]$$

with π_i symplectic for every $i \leq k-2$. A quick examination shows that we always have either $d_i < d_{k-1}$ and $\varepsilon(\pi_i) = (-1)^{n_i d_i/4}$, or $d_i > d_{k-1}$ and $d_i \equiv 0 \bmod 4$ (which only occurs for ψ_{13}, ψ_{21}, and ψ_{22}). Concretely, we see that the ε-factor $\varepsilon(\Delta_{j,k}) = (-1)^k$ is 1 for the four pairs (j,k) we are interested in, and that each time a component of the form $\Delta_w[d]$ appears in ψ, we have $d \equiv w+1 \bmod 4$. We can therefore conclude using Criterion 8.5.10.

(b) If ψ contains $\mathrm{Sym}^2\Delta_{11}$, then ψ is of the form

$$(\oplus_{i=1}^{k-2}\pi_i[d_i]) \oplus \mathrm{Sym}^2\Delta_{11} \oplus [1]$$

with π_i symplectic for every $i \leq k-2$; it therefore satisfies the assumption of Criterion 8.5.11. We therefore again conclude using this criterion, by observing that we still have either $d_i < d_{k-1}$ and $\varepsilon(\pi_i)(-1)^{n_i/2} = (-1)^{n_i d_i/4}$, or $d_i > d_{k-1}$ and $d_i \equiv 2 \bmod 4$ (which only occurs for ψ_{20} and ψ_{23}).

Thus, for every $i \leq 23$, the unique representation $\pi'_i \in \Pi(\mathrm{SO}_{24})$ such that we have $\psi(\pi'_i, \mathrm{St}) = \psi_i$ satisfies $\mathrm{m}(\pi'_i) = 1$. Finally, if we have $\psi = \psi_{20} = \Delta_{11}[12]$, Equality (8.5.1) trivially holds, and Remark 8.5.9 asserts that there exist exactly two representations $\pi'_{24}, \pi''_{24} = \mathrm{S}(\pi'_{24}) \in \Pi_{\mathrm{disc}}(\mathrm{SO}_{24})$ with standard parameter ψ_{24} and that, moreover, $\mathrm{m}(\pi'_{24}) = \mathrm{m}(\pi''_{24}) = 1$.

This discussion provides 25 distinct elements of $\Pi_{\mathrm{disc}}(\mathrm{SO}_{24})$, each of multiplicity 1. Since $\mathrm{h}(\mathrm{SO}_{24}) = |\widetilde{\mathrm{X}}_{24}| = 25$ (Corollary 4.1.9), these are exactly the $\pi \in \Pi_{\mathrm{disc}}(\mathrm{SO}_{24})$ such that $\pi_\infty = 1$. The $\mathrm{H}(\mathrm{O}_{24})$-equivariant decomposition (Sect. 4.4.4)

$$\mathrm{M}_{\mathbb{C}}(\mathrm{SO}_{24}) = \mathrm{M}_{\mathbb{C}}(\mathrm{O}_{24}) \oplus \mathrm{M}_{\det}(\mathrm{O}_{24}),$$

combined with the fact that T_2 has 24 distinct eigenvalues on $\mathrm{M}_{\mathbb{C}}(\mathrm{O}_{24})$, shows that the standard parameters of the 24 representations $\pi \in \Pi_{\mathrm{disc}}(\mathrm{O}_{24})$ such that $\pi_\infty = 1$ are exactly the ψ_i of Table 1.2. It also shows that the unique representation $\pi \in \Pi_{\mathrm{disc}}(\mathrm{O}_{24})$ such that we have $\pi_\infty = \det$ has parameter $\Delta_{11}[12]$ (in other words, we also recover Proposition 7.5.1!).

9.2.11 Second Proof of Theorem 9.2.5, *Modulo Conjecture* 8.4.22

Let us now give a "less conditional" proof of Theorem 9.2.5, which does not use Conjecture 8.4.25 and only depends on Conjecture 8.4.22. We begin with two observations.

Observation 1. Let $1 \leq g < 12$, and consider the map of Sect. 5.1

$$\vartheta_g \colon \mathbb{C}[\mathrm{X}_{24}] \longrightarrow \mathrm{M}_{12}(\mathrm{Sp}_{2g}(\mathbb{Z})) \,.$$

Suppose that $F \in \mathrm{S}_{12}(\mathrm{Sp}_{2g}(\mathbb{Z}))$ is an eigenform for $\mathrm{H}(\mathrm{Sp}_{2g})$, and denote by $\pi_F \in \Pi_{\mathrm{disc}}(\mathrm{Sp}_{2g})$ the representation it generates. By Arthur's Theorem 8.1.1, we can write

$$\psi(\pi_F, \mathrm{St}) = \oplus_{j=1}^{k} \pi_j[d_j] \in \mathcal{X}_{\mathrm{AL}}(\mathrm{SL}_{2g+1}) \,.$$

By Böcherer, F is in the image of ϑ_g if and only if we have $\mathrm{L}(12-g, \pi_F, \mathrm{St}) \neq 0$ (Remark 7.2.4). By Proposition 8.6.2, we have $\mathrm{L}(12-g, \pi_F, \mathrm{St}) \neq 0$ if and only if

$$\forall\, 1 \leq j \leq k \text{ such that } d_j \equiv 0 \bmod 2 \text{ and } d_j > 23 - 2g, \qquad (9.2.1)$$
$$\text{we have } \mathrm{L}\Big(\frac{1}{2}, \pi_j\Big) \neq 0 \,.$$

If this condition is satisfied, the representation π_F therefore admits a ϑ-correspondent π'_F in $\Pi_{\mathrm{disc}}(\mathrm{O}_{24})$ such that $(\pi'_F)_\infty = 1$ and

$$\psi(\pi'_F, \mathrm{St}) = \psi(\pi_F, \mathrm{St}) \oplus [23 - 2g] \,,$$

by Rallis (Corollary 7.1.3).

Observation 2. Consider Table C.5. We see that for every $2 \leq i \leq 23$, there exists a unique $\psi'_i \in \mathcal{X}_{\mathrm{AL}}(\mathrm{SL}_{2g_i+1})$ such that we have

$$\psi_i = \psi'_i \oplus [23 - 2g_i] \,.$$

Clearly, the eigenvalues of $(\psi'_i)_\infty$ are the $2g_i + 1$ integers $\pm 11, \pm 10, \ldots, \pm(12 - g_i)$, as well as 0. Let $\varpi_i \in \Pi(\mathrm{Sp}_{2g_i})$ be the unique representation such that $\psi(\varpi_i, \mathrm{St}) = \psi'_i$ and $(\varpi_i)_\infty \simeq \pi'_{\det^{12}}$ (the holomorphic discrete series introduced in Sect. 6.3.4). As already explained several times, the multiplicity $\mathrm{m}(\varpi_i)$ is nonzero if and only if there exists an eigenform $F_i \in \mathrm{S}_{12}(\mathrm{Sp}_{2g_i}(\mathbb{Z}))$ such that $\pi_{F_i} \simeq \varpi_i$ (Corollary 6.3.7).

These two observations suggest the following optimistic strategy for proving Theorem 9.2.5.

1. Show $\mathrm{m}(\varpi_i) \neq 0$ for every $2 \leq i \leq 23$.
2. Using Böcherer's criterion (9.2.1), verify, for every $1 \leq i \leq 23$, that if there exists an eigenform $F_i \in \mathrm{S}_{12}(\mathrm{Sp}_{2g_i}(\mathbb{Z}))$ such that $\pi_{F_i} \simeq \varpi_i$, then $F_i \in \mathrm{Im}(\vartheta_{g_i})$.

Indeed, once this is done, we deduce from this the existence, for every $2 \leq i \leq 23$, of a representation in $\Pi_{\mathrm{disc}}(\mathrm{O}_{24})$ with trivial Archimedean component and standard parameter ψ_i, namely a ϑ-correspondent of ϖ_i. The existence of a representation in $\Pi_{\mathrm{disc}}(\mathrm{SO}_{24})$ with standard parameter $\psi_1 = [23] \oplus [1]$ is clear: we can take the trivial representation (Examples 6.4.7). Finally, the existence of a representation in $\Pi_{\mathrm{disc}}(\mathrm{SO}_{24})$ with standard parameter $\psi_{24} = \Delta_{11}[12]$ was already proved in Corollary 7.3.4 (work of Ikeda and Böcherer).

Verification of Item 2. By examining the ψ_i', we see that there is nothing to check, because no integer j satisfies $d_j \equiv 0 \bmod 2$ and $d_j > 23 - 2g_i$ (Criterion (9.2.1)) unless $i \in \{13, 20, 21, 22, 23\}$, in which case the criterion can simply be written as $\mathrm{L}(1/2, \Delta_{15}) \neq 0$ for $i = 13$ and $\mathrm{L}(1/2, \Delta_{11}) \neq 0$ for $i \geq 20$. This suffices to conclude because these two values of L-functions are indeed nonzero by Remark 7.3.3.

Conjectural Verification of Item 1. The value of $\mathrm{m}(\varpi_i)$ is, of course, determined by Theorem 8.5.2. To apply this theorem, we must on the one hand, know that the morphism ν_∞ satisfies Conjecture 8.4.22 and, on the other hand, verify conditions (a) and (b) of the theorem. Condition (a) is clearly always satisfied, as can be seen by examining the ψ_i'. As far as condition (b) is concerned, we assert that it is also always satisfied. This is a miracle of the same nature at that encountered in Sect. 9.2.10, which we could verify the same way (or case by case!). This is not, in fact, necessary, because by Proposition 8.6.1 (or, more exactly, by the proof of that proposition), this verification can be deduced, formally, from that carried out in Sect. 9.2.10 if we verify that for every component of ψ_i' of the form $\pi[d]$ with $d \equiv 0 \bmod 2$ and $d > 23 - 2g_i$, we have $\varepsilon(\pi) = 1$. But a direct examination of Table C.5 shows that such a component exists only for the parameters of index $i \in \{13, 20, 21, 22, 23\}$ and that in all cases $\pi = \Delta_{11}$ or Δ_{15}, so that we indeed have $\varepsilon(\pi) = 1$. The fact that we recover exactly the exceptions we already needed to consider in the verification of item 2 is not, of course, a coincidence, as we explained in Sect. 8.6. Thus, if for a given i, Conjecture 8.4.22 is known to be true for the ν_∞ associated with ψ_i', the equality $\mathrm{m}(\varpi_i) = 1$ follows. □

Remark 9.2.12. Except for the case $i = 2$, where $\psi_i' = \mathrm{Sym}^2 \Delta_{11}$ (and where it is clear that $\mathrm{m}(\varpi_i) = 1$!), the criterion "$d_j = 1$ for every j" unfortunately never holds. On the other hand, we see that if we had the particular case of Conjecture 8.4.22 announced by Arancibia, described in Remark 8.4.23, namely the case where "$\pi_j \neq 1 \Rightarrow d_j \leq 4$," we could conclude that $\mathrm{m}(\varpi_i) = 1$ holds whenever $i \notin \{11, 12, 13, 20, 22, 23\}$. Since, furthermore, the work of Ikeda mentioned earlier [109] asserts that we have $\mathrm{m}(\varpi_i) \neq 0$ whenever $i \notin \{10, 15, 19, 21\}$, this would lead to an unconditional proof of the theorem.

9.3 Algebraic Representations of Motivic Weight at Most 22

9.3.1 A Classification Statement

The aim of this subsection is to prove the following theorem.

Theorem 9.3.2. *Let $n \geq 1$, and let $\pi \in \Pi_{\mathrm{cusp}}(\mathrm{PGL}_n)$ be algebraic of motivic weight $w \leq 22$. Then we are in one of the following cases:*

(i) *$n = 1$, $w = 0$, and π is the trivial representation;*
(ii) *$n = 2$, $w \in \{11, 15, 17, 19, 21\}$, and π is the representation Δ_w generated by the unique normalized modular cusp form of weight $w+1$ for the group $\mathrm{SL}_2(\mathbb{Z})$;*

(iii) $n = 3$, $w = 22$, *and* π *is the symmetric square of* Δ_{11}*;*
(iv) $n = 4$ *and* $\mathrm{Weights}(\pi)$ *is of the form* $\{\pm w/2, \pm v/2\}$ *with*

$$(w, v) \quad = \quad (19, 7)\,, \quad (21, 5)\,, \quad (21, 9)\,, \quad \text{or} \quad (21, 13)\,.$$

In this case, π *is the unique representation in* $\Pi_{\mathrm{alg}}(\mathrm{PGL}_4)$ *with weights* $\{\pm w/2, \pm v/2\}$*; in particular, we have* $\pi \simeq \pi^\vee$.
(v) $n = 4$, $w = 22$, *and* $\mathrm{Weights}(\pi) = \{\pm 11, \pm v\}$ *with* $v = 4$, 5, *or* 6.
In this case, the representations π *and* $\pi^\vee$ *are not isomorphic, and they are the only representations in* $\Pi_{\mathrm{alg}}(\mathrm{PGL}_4)$ *with weights* $\{\pm 11, \pm v\}$.

Moreover, if there exist representations in $\Pi_{\mathrm{alg}}(\mathrm{PGL}_4)$ *with weights* $\{\pm 21/2, \pm 9/2\}$ *and* $\{\pm 21/2, \pm 13/2\}$*, respectively, then case* (v) *does not occur.*

Let us emphasize that assertion (iv) of this theorem asserts only the uniqueness, and not the existence, of a representation $\pi \in \Pi_{\mathrm{alg}}(\mathrm{PGL}_4)$ with weights $\{\pm w/2, \pm v/2\}$, where (w, v) run through the four ordered pairs in the statement. However, in Definition-Proposition 9.1.6, we showed, using Tsushima's formula and the results of Arthur, that there indeed exists such a representation, which we denoted by $\Delta_{w,v}$. The reason why we formulate Theorem 9.3.2 this way is that its proof, as we will see, does not require the existence of these representations and, in particular, does not depend on the work of Arthur. We can therefore deduce the following theorem from it[2] (Theorem F of the introduction).

Theorem* 9.3.3. *Let* $n \geq 1$, *and let* $\pi \in \Pi_{\mathrm{cusp}}(\mathrm{PGL}_n)$ *be algebraic of motivic weight at most* 22. *Then* π *belongs to the following list of* 11 *representations:*

$$1,\ \Delta_{11},\ \Delta_{15},\ \Delta_{17},\ \Delta_{19},\ \Delta_{19,7},\ \Delta_{21},\ \Delta_{21,5},\ \Delta_{21,9},\ \Delta_{21,13},\ \text{and } \mathrm{Sym}^2\Delta_{11}\,.$$

As we already explained in the introduction, our proof of Theorem 9.3.2 depends on an analog, in the setting of automorphic L-functions, of the *explicit formulas* of Riemann and Weil in the theory of prime numbers. We refer to the surveys of Poitou [165, 166] on this subject. This analog was developed by Mestre [144] and then applied by Fermigier to the functions $\mathrm{L}(s, \pi)$ for $\pi \in \Pi_{\mathrm{cusp}}(\mathrm{PGL}_n)$ [84]. From this, the latter deduced loc. cit. the vanishing of the "cuspidal" cohomology of the group $\mathrm{SL}_n(\mathbb{Z})$ with rational coefficients for $n < 24$. This result was subsequently extended to $n < 27$ by Miller [147], inspired by work of Rudnick and Sarnak [173], by considering instead the function $\mathrm{L}(s, \pi \times \pi^\vee)$, which has several advantages regarding convergence and positivity.[3] This will also be our starting point. More generally, we will examine, in detail, the inequalities given by the explicit formula applied to the function $\mathrm{L}(s, \pi \times \pi')$ for every pair of representations $\{\pi, \pi'\}$ with $\pi \in \Pi_{\mathrm{alg}}(\mathrm{PGL}_n)$ and $\pi' \in \Pi_{\mathrm{alg}}(\mathrm{PGL}_{n'})$.

---

[2] Let us mention that, as far as we know, no one has yet proved the existence of a representation $\pi \in \Pi_{\mathrm{alg}}(\mathrm{PGL}_n)$ for $n \geq 1$ that is not self-dual.

[3] In a close context, the advantage of considering this type of L-function had already been observed by Serre [166, p. 15].

Before beginning these proofs, let us give a corollary of Theorem 9.3.2 in the particular case $\mathrm{w}(\pi) = 0$.

Proposition 9.3.4. *Suppose that* $\mathbf{L}_{\mathbb{Z}}$ *is a compact topological group satisfying Axioms* (L1) *and* (L2) *introduced on p. xi of the preface. Then* $\mathbf{L}_{\mathbb{Z}}$ *is connected.*

Proof. Let $\psi\colon \mathbf{L}_{\mathbb{Z}} \to \mathrm{GL}_n(\mathbb{C})$ be a continuous irreducible representation of $\mathbf{L}_{\mathbb{Z}}$. By Axiom (L2), there exists a unique representation π in $\Pi_{\mathrm{cusp}}(\mathrm{PGL}_n)$ satisfying $\mathrm{c}(\pi) = \mathrm{c}(\psi)$ (by convention, we have $\Pi_{\mathrm{cusp}}(\mathrm{PGL}_n) = \Pi_{\mathrm{cusp}}(\mathrm{GL}_n)$). Suppose, moreover, that the image of ψ is finite; in particular, we have $\mathrm{c}_\infty(\psi) = 0$ by the definition of Frob_∞ (Axiom (L1)). From this, we deduce $\mathrm{c}_\infty(\pi) = 0$; hence all weights of π are zero, and we have $\mathrm{w}(\pi) = 0$. Theorem 9.3.2 implies $n = 1$ and that π is the trivial representation. Axiom (L2) in the case $n = 1$ shows that ψ is also the trivial representation of $\mathbf{L}_{\mathbb{Z}}$. Since every nontrivial finite group admits a nontrivial irreducible representation, we have proved that the compact group $\mathbf{L}_{\mathbb{Z}}$ does not have a nontrivial finite quotient: it is a connected group. □

9.3.5 The Explicit Formula for the L*-Functions of Pairs*

The explicit formulas depend on the choice of a "test function." Following the analysis of Poitou and Weil [165, p. 6], by this, we mean any even function $F\colon \mathbb{R} \to \mathbb{R}$ satisfying the following conditions, where F_ϵ, for ϵ real and positive, denotes the function $\mathbb{R} \to \mathbb{R}$ defined by $x \mapsto F(x)e^{(1/2+\epsilon)x}$:

(T1) There exists an $\epsilon > 0$ such that F_ϵ is integrable over $\mathbb{R}_{>0}$.
(T2) There exists an $\epsilon > 0$ such that F_ϵ is of bounded variation on $\mathbb{R}$, equal, at each point, to the average of its left and right limits.
(T3) The function $(1/x)(F(x) - F(0))$ is of bounded variation on $\mathbb{R} - \{0\}$.

In particular, every even function of class $\mathcal{C}^2$ over $\mathbb{R}$ with compact support is a test function. This will be the case in our applications, where F will be a simple modification of Odlyzko's function. For now, however, it will be clearer if we consider arbitrary test functions.

If F is a test function and if $\epsilon > 0$ is such that F_ϵ is integrable over $\mathbb{R}$, then the integral defined by

$$\Phi_F(s) = \int_{\mathbb{R}} F(x)e^{(s-1/2)x}\,\mathrm{d}x \tag{9.3.1}$$

is absolutely convergent in the region $-\epsilon < \Re\, s < 1 + \epsilon$ in the complex plane. In particular, the function $\Phi_F(s)$ is well defined and holomorphic in this region. The parity of F implies the equality $\Phi_F(s) = \Phi_F(1-s)$. Moreover, the relation $\Phi_F(\overline{s}) = \overline{\Phi_F(s)}$ shows $\Phi_F(s) \in \mathbb{R}$ for every real s such that $0 \leq s \leq 1$.

Set

$$\Pi_{\mathrm{alg}} = \coprod_{n\geq 1} \Pi_{\mathrm{alg}}(\mathrm{PGL}_n)\ .$$

Let $\pi, \pi' \in \Pi_{\mathrm{alg}}$. As recalled in Sect. 8.2.21, according to Jacquet and Shalika [112, Theorem 5.3], we have an Euler product

$$\mathrm{L}(s, \pi \times \pi') = \prod_p \det(1 - p^{-s} \mathrm{c}_p(\pi) \otimes \mathrm{c}_p(\pi'))^{-1}$$

that is well defined and absolutely convergent for $\Re s > 1$. Moreover, the function

$$\xi(s, \pi \times \pi') := \Gamma(s, \mathrm{L}(\pi_\infty) \otimes \mathrm{L}(\pi'_\infty))\, \mathrm{L}(s, \pi \times \pi')$$

admits a meromorphic continuation to all of $\mathbb{C}$ satisfying the functional equation $\xi(s, \pi \times \pi') = \varepsilon(\pi \times \pi')\, \xi(1 - s, \pi^\vee \times \pi'^\vee)$ with

$$\varepsilon(\pi \times \pi') = \varepsilon(\mathrm{L}(\pi_\infty) \otimes \mathrm{L}(\pi'_\infty))\,.$$

In particular, all zeros of ξ are in the critical strip $0 \leq \Re s \leq 1$ (Shahidi has even proved that these zeros are in the interior of this strip). Finally, the function ξ is holomorphic on $\mathbb{C} - \{0, 1\}$, and admits a pole at $s = 1$ if and only if $\pi' = \pi^\vee$, in which case this is a simple pole, by Moeglin and Waldspurger [151, Appendice].

Proposition-Definition 9.3.6. *Let $\pi, \pi' \in \Pi_{\mathrm{alg}}$ and $\xi(s) = \xi(s, \pi^\vee \times \pi')$, let F be a test function and $T > 0$ a real number. The finite sum*

$$\sum_{\{s \in \mathbb{C}\,;\, |\Im s| < T,\ \xi(s)=0\}} \Phi_F(s)\ \mathrm{ord}_{z=s}\, \xi(z)$$

is real and admits a finite limit as $T \to +\infty$; we denote this limit by $\mathrm{Z}^F(\pi, \pi')$.

This statement is a special case of the results of Mestre [144, Sect. I], which in turn generalize, rather directly, those of Riemann, Weil, and Poitou [165]. Suppose $\pi \in \Pi_{\mathrm{alg}}(\mathrm{PGL}_n)$ and $\pi' \in \Pi_{\mathrm{alg}}(\mathrm{PGL}_{n'})$. In the notation of Mestre, we take $M = M' = nn'$, $c = 0$, $L_1(s) = \mathrm{L}(s, \pi^\vee \times \pi')$, $L_2(s) = \mathrm{L}(s, \pi \times (\pi')^\vee)$, $w = \varepsilon(\pi^\vee \times \pi')$, $\Lambda_1(s) = \xi(s, \pi^\vee \times \pi')$, and $\Lambda_2(s) = \xi(s, \pi \times (\pi')^\vee)$. By definition, we incorporate the factors that he denotes by A^s and B^s, as well as his coefficients a_i, a'_i, b_i, b'_i, in our Archimedean factors Γ, and there is no contribution from the conductor (which is 1 in this case). Having said this, the assumptions (i), (ii), and (iii) loc. cit. follow from the functional equation and the finiteness of the number of poles of the Λ_i, which have already been justified above. Assumption (iii), namely that the function ξ minus its singular parts is bounded in the entire vertical strip, is a theorem of Gelbart and Shahidi [91, Corollary 2]. Finally, only a weakened version of the last assumption (iv) of Mestre is satisfied, namely the absolute convergence of the Euler products L_1 and L_2, as well as their nonvanishing on $\Re s > 1$, but that is all he needs; see [144, pp. 213–214] and especially the argument given by Poitou [165, pp. 2–3].

The conclusion of this discussion is that all results of [144, Sect. I.2] apply. The convergence assertion in Proposition-Definition 9.3.6 is, in particular, justified by Mestre on p. 213. The fact that the finite sum that appears in the proposition-definition is real comes from the fact that the region $\{s \in \mathbb{C}\,;\ |\Im s| < T\}$ is stable under $s \mapsto 1-\overline{s}$ and from the equalities

$$\xi(1-\overline{s}, \pi^\vee \times \pi') = \varepsilon(\pi^\vee \times \pi')\, \xi(\overline{s}, \pi \times \pi'^\vee) = \varepsilon(\pi^\vee \times \pi')\, \overline{\xi(s, \pi^\vee \times \pi')}\,.$$

Mestre also establishes, loc. cit., the explicit formula we will use. It is the result of integrating the 1-form $\Phi_F(s)\,\mathrm{dlog}\,\xi(s)$ on the boundary of the rectangle $\{s \in \mathbb{C}\,;\ -\epsilon \leq \Re s \leq 1+\epsilon,\ |\Im s| \leq A\}$, where A and ϵ are suitable strictly positive real numbers, followed by passing to the limits $\epsilon \to 0$ and $A \to \infty$. In order to state it in a pleasant form, we first introduce certain "local" preliminary quantities. The convergence assertion in the following definition is justified in [144, pp. 213–214] and [165, pp. 2–3].

Proposition-Definition 9.3.7. *Let $\pi, \pi' \in \Pi_{\mathrm{alg}}$, and let F be a test function. The sum*

$$\sum_{p,k} F(k\log(p))\frac{\log(p)}{p^{k/2}}\ \overline{\mathrm{tr}\,(\mathrm{c}_p(\pi)^k)}\ \mathrm{tr}\,(\mathrm{c}_p(\pi')^k)\,,$$

taken over all pairs (p,k) with p a prime and $k \geq 1$ an integer, is absolutely convergent; we denote it by $\widetilde{\mathrm{B}}^F_f(\pi,\pi')$. We have the obvious relations $\widetilde{\mathrm{B}}^F_f(\pi,\pi') = \overline{\widetilde{\mathrm{B}}^F_f(\pi',\pi)} = \widetilde{\mathrm{B}}^F_f((\pi')^\vee, \pi^\vee)$. Finally, set

$$\mathrm{B}^F_f(\pi,\pi') := \Re\,\widetilde{\mathrm{B}}^F_f(\pi,\pi')\,.$$

Let $\mathrm{W}^{\mathrm{alg}}_{\mathbb{R}}$ be the the quotient of the Weil group $\mathrm{W}_{\mathbb{R}}$ by the connected component of its center, namely $\mathbb{R}_{>0}$. Let K_∞ be the Grothendieck ring of the category of continuous, complex, finite-dimensional representations of the group $\mathrm{W}^{\mathrm{alg}}_{\mathbb{R}}$. By the statements recalled in Sect. 8.2.12, it is the free abelian group on the (classes of the) representations

$$1\,,\quad \epsilon_{\mathbb{C}/\mathbb{R}}\,,\quad \text{and}\quad \mathrm{I}_w \ \text{for}\ w \in \mathbb{Z}_{>0}\,.$$

Note that every element of K_∞ is equal to its dual, because this is the case for the three representations above.

The map $V \mapsto \Gamma(-,V)$ introduced in Sect. 8.2.21 extends naturally to a homomorphism $\Gamma\colon \mathrm{K}_\infty \to \mathcal{M}(\mathbb{C})^\times$, where $\mathcal{M}(\mathbb{C})^\times$ denotes the multiplicative group of the field $\mathcal{M}(\mathbb{C})$ of meromorphic functions on $\mathbb{C}$. The image of this map consists of functions that have neither a pole nor a zero in the half-plane $\Re s > 0$.

Proposition-Definition 9.3.8. *If F is a test function, then the map*

$$\mathrm{J}_F\colon \mathrm{K}_\infty \to \mathbb{R}\,,\quad V \mapsto -\frac{1}{2\pi i}\int_{\Re(s)=1/2} \Phi_F(s)\frac{\Gamma'(s,V)}{\Gamma(s,V)}\mathrm{d}s$$

is well defined and $\mathbb{Z}$*-linear. Moreover, we have the identities*

(i) $\mathrm{J}_F(1) = (1/2)\log(\pi)\, F(0) + \sigma_F(1/2, 0)$,
(ii) $\mathrm{J}_F(\epsilon_{\mathbb{C}/\mathbb{R}}) = (1/2)\log(\pi)\, F(0) + \sigma_F(1/2, 1/2)$,
(iii) $\mathrm{J}_F(\mathrm{I}_w) = \log(2\pi)\, F(0) + \sigma_F(1, w/2)$ *for* $w \geq 0$,

where we have set

$$\sigma_F(a,b) = a \int_0^{+\infty} \left(F(ax)e^{-(a/2+b)x}(1-e^{-x})^{-1} - F(0)e^{-x}x^{-1} \right) \mathrm{d}x \,.$$

Finally, the map $\mathrm{K}_\infty \times \mathrm{K}_\infty \to \mathbb{R}$ *defined by* $(V, W) \mapsto \mathrm{J}_F(V^* \otimes W)$ *is bilinear; we denote it by* B^F_∞*. The obvious relation* $\mathrm{B}^F_\infty(V, W) = \mathrm{J}_F(V \otimes W)$ *for every* $V, W \in \mathrm{K}_\infty$ *shows that* B^F_∞ *is symmetric.*

Proof. Denote by $\psi \in \mathcal{M}(\mathbb{C})$ the digamma function, defined by $\psi(s) = \Gamma'(s)/\Gamma(s)$. Also set $\psi(s, V) = \Gamma'(s, V)/\Gamma(s, V)$ for $V \in \mathrm{K}_\infty$. The map $\mathrm{K}_\infty \to \mathcal{M}(\mathbb{C})$ defined by $V \mapsto \psi(-, V)$ is clearly $\mathbb{Z}$-linear. By the definitions of the $\Gamma(s, V)$-factors recalled in Sect. 8.2.21, we have the identities

$$\psi(s,1) = -\frac{1}{2}\log\pi + \frac{1}{2}\psi\Big(\frac{s}{2}\Big) \quad \text{and} \quad \psi(s,\mathrm{I}_w) = -\log(2\pi) + \psi\Big(s + \frac{w}{2}\Big)$$

for every $w \geq 0$. The function F is integrable over $\mathbb{R}$ by Condition (T1) and equal, at every point, to the average of its left and right limits by Condition (T2), and $(F(x) - F(0))/x$ is bounded in the neighborhood of 0 by Condition (T3). The Fourier inversion formula therefore holds at $x = 0$; in other words, the integral $-(1/2\pi i)\int_{\Re(s)=1/2} \Phi_F(s)\mathrm{d}s$ is convergent and has value $-F(0)$. It remains to examine the convergence of an integral of the form

$$-\frac{a}{2i\pi}\int_{\Re(s)=1/2} \Phi_F(s)\psi(as+b)\mathrm{d}s$$

with $b \in \mathbb{R}_{\geq 0}$ and $a \in \mathbb{R}_{>0}$. The (simple) convergence of this integral is verified in [165, p. 6–04] and [144, Lemma I.2.1], as is the equality of the sum with the (equally convergent) integral $\sigma_F(a, b)$ of the statement. This implies all parts of the proposition. □

We now have all ingredients necessary to state the explicit formula. For the sake of convenience, we introduce the free abelian group on the set Π_{alg}:

$$\mathrm{K} = \mathbb{Z}[\Pi_{\mathrm{alg}}] \,.$$

Let F be a test function. Each of the three functions $\Pi_{\mathrm{alg}} \times \Pi_{\mathrm{alg}} \to \mathbb{R}$ that send (π, π') onto, respectively, $\mathrm{Z}^F(\pi, \pi')$, $\mathrm{B}^F_f(\pi, \pi')$, and $\delta_{\pi,\pi'}$ (Kronecker delta), extends to a symmetric bilinear map $\mathrm{K} \times \mathrm{K} \to \mathbb{R}$ associated with F. Denote these three bilinear maps by Z^F, B^F_f, and δ, respectively. Furthermore, the map $\Pi_{\mathrm{alg}} \to \mathrm{K}_\infty$ defined by $\pi \mapsto \mathrm{L}(\pi_\infty)$ extends to a homomorphism $\mathrm{L}\colon \mathrm{K} \to \mathrm{K}_\infty$. The map $\mathrm{K} \times \mathrm{K} \to \mathbb{R}$ defined by $(V, W) \mapsto \mathrm{B}^F_\infty(\mathrm{L}(V), \mathrm{L}(W))$ is therefore also bilinear and symmetric; we allow ourselves the consistent abuse of notation of also denoting it by B_∞^F. Finally, set

$$\mathrm{B}^F = \mathrm{B}_f^F + \mathrm{B}_\infty^F : \mathrm{K} \times \mathrm{K} \to \mathbb{R} .$$

Proposition 9.3.9 ("Explicit Formula"). *For every test function F, we have the following equality between bilinear forms* $\mathrm{K} \times \mathrm{K} \to \mathbb{R}$*:*

$$\mathrm{B}^F + \frac{1}{2}\mathrm{Z}^F = \Phi_F(0)\ \delta .$$

Proof. By bilinearity, it suffices to show this equality for the ordered pair $(\pi, \pi') \in \Pi_{\mathrm{alg}} \times \Pi_{\mathrm{alg}}$. In this case, in view of the parity of the function F, the formula given by Mestre [144, Sect. I.2] can be written

$$\mathrm{B}_\infty^F(\pi,\pi') + \mathrm{B}_\infty^F(\pi',\pi) + \widetilde{\mathrm{B}}_f^F(\pi,\pi') + \widetilde{\mathrm{B}}_f^F(\pi',\pi) + \mathrm{Z}^F(\pi,\pi') \\ = \delta_{\pi,\pi'}\,(\Phi_F(0) + \Phi_F(1)) .$$

The identities $\Phi_F(1) = \Phi_F(0)$, $\mathrm{B}_\infty^F(\pi,\pi') = \mathrm{B}_\infty^F(\pi',\pi)$, and $2\,\mathrm{B}_f^F(\pi,\pi') = \widetilde{\mathrm{B}}_f^F(\pi,\pi') + \widetilde{\mathrm{B}}_f^F(\pi',\pi)$ now suffice to conclude. □

An element π of K is called *effective* if it is a finite sum of elements of Π_{alg} (in other words, a linear combination with nonnegative coefficients). If π and π' are effective, say $\pi = \sum_i \pi_i$ and $\pi' = \sum_j \pi'_j$ with $\pi_i, \pi'_j \in \Pi_{\mathrm{alg}}$, we denote by $\xi(s, \pi \times \pi')$ (resp. $\mathrm{L}(s, \pi \times \pi')$), the product of the $\xi(s, \pi_i \times \pi'_j)$ (resp. $\mathrm{L}(s, \pi_i \times \pi'_j)$). Let us furthermore consider the following condition on a test function F:

(T4) The inequality $\Re\ \Phi_F(s) \geq 0$ holds for every $s \in \mathbb{C}$ with $0 \leq \Re\, s \leq 1$.

Recall that $\Phi_F(s)$ is a real number for every real number s in $[0, 1]$; in particular, it is nonnegative if F satisfies Condition (T4). In this book, we will only use the following corollary of the explicit formula.

Corollary 9.3.10. *Let F be a test function, and let $\pi, \pi' \in \mathrm{K}$ be effective. Suppose that F satisfies Condition* (T4). *Then we have the following equality:*

$$\mathrm{B}^F(\pi,\pi') \leq \Phi_F(0)\,\delta(\pi,\pi') - \frac{1}{2}\,\Phi_F\Big(\frac{1}{2}\Big)\ \mathrm{ord}_{s=1/2}\xi(s, \pi^\vee \times \pi') .$$

Proof. Under Condition (T4) on F, we have the equality

$$\mathrm{Z}^F(\pi,\pi') = \Re\,\mathrm{Z}^F(\pi,\pi') \geq \Phi_F\Big(\frac{1}{2}\Big)\ \mathrm{ord}_{s=1/2}\xi(s, \pi^\vee \times \pi') .$$

We conclude using Proposition 9.3.9. □

Remark 9.3.11. For $V \in \mathrm{K}_\infty$, we see that the meromorphic function $\Gamma(s, V)$ is finite and nonzero at $s = 1/2$. This implies, for all effective $\pi, \pi' \in \mathrm{K}$, the equality $\mathrm{ord}_{s=1/2}\xi(s, \pi \times \pi') = \mathrm{ord}_{s=1/2}\mathrm{L}(s, \pi \times \pi')$. The corollary above therefore also holds if we replace $\xi(s, \pi^\vee \times \pi')$ by $\mathrm{L}(s, \pi^\vee \times \pi')$.

The vanishing order of the functions $\xi(s, \pi \times \pi')$ at $s = 1/2$ is known to be mysterious. It is traditionally bounded below in the following way. For $\pi, \pi' \in \Pi_{\mathrm{alg}}$, set $\mathrm{e}^{\perp}(\pi, \pi') = 1$ if π and π' are both self-dual and satisfy $\varepsilon(\pi \times \pi') = -1$; otherwise, set $\mathrm{e}^{\perp}(\pi, \pi') = 0$. The function $\mathrm{e}^{\perp}\colon \Pi_{\mathrm{alg}} \times \Pi_{\mathrm{alg}} \to \mathbb{Z}$ extends to a symmetric bilinear form $\mathrm{e}^{\perp}\colon \mathrm{K} \times \mathrm{K} \to \mathbb{Z}$. Hence, for all effective $\pi, \pi' \in \mathrm{K}$, we have the equality $\mathrm{ord}_{s=1/2}\, \xi(s, \pi \times \pi') \geq \mathrm{e}^{\perp}(\pi, \pi')$.

Corollary 9.3.12. *Let F be a test function and $\pi, \pi' \in \mathrm{K}$ be effective. Suppose that F satisfies Condition* (T4) *and $F \geq 0$. Set*

$$\mathrm{C}^F(\pi, \pi') = \Phi_F(0)\, \delta(\pi, \pi') - \frac{1}{2}\, \Phi_F\Big(\frac{1}{2}\Big)\, \mathrm{e}^{\perp}(\pi, \pi') - \mathrm{B}^F_{\infty}(\pi, \pi')\,.$$

We have the following inequalities:

(i) $\mathrm{C}^F(\pi, \pi) \geq 0$ *(in particular, we have* $\mathrm{B}^F_{\infty}(\pi, \pi) \leq \Phi_F(0)\, \delta(\pi, \pi)$*),*
(ii) $\mathrm{C}^F(\pi, \pi') + \sqrt{\mathrm{C}^F(\pi, \pi)\mathrm{C}^F(\pi', \pi')} \geq 0$.

Proof. Corollary 9.3.10 and the discussion above show that under Condition (T4) on F, we have $\mathrm{B}^F_f(\pi, \pi') \leq \mathrm{C}^F(\pi, \pi')$ for all effective $\pi, \pi' \in \mathrm{K}$. Note that the positivity assumption on the function F implies the positivity of the bilinear form B^F_f on K. Indeed, an element $\varpi \in \mathrm{K} \otimes \mathbb{R}$ can be written in the form $\varpi = \sum_i \lambda_i \pi_i$, where the π_i are in Π_{alg} and the λ_i are in $\mathbb{R}$. The identity

$$\mathrm{B}^F_f(\varpi, \varpi) = \sum_{p,k} F(k \log p)\, \frac{\log p}{p^{k/2}}\, \Big| \sum_i \lambda_i\, \mathrm{tr}\, \mathrm{c}(\pi_i)^k \Big|^2$$

therefore implies $\mathrm{B}^F_f(\varpi, \varpi) \geq 0$. Assertion (ii) follows from the Cauchy–Schwarz inequality applied to B^F_f, which gives

$$|\mathrm{B}^F_f(\pi, \pi')| \leq \sqrt{\mathrm{B}^F_f(\pi, \pi)\, \mathrm{B}^F_f(\pi', \pi')}\,,$$

and from the obvious inequality $|\mathrm{B}^F_f(\pi, \pi')| \geq -\mathrm{B}^F_f(\pi, \pi')$. (One should be aware that the bilinear form C^F is, a priori, only positive on the effective elements of K, whence the formulation of assertion (ii).) □

In [147], the inequality (i) of Corollary 9.3.12 is used in the particular case $\pi \in \Pi_{\mathrm{alg}}$. For an arbitrary effective π, this inequality implies Corollary 9.3.14 below, first observed by Olivier Taïbi. Inequality (ii) seems new; it will be of great use in the applications.

Definition 9.3.13. Let $V \in \mathrm{K}_{\infty}$. Denote by $\mathrm{m}(V)$ the number of representations $\pi \in \Pi_{\mathrm{alg}}$ that satisfy $\mathrm{L}(\pi_{\infty}) \simeq V$; by Harish-Chandra, we have $\mathrm{m}(V) < +\infty$ (Sect. 4.3.2). Moreover, denote by $\mathrm{m}^{\perp}(V)$ the number of self-dual representations $\pi \in \Pi_{\mathrm{alg}}$ that satisfy $\mathrm{L}(\pi_{\infty}) \simeq V$. We have the inequality $\mathrm{m}^{\perp}(V) \leq \mathrm{m}(V)$.

An element V of K_∞ is called effective if it is the class of a finite-dimensional, continuous representation of $\mathrm{W}_{\mathbb{R}}^{\mathrm{alg}}$ with coefficients in $\mathbb{C}$. It is clear that if $\mathrm{m}(V) \neq 0$, then V is effective.

Corollary 9.3.14 (Taïbi). *Let $V \in \mathrm{K}_\infty$ be effective and F a test function, supposed nonnegative and satisfying condition* (T4). *We have the inequality*

$$\mathrm{m}(V)\ \mathrm{B}_\infty^F(V,V) \leq \Phi_F(0)\ .$$

Proof. If $\mathrm{m}(V) = 0$, there is nothing to prove. Therefore, suppose that there exist an integer $r \geq 1$ and distinct representations $\pi_1, \ldots, \pi_r \in \Pi_{\mathrm{alg}}$ such that $\mathrm{L}((\pi_i)_\infty) \simeq V$ for every i. Apply part (i) of Corollary 9.3.12 to the element $\pi = \pi_1 + \pi_2 + \cdots + \pi_r$ of K. On the one hand, we have the equalities

$$\mathrm{B}_\infty^F(\pi,\pi)\ =\ \mathrm{B}_\infty^F(rV,rV)\ =\ r^2\,\mathrm{B}_\infty^F(V,V)$$

and, on the other hand, we have $\delta(\pi,\pi) = r$. The inequality $r^2\ \mathrm{B}_\infty^F(V,V) \leq r\,\Phi_F(0)$ follows, and therefore also $r\,\mathrm{B}_\infty^F(V,V) \leq \Phi_F(0)$. □

Note that under the additional assumption $\mathrm{B}_\infty^F(V,V) > 0$, Corollary 9.3.14 provides an explicit upper bound for $\mathrm{m}(V)$. In particular, it re-proves the result $\mathrm{m}(V) < \infty$ of Harish-Chandra (Sect. 4.3.2) mentioned above.

Corollary 9.3.15. *Let $V, V' \in \mathrm{K}_\infty$ be effective and F a nonnegative test function satisfying Condition* (T4). *Suppose $V \neq V'$ and $\mathrm{m}(V)\,\mathrm{m}(V') \neq 0$.*

(i) *If we set*

$$\mathrm{n}(V,V') = \frac{\mathrm{m}^\perp(V)\,\mathrm{m}^\perp(V')}{4\,\mathrm{m}(V)\,\mathrm{m}(V')}(1-\varepsilon(V\otimes V'))\ ,$$

we have the inequality

$$\mathrm{n}(V,V')\,\Phi_F\Big(\frac{1}{2}\Big) + \mathrm{B}_\infty^F(V,V') \leq \sqrt{\left(\frac{\Phi_F(0)}{\mathrm{m}(V)} - \mathrm{B}_\infty^F(V,V)\right)\left(\frac{\Phi_F(0)}{\mathrm{m}(V')} - \mathrm{B}_\infty^F(V',V')\right)}\ .$$

(ii) *If we, moreover, have $\mathrm{m}^\perp(V)\mathrm{m}^\perp(V') \neq 0$, then we also have the inequality*

$$\frac{1-\varepsilon(V\otimes V')}{4}\,\Phi_F\Big(\frac{1}{2}\Big) + \mathrm{B}_\infty^F(V,V') \leq \sqrt{\left(\frac{\Phi_F(0)}{\mathrm{m}^\perp(V)} - \mathrm{B}_\infty^F(V,V)\right)\left(\frac{\Phi_F(0)}{\mathrm{m}^\perp(V')} - \mathrm{B}_\infty^F(V',V')\right)}\ .$$

Proof. Let r, r' be integers ≥ 1 and $\pi_1, \ldots, \pi_r, \pi'_1, \ldots, \pi'_{r'} \in \Pi_{\mathrm{alg}}$ distinct representations such that $\mathrm{L}(\pi_i) = V$ for every i and $\mathrm{L}(\pi'_j) = V'$ for every j. Set $\pi = \sum_i \pi_i$ and $\pi' = \sum_j \pi'_j$. We have $\delta(\pi,\pi') = 0$ because $V \neq V'$. Denote by

$s \leq r$ (resp. $s' \leq r'$) the number of self-dual representations among the π_i (resp. π'_j). We have the obvious equality

$$\mathrm{e}^{\perp}(\pi, \pi') = s\, s' \, \frac{1 - \varepsilon(V \otimes V')}{2}.$$

Moreover, we have

$$\mathrm{C}^F(\pi, \pi') + \frac{\mathrm{e}^{\perp}(\pi, \pi')}{2} \Phi_F\Big(\frac{1}{2}\Big) = -\mathrm{B}^F_\infty(rV, r'V') = -rr'\,\mathrm{B}^F_\infty(V, V').$$

Likewise, it is easy to see that we have the inequalities $\mathrm{C}^F(\pi, \pi) \leq r\,\Phi_F(0) - r^2\,\mathrm{B}^F_\infty(V, V)$ and $\mathrm{C}^F(\pi', \pi') \leq r'\,\Phi_F(0) - (r')^2\,\mathrm{B}^F_\infty(V', V')$. Dividing the inequality given in part (iii) of Corollary 9.3.12 by $rr' \neq 0$ leads to the inequality

$$\frac{ss'}{4rr'}(1 - \varepsilon(V \otimes V'))\,\Phi_F\Big(\frac{1}{2}\Big) + \mathrm{B}^F_\infty(V, V') \\ \leq \sqrt{\Big(\frac{\Phi_F(0)}{r} - \mathrm{B}^F_\infty(V, V)\Big)\Big(\frac{\Phi_F(0)}{r'} - \mathrm{B}^F_\infty(V', V')\Big)}.$$

This inequality holds for all integers r and r' with $1 \leq r \leq \mathrm{m}(V)$ and $1 \leq r' \leq \mathrm{m}(V')$, where s and s' are defined as above. Assertion (i) follows from the special case $(r, r', s, s') = (\mathrm{m}(V), \mathrm{m}(V'), \mathrm{m}^{\perp}(V), \mathrm{m}^{\perp}(V'))$, and assertion (ii) from the case $(r, r', s, s') = (\mathrm{m}(V)^{\perp}, \mathrm{m}^{\perp}(V'), \mathrm{m}^{\perp}(V), \mathrm{m}^{\perp}(V'))$. □

As we will see, the corollary typically makes it possible to show that if $V \neq V'$, the existence of certain elements $\pi \in \Pi_{\mathrm{alg}}$ such that $\mathrm{L}(\pi_\infty) = V$ implies the nonexistence of elements $\pi' \in \Pi_{\mathrm{alg}}$ such that $\mathrm{L}(\pi'_\infty) = V'$. It admits several refinements which we will come back to in Sect. 9.3.29. Let us conclude with a simple well-known criterion [165] that allows the construction of test functions satisfying Condition (T4); we repeat the argument for the comfort of the reader.

Lemma 9.3.16. *Let $g\colon \mathbb{R} \to \mathbb{R}$ be an even, integrable, and square-integrable function. Its Fourier transform $\widehat{g}$ is well defined and real-valued. Suppose $\widehat{g} \geq 0$ and consider the function $F\colon \mathbb{R} \to \mathbb{R}$ defined by*

$$F(x) = \frac{g(x)}{\cosh(x/2)}.$$

Then $\Phi_F(s)$ is well defined for every $s \in \mathbb{C}$ such that $0 \leq \Re s \leq 1$ (formula (9.3.1)), and in this region, we have $\Re\,\Phi_F(s) \geq 0$.

Our convention for the Fourier transform of an integrable function g is $\widehat{g}(y) = \int_{\mathbb{R}} g(x)\, e^{-2i\pi xy}\, \mathrm{d}y$, where $y \in \mathbb{R}$.

Proof. Consider $y \in \mathbb{C}$ with $|\Im\, y| < 1/2$; we have the equalities

$$\Phi_F\Big(\frac{1}{2} + iy\Big) = 2\pi \int_{\mathbb{R}} g(2\pi x)\, \frac{e^{2i\pi xy}}{\cosh \pi x}\, \mathrm{d}x = \int_{\mathbb{R}} \frac{\widehat{g}(x/2\pi)}{\cosh \pi(x-y)}\, \mathrm{d}x\, .$$

The first is trivial for every $y \in \mathbb{C}$. The second is, for example, an application of Plancherel's formula, because the functions $x \mapsto g(2\pi x)$ and $x \mapsto e^{2i\pi xy}/\cosh\pi x$ are integrable and square-integrable over $\mathbb{R}$, and because the function $1/\cosh \pi x$ is equal to its Fourier transform. This suffices to conclude because, on the one hand, we have $\widehat{g}(z) \geq 0$ for every $z \in \mathbb{R}$ by assumption and, on the other hand, we see that we have the inequality $\Re(1/\cosh \pi z) > 0$ for every $z \in \mathbb{C}$ such that $|\Im\, z\,| < 1/2$.

9.3.17 Odlyzko's Function

Following Odlyzko [165, Sect. 3], consider the function $u\colon \mathbb{R} \to \mathbb{R}$ defined by $u(x) = \cos(\pi x)$ if $|x| \leq 1/2$ and $u(x) = 0$ otherwise, and denote by g twice the convolution of u with itself, that is $g = 2u * u$. Concretely, the function g is zero outside the segment $[-1, 1]$ and for $|x| \leq 1$, it is given by the formula

$$g(x) = (1 - |x|)\cos(\pi x) + \frac{1}{\pi}\sin(\pi|x|)\, .$$

One immediately verifies that g is an even positive $\mathcal{C}^2$-function with compact support and that we have $g(0) = 1$. Its Fourier transform, namely $2\,\widehat{u}^2$, is clearly positive, because u is real, even, and integrable. In particular, for every $\lambda \in \mathbb{R}_{>0}$, the function $\mathrm{F}_\lambda\colon \mathbb{R} \to \mathbb{R}$ defined by

$$\mathrm{F}_\lambda(x) = g(x/\lambda)/\cosh(x/2)$$

is a nonnegative test function satisfying Condition (T4) (Lemma 9.3.16).

Let us explain how to evaluate the linear form $\mathrm{J}_{\mathrm{F}_\lambda}$ on K_∞ numerically. Recall that $\psi(z) = \Gamma'(z)/\Gamma(z)$ denotes the digamma function. Also set, respectively for $z \in \mathbb{C} - \mathrm{N}$ and $z \in \mathbb{C} - \{\pm i\pi\}$,

$$\phi(z) = \frac{1}{2}\psi\Big(\frac{z+1}{2}\Big) - \frac{1}{2}\psi\Big(\frac{z}{2}\Big) \quad \text{and} \quad \mathrm{r}(z) = 2\pi^2 \frac{e^{-z}}{(z^2+\pi^2)^2}\, .$$

Note that we have the formula $\phi(z) = \sum_{n \geq 0} (-1)^n (z+n)^{-1}$. We thank Henri Cohen for putting us on the right track for the following proposition.

Proposition 9.3.18. *Let $\lambda > 0$ be a real number.*

(i) *For every integer $w \geq 0$, we have the equality*

$$\mathrm{J}_{\mathrm{F}_\lambda}(\mathrm{I}_w) = \log \pi \; - \; \Re\psi\Big(b + \frac{i\pi}{2\lambda}\Big) + \frac{1}{\pi}\Im\psi\Big(b + \frac{i\pi}{2\lambda}\Big) - \frac{1}{2\lambda}\Re\psi'\Big(b + \frac{i\pi}{2\lambda}\Big) + r_1(w,\lambda)\,,$$

with $b = 1/2 + w/4$ and $r_1(w,\lambda) = 2\lambda \sum_{n=0}^{\infty} \mathrm{r}(2\lambda(b+n))$.

(ii) *Moreover, we have the equality*

$$\mathrm{J}_{\mathrm{F}_\lambda}(1 - \epsilon_{\mathbb{C}/\mathbb{R}}) = 1 + \frac{2\pi}{\lambda}\Im\phi\Big(1 + \frac{i\pi}{\lambda}\Big) + \frac{2\pi}{\lambda^2}\Im\phi'\Big(1 + \frac{i\pi}{\lambda}\Big) + r_2(\lambda)\,,$$

with $r_2(\lambda) = 2\lambda \sum_{n=1}^{\infty} (-1)^{n+1}\, n\, \mathrm{r}(\lambda n)$.

(iii) *Finally, we have $\Phi_{\mathrm{F}_\lambda}(0) = (8/\pi^2)\lambda$ and*

$$\Phi_{\mathrm{F}_\lambda}\Big(\frac{1}{2}\Big) = 4\,\Re\phi\Big(\frac{1}{2} + \frac{i\pi}{\lambda}\Big) - \frac{4}{\pi}\Im\phi\Big(\frac{1}{2} + \frac{i\pi}{\lambda}\Big) + \frac{4}{\lambda}\Re\phi'\Big(\frac{1}{2} + \frac{i\pi}{\lambda}\Big) + r_3(\lambda)\,,$$

with $r_3(\lambda) = 4\lambda \sum_{n=0}^{\infty} (-1)^n\, \mathrm{r}(\lambda(n+1/2))$.

Proof. If $\alpha > 0$, set $h(\alpha) = \int_0^1 g(x)\, e^{-\alpha x} \mathrm{d}x$. Using, for example, the definition $g = 2\, u * u$, we first easily verify the identity $g''(x) + \pi^2 g(x) = 2\pi|\sin \pi x|$ for $|x| \leq 1$, and then the relation

$$h(\alpha) \;=\; \frac{\alpha}{\alpha^2 + \pi^2} + 2\pi^2\, \frac{1 + e^{-\alpha}}{(\alpha^2 + \pi^2)^2}\,.$$

We, moreover, have $\int_0^\infty g(x/\lambda)\, e^{-\alpha x}\, \mathrm{d}x = \lambda\, h(\lambda\, \alpha)$.

Let $w \geq 0$ be an integer; set $b = 1/2 + w/4$. Proposition 9.3.8 applied to the function $F = \mathrm{F}_\lambda$ can be written as $\mathrm{J}_\lambda(\mathrm{I}_w) = \log(2\pi) + \sigma_{\mathrm{F}_\lambda}(1, w/2)$. Elementary manipulations lead to the relation

$$\sigma_{\mathrm{F}_\lambda}\Big(1, \frac{w}{2}\Big) \;=\; \int_0^\infty \left(\frac{2e^{-2bx}}{1 - e^{-2x}} - \frac{e^{-x}}{x}\right) \mathrm{d}x + \int_0^\infty \left(g\Big(\frac{x}{\lambda}\Big) - 1\right) \frac{2e^{-2bx}}{1 - e^{-2x}}\, \mathrm{d}x\,.$$

Let us write this sum in the obvious form $\sigma_\lambda(1, w/2) = S_1 + S_2$. Gauss' formula

$$\psi(z) = -\int_0^\infty \left(\frac{e^{-zx}}{1 - e^{-x}} - \frac{e^{-x}}{x}\right) \mathrm{d}x$$

and the identity

$$\int_0^\infty \frac{e^{-\alpha x} - e^{-x}}{x} \mathrm{d}x = -\log \alpha$$

for $\alpha > 0$ imply the equality $S_1 = -\log 2 - \psi(b)$. On the other hand, the expansion

$$\frac{e^{-2bx}}{1-e^{-2x}} = \sum_{n\geq 0} e^{-2(b+n)x}$$

leads to the relation

$$S_2 = 2\lambda \sum_{n=0}^{\infty} \left(h(2\lambda(b+n)) - \frac{1}{2\lambda(b+n)} \right).$$

We have $h = h_1 + h_2 + \mathrm{r}$ with $h_1(\alpha) = \alpha(\alpha^2 + \pi^2)^{-1}$ and $h_2(\alpha) = 2\pi^2(\alpha^2+\pi^2)^{-2}$. Moreover, if u and v are nonzero real numbers, we have the identities $u/(u^2+v^2) - 1/u = \Re\,(1/(u+vi) - 1/u)$ and

$$\frac{2v}{(u^2+v^2)^2} = -\frac{1}{v^2}\Im\left(\frac{1}{u+vi} - \frac{1}{u}\right) - \frac{1}{v}\Re\frac{1}{(u+vi)^2}.$$

Let us apply these to $v = \pi/2\lambda$ and $u = b+n$ for every integer $n \geq 0$ and take the sum. The formula $\psi(b) - \psi(b+z) = \sum_{n\geq 0} 1/(b+z+n) - 1/(b+n)$ with $z = i\pi/2\lambda$ then implies

$$\begin{aligned}\sum_{n\geq 0} 2\lambda\; h_1(\,2\lambda(b+n)\,) - \frac{1}{b+n} &= \sum_{n\geq 0} \frac{(b+n)}{(b+n)^2 + (\pi/2\lambda)^2} - \frac{1}{b+n}\\ &= \psi(b) - \Re\,\psi(b+i\pi/2\lambda)\,.\end{aligned}$$

Likewise, in view of the identity $\psi'(z) = \sum_{n\geq 0} 1/(z+n)^2$, we obtain

$$\sum_{n\geq 0} 2\lambda\; h_2(\,2\lambda(b+n)\,) = \frac{1}{\pi}\Im\,\psi\left(b + \frac{i\pi}{2\lambda}\right) - \frac{1}{2\lambda}\Re\,\psi'\left(b+\frac{i\pi}{2\lambda}\right).$$

Putting all these formulas end-to-end gives assertion (i). The proof of assertion (ii) is similar. From Proposition 9.3.8, we begin by establishing the equality

$$\mathrm{J}_{\mathrm{F}_\lambda}(1-\epsilon_{\mathbb{C}/\mathbb{R}}) = 1 + \int_0^{\infty} \left(g\left(\frac{x}{2\lambda}\right) - 1\right) \frac{e^{-x/2}}{(1\;+\;e^{-x/2})^2}\,dx\,.$$

The expansion $e^{-x/2}(1+e^{-x/2})^{-2} = \sum_{n\geq 1} (-1)^{n+1}\, n\, e^{-nx/2}$ allows us to write

$$\mathrm{J}_{\mathrm{F}_\lambda}(1-\epsilon_{\mathbb{C}/\mathbb{R}}) = 1 + 2\sum_{n\geq 1} (-1)^{n+1}\,(\lambda n\, h(\lambda n) - 1)\,.$$

Moreover, if u and v are nonzero real numbers, we have the identities

$$\frac{u^2}{u^2+v^2} - 1 = v\,\Im\,\frac{1}{u+vi} \quad \text{and} \quad \frac{2\,u\,v^2}{(u^2+v^2)^2} = -v\,\Im\,\frac{1}{(u+vi)^2}.$$

By setting $u = n\lambda$ and $v = \pi/\lambda$ and noting that we have $\phi(z+1) = \sum_{n\geq 1}(-1)^{n+1}(z+n)^{-1}$, we find

$$2\sum_{n\geq 1}(-1)^{n+1}\,(\lambda n\, h_1(\lambda n)\ -\ 1) = \frac{2\pi}{\lambda}\,\Im\,\phi\Big(1+\frac{i\pi}{\lambda}\Big)\,,$$

and then $2\sum_{n\geq 1}\,(-1)^{n+1}\ \lambda n\, h_2(\lambda n) = (2\pi/\lambda^2)\,\Im\,\phi'(1+i\pi/\lambda)$. This proves part (ii). It remains to verify assertion (iii). By the definition of F_λ and g, we have the equalities $\Phi_{\mathrm{F}_\lambda}(0) = \int_{\mathbb{R}} g(x/\lambda)\mathrm{d}\,x = 2\lambda\,\widehat{u}(0)^2$; the value of $\Phi_{\mathrm{F}_\lambda}(0)$ therefore follows from the immediate relation $\widehat{u}(0) = 2/\pi$. To determine $\Phi_{\mathrm{F}_\lambda}(1/2)$, we proceed as for assertions (i) and (ii) from the immediate identities

$$\Phi_{\mathrm{F}_\lambda}\Big(\frac{1}{2}\Big) = 4\int_0^\infty g(x/\lambda)\frac{e^{-x/2}}{1+e^{-x}}\mathrm{d}x = 4\lambda\sum_{n\geq 0}(-1)^n\, h(\lambda(n+1/2))\,. \qquad \square$$

Comments on the Numerical Computations Carried Out in the Next Subsections

(1) The formulas of Proposition 9.3.18, although not very aesthetic, are very effective for evaluating $\mathrm{J}_{\mathrm{F}_\lambda}$ and $\Phi_{\mathrm{F}_\lambda}(1/2)$ numerically, and this with an arbitrary accuracy. In our applications, we will have $0 \leq w \leq 46$ and $\lambda = \log N$ with $2 \leq N \leq 100$.

(2) John L. Spouge elaborated, in 1994, a remarkable algorithm to determine the values of the gamma, ψ (digamma), and ψ' (trigamma) functions [190]. These functions are implemented in PARI [160]. However, Henri Cohen has let us know that for its computations, PARI uses the Euler–MacLaurin formula, and therefore the Bernoulli numbers, and that after computing a first value, the computation of the following ones is sped up by storing Bernoulli numbers.

(3) The three functions $r_1(w,\lambda)$, $r_2(\lambda)$, and $r_3(\lambda)$ that appear in Proposition 9.3.18 are defined as sums of series; below, we estimate the "tails" $\sum_{n=N+1}^{\infty}$ of these series.

(3.1) We have the inequality

$$0\ \leq\ r_1(w,\lambda) - 2\lambda\sum_{n=0}^{N}\mathrm{r}(2\lambda(b+n))\ \leq\ \frac{2\lambda\mathrm{r}(2\lambda(b+N+1))}{1-e^{-2\lambda}}\,;$$

this inequality follows from the fact that we have

$$0\ \leq\ \mathrm{r}(2\lambda(b+n))\ \leq\ e^{-2\lambda(n-N-1)}\,\mathrm{r}(2\lambda(b+N+1))$$

for every n with $n \geq N+1$.

(3.2) Since the function in one real variable $x \mapsto x\,\mathrm{r}(x)$ has positive values and is decreasing for $x \geq 0.773$, we have the inequality

$$0 \;\leq\; (-1)^{N}\left(r_2(\lambda) - 2\lambda\sum_{n=1}^{N}(-1)^{n+1}n\,\mathrm{r}(\lambda n)\right) \;\leq\; 2\lambda(N+1)\mathrm{r}(\lambda(N+1))$$

under the assumption $\lambda(N+1) \geq 0.773$.

(3.2) Since the function in one real variable $x \mapsto \mathrm{r}(x)$ has positive values and is decreasing, we have the inequality

$$0 \;\leq\; (-1)^{N+1}\left(r_3(\lambda) - 4\lambda\sum_{n=0}^{N}(-1)^{n}\,\mathrm{r}\Big(\lambda\Big(n+\frac{1}{2}\Big)\Big)\right) \;\leq\; 4\lambda\,\mathrm{r}\Big(\lambda\Big(N+\frac{3}{2}\Big)\Big)\,.$$

(4) The computations carried out using the formulas of Proposition 9.3.18 are confirmed, with arbitrary accuracy, by the numerical integration routines in PARI.

(5) The function $w \mapsto \mathrm{J}_{\mathrm{F}_\lambda}(\mathrm{I}_w)$, with λ fixed, is decreasing. This follows by invoking part (iii) of Proposition-Definition 9.3.8, which shows that under the assumptions $F \geq 0$ and $w' \geq w$, the difference $\mathrm{J}_F(\mathrm{I}_w) - \mathrm{J}_F(\mathrm{I}_{w'})$ is the integral over $[0, +\infty[$ of a nonnegative function.

Likewise, the function $\lambda \mapsto \mathrm{J}_{\mathrm{F}_\lambda}(\mathrm{I}_w)$, with w fixed, is increasing. Indeed, since Odlyzko's function is decreasing on $[0, +\infty[$, part (iii) of Proposition-Definition 9.3.8 shows that under the assumption $\lambda \leq \lambda'$, the difference $\mathrm{J}_{\mathrm{F}_{\lambda'}}(\mathrm{I}_w) - \mathrm{J}_{\mathrm{F}_\lambda}(\mathrm{I}_w)$ is the integral over $[0, +\infty[$ of a nonnegative function (note that we have $\mathrm{F}_\lambda(0) = 1$ for every λ).

The above implies that we have the bounds $-1.40 \leq \mathrm{J}_{\mathrm{F}_\lambda}(\mathrm{I}_w) \leq 2.63$ for $\log 2 \leq \lambda \leq \log 100$ and $0 \leq w \leq 50$. Figure 9.1 shows the graph of the function $w \mapsto \mathrm{J}_{\mathrm{F}_\lambda}(\mathrm{I}_w)$ for several values of the parameter λ.

(6) In the appendix of [144], Mestre describes a computation of $\sigma_{\mathrm{F}_\lambda}(1, 1/2)$ (in other words, $\mathrm{J}_{\mathrm{F}_\lambda}(\mathrm{I}_1) - \log 2\pi$) for $\lambda < \pi$ using a method that is completely different from ours; the restriction $\lambda < \pi$ comes from the fact that Mestre implicitly uses the expansion of the holomorphic function $z/(e^z - 1)$ as a power series in the disk $|z| < 2\pi$.

9.3.19 Beginning of the Proof of Theorem 9.3.2: The Case $w \leq 20$

For every integer $w \geq 0$, consider the subgroup of K_∞ defined as follows:

$$\mathrm{K}_\infty^{\leq w} = \begin{cases} \left(\bigoplus_{1\leq j\leq w/2}\mathbb{Z}\,\mathrm{I}_{2j}\right) \oplus \mathbb{Z}1 \oplus \mathbb{Z}\epsilon_{\mathbb{C}/\mathbb{R}} & \text{if } w \equiv 0 \bmod 2\,, \\ \bigoplus_{1\leq j\leq (w+1)/2}\mathbb{Z}\,\mathrm{I}_{2j-1} & \text{if } w \equiv 1 \bmod 2\,. \end{cases}$$

The interest of this definition is that if $\pi \in \Pi_{\mathrm{alg}}(\mathrm{PGL}_n)$ has motivic weight w, we have $\mathrm{L}(\pi_\infty) \in \mathrm{K}_\infty^{\leq w}$ (Proposition 8.2.13). We have two more simple constraints on

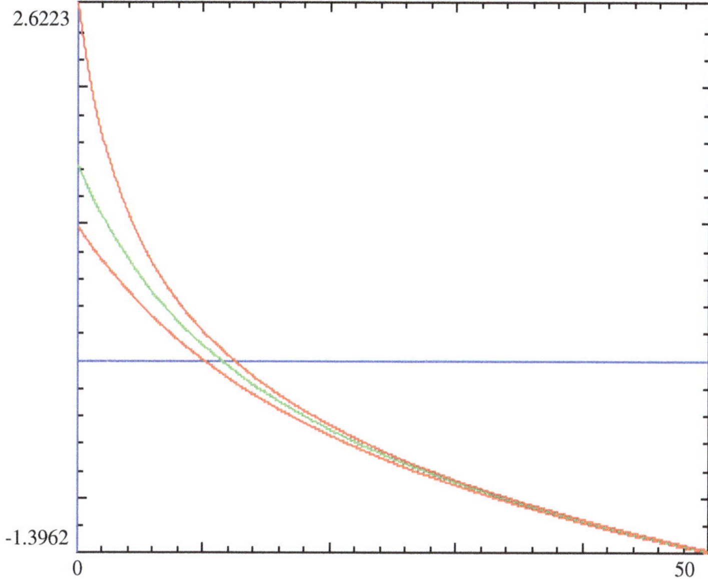

Fig. 9.1 Graph of the map $w \mapsto \mathrm{J}_{\mathrm{F}_\lambda}(\mathrm{I}_w)$ on the interval $0 \le w \le 50$, for $\lambda = \log 2$, $\log 3$ and $\log 100$

$\mathrm{L}(\pi_\infty)$. On the one hand, it is an effective element of K_∞. On the other hand, we have the relation $\det \mathrm{L}(\pi_\infty) = 1$. Recall the identity $\det \mathrm{I}_v = \epsilon_{\mathbb{C}/\mathbb{R}}^{v+1}$ for every $v \ge 0$.

The general principle of the proof will consist in showing that under a suitable assumption on π_∞, there exist no—or few—representations $\pi \in \Pi_{\mathrm{alg}}$ of motivic weight at most 22, using the inequalities given by Proposition 9.3.9 applied to the test functions introduced in Sect. 9.3.17. In this entire subsection, λ will therefore denote a positive real number, and we consider the associated test function F_λ defined in Sect. 9.3.17. In the proof, we will systematically use the bilinear form $\mathrm{B}_\infty^{\mathrm{F}_\lambda}$ on K_∞ introduced in Sect. 9.3.5. We will, in particular, need to evaluate this bilinear form explicitly, which we will do, of course, with the help of a computer, using the formulas described in Sect. 9.3.17. We refer the reader to the source code [54] for a justification of the numerical computations we carry out below. In order to acquire an intuition for this method, we begin by proving the case $n = 2$ of the theorem. Since the trivial representation is the only element of $\Pi(\mathrm{PGL}_1) = \Pi_{\mathrm{cusp}}(\mathrm{PGL}_1)$, note that there is nothing to prove for $n = 1$!

The Case $n = 2$ or $w \le 10$

Suppose given a representation $\pi \in \Pi_{\mathrm{alg}}(\mathrm{PGL}_2)$ of motivic weight w. The condition $\det \mathrm{L}(\pi_\infty) = 1$ shows that there are two cases: either w is odd and $\mathrm{L}(\pi_\infty) = \mathrm{I}_w$, or $w = 0$ and we have $\mathrm{L}(\pi_\infty) = 2 \cdot 1$ or $\mathrm{L}(\pi_\infty) = 2 \cdot \epsilon_{\mathbb{C}/\mathbb{R}}$.

If w is odd, Proposition 9.1.5 asserts that π is the representation generated by a unique normalized modular eigenform of weight $w+1$ for the group $\mathrm{SL}_2(\mathbb{Z})$; conversely, every such form obviously generates such a π. In other words, in view of Definition 9.3.13, for every odd $w \geq 1$, we have the equality

$$\mathrm{m}(\mathrm{I}_w) = \dim \mathrm{S}_{w+1}(\mathrm{SL}_2(\mathbb{Z})) .$$

Hence, the particular case of Theorem 9.3.2 where $n = 2$ and the motivic weight is nonzero is a consequence of the well-known description of $\mathrm{S}_{w+1}(\mathrm{SL}_2(\mathbb{Z}))$ [177, Chap. VII]. Since such descriptions of $\Pi_{\mathrm{alg}}(\mathrm{PGL}_n)$ do not exist in dimension $n > 2$, it is in our interest to explain how to proceed differently, if possible. In what follows, we propose to give another proof that $\mathrm{S}_{w+1}(\mathrm{SL}_2(\mathbb{Z}))$ is zero if $w < 11$ or $w = 13$, and of dimension at most 1 if $15 \leq w \leq 21$, using Corollary 9.3.10.

Since the function F_λ is continuous and with support in $[-\lambda, \lambda]$, we see that for all $\pi, \pi' \in \Pi_{\mathrm{alg}}$, we have

$$\widetilde{\mathrm{B}}_f^{\mathrm{F}_\lambda}(\pi, \pi') = \sum_{p^k < e^\lambda} \mathrm{F}_\lambda(k \log p) \frac{\log p}{p^{k/2}} \, \overline{\mathrm{tr}\,(\mathrm{c}_p(\pi)^k)} \, \mathrm{tr}\,(\mathrm{c}_p(\pi')^k) . \qquad (9.3.2)$$

In particular, we have $\mathrm{B}_f^{\mathrm{F}_\lambda} = 0$, that is, $\mathrm{B}^{\mathrm{F}_\lambda} = \mathrm{B}_\infty^{\mathrm{F}_\lambda}$, for $\lambda \leq \log 2$. Now, consider $\pi \in \Pi_{\mathrm{alg}}(\mathrm{PGL}_2)$ of odd motivic weight w and $\pi' = 1 \in \Pi_{\mathrm{alg}}(\mathrm{PGL}_1)$, in which case we have $\mathrm{B}_\infty^{\mathrm{F}_\lambda}(\pi, 1) = \mathrm{J}_{\mathrm{F}_\lambda}(\mathrm{I}_w)$. The inequality of Corollary 9.3.10 therefore implies, in particular,

$$\mathrm{J}_{\mathrm{F}_{\log 2}}(\mathrm{I}_w) \leq 0 .$$

Table 9.3 gives the numerical evaluation, up to 10^{-2}, of $\mathrm{J}_{\mathrm{F}_{\log 2}}(\mathrm{I}_w)$, for $1 \leq w \leq 21$ odd. It contradicts the inequality above for $w < 11$: the representation π does not exist. Let us emphasize that this argument is not new: it is, for example, exactly the method used by Mestre in [144]. Its accuracy is rather surprising, because we already know that we have $\mathrm{S}_{12}(\mathrm{SL}_2(\mathbb{Z})) \neq 0$. Further on, we will see many other examples of the fascinating accuracy of the explicit formulas.

Table 9.3 Values of $\mathrm{J}_{\mathrm{F}_{\log 2}}(\mathrm{I}_w)$ for $1 \leq w \leq 21$ odd, up to 10^{-2}

w	$\mathrm{J}_{\mathrm{F}_{\log 2}}(\mathrm{I}_w)$	w	$\mathrm{J}_{\mathrm{F}_{\log 2}}(\mathrm{I}_w)$	w	$\mathrm{J}_{\mathrm{F}_{\log 2}}(\mathrm{I}_w)$
1	0.85	9	0.07	17	−0.40
3	0.61	11	−0.06	19	−0.50
5	0.41	13	−0.19	21	−0.59
7	0.23	15	−0.30		

As Mestre essentially remarks [144, Remarque 1, Sect. III], the argument above has a wider reach: if $\pi \in \Pi_{\mathrm{alg}}(\mathrm{PGL}_n)$ is of motivic weight at most 10, then we have $n = 1$ and π is the trivial representation. Indeed, for $\pi \neq 1$, we have the inequality $\mathrm{B}_\infty^{\mathrm{F}_{\log 2}}(\pi, 1) \leq 0$, while we easily verify, numerically, that we have $\mathrm{B}_\infty^{\mathrm{F}_{\log 2}}(V, 1) = \mathrm{J}_{\mathrm{F}_{\log 2}}(V) > 0.002$ for $V = 1, \epsilon_{\mathbb{C}/\mathbb{R}}$, and I_w with $w < 11$.

Let us now explain how to proceed in the case $n = 2$, to begin with how to eliminate the case $w = 13$. Like every element of $\Pi(\mathrm{PGL}_2)$, the representation π is self-dual; we therefore have $\xi(s, \pi) = \varepsilon(\pi)\xi(1-s, \pi)$ with $\varepsilon(\pi) = \varepsilon(\mathrm{I}_w) = i^{w+1}$. This ε-factor is -1 if $w \equiv 1 \bmod 4$. By Corollary 9.3.10, we must therefore have

$$\mathrm{J}_{\mathrm{F}_{\log 2}}(\mathrm{I}_w) \leq -\frac{1}{2}\Phi_{\mathrm{F}_{\log 2}}\left(\frac{1}{2}\right)$$

for $w \equiv 1 \bmod 4$. A numerical evaluation shows that, up to 10^{-2}, the number $\frac{1}{2}\Phi_{\mathrm{F}_{\log 2}}(1/2)$ is 0.28. Table 9.3 therefore also shows that π does not exist for $w = 13$.

Let us now show $\mathrm{m}(\mathrm{I}_w) \leq 1$ for $w \leq 21$. We have

$$\mathrm{m}(w)\ \mathrm{B}_\infty^{\mathrm{F}_\lambda}(\mathrm{I}_w, \mathrm{I}_w)\ \leq \Phi_{\mathrm{F}_\lambda}(0) \qquad \left(\text{recall} : \Phi_{\mathrm{F}_\lambda}(0) = \frac{8}{\pi^2}\lambda\right)$$

for every $\lambda > 0$, by Corollary 9.3.14. On the other hand, we have

$$\mathrm{B}_\infty^{\mathrm{F}_\lambda}(\mathrm{I}_w, \mathrm{I}_w)\ =\ \mathrm{J}_{\mathrm{F}_\lambda}(\mathrm{I}_{2w}) + \mathrm{J}_{\mathrm{F}_\lambda}(\mathrm{I}_0)$$

by the relation $\mathrm{I}_u \otimes \mathrm{I}_v = \mathrm{I}_{u+v} + \mathrm{I}_{|u-v|}$. Experimentally, for small values of w, the inequality above turns out to be remarkably good. For example, for $\lambda = \log 8$, we verify numerically that we have $\mathrm{m}(\mathrm{I}_{11}) \leq 1.17$, $\mathrm{m}(\mathrm{I}_{15}) \leq 1.48$, $\mathrm{m}(\mathrm{I}_{17}) \leq 1.66$, $\mathrm{m}(\mathrm{I}_{19}) \leq 1.86$, and $\mathrm{m}(\mathrm{I}_{21}) \leq 2.08$. From this, we deduce the stated inequalities $\mathrm{m}(\mathrm{I}_w) \leq 1$ for $w \leq 19$, as well as $\mathrm{m}(\mathrm{I}_{21}) \leq 2$. It does not seem that we can improve this last equality by simply using other values of λ.

To conclude $\mathrm{m}(\mathrm{I}_{21}) \leq 1$, we use Corollary 9.3.15 (ii) applied to $V = 1$ and $V' = \mathrm{I}_{21}$. Indeed, we have $\mathrm{m}(1) = \mathrm{m}^\perp(1) = 1$, $\mathrm{m}(\mathrm{I}_{21}) = \mathrm{m}^\perp(\mathrm{I}_{21})$, and $\varepsilon(\mathrm{I}_{21}) = -1$. From this, we deduce, for every $\lambda > 0$, the inequality

$$\frac{1}{2}\Phi_{\mathrm{F}_\lambda}\left(\frac{1}{2}\right) + \mathrm{J}_{\mathrm{F}_\lambda}(\mathrm{I}_{21}) \leq \sqrt{(\Phi_{\mathrm{F}_\lambda}(0) - \mathrm{J}_{\mathrm{F}_\lambda}(1))\left(\frac{\Phi_{\mathrm{F}_\lambda}(0)}{\mathrm{m}(\mathrm{I}_{21})} - \mathrm{B}_\infty^{\mathrm{F}_\lambda}(\mathrm{I}_{21}, \mathrm{I}_{21})\right)}\,.$$

But for $\lambda = \log 6$, we can verify that, up to 10^{-2}, the left-hand side is 0.17, whereas the right-hand side is 0.13 if $\mathrm{m}(\mathrm{I}_{21}) = 2$ (and 0.51 if $\mathrm{m}(\mathrm{I}_{21}) = 1$!). □

The Case $w \leq 19$ Odd

Our aim in what follows is to prove that if $\pi \in \Pi_{\mathrm{alg}}(\mathrm{PGL}_n)$ is of motivic weight at most 19 and if $n > 2$, then $n = 4$ and π is the unique self-dual representation such that $\mathrm{L}(\pi_\infty) = \mathrm{I}_{19} \oplus \mathrm{I}_7$. Our starting point is the following result.

Lemma 9.3.20. *If* $\lambda = \log 9$, *the restriction of* $\mathrm{B}_\infty^{\mathrm{F}_\lambda}$ *to* $\mathrm{K}_\infty^{\leq 19}$ *is positive definite*

For $w \geq 0$, we denote by $\mathrm{Gram}(w, \lambda)$ the Gram matrix of the bilinear form $\mathrm{B}_\infty^{\mathrm{F}_\lambda}$ on $\mathrm{K}_\infty^{\leq w}$ in the natural $\mathbb{Z}$-basis defining $\mathrm{K}_\infty^{\leq w}$, namely:

- in the case where w is odd, the I_v with $1 \leq v \leq w$ and v odd;
- in the case where w is even, the representations 1, $\epsilon_{\mathbb{C}/\mathbb{R}}$, and I_v with $0 \leq v \leq w$ and v even.

Proof. Let $B = \mathrm{Gram}(19, \log 9)$. The formulas of Proposition 9.3.18 (and the computer algebra system PARI) make it possible to compute the coefficients of B, and of $\mathrm{Gram}(w, \lambda)$ in general, with a theoretically arbitrary accuracy (more than 20 significant figures in the output of the source code [54]), and we see that B is indeed positive definite. The argument we present below shows that we in fact need only very few significant figures; this argument will in particular be useful in the proof of Lemma 9.3.22, where one needs to determine the vectors $v \in \mathrm{K}_\infty^{\leq 19}$ such that $\mathrm{B}_\infty^{\mathrm{F}_\lambda}(v, v)$ is less than a certain constant.

We first observe, numerically, that all coefficients of B have absolute value in the interval $]0.01, 3.48[$. Let $A \in 10^{-4}\,\mathrm{M}_{10}(\mathbb{Z})$ be the symmetric matrix obtained by rounding the approximation of B given by the computer to the closest element of $10^{-4}\,\mathbb{Z}$, so that all coefficients of the matrix $A - B$ have absolute value at most 10^{-4}. An exact computation carried out by the computer shows that A is positive definite (it suffices to apply Sylvester's criterion). Let $||\cdot||$ be the norm on $\mathrm{M}_n(\mathbb{R})$ subordinate to the norm $\sup_i |x_i|$ on $\mathbb{R}^n$, so that $||(m_{i,j})|| = \sup_i \sum_j |m_{i,j}|$. We easily check $||A^{-1}|| \leq 3.23$ and, moreover, we have $||A - B|| \leq 10 \cdot 10^{-4} = 10^{-3}$. This implies that the spectral radius of $A^{-1}(A - B)$ is at most 0.00323, and in particular that B is positive definite by Lemma 9.3.21. □

Lemma 9.3.21. *Let* V *be a finite-dimensional* $\mathbb{R}$*-vector space,* b_1, b_2 *two symmetric bilinear forms on* V, $e = (e_1, \ldots, e_n)$ *a basis of* V, *and* M_i *the Gram matrix of* b_i *in the basis* e. *Suppose that* b_2 *is positive definite. For every* $x \in V$, *we have the inequality*

$$|b_1(x, x) - b_2(x, x)| \leq \rho(M_2^{-1}(M_1 - M_2))\, b_2(x, x)\ ,$$

where $\rho(M)$ *denotes the spectral radius of the matrix* M. *If we, moreover, have the inequality* $\rho(M_2^{-1}(M_1 - M_2)) < 1$, *then* b_1 *is positive definite, and for every* $x \in V$, *we have* $b_2(x, x) \leq (1 - \rho(M_2^{-1}(M_1 - M_2)))^{-1}\, b_1(x, x)$.

Proof. This is a classical consequence of the diagonalizability of the self-dual endomorphisms of a Euclidean space. □

Lemma 9.3.22. *Let* $V \in \mathrm{K}_\infty^{\leq 19}$ *be effective and such that* $\mathrm{B}_\infty^{\mathrm{F}_{\log 9}}(V, V) \leq 8 \log 9/\pi^2$; *then either*

(i) $V = \mathrm{I}_w$ *with* $9 \leq w \leq 19$, *or*

(ii) $V = \mathrm{I}_{19} + \mathrm{I}_v$ *with* $5 \leq v \leq 13$.

Proof. Lemma 9.3.20 asserts that there are only finitely many nonzero elements V in the "lattice" $\mathrm{K}_\infty^{\leq 19}$ such that $\mathrm{B}_\infty^{\mathrm{F}_{\log 9}}(V,V) \leq 8\log 9/\pi^2$. It remains to enumerate them, which we will do using the algorithm of Fincke and Pohst [85] implemented in `PARI` [160] (the command `qfminim`). In order not to have to justify rounding errors in the algorithm mentioned above, it is convenient to reuse the approximation $A \in 10^{-4}\,\mathrm{M}_{10}(\mathbb{Z})$ of $\mathrm{Gram}(19, \log 9)$ introduced in the proof of Lemma 9.3.20. Let $\mathrm{q}_A : \mathbb{Z}^{10} \to 10^{-4}\mathbb{Z}$ be the positive definite quadratic form $x \mapsto {}^{\mathrm{t}}xAx$. If $V \in \mathrm{K}_\infty$ satisfies $\mathrm{B}_\infty^{\mathrm{F}_{\log 9}}(V,V) \leq 8\log 9/\pi^2$, then its coordinates $(x_1, x_3, \ldots, x_{19})$ in the basis $\mathrm{I}_1, \ldots, \mathrm{I}_{19}$ satisfy

$$10^4\,\mathrm{q}_A(x_1,\ldots,x_{19}) \leq 10^4\,(1-0.00323)^{-1}\,\frac{8\log 9}{\pi^2} < 17868$$

by Lemma 9.3.21. The algorithm of Fincke and Pohst asserts that there are exactly 24 pairs of vectors $\pm(x_i) \in \mathbb{Z}^{10}$ satisfying this inequality and provides a list of them. Of these, we of course only keep the elements $x = (x_i)$ that belong to $\mathbb{N}^{10}$ (corresponding to effective V); there are 11 such elements, listed in the statement. □

Remark 9.3.23. Let us point out that 9 is not the least integer $m \geq 2$ such that $\mathrm{Gram}(19, \log m)$ is positive definite (and even less the only one!). For example, every integer $5 \leq m \leq 100$ has this property. Nevertheless, this choice $\log(9)$, obtained through trial and error, has the advantage of minimizing the size of the list obtained in the statement of Lemma 9.3.22.

Assume now that $\pi \in \Pi_{\mathrm{alg}}(\mathrm{PGL}_n)$ is of odd motivic weight $w \leq 19$, and set $V = \mathrm{L}(\pi_\infty) \in \mathrm{K}_\infty^{\leq 19}$. Corollary 9.3.12 (i) shows that we have $\mathrm{B}_\infty^{\mathrm{F}_{\log 9}}(V,V) \leq 8\log 9/\pi^2$, so that V is in the list given in Lemma 9.3.22. We may assume $n = \dim V > 2$, because the case $n = 2$ has already been treated above. It follows that we have $n = 4$, $w = 19$, and $V = \mathrm{I}_{19} + \mathrm{I}_v$ with $5 \leq v \leq 13$ and v odd. We want to show $v = 7$ and the uniqueness of the representation π. Let us first show that π is unique if it exists.

Lemma 9.3.24. *For every* $v \in \{5, 7, 9, 11, 13\}$, *we have* $\mathrm{m}(\mathrm{I}_{19} + \mathrm{I}_v) \leq 1$.

Proof. Apply Corollary 9.3.14. It suffices to see that for $V = \mathrm{I}_{19} + \mathrm{I}_v$, with v as in the statement, we have

$$\mathrm{B}_\infty^{\mathrm{F}_{\log 9}}(V,V) > \frac{1}{2}\,\frac{8\log 9}{\pi^2}\,.$$

But a numeric computation shows that, up to 10^{-2}, we have $\frac{1}{2}\,8\log 9/\pi^2 \simeq 0.89$, whereas the left-hand side of the inequality above is 1.65, 1.47, 1.42, 1.49, or 1.70 when v is, respectively, 5, 7, 9, 11, or 13. □

Lemma 9.3.25. *Let* $V \in \mathrm{K}_\infty$. *Suppose that there exists a unique representation* $\pi \in \Pi_{\mathrm{alg}}$ *such that* $\mathrm{L}(\pi_\infty) = V$. *Then* π *is self-dual. More generally, we have* $\mathrm{m}(V) \equiv \mathrm{m}^\perp(V) \bmod 2$.

Proof. If $\pi \in \Pi_{\mathrm{alg}}$, we also have $\pi^\vee \in \Pi_{\mathrm{alg}}$, as well as the equalities $\mathrm{L}((\pi^\vee)_\infty) = \mathrm{L}(\pi_\infty)^* = \mathrm{L}(\pi_\infty)$. □

To complete the proof of Theorem 9.3.2 in the case w odd and at most 19, it remains to prove that there does not exist a self-dual representation $\pi \in \Pi_{\rm alg}$ such that $\mathrm{L}(\pi_\infty) = \mathrm{I}_{19} + \mathrm{I}_v$ with $v \in \{5, 9, 11, 13\}$. A first way to proceed would be to use Proposition 9.1.4 (iii) as well as Table 9.1 (this table contains the necessary information because $19 - v \geq 6$ holds in all cases). This is, however, unnecessary, because we will see that the result follows from Corollary 9.3.15 and the existence of the representations $1, \Delta_{11}$, and Δ_{15}. For later use, let us state the following criterion.

Scholium 9.3.26. *Let $V, V' \in \mathrm{K}_\infty$, and let $\lambda > 0$ be a real number. Set*

$$\mathrm{t}(V, V', \lambda) = \sqrt{\left|\left(\Phi_{\mathrm{F}_\lambda}(0) - \mathrm{B}_\infty^{\mathrm{F}_\lambda}(V, V)\right)\left(\Phi_{\mathrm{F}_\lambda}(0) - \mathrm{B}_\infty^{\mathrm{F}_\lambda}(V', V')\right)\right|} + \frac{\varepsilon(V \otimes V') - 1}{4}\,\Phi_{\mathrm{F}_\lambda}\left(\frac{1}{2}\right) - \mathrm{B}_\infty^{\mathrm{F}_\lambda}(V, V')\,.$$

Suppose that V and V' are distinct and effective, and satisfy $\mathrm{m}^\perp(V) \geq 1$ and $\mathrm{m}^\perp(V') \geq 1$. Then we have $\mathrm{t}(V, V', \lambda) \geq 0$. In particular, if V and V' satisfy $\mathrm{m}(V) = \mathrm{m}(V') = 1$, then we have $\mathrm{t}(V, V', \lambda) \geq 0$.

Proof. The first assertion is an immediate consequence of part (ii) of Corollary 9.3.15. The second follows from the first and Lemma 9.3.25. □

Table 9.4 Values of $\mathrm{t}(\mathrm{I}_{19} + \mathrm{I}_v, \mathrm{L}(\pi'_\infty), \log 5)$, up to 10^{-3}

v \ π'	1	Δ_{11}	Δ_{15}	Δ_{17}	Δ_{19}	Δ_{21}
13	0.141	0.074	**−0.006**	0.166	0.088	0.990
11	0.697	**−0.492**	0.396	0.498	0.376	1.251
9	**−0.094**	**−0.046**	0.636	0.689	0.536	1.388
7	0.308	0.223	0.762	0.778	0.597	1.430
5	**−0.660**	0.359	0.771	0.751	0.546	1.357

We know that we have $\mathrm{m}(V) = 1$ for $V = 1$ or $V = \mathrm{L}((\Delta_w)_\infty)$ with $w \leq 21$. Consider Table 9.4. Observe that if $v \neq 7$, there always exists a representation $\pi' \in \{1, \Delta_{11}, \Delta_{15}\}$ such that $\mathrm{t}(\mathrm{I}_{19} + \mathrm{I}_v, \mathrm{L}(\pi'_\infty), \log 5) < 0$ (those negatives values in the table are printed in bold font). By Scholium 9.3.26, this proves $\mathrm{m}(\mathrm{I}_{19} + \mathrm{I}_v) \neq 1$ and therefore completes the proof of the case $w \leq 19$ odd. □

Remark 9.3.27. The parameter $\lambda = \log 5$ has been chosen through trial and error. A variation of this parameter shows that the existence of a representation in the case $v = 7$ seems rather miraculous from this point of view. The property that seems important is that the successive differences between the four weights, namely $(19 - v)/2$, v, and $(19 - v)/2$, are almost equal for $v = 7$.

THE CASE $w \leq 20$ EVEN

We proceed in a way strictly similar to that of the case $w \leq 19$ odd, which is why we give fewer details.

Lemma 9.3.28. *The restriction of* $\mathrm{B}_{\infty}^{\mathrm{F}_{\log 9}}$ *to* $\mathrm{K}_{\infty}^{\leq 20}$ *is positive definite. Let* $V \in \mathrm{K}_{\infty}^{\leq 20}$ *be effective, of determinant* 1*, and satisfying* $\mathrm{B}_{\infty}^{\mathrm{F}_{\log 9}}(V,V) \leq \Phi_{\mathrm{F}_{\log 9}}(0)$*; then we are in exactly one of the following cases:*

(i) $V = 1$,

(ii) $V = \mathrm{I}_w + \epsilon_{\mathbb{C}/\mathbb{R}}$ *with* $w = 18$ *or* 20,

(iii) $V = \mathrm{I}_w + \mathrm{I}_v$ *with either* $w = 18$ *and* $8 \leq v \leq 10$*, or* $w = 20$ *and* $4 \leq v \leq 14$,

(iv) $V = \mathrm{I}_{20} + \mathrm{I}_v + 1$ *with* $10 \leq v \leq 16$.

Moreover, we have $\mathrm{m}(V) \leq 1$.

Proof. The first two assertions follow by studying the matrix $\mathrm{Gram}(20, \log 9)$, which is analogous to that leading to the proofs of Lemmas 9.3.20 and 9.3.22; see the file [54]. For the 15 elements V of the statement, we see that we have $\mathrm{m}(V) \leq 1.6$ by using Corollary 9.3.14 applied to $\lambda = \log 9$, whence the last assertion. □

It follows that if $\pi \in \Pi_{\mathrm{alg}}$ is of even motivic weight $w \leq 20$, then $V = \mathrm{L}(\pi_\infty)$ is in the list above, we have $\mathrm{m}(V) \leq 1$, and π is self-dual. We eliminate the possibilities

$$\mathrm{I}_{18} + \mathrm{I}_{10}\,, \quad \mathrm{I}_{20} + \mathrm{I}_{10} + 1\,, \quad \text{and} \quad \mathrm{I}_{20} + \mathrm{I}_{16} + 1$$

for V using the criterion $\mathrm{B}_{\infty}^{\mathrm{F}_\lambda}(V,V) \leq \Phi_{\mathrm{F}_\lambda}(0)$ taking λ equal to, respectively, $\log 10$, $\log 16$, and $\log 16$. Note that for $V = \mathrm{I}_{18} + \mathrm{I}_{10}$ and $\lambda = \log 10$, the quantity $\Phi_{\mathrm{F}_\lambda}(0) - \mathrm{B}_{\infty}^{\mathrm{F}_\lambda}(V,V)$ has value -0.00012 (up to 10^{-5}), which just barely passes!

It remains to show that π does not exist if $V \neq 1$. By chance, we achieve this using Scholium 9.3.26 applied to the particular cases indicated in Table 9.5. This concludes the proof of Theorem 9.3.2 when the motivic weight w is even and at most 20.

Table 9.5 Pairs (V, π') satisfying $\mathrm{t}(V, \mathrm{L}(\pi'_\infty), \log 5) < -0.05$

V	π'	V	π'	V	π'
$\mathrm{I}_{18} + \epsilon_{\mathbb{C}/\mathbb{R}}$	1	$\mathrm{I}_{20} + \mathrm{I}_{12}$	Δ_{15}	$\mathrm{I}_{20} + \mathrm{I}_4$	1
$\mathrm{I}_{20} + \epsilon_{\mathbb{C}/\mathbb{R}}$	1	$\mathrm{I}_{20} + \mathrm{I}_{10}$	1	$\mathrm{I}_{20} + \mathrm{I}_{14} + 1$	Δ_{15}
$\mathrm{I}_{18} + \mathrm{I}_8$	Δ_{11}	$\mathrm{I}_{20} + \mathrm{I}_8$	1	$\mathrm{I}_{20} + \mathrm{I}_{12} + 1$	1
$\mathrm{I}_{20} + \mathrm{I}_{14}$	Δ_{15}	$\mathrm{I}_{20} + \mathrm{I}_6$	Δ_{11}		

9.3.29 Intermezzo: A Geometric Criterion

Lemma 9.3.30. *Let E be a Euclidean space, $m \geq 1$ an integer, $x_0, x_1, \dots, x_m$ elements of E, and $C_0, C_1, \dots, C_m$ real numbers. Suppose that for every $i = 0, \dots, m$, we have the inequality $x_0 \cdot x_i \leq C_i$.*

(i) *We have $C_0 \geq 0$ and $C_i + \sqrt{C_0\,(x_i \cdot x_i)} \geq 0$ for every $i = 1, \dots, m$.*

Let $G = (x_i \cdot x_j)_{1 \leq i,j, \leq m} \in \mathrm{M}_m(\mathbb{R})$ be the Gram matrix of the vectors $x_1, \dots, x_m$ and C the column vector $(C_i)_{1 \leq i \leq m} \in \mathbb{R}^m$. Suppose $\det(G) \neq 0$, that is, that the vectors $x_1, \dots, x_m$ are linearly independent. Then one of the two following assertions holds:

(ii) *At least one of the coordinates of the vector $G^{-1}C$ is strictly positive.*
(ii)' *We have the equality (of real numbers) ${}^{\mathrm{t}}C\,G^{-1}\,C \leq C_0$.*

Proof. The inequality $C_0 \geq 0$ is obvious. Moreover, for $i = 1, \dots, m$, the Cauchy–Schwarz inequality gives

$$C_i \geq x_0 \cdot x_i \geq -|x_0 \cdot x_i| \geq -\sqrt{(x_0 \cdot x_0)(x_i \cdot x_i)} \geq -\sqrt{C_0\ \ x_i \cdot x_i}\,.$$

Let us verify the second assertion of the lemma. Let $(x_i^*)_{1 \leq i \leq m}$ be the dual basis of $(x_i)_{1 \leq i \leq m}$ in the Euclidean space $F = \mathrm{Vect}_{\mathbb{R}}(x_1, \dots, x_m)$. Let $H = (h_{i,j}) \in \mathrm{M}_m(\mathbb{R})$ be the matrix defined by the equalities $x_j^* = \sum_{i=1}^m h_{i,j} x_i$ for $j = 1, \dots, m$. By the definition of the dual basis, H is also the Gram matrix $(x_i^* \cdot x_j^*)$ and we have the relation $H = G^{-1}$. Hence, the coefficients of the vector $G^{-1}\,C = H\,C$ are none other than the inner products $x_i^* \cdot v$ for $i = 1, \dots, m$, with

$$v := \sum_{j=1}^{m} C_j\, x_j^*\,.$$

Suppose that assertion (ii) does not hold, that is, $x_i^* \cdot v \leq 0$ for every $i = 1, \dots, m$. We will see that we have $x_0 \cdot x_0 \geq v \cdot v$, which is assertion (ii)'. The vector x_0 can be written uniquely as $x_0 = v - \sum_{i=1}^m t_i x_i^* + w$ with $w \in F^\perp$ and $t_i \in \mathbb{R}$ for $i = 1, \dots, m$. For $i = 1, \dots, m$, the condition $x_0 \cdot x_i \leq C_i$ is equivalent to $t_i \geq 0$. Set $\|x\|^2 = x \cdot x$ for $x \in E$. We conclude by considering the following equality:

$$\|x_0\|^2 \;=\; \|v\|^2 \;-\; 2\sum_{i=1}^{m} t_i\; x_i^* \cdot v \;+\; \|w - \sum_{i=1}^{m} t_i\, x_i^*\|^2\,. \qquad \square$$

Remark 9.3.31. The geometric interpretation of Lemma 9.3.30 is as follows. By assumption, the point x_0 is in the intersection of the ball B with center 0 and radius $\sqrt{C_0}$ and the "polyhedron" P that is the intersection of the m half-spaces $x \cdot x_i \leq C_i$ for $i = 1, \dots, m$. Assertion (i) asserts that each of these half-spaces meets B, an obvious necessary condition! The quantity ${}^{\mathrm{t}}C\,G^{-1}\,C$ of the statement is the square of the distance to the origin of the affine space $\{x \in V\,;\, x \cdot x_i = C_i,\ \ i = 1 \dots m\}$ or, equivalently, $v + F^\perp$ in the notation of the proof. If (and only if) condition (ii) does not hold, this distance is also the distance from P to the origin, whence the result.

Corollary 9.3.10 shows that the Satake parameters of a representation $\pi \in \Pi_{\mathrm{alg}}$ such that we have $\mathrm{L}(\pi_\infty) = V$ are subject to a set of constraints; these contraints may be expressed in the setting of Lemma 9.3.30. To set this up, it will be convenient to denote by $\mathrm{Q} \subset \mathbb{N}$ the subset consisting of the powers of the prime numbers; every $q \in \mathrm{Q}$ can thus be written uniquely as $q = p^k$ with p prime and $k \geq 1$ an integer. To $\pi \in \Pi(\mathrm{PGL}_n)$ and $q \in \mathrm{Q}$ corresponds the complex number

$$x_q(\pi) = \mathrm{tr}\ \mathrm{c}_p(\pi)^k \in \mathbb{C}\ ,$$

where we have written $q = p^k$ with p prime and $k \geq 1$. If π is self-dual and in Π_{alg}, we even have $x_q(\pi) \in \mathbb{R}$ by Proposition 8.2.2. Fix a real number $\lambda > 0$, and set

$$\mathrm{Q}_\lambda = \{q \in \mathrm{Q},\ \ q < e^\lambda\} \quad \text{and} \quad \mathrm{E}_\lambda = \prod_{q \in \mathrm{Q}_\lambda} \mathbb{C}\ .$$

We endow the $\mathbb{R}$-vector space underlying E_λ with the structure of a Euclidean space via the inner product

$$(x_q) \cdot (y_q) = \sum_{q \in \mathrm{Q}_\lambda} \mathrm{F}_\lambda(\log q)\ \frac{\log p}{\sqrt{q}}\ \Re\, \overline{x}_q\, y_q\ ,$$

where p denotes the prime divisor of q. For every integer $n \geq 1$ and every $\pi \in \Pi(\mathrm{PGL}_n)$, we have a vector

$$x_\lambda(\pi) := (x_q(\pi))_q \in \mathrm{E}_\lambda\ .$$

By the definition of $\mathrm{B}_f^{\mathrm{F}_\lambda}$ (Proposition-Definition 9.3.7), we have

$$\mathrm{B}_f^{\mathrm{F}_\lambda}(\pi, \pi') = x_\lambda(\pi) \cdot x_\lambda(\pi')$$

for all $\pi, \pi' \in \Pi_{\mathrm{alg}}$.

Denote by $\Pi_{\mathrm{alg}}^{\perp} \subset \Pi_{\mathrm{alg}}$ the subset of self-dual representations. To avoid multiplying the statements, and given the applications we have in mind, we restrict the analysis below to the elements of $\Pi_{\mathrm{alg}}^{\perp}$. Fix $\pi_0 \in \Pi_{\mathrm{alg}}^{\perp}$ and set $V_0 = \mathrm{L}((\pi_0)_\infty)$. Proposition 9.3.10 leads to the following system of inequalities:

$$\begin{cases} x_\lambda(\pi_0) \cdot x_\lambda(\pi_0) \leq \Phi_{\mathrm{F}_\lambda}(0)\ -\ \mathrm{B}_\infty^{\mathrm{F}_\lambda}(V_0, V_0)\ , \\ \qquad\quad \text{and} \quad \forall \pi \in \Pi_{\mathrm{alg}}^{\perp} - \{\pi_0\}\ , \\ x_\lambda(\pi_0) \cdot x_\lambda(\pi)\ \leq\ -\ \frac{1-\varepsilon(V_0 \otimes \mathrm{L}(\pi_\infty))}{4}\ \Phi_{\mathrm{F}_\lambda}\left(\frac{1}{2}\right)\ -\ \mathrm{B}_\infty^{\mathrm{F}_\lambda}(V_0, \mathrm{L}(\pi_\infty))\ . \end{cases} \tag{9.3.3}$$

The assumptions of Lemma 9.3.30 clearly hold, which immediately implies the following scholium.

Scholium 9.3.32. *Let $V_0 \in \mathrm{K}_\infty$, let $\lambda > 0$ be a real number, $m \geq 1$ an integer, and $\pi_1, \ldots, \pi_m$ distinct elements of $\Pi_{\mathrm{alg}}^{\perp}$. Set $C_0 = \Phi_{\mathrm{F}_\lambda}(0) - \mathrm{B}_\infty^{\mathrm{F}_\lambda}(V_0, V_0)$,*

$$C_i = -\frac{1 - \varepsilon(V_0 \otimes \mathrm{L}((\pi_i)_\infty))}{4} \Phi_\lambda\left(\frac{1}{2}\right) - \mathrm{B}_\infty^{\mathrm{F}_\lambda}(V_0, \mathrm{L}((\pi_i)_\infty))$$

for $i = 1, \ldots, m$, and $C = (C_i) \in \mathbb{R}^m$. Suppose that the matrix of $\mathrm{M}_m(\mathbb{R})$

$$G = (x_\lambda(\pi_i) \cdot x_\lambda(\pi_j))_{1 \leq i,j \leq m}$$

is invertible, that the coordinates of the vector $G^{-1}C$ are all strictly negative, and that the real number $C_0 - C^t G^{-1} C$ is strictly negative.

If $\pi \in \Pi_{\mathrm{alg}}^{\perp}$ satisfies $\mathrm{L}(\pi_\infty) = V_0$, then there exists $1 \leq i \leq m$ such that $\pi = \pi_i$. In particular, if $\mathrm{L}((\pi_i)_\infty) \neq V_0$ for every i, then $\mathrm{m}^\perp(V_0) = 0$.

In what follows, we apply this criterion to the elements π_i of the set

$$\mathcal{R} = \{1, \Delta_{11}, \Delta_{15}, \Delta_{17}, \Delta_{19}, \Delta_{21}, \mathrm{Sym}^2\Delta_{11}\} \subset \Pi_{\mathrm{alg}}^{\perp}.$$

The vectors $x_\lambda(\pi)$, for $\pi \in \mathcal{R}$ and λ reasonable, are considered known. For example, we have $x_\lambda(1) = (1, 1, 1, \ldots)$ and

$$\begin{aligned} x_\lambda(\Delta_{11}) &= (\tau(2)\, 2^{-11/2},\ \tau(3)\, 3^{-11/2},\ (\tau(4) - 2^{11})\, 4^{-11/2},\ \ldots) \\ &\simeq (-0.530,\ 0.599,\ -1.719,\ \ldots). \end{aligned}$$

By way of application, let us prove the following result, which we will need further on.

Lemma 9.3.33. *Suppose that $V \in \mathrm{K}_\infty$ belongs to the list of eight elements*

$$\begin{gathered} \mathrm{I}_{21} + \mathrm{I}_{17} + \mathrm{I}_7,\ \ \mathrm{I}_{22} + \mathrm{I}_4,\ \ \mathrm{I}_{22} + \mathrm{I}_{12},\ \ \mathrm{I}_{22} + \mathrm{I}_{16} + 1,\ \ \mathrm{I}_{22} + \mathrm{I}_{12} + 1, \\ \mathrm{I}_{22} + \mathrm{I}_{16} + \mathrm{I}_{10} + \epsilon_{\mathbb{C}/\mathbb{R}},\ \ \mathrm{I}_{22} + \mathrm{I}_{20} + \mathrm{I}_{10} + \epsilon_{\mathbb{C}/\mathbb{R}},\ \ \mathrm{I}_{22} + \mathrm{I}_{20} + \mathrm{I}_{14} + \mathrm{I}_4. \end{gathered}$$

Then we have $\mathrm{m}^\perp(V) = 0$.

Proof. Apply Scholium 9.3.32 with $m = 2$, taking for V_0, $\{\pi_1, \pi_2\}$, λ each of the triples in Table 9.6. We easily verify case by case that the assumptions of the scholium are indeed satisfied (see the source code [54]). □

Table 9.6 Triples $(V_0, \{\pi_1, \pi_2\}, \lambda)$ satisfying the assumptions of Scholium 9.3.32

V_0	$\{\pi_1, \pi_2\}$	λ
$\mathrm{I}_{21} + \mathrm{I}_{17} + \mathrm{I}_7$	$\{\Delta_{15}, \Delta_{17}\}$	$\log 14$
$\mathrm{I}_{22} + \mathrm{I}_{12}$	$\{\Delta_{11}, \Delta_{15}\}$	$\log 5$
$\mathrm{I}_{22} + \mathrm{I}_{12} + 1$	$\{1, \Delta_{11}\}$	$\log 5$
$\mathrm{I}_{22} + \mathrm{I}_{20} + \mathrm{I}_{10} + \epsilon_{\mathbb{C}/\mathbb{R}}$	$\{\Delta_{19}, \mathrm{Sym}^2\,\Delta_{11}\}$	$\log 38$
$\mathrm{I}_{22} + \mathrm{I}_4$	$\{1, \Delta_{21}\}$	$\log 7$
$\mathrm{I}_{22} + \mathrm{I}_{16} + 1$	$\{1, \Delta_{17}\}$	$\log 8$
$\mathrm{I}_{22} + \mathrm{I}_{16} + \mathrm{I}_{10} + \epsilon_{\mathbb{C}/\mathbb{R}}$	$\{\Delta_{11}, \Delta_{15}\}$	$\log 9$
$\mathrm{I}_{22} + \mathrm{I}_{20} + \mathrm{I}_{14} + \mathrm{I}_4$	$\{1, \Delta_{21}\}$	$\log 40$

9.3.34 End of the Proof of Theorem 9.3.2: The Case of Motivic Weights 21 and 22

Lemma 9.3.35. *Let $V \in \mathrm{K}_\infty^{\leq 21}$ be such that $V - \mathrm{I}_{21}$ is effective, nonzero, and such that $\mathrm{B}_\infty^{\mathrm{F}_\lambda}(V, V) \leq \Phi_\lambda(0)$ for $\lambda = \log 28$. Then V is one of the following 26 elements:*

(i) $\mathrm{I}_{21} + \mathrm{I}_v$ *with* $17 \geq v \geq 3$,
(ii) $\mathrm{I}_{21} + \mathrm{I}_v + \mathrm{I}_u$ *with* $19 \geq v \geq 13$, $9 \geq u \geq 3$, *and* $(v, u) \neq (13, 9)$, *or with* $(v, u) = (17, 11)$,
(iii) $\mathrm{I}_{21} + \mathrm{I}_{19} + \mathrm{I}_{13} + \mathrm{I}_v$ *with* $5 \geq v \geq 3$.

Moreover, we have $\mathrm{m}(V) \leq 1$.

Proof. The first assertion follows by studying the (positive definite!) matrix $\mathrm{Gram}(21, \log 28)$, which is analogous to that leading to the proofs of Lemmas 9.3.20 and 9.3.22; see [54]. For the 26 elements V of the statement, use Corollary 9.3.14 applied to $\lambda = \log 28$ to check that we have $\mathrm{m}(V) < 1.8$, whence the last assertion. □

Lemma 9.3.36. *Let $V \in \mathrm{K}_\infty$ be one of the 26 elements listed in the statement of Lemma 9.3.35. Suppose that we have $\mathrm{t}(V, \mathrm{L}(\pi_\infty), \log 27) > 0$ for every $\pi \in \mathcal{R}$* (see Scholium 9.3.26 and Sect. 9.3.29). *Then V is one of the elements $\mathrm{I}_{21} + \mathrm{I}_5$, $\mathrm{I}_{21} + \mathrm{I}_9$, $\mathrm{I}_{21} + \mathrm{I}_{13}$, and $\mathrm{I}_{21} + \mathrm{I}_{17} + \mathrm{I}_7$.*

Proof. This is a simple numerical computation for which we refer to [54]. □

Proof of Theorem 9.3.2 in the Case $w = 21$, End. To prove the case $w = 21$ of Theorem 9.3.2, it therefore only remains to show that we have $\mathrm{m}^\perp(\mathrm{I}_{21} + \mathrm{I}_{17} + \mathrm{I}_7) = 0$. But this has already been proved in Lemma 9.3.33, so we are done. □

Lemma 9.3.37. *Let* $V \in \mathrm{K}_\infty^{\leq 22}$. *Suppose that* $V - \mathrm{I}_{22}$ *is effective and that the inequality* $\mathrm{B}_\infty^{\mathrm{F}_\lambda}(V,V) \leq \Phi_\lambda(0)$ *holds for* $\lambda = \log 80$.

(i) *We have* $\mathrm{m}(V) \leq 1$, *unless* V *is one of the following elements:*

$$\mathrm{I}_{22} + \mathrm{I}_{12}\ , \quad \mathrm{I}_{22} + \mathrm{I}_{10}\ , \quad \mathrm{I}_{22} + \mathrm{I}_{8}\ ,$$

in which case we only have $\mathrm{m}(V) \leq 2$.

(ii) *Suppose, moreover,* $V \neq \mathrm{I}_{22} + \epsilon_{\mathbb{C}/\mathbb{R}}$ *and that we have the inequality* $\mathrm{t}(V, \mathrm{L}(\pi_\infty), \log 77) > 0$ *for every* $\pi \in \mathcal{R}$. *Then* V *belongs to the following list of eight representations:*

$$\begin{cases} \mathrm{I}_{22} + \mathrm{I}_{12},\ \ \mathrm{I}_{22} + \mathrm{I}_{8},\ \ \mathrm{I}_{22} + \mathrm{I}_{4},\ \ \mathrm{I}_{22} + \mathrm{I}_{16} + 1\ , \\ \mathrm{I}_{22} + \mathrm{I}_{12} + 1,\ \ \mathrm{I}_{22} + \mathrm{I}_{16} + \mathrm{I}_{10} + \epsilon_{\mathbb{C}/\mathbb{R}}\ , \\ \mathrm{I}_{22} + \mathrm{I}_{20} + \mathrm{I}_{10} + \epsilon_{\mathbb{C}/\mathbb{R}},\ \ \mathrm{I}_{22} + \mathrm{I}_{20} + \mathrm{I}_{14} + \mathrm{I}_{4}\ . \end{cases} \tag{9.3.4}$$

Proof. We first verify that the matrix $B = \mathrm{Gram}(22, \log 80)$ is positive definite, using the same method as in the proof of Lemma 9.3.20; see [54], in which we study the symmetric matrix $10^6\,A$ obtained by rounding all coefficients of the matrix $10^6\,B$ to the nearest integer.

We then proceed as in the proof of Lemma 9.3.22. The algorithm `qfminim` of `PARI` applied to $10^6 A \in \mathrm{M}_{12}(\mathbb{Z})$ returns a set of 701 pairs $\pm V$ containing all the elements $V \in \mathrm{K}_\infty^{\leq 22}$ satisfying $\mathrm{B}_\infty^{\mathrm{F}_\lambda}(V,V) \leq \Phi_\lambda(0)$ for $\lambda = \log 80$. If, from this set, we only retain the subset $\mathcal{L}$ consisting of the V such that $V - \mathrm{I}_{22}$ is effective and satisfies $\det V = 1$, there "only" remain 158 possibilities for V; in other words, we have $|\mathcal{L}| = 158$.

Next, we determine the subset consisting of the $V \in \mathcal{L}$ that moreover satisfy $0 \leq 2\,\mathrm{B}_\infty^{\mathrm{F}_\lambda}(V,V) \leq \Phi_\lambda(0)$ for $\lambda = \log 77$: we see that there are only three elements left, of the form $\mathrm{I}_{22} + \mathrm{I}_v$ with $v = 12$, 10, or 8. Since in each of these cases, we also have the inequality $3\,\mathrm{B}_\infty^{\mathrm{F}_\lambda}(V,V) > \Phi_\lambda(0)$ (still for $\lambda = \log 77$), Corollary 9.3.14 proves the first assertion.

To prove the second assertion, we simply compute the $|\mathcal{L}| \cdot |\mathcal{R}| = 1106$ quantities $\mathrm{t}(V, \mathrm{L}(\pi_\infty), \log 77)$, with $V \in \mathcal{L}$ and $\pi \in \mathcal{R}$. We refer to [54] for the justification of the results. □

Proof of Theorem 9.3.2, *End.* Let $\pi \in \Pi_{\mathrm{alg}}$ be of motivic weight 22. Set $V = \mathrm{L}(\pi_\infty)$. The element $V - \mathrm{I}_{22}$ is effective, and we have $\mathrm{B}_\infty^{\mathrm{F}_\lambda}(V,V) \leq \Phi_{\mathrm{F}_\lambda}(0)$ for every $\lambda > 0$ by Corollary 9.3.12 (i).

First, suppose that π is self-dual; in particular, we have $\mathrm{m}^\perp(V) \geq 1$. Scholium 9.3.26 and Lemma 9.3.37 (ii) show that either we have $V = \mathrm{I}_{22} + \epsilon_{\mathbb{C}/\mathbb{R}}$, or V is in the list (9.3.4) above. Moreover, it is easy to check that we have the inequality

$$\mathrm{t}(\mathrm{I}_{22} + \mathrm{I}_8, \mathrm{I}_{11}, \log 5) < 0\ ,$$

so that we also have $V \neq \mathrm{I}_{22}+\mathrm{I}_8$. By Lemma 9.3.33, for the seven remaining elements W of the list (9.3.4), we have $\mathrm{m}^{\perp}(W) = 0$. To conclude, we have $V = \mathrm{I}_{22}+\epsilon_{\mathbb{C}/\mathbb{R}}$. But Lemma 9.3.37 (i) implies $\mathrm{m}(V) \leq 1$, and therefore $\mathrm{m}^{\perp}(V) = 1$ and $\pi = \mathrm{Sym}^2\Delta_{11}$.

Next, suppose that π is not self-dual. By Lemma 9.3.37 (i), we therefore have $\mathrm{m}(V) = 2$ and $V = \mathrm{I}_{22} + \mathrm{I}_v$ with $v \in \{8, 10, 12\}$. In particular, the two representations $\varpi \in \Pi_{\mathrm{alg}}$ that satisfy $\mathrm{L}(\varpi_\infty) = V$ are π and $\pi^\vee$.

To conclude, it suffices to prove that the equality $\mathrm{m}(\mathrm{I}_{22}+\mathrm{I}_v) = 2$ for $v = 8$ (resp. $10, 12$) implies $\mathrm{m}(\mathrm{I}_{21}+\mathrm{I}_u) = 0$ for $u = 9$ (resp. $9, 13$). Let (v, u) be one of the three ordered pairs $(8, 9)$, $(10, 9)$, and $(12, 13)$, and let $V = \mathrm{I}_{22} + \mathrm{I}_v$ and $V' = \mathrm{I}_{21} + \mathrm{I}_u$. By Corollary 9.3.15 (i), it suffices to verify that there exists a $\lambda > 0$ such that we have the inequality

$$\sqrt{\left(\frac{\Phi_{\mathrm{F}_\lambda}(0)}{2} - \mathrm{B}_\infty^{\mathrm{F}_\lambda}(V,V)\right)\left(\Phi_{\mathrm{F}_\lambda}(0) - \mathrm{B}_\infty^{\mathrm{F}_\lambda}(V',V')\right)} \;-\; \mathrm{B}_\infty^{\mathrm{F}_\lambda}(V,V') < 0\,.$$

But we easily check that for $\lambda = \log 22$, and say up to 10^{-2}, the left-hand side has value -0.14, -0.03, and -0.23 when (v, u) equals $(8, 9)$, $(10, 9)$, and $(12, 13)$, respectively. □

9.3.38 Complements

The first complement concerns the vanishing order of $\mathrm{L}(s, \pi)$ at $s = 1/2$ when $\pi \in \Pi_{\mathrm{alg}}$ is of motivic weight at most 22 (compare with Remark 7.3.3).

Proposition 9.3.39. *Let $\pi \in \Pi_{\mathrm{alg}}^{\perp}$ be of motivic weight at most 22. We have*

$$\mathrm{ord}_{s=1/2}\,\mathrm{L}(s,\pi) \;=\; \begin{cases} 0 \text{ if } \varepsilon(\pi) = 1, \\ 1 \text{ otherwise.} \end{cases} \tag{9.3.5}$$

Moreover, we have $\varepsilon(\pi) = -1$ if and only if $\pi = \Delta_{17}$ or $\pi = \Delta_{21}$.

Proof. Set $r = \mathrm{ord}_{s=\frac{1}{2}}\,\mathrm{L}(s,\pi)$; we also have $r = \mathrm{ord}_{s=\frac{1}{2}}\,\xi(s,\pi)$ by the remark following Corollary 9.3.10. The functional equation

$$\xi(s,\pi) \;=\; \varepsilon(\pi)\,\xi(1-s,\pi)$$

shows that r is even if $\varepsilon(\pi) = 1$ and odd otherwise. It therefore suffices to show $r < 2$. We may assume $\pi \neq 1$ because we have $\varepsilon(1) = 1$ and $\zeta(1/2) \neq 0$. An argument similar to that given in the proof of Corollary 9.3.12 (ii), applied to π and 1, shows that under the assumption $r \geq 2$, we have the inequality

$$-\Phi_{\mathrm{F}_\lambda}\!\left(\frac{1}{2}\right) - \mathrm{J}_{\mathrm{F}_\lambda}(V) + \sqrt{\left(\Phi_{\mathrm{F}_\lambda}(0) - \mathrm{J}_{\mathrm{F}_\lambda}(1)\right)\left(\Phi_{\mathrm{F}_\lambda}(0) - \mathrm{B}_\infty^{\mathrm{F}_\lambda}(V,V)\right)} \geq 0 \tag{9.3.6}$$

for every $\lambda > 0$: it suffices to bound $\mathrm{ord}_{s=1/2}\,\xi(s,\pi)$ from below by 2 rather than by $\mathrm{e}^{\perp}(\pi,1)$. But when V is, respectively,

$$\mathrm{I}_{11},\ \mathrm{I}_{15},\ \mathrm{I}_{17},\ \mathrm{I}_{19},\ \mathrm{I}_{21},\ \mathrm{I}_{19}+\mathrm{I}_{7},\ \mathrm{I}_{21}+\mathrm{I}_{5},\ \mathrm{I}_{21}+\mathrm{I}_{9},\ \mathrm{I}_{21}+\mathrm{I}_{13},\ \mathrm{I}_{22}+\epsilon_{\mathbb{C}/\mathbb{R}}\ ,$$

and, say, $\lambda = \log 4$, we see that, up to 10^{-2}, the left-hand side of (9.3.6) has value -1.07, -0.64, -0.49, -0.35, -0.23, -0.79, -0.86, -0.35, -0.05, -0.82. The first assertion of the proposition therefore follows from Theorem 9.3.2. Note that in the list of 10 elements V above, we have $\varepsilon(V) = 1$ except for $V = \mathrm{I}_{17}$ and I_{21}; this concludes the proof. □

A very simple, but surprising, consequence of Theorem 9.3.2 is the fact that there exist only finitely many $\pi \in \Pi_{\mathrm{alg}}$ such that $\mathrm{w}(\pi) \leq 22$. Returning to our proof, we see that this finiteness assertion, which is in fact our starting point, is a consequence of the following property: if we have $w \leq 22$, then there exist real numbers $\lambda > 0$ such that the restriction of the symmetric bilinear form $\mathrm{B}_{\infty}^{\mathrm{F}_\lambda}$ to $\mathrm{K}_{\infty}^{\leq w}$ is positive definite (Lemmas 9.3.20, 9.3.28, 9.3.35, and 9.3.37). It turns out that this property still holds for $w = 23$ (but no longer does for $w > 23$!).

Proposition 9.3.40. *There are only finitely many representations in Π_{alg} of motivic weight $w \leq 23$.*

Proof. A simple computation indeed shows that $\mathrm{Gram}(23, 9.74)$ is positive definite. □

We defer to a later work the detailed study of the representations in Π_{alg} of motivic weight $w \geq 23$, which would lead us too far from our current preoccupations. Let us, however, mention two works in relation to these problems. In [55], assuming that Conjecture 8.4.25 holds, the authors prove an explicit and computable formula for $\mathrm{m}^{\perp}(V)$ when

- V is *without multiplicities*; that is, its coefficients in the basis 1, $\epsilon_{\mathbb{C}/\mathbb{R}}$, $\{\mathrm{I}_w, w > 0\}$ are all in $\{0,1\}$;
- $\dim V \leq 8$, with only partial results when $\dim V = 7$.

In a remarkable tour de force [195], Taïbi then re-proved these formulas by assuming only Conjecture 8.4.22 and extended them to the more general case $\dim V \leq 14$. His results are even independent of any conjecture if the weights of V are "sufficiently spread out."[4]

Let us return to the case of motivic weight 23 and admit Conjecture 8.4.22. The theory of modular forms of course leads to the equality $\mathrm{m}(\mathrm{I}_{23}) = 2$. Next, consider $V \in \mathrm{K}_{\infty}$ without multiplicities and such that $V - \mathrm{I}_{23}$ is effective and satisfies $\mathrm{m}^{\perp}(V) \geq 1$. The theory of Siegel forms of genus 2 and Tsushima's formula

[4] Let us be more precise about this notion. Suppose that V is effective, of dimension $n = \dim V$, and denote by $\lambda_1, \ldots, \lambda_n$ the multi-set of n complex numbers associated with $V_{|\mathbb{C}^*}$ as in the assertion of the compatibility of the Langlands parametrization with the infinitesimal character (Sect. 8.2.12 (iii)). We may assume $\lambda_i \in \frac{1}{2}\mathbb{Z}$ for all i and $\lambda_i - \lambda_j \in \mathbb{Z}$ for every i, j. We say that the weights of V are sufficiently spread out if we have the inequality $|\lambda_i - \lambda_j| \neq 1$ for every $1 \leq i, j \leq n$.

(see Remark 9.3.41) show that if $\dim V = 2$, we have $\mathrm{m}(V) = 1$ and $V - \mathrm{I}_{23} \in \{I_7, \mathrm{I}_9, \mathrm{I}_{13}\}$. The computations of [55, Corollary I.1.5], based on the study of the invariant polynomials of the orthogonal group of the lattice E_7, moreover show that if $6 \leq \dim V \leq 8$, then we have $\dim V = 6$ and $\mathrm{m}^{\perp}(V) = 1$, and $V - \mathrm{I}_{23}$ runs through the following list:

$$\mathrm{I}_{13} + \mathrm{I}_5,\ \mathrm{I}_{15} + \mathrm{I}_3,\ \mathrm{I}_{15} + \mathrm{I}_7,\ \mathrm{I}_{17} + \mathrm{I}_5,\ \mathrm{I}_{17} + \mathrm{I}_9,\ \mathrm{I}_{19} + \mathrm{I}_3,\ \mathrm{I}_{19} + \mathrm{I}_{11}\ .$$

Finally, the results of Taïbi mentioned above show that if $8 < \dim V \leq 14$, we have $\dim V = 10$, $\mathrm{m}^{\perp}(V) = 1$, and $V = \mathrm{I}_{23} + \mathrm{I}_{21} + \mathrm{I}_{17} + \mathrm{I}_{11} + \mathrm{I}_3$. In all, this gives 13 representations in Π_{alg} of motivic weight 23; the remaining question is whether there are any others. Let us also mention that we know only three $\pi \in \Pi_{\mathrm{alg}}$ of motivic weight 24, of respective dimensions 7, 8, and 8. That of dimension 7 satisfies $\mathrm{L}(\pi_\infty) = \epsilon_{\mathbb{C}/\mathbb{R}} \oplus \bigoplus_{i=1}^{3} \mathrm{I}_{8i}$ and is related to the triality and to the group G_2 [55, Corollary I.1.10].

Remark 9.3.41. A very simple special case of Taïbi's formulas is that the formula given by Tsushima [199] to compute $\dim \mathrm{S}_{j,k}$ when $k \geq 5$ holds more generally for $k \geq 3$ (Sect. 9.1), except for the case $(j, k) = (0, 3)$. An examination of its values shows, in particular, the vanishing $\dim \mathrm{S}_{j,k} = 0$ when $k = 3, 4$ and $j + 2k - 3 \leq 21$. Another proof of this vanishing is given by Theorem 9.3.2 and Proposition 9.1.4.

9.4 Proof of Theorem E

9.4.1 A New Proof of Theorem A

Let us begin by giving a new proof of Theorem A, using a method that will generalize to dimension 24. Recall that we have $|\mathrm{X}_{16}| = 2$ and that the Hecke operator T_2 admits distinct eigenvalues on $\mathbb{Z}[\mathrm{X}_{16}]$ (Corollary 2.3.6, Sect. 3.3.1). In particular, there exist exactly two representations $\pi \in \Pi_{\mathrm{disc}}(\mathrm{O}_{16})$ such that π_∞ is the trivial representation of $\mathrm{O}_{16}(\mathbb{R})$. We already explained in Sect. 5.2 that the following proposition implies Theorem A; in fact, we already gave a first proof of this proposition in Corollary 7.2.6 (ii).

Proposition* 9.4.2. *The standard parameters* $\psi(\pi, \mathrm{St})$ *of the two representations* $\pi \in \Pi_{\mathrm{disc}}(\mathrm{O}_{16})$ *such that* π_∞ *is the trivial representation are* $[15] \oplus [1]$ *and* $\Delta_{11}[4] \oplus [7] \oplus [1]$.

Proof. By the example given at the end of Sect. 6.4.7, we know that the trivial representation $1 \in \Pi_{\mathrm{disc}}(\mathrm{O}_{16})$ satisfies

$$\psi(1, \mathrm{St}) = [15] \oplus [1]\ .$$

We must therefore show that if π denotes the nontrivial representation of $\Pi_{\mathrm{disc}}(\mathrm{O}_{16})$ such that π_∞ is trivial, then $\psi(\pi, \mathrm{St}) = \Delta_{11}[4] \oplus [7] \oplus [1]$. Let us add that we have

$\psi(\pi, \mathrm{St}) \neq \psi(1, \mathrm{St})$. There is a general reason for this, but one way to see it here is to use that the two eigenvalues of T_2 acting on $\mathbb{Z}[\mathrm{X}_{16}]$, namely $2^7 \operatorname{tr} \psi(\pi, \mathrm{St})_2$ and $2^7 \operatorname{tr} \psi(1, \mathrm{St})_2$ (formula (6.2.5)), are distinct.

By Proposition 5.2.1, the representation π admits a ϑ-correspondent in genus $1 \leq g \leq 4$. Given the inequality $16 > 2g$, Arthur's Theorem 8.1.1 and Corollary 7.1.3 show that the pair (π, St) satisfies the Arthur–Langlands conjecture. In other words, there exist an integer $k \geq 1$ and, for $i = 1, \ldots, k$, representations $\pi_i \in \Pi_{\mathrm{cusp}}(\mathrm{PGL}_{n_i})$ and integers $d_i \geq 1$, such that

$$\psi(\pi, \mathrm{St}) = \oplus_{i=1}^{k} \pi_i[d_i] .$$

The assumption on π_∞ forces the eigenvalues of $\mathrm{St}\, \mathrm{c}_\infty(\pi)$ to be the 14 integers $\pm 7, \pm 6, \ldots, \pm 1$, and 0 with multiplicity 2. It follows that the π_i are algebraic (Proposition 8.2.8), of motivic weight satisfying $(\mathrm{w}(\pi_i) + d_i - 1)/2 \leq 7$. In particular, we have $\mathrm{w}(\pi_i) \leq 14$ for every i.

By Theorem 9.3.2, for every $i = 1, \ldots, k$, we have $\pi_i = 1$ or $\pi_i = \Delta_{11}$. Moreover, the weights of Δ_{11} are $\pm 11/2$. By considering the eigenvalue ± 7, which can only "belong" to a component of the form $\Delta_{11}[4]$ or $[15]$, we see that the only possibilities for $\psi(\pi, \mathrm{St})$ are the two in the statement. At this point, we can also invoke the equality $\Psi_{16} = \{\, [15] \oplus [1], \Delta_{11}[4] \oplus [7] \oplus [1] \,\}$, which was verified during the proof of Proposition 9.2.2. This concludes the proof (and re-proves the Witt conjecture $g = 4$!). □

9.4.3 *Proof of Theorem* E

Theorem* 9.4.4. *The elements ψ in $\mathcal{X}_{\mathrm{AL}}(\mathrm{SL}_{24})$ such that the eigenvalues of ψ_∞ are the integers $\pm 11, \pm 10, \ldots, \pm 1$, as well as the integer* 0 *with multiplicity* 2, *are exactly the* 24 *parameters of Table* 1.2.

Proof. Let $\psi \in \mathcal{X}_{\mathrm{AL}}(\mathrm{SL}_{24})$ be such that ψ_∞ satisfies the property of the theorem. Write $\psi = \oplus_{i=1}^{k} \pi_i[d_i]$ with $\pi_i \in \Pi_{\mathrm{alg}}(\mathrm{PGL}_{n_i})$ for every integer $1 \leq i \leq k$. Let i be such an integer. The condition on ψ_∞ implies $(\mathrm{w}(\pi_i) + d_i - 1)/2 \leq 11$, and then the inequality $\mathrm{w}(\pi_i) \leq 22$. By Theorem 9.3.2, for every integer i, the representation π_i is therefore in the set Π introduced just before Proposition 9.2.2. An application of that proposition concludes the proof. □

Proof of Theorem E. We use the notation of Sect. 9.2.4: in particular, we have 24 elements $\psi_i \in \mathcal{X}(\mathrm{SO}_{24})$, for $i = 1, \ldots, 24$, which are distinct by Nebe and Venkov, and we must prove that they are the elements of Table 1.2. Since this table also has exactly 24 elements, and given Theorem 9.4.4, it remains to prove that we have $\psi_i \in \mathcal{X}_{\mathrm{AL}}(\mathrm{SL}_{24})$ for every i.

Following Ikeda and Böcherer, we already know that we have $\psi_{24} = \Delta_{11}[12]$ (Corollary 7.3.4; see also the beginning of the proof of Proposition 7.5.1). Let $1 \leq i \leq 23$, and let $\psi_i' \in \mathcal{X}(\mathrm{SL}_{2g_i+1})$ be the standard parameter of the ϑ-correspondent of

π_i in $\Pi_{\text{cusp}}(\text{Sp}_{2g_i})$ (the notation π_i and g_i is recalled in Sect. 9.2.4). By Lemma 9.2.9, we have $g_i \leq 11$, so that by Rallis (Corollary 7.1.3) we have the identity

$$\psi_i = \psi_i' \oplus [23 - 2g_i] \, .$$

By Arthur's Theorem 8.1.1, we have $\psi_i' \in \mathcal{X}_{\text{AL}}(\text{SL}_{2g_i+1})$. This implies $\psi_i \in \mathcal{X}_{\text{AL}}(\text{SL}_{24})$ and concludes the proof. □

Remark 9.4.5. In the proof above, it was convenient to treat the case of the parameter $\Delta_{11}[12]$ separately, which was possible thanks to [108] and [31]. An examination of the proof we will give for Theorem 9.5.2 in fact shows that we could have done without these two references and simply used Erokhin's result [80].

9.5 Siegel Modular Forms of Weight at Most 12

The aim of this section is to study the space $\text{S}_k(\text{Sp}_{2g}(\mathbb{Z}))$, with $g \leq k \leq 12$. We wish to first determine its dimension, and then describe, for every eigenform $F \in \text{S}_k(\text{Sp}_{2g}(\mathbb{Z}))$ for the action of $\text{H}(\text{Sp}_{2g})$, the standard parameter $\psi(\pi_F, \text{St})$ of the representation $\pi_F \in \Pi_{\text{cusp}}(\text{Sp}_{2g})$ generated by F. For the sake of brevity, we simply say that $\psi(\pi_F, \text{St})$ is *the standard parameter of the eigenform* F.

The problem of determining the dimension of $\text{S}_k(\text{Sp}_{2g}(\mathbb{Z}))$ has been the object of works by many authors, in different particular cases. We refer, for example, to the articles of Poor and Yuen [167, 168], and to that of Nebe and Venkov [156], for a discussion of the dimensions known before this work and the recent article [195]. Recall that the case $g = 1$ is classical, and that in the cases $g = 2$ and 3, a formula for $\dim \text{S}_k(\text{Sp}_{2g}(\mathbb{Z}))$ valid for every k was proved by, respectively, Igusa [105] and Tsuyumine [200]. The situation for $g > 3$ has long remained very partial, in the sense that $\dim \text{S}_k(\text{Sp}_{2g}(\mathbb{Z}))$ had only been determined for a (finite) small number of pairs (g, k) with $g > 3$, $k \geq 0$, and $gk \equiv 0 \bmod 2$. It has recently evolved substantially with the algorithm of Taïbi [195], which has led to a concrete formula for $\dim \text{S}_k(\text{Sp}_{2g}(\mathbb{Z}))$ valid for every $k > g$ and every $g \leq 7$; this formula is, however, still conditional on Conjecture 8.4.22, at least at the time when we write this!

The method we will use is to a great extent independent of those of the authors mentioned above. In particular, it uses none of the computations mentioned above in genus $g > 2$ and therefore gives new proofs of the previously known cases in these genera. In spirit, it is close to the proof by Duke and Imamoḡlu [77] of the vanishing $\text{S}_k(\text{Sp}_{2g}(\mathbb{Z})) = 0$ for every $g \geq 1$ and every $k \leq 6$. Indeed, our proof and theirs have in common the use of "explicit formulas" (in the sense of Sect. 9.3). Duke and Imamoḡlu apply them to the standard L-function of a Siegel eigenform, basing themselves on the work of Böcherer and Mizumoto recalled in Sect. 8.7, while we applied them to the L-functions of pairs of cuspidal automorphic representations of linear groups (Jacquet, Piatetski-Shapiro, Shalika). Our approach is, of course, permitted by Arthur's Theorem 8.1.1. Viewed from this perspective, the crucial ingredient of the proofs that will follow becomes Theorem 9.3.2.

9.5.1 Forms of Weight 12 *and a Proof of Theorem* D *of the Introduction*

Recall that for every $g \geq 1$, we have a linear map $\vartheta_g \colon \mathbb{C}[\mathrm{X}_{24}] \to \mathrm{M}_{12}(\mathrm{Sp}_{2g}(\mathbb{Z}))$, as well as $\vartheta_0 \colon \mathbb{C}[\mathrm{X}_{24}] \to \mathbb{C}$ (Sect. 5.1).

Theorem* 9.5.2. (i) *The dimension of the spaces* $\mathrm{S}_{12}(\mathrm{Sp}_{2g}(\mathbb{Z}))$ *for* $g \leq 12$ *is given by the following table:*

g	1	2	3	4	5	6	7	8	9	10	11	12
$\dim \mathrm{S}_{12}(\mathrm{Sp}_{2g}(\mathbb{Z}))$	1	1	1	2	2	3	3	4	2	2	1	1

In particular, $\bigoplus_{1\leq g\leq 12} \mathrm{S}_{12}(\mathrm{Sp}_{2g}(\mathbb{Z}))$ *is of dimension* 23.

(ii) *For every* $1 \leq g \leq 12$, *the map* $\vartheta_g \colon \mathbb{C}[\mathrm{X}_{24}] \to \mathrm{M}_{12}(\mathrm{Sp}_{2g}(\mathbb{Z}))$ *induces an isomorphism* $\mathrm{Ker}\, \vartheta_{g-1}/\mathrm{Ker}\, \vartheta_g \xrightarrow{\sim} \mathrm{S}_{12}(\mathrm{Sp}_{2g}(\mathbb{Z}))$.

(iii) *There exist exactly* 23 *representations in the* $\Pi_{\mathrm{cusp}}(\mathrm{Sp}_{2g}(\mathbb{Z}))$ *with* $1 \leq g \leq 12$, *that are generated by a Siegel eigenform of weight* 12. *Their standard parameters are those of Table* C.1.

Proof. We begin by establishing a preliminary lemma that explains the contents of Table C.1. Recall the set

$$\Pi = \{\mathrm{Sym}^2\Delta_{11}\,,\ \Delta_{21,13}\,,\ \Delta_{21,9}\,,\ \Delta_{21,5}\,,\ \Delta_{21}\,,\ \Delta_{19,7}\,,\ \Delta_{19}\,,\ \Delta_{17}\,,\ \Delta_{15}, \Delta_{11}\,,\ 1\} \tag{9.5.1}$$

introduced before Proposition 9.2.2. Consider, for $1 \leq g \leq 12$, the set Φ_g of all elements $\phi \in \mathcal{X}(\mathrm{SL}_{2g+1})$ such that

(a) the eigenvalues of ϕ_∞ are the $2g+1$ integers 0 and $\pm(12-j)$ with $j = 1, \ldots, g$;
(b) there exist $r \geq 1$, integers $d_1, \ldots, d_r \geq 1$, and representations $\pi_1, \ldots, \pi_r \in \Pi$, such that $\phi = \oplus_{i=1}^r \pi_i[d_i]$.

If $1 \leq g \leq 12$ and $\phi \in \Phi_g$, we say that ϕ *satisfies Condition* (C) if in its decomposition as in part (b) above, there does not exist an integer $1 \leq i \leq r$ such that $\pi_i = 1$ and $d_i > 1$, and if there always exists at most one integer i such that $\pi_i = 1$. Condition (C) is always satisfied if $g < 11$, because in this case 1 is not an eigenvalue of ϕ_∞ and 0 is a simple eigenvalue of it.

Lemma 9.5.3. *The set of elements of* $\coprod_{1\leq g\leq 12} \Phi_g$ *satisfying Condition* (C) *is exactly the set of parameters gathered in Table* C.1.

Proof. This is an exercise in combinatorics of the same nature as the one carried out in the proof of Proposition 9.2.2. We can deduce it from this proposition as follows.

Let $\phi \in \Phi_g$ with $g \leq 11$. By Proposition 9.2.2, $\phi \oplus [23-2g]$ is an element of Table 1.2. The property $\phi \in \mathcal{X}_{\mathrm{AL}}(\mathrm{SL}_{2g+1})$, Condition (C), and the Jacquet–Shalika theorem (Proposition 6.4.5) then determine ϕ uniquely. This is, in fact, how we defined the 22 elements of Table C.1 corresponding to the genera $g < 12$.

It remains to see that the only element $\phi \in \Phi_{12}$ satisfying Condition (C) is $\Delta_{11}[12] \oplus [1]$. For this, write $\phi = \oplus_{i=1}^{r} \pi_i[d_i]$ with $\pi_i \in \Pi$ for every i. By Condition (C), there exists at most one i such that $\pi_i = 1$. Since the eigenvalue 11 of ϕ_∞ is simple, there also exists at most one integer i such that $\pi_i = \mathrm{Sym}^2\Delta_{11}$. Since 0 is a triple eigenvalue of ϕ_∞, Lemma 9.2.3 (i) shows that there exists an integer i such that $\pi_i = \Delta_{11}$ and $d_i = 12$. The only possibility is then $\phi = \Delta_{11}[12] \oplus [1]$. □

Having established this preliminary lemma, consider an eigenform $F \in \mathrm{S}_{12}(\mathrm{Sp}_{2g}(\mathbb{Z}))$ for $\mathrm{H}(\mathrm{Sp}_{2g})$ with $1 \leq g \leq 12$ and its standard parameter $\psi \in \mathcal{X}(\mathrm{SL}_{2g+1})$. By Corollary 6.3.6, the eigenvalues of ψ_∞ are the $2g+1$ integers $\pm 11, \pm 10, \ldots, \pm(12-g)$ and 0. All these eigenvalues are simple, except for the eigenvalue 0 for $g = 12$, which has multiplicity 3. By Arthur's Theorem 8.1.1, we can write

$$\psi = \oplus_{i=1}^{r} \pi_i[d_i] \tag{9.5.2}$$

with $d_i \geq 1$ and $\pi_i \in \Pi_{\mathrm{cusp}}(\mathrm{PGL}_{n_i})$ for $i = 1, \ldots, r$. We see that the π_i are algebraic of motivic weight at most 22. By Theorem 9.3.2, it follows that for every i, the representation π_i is in the set Π. In other words, we have $\psi \in \Phi_g$.

Lemma⋆ 9.5.4. *Let* $1 \leq g \leq 12$, *let* $F \in \mathrm{S}_{12}(\mathrm{Sp}_{2g}(\mathbb{Z}))$ *be an eigenform and* ψ *the standard parameter of* F. *Then* ψ *is in* Φ_g *and satisfies Condition* (C).

Proof. We have just verified $\psi \in \Phi_g$. Since condition (C) automatically holds for $g \leq 10$, we may assume $g \geq 11$. Write $\psi = \oplus_{i=1}^{r} \pi_i[d_i]$ with $\pi_i \in \Pi$ for every i.

THE CASE $g = 11$. If $g = 11$, then 0 is a simple eigenvalue of ψ_∞. We may therefore assume, after reindexing the π_i if necessary, that we have[5] $\pi_r = 1$ and $d_r > 1$. Set

$$g' = \frac{23 - d_r}{2}.$$

This is an integer satisfying $0 \leq g' \leq 10$ because d_r is odd and greater than 1. Hence, either we have $\psi = [23]$, or we have $d_r < 23$ and ψ can be written as $\psi' \oplus [d_r]$ with $\psi' \oplus [1] \in \Phi_{g'}$. The inequality $g' \leq 10$ ensures that $\psi' \oplus [1]$ satisfies Condition (C): it is one of the 12 elements of Table C.1 containing [1] and satisfying

[5] The reader who has digested the considerations of Chap. 8 will note that this assumption is in contradiction with Conjecture 8.4.22, for example by Theorem 8.5.2; we will, indeed, end up with a contradiction, but by using instead the results of Sect. 8.7.

$g' \leq 10$ (Lemma 9.5.3). Consequently, in all, there are 13 possibilities for ψ, and it remains to prove that none of them is possible. By Proposition 8.7.1 applied to the form F (the case $k = g+1$), for this it suffices to verify that we have $\delta(\pi_F, (d_r+1)/2) = 0$ in each case, in the notation loc. cit. (see formula (8.7.1)). This is obvious if $\psi = [23]$. In the other cases, we conclude by Lemma 9.5.5 below and the relation $d_r + 1 = 24 - 2g'$.

THE CASE $g = 12$. We proceed similarly in the case of genus $g = 12$. Suppose that ψ does not satisfy Condition (C). We assert that one of the π_i equals $\mathrm{Sym}^2 \Delta_{11}$ and that two of the π_i equal 1. Indeed, ψ_∞ admits 0 as triple eigenvalue and the integers $\pm 1, \ldots, \pm 11$ as simple eigenvalues. If $(\pi_i[d_i])_\infty$, for $i = 1, \ldots, r$, admits the eigenvalue 0, Lemma 9.2.3 (i) shows that we are in one of the following cases: $\pi_i = \mathrm{Sym}^2 \Delta_{11}$ and $d_i = 1$, or $\pi_i = \Delta_{11}$ and $d_i = 12$, or $\pi_i = 1$. The second case is excluded because it implies $\psi = \Delta_{11}[12] \oplus [1]$, which satisfies Condition (C). Since the eigenvalue 11 of ψ_∞ is simple, this proves the affirmation above.

So, we have $r \geq 3$ and, after reindexing the π_i if necessary, we may assume that we have $\pi_1 = \mathrm{Sym}^2 \Delta_{11}$, $\pi_r = \pi_{r-1} = 1$, $d_{r-1} = 1$, and $d_r > 1$ (following Arthur's Theorem 8.2.4, recall that we cannot have $d_r = d_{r-1} = 1$). In particular, if we once again set $g' = (23 - d_r)/2$, we have

$$\psi = \psi' \oplus [1] \oplus [d_r]$$

with $\psi' \in \Phi_{g'}$ containing $\mathrm{Sym}^2 \Delta_{11}$. The inequality $d_r \geq 3$, that is, $g' \leq 10$, shows that ψ' satisfies Condition (C): it is in Table C.1 by Lemma 9.5.3. By examining the table, we see that there are nine possibilities for ψ', and hence for ψ.

We then exclude each of these nine possibilities by using Proposition 8.7.1 applied to F (the case $k = g$). This proposition concludes the proof in these cases because of Lemma 9.5.5, which contradicts the inequality $\delta(\pi_F, (d_r+1)/2) > 0$ (note the relation $d_r + 1 = 24 - 2g'$). This concludes the proof of Lemma 9.5.4. □

Lemma 9.5.5. *Let* $1 \leq g' \leq 10$ *and* $\phi \in \Phi_{g'}$. *Write* $\phi = \oplus_{i=1}^{s} \varpi_i[q_i]$. *Then we have* $\mathrm{L}(1/2, \varpi_i) \neq 0$ *for every* $1 \leq i \leq s$ *such that* $q_i \geq 24 - 2g'$ *and* $\pi_i \neq 1$.

Proof. Only the ϕ of Table C.1 containing a factor of the form $\Delta_{17}[d]$ (resp. $\Delta_{21}[2]$) deserve attention, by Proposition 9.3.39. Examining them shows that we always have $24 - 2g' > d$ (resp. $g' \leq 10$), concluding the proof. □

Let us finish the proof of the theorem. For this, we first verify assertion (ii), namely that for every $g = 1, \ldots, 12$, the injection

$$\mathrm{Ker}\, \vartheta_{g-1} / \mathrm{Ker}\, \vartheta_g \to \mathrm{S}_{12}(\mathrm{Sp}_{2g}(\mathbb{Z}))$$

induced by ϑ_g is surjective. We need to see that if $g \leq 12$, every eigenform $F \in \mathrm{S}_{12}(\mathrm{Sp}_{2g}(\mathbb{Z}))$ is in the image of ϑ_g. The two lemmas above assert that its standard parameter $\psi(\pi_F, \mathrm{St})$ is in Table C.1. We conclude by observing that in all cases, Böcherer's criterion applies: this has, in fact, already been justified during the verification of part 2 in Sect. 9.2.11. Assertion (ii) is therefore proved.

Next, observe that if $g \le 12$ and if $G, H \in \mathrm{S}_{12}(\mathrm{Sp}_{2g}(\mathbb{Z}))$ are two eigenforms for $\mathrm{H}(\mathrm{Sp}_{2g})$ such that $\psi(\pi_G, \mathrm{St}) = \psi(\pi_H, \mathrm{St})$, then G and H are proportional. Indeed, the previous observation ensures that there exist $G', H' \in \mathbb{C}[\mathrm{X}_{24}]$ such that $\vartheta_g(G') = G$ and $\vartheta_g(H') = H$. Since G and H are eigenforms, the Eichler commutation relations (Proposition 5.1.1) assert that we may assume that G' and H' are eigenforms for T_2. This same relation and the identity $\psi(\pi_G, \mathrm{St}) = \psi(\pi_H, \mathrm{St})$ force G' and H' to have the same eigenvalue for T_2: they are therefore proportional by the computation of Nebe and Venkov, and therefore G and H are proportional.

Let Φ be the set of parameters ϕ of Table C.1 such that there exists an eigenform $G \in \mathrm{S}_{12}(\mathrm{Sp}_{2g}(\mathbb{Z}))$ satisfying $\phi = \psi(\pi_G, \mathrm{St})$ (where the integer g is, of course, uniquely determined by ϕ). The previous subsection proves $|\Phi| = \sum_{g=1}^{12} \dim \mathrm{S}_{12}(\mathrm{Sp}_{2g}(\mathbb{Z}))$. Assertion (ii) of the theorem then implies

$$|\Phi| = \sum_{g=1}^{12} \dim(\mathrm{Ker}\, \vartheta_{g-1}/\mathrm{Ker}\, \vartheta_g) = \dim(\mathrm{Ker}\, \vartheta_0) - \dim(\mathrm{Ker}\, \vartheta_{12}) = 23\,,$$

where the last equality comes from Erokin's result $\mathrm{Ker}\, \vartheta_{12} = 0$ [80]. Since there are only 23 parameters in Table C.1, Φ is the set of all parameters of this table, which proves assertions (i) and (iii) of the theorem (and justifies Remark 9.4.5). □

Corollary 9.5.6. *Let $1 \le g \le 12$. The map $\vartheta_g\colon \mathbb{C}[\mathrm{X}_{24}] \to \mathrm{M}_{12}(\mathrm{Sp}_{2g}(\mathbb{Z}))$ is surjective, and the dimension of $\mathrm{M}_{12}(\mathrm{Sp}_{2g}(\mathbb{Z}))$ is given by the following table:*

g	1	2	3	4	5	6	7	8	9	10	11	12
$\dim \mathrm{M}_{12}(\mathrm{Sp}_{2g}(\mathbb{Z}))$	2	3	4	6	8	11	14	18	20	22	23	24

Proof. We proceed by induction on the integer g. The result is well known if $g = 1$. If $g > 1$, recall that we have the Siegel operator $\Phi_g\colon \mathrm{M}_{12}(\mathrm{Sp}_{2g}(\mathbb{Z})) \to \mathrm{M}_{12}(\mathrm{Sp}_{2g-2}(\mathbb{Z}))$; it satisfies the relation $\Phi_g \circ \vartheta_g = \vartheta_{g-1}$. By the induction hypothesis, the map $\Phi_g \circ \vartheta_g$ is therefore surjective. But the map $\vartheta_g\colon \mathrm{Ker}\, \vartheta_{g-1} \to \mathrm{S}_{12}(\mathrm{Sp}_{2g}(\mathbb{Z}))$ is also surjective if $g \le 12$, by assertion (ii) of Theorem 9.5.2. This proves the the surjectivity of ϑ_g and Φ_g. □

Remark 9.5.7. It is not true that every Siegel cusp form of weight 16 for $\mathrm{Sp}_{2g}(\mathbb{Z})$ is a linear combination of theta series of elements of X_{32}, as follows from Corollary 7.3.5 (the given counterexample is in genus $g = 14$).

9.5.8 Forms of Weight at Most 11

Theorem⋆ 9.5.9. *Let $k, g \in \mathbb{Z}$ be such that $g \le k \le 11$. Then we have $\mathrm{S}_k(\mathrm{Sp}_{2g}(\mathbb{Z})) = 0$ unless we are in one of the following cases:*

(i) *$k = 8$ and $g = 4$: In this case, $\mathrm{S}_8(\mathrm{Sp}_8(\mathbb{Z}))$ is of dimension 1, generated by the Schottky form, with standard parameter $\Delta_{11}[4] \oplus [1]$.*

(ii) $k = 10$ *and* $g = 2$*: In this case,* $\mathrm{S}_{10}(\mathrm{Sp}_4(\mathbb{Z}))$ *is of dimension* 1*, generated by the Saito–Kurorawa form* F_{10}*, with standard parameter* $\Delta_{17}[2] \oplus [1]$.
(iii) $k = 10$ *and* $g = 4$*: In this case,* $\mathrm{S}_{10}(\mathrm{Sp}_8(\mathbb{Z}))$ *is of dimension* 1*, generated by the Ikeda form with standard parameter* $\Delta_{15}[4] \oplus [1]$.
(iv) $k = 10$ *and* $g = 6$*: In this case,* $\mathrm{S}_{10}(\mathrm{Sp}_{12}(\mathbb{Z}))$ *is of dimension* 1*, generated by a form with standard parameter* $\Delta_{17}[2] \oplus \Delta_{11}[4] \oplus [1]$.
(v) $k = 10$ *and* $g = 8$*: In this case,* $\mathrm{S}_{10}(\mathrm{Sp}_{16}(\mathbb{Z}))$ *is of dimension* 1*, generated by the Ikeda form with standard parameter* $\Delta_{11}[8] \oplus [1]$.
(vi) $k = 11$ *and* $g = 6$*: In this case, every eigenform of* $\mathrm{S}_{11}(\mathrm{Sp}_{12}(\mathbb{Z}))$ *has standard parameter* $\Delta_{17}[4] \oplus \Delta_{11}[2] \oplus [1]$*. Moreover, if we admit Conjecture* 8.4.22*, we have* $\mathrm{S}_{11}(\mathrm{Sp}_{12}(\mathbb{Z})) = 0$.

Note that the particular case $g = 2$ of Theorem 9.5.9 follows from the work of Igusa [105], who shows that $\mathrm{S}_k(\mathrm{Sp}_4(\mathbb{Z}))$ is zero if $k \leq 11$, and of dimension 1 and generated by F_{10} if $k = 10$; see Sect. 9.1. Also recall that the vanishing $\mathrm{S}_k(\mathrm{Sp}_{2g}(\mathbb{Z})) = 0$ for $k \leq 6$ is due to Duke and Imamoḡlu [77].

Proof. Let $g \leq k \leq 11$, and let $F \in \mathrm{S}_k(\mathrm{Sp}_{2g}(\mathbb{Z}))$ be an eigenform for $\mathrm{H}(\mathrm{Sp}_{2g})$, with generated representation $\pi_F \in \Pi_{\mathrm{disc}}(\mathrm{Sp}_{2g})$ and standard parameter $\psi = \psi(\pi_F, \mathrm{St}) = \oplus_{i=1}^{r} \pi_i[d_i]$, by Arthur's results. Recall that the eigenvalues of $\psi(\pi_F, \mathrm{St})_\infty$ are the $2g+1$ integers 0 and $\pm(k-j)$ for $j = 1, \ldots, g$, by Corollary 6.3.6. The representations π_i are therefore algebraic of motivic weight $w \leq 20$. Theorem 9.3.2 asserts that the π_i are in the set

$$\Pi = \{1, \Delta_{11}, \Delta_{15}, \Delta_{17}, \Delta_{19}, \Delta_{19,7}\} .$$

It follows that ψ is a sum, without multiplicities by Arthur [13, Theorem 1.5.2], of elements of the form $[m]$ with $m \geq 1$ odd, or of the set

$$\Psi := \{\Delta_w[d] \mid d \equiv 0 \bmod 2, w + d - 1 \leq 20\} \coprod \{\Delta_{19,7}[2]\} .$$

Lemma⋆ 9.5.10. *If* $F \in \mathrm{S}_k(\mathrm{Sp}_{2g}(\mathbb{Z}))$ *is an eigenform with* $g \leq k \leq 11$*, then we have* $k > g$*. Moreover, if* $\psi = \oplus_{i=1}^{k} \pi_i[d_i]$ *denotes the standard parameter of* F*, then there exists a unique* $i \in \{1, \ldots, r\}$ *such that* $\pi_i = 1$ *and we then have* $d_i = 1$.

Let us temporarily admit this lemma and continue the proof of Theorem 9.5.9. Note that its statement is obvious under the assumption $k > g + 1$, which is already a sufficiently interesting case. We therefore have $k > g$ and ψ is a sum without multiplicities of $[1]$ and elements of Ψ. We are thus led to study a rather small number of possibilities, which we will do case by case.

THE CASE $k \leq 6$. In this case, we have $\mathrm{w}(\pi_i) \leq 10$ for every i, and therefore $\psi = [1]$ and the form F does not exist!

THE CASE $k = 7$. In this case, we have $\mathrm{w}(\pi_i) \leq 11$, and therefore $\pi_i \in \{\Delta_{11}, 1\}$ for every i. The only possibility is $\psi = \Delta_{11}[2] \oplus [1]$, and in particular $g = 2$, a case we have already treated: the form F also does not exist because according to Igusa, we have the vanishing $\dim \mathrm{S}_7(\mathrm{Sp}_4(\mathbb{Z})) = 0$.

THE CASE $k = 8$. The only possibility is $\psi = \Delta_{11}[4] \oplus [1]$. In particular, $g = 4$ and $F \in \mathrm{S}_8(\mathrm{Sp}_8(\mathbb{Z}))$. Incidentally, note that if we apply our reasoning to the case where $F = J = \vartheta_4(\mathrm{E}_8 \oplus \mathrm{E}_8) - \vartheta_4(\mathrm{E}_{16})$ is the Schottky form (see Sect. 5.2), we obtain a new proof of the fact that the standard parameter of the representation generated by J is $\Delta_{11}[4] \oplus [1]$ (Corollary 7.2.6 (i)), because it is the only possible parameter.

To conclude the proof of Theorem 9.5.9 for $k = 8$, it suffices to invoke the fact that $\mathrm{S}_8(\mathrm{Sp}_8(\mathbb{Z}))$ is of dimension 1 (and generated by the Schottky form) by a result of Poor and Yuen [167] already mentioned in Sect. 5.2. Let us give another argument. Note that F is in the image of the linear map $\vartheta_4 \colon \mathbb{C}[\mathrm{X}_{16}] \to \mathrm{S}_8(\mathrm{Sp}_8(\mathbb{Z}))$. Indeed, Böcherer's criterion applies because the function

$$\mathrm{L}(s, \pi_F, \mathrm{St}) = \zeta(s) \prod_{i=0}^{3} \mathrm{L}\Big(s + i - \frac{3}{2}, \Delta_{11}\Big)$$

does not vanish at $s = 4$ (Sect. 7.2.4). Since $\mathrm{S}_k(\mathrm{Sp}_{2g}(\mathbb{Z}))$ is generated by eigenforms, this shows $\mathrm{S}_8(\mathrm{Sp}_8(\mathbb{Z})) \subset \mathrm{Im}\,\vartheta_4$. This finishes the proof because it is obvious that $\mathrm{S}_8(\mathrm{Sp}_8(\mathbb{Z})) \cap \mathrm{Im}\,\vartheta_4$ is generated by J (Sect. 5.2). Let us emphasize that this argument for proving $\dim \mathrm{S}_8(\mathrm{Sp}_8(\mathbb{Z})) = 1$ is not new: it was already observed by Duke and Imamoğlu in [77]; they were able to prove that $\mathrm{L}(s, \pi_F, \mathrm{St})$ necessarily has a simple pole at $s = 1$, without, however, being able to directly deduce from this the presumed exact form of the parameter ψ.

Finally, let us give a third argument to prove that $\mathrm{S}_8(\mathrm{Sp}_8(\mathbb{Z}))$ is of dimension 1. Indeed, it is an immediate consequence of our analysis above and the following general result of Ikeda refining his own Theorem 7.3.1, which will turn out to be quite useful in the remainder of this proof (see also Example 8.5.3).

Lemma 9.5.11 ([110, Theorem 7.1, Sect. 15]). *Let m and g be even integers, and let $\pi \in \Pi_{\mathrm{cusp}}(\mathrm{PGL}_2)$ be the representation generated by an eigenform of weight m for $\mathrm{SL}_2(\mathbb{Z})$. There exists an eigenform $G \in \mathrm{S}_{(m+g)/2}(\mathrm{Sp}_{2g}(\mathbb{Z}))$ for $\mathrm{H}(\mathrm{Sp}_{2g})$ such that $\psi(\pi_G, \mathrm{St}) = \pi[g] \oplus [1]$ if and only if $m \equiv g \bmod 4$. Moreover, if this condition is satisfied, then the form G is unique up to a scalar.*

THE CASE $k = 9$. This time, there are three possibilities for ψ, corresponding, respectively, to genus 2, 4, and 6. That of genus 2 is $\psi = \Delta_{15}[2] \oplus [1]$, which does not occur because $\dim \mathrm{S}_9(\mathrm{Sp}_4(\mathbb{Z})) = 0$ by Igusa. The second is

$$\psi = \Delta_{15}[2] \oplus \Delta_{11}[2] \oplus [1] ,$$

for which $g = 4$. Consider the linear map (Sect. 5.4.1)

$$\vartheta_{5,4} \colon \mathrm{M}_{\mathrm{H}_{5,4}(\mathbb{R}^8)}(\mathrm{O}_8) \to \mathrm{S}_9(\mathrm{Sp}_8(\mathbb{Z})) .$$

Since we have $\mathrm{L}(1/2, \Delta_w) \neq 0$ for $w = 11, 15$, the product $\zeta(s) \prod_{i=0}^{1} \mathrm{L}(s + i - 1/2, \Delta_{11}) \mathrm{L}(s + i - 1/2, \Delta_{15})$ has a simple pole at $s = 1$, so that Böcherer's criterion applies and shows that $\vartheta_{5,4}$ is surjective. To eliminate this second case, it therefore suffices to show $\mathrm{M}_{\mathrm{H}_{5,4}(\mathbb{R}^8)}(\mathrm{O}_8) = 0$. But this vanishing follows from the tables of

[55, Sect. 2]; we have, in fact, already come across this property in Sect. 7.4. A more direct way to obtain it is to note that by triality (Sect. 5.4.14), we have, for every even integer $d \geq 0$, the equality

$$\dim \mathrm{M}_{\mathrm{H}_{d,1}(\mathbb{R}^8)}(\mathrm{SO}_8) = \dim \mathrm{M}_{\mathrm{H}_{d/2,4}(\mathbb{R}^8)}(\mathrm{O}_8)\ . \tag{9.5.3}$$

This suffices to conclude the proof because we have $\mathrm{M}_{\mathrm{H}_{10,1}(\mathbb{R}^8)}(\mathrm{SO}_8) = 0$ by Lemma 5.4.2.

To reassure ourselves, let us verify that this last reasoning is coherent with the formula of Theorem 8.5.2, in other words, that Arthur's multiplicity formula indeed suggests that $\Delta_{15}[2] \oplus \Delta_{11}[2] \oplus [1]$ is not the standard parameter of a Siegel form. But this follows from the fact that in the notation of that theorem, we have $\chi(s_2) = -1$ and $\varepsilon(\Delta_{15} \times \Delta_{11}) = \varepsilon(\Delta_{11}) = 1$.

The last possibility is $\psi = \Delta_{11}[6] \oplus [1]$, for which $g = 6$. According to Ikeda, this does not occur, because we have $6 \not\equiv 12 \bmod 4$ (Lemma 9.5.11).

The Case $k = 10$. This time, there are four possibilities for ψ, corresponding, respectively, to genus 2, 4, 6, and 8, namely

$$\Delta_{17}[2] \oplus [1]\ ,\ \Delta_{15}[4] \oplus [1]\ ,\ \Delta_{17}[2] \oplus \Delta_{11}[4] \oplus [1]\ ,\ \Delta_{11}[8] \oplus [1]\ .$$

They are all treated by Lemma 9.5.11, except for that of genus 6, namely $\psi = \Delta_{17}[2] \oplus \Delta_{11}[4] \oplus [1]$. This shows the assertions (ii)–(v) of the theorem, except for the fact that $\dim \mathrm{S}_{10}(\mathrm{Sp}_{12}(\mathbb{Z})) = 1$ in part (iv).

Let us now prove this assertion. We have seen, above, that every eigenform of $\mathrm{S}_{10}(\mathrm{Sp}_{12}(\mathbb{Z}))$ has standard parameter $\Delta_{17}[2] \oplus \Delta_{11}[4] \oplus [1]$. An examination of Arthur's multiplicity formula suggests that such a form exists and has multiplicity 1. Let us take a different approach. Combined with Böcherer's criterion, this property also shows that the linear map

$$\vartheta_{2,6} \colon \mathrm{M}_{\mathrm{H}_{2,6}(\mathbb{R}^{16})}(\mathrm{O}_{16}) \to \mathrm{S}_{10}(\mathrm{Sp}_{12}(\mathbb{Z}))$$

is surjective. To conclude, it therefore suffices to show $\dim \mathrm{M}_{\mathrm{H}_{2,6}(\mathbb{R}^{16})}(\mathrm{O}_{16}) = 1$ and $\vartheta_{2,6} \neq 0$. But the assertion on the dimension follows from part (i) of Corollary 9.5.13. The verification of the nonvanishing of $\vartheta_{2,6}$ will be pushed back to the end of the section so as not to interrupt the flow of this proof.

The Case $k = 11$. This case is already rather tedious. This time, we find eight possible parameters ψ: one in genus 2, two in genus 4, two in genus 6, two in genus 8, and one in genus 10.

The two parameters of genus 4 are $\Delta_{17}[4] \oplus [1]$ and $\Delta_{19}[2] \oplus \Delta_{15}[2] \oplus [1]$. The first does not occur according to Ikeda (Lemma 9.5.11). Consider the map $\vartheta_{7,4} \colon \mathrm{M}_{\mathrm{H}_{7,4}(\mathbb{R}^8)}(\mathrm{O}_8) \to \mathrm{S}_{11}(\mathrm{Sp}_8(\mathbb{Z}))$. Böcherer's criterion then shows that $\vartheta_{7,4}$ is surjective. We will see that it is zero, which shows that the case $\psi = \Delta_{19}[2] \oplus \Delta_{15}[2] \oplus [1]$ does not occur either. For this, note that the relation (9.5.3) ensures $\dim \mathrm{M}_{\mathrm{H}_{7,4}(\mathbb{R}^8)}(\mathrm{O}_8) = 1$, given that we have $\dim \mathrm{M}_{\mathrm{H}_{14,1}(\mathbb{R}^8)}(\mathrm{SO}_8) = 1$ (Lemma 5.4.2). By Theorem 7.2.1, we know that there exists an eigenform

$G \in \mathrm{M}_{\mathrm{H}_{7,4}(\mathbb{R}^8)}(\mathrm{O}_8)$ that generates a representation in $\Pi_{\mathrm{disc}}(\mathrm{O}_8)$ with standard parameter $\Delta_{17}[4]$; we therefore have $\mathrm{M}_{\mathrm{H}_{7,4}(\mathbb{R}^8)}(\mathrm{O}_8) = \mathbb{C}G$. It then suffices to prove $\vartheta_{7,4}(G) = 0$. If this form is nonzero, the Eichler–Rallis relations show that $\vartheta_{7,4}(G)$ is an eigenform with standard parameter $\Delta_{17}[4] \oplus [1]$ (Corollary 7.1.3), which contradicts Lemma 9.5.11 (as well as Böcherer's criterion!).

The two possible parameters of genus 6 are $\Delta_{15}[6] \oplus [1]$ and $\Delta_{17}[4] \oplus \Delta_{11}[2] \oplus [1]$. The first is again excluded by Ikeda (Lemma 9.5.11). The second should not occur, by Arthur's multiplicity formula. Indeed, in the notation of Theorem 8.5.2, we have $\chi(s_2) = -1$ and $\varepsilon(\Delta_{11} \times \Delta_{17})^2 \varepsilon(\Delta_{11}) = 1$. However, we do not see how to eliminate it directly, which explains part (vi) of the statement of the theorem. Note that this time, we cannot deduce anything from the map $\vartheta_{3,6}\colon \mathrm{M}_{\mathrm{H}_{3,6}(\mathbb{R}^{16})}(\mathrm{O}_{16}) \to \mathrm{S}_{11}(\mathrm{Sp}_{12}(\mathbb{Z}))$, because Böcherer's criterion shows $\vartheta_{3,6} = 0$ (in fact, we have $\mathrm{M}_{\mathrm{H}_{3,6}(\mathbb{R}^{16})}(\mathrm{O}_{16}) = 0$ by Lemma 9.5.13).

The two possible parameters of genus 8 are $\Delta_{19,7}[2] \oplus \Delta_{15}[2] \oplus \Delta_{11}[2] \oplus [1]$ and $\Delta_{19}[2] \oplus \Delta_{11}[6] \oplus [1]$. Böcherer's criterion then shows that the map

$$\vartheta_{3,8}\colon \mathrm{M}_{\mathrm{H}_{3,8}(\mathbb{R}^{16})}(\mathrm{O}_{16}) \to \mathrm{S}_{11}(\mathrm{Sp}_{16}(\mathbb{Z}))$$

is surjective, as one can immediately verify using Proposition 9.3.39. Consequently, we have $\mathrm{S}_{11}(\mathrm{Sp}_{16}(\mathbb{Z})) = 0$ because we have $\mathrm{M}_{\mathrm{H}_{3,8}(\mathbb{R}^{16})}(\mathrm{O}_{16}) = 0$ by Lemma 9.5.13 (ii).

The last parameter, of genus 10, is $\Delta_{11}[10] \oplus [1]$, but it does not occur, by Lemma 9.5.11, which concludes the proof of the theorem. It only remains to prove Lemma 9.5.10. □

Proof of Lemma 9.5.10. The analysis carried out before Lemma 9.5.10 shows $\psi = \oplus_{i=1}^{r} \pi_i[d_i]$, where for every integer i, either $\pi_i[d_i]$ belongs to the set Ψ introduced loc. cit., or $\pi_i = 1$. Set $I = \{i, \pi_i = 1\}$; recall that the integers d_i for i in I are odd, and distinct by Arthur.

Assume first we have $g = k$. As the eigenvalue 0 of ψ_∞ has multiplicity 3, and as the other eigenvalues are ≤ 10 in absolute value, Lemma 9.2.3 shows $|I| = 3$. On the other hand, the eigenvalue 1 has multiplicity ≤ 1, so we have $d_i = 1$ for at least two $i \in I$: a contradiction. We have thus proved $k > g$. As the eigenvalue 0 of ψ_∞ has now multiplicity 1, Lemma 9.2.3 shows $|I| = 1$, say $I = \{i\}$.

Assume we have $d_i > 1$. This forces $k = g + 1$, and Proposition 8.7.1 implies

$$\delta\Big(\pi_F, \frac{d_i + 1}{2}\Big) > 0 \,. \tag{9.5.4}$$

By Proposition 9.3.39, the quantity $\delta(\pi_F, a)$ is the number of components of ψ of the form $\Delta_{17}[d']$ with $d' \geq 2a$. For such a component we necessarily have $d' \in \{2, 4\}$. This shows $d_i \leq 3$, and thus $d_i = 3$. Note that $\Delta_{17}[d']_\infty$ has the eigenvalue 9 (in particular, we have $k = 10$ or $k = 11$). On the other hand, if ϕ is in Ψ and if 2 is an eigenvalue of ϕ_∞, we must have $\phi = \Delta_{11}[8]$ or $\Delta_{11}[10]$, so 9 is an eigenvalue of ϕ: a contradiction as the eigenvalue 9 has multiplicity 1 in ψ_∞. □

Proposition 9.5.12. *Let $L \subset \mathbb{R}^{16}$ be an even unimodular lattice. Denote by V_λ the irreducible representation of* $\mathrm{SO}(\mathbb{R}^{16})$ *of highest weight*

$$\lambda = m_1\varepsilon_1 + m_2\varepsilon_2 + \cdots + m_8\varepsilon_8 \quad \textit{with} \quad m_1 \geq m_2 \geq \cdots \geq m_8 \geq 0$$

(Sect. 6.4.3*), and by $V_\lambda^{\mathrm{SO}(L)} \subset V_\lambda$ the subspace of invariants under the finite group* $\mathrm{SO}(L) \subset \mathrm{SO}(\mathbb{R}^{16})$. *Suppose $m_1 \leq 4$.*

(i) *If $L \simeq \mathrm{E}_{16}$, then $V_\lambda^{\mathrm{SO}(L)} = 0$, unless λ is equal to 0 or of the form $4(\sum_{i=1}^{k} \varepsilon_i)$ with $1 \leq k \leq 8$, in which case $\dim V_\lambda^{\mathrm{SO}(L)} = 1$.*

(ii) *If $L \simeq \mathrm{E}_8 \oplus \mathrm{E}_8$, then the pairs $(\lambda, \dim V_\lambda^{\mathrm{SO}(L)})$ such that $V_\lambda^{\mathrm{SO}(L)} \neq 0$ are given by Table* C.6.

Proof. This is a computation based on Weyl's character formula, in the manner of those carried out in [55, Sect. 2]. We thank Olivier Taïbi for having let us benefit from his own algorithm, which is faster than that used loc. cit., for the final evaluation. It requires, as preliminary work, an enumeration of the characteristic polynomials of the elements of $\mathrm{SO}(L)$, as well as their multiplicities. We refer to the output of the source code [54] for a justification of the affirmations that follow.

For the lattice $L = \mathrm{E}_{16}$, we have $\mathrm{O}(L) = \mathrm{W}(\mathbf{D}_{16}) \simeq \{\pm 1\}^{15} \ltimes \mathfrak{S}_{16}$, and the enumeration does not pose any difficulty if we use a computer (if we restrict ourselves to elements of determinant 1, for example, we find 823 polynomials). In the case of the lattice $L = \mathrm{E}_8 \oplus \mathrm{E}_8$, the group $\mathrm{O}(L)$ is the semidirect product of $\mathbb{Z}/2\mathbb{Z}$ and $\mathrm{O}(\mathrm{E}_8)^2$, where $\mathbb{Z}/2\mathbb{Z}$ acts on $\mathrm{O}(\mathrm{E}_8)^2$ by interchanging the two factors. The characteristic polynomials of the conjugacy classes of elements of $\mathrm{O}(\mathrm{E}_8) = \mathrm{W}(\mathbf{E}_8)$, as well as the cardinalities of these classes, have been determined by Carter [47, Table 11]. This allows us to conclude by observing that the determinant of a block matrix of the form $\left[\begin{smallmatrix} X\,\mathrm{I}_m & g \\ h & X\,\mathrm{I}_m \end{smallmatrix}\right]$, with $g, h \in \mathrm{M}_m$ and X an indeterminate, is $\det(X^2\mathrm{I}_m - gh)$. For example, after a computation, we find 1544 characteristic polynomials for $\mathrm{SO}(\mathrm{E}_8 \oplus \mathrm{E}_8)$.

Let us mention that with a little patience, which at this point the authors lack, it should also be possible to prove the proposition "by hand" !

Corollary 9.5.13. (i) *We have the equalities*

$$\dim \mathrm{M}_{\mathrm{H}_{2,6}(\mathbb{R}^{16})}(\mathrm{SO}_{16}) = \dim \mathrm{M}_{\mathrm{H}_{2,8}(\mathbb{R}^{16})}(\mathrm{O}_{16}) = 1\ .$$

(ii) *For every integer $1 \leq g \leq 8$, we have $\mathrm{M}_{\mathrm{H}_{3,g}(\mathbb{R}^{16})}(\mathrm{O}_{16}) = 0$.*

Proof. Recall that if U is a representation of $\mathrm{O}_n(\mathbb{R})$, whose restriction to $\mathrm{SO}_n(\mathbb{R})$ we denote by U', then $\mathrm{M}_U(\mathrm{O}_n)$ is a subspace of $\mathrm{M}_{U'}(\mathrm{SO}_n)$ (it is the map res introduced in Sect. 4.4.4). Moreover, formula (4.4.1) implies the equality

$$\dim \mathrm{M}_{U'}(\mathrm{SO}_n) = \dim(U'^{\mathrm{SO}(\mathrm{E}_{16})}) + \dim(U'^{\mathrm{SO}(\mathrm{E}_8 \oplus \mathrm{E}_8)})\ .$$

If $g < r$, the irreducible representation $\mathrm{H}_{d,g}(\mathbb{R}^{2r})$ of $\mathrm{O}(\mathbb{R}^{2r})$ remains irreducible when restricted to $\mathrm{SO}(\mathbb{R}^{2r})$; if $g = r$, it decomposes into a sum of nonisomorphic representations $\mathrm{H}_{d,g}(\mathbb{R}^{2r})^{\pm}$ that are conjugate under the outer action of $\mathrm{O}(\mathbb{R}^{2r})$ (Sect. 5.4.14). Recall that by formula (5.4.1), the highest weight of $\mathrm{H}_{d,g}(\mathbb{R}^{2r})$ is $d\sum_{i=1}^{g}\varepsilon_i$ if $g<r$, and that those of $\mathrm{H}_{d,g}(\mathbb{R}^{2g})^{\pm}$ are $d(\pm\varepsilon_g+\sum_{i=1}^{g-1}\varepsilon_i)$. Part (ii) then follows from Proposition 9.5.12, because the two dimensions in question are zero for the weights of the form $3(\sum_{i=1}^{g}\varepsilon_i)$, for $g\geq 1$. Part (i) is proved similarly, the nonzero invariants now only arising from $\mathrm{E}_8\oplus\mathrm{E}_8$, by observing the isomorphism

$$\mathrm{M}_{\mathrm{H}_{d,8}(\mathbb{R}^{16})}(\mathrm{O}_{16})\simeq\mathrm{M}_{\mathrm{H}_{d,8}(\mathbb{R}^{16})^{\pm}}(\mathrm{SO}_{16})$$

(this is the map ind of Sect. 4.4.4). □

To conclude this subsection, let us say a few words about the construction of an element of $\mathrm{S}_{10}(\mathrm{Sp}_{12}(\mathbb{Z}))$ using theta series. We begin with the lattice $L=\mathrm{E}_8\oplus\mathrm{E}_8$ of $\mathbb{R}^8\oplus\mathbb{R}^8$. Let $e=(e_1,\ldots,e_6)$ be a 6-tuple of elements of $L\otimes\mathbb{C}$ generating an isotropic subspace of dimension 6, and let $P_e(v_1,\ldots,v_6)=\det\left[e_i\cdot v_j\right]_{1\leq i,j\leq 6}$, so that $P_e^2\in\mathrm{H}_{2,6}(L\otimes\mathbb{R})$. Let $Q\subset L$ be a lattice of rank 6 and $v_1,\ldots,v_6$ a $\mathbb{Z}$-basis of Q. For reasons similar to those mentioned in Sect. 5.4.21, the element $P_e(v_1,\ldots,v_6)^2$ does not depend on the choice of the $\mathbb{Z}$-basis v_i of Q, and may therefore be denoted by $P_e(Q)^2$. The Fourier coefficient of $\vartheta_{2,6}(\mathrm{E}_8\oplus\mathrm{E}_8,P_e^2)$ corresponding to the Gram matrix of a $\mathbb{Z}$-basis of Q is then given by the formula

$$\mathrm{c}_Q(P_e^2)=|\mathrm{O}(Q)|\sum_{M\subset\mathrm{E}_8\oplus\mathrm{E}_8}P_e(M)^2\ ,$$

where the sum is taken over the sublattices M isometric to Q. We will apply this to the particular case where $Q\simeq\mathrm{Q}(R)$ with $R=\mathbf{D}_6$ or $R=\mathbf{E}_6$. Set $\mathrm{E}_6:=\mathrm{Q}(\mathbf{E}_6)$. First note that if such a sublattice is in $\mathrm{E}_8\oplus\mathrm{E}_8$, it is necessarily included in one of the two factors E_8. Moreover, it is not difficult to prove that the sublattices of E_8 isometric to D_6 (resp. E_6) are exactly the orthogonal complements of the sublattices of E_8 isometric to $\mathrm{A}_1\oplus\mathrm{A}_1$ (resp. A_2); they can therefore easily be enumerated using a computer (see [54]).

Let us conclude with a numerical application. Denote by $\varepsilon_1,\ldots,\varepsilon_8$ the canonical basis of the first factor $\mathbb{R}^8$, by $\varepsilon'_1,\ldots,\varepsilon'_8$ that of the second, and let $e=(\varepsilon_j+i\varepsilon'_j)_{1\leq j\leq 6}$. The first observation above ensures that if $Q\simeq\mathrm{D}_6,\mathrm{E}_6$, we have $\mathrm{c}_Q(P_e^2)=2|\mathrm{O}(Q)|\sum_{M\subset\mathrm{E}_8\oplus 0}P_e(M)^2$, where, this time, the sum is taken over the sublattices of the first factor E_8 isometric to Q. The second observation allows the evaluation of this sum. The computer gives [54]

$$\frac{\mathrm{c}_{\mathrm{E}_6}(P_e^2)}{2|\mathrm{O}(\mathrm{E}_6)|}=120\quad\text{and}\quad\frac{\mathrm{c}_{\mathrm{D}_6}(P_e^2)}{2|\mathrm{O}(\mathrm{D}_6)|}=540\ .$$

In particular, these two coefficients are nonzero, and it is easy to see that we have

$$\frac{\mathrm{c}_{\mathrm{D}_6}(P_e^2)}{\mathrm{c}_{\mathrm{E}_6}(P_e^2)}=2\ .$$

To conclude, the following corollary gives a concrete process for constructing all Siegel cusp forms of weight at most 11 using theta series.

Corollary⋆ 9.5.14. *The maps $\vartheta_{d,g}$ induce isomorphisms between spaces of dimension* 1*:*

$$\vartheta_{4,4}\colon \mathrm{M}_{\mathrm{H}_{4,4}(\mathbb{R}^8)}(\mathrm{O}_8) \xrightarrow{\sim} \mathrm{S}_8(\mathrm{Sp}_8(\mathbb{Z})),\ \vartheta_{6,2}\colon \mathrm{M}_{\mathrm{H}_{6,2}(\mathbb{R}^8)}(\mathrm{O}_8) \xrightarrow{\sim} \mathrm{S}_{10}(\mathrm{Sp}_4(\mathbb{Z}))\ ,$$
$$\vartheta_{6,4}\colon \mathrm{M}_{\mathrm{H}_{6,4}(\mathbb{R}^8)}(\mathrm{O}_8) \xrightarrow{\sim} \mathrm{S}_{10}(\mathrm{Sp}_8(\mathbb{Z})),\ \vartheta_{2,6}\colon \mathrm{M}_{\mathrm{H}_{2,6}(\mathbb{R}^{16})}(\mathrm{O}_{16}) \xrightarrow{\sim} \mathrm{S}_{10}(\mathrm{Sp}_{12}(\mathbb{Z}))\ ,$$
$$\vartheta_{2,8}\colon \mathrm{M}_{\mathrm{H}_{2,8}(\mathbb{R}^{16})}(\mathrm{O}_{16}) \xrightarrow{\sim} \mathrm{S}_{10}(\mathrm{Sp}_{16}(\mathbb{Z}))\ .$$

Proof. The assertion on $\vartheta_{4,4}$ is Proposition 5.4.22, and that on $\vartheta_{6,4}$ is likewise contained in Table 5.3. The assertion on $\vartheta_{6,2}$ is a particular case of Proposition 9.1.2. That on $\vartheta_{2,6}$ follows from the discussion preceding the corollary. Finally, the assertion on $\vartheta_{2,8}$ follows from $\dim \mathrm{M}_{\mathrm{H}_{2,8}(\mathbb{R}^{16})}(\mathrm{O}_{16}) = 1$ (Lemma 9.5.13 (i)) and Böcherer's criterion because we have $\mathrm{L}(1/2, \Delta_{11}) \neq 0$. □

9.6 Toward a New Proof of the Equality $|\mathrm{X}_{24}| = 24$

The interest of the following theorem resides in the fact that its proof does not use any computation from the theory of unimodular lattices. In particular, it naturally does not use Niemeier's determination of X_{24}. Instead, it relies on Arthur's theory and Theorem 9.3.2. Recall that the sets X_n and $\widetilde{\mathrm{X}}_n$ were introduced in Sect. 4.1.2.

Theorem 9.6.1. *Admitting Conjectures* 8.1.2 *and* 8.4.25*, we have the equalities*

$$|\mathrm{X}_{24}| = 24 \quad \textit{and} \quad |\widetilde{\mathrm{X}}_{24}| = 25\ .$$

Proof. Let us first prove the equality $|\widetilde{\mathrm{X}}_{24}| = 25$. Let π be an element of $\Pi_{\mathrm{disc}}(\mathrm{SO}_{24})$. By Conjecture 8.1.2, we have $\psi(\pi, \mathrm{St}) \in \mathcal{X}_{\mathrm{AL}}(\mathrm{SL}_{24})$. If we, moreover, suppose that the representation π_∞ is trivial, then Theorem 9.4.4 implies that $\psi(\pi, \mathrm{St})$ is one of the parameters listed in Table 1.2. The relation

$$|\widetilde{\mathrm{X}}_{24}| \;=\; \dim \mathrm{M}_{\mathbb{C}}(\mathrm{SO}_{24})$$

shows that it only remains to determine, for each of the parameters ψ of this table, the sum m_ψ of the multiplicities of the representations $\pi' \in \Pi(\mathrm{SO}_{24})$ such that $\pi'_\infty = 1$ and $\psi(\pi', \mathrm{St}) = \psi$. By Conjecture 8.4.25, we always have $m_\psi \in \{0, 1, 2\}$. The exact value of m_ψ can be determined using the formulas given in Sect. 8.5; as it happens, here we use Theorem 8.5.8. This determination of the 24 integers m_ψ, essentially done "case by case," has already been carried out in Sect. 9.2.10. The conclusion of the analysis loc. cit. is the following: if $\psi \in \mathcal{X}_{\mathrm{AL}}(\mathrm{SL}_{24})$ is one of the elements listed in Table 1.2, then either

(a) $\psi \neq \Delta_{11}[12]$ and we have $m_\psi = 1$, or
(b) $\psi = \Delta_{11}[12]$ and we have $m_\psi = 2$.

In the second case, there exist exactly two distinct representations π', π'' in $\Pi_{\mathrm{disc}}(\mathrm{SO}_{24})$ satisfying $\psi(\pi', \mathrm{St}) = \psi(\pi'', \mathrm{St}) = \psi$, and we have $\mathrm{m}(\pi') = \mathrm{m}(\pi'') = 1$ and $\pi'' = S(\pi')$ in the sense of Remark 8.5.9.

We have indeed proved $|\widetilde{\mathrm{X}}_{24}| = 23 + 2 = 25$. Finally, let us show $|\mathrm{X}_{24}| = 24$. We begin with the equality $\dim \mathrm{M}_{\mathbb{C}}(\mathrm{O}_{24}) = |\mathrm{X}_{24}|$. Recall that the group with two elements $\mathrm{O}_{24}(\mathbb{Q})/\mathrm{SO}_{24}(\mathbb{Q}) = \langle s \rangle$ acts on $\mathrm{M}_{\mathbb{C}}(\mathrm{SO}_{24})$ with fixed subspace $\mathrm{M}_{\mathbb{C}}(\mathrm{O}_{24})$ (see Example 4.4.6). Let $f' \in \mathrm{M}_{\mathbb{C}}(\mathrm{SO}_{24})$ be an eigenform under $\mathrm{H}(\mathrm{SO}_{24})$ generating the representation π' introduced in part (b) above. The assertion $S(\pi') \neq \pi'$ shows that the line $\mathbb{C}f'$ is not stable under the action of the element s and implies that the linear subspace

$$V := \langle f, sf \rangle \subset \mathrm{M}_{\mathbb{C}}(\mathrm{SO}_{24})$$

is of dimension 2 and satisfies $\dim(\mathrm{M}_{\mathbb{C}}(\mathrm{O}_{24}) \cap V) = 1$ (by the way, this fact has already been exploited in the proof of Proposition 7.5.1). From this follows, on the one hand, the inequality $|\mathrm{X}_{24}| < 25$ and, on the other hand, the existence of a representation $\pi \in \Pi_{\mathrm{disc}}(\mathrm{O}_{24})$ such that $\psi(\pi, \mathrm{St}) = \Delta_{11}[12]$.

To conclude $|\mathrm{X}_{24}| = 24$, it suffices to show that for each of the 23 elements $\psi \neq \Delta_{11}[12]$ listed in Table 1.2, there exists a representation $\pi \in \Pi_{\mathrm{disc}}(\mathrm{O}_{24})$ such that $\psi(\pi, \mathrm{St}) = \psi$. This is obvious in the special case $\psi = [23] \oplus [1]$, which corresponds to the trivial representation (Examples 6.4.7). For the other cases, the argument we propose is rather indirect. It consists in applying verbatim the method of Sect. 9.2.11. It is summarized as follows: first observe that ψ can be written uniquely in the form $\psi' \oplus [23 - 2g]$ with $1 \leq g < 12$ and $\psi' \in \mathcal{X}_{\mathrm{AL}}(\mathrm{SL}_{2g+1})$, then use Conjecture 8.4.25 to verify (case by case!) that ψ' is the standard parameter of a Siegel cuspidal modular eigenform of weight 12 for $\mathrm{Sp}_{2g}(\mathbb{Z})$, and finally, verify that this modular form is a linear combination of theta series of elements of X_{24} using Böcherer's criterion. None of these arguments use the determination of X_{24}. This concludes the proof. □

9.7 A Few Elements of $\Pi_{\mathrm{disc}}(\mathrm{SO}_n)$ for $n = 15, 17$ and 23

Theorem* 9.7.1. *Let $n \geq 1$ be an odd integer. The elements $\psi \in \mathcal{X}_{\mathrm{AL}}(\mathrm{SL}_{n-1})$ such that the conjugacy class ψ_∞ has as eigenvalues the $n-1$ half-integers $\pm(n-2)/2$, $\pm(n-4)/2$, ..., $\pm 1/2$ are the following:*

(i) *the unique element $[n-1]$ if $n \leq 11$,*
(ii) $[12]$ *and* $\Delta_{11} \oplus [10]$ *if $n = 13$,*
(iii) $[14]$ *and* $\Delta_{11}[3] \oplus [8]$ *if $n = 15$,*
(iv) $[16]$, $\Delta_{15} \oplus [14]$, $\Delta_{15} \oplus \Delta_{11}[3] \oplus [8]$, *and* $\Delta_{11}[5] \oplus [6]$ *if $n = 17$,*
(v) *the 32 parameters listed in Table* C.7 *if $n = 23$.*

Proof. This is a consequence of Theorem 9.3.2, whose proof is similar to those of Theorem 9.4.4 and Proposition 9.2.2. Let us mention that the cases $n = 19$ and 21, although not explicit in the statement, also immediately follow from this method, and even from Table C.7. □

This theorem has consequences for the classification of the $\pi \in \Pi_{\mathrm{disc}}(\mathrm{SO}_n)$ satisfying $\pi_\infty = 1$ if n is odd, in which case the eigenvalues of $\mathrm{St}\,\mathrm{c}_\infty(\pi) \subset \mathfrak{sl}_{n-1}(\mathbb{C})$ are the $n-1$ half-integers $\pm(n-2)/2, \pm(n-4)/2, \ldots, \pm 1/2$.

Theorem 9.7.2. *Assuming Conjecture* 8.1.2, *the standard parameters* $\psi(\pi, \mathrm{St})$ *of the representations* $\pi \in \Pi_{\mathrm{disc}}(\mathrm{SO}_n)$ *such that* $\pi_\infty = 1$ *are*

(i) $[14]$ *and* $\Delta_{11}[3] \oplus [8]$ *if* $n = 15$,
(ii) $[16]$, $\Delta_{15} \oplus [14]$, $\Delta_{15} \oplus \Delta_{11}[3] \oplus [8]$, *and* $\Delta_{11}[5] \oplus [6]$ *if* $n = 17$.

Proof. Recall that if $n = 15$ or $n = 17$, the dimension $\dim \mathrm{M}_{\mathbb{C}}(\mathrm{SO}_n) = |\mathrm{X}_n|$ is, respectively, 2 or 4 by Corollary 4.1.11. The operator T_2 of $\mathbb{C}[\mathrm{X}_n]$ is determined in Sect. B.5. In both cases, its eigenvalues are indeed compatible with the theorem above; what will matter here is that they are distinct. Thus, if $n = 15$ (resp. $n = 17$), there exist exactly two (resp. four) elements of $\Pi_{\mathrm{disc}}(\mathrm{SO}_n)$ satisfying $\pi_\infty = 1$, each with multiplicity 1. Since the integer n is odd, the standard parameters $\psi(\pi, \mathrm{St})$ of these elements are obviously distinct. By Conjecture 8.1.2, these parameters satisfy the assumptions of Theorem 9.7.1. But by the conclusions of that theorem, there exist only two possible parameters for $n = 15$, and four for $n = 17$, which concludes the proof. □

Theorem 9.7.3. *Assuming Conjecture* 8.4.25, *the standard parameters* $\psi(\pi, \mathrm{St})$ *of the representations* $\pi \in \Pi_{\mathrm{disc}}(\mathrm{SO}_{23})$ *such that* $\pi_\infty = 1$ *are the* 32 *elements listed in Table* C.7 *if* $n = 23$.

Proof. Recall the equality $|\mathrm{X}_{23}| = 32$ (Corollary 4.1.11). Let ψ be one of the 32 elements of Table C.7 and $\pi \in \Pi(\mathrm{SO}_{23})$ the unique representation such that $\psi(\pi, \mathrm{St}) = \psi$. We must prove that the multiplicity of π is nonzero. By Conjecture 8.4.25, for this, it suffices to apply Arthur's multiplicity formula in the form given by Theorem 8.5.6. We argue using a case-by-case analysis, which we can simplify slightly by using criteria of the same type as those given in Sect. 8.5.7 in the case $n \equiv 0 \bmod 8$. We leave this as an exercise for the reader because the reasoning is very similar to that studied in detail loc. cit. To conclude, we note, oh miracle, that we indeed have $\mathrm{m}(\pi) = 1$, regardless of the element ψ chosen initially. □

Remark 9.7.4. (i) Theorem 9.7.3 was extended to dimension 25 in [55, Theorem 1.14]. The 121 parameters in question involve, in particular, the seven representations in $\Pi_{\mathrm{alg}}(\mathrm{PGL}_6)$ of motivic weight 23 mentioned in Sect. 9.3.38.

(ii) An argument similar to that of the proof of Theorem 9.6.1 allows one to reprove the equalities $|\mathrm{X}_7| = |\mathrm{X}_9| = 1$, $|\mathrm{X}_{15}| = 2$, $|\mathrm{X}_{17}| = 4$, and $|\mathrm{X}_{23}| = 32$, using Theorem 9.3.2 and conditionally on Conjectures 8.1.2 and 8.4.25, but without any computation from the theory of Euclidean lattices.

Let us conclude with a curious observation. For every integer $n \leq 24$ such that $n \equiv -1, 0, 1 \bmod 8$, we have described, in this book, the subset

$$\Phi_n \subset \mathcal{X}(\mathrm{SL}_{2[n/2]})$$

consisting of the elements of the form $\psi(\pi, \mathrm{St})$ with $\pi \in \Pi_{\mathrm{disc}}(\mathrm{SO}_n)$ such that $\pi_\infty = 1$. For example, Φ_{24} and Φ_{25} are, respectively, given by Tables 1.2 and C.7. This description is still conditional when n is odd, in which case it even extends to $n = 25$ by Remark 9.7.4 (i), but that is not the problem in this discussion, where we would gladly admit Conjecture 8.4.25. Consider the second problem, which appears rather different, of determining the subset

$$\Phi'_n \subset \mathcal{X}_{\mathrm{AL}}(\mathrm{SL}_{2[n/2]})$$

consisting of all the ψ such that ψ_∞ is the infinitesimal character of the trivial representation of $\mathrm{SO}_n(\mathbb{R})$ (a simple condition on its eigenvalues). Arthur's theory first asserts $\Phi_n \subset \Phi'_n$; it also gives an explicit criterion, "Arthur's multiplicity formula," that allows one to determine whether a given element $\psi \in \Phi'_n$ is in Φ_n: these are the formulas in Theorems 8.5.6 and 8.5.8. But Theorem 9.3.2 allows one to determine Φ'_n for every $n \leq 24$. The miraculous property, satisfied in all cases, is then the equality

$$\Phi_n = \Phi'_n \quad \forall n \leq 24\,. \tag{9.7.1}$$

It is even conceivable that this equality extend to $n = 25$. Concretely, this means that for $n \leq 24$ and every $\psi \in \Phi'_n$, Arthur's multiplicity formula applied to ψ always leads to a nonzero multiplicity. It would be interesting to find a deeper reason for this phenomenon.

One might hope that Equality (9.7.1) holds for every $n \equiv -1, 0, 1 \bmod 8$, at least if we replace Φ'_n by its subset consisting of the ψ such that $\psi_p = \psi_p^{-1}$ for every p. This is not so. Indeed, Arthur's multiplicity formula shows that if $n = 32$, the parameter $\Delta_{17}[14] \oplus [3] \oplus [1]$ of Φ'_{32} does not belong to Φ_{32}. Likewise, the element $\Delta_{17}[13] \oplus [4]$ of Φ'_{31} must not belong to Φ_{31}. If $\mathrm{S}_{14}(\mathrm{SL}_2(\mathbb{Z}))$ were not zero, such examples would also exist in dimensions 23 and 24.

Chapter 10
Applications

10.1 24 ℓ-Adic Galois Representations

Recall that the vector space $\mathbb{Q}[\mathrm{X}_{24}]$ admits a $\mathbb{Q}$-basis $\mathrm{v}_1, \ldots, \mathrm{v}_{24}$ consisting of eigenvectors common to all elements of $\mathrm{H}(\mathrm{O}_{24})$ (Sect. 9.2). Each of these vectors v_i generates an automorphic representation $\pi_i \in \Pi_{\mathrm{disc}}(\mathrm{O}_{24})$ whose standard parameter $\psi(\pi_i, \mathrm{St})$ is determined by Theorem* 9.2.5 (Table C.5). In particular, the pair (π_i, St) satisfies the Arthur–Langlands conjecture, so that Corollary 8.2.19 applies and associates with it the 24-dimensional ℓ-adic representations of the group $\mathrm{Gal}(\overline{\mathbb{Q}}/\mathbb{Q})$. In what follows, we specify the resulting statement.

For a prime p and $1 \leq i \leq 24$, denote by $\lambda_i(p) \in \mathbb{Z}$ the eigenvalue of T_p on the vector v_i. By formula (6.2.5), we have $\lambda_i(p) = p^{11}\,\mathrm{trace}\,\mathrm{St}(\mathrm{c}_p(\pi_i))$. More generally, the following lemma shows that the polynomial $\det(t - p^{11}\,\mathrm{St}(\mathrm{c}_p(\pi_i)))$ (in the indeterminate t) has integral coefficients, each of which can be seen as the eigenvalue of a well-chosen Hecke operator in $\mathrm{H}(\mathrm{O}_{24})$.

Lemma 10.1.1. *Let G be a split semisimple $\mathbb{Z}_p$-group, λ a dominant weight of $\widehat{G}$, V_λ the associated irreducible representation of $\widehat{G}$, and ρ the half-sum of the positive roots of $G_{\mathbb{C}}$. For every integer $m \geq 1$, there exists a unique element $T \in \mathrm{H}(G)$ such that for every $c \in \widehat{G}_{\mathrm{ss}}$, we have*

$$p^{m\langle\lambda,\rho\rangle}\,\mathrm{trace}(c \,|\, \Lambda^m V_\lambda) = \mathrm{tr}(c)(\mathrm{Sat}(T))\ .$$

Moreover, if $G = \mathrm{SO}_n$ and if the representation V_λ extends to $\mathrm{O}_n(\mathbb{C})$, for example if $V = \mathrm{St}$, then T belongs to the subring $\mathrm{H}(\mathrm{O}_n) \subset \mathrm{H}(\mathrm{SO}_n)$.

Proof. The existence and uniqueness of an element $T \in \mathrm{H}(G) \otimes \mathbb{Z}[p^{-1/2}]$ with $\mathrm{Sat}(T) = p^{m\langle\lambda,\rho\rangle}[\Lambda^m V_\lambda]$ is an immediate consequence of the Satake isomorphism

G. Chenevier, J. Lannes, *Automorphic Forms and Even Unimodular Lattices*,
Ergebnisse der Mathematik und ihrer Grenzgebiete. 3. Folge / A Series of Modern Surveys in Mathematics 69, https://doi.org/10.1007/978-3-319-95891-0_10

(Sect. 6.2.1). We must therefore show that we have $T \in \mathrm{H}(G)$. Recall that we have

$$p^{\langle \rho, \mu \rangle}[V_\mu] \in \mathrm{Sat}(\mathrm{H}(G))$$

for every dominant weight μ of $\widehat{G}$, by formula (6.2.4). The case $m = 1$ immediately follows. For an arbitrary $m \geq 1$, observe that the irreducible components of $\Lambda^m V_\lambda$, which are of the form V_μ, with μ a dominant weight of $\widehat{G}$, satisfy $\mu \leq m\lambda$ (Sect. 6.1.4). Indeed, this inequality holds more generally for all weights μ of $V_\lambda^{\otimes m}$, and $\Lambda^m V_\lambda$ is a quotient of the latter. The first assertion of the lemma follows because for such a μ, $\langle \rho, m\lambda - \mu \rangle$ is a nonnegative integer.

To verify the second assertion, concerning $G = \mathrm{SO}_n$, it suffices to note that if V_λ extends to $\mathrm{O}_n(\mathbb{C}) \supset \widehat{G} = \mathrm{SO}_n(\mathbb{C})$, then the same holds for $\Lambda^m V_\lambda$. Relation (6.2.2) then shows that the element $T \in \mathrm{H}(\mathrm{SO}_n)$ defined above belongs to $\mathrm{H}(\mathrm{O}_n)$. □

Remark 10.1.2. In general, it is difficult to explicitly determine the operator T given by Lemma 10.1.1, say in terms of the $\mathbb{Z}$-basis consisting of the c_μ (Sect. 6.2.5), even in the particular case of the group $G = \mathrm{SO}_n$ and the standard representation $V_\lambda = \mathrm{St}$ of $\widehat{G}$. In this case, however, we already noted that we have $T = \mathrm{T}_p$ for $m = 1$, and, furthermore, we have $T = p\,\mathrm{T}_{p,p} + p^{n/2-1} + \sum_{i=0}^{n/2-2} p^{2i+1}$ for $m = 2$ (formula (6.2.6)).

In Sect. 8.2, we recalled the existence and several properties of the Galois representations $\rho_{\pi,\iota}$ associated with an algebraic, self-dual, regular automorphic representation $\pi \in \Pi_{\mathrm{cusp}}(\mathrm{PGL}_n)$ and an embedding $\iota \colon \overline{\mathbb{Q}} \to \overline{\mathbb{Q}}_\ell$, where $\overline{\mathbb{Q}}_\ell$ denotes an algebraic closure of $\mathbb{Q}_\ell$. The Galois representation $\rho_{\pi,\iota}$ is continuous, semisimple, and unramified outside ℓ, and its isomorphism class is uniquely determined by the relation

$$\det(t - \rho_{\pi,\iota}(\mathrm{Frob}_p)) = \iota\big(\det\big(t - \mathrm{c}_p(\pi)p^{\mathrm{w}(\pi)/2}\big)\big) \tag{10.1.1}$$

for every prime $p \neq \ell$ (the uniqueness follows from the Chebotarev density theorem). If the polynomial $\det(t - \mathrm{c}_p(\pi)p^{\mathrm{w}(\pi)/2})$ is in $\mathbb{Q}[t]$ for every prime p, then the Galois representation $\rho_{\pi,\iota}$ depends only on ℓ and not on the choice of ι; we then denote it simply by $\rho_{\pi,\ell}$.

The automorphic representations π of interest to us here will be the Δ_w and the $\Delta_{w,v}$. In these cases, the Galois representations $\rho_{\pi,\iota}$ (satisfying conditions (i) and (ii) of Theorem 8.2.17) have been constructed by, respectively, Deligne [70] (generalizing a prior construction of Eichler, Shimura, Kuga, and Sato [131]) and Weissauer [212] (see also prior work of Chai–Faltings [57] and Taylor [197]).

When π is generated by a modular form $f = q + a_2\,q^2 + \ldots$ in $\mathrm{S}_k(\mathrm{SL}_2(\mathbb{Z}))$ that is an eigenform for the Hecke operators, the relation (10.1.1) reduces to the well-known relation $\det(t - \rho_{\pi,\iota}(\mathrm{Frob}_p)) = t^2 - \iota(a_p)t + p^{k-1}$. This polynomial has integral coefficients if $k \leq 22$ (because the form f does); it is, moreover, well-known that $\rho_{\pi,\ell}$ can in this case be chosen with coefficients in the field $\mathbb{Q}_\ell$.

Recall that when π is one of the four representations $\Delta_{w,v}$ defined in Sect. 9.1, the right-hand side of Equality (10.1.1) can also be written as

$$t^4 - \tau_{j,k}(p)\, t^3 + \frac{\tau_{j,k}(p)^2 - \tau_{j,k}(p^2)}{2}\, t^2 - \tau_{j,k}(p)\, p^{j+2k-3}\, t + p^{2j+4k-6},$$

where $(j,k) = (v-1, (w-v)/2+2)$ (formula (9.1.3)). Here again, this polynomial has rational coefficients, and even integral ones by Proposition 9.1.9.

Theorem* 10.1.3. *Let $i = 1, \dots, 24$; let ℓ be a prime and $\overline{\mathbb{Q}}_\ell$ an algebraic closure of $\mathbb{Q}_\ell$. There exists a continuous semisimple representation $\rho_{i,\ell}\colon \mathrm{Gal}(\overline{\mathbb{Q}}/\mathbb{Q}) \longrightarrow \mathrm{GL}_{24}(\overline{\mathbb{Q}}_\ell)$, unique up to isomorphism, that is unramified outside ℓ and such that for every prime $p \neq \ell$, we have the following equality in $\mathbb{Z}[t]$:*

$$\det(t - \rho_{i,\ell}(\mathrm{Frob}_p)) = \det(t - p^{11}\, \mathrm{St}(\mathrm{c}_p(\pi_i)))\,. \tag{10.1.2}$$

In particular, we have $\lambda_i(p) = \mathrm{trace}\, \rho_{i,\ell}(\mathrm{Frob}_p)$ for every prime $p \neq \ell$.

Proof. Let $i = 1, \dots, 24$. Theorem 9.2.5 asserts that $\psi(\pi_i, \mathrm{St})$ is of the form $\oplus_{j=1}^{k} \varpi_j[d_j]$, where the ϖ_j are among the automorphic representations 1, Δ_w, $\Delta_{w,v}$, and $\mathrm{Sym}^2\Delta_{11}$. The existence of $\rho_{i,\ell}$ follows by setting

$$\rho_{i,\ell} = \bigoplus_{j=1}^{k} \rho_{\varpi_j,\ell} \otimes (\oplus_{m=0}^{d_j-1} \omega_\ell^m) \otimes \omega_\ell^{(22-\mathrm{w}(\varpi_j)+1-d_j)/2}, \tag{10.1.3}$$

using the notation $\rho_{\varpi_j,\ell}$ introduced after Remark 10.1.2. The uniqueness follows from the Chebotarev density theorem. Naturally, when ϖ_j is the trivial automorphic representation $\Pi_{\mathrm{cusp}}(\mathrm{PGL}_1)$, $\rho_{\varpi_j,\ell}$ denotes the trivial Galois representation (of dimension 1). Moreover, we can take the representation $\mathrm{Sym}^2\rho_{\Delta_{11},\ell}$ for $\rho_{\mathrm{Sym}^2\Delta_{11},\ell}$.

Remark 10.1.4. Following the construction of Weissauer [212], it should be possible to show that the representations $\rho_{\Delta_{w,v},\ell}$, hence also the $\rho_{i,\ell}$, are defined over $\mathbb{Q}_\ell$, because we have $\dim \mathrm{S}_{j,k} = 1$ for the four corresponding pairs (j,k) (Sect. 9.1.1).

It would be interesting to study in detail the images of the Galois representations $\rho_{\Delta_{w,v},\ell}$, in the way done by Serre and Swinnerton-Dyer in their work on the representations $\rho_{\Delta_w,\ell}$ [194]. We will content ourselves, in Sect. 10.4, with proving several congruences satisfied by these representations, in the spirit of the Ramanujan congruence.

Corollary* 10.1.5. *Let $i = 1, \dots, 24$, and let ℓ be a prime. There exists a continuous semisimple representation $\overline{\rho}_{i,\ell}\colon \mathrm{Gal}(\overline{\mathbb{Q}}/\mathbb{Q}) \longrightarrow \mathrm{GL}_{24}(\mathbb{F}_\ell)$, unique up to isomorphism, that is unramified outside ℓ and such that for every prime $p \neq \ell$, we have the congruence*

$$\det(t - \overline{\rho}_{i,\ell}(\mathrm{Frob}_p)) \equiv \det(t - p^{11}\, \mathrm{St}(\mathrm{c}_p(\pi_i))) \bmod \ell\,. \tag{10.1.4}$$

In particular, we have $\lambda_i(p) \equiv \mathrm{trace}\, \overline{\rho}_{i,\ell}(\mathrm{Frob}_p) \bmod \ell$ for every prime $p \neq \ell$.

The uniqueness follows from the Chebotarev density theorem and a classical result of Brauer–Nesbitt.[1] The existence of $\overline{\rho}_{i,\ell}$ follows from that of $\rho_{i,\ell}$ by a standard general procedure, recalled below.

Fix an algebraic closure $\overline{\mathbb{Q}}_\ell$ (resp. $\overline{\mathbb{F}}_\ell$) of $\mathbb{Q}_\ell$ (resp. $\mathbb{F}_\ell$), as well as a ring homomorphism $\mathcal{O} \to \overline{\mathbb{F}}_\ell$, where $\mathcal{O} \subset \overline{\mathbb{Q}}_\ell$ is the integral closure of $\mathbb{Z}_\ell$. Let G be a profinite group and $\rho\colon G \to \mathrm{GL}_n(\overline{\mathbb{Q}}_\ell)$ a continuous representation. It is shown that the following hold:

(i) The polynomial $\det(t - \rho(g))$ belongs to $\mathcal{O}[t]$ for every g in G.
(ii) There exists a continuous semisimple representation $\overline{\rho}\colon G \to \mathrm{GL}_n(\overline{\mathbb{F}}_\ell)$, unique up to isomorphism, such that for every g in G, the characteristic polynomial $\det(t - \overline{\rho}(g))$ is the image of $\det(t - \rho(g)) \in \mathcal{O}[t]$ in $\overline{\mathbb{F}}_\ell[t]$.
(iii) If, moreover, $\det(t - \rho(g))$ belongs to $\mathbb{Q}_\ell[t]$ for every g in G, then $\overline{\rho}$ can be chosen with coefficients in $\mathbb{F}_\ell$ and its isomorphism class does not depend on the choice of the morphism $\mathcal{O} \to \overline{\mathbb{F}}_\ell$.

The representation $\overline{\rho}$ is then called "the" residual representation of ρ. This construction (including Assertion (iii)) applies to the $\rho_{i,\ell}$, as well as to the representations of the form $\rho_{\pi,\ell}$ introduced earlier; it leads to residual representations $\overline{\rho}_{i,\ell}$ and $\overline{\rho}_{\pi,\ell}$ with coefficients in $\mathbb{F}_\ell$. Naturally, the relation (10.1.3) induces a similar decomposition

$$\overline{\rho}_{i,\ell} \simeq \bigoplus_{j=1}^{k} \overline{\rho}_{\varpi_j,\ell} \otimes (\oplus_{m=0}^{d_j-1} \overline{\omega_\ell}^{\,m}) \otimes \overline{\omega_\ell}^{\,(22-\mathrm{w}(\varpi_j)+1-d_j)/2}. \tag{10.1.5}$$

Let us briefly indicate how to prove Assertions (i), (ii), and (iii). First, a classical application of the Baire category theorem asserts that we have $\rho(G) \subset \mathrm{GL}_n(F)$, where $F \subset \overline{\mathbb{Q}}_\ell$ is a finite extension of $\mathbb{Q}_\ell$. After conjugating ρ by an element of $\mathrm{GL}_n(F)$ if necessary, the compactness of $\rho(G)$ allows us to assume that we have $\rho(G) \subset \mathrm{GL}_n(\mathcal{O}_F)$, where $\mathcal{O}_F = \mathcal{O} \cap F$. Assertion (i) follows. Composing with the ring homomorphism $\mathcal{O} \to \overline{\mathbb{F}}_\ell$ gives a continuous representation $G \to \mathrm{GL}_n(\overline{\mathbb{F}}_\ell)$. To prove part (ii), it suffices to take a semisimplification of this representation for $\overline{\rho}$. The uniqueness, as well as the last assertion of part (iii), follows from the result of Brauer–Nesbitt. Since Schur's obstruction is trivial over finite fields, the first assertion of part (iii) follows (see [197, Lemma 2]).

Remark 10.1.6. Let $\rho = \rho_{\Delta_{w,v},\ell}$. In Sect. 8.2.16, formula (8.2.1), we already observed that we have an isomorphism $\rho^* \simeq \rho \otimes \omega_\ell^{-w}$. Even better, by Bellaïche and Chenevier [20, Corollary 1.3], there exists a nondegenerate, alternating, and $\mathrm{Gal}(\overline{\mathbb{Q}}/\mathbb{Q})$-equivariant pairing $\rho \otimes \rho \to \omega_\ell^w$ (see also [212]). It is not difficult to deduce from this the existence of a nondegenerate, alternating, and $\mathrm{Gal}(\overline{\mathbb{Q}}/\mathbb{Q})$-equivariant pairing $\overline{\rho} \otimes \overline{\rho} \to \overline{\omega_\ell}^{\,w}$.

[1] Let us recall the statement of the latter. Let G be a group, k a field, and $\rho_1, \rho_2\colon G \to \mathrm{GL}_m(k)$ two semisimple representations. The representations ρ_1 and ρ_2 are isomorphic if and only if we have $\det(t - \rho_1(g)) = \det(t - \rho_2(g))$ for every $g \in G$.

10.2 Back to p-Neighbors of Niemeier Lattices

We number the 24 Niemeier lattices (or, rather, their isomorphism classes) $\mathrm{L}_1, \mathrm{L}_2, \ldots, \mathrm{L}_{24}$, following the convention used by Conway and Sloane [68, Chap. 16, Table 16.1] (the Greek letters $\alpha, \beta, \ldots, \omega$ of this reference being replaced with the integers $1, 2, \ldots, 24$). We therefore have $\mathrm{R}(\mathrm{L}_i) = \mathbf{R}_i$ for $i \leq 23$ (see Sect. 2.3), and L_{24} is the Leech lattice that we also denoted by Leech in Sect. 3.4.

Let p be a prime; we denote by T_p the matrix of the Hecke operator $\mathrm{T}_p \colon \mathbb{Z}[\mathrm{X}_{24}] \to \mathbb{Z}[\mathrm{X}_{24}]$ in the basis $(\mathrm{L}_1, \mathrm{L}_2, \ldots, \mathrm{L}_{24})$.

As mentioned before, Nebe and Venkov determined T_2 and deduced that the eigenvalues of this operator are integral and distinct. Denote them by

$$\lambda_1 > \lambda_2 > \ldots > \lambda_{24} \, .$$

Denote by v_j, for $1 \leq j \leq 24$, an eigenvector associated with λ_j whose coordinates, in the bases mentioned above, are integral and pairwise relatively prime (such an eigenvector is determined up to a sign); denote by V the 24×24 matrix whose jth column is the column vector v_j.

Denote by $\lambda_j(p)$ the integer defined by the equality $\mathrm{T}_p \mathrm{v}_j = \lambda_j(p) \mathrm{v}_j$ (recall that T_2 and T_p commute); we therefore have $\lambda_j(2) = \lambda_j$, by definition.

Set $\theta_1(p) = \tau_{12}(p) (= \tau(p))$, $\theta_2(p) = \tau_{16}(p)$, $\theta_3(p) = \tau_{18}(p)$, $\theta_4(p) = \tau_{20}(p)$, and $\theta_5(p) = \tau_{22}(p)$; in other words, denote by $\theta_r(p)$ the pth Fourier coefficient of the normalized cusp form (for $\mathrm{SL}_2(\mathbb{Z})$) of respective weight $12, 16, 18, 20, 22$ for $r = 1, 2, 3, 4, 5$. Set $\theta_6(p) = (\theta_1(p))^2 - p^{11}$ (we have $\theta_6(p) = p^{11} \operatorname{tr}(\mathrm{Sym}^2 \mathrm{c}_p(\Delta_{11}))$). Finally, set $\theta_7(p) = \tau_{6,8}(p)$, $\theta_8(p) = \tau_{8,8}(p)$, $\theta_9(p) = \tau_{12,6}(p)$, and $\theta_{10}(p) = \tau_{4,10}(p)$.

By Theorem 9.2.5 and formula (6.2.5), there exist uniquely determined polynomials $\mathrm{C}_{j,r}$ in $\mathbb{Z}[X]$, for $1 \leq j \leq 24$ and $0 \leq r \leq 10$, such that we have

$$\lambda_j(p) \;=\; \mathrm{C}_{j,0}(p) + \sum_{r=1}^{10} \mathrm{C}_{j,r}(p)\, \theta_r(p) \tag{10.2.1}$$

for every prime p.

Let us recall the value of some of these polynomials $\mathrm{C}_{j,r}$.

We have $\mathrm{C}_{1,0} = \sum_{k=0}^{k=22} X^k + X^{11}$ and $\mathrm{C}_{1,r} = 0$ for $r \geq 1$; in other words, we have $\lambda_1(p) = \mathrm{c}_{24}(p) := \sum_{k=0}^{k=22} p^k + p^{11}$ (see Proposition 3.2.4 and Proposition-Definition 3.2.1).

We have $\mathrm{C}_{2,0} = \sum_{k=1}^{k=21} X^k$, $\mathrm{C}_{2,6} = 1$, and $\mathrm{C}_{2,r} = 0$ for $r \neq 0, 6$.

For $r \geq 7$, the polynomials $\mathrm{C}_{j,r}$ are as follows:

- $\mathrm{C}_{j,7} = 0$ for $j \neq 19$ and $\mathrm{C}_{19,7} = X(X+1)$;
- $\mathrm{C}_{j,8} = 0$ for $j \neq 15$ and $\mathrm{C}_{15,8} = X + 1$;
- $\mathrm{C}_{j,9} = 0$ for $j \neq 10$ and $\mathrm{C}_{10,9} = X + 1$;
- $\mathrm{C}_{j,10} = 0$ for $j \neq 21$ and $\mathrm{C}_{21,10} = X + 1$.

Considering the formula

$$\mathrm{T}_p \;=\; \mathrm{V}\ \mathrm{diag}(\lambda_1(p), \lambda_2(p), \ldots, \lambda_{24}(p))\ \mathrm{V}^{-1} \tag{10.2.2}$$

(here, $\mathrm{diag}(\lambda_1(p), \lambda_2(p), \ldots, \lambda_{24}(p))$ denotes the diagonal matrix with diagonal entries the $\lambda_j(p)$), leads to the following statement.

Theorem 10.2.1. *Let L and L' be two even unimodular lattices of dimension 24. There exist polynomials $\mathrm{P}_r(L, L'; X)$ in $\mathbb{Q}[X]$, for $0 \leq r \leq 10$, uniquely determined in terms of the isomorphism classes of L and L', such that we have*

$$\mathrm{N}_p(L, L') \;=\; \mathrm{P}_0(L, L'; p) + \sum_{r=1}^{10} \mathrm{P}_r(L, L'; p)\, \theta_r(p)$$

for every prime p.

Remark. By definition, for every r with $0 \leq r \leq 10$, we have the following equality of 24×24 matrices with coefficients in $\mathbb{Q}[X]$:

$$[\,\mathrm{P}_r(\mathrm{L}_j, \mathrm{L}_i; X)\,] \;=\; \mathrm{V}\ \mathrm{diag}(\mathrm{C}_{1,r}(X), \mathrm{C}_{2,r}(X), \ldots, \mathrm{C}_{24,r}(X))\ \mathrm{V}^{-1}\,.$$

Since the columns of V are pairwise orthogonal for the inner product with matrix $\mathrm{diag}(|\mathrm{O}(\mathrm{L}_1)|, |\mathrm{O}(\mathrm{L}_2)|, \ldots, |\mathrm{O}(\mathrm{L}_{24})|)$ (Proposition 3.2.3), this equality shows that we have

$$\frac{1}{|\mathrm{O}(L)|}\ \mathrm{P}_r(L, L'; X) \;=\; \frac{1}{|\mathrm{O}(L')|}\ \mathrm{P}_r(L', L; X)$$

for every r, L, and L' (which is, of course, compatible with Scholium 3.1.7).

Denote by Proj_1 the orthogonal projection for the inner product introduced in Proposition 3.2.3, from $\mathbb{Q}[\mathrm{X}_{24}]$ onto the line generated by v_1. Denote by w the vector $\sum_{x \in \mathrm{X}_{24}} (1/|\mathrm{O}(x)|)\, x$ in $\mathbb{Q}[\mathrm{X}_{24}]$; it follows from Proposition 3.2.4 that the vector v_1 is collinear to the vector w. Let y be an element of X_{24}; the equalities $\mathrm{w}.y = 1$ and $\mathrm{w}.\mathrm{w} = \sum_{x \in \mathrm{X}_{24}} 1/|\mathrm{O}(x)|$ imply

$$\mathrm{Proj}_1(y) \;=\; \frac{\mathrm{w}}{\sum_{x \in \mathrm{X}_{24}} 1/|\mathrm{O}(x)|} \;=\; \sum_{x \in \mathrm{X}_{24}} \mu(x)\, x$$

with

$$\mu(x) \;=\; \frac{1/|\mathrm{O}(x)|}{\sum_{x \in \mathrm{X}_{24}} 1/|\mathrm{O}(x)|}\,.$$

Note that $\mu(x)$ is the quotient of the mass of x and the mass of the genus of the even unimodular lattices of dimension 24, masses in the sense of Minkowski–Siegel: $\sum_{x \in \mathrm{X}_{24}} \mu(x)\, \delta_x$ is *the probability measure of Minkowski–Siegel* on the set X_{24}.

Theorem 10.2.2. *As the prime p tends to infinity, we have*

$$\mathrm{T}_p \;=\; p^{22}\ \mathrm{Proj}_1 \;+\; \mathrm{O}(p^{21})$$

(the notation $\mathrm{O}(-)$ *is the notation of Landau... and has nothing to do with orthogonal groups!).*

Proof. In view of Proposition 3.2.3, we have $\mathrm{T}_p = \sum_{j=1}^{24} \lambda_j(p)\mathrm{Proj}_j$, where Proj_j is the orthogonal projection from $\mathbb{Q}[\mathrm{X}_{24}]$ onto the line generated by v_j. But we have $\lambda_1(p) = p^{22} + \mathrm{O}(p^{21})$ and the Ramanujan inequalities for the θ_r imply $\lambda_j(p) = \mathrm{O}(p^{21})$ for $j \geq 2$. Let us recall these inequalities. We have $|\tau_k(p)| \leq 2p^{(k-1)/2}$ for $k = 12, 16, 18, 20, 22$, $|\tau(p)^2 - p^{11}| \leq 3p^{11}$, $|\tau_{6,8}(p)| \leq 4p^{19/2}$, and $\tau_{j,k}(p)| \leq 4p^{21/2}$ for $(j,k) = (8,8), (12,6), (4,10)$. The first five inequalities are due to Deligne [70], the last four to Weissauer [212] (the sixth is a consequence of the first!); for a general discussion concerning this type of inequality, see Sect. 8.2.16. □

Scholium 10.2.3. *Let x and y be two elements of* X_{24}*; we have*

$$\frac{\mathrm{N}_p(x,y)}{\mathrm{c}_{24}(p)} \;=\; \mu(y) \;+\; \mathrm{O}\left(\frac{1}{p}\right)$$

as p tends to infinity.

Comments. Let L be an even unimodular lattice of dimension 24 and y an element of X_{24}. The quotient

$$\frac{\mathrm{N}_p([L],y)}{\mathrm{c}_{24}(p)} \;=\; \frac{\mathrm{N}_p([L],y)}{|\mathrm{C}_L(\mathbb{F}_p)|}$$

is the proportion of the points c of the quadric $\mathrm{C}_L(\mathbb{F}_p)$ such that the isomorphism class of the p-neighbor of L associated with c (see Proposition 2.1.5) is y. Scholium 10.2.3 says that this proportion tends to $\mu(y)$ as p tends to infinity (with convergence rate $1/p$).

Remark. The Minkowski–Siegel probability measure on X_{24} is very far from being uniform. For example, we have $\mu([\mathrm{E}_{24}]) \approx 2.42 \times 10^{-17}$ (this is the minimum of the function μ on X_{24}), $\mu([\mathrm{L}_{21}]) \approx 0.426$ (this is the maximum), and $\mu([\mathrm{L}_{20}]) + \mu([\mathrm{L}_{21}]) + \mu([\mathrm{L}_{22}]) \approx 0.906$.

Remark. By examining the equalities (10.2.1), we see that the Ramanujan inequalities give $\lambda_j(p) = \mathrm{C}_{0,j}(p) + \mathrm{O}(p^{33/2})$ for every j. As before, this leads to estimates of the form

$$\frac{\mathrm{N}_p(x,y)}{\mathrm{c}_{24}(p)} \;=\; \mu(y)\left(1 + \sum_{n=1}^{5} \kappa_n(x,y)\,\frac{1}{p^n} \;+\; \mathrm{O}\left(p^{-11/2}\right)\right);$$

the relation between the eigenvector v_2 and the theta series of genus 1 of the even unimodular lattices of dimension 24 make it possible to give $\kappa_1(x,y)$ explicitly:

$$\kappa_1(x,y) \;=\; \frac{37092156523}{34673184000}\left(\mathrm{h}(x) - \frac{2730}{691}\right)\left(\mathrm{h}(y) - \frac{2730}{691}\right).$$

Let us give some details on the relation between v_2 and the theta series of genus 1 invoked above; these details are in fact an expansion of the second comment after Table 5.1. On the one hand, Theorem 5.5.1 shows that the diagram

$$\begin{array}{ccc} \mathbb{C}[\mathrm{X}_{24}] & \xrightarrow{\vartheta_1} & \mathrm{M}_{12}(\mathrm{SL}_2(\mathbb{Z})) \\ \mathrm{T}_p \downarrow & & \downarrow p\frac{p^{21}-1}{p-1}+\mathrm{T}(p^2) \\ \mathbb{C}[\mathrm{X}_{24}] & \xrightarrow{\vartheta_1} & \mathrm{M}_{12}(\mathrm{SL}_2(\mathbb{Z})) \end{array}$$

is commutative (the Eichler commutation relations). On the other hand, $\{\mathbb{E}_{12}, \Delta\}$ is a basis of the $\mathbb{C}$-vector space $\mathrm{M}_{12}(\mathrm{SL}_2(\mathbb{Z}))$ that consists of eigenvectors for the Hecke operators; in particular, we have $\mathrm{T}(p^2)(\Delta) = \tau(p^2)\,\Delta = (\tau(p)^2 - p^{11})\,\Delta$. Let $\mathrm{coord}_\Delta : \mathrm{M}_{12}(\mathrm{SL}_2(\mathbb{Z})) \to \mathbb{C}$ be the linear form giving the coordinate "of index Δ" defined by this basis and $\eta\colon \mathbb{C}[\mathrm{X}_{24}] \to \mathbb{C}$ the composed linear form $\mathrm{coord}_\Delta \circ \vartheta_1$. It follows from the above that η is an eigenvector of the endomorphism T_p^* of $(\mathbb{C}[\mathrm{X}_{24}])^*$, for the eigenvalue $\sum_{k=1}^{21} p^k + (\tau(p)^2 - p^{11}) = \lambda_2(p)$. We easily verify that we have

$$\eta(x) \;=\; |\mathrm{R}(x)| - \frac{65520}{691} \;=\; 24\,\left(\mathrm{h}(x) - \frac{2730}{691}\right)$$

for every x in X_{24} (see, for example, [177, Sect. 6.6, formule (108)]). As in Sect. 2.2 (Proposition 3.2.4), it follows that the vector

$$\sum_{x\in \mathrm{X}_{24}} \frac{1}{|\mathrm{O}(x)|}\left(\mathrm{h}(x) - \frac{2730}{691}\right) x$$

is an eigenvector of the endomorphism T_p of $\mathbb{C}[\mathrm{X}_{24}]$, for the eigenvalue $\lambda_2(p)$.

On the Diameter of the Graph of the p-Neighbors in Dimension 24

The formula of Theorem 10.2.1 shows that if we have $\mathrm{N}_p(L, L') = 0$, then we have

$$\mathrm{P}_0(L, L'; p)^{\,2} \;=\; \left(\sum_{r=1}^{10} \mathrm{P}_r(L, L'; p)\,\theta_r(p)\right)^2 .$$

By the Schwarz inequality, we have

$$\left(\sum_{r=1}^{10} \mathrm{P}_r(L, L'; p)\,\theta_r(p)\right)^2 \;\leq\; \left(\sum_{r=1}^{10} \gamma_r\,\mathrm{P}_r(L, L'; p)^2\right)\left(\sum_{r=1}^{10} \gamma_r^{-1}\,\theta_r(p)^2\right)$$

for every 10-tuple $(\gamma_1, \gamma_2, \ldots, \gamma_{10})$ of strictly positive real numbers. By taking

$$(4p^{11}, 4p^{15}, 4p^{17}, 4p^{19}, 4p^{21}, 9p^{22}, 16p^{19}, 16p^{21}, 16p^{21}, 16p^{21})$$

for $(\gamma_1, \gamma_2, \ldots, \gamma_{10})$, in view of the Ramanujan inequalities, we obtain the inequality

$$\left(\sum_{r=1}^{10} \mathrm{P}_r(L, L'; p)\,\theta_r(p)\right)^2 \;\leq\; 10\left(\sum_{r=1}^{10} \gamma_r\,\mathrm{P}_r(L, L'; p)^2\right).$$

Set

$$(\Gamma_1(X), \Gamma_2(X), \ldots, \Gamma_{10}(X)) = (4X^{11}, 4X^{15}, \ldots, 16X^{21})$$

and

$$\mathrm{Q}(L, L'; X) = \mathrm{P}_0(L, L'; X)^2 - 10\left(\sum_{r=1}^{10} \Gamma_r(X)\,\mathrm{P}_r(L, L'; X)^2\right);$$

we see that $\mathrm{Q}(L, L'; X)$ is a polynomial in $\mathbb{Q}[X]$ whose monomial of highest degree is $\mu(L')^2 X^{44}$. Note that the remark following Theorem 10.2.1 implies the equality $\mu(L)^2\,\mathrm{Q}(L, L'; X) = \mu(L')^2\,\mathrm{Q}(L', L; X)$. Denote by $\rho(L, L')$ the greatest real root of the polynomial $\mathrm{Q}(L, L'; X)$ (we could agree to set $\rho(L, L') = -\infty$ if $\mathrm{Q}(L, L'; X)$ does not have any real roots, but this polynomial in fact always has real roots); finally, denote by $\mathrm{p}(L, L')$ the least prime strictly greater than $\rho(L, L')$. We did the necessary to ensure $\mathrm{N}_p(L, L') > 0$ for $p \geq \mathrm{p}(L, L')$.

Example. The 24-tuple $(\rho(\mathrm{L}_i, \mathrm{Leech}))_{1 \leq i \leq 24}$ is approximately as follows:

$$(46.77, 30.11, 30.88, 23.97, 21.71, 17.80, 17.59, 15.63, 13.72, 12.00, 11.27, 12.14, 9.36, 9.58, 8.48, 7.03, 6.19, 5.21, 5.86, 4.12, 3.10, 2.13, 1.37, 1.68)\,.$$

On a case-by-case basis, we check that for every Niemeier lattice L, the prime $\mathrm{p}(L, \mathrm{Leech})$ is the least prime greater than or equal to the Coxeter number $\mathrm{h}(L)$. This checking and Proposition 3.4.1.1 show that the statement "$\mathrm{N}_p(L, L') > 0$ for $p \geq \mathrm{p}(L, L')$" is optimal for $L' = \mathrm{Leech}$.

Again on a case-by-case basis, we check that we have

$$\rho(L, L') \leq \rho(\mathrm{E}_{24}, \mathrm{Leech}) < 47$$

for all Niemeier lattices L and L'. We therefore see that the graph of the p-neighbors in dimension 24 is the complete graph with set of vertices X_{24} for $p \geq 47$. Since we have computed the $\tau_{j,k}(p)$ for $p \leq 43$ (and even $p \leq 113$, see the next subsection), we are now able to determine the diameter of the graph of the p-neighbors in dimension 24; here is the result.

Theorem 10.2.4. *Let* p *be a prime. The diameter of the graph* $\mathrm{K}_{24}(\mathrm{p})$ *is as follows:* 5 *for* $\mathrm{p} = 2$, 4 *for* $\mathrm{p} = 3$, 3 *for* $\mathrm{p} = 5$, 2 *for* $7 \leq \mathrm{p} \leq 43$, *and* 1 *for* $\mathrm{p} \geq 47$.

10.3 Determination of the $\tau_{j,k}(q)$ for Small Values of q

The $\tau_{j,k}(q)$ in question, where $q = p^n$ with p prime and $n \geq 1$ integral, are defined in Sect. 9.1.7; we have seen that the determination of the $\tau_{j,k}(p^n)$ for $n > 2$ reduces to that of the $\tau_{j,k}(p^n)$ for $n = 1, 2$. The values of the $\tau_{j,k}(p)$ for $p \leq 113$ are gathered in Table C.3, those of the $\tau_{j,k}(p^2)$ for $p \leq 29$ in Table C.4.

10.3.1 Determination of the $\tau_{j,k}(p)$ for $p \le 113$

The integers $\theta_r(p)$ for $r \le 6$ are not difficult to compute (at least for a prime of reasonable size!). On the other hand, as already mentioned, the tables for the $\theta_r(p)$ for $r \ge 7$ are quite short. We propose to show that the theory developed in this book and the information we collected in Sect. 3.4 on the last column of T_p (that is, on the number of neighbors $\mathrm{N}_p(L, \mathrm{Leech})$ for L an even unimodular lattice of dimension 24 representing 2) allow us to determine these $\theta_r(p)$ for $p \le 113$. Formulas (10.2.1) and (10.2.2) (or, equivalently, Theorem 10.2.1) and the determination of $\tau_{6,8}(p)$, $\tau_{8,8}(p)$, $\tau_{12,6}(p)$, and $\tau_{4,10}(p)$ for $p \le 113$ make it possible to compute explicitly the Hecke operator $\mathrm{T}_p \colon \mathbb{Z}[\mathrm{X}_{24}] \to \mathbb{Z}[\mathrm{X}_{24}]$ for $p \le 113$.

Write $\mathrm{N}_p(L, L') = \mathrm{N}_p^1(L, L') + \mathrm{N}_p^2(L, L')$ with

$$\mathrm{N}_p^1(L, L') = \mathrm{P}_0(L, L'; p) + \sum_{r=1}^{6} \mathrm{P}_r(L, L'; p)\, \theta_r(p)$$

(this is the "easily computable" term) and

$$\mathrm{N}_p^2(L, L') = \sum_{r=7}^{10} \mathrm{P}_r(L, L'; p)\, \theta_r(p)$$

(this is the "mysterious" term). In view of what we recalled earlier on the polynomials $\mathrm{C}_{j,r}$, with $j \ge 1, r \ge 7$, that occur in the expression of the $\lambda_j(p)$, we have

$$\mathrm{N}_p^2(L, L') = \mathrm{c}_7(L, L')\, p(p+1)\, \theta_7(p) + \sum_{r=8}^{10} \mathrm{c}_r(L, L')\, (p+1)\, \theta_r(p)\,, \quad (10.3.1)$$

where the $\mathrm{c}_r(L, L')$, for $r \ge 7$, are rational numbers determined in terms of the isomorphism classes of the lattices L and L'.

We therefore see that if we know the integers $\mathrm{N}_p(L, L')$ for four ordered pairs (L, L') (whose orbits under the action of $\mathfrak{S}_2$ by interchanging the factors are distinct), then we may hope to determine the $\theta_r(p)$ for $r \ge 7$ by solving a linear system.

From here on, we take the Leech lattice for L', and for L the four Niemeier lattices with greatest Coxeter number: $\mathrm{L}_1 := \mathrm{E}_{24}$ ($h = 46$), $\mathrm{L}_2 := \mathrm{E}_{16} \oplus \mathrm{E}_8$ ($h = 30$), $\mathrm{L}_3 := \mathrm{E}_8 \oplus \mathrm{E}_8 \oplus \mathrm{E}_8$ ($h = 30$), and $\mathrm{L}_4 := \mathrm{A}_{24}^+$ ($h = 25$).

The four equalities $\mathrm{N}_p^2(\mathrm{L}_i, \mathrm{Leech}) = \sum_{r=7}^{10} \mathrm{P}_r(\mathrm{L}_i, \mathrm{Leech}; p)\, \theta_r(p)$, for $i = 1, 2, 3, 4$, can also be written as

$$\mathrm{A}(p) \begin{bmatrix} \theta_7(p) \\ \theta_8(p) \\ \theta_9(p) \\ \theta_{10}(p) \end{bmatrix} = \mathrm{B}(p)\,, \quad (10.3.2)$$

where $\mathrm{A}(p)$ is the 4×4 matrix obtained by taking the product of the matrix

$$\mathrm{a} := \begin{bmatrix} \frac{-20360704}{297} & \frac{31085824}{1495} & \frac{210852224}{15795} & \frac{182174720}{9963} \\ \frac{-1048320}{2057} & \frac{110194560}{116909} & \frac{901568}{2223} & \frac{-15608320}{77121} \\ \frac{16329600}{22627} & \frac{16092820800}{16717987} & \frac{12615840}{35321} & \frac{27014400}{94259} \\ \frac{994175}{4752} & \frac{-36575}{208} & \frac{37053115}{202176} & \frac{-4447625}{79704} \end{bmatrix} \tag{10.3.3}$$

and the diagonal matrix $\mathrm{diag}(p(p+1), p+1, p+1, p+1)$, and $\mathrm{B}(p)$ is the column matrix

$$\begin{bmatrix} \mathrm{N}_p(\mathrm{E}_{24}, \mathrm{Leech}) - \mathrm{N}_p^1(\mathrm{E}_{24}, \mathrm{Leech}) \\ \mathrm{N}_p(\mathrm{E}_{16} \oplus \mathrm{E}_8, \mathrm{Leech}) - \mathrm{N}_p^1(\mathrm{E}_{16} \oplus \mathrm{E}_8, \mathrm{Leech}) \\ \mathrm{N}_p(\mathrm{E}_8 \oplus \mathrm{E}_8 \oplus \mathrm{E}_8, \mathrm{Leech}) - \mathrm{N}_p^1(\mathrm{E}_8 \oplus \mathrm{E}_8 \oplus \mathrm{E}_8, \mathrm{Leech}) \\ \mathrm{N}_p(\mathrm{A}_{24}^+, \mathrm{Leech}) - \mathrm{N}_p^1(\mathrm{A}_{24}^+, \mathrm{Leech}) \end{bmatrix}.$$

We see that we have $\det \mathrm{a} \neq 0$ (thanks, PARI); the linear system (10.3.2) therefore makes it possible to determine the $\theta_r(p)$ for $r \geq 7$ if we know the integers $\mathrm{N}_p(\mathrm{L}_i, \mathrm{Leech})$ for $i = 1, 2, 3, 4$.

The Cases $p \leq 23$

Let L be an even unimodular lattice of dimension 24 with roots. Proposition 3.4.1.1 says that we have $\mathrm{N}_p(L, \mathrm{Leech}) = 0$ for $p < \mathrm{h}(L)$. In particular, we therefore have $\mathrm{N}_p(\mathrm{L}_i, \mathrm{Leech}) = 0$ for $i = 1, 2, 3, 4$: the linear system (10.3.2) makes it possible to determine the $\theta_r(p)$ with $r \geq 7$ for $p \leq 23$.

The Cases $p = 29$ and $p = 31$

We computed $\mathrm{N}_p(\mathrm{A}_{24}^+, \mathrm{Leech})$ in Sect. 3.4.3 for $p = 29$ and $p = 31$. By Proposition 3.4.1.1, we have $\mathrm{N}_{29}(L, \mathrm{Leech}) = 0$ for $L = \mathrm{E}_{24}, \mathrm{E}_{16} \oplus \mathrm{E}_8, \mathrm{E}_8 \oplus \mathrm{E}_8 \oplus \mathrm{E}_8$ and $\mathrm{N}_{31}(L, \mathrm{Leech}) = 0$ for $L = \mathrm{E}_{24}$; on the other hand, part (d) of Theorem 3.4.2.10 gives the value of $\mathrm{N}_{31}(L, \mathrm{Leech})$ for $L = \mathrm{E}_{16} \oplus \mathrm{E}_8, \mathrm{E}_8 \oplus \mathrm{E}_8 \oplus \mathrm{E}_8$. The linear system (10.3.2) also makes it possible to determine the $\theta_r(p)$ with $r \geq 7$ for $p = 29$ and $p = 31$.

The Cases $p = 3$ and $7 \leq p \leq 59$

Although elementary, the computation of $\mathrm{N}_{31}(\mathrm{A}_{24}^{+}, \mathrm{Leech})$ evoked above is quite acrobatic. Below, we propose a method, which, while far from being as elementary, is decidedly more effective, to determine the integers $\mathrm{N}_p(L, \mathrm{Leech})$ for L the Niemeier lattice with roots, when p is "not too great in terms of L," at least if p does not divide the index of the submodule of L generated by its roots; recall that we introduced the notation $\mathrm{g}(L)$ for this index in Sect. 3.4. This method is based on the following two observations:

- Let p be a prime that does not divide $\mathrm{g}(L)$. Scholium-Definition 3.4.3.3 says that $\mathrm{N}_p(L, \mathrm{Leech})$ belongs to an arithmetic sequence (containing 0) with common difference
$$\mathrm{pas}(L;p) \ := \ \frac{|\mathrm{W}(L)|}{\gcd(p-1, 24\,\mathrm{h}(L), |\mathrm{W}(L)|)} \ .$$
- Let L and L' be two even unimodular lattices of dimension 24. The Ramanujan inequalities for the $\theta_r(p)$ for $r \geq 7$ provide a lower and upper bounds
$$\mathrm{N}_p^{\mathrm{inf}}(L, L') \ \leq \ \mathrm{N}_p(L, L') \ \leq \ \mathrm{N}_p^{\mathrm{sup}}(L, L')$$
such that the difference $\mathrm{N}_p^{\mathrm{sup}}(L, L') - \mathrm{N}_p^{\mathrm{inf}}(L, L')$ is $2\mathrm{K}(L, L')\,(p+1)\,p^{21/2}$ with $\mathrm{K}(L, L') := 4\sum_{r=7}^{10} |\mathrm{c}_r(L, L')|$ (notation of (10.3.1)).

For L, we take a Niemeier lattice with roots with $\mathrm{g}(L)$ not divisible by p, and for L', we take the Leech lattice. If the difference in question is strictly less than $\mathrm{pas}(L;p)$, then $\mathrm{N}_p(L, \mathrm{Leech})$ is uniquely determined.

Let us be more precise. We have $|\theta_7(p)| \leq 4p^{19/2}$ and $|\theta_r(p)| \leq 4p^{21/2}$ for $r = 8, 9, 10$. We consequently have the inequality
$$|\mathrm{N}_p^2(L, L')| \ \leq \ \mathrm{K}(L, L')\,(p+1)\,p^{21/2} \ .$$

We set
$$\mathrm{N}_p^{\mathrm{inf}}(L, L') \ = \ \mathrm{N}_p^1(L, L') - \mathrm{K}(L, L')\,(p+1)\,p^{21/2} \ ,$$
$$\mathrm{N}_p^{\mathrm{sup}}(L, L') \ = \ \mathrm{N}_p^1(L, L') + \mathrm{K}(L, L')\,(p+1)\,p^{21/2} \ .$$

Recall that $\mathrm{n}_p(L)$ is the integer defined by the equality $\mathrm{N}_p(L, \mathrm{Leech}) = \mathrm{n}_p(L)\,\mathrm{pas}(L;p)$. Set
$$\nu_p^{\mathrm{inf}}(L) \ = \ \frac{\mathrm{N}_p^{\mathrm{inf}}(L, \mathrm{Leech})}{\mathrm{pas}(L;p)} \ , \quad \nu_p^{\mathrm{sup}}(L) \ = \ \frac{\mathrm{N}_p^{\mathrm{sup}}(L, \mathrm{Leech})}{\mathrm{pas}(L;p)} \ ;$$
we therefore have the bounds $\nu_p^{\mathrm{inf}}(L) \leq \mathrm{n}_p(L) \leq \nu_p^{\mathrm{sup}}(L)$.

Examples. Let us illustrate the effectiveness of these bounds through a few examples:

- We have $\nu_3^{\rm inf}(\mathrm{L}_{23}) \approx 0.99953$ and $\nu_3^{\rm sup}(\mathrm{L}_{23}) \approx 1.00041$, from which follow $\mathrm{n}_3(\mathrm{L}_{23}) = 1$ and $\mathrm{N}_3(\mathrm{L}_{23}, \mathrm{Leech}) = 8388608$, which agrees with part (d) of Theorem 3.4.2.10.
- We have $\nu_{31}^{\rm inf}(\mathrm{A}_{24}^{+}) \approx 275.99920$ and $\nu_{31}^{\rm sup}(\mathrm{A}_{24}^{+}) \approx 276.00061$, from which follow $\mathrm{n}_{31}(\mathrm{A}_{24}^{+}) = 276$ and $\mathrm{N}_{31}(\mathrm{A}_{24}^{+}, \mathrm{Leech}) = 1427031323986450710528000000$, which agrees with the computation carried out in Sect. 3.4.3.
- We check that we have $\nu_p^{\rm sup}(\mathrm{E}_8 \oplus \mathrm{E}_8 \oplus \mathrm{E}_8) < 8 \cdot 10^{-6}$ for $p \leq 29$, from which follows $\mathrm{N}_p(\mathrm{E}_8 \oplus \mathrm{E}_8 \oplus \mathrm{E}_8, \mathrm{Leech}) = 0$ for $p \leq 29$, which agrees with Proposition 3.4.1.1.
- We have $\nu_{47}^{\rm inf}(\mathrm{E}_{24}) \approx 0.99992$ and $\nu_{47}^{\rm sup}(\mathrm{E}_{24}) \approx 1.00006$, from which follow $\mathrm{n}_{47}(\mathrm{E}_{24}) = 1$ and $\mathrm{N}_{47}(\mathrm{E}_{24}, \mathrm{Leech}) = 1131456179644927440637132800000$, which agrees with part (d) of Theorem 3.4.2.10.

Set $\mathrm{n}_p^{\rm inf}(L) = \lceil \nu_p^{\rm inf}(L) \rceil$ and $\mathrm{n}_p^{\rm sup}(L) = \lfloor \nu_p^{\rm sup}(L) \rfloor$. Let us recall the notation: for a real number ν, $\lceil \nu \rceil$ is the least integer n with $\nu \leq n$ and $\lfloor \nu \rfloor$ is the greatest integer n with $n \leq \nu$. By definition, we therefore have the bounds

$$\mathrm{n}_p^{\rm inf}(L) \ \leq \ \mathrm{n}_p(L) \ \leq \ \mathrm{n}_p^{\rm sup}(L) \,.$$

Denote by $\mathrm{e}_p(L)$ the nonnegative integer $\mathrm{n}_p^{\rm sup}(L) - \mathrm{n}_p^{\rm inf}(L)$. If we have $\mathrm{e}_p(L) = 0$, then $\mathrm{N}_p(L, \mathrm{Leech})$ is uniquely determined: $\mathrm{N}_p(L, \mathrm{Leech}) = \mathrm{n}_p^{\rm inf}(L)\, \mathrm{pas}(L; p)$.

Example. Let $L = \mathrm{E}_{24}$. Since we have $\mathrm{g}(\mathrm{E}_{24}) = 2$, we must assume $p \geq 3$ for the theory above to apply. PARI tells us that $\mathrm{e}_p(\mathrm{E}_{24})$ is zero for $3 \leq p \leq 131$. The integer $\mathrm{N}_p(\mathrm{E}_{24}, \mathrm{Leech})$ has been determined for these primes; for example, we have (for the childish pleasure of writing a very large integer!):

$$\mathrm{N}_{131}(\mathrm{E}_{24}, \mathrm{Leech}) \ = \ 123625448053001992116952381878687498240000 \,.$$

Now, we consider the quadruple

$$\underline{\mathrm{e}}_p \ := \ (\mathrm{e}_p(\mathrm{E}_{24}), \mathrm{e}_p(\mathrm{E}_{16} \oplus \mathrm{E}_8), \mathrm{e}_p(\mathrm{E}_8 \oplus \mathrm{E}_8 \oplus \mathrm{E}_8), \mathrm{e}_p(\mathrm{A}_{24}^{+}))\,;$$

since we have $(\mathrm{g}(\mathrm{E}_{24}), \mathrm{g}(\mathrm{E}_{16} \oplus \mathrm{E}_8), \mathrm{g}(\mathrm{E}_8 \oplus \mathrm{E}_8 \oplus \mathrm{E}_8), \mathrm{g}(\mathrm{A}_{24}^{+})) = (2, 2, 1, 5)$, we assume $p \neq 2, 5$. PARI tells us that we have $\underline{\mathrm{e}}_p = (0, 0, 0, 0)$ for $p = 3$ and $7 \leq p \leq 59$. The integers $\mathrm{N}_p(\mathrm{L}_i, \mathrm{Leech})$ have been determined for these primes and $i = 1, 2, 3, 4$; consequently, the linear system (10.3.2) allows the computation of the $\theta_r(p)$ with $r \geq 7$ for the primes in question.

The Cases $61 \leq p \leq 107$

For $61 \leq p \leq 107$, we no longer have $\underline{\mathrm{e}}_p = (0, 0, 0, 0)$; we will, however, show that we can still determine the $\theta_r(p)$ for $r \geq 7$. The method is described below. Set $\mathrm{n}_p(\mathrm{L}_i) = \mathrm{n}_p^{\rm inf}(\mathrm{L}_i) + x_i$ for $1 \leq i \leq 4$; (x_1, x_2, x_3, x_4) is therefore a quadruple of nonnegative integers that need to be determined. This quadruple is subject to the following constraints, numbered (1)–(4):

(1) We have the inequalities $0 \leq x_i \leq \mathrm{e}_p(\mathrm{L}_i)$ for $1 \leq i \leq 4$.

(2) The quadruple (x_1, x_2, x_3, x_4) satisfies linear congruences modulo p or modulo divisors of $p+1$. Let us be more precise: the relation (10.3.2) can be written in the following form:

$$\begin{bmatrix} p(p+1)\,\theta_7(p) \\ (p+1)\,\theta_8(p) \\ (p+1)\,\theta_9(p) \\ (p+1)\,\theta_{10}(p) \end{bmatrix} = \mathrm{F}(p) \begin{bmatrix} x_1 \\ x_2 \\ x_3 \\ x_4 \end{bmatrix} + \mathrm{G}(p)\ , \tag{10.3.4}$$

where $\mathrm{F}(p)$ is the square matrix $a^{-1}\mathrm{diag}(\mathrm{pas}(\mathrm{L}_1;p), \ldots, \mathrm{pas}(\mathrm{L}_4;p))$ and $\mathrm{G}(p)$ is the column matrix $a^{-1}\mathrm{B}^{\mathrm{inf}}(p)$ with

$$\mathrm{B}^{\mathrm{inf}}(p) \ := \begin{bmatrix} \mathrm{n}_p^{\mathrm{inf}}(\mathrm{L}_1)\,\mathrm{pas}(\mathrm{L}_1;p) - \mathrm{N}_p^1(\mathrm{L}_1, \mathrm{Leech}) \\ \mathrm{n}_p^{\mathrm{inf}}(\mathrm{L}_2)\,\mathrm{pas}(\mathrm{L}_2;p) - \mathrm{N}_p^1(\mathrm{L}_2, \mathrm{Leech}) \\ \mathrm{n}_p^{\mathrm{inf}}(\mathrm{L}_3)\,\mathrm{pas}(\mathrm{L}_3;p) - \mathrm{N}_p^1(\mathrm{L}_3, \mathrm{Leech}) \\ \mathrm{n}_p^{\mathrm{inf}}(\mathrm{L}_4)\,\mathrm{pas}(\mathrm{L}_4;p) - \mathrm{N}_p^1(\mathrm{L}_4, \mathrm{Leech}) \end{bmatrix} .$$

The matrix $\mathrm{F}(p)$ has integral coefficients for $p \not\equiv 1 \bmod 23$. Let us explain why. As in the remark following Scholium-Definition 3.4.3.3, we write $\mathrm{pas}(\mathrm{L}_i;p) = \mathrm{pas}_1(\mathrm{L}_i)\,\mathrm{pas}_2(\mathrm{L}_i;p)$. We see that the matrix

$$a^{-1}\,\mathrm{diag}(\mathrm{pas}_1(\mathrm{L}_1), \ldots, \mathrm{pas}_1(\mathrm{L}_4))\,\mathrm{diag}(23, 1, 1, 1)$$

(which is independent of p) has integral coefficients; now, by its very definition, 23 divides the integer $\mathrm{pas}_2(\mathrm{E}_{24};p)$ for $p \not\equiv 1 \bmod 23$. Note that if $\mathrm{F}(p)$ has integral coefficients, then in view of (10.3.4), the same holds for $\mathrm{G}(p)$. The fact that $\mathrm{F}(p)$ and $\mathrm{G}(p)$ have integral coefficients and that the $\theta_r(p)$ with $r \geq 7$ are integral leads to the congruences mentioned above.

(3) Let $\theta_r(p; X_1, X_2, X_3, X_4)$, for $7 \leq r \leq 10$, be the four linear polynomials in $\mathbb{Q}[X_1, X_2, X_3, X_4]$ defined by

$$\begin{bmatrix} \theta_7(p; X_1, X_2, X_3, X_4) \\ \theta_8(p; X_1, X_2, X_3, X_4) \\ \theta_9(p; X_1, X_2, X_3, X_4) \\ \theta_{10}(p; X_1, X_2, X_3, X_4) \end{bmatrix} := \mathrm{diag}(p(p+1), p+1, p+1, p+1)^{-1} \left(\mathrm{F}(p) \begin{bmatrix} X_1 \\ X_2 \\ X_3 \\ X_4 \end{bmatrix} + \mathrm{G}(p) \right)$$

(the notation is strange... but transparent; note that Condition (2) is equivalent to the fact that $\theta_r(p; x_1, x_2, x_3, x_4)$ is integral for $7 \le r \le 10$). Let $\mathrm{T}_p(X_1, X_2, X_3, X_4)$ be the matrix obtained by substituting in (10.2.2) the $\theta_r(p; X_1, X_2, X_3, X_4)$ for the $\theta_r(p)$, when we have $r \ge 7$; $\mathrm{T}_p(X_1, X_2, X_3, X_4)$ is therefore a 24×24 matrix whose coefficients are linear polynomials in $\mathbb{Q}[X_1, X_2, X_3, X_4]$, such that we have $\mathrm{T}_p(x_1, x_2, x_3, x_4) = \mathrm{T}_p$. Denote by $\mathrm{N}_p(\mathrm{L}_i, \mathrm{L}_j; X_1, X_2, X_3, X_4)$ the coefficient of index (j, i) of $\mathrm{T}_p(X_1, X_2, X_3, X_4)$ and set, for $i \le 23$,

$$\mathrm{n}_p(\mathrm{L}_i; X_1, X_2, X_3, X_4) \ := \ \frac{\mathrm{N}_p(\mathrm{L}_i, \mathrm{Leech}; X_1, X_2, X_3, X_4)}{\mathrm{pas}(\mathrm{L}_i; p)}$$

(hence, by construction, we have $\mathrm{n}_p(\mathrm{L}_i; X_1, X_2, X_3, X_4) = X_i + \mathrm{n}_p^{\mathrm{inf}}(\mathrm{L}_i)$ for $i \le 4$). For example, we have

$$\mathrm{n}_p(\mathrm{L}_5; X_1, X_2, X_3, X_4) \ = \ \sum_{i=1}^{4} \alpha_i \, \gamma_i(p) \, X_i + \beta(p)$$

with

$$(\alpha_1, \alpha_2, \alpha_3, \alpha_4) = \Big(-1472, -\frac{41}{8}, \frac{119}{16}, \frac{281}{256} \Big)$$

$$\gamma_i(p) = \frac{\gcd(p-1, 24\,\mathrm{h}(\mathrm{L}_5))}{\gcd(p-1, 24\,\mathrm{h}(\mathrm{L}_i))}, \quad \beta(p) = \frac{\mathrm{N}_p^1(\mathrm{L}_5, \mathrm{Leech})}{\mathrm{pas}(\mathrm{L}_i; p)} \, .$$

By Scholium-Definition 3.4.3.3, $\mathrm{n}_p(\mathrm{L}_5; x_1, x_2, x_3, x_4) = \mathrm{n}_p(\mathrm{L}_5)$ is integral; it follows that the quadruple (x_1, x_2, x_3, x_4) satisfies a linear congruence modulo an explicit integer depending on p, which we denote by $\mathrm{m}(p)$. Let us be more precise. We see that the common denominator of the rational numbers $\alpha_i \, \gamma_i(p)$ (in irreducible form) is $\mathrm{m}(p)$; for $61 \le p \le 113$, the function $p \mapsto \mathrm{m}(p)$ is given by the following table (we have added the values of $\mathrm{m}(p)$ for $p = 109$ and $p = 113$ in view of a later application):

p	61	67	71	73	79	83	89	97	101	103	107	109	113
m	1280	256	1280	768	256	256	256	128	6400	256	256	768	128

From this, we deduce that $\mathrm{m}(p)\beta(p)$ is integral and also deduce the congruence mentioned above.

Note that the 2-adic valuation of $\mathrm{v}_2(\gamma_i(p))$ is zero for $i = 1, 2, 3$ and that we have $\mathrm{v}_2(\gamma_4(p)) = 0$ when $\mathrm{v}_2(p-1) \le 3$ and $\mathrm{v}_2(\gamma_4(p)) = 1$ when $\mathrm{v}_2(p-1) \ge 4$. By considering the quadruple $(\alpha_1, \alpha_2, \alpha_3, \alpha_4)$, it follows that the congruence we have just deduced determines the class of x_4 modulo 16 for $p \ne 97, 113$ (and

$61 \le p \le 113$) and modulo 8 for $p = 97, 113$. The value of x_4 modulo 16 for $61 \le p \le 113$ is given by the following table:

p	61	67	71	73	79	83	89	97	101	103	107	109	113
$x_4 \bmod 16$	3	0	2	11	7	3	2	0, 8	5	4	5	0	0, 8

From now on, we will denote this weak form of Constraint (3) by (3_4).

(4) We must have the inequalities $|\theta_7(p; x_1, x_2, x_3, x_4)| \le 4p^{19/2}$ and $|\theta_r(p; x_1, x_2, x_3, x_4)| \le 4p^{21/2}$ for $r = 8, 9, 10$.

The Case $p = 61$. We have $\underline{e}_{61} = (0, 0, 0, 5)$. Constraint ($3_4$) determines the desired quadruple: $(x_1, x_2, x_3, x_4) = (0, 0, 0, 3)$.

The Case $p = 67$. We have $\underline{e}_{67} = (0, 0, 0, 1)$. Constraint ($3_4$) determines the desired quadruple: $(x_1, x_2, x_3, x_4) = (0, 0, 0, 0)$.

The Case $p = 71$. We have $\underline{e}_{71} = (0, 0, 0, 6)$. Constraint ($3_4$) determines the desired quadruple: $(x_1, x_2, x_3, x_4) = (0, 0, 0, 2)$.

The Case $p = 73$. We have $\underline{e}_{73} = (0, 6, 9, 20)$. Constraint (2) gives, in particular, $(x_1, x_2, x_3, x_4) \equiv (0, 3, 5, 11) \bmod 37$. The desired quadruple is $(0, 3, 5, 11)$.

The Case $p = 79$. We have $\underline{e}_{79} = (0, 0, 1, 12)$. Constraint (2) gives $2x_3 + x_4 \equiv 9 \bmod 79$; the bounds $0 \le 2x_3 + x_4 \le 14$ then imply the equality $2x_3 + x_4 = 9$. Constraint (3_4) then determines the desired quadruple: $(x_1, x_2, x_3, x_4) = (0, 0, 1, 7)$.

The Case $p = 83$. We have $\underline{e}_{83} = (0, 0, 0, 6)$. Constraint ($3_4$) determines the desired quadruple: $(x_1, x_2, x_3, x_4) = (0, 0, 0, 3)$.

The Case $p = 89$. We have $\underline{e}_{89} = (0, 6, 10, 67)$. Constraint (2) gives $-2x_2 - 7x_3 + x_4 + 7 \equiv 0 \bmod 89$; the bounds $-75 \le -2x_2 - 7x_3 + x_4 - 7 \le 74$ then imply the equality $x_4 = 2x_2 + 7x_3 - 7$. Constraint (3) gives $10x_2 + 3x_3 - 45 \equiv 0 \bmod 256$; the same argument as before shows that we have $10x_2 + 3x_3 - 45 = 0$. In particular, x_2 is divisible by 3. The bounds $0 \le x_3 \le 10$ then imply the equality $x_2 = 3$. The desired quadruple is $(0, 3, 5, 34)$.

The Case $p = 97$. We have $\underline{e}_{97} = (0, 117, 187, 548)$. The computer says that the quadruples satisfying Constraints (1)–(3) are $(0, 22, 63, 432)$ and $(0, 71, 105, 272)$ (we can help the computer by observing that Constraint (2) implies $(x_2, x_3) \equiv (1, 0) \bmod 7$ and that Constraint (3_4) says that x_4 is divisible by 8). The first does not pass the "Ramanujan test" (Constraint (4)): $(x_1, x_2, x_3, x_4) = (0, 71, 105, 272)$.

The Case $p = 101$. We have $\underline{e}_{101} = (0, 78, 124, 3643)$. The computer says that the only quadruple satisfying Constraints (1)–(3) is $(0, 30, 63, 2149)$.

The Case $p = 103$. We have $\underline{e}_{103} = (0, 29, 46, 273)$. The computer says that the quadruples satisfying Constraints (1)–(3) are $(0, 7, 46, 196)$, $(0, 15, 27, 148)$, and $(0, 23, 8, 100)$. The first and third do not pass the "Ramanujan test": $(x_1, x_2, x_3, x_4) = (0, 15, 27, 148)$.

The Case $p = 107$. We have $\underline{e}_{107} = (0, 14, 23, 141)$. The computer says that the only quadruple satisfying Constraints (1)–(3) is $(0, 7, 10, 53)$.

The Cases $p = 109$ and $p = 113$

An unexpected consequence of Theorem 10.4.4, which we will prove in the next subsection by invoking the theory of Galois representations, is that the quadruple (x_1, x_2, x_3, x_4) introduced while studying the cases $61 \leq p \leq 107$ satisfies linear congruences modulo divisors of $p + 1$, of which some may be "independent" of the congruences of Constraint (2). These additional constraints allow the determination of the quadruple (x_1, x_2, x_3, x_4) for $p = 109$ and $p = 113$.

The Case $p = 109$. We have $\underline{e}_{109} = (0, 337, 538, 1049)$. The computer says that there exist 208 quadruples satisfying Constraints (1)–(3) and that of these, 12 satisfy Constraint (4). Denote these 12 quadruples of nonnegative integers by $\underline{x}^{(1)}, \underline{x}^{(2)}, \ldots, \underline{x}^{(12)}$; suppose, to fix ideas, that we have $\underline{x}^{(1)} < \underline{x}^{(2)} < \ldots < \underline{x}^{(12)}$ for the lexicographical order. For $k = 1, 2, \ldots, 12$, we easily verify that $\mathrm{T}_{109}(\underline{x}^{(k)})$ has integral coefficients and that $\mathrm{n}_{109}(\mathrm{L}_i; \underline{x}^{(k)})$ is also integral for $6 \leq i \leq 23$ (we have done the necessary to ensure that $\mathrm{n}_{109}(\mathrm{L}_i; \underline{x}^{(k)})$ is integral for $1 \leq i \leq 5$).

The above shows that the method that made it possible to determine $\tau_{6,8}(p)$, $\tau_{8,8}(p)$, $\tau_{12,6}(p)$, and $\tau_{4,10}(p)$ for $p \leq 107$ does not work for $p = 109$. However, we succeed in removing this indetermination as follows. Consider the sequence of quadruples of integers

$$\left(\theta_7(109; \underline{x}^{(k)})\,,\ \theta_8(109; \underline{x}^{(k)})\,,\ \theta_9(109; \underline{x}^{(k)})\,,\ \theta_{10}(109; \underline{x}^{(k)})\right)_{k=1,2,\ldots,12}\,.$$

Luckily, only one of these quadruples, namely the fifth, verifies the congruence imposed by item (12) of Theorem 10.4.4 (incidentally, the quadruple in question is the one that was the most probable "in the sense of Sato–Tate"). Note that item (12) of Theorem 10.4.4 is a congruence modulo 11 and that 11 divides $109 + 1$!

We can paraphrase the above as follows. In addition to Constraints (1)–(4), for $i = 1, 2, 3, 4$, the nonnegative numbers $x_i := \mathrm{n}_{109}(\mathrm{L}_i) - \mathrm{n}_{109}^{\mathrm{inf}}(\mathrm{L}_i)$ are subject to the constraint, which we call (2-supp), imposed by item (12) of Theorem 10.4.4. It is not difficult to give (2-supp) explicitly: it is the congruence

$$x_1 + x_2 + 6x_3 + x_4 + 2 \equiv 0 \mod 11\,.$$

Only one quadruple (x_1, x_2, x_3, x_4) satisfies Constraints (1), (2), (2-supp), (3), and (4): the quadruple $(0, 138, 284, 576)$.

The Case $p = 113$. We have $\underline{e}_{113} = (0, 227, 361, 1058)$. This time, consider items (3) and (4) of Theorem 10.4.4 (because 19 divides $113 + 1$). These two items impose one constraint on the quadruple (x_1, x_2, x_3, x_4), which we denote by (2-supp), consisting of two linear congruences modulo 19; we easily see that these are independent of the two linear congruences modulo 19 that appear in Constraint (2).

Consequently, the quadruple (x_1, x_2, x_3, x_4) is fixed modulo 19; we find

$$(x_1, x_2, x_3, x_4) \equiv (0, 6, 3, 16) \mod 19$$

(the presence of the 0 in the first position on the right-hand side is reassuring!). The computer then shows that there exist only two quadruples satisfying Constraints (1), (2), (2-supp), and (3): $(0, 120, 155, 396)$ and $(0, 177, 326, 244)$. The second does not pass the Ramanujan test. The integers $\tau_{6,8}(113)$, $\tau_{8,8}(113)$, $\tau_{12,6}(113)$, and $\tau_{4,10}(113)$ and the endomorphism T_{113} of $\mathbb{Z}[\mathrm{X}_{24}]$ are thus determined.

For example, we find

$$\mathrm{N}_{113}(\mathrm{L}_{12}, \mathrm{L}_{21}) = 63332383852347806963662416686287375220736000\,0$$

(this is the greatest number of p-neighbors for p prime among those we were able to compute).

The Case $p = 127$

The methods that have allowed us to determine $\tau_{6,8}(p)$, $\tau_{8,8}(p)$, $\tau_{12,6}(p)$, and $\tau_{4,10}(p)$ for $p \leq 113$ do not work for $p = 127$. Let us explain why below (in a smaller font).

- The computer says that there exist 3329 quadruples that satisfy Constraints (1), (2), (3), and (4), which we denote by $\underline{x}^{(1)}, \underline{x}^{(2)}, \ldots, \underline{x}^{(3329)}$. For $k = 1, 2, \ldots, 3329$, we check that $\mathrm{T}_{127}(\underline{x}^{(k)})$ has integral coefficients and that $\mathrm{n}_{127}(\mathrm{L}_i; \underline{x}^{(k)})$ is also integral for $6 \leq i \leq 23$.
- The only prime that divides $127 + 1$ is 2, and we easily see that the integers $\theta_r(127; \underline{x}^{(k)})$, for $7 \leq r \leq 10$ and $1 \leq k \leq 3329$, are all even, which agrees with congruence (12) of Theorem 10.4.4. In fact, by considering the four polynomials $\theta_r(127; X_1, X_2, X_3, X_4)$, for $7 \leq r \leq 10$, we see that we have $\theta_7(127; \underline{x}) \equiv 134400 \mod 2^{18}$, $\theta_8(127; \underline{x}) \equiv 3840 \mod 2^{13}$, $\theta_9(127; \underline{x}) \equiv -3840 \mod 2^{13}$, and $\theta_{10}(127; \underline{x}) \equiv 256 \mod 2^{10}$ for every $\underline{x}$ in $\mathbb{Z}^4$.

10.3.2 Determination of the $\tau_{j,k}(p^2)$ for $p \leq 29$

Below, we denote by $\varpi_1, \varpi_2, \ldots, \varpi_{10}$, respectively, the automorphic representations $\Delta_{11}, \Delta_{15}, \Delta_{17}, \Delta_{19}, \Delta_{21}, \mathrm{Sym}^2\Delta_{11}, \Delta_{19,7}, \Delta_{21,9}, \Delta_{21,13}$, and $\Delta_{21,5}$ (see Sects. 7.3 and 6.4.7 and Definition 9.1.6). The representation ϖ_r is therefore in $\Pi_{\mathrm{cusp}}(\mathrm{PGL}_2)$ for $r \leq 5$, in $\Pi_{\mathrm{cusp}}(\mathrm{PGL}_3)$ for $r = 6$, and in $\Pi_{\mathrm{cusp}}(\mathrm{PGL}_4)$ for $r \geq 7$.

Let p be a prime. By definition, the integers $\theta_r(p)$, for $1 \leq r \leq 10$, that we introduced in Sect. 10.2 satisfy the relation

$$\theta_r(p) = p^{\mathrm{w}(\varpi_r)/2} \operatorname{trace} \mathrm{c}_p(\varpi_r) .$$

(See Sects. 6.4.1 and 8.2.6; recall that $\mathrm{w}(\pi)$ denotes the motivic weight of an algebraic automorphic representation π in $\Pi_{\mathrm{cusp}}(\mathrm{PGL}_n)$. Here we have $\mathrm{w}(\Delta_w) = w$ and $\mathrm{w}(\Delta_{w,v}) = w$.)

Likewise, set

$$\theta_r(p^2) \;:=\; p^{\mathrm{w}(\varpi_r)}\,\mathrm{trace}\,(\mathrm{c}_p(\varpi_r)^2)\,.$$

For $r \geq 7$, the definition above agrees with Sect. 9.1.7: $\theta_7(p^2) = \tau_{6,8}(p^2)$, $\theta_8(p^2) = \tau_{8,8}(p^2)$, $\theta_9(p^2) = \tau_{12,6}(p^2)$, and $\theta_9(p^2) = \tau_{4,10}(p^2)$.

On the other hand, let us stress that for $r = 1, 2, 3, 4, 5$, $\theta_i(p^2)$ is not the value in p^2 of the arithmetic functions τ_{12}, τ_{16}, τ_{18}, τ_{20}, τ_{22}, respectively; we in fact have $\theta_1(p^2) = \tau_{12}(p)^2 - 2p^{11}$, $\theta_2(p^2) = \tau_{16}(p)^2 - 2p^{15}$, $\theta_3(p^2) = \tau_{18}(p)^2 - 2p^{17}$, $\theta_4(p^2) = \tau_{20}(p)^2 - 2p^{19}$, and $\theta_5(p^2) = \tau_{22}(p)^2 - 2p^{21}$. Finally, it is easy to verify that we have $\theta_6(p^2) = \tau_{12}(p)^4 - 4p^{11}\tau_{12}(p)^2 + 3p^{22}$.

Let $\mathrm{V_{St}}$ be the standard representation of $\mathrm{SO}_n(\mathbb{C})$. For a prime p, denote by $\mathrm{T}_p^{\psi^2}$ the Hecke operator in $\mathrm{H}_p(\mathrm{O}_n)$ defined, via the Satake isomorphism, by the formula

$$p^{2-n}\;\mathrm{Sat}(\mathrm{T}_p^{\psi^2}) \;=\; \psi^2\,[\mathrm{V_{St}}] \;:=\; [\mathrm{V_{St}} \otimes \mathrm{V_{St}}] - 2\,[\Lambda^2\,\mathrm{V_{St}}]$$

(see Sect. 6.2.1). This operator, which is a priori in $\mathrm{H}_p(\mathrm{O}_n)[p^{-1/2}]$, in fact belongs to $\mathrm{H}_p(\mathrm{O}_n)$ by Lemma 10.1.1; actually, in the present case, we have

$$\mathrm{T}_p^{\psi^2} \;=\; \mathrm{T}_p^2 - 2p\mathrm{T}_{p,p} - 2p\left(\sum_{i=0}^{n/2-2} p^{2i} + p^{n/2-2}\right) \tag{10.3.5}$$

by formulas (6.2.5) and (6.2.6).

Now, suppose $n = 24$, and for $1 \leq j \leq 24$ denote by $\lambda_j^{\psi^2}(p)$ the eigenvalue of $\mathrm{T}_p^{\psi^2}$ on the vector v_j of $\mathbb{Z}[\mathrm{X}_{24}]$; we have done what we can to obtain

$$\lambda_j^{\psi^2}(p) \;=\; p^{22}\;\mathrm{trace}\;\mathrm{St}(\mathrm{c}_p(\pi_j)^2) \;=\; \mathrm{C}_{j,0}(p^2) + \sum_{r=1}^{10}\mathrm{C}_{j,r}(p^2)\,\theta_r(p^2)\,,$$

where the polynomials $\mathrm{C}_{j,r}$ in $\mathbb{Z}[X]$ are those introduced in Sect. 10.2.

Let us return to the general case. The formula

$$(p+1)\,\mathrm{T}_{p,p} \;=\; \mathrm{T}_p^2 - \mathrm{T}_{p^2} - \frac{(p^{n/2}-1)(p^{n/2-1}+1)}{(p-1)}$$

of Example 6.2.11 and formula (10.3.5) shows that the operator T_{p^2} can be expressed in terms of $\mathrm{T}_p^{\psi^2}$ and T_p^2:

$$\mathrm{T}_{p^2} \;=\; \frac{p+1}{2p}\,\mathrm{T}_p^{\psi^2} + \frac{p-1}{2p}\,\mathrm{T}_p^2\; - p^{n-2} + p^{n/2-2}\,. \tag{10.3.6}$$

Since we determined the $\tau_{j,k}(p)$ for $p \leq 113$ in Sect. 10.3.1, the considerations above show that determining the $\tau_{j,k}(p^2)$ with $p \leq 29$ reduces to determining the number of neighbors $\mathrm{N}_{p^2}(\mathrm{L}_i, \mathrm{Leech})$ for $i \leq 4$ and $p \leq 29$ (recall that we have set $\mathrm{E}_{24} = \mathrm{L}_1$, $\mathrm{E}_{16} \oplus \mathrm{E}_8 = \mathrm{L}_2$, $\mathrm{E}_8 \oplus \mathrm{E}_8 \oplus \mathrm{E}_8 = \mathrm{L}_3$, and $\mathrm{A}_{24}^+ = \mathrm{L}_4$). Indeed, the system that expresses the $\mathrm{N}_{p^2}(\mathrm{L}_i, \mathrm{Leech})$ in terms of the “unknowns” $\tau_{j,k}(p^2)$ is still a system with as many equations as variables and a unique solution, because the matrix a of (10.3.3) is invertible. We determine these numbers of neighbors below.

The Case $p \leq 3$. We have $\mathrm{N}_{p^2}(\mathrm{L}_i, \mathrm{Leech}) = 0$ for $i \leq 4$ and $p = 2, 3$ by Proposition 3.4.1.1.

The Case $p = 5$. We again have $\mathrm{N}_{25}(\mathrm{L}_i, \mathrm{Leech}) = 0$ for $i \leq 3$, by Proposition 3.4.1.1; part (d) of Theorem 3.4.2.10 gives the value of $\mathrm{N}_{25}(\mathrm{L}_4, \mathrm{Leech})$.

The Case $p = 7$. We adapt the method used previously to determine $\mathrm{N}_p(\mathrm{L}_i, \mathrm{Leech})$ for $i = 1, 2, 3, 4$, in the cases $7 \leq p \leq 59$.

Proposition 3.4.3.1, whose notation we use, must be modified as follows.

Proposition 10.3.2.1. *Let L be a Niemeier lattice with roots and p a prime that does not divide the index of Q in L; let S be the stabilizer, for the action of W, of an element of $\mathrm{P}_L^{\mathrm{reg}}(\mathbb{Z}/p^2)$.*

(a) *The group S can be identified with a subgroup of $(\mathbb{Z}/p^2)^\times$.*
(b) *Let S^p be the image of S by the endomorphism $x \mapsto x^p$ of the group $(\mathbb{Z}/p^2)^\times$; the action of S^p on R (induced by that of W) is free.*

Consequently, Scholium-Definition 3.4.3.3 must be modified as follows.

Scholium-Definition 10.3.2.2. *Let L be a Niemeier lattice with roots and p a prime; denote by $\mathrm{pas}(L; p^2)$ the integer defined by*

$$\mathrm{pas}(L; p^2) \ := \ \frac{|\mathrm{W}(L)|}{\gcd(p(p-1), 24p\mathrm{h}(L), |\mathrm{W}(L)|)} \ .$$

If p does not divide the index of Q in L, then $\mathrm{N}_{p^2}(L, \mathrm{Leech})$ is divisible by $\mathrm{pas}(L; p^2)$. In this case, we denote by $\mathrm{n}_{p^2}(L)$ the integer defined by

$$\mathrm{N}_{p^2}(L, \mathrm{Leech}) \ = \ \mathrm{n}_{p^2}(L) \ \mathrm{pas}(L; p^2) \ .$$

Remark. As we have $9 = \mathrm{h}(\mathrm{L}_{16}) + 1$, item (d) of Theorem 3.4.2.10 gives us the value of $\mathrm{N}_9(\mathrm{L}_{16}, \mathrm{Leech})$ and therefore that of $\mathrm{n}_9(\mathrm{L}_{16})$. We find $\mathrm{n}_9(\mathrm{L}_{16}) = 1$; as we have $\mathrm{g}(\mathrm{L}_{16}) = 64$, this shows that Scholium-Definition 10.3.2.2 is, in a sense, optimal.

We then proceed as when we determined $\mathrm{N}_p(\mathrm{L}_i, \mathrm{Leech})$ for $i = 1, 2, 3, 4$ and $7 \leq p \leq 59$. Mutatis mutandis, we define "easily computable" integers $\mathrm{n}_{p^2}^{\inf}(L)$ and $\mathrm{n}_{p^2}^{\sup}(L)$ such that we have the bounds

$$\mathrm{n}_{p^2}^{\inf}(L) \ \leq \ \mathrm{n}_{p^2}(L) \ \leq \ \mathrm{n}_{p^2}^{\sup}(L) \ .$$

Let us be more precise. To obtain these bounds, consider the expression (10.3.6) of T_{p^2} in terms of T_p and $\mathrm{T}_p^{\psi^2}$ obtained earlier, and use the determination of T_p for $p \leq 113$ and the Ramanujan inequalities for the $\tau_{j,k}(p^2)$, namely $|\tau_{6,8}(p^2)| \leq 4\,p^{19}$,

$|\tau_{8,8}(p^2)| \leq 4\,p^{21}$, $|\tau_{12,6}(p^2)| \leq 4\,p^{21}$, and $|\tau_{4,10}(p^2)| \leq 4\,p^{21}$ (for finer Ramanujan inequalities, see (10.3.8)).

We find $\mathrm{n}_{49}^{\mathrm{inf}}(\mathrm{L}_i) = \mathrm{n}_{49}^{\mathrm{sup}}(\mathrm{L}_i)$ for $i = 1, 2, 3, 4$; from this we deduce the determination of $\mathrm{N}_{49}(\mathrm{L}_i, \mathrm{Leech})$ for $i = 1, 2, 3, 4$.

The Cases $11 \leq p \leq 29$

This time, we adapt the method used to determine $\mathrm{N}_p(\mathrm{L}_i, \mathrm{Leech})$ for $i = 1, 2, 3, 4$ in the cases $61 \leq p \leq 107$.

Set $x_i = \mathrm{n}_{p^2}(\mathrm{L}_i) - \mathrm{n}_{p^2}^{\mathrm{inf}}(\mathrm{L}_i)$; we must once again determine the quadruple of nonnegative integers (x_1, x_2, x_3, x_4).

Set $\mathrm{e}_{p^2}(\mathrm{L}_i) := \mathrm{n}_{p^2}^{\mathrm{sup}}(\mathrm{L}_i) - \mathrm{n}_{p^2}^{\mathrm{inf}}(\mathrm{L}_i)$ for $1 \leq i \leq 23$ and

$$\underline{\mathrm{e}}_{p^2} \;:=\; (\mathrm{e}_{p^2}(\mathrm{L}_1), \mathrm{e}_{p^2}(\mathrm{L}_2), \mathrm{e}_{p^2}(\mathrm{L}_3), \mathrm{e}_{p^2}(\mathrm{L}_4))\ .$$

By definition, we then have the inequalities

$$(1) \qquad x_1 \leq \mathrm{e}_{p^2}(\mathrm{L}_1)\,, \quad x_2 \leq \mathrm{e}_{p^2}(\mathrm{L}_2)\,, \quad x_3 \leq \mathrm{e}_{p^2}(\mathrm{L}_3)\,, \quad x_4 \leq \mathrm{e}_{p^2}(\mathrm{L}_4)\ .$$

The Case $p = 11$. First, compute $\underline{\mathrm{e}}_{11^2}$; this gives $\underline{\mathrm{e}}_{11^2} = (1, 1868, 270, 17436)$.

Then, express the integers $\tau_{j,k}(11^2)$ in terms of the "unknowns" x_1, x_2, x_3, x_4; this gives (the computation uses the determination of T_{11} that we carried out earlier):

$$\begin{bmatrix} \tau_{6,8}(11^2) \\ \tau_{8,8}(11^2) \\ \tau_{12,6}(11^2) \\ \tau_{4,10}(11^2) \end{bmatrix} \;=\; \frac{1}{61}\,A \begin{bmatrix} x_1 \\ x_2 \\ x_3 \\ x_4 \end{bmatrix} \;+\; \frac{1}{61}\,B\,,$$

where A and B are two explicit matrices with integral coefficients, of respective sizes 4×4 and 4×1 (the occurrence of the prime 61 is due to the fact that 61 divides $11^2 + 1$).

Note that the reduction modulo 61 of the matrix A is invertible; we can therefore compute the reduction modulo 61 of the quadruple (x_1, x_2, x_3, x_4). This gives

$$(2) \qquad (x_1, x_2, x_3, x_4) \;\equiv\; (1, 52, 25, 15) \mod 61$$

(note that this congruence already implies the equality $x_1 = 1$).

Next, also express the integers $\mathrm{n}_{11^2}(\mathrm{L}_i)$ for $5 \leq i \leq 23$ in terms of x_1, x_2, x_3, x_4; this gives

$$\mathrm{n}_{11^2}(\mathrm{L}_i) \;=\; a_{i,1}\,x_1 + a_{i,2}\,x_2 + a_{i,3}\,x_3 + a_{i,4}\,x_4 + b_i\ ,$$

where $a_{i,1}$, $a_{i,2}$, $a_{i,3}$, $a_{i,4}$, b_i are rational numbers. This equality shows that the quadruple (x_1, x_2, x_3, x_4) satisfies a certain linear congruence (that may be trivial) modulo the gcd of the denominators of the $a_{i,j}$ for $1 \leq j \leq 4$; we denote by (3) the set of these new congruences.

The computer says that the only quadruples satisfying (1), (2), and (3) are

$$(1, 662, 269, 6481)\,, \quad (1, 1333, 147, 6481)\,, \quad (1, 1333, 208, 17217)\,.$$

Finally, the first and third quadruple do not pass the Ramanujan test. This concludes the determination of the $\tau_{j,k}(11^2)$.

Alternative Method. By Proposition 9.1.9, the coefficient of t^2 in the characteristic polynomial $\det(t - p^{\mathrm{w}(\varpi_i)/2}\,\mathrm{c}_p(\varpi_i))$, for $i = 7, 8, 9, 10$, is divisible by p^6, p^6, p^4, p^8, respectively; that is, we have the congruence

$$\tau_{j,k}(p)^2 \;\equiv\; \tau_{j,k}(p^2) \quad \bmod 2p^{k-2}\,. \tag{10.3.7}$$

Set $\varepsilon_{j,k}(p) := \frac{1}{2}(\tau_{j,k}(p)^2 - \tau_{j,k}(p^2))$ and express these $\varepsilon_{j,k}(p)$ in terms of the quadruple of nonnegative integers $(x_i)_{1\leq i\leq 4} := (\mathrm{n}_{p^2}(\mathrm{L}_i) - \mathrm{n}^{\mathrm{inf}}_{p^2}(\mathrm{L}_i))$ (assume that the $\tau_{j,k}(p)$ are known, which is the case for all primes we are interested in here). We obtain an expression of the form

$$\begin{bmatrix} \varepsilon_{6,8}(p) \\ \varepsilon_{8,8}(p) \\ \varepsilon_{12,6}(p) \\ \varepsilon_{4,10}(p) \end{bmatrix} = \mathrm{E}(p) \begin{bmatrix} x_1 \\ x_2 \\ x_3 \\ x_4 \end{bmatrix} + \mathrm{H}(p)\,,$$

where $\mathrm{E}(p)$ and $\mathrm{H}(p)$ are two matrices with rational coefficients of respective size 4×4 and 4×1. The fact that the $\varepsilon_{j,k}(p)$ are integers and satisfy the congruences modulo p^{k-2} provided by (10.3.7) imposes a constraint on the quadruple (x_1, x_2, x_3, x_4), which we denote by (2-bis).

For $p = 11$, we easily see that there exists a single quadruple that satisfies Constraints (1) and (2-bis), namely $(1, 1333, 147, 6481)$.

The Cases $p = 13$, $p = 17$ *and* $p = 19$. We have

$$\underline{\mathrm{e}}_{13^2} = (655, 121728, 14943, 1135678)\,;$$
$$\underline{\mathrm{e}}_{17^2} = (536541, 5855913, 9346120, 46438144)\,;$$
$$\underline{\mathrm{e}}_{19^2} = (2884703, 84510145, 134879385, 4993470088)\,.$$

We see that in all three cases, there exists a single quadruple (x_1, x_2, x_3, x_4) that satisfies Constraints (1) and (2-bis), namely

$$(453, 50943, 3642, 439453) \text{ for } p = 13\,,$$
$$(217661, 1571118, 3271290, 261210371) \text{ for } p = 17\,,$$
$$(964326, 29790571, 55543719, 3055506804) \text{ for } p = 19\,.$$

The Case $p = 23$. We have

$$\underline{\mathrm{e}}_{23^2} = (93365728, 753181406, 1202088152, 161617609778)\,.$$

This time, there exist two quadruples that satisfy Constraints (1) and (2-bis). The only one to pass the Ramanujan test is

$$(52157635, 398996852, 418588772, 78467649933)\,.$$

The Case $p = 29$. We have

$$\underline{e}_{29^2} = (1662796593, 308516971151, 492397438725, 2878328193860)\,.$$

There exist 156 quadruples that satisfy Constraints (1) and (2-bis). Of these, only six that pass the Ramanujan test.

To carry out a final selection, proceed as in the case $p = 11$, first method. Express $\mathrm{n}_{29^2}(\mathrm{L}_5)$ in terms of (x_1, x_2, x_3, x_4) and observe that the integrality of $\mathrm{n}_{29^2}(\mathrm{L}_5)$ implies that the quadruple (x_1, x_2, x_3, x_4) satisfies a certain congruence modulo 256, which we denote by (3). We easily check that this constraint is satisfied for only one of the six quadruples above, namely

$$(773950187, 87165709281, 106617389411, 1454026724829)$$

(in fact, this quadruple is the only one to satisfy both Constraint (1) and Constraint (3)).

Remark. The determination of the $\tau_{j,k}(p^2)$ for $p \leq 29$ allows us to compute explicitly the Hecke operator $\mathrm{T}_{p^2} : \mathbb{Z}[\mathrm{X}_{24}] \to \mathbb{Z}[\mathrm{X}_{24}]$ for $p \leq 29$. For example, we find

$$\mathrm{N}_{29^2}(\mathrm{E}_{24}, \mathrm{L}_{21}) = 9787847431870605615736000813350868753051894303124387738419200000$$

(approximately 0.98×10^{64}, a new record!).

The Case $p = 31$

The method we used for $11 \leq p \leq 29$ does not work for $p = 31$. Let us explain why (in a smaller font).

This method can be described as follows. Set $\mathrm{n}_{p^2}(\mathrm{L}_i) = \mathrm{n}_{p^2}^{\mathrm{inf}}(\mathrm{L}_i) + x_i$, with $x_i \in \mathbb{Z}$, for $i = 1, 2, 3, 4$. Set $\underline{x} = (x_1, x_2, x_3, x_4)$, so $\underline{x}$ is a priori an element of $\mathbb{Z}^4 \subset \mathbb{R}^4$.

The Ramanujan inequalities satisfied by the $\tau_{j,k}(p^2)$ say that $\underline{x}$ belongs to a parallelotope, which we denote by Par_p, in the affine space $\mathbb{R}^4$; the definition of the integers $\mathrm{n}_{p^2}^{\mathrm{inf}}(\mathrm{L}_i)$ and $\mathrm{n}_{p^2}^{\mathrm{sup}}(\mathrm{L}_i)$ is such that the condition $\underline{x} \in \mathrm{Par}_p$ implies the bounds $0 \leq x_i \leq \mathrm{e}_{p^2}(\mathrm{L}_i)$. The fact that the $\varepsilon_{j,k}(p)$ are integers, that these integers satisfy the congruences modulo p^{k-2} provided by (10.3.7), and finally that the $\mathrm{n}_{p^2}(\mathrm{L}_i)$ are integers for $5 \leq i \leq 23$ (these $\mathrm{n}_{p^2}(\mathrm{L}_i)$ can be expressed as linear functions of $\underline{x}$ with rational coefficients) mean that $\underline{x}$ belongs to a translate Γ_p^{aff} of a lattice Γ_p in the vector space $\mathbb{R}^4$, containing $\mathbb{Z}^4$.

Consider the quotient

$$\phi(p) := \frac{\text{volume}\ (\mathrm{Par}_p)}{\text{covolume}\ (\Gamma_p)}$$

(volume and covolume for the Lebesgue measure). The essential difference between the cases $p = 29$ and $p = 31$ is the following: we have $\phi(29) \approx 0.02409$ and $\phi(31) \approx 31918.2436$. In the first case, we have been able to show that the intersection $\Gamma_{29}^{\mathrm{aff}} \cap \mathrm{Par}_{29}$ contains a single point and to determine this point. In the second case, the computation of $\phi(31)$ indicates, heuristically, that the number of points of $\Gamma_{31}^{\mathrm{aff}} \cap \mathrm{Par}_{31}$ is approximately 32000; in fact, this number is 31995.

The diligent reader will object that we have been a bit lazy regarding the Ramanujan inequalities. Indeed, we have simply used the fact that for $7 \leq r \leq 10$, $\theta_r(p^2)$ is the sum of four complex numbers of absolute value $p^{\mathrm{w}(\varpi_r)}$ (recall that $\theta_7, \theta_8, \theta_9, \theta_{10}$ is an alternative notation for $\tau_{6,8}, \tau_{8,8}, \tau_{12,6}, \tau_{4,10}$). But here we know the $\theta_r(p)$ for $7 \leq r \leq 10$; the fact that the roots in $\mathbb{C}$ of the characteristic polynomial $\det(t - p^{\mathrm{w}(\varpi_r)/2}\,\mathrm{c}_p(\varpi_r))$ have absolute value $p^{\mathrm{w}(\varpi_r)/2}$ is equivalent to the inequalities

$$- 4\, p^{\mathrm{w}(\varpi_r)} + \frac{\theta_r(p)^2}{2} \;\leq\; \theta_r(p^2) \;\leq\; (2\, p^{\mathrm{w}(\varpi_r)/2} - |\theta_r(p)|\,)^{\,2}\,. \tag{10.3.8}$$

These inequalities show that the point $\underline{x}$ belongs to a parallelotope $\mathrm{Par}_p^{\mathrm{slim}}$ contained in Par_p. We have

$$\frac{\mathrm{volume}\,(\,\mathrm{Par}_p^{\mathrm{slim}}\,)}{\mathrm{volume}\,(\,\mathrm{Par}_p\,)} \;=\; \prod_{r=7}^{10} \left(1 - \frac{|\theta_r(p)|}{4p^{\mathrm{w}(\varpi_r)/2}}\,\right)^{2}.$$

For $p = 31$, this ratio is approximately 0.2115 and the cardinality of $\Gamma_{31}^{\mathrm{aff}} \cap \mathrm{Par}_{31}^{\mathrm{slim}}$ is 6735.

10.4 Harder-Type Congruences

This section consists of three parts.

In the first, very elementary one, we exploit the following observation: the very fact that the endomorphisms T_p of $\mathbb{Z}[\mathrm{X}_{24}]$ for p prime have (after extension of scalars to $\mathbb{Q}$) a basis of common eigenvectors, namely that of T_2, implies that the $\lambda_j(p)$ satisfy numerous congruences. For example, we obtain the following congruence:

$$(p+1)\,(\tau_{4,10}(p) - \tau_{22}(p) - p^{13} - p^{8}) \;\equiv\; 0 \quad \bmod 41\;.$$

In the second, more subtle part, we "divide by $p+1$" some of the congruences obtained in the first part by, in particular, invoking the theory of Galois representations. For example, we prove that we have the congruence

$$\tau_{4,10}(p) \;\equiv\; \tau_{22}(p) + p^{13} + p^{8} \quad \bmod 41$$

conjectured by Günter Harder [100]. In the third part, we analyze the form that can a priori be taken by a decomposition into irreducible components of the ℓ-adic residual representation associated with a $\tau_{j,k}$. From this analysis and the computations carried out in Sect. 10.3, we deduce that some of these representations are irreducible, which explains why the corresponding $\tau_{j,k}$ do not appear in the congruences stated in the second part of this section.

On Certain Congruences Satisfied by the $\lambda_j(p)$

We again consider formula (10.2.2):

$$\mathrm{T}_p \;=\; \mathrm{V}\;\mathrm{diag}(\lambda_1(p), \lambda_2(p), \ldots, \lambda_{24}(p))\;\mathrm{V}^{-1}\,.$$

The matrix V has integral coefficients, but this is not the case for the matrix V^{-1}. Indeed, PARI tells us that we have

$$\begin{aligned}|\det \mathrm{V}| &= 2^{220}\cdot 3^{85}\cdot 5^{35}\cdot 7^{23}\cdot 11^{9}\cdot 13^{10}\cdot 17^{3}\cdot 19^{2}\cdot 23^{2}\cdot 41\\ &\quad\cdot 131\cdot 283^{2}\cdot 593\cdot 617^{2}\cdot 691^{10}\cdot 3617^{4}\cdot 43867^{3}\,;\end{aligned}$$

PARI also tells us that the least integer $d>0$ such that $d\,\mathrm{V}^{-1}$ has integral coefficients is

$$\begin{aligned}\mathrm{D} &:= 2^{21}\cdot 3^{10}\cdot 5^{5}\cdot 7^{2}\cdot 11^{3}\cdot 13^{2}\cdot 17\cdot 19\cdot 23^{2}\cdot 41\\ &\quad\cdot 131\cdot 283\cdot 593\cdot 617\cdot 691^{2}\cdot 3617\cdot 43867\,.\end{aligned}$$

We therefore see that for the matrix T_p to have integral coefficients, numerous congruences modulo the divisors of D, which imply the eigenvalues of $\lambda_j(p)$, must be satisfied. Now that we have the matrix V (thanks to Nebe–Venkov), obtaining these congruences involves only the theory of modules over principal ideal domains. The expression (10.2.1) of the $\lambda_j(p)$ in terms of $\tau_{12}(p)$, $\tau_{16}(p)$, $\tau_{18}(p)$, $\tau_{20}(p)$, $\tau_{22}(p)$, $\tau_{6,8}(p)$, $\tau_{8,8}(p)$, $\tau_{12,6}(p)$, and $\tau_{4,10}(p)$ then gives congruences concerning these arithmetic functions.

The theory of modules over principal ideal domains tells us that there exist two matrices R and S in $\mathrm{GL}_{24}(\mathbb{Z})$ and strictly positive integers $\mathrm{d}_1,\mathrm{d}_2,\ldots,\mathrm{d}_{24}$ with d_j dividing d_i for $j>i$, such that we have

$$\mathrm{V} = \mathrm{R}\ \mathrm{diag}(\mathrm{d}_1,\mathrm{d}_2,\ldots,\mathrm{d}_{24})\ \mathrm{S}^{-1}$$

(note that we have $\mathrm{d}_1=\mathrm{D}$ and $\prod_i \mathrm{d}_i = |\det \mathrm{V}|$).

The following conditions are equivalent:

- The matrix $\mathrm{V}\,\mathrm{diag}(\lambda_1(p),\lambda_2(p),\ldots,\lambda_{24}(p))\,\mathrm{V}^{-1}$ has integral coefficients.
- The matrix

  $$\mathrm{diag}(\mathrm{d}_1,\ldots,\mathrm{d}_{24})\,\mathrm{S}^{-1}\,\mathrm{diag}(\lambda_1(p),\lambda_2(p),\ldots,\lambda_{24}(p))\,\mathrm{S}\,\mathrm{diag}(\mathrm{d}_1^{-1},\ldots,\mathrm{d}_{24}^{-1})$$

  has integral coefficients.

Let k be an integer with $1\leq k\leq 24$; set

$$\mathrm{E}_k := \mathrm{S}^{-1}\,\mathrm{diag}(\delta_{k,1},\delta_{k,2},\ldots,\delta_{k,24})\,\mathrm{S}$$

($\delta_{-,-}$ is the Kronecker delta), and denote by $\mathrm{e}_{i,j,k}$ the coefficient of index (i,j) of the matrix E_k (the $\mathrm{e}_{i,j,k}$ are "universal" integers, determined by V and a choice of the ordered pair (R, S)). The second condition above is also equivalent to the following:

- For every ordered pair (i,j) with $i>j$, we have the congruence

$$\sum_{k=1}^{24} \mathrm{e}_{i,j,k}\,\lambda_k(p) \equiv 0 \mod \frac{\mathrm{d}_j}{\mathrm{d}_i}\,. \tag{10.4.1}$$

Let us conceptualize the above a little.

First of all, observe that the only property of T_p that we have used above is that for $1 \leq j \leq 24$, the v_j are all eigenvectors of T_p. Let U be an endomorphism of $\mathbb{Z}[\mathrm{X}_{24}]$ satisfying this property, and for $1 \leq j \leq 24$, let $\lambda_j(U)$ be the integer defined by the equality $U(\mathrm{v}_j) = \lambda_j(U)\,\mathrm{v}_j$; then for every ordered pair (i,j) with $i > j$, we have the congruence

$$\sum_{k=1}^{24} \mathrm{e}_{i,j,k}\,\lambda_k(U) \equiv 0 \mod \frac{\mathrm{d}_j}{\mathrm{d}_i}\,. \tag{10.4.2}$$

Denote by C the subring of $\mathrm{End}_{\mathbb{Z}}(\mathbb{Z}[\mathrm{X}_{24}])$ consisting of the endomorphisms U considered above; the maps $\lambda_j \colon \mathrm{C} \to \mathbb{Z}$ given by $U \mapsto \lambda_j(U)$ are ring homomorphisms whose product

$$\underline{\lambda} \colon \mathrm{C} \to \mathbb{Z}^{24}\,, \quad U \mapsto (\lambda_1(U), \lambda_2(U), \ldots, \lambda_{24}(U))$$

is an injective ring homomorphism (which, in particular, shows that the ring C is commutative). The image of $\underline{\lambda}$ is the subring of $\mathbb{Z}^{24}$ consisting of the 24-tuples $(x_1, x_2, \ldots, x_{24})$ satisfying

$$\sum_{k=1}^{24} \mathrm{e}_{i,j,k}\,x_k \equiv 0 \mod \frac{\mathrm{d}_j}{\mathrm{d}_i}$$

for every ordered pair (i,j) with $i > j$.

Remarks.

(1) Since the coefficients of V are pairwise relatively prime, we have $\mathrm{d}_{24} = 1$ (in fact, we have $\mathrm{d}_j = 1$ for $j \geq 21$).

(2) Set $\underline{\mathrm{d}} := (\mathrm{d}_1, \mathrm{d}_2, \ldots, \mathrm{d}_{24})$, and denote by $\Gamma(\underline{\mathrm{d}})$ the subgroup of $\mathrm{GL}_{24}(\mathbb{Z})$ defined as the intersection of $\mathrm{GL}_{24}(\mathbb{Z})$ and $\mathrm{diag}(\underline{\mathrm{d}})\,\mathrm{GL}_{24}(\mathbb{Z})\,\mathrm{diag}(\underline{\mathrm{d}})^{-1}$ in $\mathrm{GL}_{24}(\mathbb{Q})$. We see that the class of S in the finite set $\mathrm{GL}_{24}(\mathbb{Z})/\Gamma(\underline{\mathrm{d}})$ depends only on V and that we can define the subring C of $\mathbb{Z}^{24}$ in terms of the 24-tuple $\underline{\mathrm{d}}$ and this class.

(3) Let ℓ be a prime; then the ring homomorphisms $\mathrm{C} \to \mathbb{Z}/\ell$, viewed as elements of the $\mathbb{Z}/\ell$-vector space $\mathrm{Hom}_{\mathbb{Z}}(\mathrm{C}, \mathbb{Z}/\ell)$, are linearly independent ("independence of the characters"). If the ℓ-adic valuation of D is 1, in other words, if ℓ appears in the list

$$\{17, 19, 41, 131, 283, 593, 617, 3617, 43867\}\,,$$

then $\mathbb{Z}_{(\ell)} \otimes_{\mathbb{Z}} \mathrm{coker}\,\underline{\lambda}$ is annihilated by the multiplication by ℓ. In this case, the previous observation shows that there exists a uniquely determined equivalence relation on $\{1, 2, \ldots, 24\}$, which we denote by $\mathcal{R}_\ell$, such that $\mathbb{Z}_{(\ell)} \otimes_{\mathbb{Z}} \mathrm{C}$ is the subring of $\mathbb{Z}_{(\ell)}^{24}$ consisting of the 24-tuples $(x_1, x_2, \ldots, x_{24})$ satisfying the congruences $x_i \equiv x_j \bmod \ell$ for $i\,\mathcal{R}_\ell\,j$. For example, we will see further on that $\mathbb{Z}_{(41)} \otimes_{\mathbb{Z}} \mathrm{C}$ is the subring of $\mathbb{Z}_{(41)}^{24}$ consisting of the 24-tuples $(x_1, x_2, \ldots, x_{24})$ satisfying $x_{18} \equiv x_{21} \bmod 41$.

In general, we can determine the isomorphism class of the $(\mathbb{Z}/\mathrm{D})$-module coker $\underline{\lambda}$ using the linear algebra "routines" of `PARI` (`mathnf`, `mathnfmod`, and `matsnf`). We have carried out the computation, viewing C as the submodule of $\mathbb{Z}^{24}$ consisting of the 24-tuples $(x_1, x_2, \ldots, x_{24})$ such that the matrix $\sum_j x_j \operatorname{Proj}_j$ has integral coefficients (the notation Proj_j was introduced in the proof of Theorem 10.2.2). We give the result below.

Proposition 10.4.1. *Let ℓ be a prime that divides* D, *in other words, an element of the list*

$$\{2, 3, 5, 7, 11, 13, 17, 19, 23, 41, 131, 283, 593, 617, 691, 3617, 43867\}.$$

We have isomorphisms of the form

$$\mathbb{Z}_{(\ell)} \otimes_{\mathbb{Z}} \operatorname{coker}(\underline{\lambda} : \mathrm{C} \to \mathbb{Z}^{24}) \;\simeq\; \mathbb{Z}/\ell^{\mathrm{e}_{\ell,1}} \times \mathbb{Z}/\ell^{\mathrm{e}_{\ell,1}} \times \ldots \times \mathbb{Z}/\ell^{\mathrm{e}_{\ell,\mathrm{r}_\ell}} ,$$

where $\mathrm{e}_\ell := (\mathrm{e}_{\ell,1}, \mathrm{e}_{\ell,2}, \ldots, \mathrm{e}_{\ell,\mathrm{r}_\ell})$ denotes the finite decreasing sequence of strictly positive integers given explicitly below:

$$\begin{aligned}
\mathrm{e}_2 &= (21, 19, 17, 17, 15, 15, 14, 14, 12, 12, 11, 10, 9, 9, 9, 8, 8, 7, 6, 6, 3, 1) ,\\
\mathrm{e}_3 &= (10, 9, 9, 7, 7, 6, 5, 5, 5, 5, 5, 5, 5, 5, 4, 4, 4, 3, 3, 3, 2, 1) ,\\
\mathrm{e}_5 &= (5, 5, 3, 3, 3, 2, 2, 2, 2, 2, 2, 2, 2, 2, 2, 2, 2, 1, 1, 1, 1) ,\\
\mathrm{e}_7 &= (2, 2, 2, 2, 2, 2, 2, 1, 1, 1, 1, 1, 1, 1, 1, 1, 1, 1) ,\\
\mathrm{e}_{11} &= (3, 2, 1, 1, 1, 1, 1) ,\\
\mathrm{e}_{13} &= (2, 2, 1, 1, 1, 1, 1, 1, 1) ,\\
\mathrm{e}_{17} &= (1, 1, 1) ,\\
\mathrm{e}_{19} &= (1, 1) ,\\
\mathrm{e}_{23} &= (2) ,\\
\mathrm{e}_{41} &= (1) ,\\
\mathrm{e}_{131} &= (1) ,\\
\mathrm{e}_{283} &= (1, 1) ,\\
\mathrm{e}_{593} &= (1) ,\\
\mathrm{e}_{617} &= (1, 1) ,\\
\mathrm{e}_{691} &= (2, 1, 1, 1, 1, 1, 1, 1, 1) ,\\
\mathrm{e}_{3617} &= (1, 1, 1, 1) ,\\
\mathrm{e}_{43867} &= (1, 1, 1) .
\end{aligned}$$

Let us now describe two generalizations of the congruences (10.4.2) (and (10.4.1)).

(1) A priori, the congruences (10.4.2) involve the 24 eigenvalues $\lambda_j(U)$ simultaneously. Let J be an arbitrary subset of $\{1, 2, \ldots, 24\}$; below, we describe an

algorithm, similar to the one that leads to (10.4.2), to obtain congruences that involve only the $\lambda_j(U)$ with j in J.

Denote by M_J and L_J, respectively, the submodule of $\mathbb{Z}[\mathrm{X}_{24}]$ generated by the v_j with $j \in J$ and the intersection of $\mathbb{Q} \otimes_{\mathbb{Z}} \mathrm{M}_J$ and $\mathbb{Z}[\mathrm{X}_{24}]$ in $\mathbb{Q}[\mathrm{X}_{24}]$. Still by the theory of modules over principal ideal domains, there exist

- a $J \times J$ matrix $\mathrm{S}_J = [\mathrm{s}_{J,i,j}]_{(i,j)\in J\times J}$ with integral coefficients that is invertible and whose inverse S_J^{-1} also has integral coefficients,
- strictly positive integers $\mathrm{d}_{J,i}$, for $i \in J$, with $\mathrm{d}_{J,j}$ dividing $\mathrm{d}_{J,i}$ for $i < j$,

such that the set

$$\left\{ \frac{1}{d_{J,j}} \sum_{i\in J} \mathrm{s}_{J,i,j}\, \mathrm{v}_i \right\}_{j\in J}$$

is a basis of L_J (so that the quotient $\mathrm{L}_J/\mathrm{M}_J$ is isomorphic to a direct sum $\bigoplus_{j\in J} \mathbb{Z}/\mathrm{d}_{J,j}$).

Let (i,j,k) be an element of $J \times J \times J$; denote by $\mathrm{e}_{J,i,j,k}$ the coefficient of index (i,j) of the matrix $\mathrm{S}_J^{-1}\,\mathrm{diag}((\delta_{k,i})_{i\in J})\,\mathrm{S}_J$. For every ordered pair (i,j) with $i > j$, we have the congruence

$$\sum_{k\in J} \mathrm{e}_{J,i,j,k}\, \lambda_k(U) \;\equiv\; 0 \mod \frac{\mathrm{d}_{J,j}}{\mathrm{d}_{J,i}} \tag{10.4.3}$$

(note that the sum $\sum_{k\in J} \mathrm{e}_{J,i,j,k}$ is zero).

(2) Let m be a divisor of D; denote by $\mathbb{Z}_{(m)}$ the localization of $\mathbb{Z}$ obtained by inverting the elements prime to m. We obtain congruences modulo divisors of a power of m (and of D) by replacing the principal ideal domain $\mathbb{Z}$ in item 1 above with the principal ideal domain $\mathbb{Z}_{(m)}$.

Finally, consider a particular case of the above. Set $\mathcal{V} = \{\mathrm{v}_1, \mathrm{v}_2, \dots, \mathrm{v}_{24}\}$. Let $\mathcal{W}$ be a subset of $\mathcal{V}$; denote by $\mathrm{J}(\mathcal{W})$ the subset of $\{1, 2, \dots, 24\}$ consisting of the j with $\mathrm{v}_j \in \mathcal{W}$.

Let m be a divisor of D and $\rho_m \colon \mathbb{Z}[\mathrm{X}_{24}] \to (\mathbb{Z}/m)[\mathrm{X}_{24}]$ the homomorphism defined by reduction modulo m. Let $\mathcal{W}$ be a subset of $\mathcal{V}$. First, suppose that m is prime. If the set of vectors $\rho_m(\mathcal{W})$ is linearly dependent and if $\mathcal{W}$ is minimal among the subsets of $\mathcal{V}$ with this property, then for every v in $\mathcal{W}$, the image $\rho_m(\mathcal{W} - \{v\})$ is a basis of the linear subspace generated by $\rho_m(\mathcal{W})$. Next, suppose that m is an arbitrary divisor of D; we say, more generally, that $\mathcal{W}$ is a *minimal m-unfree* set if the submodule of $(\mathbb{Z}/m)[\mathrm{X}_{24}]$ generated by $\rho_m(\mathcal{W})$ is a free $(\mathbb{Z}/m)$-module and if $\rho_m(\mathcal{W} - \{v\})$ is a basis of this space for every v in $\mathcal{W}$. If $\mathcal{W}$ is minimal m-unfree, then the $\mathbb{Z}_{(m)}$-module $\mathbb{Z}_{(m)} \otimes_{\mathbb{Z}} (\mathrm{L}_{\mathrm{J}(\mathcal{W})}/\mathrm{M}_{\mathrm{J}(\mathcal{W})})$ is isomorphic to $\mathbb{Z}/\widetilde{m}$, with $\widetilde{m}$ a multiple of m that divides a power of m. Item 2 above provides congruences modulo $\widetilde{m}$ and a fortiori modulo m.

Proposition 10.4.2. *Let U be an endomorphism of $\mathbb{Z}[\mathrm{X}_{24}]$ that has the v_j, for $1 \leq j \leq 24$, as eigenvectors, with respective eigenvalues $\lambda_j(U)$. Let m be a divisor of* D *and* $\mathcal{W}$ *a minimal* m*-unfree subset of* $\mathcal{V}$*. Then we have the congruences*

$$\lambda_i(U) \equiv \lambda_j(U) \mod m$$

for all i *and* j *in* $\mathrm{J}(\mathcal{W})$.

For the comfort of the reader, we give a proof of Proposition 10.4.2 ab initio.

Proof. By definition, we have at our disposal a dependence relation of the form $\sum_{v\in\mathcal{W}} \mu_v\, \rho_m(v) = 0$ with $\mu_v \in (\mathbb{Z}/m)^\times$, and if $\sum_{v\in\mathcal{W}} \mu'_v\, \rho_m(v) = 0$ is another dependence relation, then we have $\mu'_v/\mu_v = \mu'_w/\mu_w$ for all v and w in $\mathcal{W}$. The proposition follows by considering the dependence relation $U(\sum_{v\in\mathcal{W}} \mu_v\, \rho_m(v))$ $= 0$. □

It is clear that the cardinality of a minimal m-unfree subset of $\mathcal{V}$ is greater than or equal to 2. The following proposition, which we verify on a case-by-case basis says that we often have equality, at least if m is prime.

Proposition 10.4.3. *Let* ℓ *be a prime divisor of* D *other than* $3, 5, 7, 11$ *and* $\mathcal{W}$ *a minimal* ℓ*-unfree subset of* $\mathcal{V}$*. Then the cardinality of* $\mathcal{W}$ *is* 2.

Remark. Let $\mathrm{P}_{\mathrm{X}_{24}}$ be the $\mathbb{Z}$-scheme whose A-points (A a commutative ring with unit) are the direct factors of rank 1 of $A[\mathrm{X}_{24}]$ ($\mathrm{P}_{\mathrm{X}_{24}}$ is therefore an avatar of the projective space P^{23}). There is a canonical subset of $\mathrm{P}_{\mathrm{X}_{24}}(\mathbb{Q}) = \mathrm{P}_{\mathrm{X}_{24}}(\mathbb{Z})$, namely the set consisting of the classes of the v_j; denote it by $[\mathcal{V}]$. The proposition above says that this set of 24 elements is far from being "generic." Indeed, it shows that for the ℓ in its statement, the points of $\rho_\ell([\mathcal{V}])$ are "projectively independent."

Examples

The Case $m = 43867$. The minimal m-unfree subsets of $\mathcal{V}$ are $\{\mathrm{v}_1, \mathrm{v}_{11}\}$, $\{\mathrm{v}_2, \mathrm{v}_8\}$, and $\{\mathrm{v}_3, \mathrm{v}_6\}$. By taking $U = \mathrm{T}_p$ with p prime in Proposition 10.4.2, we obtain the following congruences modulo 43867:

$$\lambda_1(p) \equiv \lambda_{11}(p)\,, \quad \lambda_2(p) \equiv \lambda_8(p)\,, \quad \lambda_3(p) \equiv \lambda_6(p)\,. \qquad (10.4.4)$$

We see that we have

$$\begin{aligned}
\lambda_{11}(p) - \lambda_1(p) &= (p^5 + p^4 + p^3 + p^2 + p + 1)\,(\tau_{18}(p) - p^{17} - 1)\,, \\
\lambda_8(p) - \lambda_2(p) &= (p^4 + p^3 + p^2 + p)\,(\tau_{18}(p) - p^{17} - 1)\,, \\
\lambda_6(p) - \lambda_3(p) &= (p^3 + p^2)\,(\tau_{18}(p) - p^{17} - 1)\,.
\end{aligned}$$

Since the gcd of the polynomials $X^5+X^4+X^3+X^2+X+1$, $X^4+X^3+X^2+X$, and $X^3 + X^2$ is $X + 1$, we see that (10.4.4) implies the congruence

$$(p+1)\,(\tau_{18}(p) - p^{17} - 1) \equiv 0 \mod 43867\,. \qquad (10.4.5)$$

This congruence is weaker than the well-known congruence (see, for example, [194])

$$\tau_{18}(p) - p^{17} - 1 \equiv 0 \mod 43867 \tag{10.4.6}$$

(nonetheless, note that (10.4.5) implies (10.4.6) for $p \not\equiv -1 \bmod 43867$!). Nevertheless, during the proof of Theorem 10.4.4, we will explain how the intervention of the theory of Galois representations and a more elaborate version of Proposition 10.4.2 (Proposition 10.4.5) make it possible to obtain (10.4.6) (which gives a quite complicated proof of this congruence!).

Remark. By Proposition 10.4.1, we have $\mathbb{Z}_{(43867)} \otimes_{\mathbb{Z}} \operatorname{coker} \underline{\lambda} \approx (\mathbb{Z}/43867)^3$; the above in fact shows that $\mathbb{Z}_{(43867)} \otimes_{\mathbb{Z}} \mathrm{C}$ is the subring of $\mathbb{Z}^{24}_{(43867)}$ consisting of the 24-tuples $(x_1, x_2, \dots, x_{24})$ satisfying $x_1 \equiv x_{11}$, $x_2 \equiv x_8$, and $x_3 \equiv x_6$ modulo 43867.

The Case $m = 3617$. The minimal m-unfree subsets of $\mathcal{V}$ are $\{\mathrm{v}_1, \mathrm{v}_{13}\}$, $\{\mathrm{v}_2, \mathrm{v}_{12}\}$, $\{\mathrm{v}_3, \mathrm{v}_9\}$, and $\{\mathrm{v}_4, \mathrm{v}_7\}$. This time, we obtain the following congruences modulo 3617:

$$\lambda_1(p) \equiv \lambda_{13}(p)\,, \quad \lambda_2(p) \equiv \lambda_{12}(p)\,, \quad \lambda_3(p) \equiv \lambda_9(p)\,, \quad \lambda_4(p) \equiv \lambda_7(p)\,.$$

As before, we see that these congruences imply the congruence

$$(p+1)(\tau_{16}(p) - p^{15} - 1) \equiv 0 \mod 3617\,.$$

The Case $m = 691$. The minimal m-unfree subsets have cardinality 2, and there are 12 of them. Considering the two m-unfree sets $\{\mathrm{v}_1, \mathrm{v}_{24}\}$ and $\{\mathrm{v}_2, \mathrm{v}_{23}\}$ leads to the congruence

$$(p+1)(\tau_{12}(p) - p^{11} - 1) \equiv 0 \mod 691\,.$$

The Case $m = 283 \cdot 617$. The minimal m-unfree subsets are $\{\mathrm{v}_1, \mathrm{v}_5\}$ and $\{\mathrm{v}_2, \mathrm{v}_4\}$. This time, we obtain the congruence

$$(p+1)(\tau_{20}(p) - p^{19} - 1) \equiv 0 \mod 283 \cdot 617\,.$$

The Case $m = 131 \cdot 593$. The only minimal m-unfree subset is $\{\mathrm{v}_1, \mathrm{v}_3\}$. Since we have the equality $\lambda_3(p) - \lambda_1(p) = (p+1)(\tau_{22}(p) - p^{21} - 1)$, we obtain the congruence

$$(p+1)(\tau_{22}(p) - p^{21} - 1) \equiv 0 \mod 131 \cdot 593\,.$$

The Case $m = 41$. In this case, the only minimal m-unfree subset is $\{\mathrm{v}_{18}, \mathrm{v}_{21}\}$. Since we have the equality $\lambda_{21}(p) - \lambda_{18}(p) = (p+1)(\tau_{4,10}(p) - \tau_{22}(p) - p^{13} - p^8)$, we obtain the congruence

$$(p+1)(\tau_{4,10}(p) - \tau_{22}(p) - p^{13} - p^8) \equiv 0 \mod 41\,. \tag{10.4.7}$$

As already mentioned, we will see that the theory of Galois representations makes it possible to show that, in fact, we have

$$\tau_{4,10}(p) - \tau_{22}(p) - p^{13} - p^{8} \equiv 0 \mod 41 , \qquad (10.4.8)$$

a congruence conjectured by G. Harder [100].

The subset $\{v_{18}, v_{21}\}$ is minimal m-unfree for $m = 2^4 \cdot 3 \cdot 41$, so that the congruence (10.4.7) refines to

$$(p+1)(\tau_{4,10}(p) - \tau_{22}(p) - p^{13} - p^{8}) \equiv 0 \mod 2^4 \cdot 3 \cdot 41 .$$

Remark. In view of Proposition 10.4.1, the above shows that $\mathbb{Z}_{(41)} \otimes_{\mathbb{Z}} \mathrm{C}$ is the subring of $\mathbb{Z}_{(41)}^{24}$ consisting of the 24-tuples $(x_1, x_2, \ldots, x_{24})$ satisfying $x_{18} \equiv x_{21} \mod 41$.

The Case $m = 23$. In this case, the only minimal m-unfree subset is $\{v_{13}, v_{15}\}$; in fact, $\{v_{13}, v_{15}\}$ is minimal $\widetilde{m}$-unfree with $\widetilde{m} = 23^2$. This time, we obtain the congruence

$$(p+1)(\,\tau_{8,8}(p) - (p^6+1)\,\tau_{16}(p)\,) \equiv 0 \mod 23^2 . \qquad (10.4.9)$$

Here, too, we will see further on that we in fact have

$$\tau_{8,8}(p) - (p^6+1)\,\tau_{16}(p) \equiv 0 \mod 23^2 .$$

By the same argument as before, the congruence (10.4.9) refines to

$$(p+1)(\,\tau_{8,8}(p) - (p^6+1)\,\tau_{16}(p)\,) \equiv 0 \mod 2^3 \cdot 3^2 \cdot 23^2 .$$

Remark. In view of Proposition 10.4.1, the above shows that $\mathbb{Z}_{(23)} \otimes_{\mathbb{Z}} \mathrm{C}$ is the subring of $\mathbb{Z}_{(23)}^{24}$ consisting of the 24-tuples $(x_1, x_2, \ldots, x_{24})$ satisfying $x_{13} \equiv x_{15} \mod 23^2$.

The Case $m = 19$. The minimal m-unfree subsets are $\{v_9, v_{10}\}$ and $\{v_{21}, v_{22}\}$.

Considering $\{v_9, v_{10}\}$ leads to

$$(p+1)(\,\tau_{12,6}(p) - (p^4+p^2)\,\tau_{16}(p) + p^2\,\tau_{18}(p) - \tau_{22}(p)\,) \equiv 0 \mod 19 . \quad (10.4.10)$$

Since $\{v_9, v_{10}\}$ is minimal m-unfree for $m = 2^4 \cdot 19$, we also have

$$(p+1)(\,\tau_{12,6}(p) - (p^4+p^2)\,\tau_{16}(p) + p^2\,\tau_{18}(p) - \tau_{22}(p)\,) \equiv 0 \mod 2^4 \cdot 19 .$$

Considering $\{v_{21}, v_{22}\}$ leads to

$$(p+1)(\,\tau_{4,10}(p) - (p^8+p^2)\,\tau_{12}(p) + p^2\,\tau_{18}(p) - \tau_{22}(p)\,) \equiv 0 \mod 19 .$$

Since $\{v_{21}, v_{22}\}$ is minimal m-unfree for $m = 2^4 \cdot 3^2 \cdot 19$, we also have

$$(p+1)(\,\tau_{4,10}(p) - (p^8+p^2)\,\tau_{12}(p) + p^2\,\tau_{18}(p) - \tau_{22}(p)\,) \equiv 0 \mod 2^4 \cdot 3^2 \cdot 19 .$$

The Case $m = 17$. The minimal m-unfree subsets are $\{v_5, v_9\}$, $\{v_{15}, v_{17}\}$, and $\{v_{19}, v_{20}\}$. The congruences associated with these unordered pairs are, respectively,

$$(p+1)((p^4+p^2)\tau_{16}(p) - (p^2+1)\tau_{20}(p) + \tau_{22}(p) - p^{17} - p^4) \equiv 0 \mod 17\,, \tag{10.4.11}$$

$$(p+1)(\tau_{8,8}(p) - (p^6+p^4)\tau_{12}(p) + (p^4+p^2)\tau_{16}(p) - (p^2+1)\tau_{20}(p)) \equiv 0 \mod 17\,, \tag{10.4.12}$$

$$(p+1)(\tau_{6,8}(p) - (p^6+p^2)\tau_{12}(p) + p^2\tau_{16}(p) - \tau_{20}(p)) \equiv 0 \mod 17\,. \tag{10.4.13}$$

The Case $m = 13$. The minimal m-unfree subsets have cardinality 2, and there are 12 of them. Considering the minimal m-unfree sets $\{v_6, v_{10}\}$, $\{v_9, v_{15}\}$, $\{v_{10}, v_{11}\}$, $\{v_{15}, v_{17}\}$, and $\{v_{15}, v_{18}\}$ leads, respectively, to the congruences

$$(p+1)(\tau_{12,6}(p) - \tau_{22}(p) - p^{17} - p^4) \equiv 0 \mod 13\,, \tag{10.4.14}$$

$$(p+1)(\tau_{8,8}(p) - \tau_{22}(p) - p^{15} - p^6) \equiv 0 \mod 13\,, \tag{10.4.15}$$

$$(p+1)(\tau_{12,6}(p) - (p^4+1)\tau_{18}(p)) \equiv 0 \mod 13\,, \tag{10.4.16}$$

$$(p+1)(\tau_{8,8}(p) - (p^6+p^4)\tau_{12}(p) + (p^4+p^2)\tau_{16}(p) - (p^2+1)\tau_{20}(p)) \equiv 0 \mod 13\,,$$

$$(p+1)(\tau_{8,8}(p) - (p^6+p^4)\tau_{12}(p) + (p^4+p^2)\tau_{16}(p) - p^2\tau_{18}(p) - \tau_{22}(p)) \equiv 0 \mod 13\,.$$

In fact, $\{v_{10}, v_{11}\}$ is minimal m-unfree for $m = 2^5 \cdot 7 \cdot 13$, so that the congruence (10.4.16) refines to

$$(p+1)(\tau_{12,6}(p) - (p^4+1)\tau_{18}(p)) \equiv 0 \mod 2^5 \cdot 7 \cdot 13\,. \tag{10.4.17}$$

The Case $m = 11$. The minimal m-unfree subsets are $\{v_5, v_{13}\}$, $\{v_{10}, v_{15}\}$, $\{v_{14}, v_{16}\}$, $\{v_{14}, v_{19}\}$, $\{v_{16}, v_{19}\}$, $\{v_{17}, v_{21}\}$, and $\{v_7, v_8, v_{12}\}$ (note that the last subset has three elements!).

By taking $\mathcal{W} = \{v_7, v_8, v_{12}\}$ in Proposition 10.4.2, we obtain

$$\lambda_7(p) \equiv \lambda_8(p) \equiv \lambda_{12}(p) \mod 11\,.$$

By taking $\mathcal{W} = \{v_{14}, v_{16}\}$, $\mathcal{W} = \{v_{14}, v_{19}\}$, and $\mathcal{W} = \{v_{17}, v_{21}\}$ in the same proposition, we obtain, respectively,

$$\lambda_{14}(p) \equiv \lambda_{16}(p) \mod 11\,,\ \lambda_{14}(p) \equiv \lambda_{19}(p) \mod 11^2\,,\ \lambda_{17}(p) \equiv \lambda_{21}(p) \mod 11^2\,.$$

The second congruence can also be written as

$$p(p+1)\,(\tau_{6,8}(p) - \tau_{20}(p) - p^{13} - p^{6}) \;\equiv\; 0 \quad \bmod 11^2\,,$$

a congruence that, by considering the case $p = 11$, implies the following:

$$(p+1)\,(\tau_{6,8}(p) - \tau_{20}(p) - p^{13} - p^{6}) \;\equiv\; 0 \quad \bmod 11^2\,. \tag{10.4.18}$$

The third can also be written as

$$(p+1)\,(\tau_{4,10}(p) - (p^2+1)\,\tau_{20}(p) + p^2\,\tau_{18}(p) - p^{13} - p^{8}) \equiv 0 \bmod 11^2\,. \tag{10.4.19}$$

Examples of Specializations of Congruence (10.4.3) *that Escape Proposition* 10.4.2

- Since the three sets $\{v_9, v_{15}\}$, $\{v_{15}, v_{17}\}$, and $\{v_{17}, v_{18}\}$ are minimal 13-unfree, we have $\lambda_9(p) \equiv \lambda_{15}(p) \bmod 13$, $\lambda_{15}(p) \equiv \lambda_{17}(p) \bmod 13$, and $\lambda_{17}(p) \equiv \lambda_{18}(p) \bmod 13$ (the first two congruences were used above, the third was not because it does not involve the $\tau_{j,k}$). Let us analyze what the congruence (10.4.3) gives for $J = \{9, 15, 17, 18\}$ (and $(i,j) = (4,1)$).

  We obtain (thanks, PARI)

$$2407302\,\lambda_9(p) - 513085\,\lambda_{15}(p) - 482792\,\lambda_{17}(p) - 1411425\,\lambda_{18}(p) \equiv 0 \quad \bmod 2^8 \cdot 3^2 \cdot 13^2 \cdot 17$$

and a fortiori

$$\lambda_9(p) + 64\,\lambda_{15}(p) - 89\,\lambda_{17}(p) + 24\,\lambda_{18}(p) \;\equiv\; 0 \quad \bmod 13^2\,.$$

  We have

$$\begin{aligned}\lambda_9(p) + 64\,\lambda_{15}(p) - 89\,\lambda_{17}(p) + 24\,\lambda_{18}(p) \;=\;& \\ \lambda_9(p) - \lambda_{15}(p) + 2\,\lambda_{17}(p) - 2\,\lambda_{18}(p)& \\ - 65\,(\lambda_{17}(p) - \lambda_{15}(p)) + 26\,(\lambda_{18}(p) - \lambda_{17}(p))&\,.\end{aligned}$$

Since the two differences $\lambda_{17}(p) - \lambda_{15}(p)$ and $\lambda_{18}(p) - \lambda_{17}(p)$ are divisible by 13, we end up with the congruence

$$\lambda_9(p) - \lambda_{15}(p) + 2\,\lambda_{17}(p) - 2\,\lambda_{18}(p) \;\equiv\; 0 \quad \bmod 13^2$$

or, equivalently,

$$(p+1)\,(\tau_{8,8}(p) + 2p^2\,\tau_{18}(p) - 2\,(p^2+1)\,\tau_{20}(p) + \tau_{22}(p) - p^{15} - p^{6}) \equiv 0 \bmod 13^2\,.$$

Remark. The computer says that we in fact have

$$\lambda_9(p) - \lambda_{15}(p) + 2\,\lambda_{17}(p) - 2\,\lambda_{18}(p) \equiv 0 \mod 2^4 \cdot 3^2 \cdot 5 \cdot 13^2$$

for $p \leq 113$ (recall that we computed the $\tau_{j,k}(p)$ for $p \leq 113$). We can show that we have

$$\lambda_9(p) - \lambda_{15}(p) + 2\,\lambda_{17}(p) - 2\,\lambda_{18}(p) \equiv 0 \mod 5$$

for every p, as follows. We verify that $\{\mathrm{v}_9, \mathrm{v}_{14}, \mathrm{v}_{15}\}$ and $\{\mathrm{v}_6, \mathrm{v}_{17}, \mathrm{v}_{18}\}$ are, respectively, minimal 25-unfree and minimal 5-unfree; in particular, we have $\lambda_9(p) \equiv \lambda_{15}(p) \bmod 25$ and $\lambda_{17}(p) \equiv \lambda_{18}(p) \bmod 5$.

- Earlier, we saw that we have the congruences $\lambda_{14}(p) \equiv \lambda_{19}(p) \bmod 11^2$ and $\lambda_{14}(p) \equiv \lambda_{16}(p) \bmod 11$; we therefore also have

  $$\lambda_{19}(p) - \lambda_{14}(p) + 22\,(\lambda_{16}(p) - \lambda_{14}(p)) \equiv 0 \mod 11^2\,.$$

  The congruence (10.4.3) for $J = \{14, 16, 19\}$ (and $(i, j) = (3, 1)$) makes it possible to show, using the same method as above, that we in fact have

  $$\lambda_{19}(p) - \lambda_{14}(p) + 22\,(\lambda_{16}(p) - \lambda_{14}(p)) \equiv 0 \mod 11^3\,.$$

- If we take $J = \{1, 2, 23, 24\}$ (and $(i, j) = (4, 1)$), we obtain

  $$\lambda_1(p) - \lambda_2(p) + 2\,\lambda_{23}(p) - 2\,\lambda_{24}(p) \equiv 0 \mod 691^2\,;$$

  this congruence is not very surprising because the left-hand side is equal to $(\tau(p) - p^{11} - 1)^2$!

Where We Explain How the Theory of Galois Representations Allows the "Division by $p + 1$" of Certain of the Previous Congruences

Theorem* 10.4.4. *For every prime p, the following congruences hold:*

$$\begin{aligned}
&(1)\quad \tau_{4,10}(p) \equiv \tau_{22}(p) + p^{13} + p^8 \mod 41 \text{ (Harder conjecture [100])},\\
&(2)\quad \tau_{8,8}(p) \equiv (p^6 + 1)\tau_{16}(p) \mod 23^2\,,\\
&(3)\quad \tau_{12,6}(p) \equiv (p^4 + p^2)\tau_{16}(p) \mod 19\,,\\
&(4)\quad \tau_{4,10}(p) \equiv (p^8 + p^2)\tau_{12}(p) \mod 19\,,\\
&(5)\quad \tau_{6,8}(p) \equiv (p^6 + p^2)\tau_{12}(p) \mod 17\,,\\
&(6)\quad \tau_{8,8}(p) \equiv (p^6 + p^4)\tau_{12}(p) \mod 17\,,\\
&(7)\quad \tau_{8,8}(p) \equiv p^8 + p^6 + p^3 + p \mod 13\,,\\
&(8)\quad \tau_{12,6}(p) \equiv p^8 + p^5 + p^4 + p \mod 13\,,\\
&(9)\quad \tau_{6,8}(p) \equiv p^8 + p^6 + p^3 + p \mod 11\,,
\end{aligned}$$

$$\begin{aligned}
&(10)\quad \tau_{6,8}(p) \equiv \tau_{20}(p) + p^{13} + p^{6} \mod 11^2\,,\\
&(11)\quad \tau_{4,10}(p) \equiv p^{10} + p^{8} + p^{3} + p \mod 11\,,\\
&(12)\quad \tau_{8,8}(p) \equiv \tau_{12,6}(p) \mod 11\,,\\
&(13)\quad \tau_{12,6}(p) \equiv p^{5} + p^{4} + p^{2} + p \mod 7\,,\\
&(14)\quad p\tau_{6,8}(p) \equiv \tau_{8,8}(p) \equiv \tau_{4,10}(p) \mod 7\,,\\
&(15)\quad \tau_{8,8}(p) \equiv 2(p^{3} + p^{2}) \mod 5\,,\\
&(16)\quad \tau_{6,8,}(p) \equiv \tau_{12,6}(p) \equiv \tau_{4,10}(p) \equiv p^{4} + p^{3} + p^{2} + p \mod 5\,,\\
&(17)\quad \tau_{j,k}(p) \equiv 2(p^{2} + p) \mod 3\,,\\
&(18)\quad \tau_{j,k}(p) \equiv 0 \mod 2\,.
\end{aligned}$$

Proof of Item (1). Earlier, we showed that we have the congruence $\lambda_{18}(p) \equiv \lambda_{21}(p)$ modulo 41, using Proposition 10.4.2, and by invoking Theorem 9.2.5 (the principal result of this book!), we saw that we have

$$\lambda_{21}(p) - \lambda_{18}(p) = (p+1)\,(\,\tau_{4,10}(p) - (\tau_{22}(p) + p^{13} + p^{8})\,)\,.$$

From this, we deduced the congruence (10.4.7) that we now need to "divide by $p+1$." To do this, we will involve the 24 ℓ-adic Galois representations $\rho_{i,\ell}\colon \mathrm{Gal}(\overline{\mathbb{Q}}/\mathbb{Q}) \to \mathrm{GL}_{24}(\overline{\mathbb{Q}}_\ell)$ introduced in Sect. 10.1. These are semisimple and unramified outside ℓ (with, in this case, $\ell = 41$) and characterized by the equalities $\lambda_i(p) = \mathrm{trace}\,\rho_{i,\ell}(\mathrm{Frob}_p)$ for every $p \neq \ell$.

In Sect. 10.1, we also saw that the characteristic polynomial of $\rho_{i,\ell}(\mathrm{Frob}_p)$ has integral coefficients (that are, moreover, independent of ℓ) and that there exists a continuous, semisimple representation $\overline{\rho}_{i,\ell}\colon \mathrm{Gal}(\overline{\mathbb{Q}}/\mathbb{Q}) \to \mathrm{GL}_{24}(\mathbb{F}_\ell)$, unique up to isomorphism, that is unramified outside ℓ and such that the characteristic polynomial of $\overline{\rho}_{i,\ell}(\mathrm{Frob}_p)$ is the reduction modulo ℓ of the characteristic polynomial of $\rho_{i,\ell}(\mathrm{Frob}_p)$. Lemma 10.1.1 and Proposition 10.4.2 imply the following statement.

Proposition 10.4.5. *Let m be a divisor of D, ℓ a prime divisor of m, and $\mathcal{W}$ a minimal m-unfree subset of $\mathcal{V}$. Then we have the congruence*

$$\det(t - \rho_{i,\ell}(\gamma)) \equiv \det(t - \rho_{j,\ell}(\gamma)) \mod m\,\mathbb{Z}_\ell$$

for all i, j with $\mathrm{v}_i, \mathrm{v}_j$ in $\mathcal{W}$ and all γ in $\mathrm{Gal}(\overline{\mathbb{Q}}/\mathbb{Q})$. In particular, the representations $\overline{\rho}_{i,\ell}$ and $\overline{\rho}_{j,\ell}$ are isomorphic.

By taking $m = \ell = 41$ and $\mathcal{W} = \{\mathrm{v}_{18}, \mathrm{v}_{21}\}$ in Proposition 10.4.5, we obtain $\overline{\rho}_{18,41} \simeq \overline{\rho}_{21,41}$. To obtain an equation of a form similar to that of (10.4.7), we introduce the following formalism.

Let ℓ be a prime; denote by A_ℓ the Grothendieck ring of finite-dimensional continuous representations of $\mathrm{Gal}(\overline{\mathbb{Q}}/\mathbb{Q})$ with coefficients in $\mathbb{F}_\ell$, which we assume to be unramified outside ℓ. Therefore, by its very definition, we have $\overline{\rho}_{21,41} - \overline{\rho}_{18,41} = 0$ in A_{41}.

The ℓ-adic representations denoted by $\rho_{\Delta_w,\ell}$ for $w = 11, 15, 17, 19, 21$ and $\rho_{\Delta_{w,v},\ell}$ for $(w,v) = (19,7), (21,9), (21,13), (21,5)$ in the proof of Theorem 10.1.3

will, here, be respectively denoted by $\mathrm{r}_{i;\ell}$ with $i = 12, 16, 18, 20, 22$, and $\mathrm{r}_{j,k;\ell}$ with $(j,k) = (6,8), (8,8), (12,6), (4,10)$. With this notation, we have $\tau_i(p) = \operatorname{trace} \mathrm{r}_{i;\ell}(\mathrm{Frob}_p)$ and $\tau_{j,k}(p) = \operatorname{trace} \mathrm{r}_{j,k;\ell}(\mathrm{Frob}_p)$ for every $p \neq \ell$.

A few reminders:

- The representations $\mathrm{r}_{i;\ell}$ and $\mathrm{r}_{j,k;\ell}$ are of dimension 2 and 4, respectively.
- The representations $\mathrm{r}_{i;\ell}$ can be defined over $\mathbb{Z}_\ell$. It is probable that this also holds for the representations $\mathrm{r}_{j,k;\ell}$ (see Remark 10.1.4); in what follows, we will use that they can be defined over the integral closure of $\mathbb{Z}_\ell$ in a finite extension of $\mathbb{Q}_\ell$ (see the proof of Corollary 10.1.5).
- We denote by $\omega_\ell \colon \mathrm{Gal}(\overline{\mathbb{Q}}/\mathbb{Q}) \to \mathbb{Z}_\ell^\times$ the homomorphism (the ℓ-adic representation of dimension 1) defined by the action of $\mathrm{Gal}(\overline{\mathbb{Q}}/\mathbb{Q})$ on the ℓ^αth roots of unity, with $\alpha \geq 1$.
- We have $\det \mathrm{r}_{i;\ell} = \omega_\ell^{i-1}$, $\det \mathrm{r}_{6,8;\ell} = \omega_\ell^{38}$, and $\det \mathrm{r}_{j,k;\ell} = \omega_\ell^{42}$ for $(j,k) = (8,8), (12,6), (4,10)$, respectively.
- We denote by $\overline{\mathrm{r}}_{i;\ell} \colon \mathrm{Gal}(\overline{\mathbb{Q}}/\mathbb{Q}) \to \mathrm{GL}_2(\mathbb{F}_\ell)$, $\overline{\mathrm{r}}_{j,k;\ell} \colon \mathrm{Gal}(\overline{\mathbb{Q}}/\mathbb{Q}) \to \mathrm{GL}_4(\mathbb{F}_\ell)$, and $\overline{\omega}_\ell \colon \mathrm{Gal}(\overline{\mathbb{Q}}/\mathbb{Q}) \to \mathbb{F}_\ell^\times$ the respective residual representations associated with the ℓ-adic representations $\mathrm{r}_{i;\ell}$, $\mathrm{r}_{j,k;\ell}$ (see Corollary 10.1.5), and ω_ℓ; the representation $\overline{\omega}_\ell$ can be identified with the homomorphism defined by the action of $\mathrm{Gal}(\overline{\mathbb{Q}}/\mathbb{Q})$ on the ℓth roots of unity, a homomorphism that we will also denote by χ_ℓ.

In what follows, the prime ℓ will be fixed, hence we will leave the index ℓ out of the notation except in the case of A_ℓ.

Let us now return to the proof of the congruence in item (1). Take $\ell = 41$. The equalities

$$\begin{aligned}\rho_{18} \;=\; & (\omega^{14} \oplus \omega^{13} \oplus \omega^{12} \oplus 2\,\omega^{11} \oplus \omega^{10} \oplus \omega^{9} \oplus \omega^{8}) \;\oplus \\ & (\omega^{7} \oplus \omega^{6} \oplus \omega^{5} \oplus \omega^{4}) \otimes \mathrm{r}_{12} \;\oplus\; (\omega^{3} \oplus \omega^{2}) \otimes \mathrm{r}_{18} \;\oplus\; (\omega \oplus 1) \otimes \mathrm{r}_{18}\end{aligned}$$

and

$$\begin{aligned}\rho_{21} \;=\; & (\omega^{12} \oplus 2\,\omega^{11} \oplus \omega^{10}) \;\oplus \\ & (\omega^{7} \oplus \omega^{6} \oplus \omega^{5} \oplus \omega^{4}) \otimes \mathrm{r}_{12} \;\oplus\; (\omega^{3} \oplus \omega^{2}) \otimes \mathrm{r}_{18} \;\oplus\; (\omega \oplus 1) \otimes \mathrm{r}_{4,10}\end{aligned}$$

imply that in the Grothendieck ring A_{41}, we have the equality

$$\overline{\rho}_{21} - \overline{\rho}_{18} \;=\; (\chi + 1)\,\left(\,\overline{\mathrm{r}}_{4,10} - (\overline{\mathrm{r}}_{18} + \chi^{13} + \chi^{8})\,\right)$$

and the promised equation

$$(\chi + 1)\,\left(\,\overline{\mathrm{r}}_{4,10} - (\overline{\mathrm{r}}_{18} + \chi^{13} + \chi^{8})\,\right) \;=\; 0\,, \qquad (10.4.20)$$

which is the "Galois counterpart" of (10.4.7). To "divide this equation by $\chi + 1$," we use Proposition 10.4.6 below. Before stating this proposition, we will need to make a few observations and introduce some more notation.

Let ℓ be a prime and ρ a finite-dimensional continuous representation of $\mathrm{Gal}(\overline{\mathbb{Q}}/\mathbb{Q})$ with coefficients in $\mathbb{F}_\ell$ that is unramified outside ℓ. The map $\rho \mapsto \dim \rho$ induces a ring homomorphism that we also denote by $\dim\colon \mathrm{A}_\ell \to \mathbb{Z}$. The Knonecker–Weber theorem shows that the determinant of ρ is a power of χ. The map $\rho \mapsto \det \rho$ induces a map $\mathrm{A}_\ell \to \mathrm{C}_\chi$, where C_χ denotes the subgroup of $\mathrm{A}_\ell^\times$ generated by χ; this map is also denoted by det. Since χ is of order $\ell - 1$, the group C_χ is canonically isomorphic to $\mathbb{Z}/(\ell-1)$. We easily verify that we have $\det(x+y) = \det(x)\det(y)$ and $\det(xy) = \det(x)^{\dim y}\det(y)^{\dim x}$ for all x and y in A_ℓ.

The abelian group underlying the commutative ring A_ℓ is the free abelian group generated by the set $\mathcal{S}$ of isomorphism classes of the simple representations. Let $H = \sum_{S\in\mathcal{S}} n_S\, S$, with $n_S \in \mathbb{Z}$, be an element of A_ℓ; set

$$\|H\| \;=\; \sum_{S\in\mathcal{S}} |n_S| \dim S\ .$$

The map $\mathrm{A}_\ell \to \mathbb{N}$ defined by $H \mapsto \|H\|$ is a "norm"; in other words, it has the following properties:

- $H = 0 \iff \|H\| = 0$;
- $\|n\,H\| = |n|\,\|H\|$ for every n in $\mathbb{Z}$;
- $\|H_1 + H_2\| \le \|H_1\| + \|H_2\|$ for all H_1 and H_2 in A_ℓ.

Let ρ_+ and ρ_- be two representations of $\mathrm{Gal}(\overline{\mathbb{Q}}/\mathbb{Q})$ with coefficients in $\mathbb{F}_\ell$; we observe that we have the equalities $\|\rho_+\| = \dim\rho_+$ and $\|\rho_-\| = \dim\rho_-$ and the inequality $\|\rho_+ - \rho_+\| \le \dim\rho_+ + \dim\rho_-$.

We finally reach the statement we had in mind.

Proposition 10.4.6. *Let $\ell \neq 2$ be a prime, and let H be an element of A_ℓ. If we have $(\chi+1)H = 0$, then the integer $\|H\|$ is divisible $\ell - 1$. If we moreover have $\det H = 1$, then the integer $\|H\|$ is divisible by $2(\ell-1)$.*

Proof. The obvious action of the group C_χ on the abelian group underlying A_ℓ, given by $(\chi^k, x) \mapsto \chi^k x$, preserves the subset $\mathcal{S}$ introduced above. Let S be an element of $\mathcal{S}$; denote by $\Omega(S)$ the orbit of S under the action of C_χ, by $\mathrm{m}(S)$ the cardinality of this orbit (in other words, $\mathrm{m}(S)$ is the least integer $k \ge 1$ such that we have $\chi^k S = S$), and by $\mathbb{Z}[\Omega(S)]$ the (free abelian) subgroup of the abelian group underlying A_ℓ generated by S. We therefore have a decomposition of the abelian group underlying A_ℓ into a direct sum:

$$\mathrm{A}_\ell \;=\; \bigoplus_{S\in\mathcal{S}_0} \mathbb{Z}[\Omega(S)]\ ,$$

where $\mathcal{S}_0 \subset \mathcal{S}$ is a system of representatives for the quotient set $\mathrm{C}_\chi\backslash\mathcal{S}$. This decomposition is compatible with the action of C_χ; in particular, each factor is sent to itself by the multiplication by $\chi + 1$.

Proposition 10.4.7. *Let S be an element of $\mathcal{S}$. Let $\Omega(S)$ be the orbit of S under the action of C_χ and $\mathrm{m}(S)$ the cardinality of this orbit, that is, the least integer $k \geq 1$ such that we have $\chi^k S = S$.*

(a) *The integer $\mathrm{m}(S)$ divides $\ell - 1$ and $\ell - 1$ divides $\mathrm{m}(S) \dim S$.*
(b) *The kernel of the endomorphism of the abelian group $\mathbb{Z}[\Omega(S)]$ induced by the multiplication by $\chi + 1$ is trivial if $\mathrm{m}(S)$ is odd and is generated (as an abelian group) by*

$$(1 - \chi + \chi^2 - \chi^3 + \ldots - \chi^{\mathrm{m}(S)-1})\, S$$

if $\mathrm{m}(S)$ is even.

Proof. The only part that is not completely obvious is the second part of item (a). To see that it is true, note that we have $\chi^{\mathrm{m}(S)} S = S$ and $\det(\chi^{\mathrm{m}(S)} S) = \chi^{\mathrm{m}(S) \dim S} \det S$. □

Proof of Proposition 10.4.6, *Continued.* Let $\mathcal{S}_{0,0}$ be the subset of $\mathcal{S}_0$ consisting of the S with $\mathrm{m}(S)$ even; Proposition 10.4.7 shows that if we have the equality $(\chi + 1)H = 0$, then there exist integers n_S, where S runs through $\mathcal{S}_{0,0}$, such that we have

$$H = \sum_{S \in \mathcal{S}_{0,0}} n_S\, (1 - \chi + \chi^2 - \chi^3 + \ldots - \chi^{\mathrm{m}(S)-1})\, S\,. \tag{10.4.21}$$

By the very definition of $\|H\|$, we have

$$\|H\| = \sum_{S \in \mathcal{S}_{0,0}} |n_S|\, \mathrm{m}(S) \dim S\,.$$

The second part of item (a) of Proposition 10.4.7 says that all the products $\mathrm{m}(S) \dim S$ in the sum are divisible by $\ell - 1$, which proves the first part of Proposition 10.4.6.

Let us turn to the proof of the second part of Proposition 10.4.6. Equality (10.4.21) implies

$$\det H = \chi^{-\frac{1}{2} \sum_{S \in \mathcal{S}_{0,0}} n_S\, \mathrm{m}(S) \dim S}$$

(note that we have $\dim(1 - \chi + \chi^2 - \chi^3 + \ldots - \chi^{\mathrm{m}(S)-1}) = 0$ and $\det(1 - \chi + \chi^2 - \chi^3 + \ldots - \chi^{\mathrm{m}(S)-1}) = \chi^{-\mathrm{m}(S)/2}$), so that the equality $\det H = 1$ is equivalent to the congruence

$$\sum_{S \in \mathcal{S}_{0,0}} n_S\, \mathrm{m}(S) \dim S \equiv 0 \mod 2\,(\ell - 1)\,.$$

Since $|n_S|$ and n_S have the same parity and all the $\mathrm{m}(S) \dim S$ are divisible by $\ell - 1$, we see that the equalities $(\chi + 1)H = 0$ and $\det H = 1$ indeed imply that $\|H\|$ is divisible by $2\,(\ell - 1)$. □

Remark. The equality $(\chi + 1)H = 0$ implies $\dim H = 0$ and $(\det H)^2 = 1$. This implication is "optimal" (for $\ell \neq 2$). To see this, take $H = H_0 := 1 - \chi + \chi^2 -$

$\chi^3 + \ldots - \chi^{\ell-2}$ and check that we have $\det H_0 = \chi^{(\ell-1)/2}$. We also see that we have $\|H_0\| = \ell - 1$, which shows that the first part of Proposition 10.4.6 is optimal; moreover, we have $(\chi+1)(2H_0) = 0$, $\det(2H_0) = 1$, and $\|2H_0\| = 2(\ell-1)$, which shows that the second part of this proposition is also optimal.

Proof of Item (1) *of Theorem* 10.4.4 *Using* 10.4.6. Take $\ell = 41$ and $H = \bar{\mathrm{r}}_{4,10} - (\bar{\mathrm{r}}_{18} + \chi^{13} + \chi^8)$. We have $\|H\| \le 8$. Since we have $(\chi+1)H = 0$ (Eq. (10.4.20)), Proposition 10.4.6 says that $\|H\|$ is divisible by 40 (and even 80 because we have $\det H = 1$). From this, we deduce $\|H\| = 0$ and $H = 0$. By evaluating the representations $\bar{\mathrm{r}}_{4,10}$ and $\bar{\mathrm{r}}_{18} \oplus \chi^{13} \oplus \chi^8$ "at the conjugation class Frob_p," we obtain item (1) for $p \neq 41$. Moreover, the congruence (10.4.7) trivially implies item (1) for $p = 41$. □

Proof of Item (2) *of Theorem* 10.4.4. Fix $\ell = 23$.

By taking $m = \ell = 23$ and $\mathcal{W} = \{\mathrm{v}_{13}, \mathrm{v}_{15}\}$ in Proposition 10.4.5, we obtain, as before, the isomorphism of Galois representations

$$\bar{\mathrm{r}}_{8,8} \simeq (\chi^6 \oplus 1) \otimes \bar{\mathrm{r}}_{16} \tag{10.4.22}$$

and the congruence

$$\tau_{8,8}(p) \equiv (p^6 + 1)\,\tau_{16}(p) \mod 23\,.$$

To obtain the congruence (2) (which refines both this congruence and the congruence (10.4.9)), we use the lemma below.

Lemma 10.4.8. *Let B be an Artinian local ring with residue field k, G a group, and V_1, V_2, W_1, W_2, $B[G]$-modules that we assume to be free of finite dimension as B-modules. We make the following assumptions:*

(i) *For $i = 1, 2$, the semisimplifications of the $k[G]$-modules $k \otimes_B V_i$ and $k \otimes_B W_i$ are isomorphic.*
(ii) *The $k[G]$-modules $k \otimes_B V_1$ and $k \otimes_B V_2$ have no common Jordan–Hölder factor.*
(iii) *For every g in G, we have* $\det(t - g_{|V_1 \oplus V_2}) = \det(t - g_{|W_1 \oplus W_2})$.

Then for $i = 1, 2$ and for every g, we have $\det(t - g_{|V_i}) = \det(t - g_{|W_i})$.

Proof. Let U be the $B[G]$-module $V_1 \oplus V_2 \oplus W_1 \oplus W_2$ and R the B-algebra that is the image of $B[G]$ in $\mathrm{End}_B(U)$. Let J be the Jacobson radical of R. Since the B-module underlying R is of finite type, J is the greatest nilpotent bilateral ideal of R. It is also the kernel of the natural homomorphism from R to the endomorphisms of the semisimplification of $k \otimes_B U$. In particular, we have $\mathfrak{m}R \subset J$, where $\mathfrak{m}$ denotes the maximal ideal of B and R/J is a finite-dimensional semisimple k-algebra.

Artin–Wedderburn theory applied to R/J and assumption (ii) show that we can find an idempotent f in R/J such that f acts by the identity on the semisimplification of $k \otimes_B V_1$ and by 0 on that of $k \otimes_B V_2$. Since J is nilpotent, this idempotent lifts to an idempotent e in R. This ensures that e acts by 0 on V_2 (because it acts that way on all its Jordan–Hölder factors) and by the identity on V_1 and W_1 (for the same reason).

The classical Amitsur identity, which expresses the coefficients of the characteristic polynomial of a sum of two matrices as a universal function of the coefficients of the characteristic polynomials of these two matrices, shows that the equality of the determinants in assumption (iii) more generally implies $\det(t - r_{|V_1 \oplus V_2}) = \det(t - r_{|W_1 \oplus W_2})$ for every r in R. The lemma follows: let g be an element of G; for $i = 1$ (resp. $i = 2$), we specialize this identity to $r = ge$ (resp. $r = g(1-e)$). □

Proof of Item (2) *of Theorem* 10.4.4 *Using* 10.4.8. We specialize the lemma in question.

(Recall: the prime ℓ is fixed, equal to 23, the notation $\mathrm{r}_{8,8}$, r_{16}, ω, ρ_{13}, ρ_{15}, and χ that appears below is the abbreviation of $\mathrm{r}_{8,8;23}$, $\mathrm{r}_{16;23}$, ω_{23}, $\rho_{13,23}$, $\rho_{15,23}$, and χ_{23}, respectively.)

The representation $\mathrm{r}_{8,8}$ can be defined over the integral closure of $\mathbb{Z}_{23}$ in a finite extension of $\mathbb{Q}_{23}$, which we denote by $\mathcal{D}$. The representation r_{16} can be defined over $\mathbb{Z}_{23}$ and a fortiori over $\mathcal{D}$; likewise, the representation ω is defined over $\mathbb{Z}_{23}$ and a fortiori over $\mathcal{D}$.

For B, take the quotient ring $\mathcal{D}/23^2$. The ring B is local, its residue field k is a finite field of characteristic 23; B is Artinian (it is finite!).

For G, take the Galois group $\mathrm{Gal}(\overline{\mathbb{Q}}/\mathbb{Q})$.

For V_1, V_2, W_1, and W_2, take the B-module B^4 endowed with the linear action of the group G given by the representations $(\omega^6 \oplus 1) \otimes \mathrm{r}_{16}$, $\omega \otimes (\omega^6 \oplus 1) \otimes \mathrm{r}_{16}$, $\mathrm{r}_{8,8}$, and $\omega \otimes \mathrm{r}_{8,8}$, respectively.

Assumption (i) of Lemma 10.4.8 follows from the isomorphism (10.4.22) and the extension of scalars of $\mathbb{F}_{23}$ to k.

It is not difficult to verify assumption (ii). The (residual) representation $\overline{\mathrm{r}}_{16}$ (modulo 23) is simple [194] (to see this ab initio, note that we have $47 \equiv 1 \bmod 23$ and $\tau_{16}(47) \not\equiv 2 \bmod 23$); it follows, again by Kronecker–Weber, that $k \otimes_{\mathbb{F}_{23}} \overline{\mathrm{r}}_{16}$ is also simple. The Jordan–Hölder factors of $k \otimes_B V_1$ (resp. $k \otimes_B V_2$) are therefore $k \otimes_{\mathbb{F}_{23}} \overline{\mathrm{r}}_{16}$ and $k \otimes_{\mathbb{F}_{23}} (\chi^6\, \overline{\mathrm{r}}_{16})$ (resp. $k \otimes_{\mathbb{F}_{23}} (\chi\, \overline{\mathrm{r}}_{16})$ and $k \otimes_{\mathbb{F}_{23}} (\chi^7\, \overline{\mathrm{r}}_{16})$). We conclude by observing that the determinant of $\overline{\mathrm{r}}_{16}$, $\chi^6\, \overline{\mathrm{r}}_{16}$, $\chi\, \overline{\mathrm{r}}_{16}$, and $\chi^7\, \overline{\mathrm{r}}_{16}$ is, respectively, χ^{15}, χ^5, χ^{17}, and χ^7.

Assumption (iii) of Lemma 10.4.8 is implied by Proposition 10.4.5 and the fact that we have $\rho_{13} = v \oplus \sigma$ and $\rho_{15} = w \oplus \sigma$, with $v = (\omega \oplus 1) \otimes (\omega^6 \oplus 1) \otimes \mathrm{r}_{16}$, $w = (\omega \oplus 1) \otimes \mathrm{r}_{8,8}$, and σ a 23-adic representation of dimension 16.

The conclusion of the lemma says that "the characteristic polynomials at Frob_p," for $p \neq 23$, of the 23-adic representations $(\omega^6 \oplus 1) \otimes \mathrm{r}_{16}$ and $\mathrm{r}_{8,8}$ are congruent modulo 23^2. A fortiori, "the traces in Frob_p," for $p \neq 23$, are congruent modulo 23^2; in other words, the congruence in item (2) is satisfied for $p \neq 23$. The case $p = 23$ trivially follows from the congruence (10.4.9). □

Proof of Item (3) *of Theorem* 10.4.4. Fix $\ell = 19$.

By taking $m = \ell = 19$ and $\mathcal{W} = \{v_9, v_{10}\}$ in Proposition 10.4.5 and using Proposition 10.4.6, we obtain the equality

$$\bar{r}_{12,6} - (\chi^4 + \chi^2)\,\bar{r}_{16} + \chi^2\,\bar{r}_{18} - \bar{r}_{22} \;=\; 0$$

in the Grothendieck ring A_{19} or, equivalently, the isomorphism of representations

$$\bar{r}_{12,6} \oplus \chi^2\,\bar{r}_{18} \;\simeq\; (\chi^4 \oplus \chi^2)\,\bar{r}_{16} \oplus \bar{r}_{22}\,. \qquad (10.4.23)$$

Since the representations $\bar{r}_{16}$ and $\bar{r}_{22}$ are simple (we can see this by observing that we have $\tau_{16}(5) \not\equiv \tau_{16}(43) \bmod 19$ and $\tau_{22}(5) \not\equiv \tau_{22}(43) \bmod 19$, while we have $5 \equiv 43 \bmod 19$), the representation $\chi^2\,\bar{r}_{18}$ is necessarily isomorphic to one of the representations $\chi^4\,\bar{r}_{16}$, $\chi^2\,\bar{r}_{16}$, or $\bar{r}_{22}$. Computing the determinants shows that the only possibility is

$$\chi^2\,\bar{r}_{18} \;\simeq\; \bar{r}_{22}\,. \qquad (10.4.24)$$

Since the representations $\bar{r}_{12,6}$ and $(\chi^4 \oplus \chi^2)\,\bar{r}_{16}$ are semisimple, the isomorphisms (10.4.23) and (10.4.24) imply

$$\bar{r}_{12,6} \;\simeq\; (\chi^4 \oplus \chi^2)\,\bar{r}_{16}\,.$$

This isomorphism implies item (3) for $p \neq 19$. We easily verify that this congruence also holds for $p = 19$ (the use of the computation of $\tau_{12,6}(19)$ can be avoided by observing that the congruences $19^2\tau_{18}(19) \equiv \tau_{22}(19)$ and (10.4.10) imply item (3) for $p = 19$). □

Proof of Items (4), (5), *and* (6) *of Theorem* 10.4.4. The proof of item (4) is the same as that of item (3); it is, moreover, quicker if we use (10.4.24).

Let us move on to items (5) and (6). Obviously, set $\ell = 17$.

By using the “Galois counterpart” of (10.4.13) and copying the proof of item (3), we obtain the isomorphisms of representations

$$\chi^2\,\bar{r}_{16} \;\simeq\; \bar{r}_{20} \qquad (10.4.25)$$

and

$$\bar{r}_{6,8} \;\simeq\; (\chi^6 \oplus \chi^2) \otimes \bar{r}_{12}\,.$$

This isomorphism implies item (5) for $p \neq 17$; the case $p = 17$ can be taken care of as in the case of item (3).

The Galois counterpart of (10.4.12) is the following equation in A_{17}:

$$(\chi + 1)\,(\bar{r}_{8,8} - (\chi^6 + \chi^4)\bar{r}_{12} + (\chi^4 + \chi^2)\bar{r}_{16} - (\chi^2 + 1)\bar{r}_{20}) \;=\; 0\,.$$

This equation and the isomorphism (10.4.25) imply

$$(\chi + 1)\,(\bar{r}_{8,8} - (\chi^6 + \chi^4)\bar{r}_{12}) \;=\; 0\,.$$

By invoking Proposition 10.4.6, we obtain the equation

$$\bar{\mathrm{r}}_{8,8} - (\chi^6 + \chi^4)\bar{\mathrm{r}}_{12} \;=\; 0$$

or, equivalently, the isomorphism of representations

$$\bar{\mathrm{r}}_{8,8} \;\simeq\; (\chi^6 \oplus \chi^4) \otimes \bar{\mathrm{r}}_{12}$$

(the two sides are semisimple). This isomorphism implies item (6) for $p \neq 17$; the case $p = 17$ can be taken care of as before. □

Remark. The Galois counterpart of (10.4.11) is the following equation in A_{17}:

$$(\chi + 1)\,((\chi^4 + \chi^2)\bar{\mathrm{r}}_{16} - (\chi^2 + 1)\bar{\mathrm{r}}_{20} + \bar{\mathrm{r}}_{22} - \chi^{17} - \chi^4) \;=\; 0\,.$$

In view of (10.4.25), this gives the equation

$$(\chi + 1)\,(\bar{\mathrm{r}}_{22} - \chi^{17} - \chi^4) \;=\; 0\,.$$

By once again invoking Proposition 10.4.6, we obtain the equation

$$\bar{\mathrm{r}}_{22} - \chi^{17} - \chi^4 \;=\; 0$$

or, equivalently, the isomorphism of representations

$$\bar{\mathrm{r}}_{22} \;\simeq\; \chi^{17} \oplus \chi^4 \;=\; \chi \oplus \chi^4\,.$$

This isomorphism $\bar{\mathrm{r}}_{22} \simeq \chi \oplus \chi^4$ is one of the isomorphisms given by Swinnerton-Dyer in [194]; from now, we systematically use this type of isomorphism.

Proof of Items (7), (8), (9), (11), *and* (13) *of Theorem* 10.4.4. The starting point of these proofs is, respectively,

- the equation in A_{13} that is the Galois counterpart of the congruence (10.4.15),
- the equation in A_{13} that is the Galois counterpart of the congruence (10.4.14) or the congruence (10.4.16),
- the equation in A_{11} that is the Galois counterpart of the reduction modulo 11 of the congruence (10.4.18) (that is a congruence modulo 11^2),
- the equation in A_{11} that is the Galois counterpart of the reduction modulo 11 of the congruence (10.4.19) (that is a congruence modulo 11^2),
- the equation in A_7 that is the Galois counterpart of the reduction modulo 7 of the congruence (10.4.17) (that is a congruence modulo $2^5 \cdot 7 \cdot 13$ whose reduction modulo 13 is the congruence (10.4.16) mentioned above).

By the method we repeatedly used above, we express the $\bar{\mathrm{r}}_{j,k}$ that interest us in terms of certain $\bar{\mathrm{r}}_i$ and χ. All the $\bar{\mathrm{r}}_i$ that appear can, in turn, be expressed in terms of χ thanks to the Swinnerton-Dyer isomorphisms [194]. In the end, we obtain isomorphisms of the form $\bar{\mathrm{r}}_{j,k} \simeq \chi^{a_1} \oplus \chi^{a_2} \oplus \chi^{a_3} \oplus \chi^{a_4}$ that lead to items (7), (8), (9), (11), and (13).

Let us, for example, treat the case of the congruence in item (13).

We fix $\ell = 7$. By taking $m = \ell = 7$ and $\mathcal{W} = \{\mathrm{v}_{10}, \mathrm{v}_{11}\}$ in Proposition 10.4.5, we obtain the following equation in A_7:

$$(\chi + 1)(\bar{\mathrm{r}}_{12,6} - (\chi^4 + 1)\bar{\mathrm{r}}_{18}) \ = \ 0 \, .$$

Since the representations $\bar{\mathrm{r}}_{12,6}$ and $(\chi^4 \oplus 1) \otimes \bar{\mathrm{r}}_{18}$ have the same determinant (namely χ^{42}) and since we have the inequality $\|\bar{\mathrm{r}}_{12,6} - (\chi^4 + 1)\bar{\mathrm{r}}_{18}\| \leq 8$, the second part of Proposition 10.4.6 shows that we in fact have the equation

$$\bar{\mathrm{r}}_{12,6} - (\chi^4 + 1)\bar{\mathrm{r}}_{18} \ = \ 0$$

or, equivalently, the isomorphism of representations

$$\bar{\mathrm{r}}_{12,6} \ \simeq \ (\chi^4 \oplus 1) \otimes \bar{\mathrm{r}}_{18} \, .$$

But Swinnerton-Dyer tells us that we have $\bar{\mathrm{r}}_{18} \simeq \chi \oplus \chi^4$, so that in the end, we obtain

$$\bar{\mathrm{r}}_{12,6} \ \simeq \ \chi^5 \oplus \chi^4 \oplus \chi^2 \oplus \chi \, .$$

This isomorphism gives item (13) for $p \neq 7$; the case $p = 7$ follows from (10.4.17).

□

Proof of Item (10) *of Theorem* 10.4.4. This is similar to that of item (2).

Fix $\ell = 11$.

By taking $m = \ell = 11$ and $\mathcal{W} = \{\mathrm{v}_{14}, \mathrm{v}_{19}\}$ in Proposition 10.4.5, we obtain the following equation in A_{11}:

$$\chi(\chi + 1)\,(\bar{\mathrm{r}}_{6,8} - \bar{\mathrm{r}}_{20} - \chi^{13} - \chi^6) \, .$$

We "divide this equation by $\chi(\chi + 1)$," observing that χ is invertible and using Proposition 10.4.6; we thus obtain the isomorphism of Galois representations

$$\bar{\mathrm{r}}_{6,8} \ \simeq \ \bar{\mathrm{r}}_{20} \oplus \chi^{13} \oplus \chi^6$$

and the congruence

$$\tau_{6,8}(p) \ \equiv \ \tau_{20}(p) + p^{13} + p^6 \mod 11 \, . \tag{10.4.26}$$

This congruence transforms into that in item (9) using the congruence $\tau_{20}(p) \equiv p^8 + p \bmod 11$ of [194].

But in the first part of this section, we saw that for every prime p, we have the congruence (10.4.18):

$$(p + 1)(\tau_{6,8}(p) - \tau_{20}(p) - p^{13} - p^6) \ \equiv \ 0 \mod 11^2 \, .$$

To "divide this congruence by $p+1$," we use the method that has allowed us to obtain item (2) from the congruence (10.4.9) (namely, the application of Lemma 10.4.8). □

Remark. We cannot have a congruence of the form

$$\tau_{6,8}(p) \equiv p^{a_1} + p^{a_2} + p^{a_3} + p^{a_4} \mod 11^2$$

for every prime p, where we assume $p \neq 11$. Indeed, in view of item (10), we would have $\tau_{20}(p) \equiv p^{a_1} + p^{a_2} + p^{a_3} + p^{a_4} - p^{13} - p^6 \mod 11^2$. But this congruence does not hold for the least prime p with $p \equiv 1 \mod 11^2$, namely $p = 727$: $\tau_{20}(727) \equiv 68 \not\equiv 2 \mod 11^2$.

Remark. In the first part of this section, we saw that for every prime p, we have the congruence (10.4.19):

$$(p+1)\,(\tau_{4,10}(p) - (p^2+1)\,\tau_{20}(p) + p^2\,\tau_{18}(p) - p^{13} - p^8) \equiv 0 \mod 11^2 .$$

The Galois counterpart of the reduction modulo 11 of this congruence is the following equation in A_{11}:

$$(\chi+1)\,(\bar{\mathrm{r}}_{4,10} - (\chi^2+1)\,\bar{\mathrm{r}}_{20} + \chi^2\,\bar{\mathrm{r}}_{18} - \chi^{13} - \chi^8) = 0 .$$

Using the isomorphisms $\bar{\mathrm{r}}_{20} \simeq \chi^8 \oplus \chi$ and $\bar{\mathrm{r}}_{18} \simeq \chi^6 \oplus \chi$ of [194], we obtain the equation

$$(\chi+1)\,(\bar{\mathrm{r}}_{4,10} - \chi^8 - \chi^3 - \chi - 1) = 0$$

that, after "dividing by $p+1$," leads to the congruence in item (11). But this time, Lemma 10.4.8 does not allow "dividing the congruence (10.4.19) by $p+1$" because the assumption on the Jordan–Hölder factors is not satisfied. However, we see that the congruence

$$\tau_{4,10}(p) \equiv (p^2+1)\,\tau_{20}(p) - p^2\,\tau_{18}(p) + p^{13} + p^8 \mod 11^2$$

holds for $p \leq 113$ (recall that we have determined the $\tau_{j,k}(p)$ for $p \leq 113$); note that this is truly information only for the primes $p \leq 113$ with $p+1 \equiv 0 \mod 11$, namely 43 and 109!

Proof of Items (12) *and* (14) *of Theorem* 10.4.4. The proofs of these items are of the same type (which is why we combined them). Let us prove, for example, that for every prime p, we have the congruence

$$p\,\tau_{6,8}(p) \equiv \tau_{8,8}(p) \mod 7 . \tag{10.4.27}$$

Fix $\ell = 7$.

By taking $m = \ell = 7$ and $\mathcal{W} = \{\mathrm{v}_{15}, \mathrm{v}_{19}\}$ in Proposition 10.4.5, we obtain the following equation in A_7:

$$(\chi+1)\,\bar{\mathrm{r}}_{8,8} - \chi(\chi+1)\,\bar{\mathrm{r}}_{6,8} + (\chi^5+\chi^2)\,\bar{\mathrm{r}}_{16} - (\chi^5+1)\,\bar{\mathrm{r}}_{12} - \mathrm{Sym}^2\,\bar{\mathrm{r}}_{12} + \chi^5 + 2\chi^2 \;=\; 0\,. \quad (10.4.28)$$

We see that the isomorphisms $\bar{\mathrm{r}}_{16} \simeq \chi^2 + \chi$ and $\bar{\mathrm{r}}_{12} \simeq \chi^4 + \chi$ of [194] imply

$$(\chi^5+\chi^2)\,\bar{\mathrm{r}}_{16} - (\chi^5+1)\,\bar{\mathrm{r}}_{12} - \mathrm{Sym}^2\,\bar{\mathrm{r}}_{12} + \chi^5 + 2\chi^2 \;=\; 0\,,$$

so that Eq. (10.4.28) simplifies to the following:

$$(\chi+1)\,\bar{\mathrm{r}}_{8,8} - \chi(\chi+1)\,\bar{\mathrm{r}}_{6,8} \;=\; 0\,. \quad (10.4.29)$$

Set $H = \bar{\mathrm{r}}_{8,8} - \chi\,\bar{\mathrm{r}}_{6,8}$; Eq. (10.4.29) says that we have $(\chi+1)\,H = 0$, and we see that we have $\det H = 1$. We can therefore apply the second part of Proposition 10.4.6: $\|H\|$ is divisible by 12. Since, a priori, we have $\|H\| \leq 8$, it follows that we have $H = 0$, the isomorphism of representations

$$\chi\,\bar{\mathrm{r}}_{6,8} \;\simeq\; \bar{\mathrm{r}}_{8,8}\,, \quad (10.4.30)$$

and the congruence (10.4.27) for $p \neq 7$. The case $p = 7$ is left to the reader. □

Proof of Items (15), (16), (17), *and* (18) *of Theorem* 10.4.4. As before, the proofs of these items are all of the same type. We give a few details on the proof of the congruences in item (16), and we restrict ourselves to indicating the essential modifications needed to obtain the proofs of items (15), (17), and (18).

Fix $\ell = 5$.

Consider the subset $\mathcal{W} := \{\mathrm{v}_{10}, \mathrm{v}_{17}, \mathrm{v}_{19}, \mathrm{v}_{21}\}$ of $\mathcal{V}$; we easily verify that $\mathcal{W}$ is minimal 5-unfree. By Proposition 10.4.5, it follow that the four representations $\overline{\rho}_{10}$, $\overline{\rho}_{17}$, $\overline{\rho}_{19}$, and $\overline{\rho}_{21}$ are pairwise isomorphic. The isomorphism $\overline{\rho}_{10} \simeq \overline{\rho}_{17}$ gives the following equation in A_5:

$$(\chi+1)\,\bar{\mathrm{r}}_{12,6} \;=\; (\chi^3+\chi^2+\chi+1)\,\bar{\mathrm{r}}_{20} - (\chi^3+\chi^2)\,\bar{\mathrm{r}}_{18} + (\chi^3+\chi^2+\chi+1)\,\bar{\mathrm{r}}_{12} - 2\chi^3 - \chi^2 - 1\,. \quad (10.4.31)$$

In view of [194], this equation becomes

$$(1+\chi)\,\bar{\mathrm{r}}_{12,6} \;=\; 2(1+\chi+\chi^2+\chi^3)\,. \quad (10.4.32)$$

This equation shows that the (semisimple) representation $(1 \oplus \chi) \otimes \bar{\mathrm{r}}_{12,6}$ is a direct sum of powers of χ; it follows that the same holds for $\bar{\mathrm{r}}_{12,6}$. Consequently, in A_5 we have an equation of the form

$$\bar{\mathrm{r}}_{12,6} \;=\; a_0 + a_1\,\chi + a_2\,\chi^2 + a_3\,\chi^3 \quad (10.4.33)$$

with a_k, for $k = 0, 1, 2, 3$, integers satisfying $a_k \geq 0$ and $a_0 + a_1 + a_2 + a_3 = 1$. Equation (10.4.32) can be rewritten as follows:

$$(1+\chi)\,(\bar{\mathrm{r}}_{12,6} - (1+\chi+\chi^2+\chi^3)) \;=\; 0\,;$$

this form and part (b) of Proposition 10.4.7 (take $S = \chi$) show that there exists an integer n such that we have

$$\bar{\mathrm{r}}_{12,6} \;=\; 1+\chi+\chi^2+\chi^3+n\,(1-\chi+\chi^2-\chi^3)\,.$$

The inequalities $a_k \geq 0$ show that we have $|n| \leq 1$. The computation of the determinant of the two sides shows that we have $n \equiv 0 \bmod 2$. We therefore have $n = 0$ and an isomorphism of representations

$$\bar{\mathrm{r}}_{12,6} \;\simeq\; 1 \oplus \chi \oplus \chi^2 \oplus \chi^3\,.$$

This isomorphism implies the congruence $\tau_{12,6}(p) \equiv 1 + p + p^2 + p^3 \bmod 5$ or, equivalently, $\tau_{12,6}(p) \equiv p^4 + p^3 + p^2 + p \bmod 5$, for $p \neq 5$. The case $p = 5$ is left to the reader.

Likewise, the isomorphisms $\overline{\rho}_{19} \simeq \overline{\rho}_{17}$ and $\overline{\rho}_{21} \simeq \overline{\rho}_{17}$ lead to the congruences $\tau_{6,8}(p) \equiv p^4 + p^3 + p^2 + p \bmod 5$ and $\tau_{4,10}(p) \equiv p^4 + p^3 + p^2 + p \bmod 5$. (Note that the isomorphism $\overline{\rho}_{19} \simeq \overline{\rho}_{17}$ naturally leads to the isomorphism $\chi\bar{\mathrm{r}}_{6,8} \simeq 1 \oplus \chi \oplus \chi^2 \oplus \chi^3$ but that we have $\chi^{-1}(1 \oplus \chi \oplus \chi^2 \oplus \chi^3) \cong 1 \oplus \chi \oplus \chi^2 \oplus \chi^3$.)

The congruence in item (15), in turn, follows by taking, for example, $\mathcal{W} = \{\mathrm{v}_3, \mathrm{v}_{15}\}$.

Let us finally move on to the congruences in items (17) and (18). We can prove them by taking $\ell = 2, 3$, $m = 6$, and $\mathcal{W} = \{\mathrm{v}_6, \mathrm{v}_{10}\}, \{\mathrm{v}_6, \mathrm{v}_{15}\}, \{\mathrm{v}_6, \mathrm{v}_{19}\}, \{\mathrm{v}_6, \mathrm{v}_{21}\}$ in Proposition 10.4.5.

Recall that when the prime ℓ is 2 or 3, the isomorphisms of [194] that concern us take on a particularly simple form: $\bar{\mathrm{r}}_i \simeq 1 \oplus 1$ for $\ell = 2$ and $\bar{\mathrm{r}}_i \simeq 1 \oplus \chi$ for $\ell = 3$.

For $\ell = 2$, the previous methods give

$$\bar{\mathrm{r}}_{j,k} \;\simeq\; 1 \oplus 1 \oplus 1 \oplus 1\,. \tag{10.4.34}$$

For $\ell = 3$, we find that the representations $\bar{\mathrm{r}}_{j,k}$ are each isomorphic to one of the following three representations: $1 \oplus 1 \oplus 1 \oplus 1$, $1 \oplus 1 \oplus \chi \oplus \chi$, or $\chi \oplus \chi \oplus \chi \oplus \chi$. We remove the ambiguity by using the fact that the dual representation $\bar{\mathrm{r}}^*_{j,k}$ is isomorphic to the representation $\chi\bar{\mathrm{r}}_{j,k}$ (see the beginning of Remark 10.1.6):

$$\bar{\mathrm{r}}_{j,k} \;\simeq\; 1 \oplus 1 \oplus \chi \oplus \chi\,. \tag{10.4.35}$$

□

On the Decomposition of the $\bar{\mathrm{r}}_{j,k;\ell}$ into Irreducible Factors

As we have just seen, items (12) and (14) of Theorem 10.4.4 are consequences of isomorphisms between certain representations of the form $\bar{\mathrm{r}}_{j,k;\ell}$ or $\chi \otimes \bar{\mathrm{r}}_{j,k;\ell}$.

Each of the other congruences of this theorem, with the exception of those in items (2) and (10) (however, their reductions modulo 23 and 11, respectively, are no exception), is the manifestation of a reducibility property of a representation $\overline{\mathrm{r}}_{j,k;\ell}$. Proposition 10.4.9 below, which is probably well known, describes exhaustively the different possibilities for a reduction of a representation of this type; it explains, in part, the structure of the congruences we have stated.

Let ℓ be a prime, and let κ be an integer. Denote by $\mathrm{R}_{\kappa,\ell}$ the (finite) set of the isomorphism classes of irreducible representations $\mathrm{Gal}(\overline{\mathbb{Q}}/\mathbb{Q}) \to \mathrm{GL}_2(\overline{\mathbb{F}}_\ell)$ of the form $\overline{\rho}_{\pi,\iota}$, where $\pi \in \Pi_{\mathrm{cusp}}(\mathrm{PGL}_2)$ is the automorphic representation generated by an element of $\mathrm{S}_\kappa(\mathrm{SL}_2(\mathbb{Z}))$ that is an eigenform for the Hecke operators. Recall that we denote by $\nu\colon \mathrm{GSp}_{2g} \to \mathbb{G}_{\mathrm{m}}$ the "similitude factor" homomorphism (see Sect. 2.1). Finally, if S is a finite-dimensional irreducible $\overline{\mathbb{F}}_\ell$-representation of $\mathrm{Gal}(\overline{\mathbb{Q}}/\mathbb{Q})$, denote by $\mathrm{m}(S)$ the least integer $m \geq 1$ such that we have $\chi^m \otimes S \simeq S$ (recall that we have set $\chi = \overline{\omega}_\ell$); this notation agrees with the one introduced in Proposition 10.4.7.

Proposition 10.4.9. *Let ℓ be an odd prime and*

$$r : \mathrm{Gal}(\overline{\mathbb{Q}}/\mathbb{Q}) \to \mathrm{GSp}_4(\overline{\mathbb{F}}_\ell)$$

a continuous semisimple representation that is unramified outside ℓ. Denote by w the element of $\mathbb{Z}/(\ell-1)$ such that we have $\nu \circ r = \chi^w$, and suppose $w \equiv 1 \bmod 2$.

Then we are in one, and only one, of the following cases:

(i) *There exist a and b in $\mathbb{Z}/(\ell-1)$ such that we have $r \simeq \chi^a \oplus \chi^b \oplus \chi^{w-a} \oplus \chi^{w-b}$.*
(ii) *There exist $\kappa \leq \ell+1$, a representation ρ in $\mathrm{R}_{\kappa;\ell}$, and a, b in $\mathbb{Z}/(\ell-1)$ with $2a+\kappa-1 \equiv w \bmod \mathrm{m}(\rho)$, such that we have $r \simeq (\chi^a \otimes \rho) \oplus \chi^b \oplus \chi^{w-b}$.*
(iii)$_1$ *There exist $\kappa \leq \ell+1$, a representation ρ in $\mathrm{R}_{\kappa,\ell}$, and a in $\mathbb{Z}/(\ell-1)$ with $2a+\kappa-1 \not\equiv w \bmod \mathrm{m}(\rho)$, such that we have $r \simeq (\chi^a \oplus \chi^{w-a-\kappa+1}) \otimes \rho$.*
(iii)$_2$ *There exists a $\overline{\mathbb{F}}_\ell$-representation ρ, irreducible of dimension 2, of determinant $\det \rho = \chi^a$ with $a \in 2\mathbb{Z}$ and $a \not\equiv w \bmod \mathrm{m}(\rho)$, such that we have $r \simeq (1 \oplus \chi^{w-a}) \otimes \rho$.*
(iv) *For $i = 1, 2$, there exist $\kappa_i \leq \ell+1$, a representation ρ_i in $\mathrm{R}_{\kappa_i;\ell}$, and a_i in $\mathbb{Z}/(\ell-1)$ with $2a_i+\kappa_i-1 \equiv w \bmod \mathrm{m}(\rho_i)$, such that we have $r \simeq (\chi^{a_1} \otimes \rho_1) \oplus (\chi^{a_2} \otimes \rho_2)$.*
(v) *The representation r is irreducible.*

Proof. Let V be a finite-dimensional $\overline{\mathbb{F}}_\ell$-representation of $G = \mathrm{Gal}(\overline{\mathbb{Q}}/\mathbb{Q})$; denote by $\mathrm{i}(V)$ the representation $V^* \otimes \chi^w$. It is clear that we have $V \cong \mathrm{i}(\mathrm{i}(V))$ and that the map $V \mapsto \mathrm{i}(V)$ defines an auto-equivalence, which is exact and contravariant, of the category of finite-dimensional $\overline{\mathbb{F}}_\ell$-representations of G. In particular, the finite set $\mathrm{J}(r)$ of Jordan–Hölder factors of the representation r is stable under the involution i.

Note that the elements of $\mathrm{J}(r)$ are of dimension 1 or 2. Indeed, suppose $r \simeq V \oplus W$ with V irreducible of dimension 3 (and W of dimension 1). The nondegenerate alternating bilinear form associated, by definition, with the $\overline{\mathbb{F}}_\ell$-vector space underlying r provides a natural G-equivariant isomorphism $r \to \mathrm{i}(r)$ that induces an isomorphism $V \to \mathrm{i}(V)$. It follows, in particular, that the restriction to V of the alternating bilinear form mentioned above is nondegenerate, which is absurd because the dimension of V is odd.

As mentioned before, the Kronecker–Weber theorem asserts that the only homomorphisms $G \to \overline{\mathbb{F}}_\ell^\times$, assumed continuous and unramified outside ℓ, are the powers of χ. Since the integers ℓ and w are odd, we have $\mathrm{i}(\chi^a) = \chi^{w-a} \neq \chi^a$ for every a in $\mathbb{Z}$, so that no representation of dimension 1 of G is "fixed" by i. In particular, r is the sum of four characters if and only if we are in case (i) of the proposition.

Let V be an irreducible $\overline{\mathbb{F}}_\ell$-representation of dimension 2 of G assumed to be continuous and unramified outside ℓ. Recall that V is called *odd* if the conjugacy class of G consisting of the complex conjugations admits the eigenvalues 1 and -1 in the representation V. This is equivalent to saying that we have $\det V = \chi^s$ with $s \equiv 1 \bmod 2$. If V is odd, then the "level 1 case" of Serre's conjecture, proved by Khare [117], asserts that there exist a in $\mathbb{Z}/(\ell - 1)$, an integer $\kappa \leq \ell + 1$, and a representation ρ in $\mathrm{R}_{\kappa;\ell}$, such that we have $V \simeq \chi^a \otimes \rho$. The equality $\det \rho = \chi^{\kappa-1}$ moreover shows that we have $V \simeq \mathrm{i}(V)$ if and only if we have $2a + \kappa - 1 \equiv w \bmod \mathrm{m}(\rho)$.

Suppose that there exists a V in $\mathrm{J}(r)$ of dimension 2 with $\mathrm{i}(V) \not\simeq V$. In this case, we have $r \simeq V \oplus \mathrm{i}(V)$ and $\mathrm{i}(V) \simeq V \otimes \chi^w (\det V)^{-1}$. We are therefore in case $(\mathrm{iii})_1$ if V is odd, and in case $(\mathrm{iii})_2$ otherwise. Note that V is odd if and only if $\mathrm{i}(V)$ is, so that cases $(\mathrm{iii})_1$ and $(\mathrm{iii})_2$ are mutually exclusive.

We may therefore assume that every representation V of $\mathrm{J}(r)$ is either of dimension 1, or of dimension 2 with $\mathrm{i}(V) \simeq V$. Note that in the latter case, V is automatically odd (because $w \equiv 1 \bmod 2$). By Khare's theorem, V is therefore of the form $\chi^a \otimes \rho$ with $\rho \in \mathrm{R}_{\kappa;\ell}$ and $\kappa \leq \ell + 1$, and, moreover, we have $2a + \kappa - 1 \equiv w \bmod \mathrm{m}(\rho)$. We are therefore in case (ii) or (iv), according to whether $\mathrm{J}(r)$ contains a representation of dimension 1 or two representations of dimension 2, respectively. In the remaining case, r is irreducible. □

This result and Table C.3, in turn, make it possible to prove the nonexistence of certain congruences. Let us give a few examples to conclude.

Proposition* 10.4.10. *The representation $\overline{\mathrm{r}}_{j,k;\ell}$ is irreducible (over $\overline{\mathbb{F}}_\ell$) in each of the following cases:*

$(j, k) = (6, 8)$ *and* $\ell = 7, 13, 19,$
$(j, k) = (8, 8)$ *and* $\ell = 7, 11, 19,$
$(j, k) = (12, 6)$ *and* $\ell = 11, 17,$
$(j, k) = (4, 10)$ *and* $\ell = 7, 13, 17.$

Moreover, in each of these cases, we have $\mathrm{m}(\overline{\mathrm{r}}_{j,k;\ell}) = \ell - 1$, *except when* $(j, k) = (6, 8)$ *and* $\ell = 13$, *in which case we merely have* $\mathrm{m}(\overline{\mathrm{r}}_{6,8;13}) \equiv 0 \bmod 6$.

Proof. The representation $\overline{\mathrm{r}}_{j,k;\ell}$, viewed over $\overline{\mathbb{F}}_\ell$, satisfies the assumptions of Proposition 10.4.9 (with $w = j + 2k - 3$), by Remark 10.1.6. Applying this proposition, we must therefore exclude, for each of the triples $(j, k; \ell)$ in the statement above, the possibility of a decomposition of the form (i)–(iv). Table C.3 is sufficiently stocked to allow us several ways to proceed. Let us give a few simple recipes.

Criterion 1. If $\bar{\mathrm{r}}_{j,k;\ell}$ is in one of the cases (i), (ii), (iii)$_1$, *and* (iv), *and if p is a prime with $p \equiv 1 \bmod \ell$, we have, respectively,*

$\tau_{j,k}(p) \equiv 4 \bmod \ell$ *in case* (i),

$\tau_{j,k}(p) \equiv \tau_\kappa(p) + 2 \bmod \ell$ *in case* (ii),

$\tau_{j,k}(p) \equiv 2\tau_\kappa(p) \bmod \ell$ *in case* (iii)$_1$, *and*

$\tau_{j,k}(p) \equiv \tau_{\kappa_1}(p) + \tau_{\kappa_2}(p) \bmod \ell$ *in case* (iv).

Criterion 2. If $\bar{\mathrm{r}}_{j,k;\ell}$ is in case (iii)$_2$, *then $\tau_{j,k}(p) \equiv 0 \bmod \ell$ for every prime p with $p \equiv -1 \bmod \ell$.*

First, suppose $\ell = 7$. We have $\mathrm{S}_\kappa(\mathrm{SL}_2(\mathbb{Z})) = 0$ for every $\kappa \leq \ell + 1$, and $\mathrm{R}_{\kappa,7} = \emptyset$ for $\kappa \leq 8$. It therefore suffices to eliminate the cases (i) and (iii)$_2$. We have $29 \equiv 1 \bmod 7$, and from Table C.3 we extract the congruences

$$\tau_{6,8}(29) \equiv \tau_{4,10}(29) \equiv \tau_{8,8}(29) \equiv 0 \bmod 7 ,$$

which eliminates case (i) by Criterion 1. We eliminate case (iii)$_2$ likewise, using Criterion 2: we have $13 \equiv -1 \bmod 7$, $\tau_{6,8}(13) \equiv 6 \bmod 7$, and $\tau_{4,10}(13) \equiv \tau_{8,8}(13) \equiv 1 \bmod 7$.

Suppose $\ell = 11$. We eliminate cases (i) and (iii)$_2$ as before, by observing first that we have $23 \equiv 1 \bmod 11$ while we have $\tau_{8,8}(23) \equiv \tau_{12,6}(23) \equiv 0 \bmod 11$, then that we have $43 \equiv -1 \bmod 11$ while we have $\tau_{8,8}(43) \equiv \tau_{12,6}(23) \equiv 6 \bmod 11$. The unique integer $\kappa \leq \ell + 1$ such that we have $\mathrm{S}_\kappa(\mathrm{SL}_2(\mathbb{Z})) \neq 0$ is $\kappa = 12$, and we have $\tau_{12}(23) \equiv -1 \bmod 11$. The irreducibility of $\bar{\mathrm{r}}_{j,k;11}$ for $(j,k) = (8,8)$ and $(12,6)$ follows by observing that we have $\tau_{j,k}(23) \not\equiv 1, -2 \bmod 11$ (Criterion 1).

The case $\ell = 13$ is similar because we have $\mathrm{S}_{14}(\mathrm{SL}_2(\mathbb{Z})) = 0$. Criterion 1 applies because we have $53 \equiv 1 \bmod 13$, $\tau_{12}(53) \equiv -3 \bmod 13$, and $\tau_{6,8}(53) \equiv \tau_{4,10}(53) \equiv 3 \bmod 13$. Criterion 2 also applies because we have $103 \equiv -1 \bmod 13$, $\tau_{6,8}(103) \equiv 11 \bmod 13$, and $\tau_{4,10}(103) \equiv 5 \bmod 13$.

In the case $\ell = 17$, we again conclude using Criteria 1 and 2, thanks to the following congruences: $103 \equiv 1 \bmod 17$, $\tau_{12}(103) \equiv 2 \bmod 17$, $\tau_{16}(103) \equiv 6 \bmod 17$, $\tau_{18}(103) \equiv 8 \bmod 17$, and $\tau_{4,10}(103) \equiv \tau_{12,6}(103) \equiv 1 \bmod 17$; $67 \equiv -1 \bmod 17$, $\tau_{4,10}(67) \equiv 8 \bmod 67$, and $\tau_{12,6}(67) \equiv 12 \bmod 67$.

In the case $\ell = 19$, the least prime p with $p \equiv 1 \bmod 19$ is $191 > 113$, which falls outside of Table C.3. On the other hand, Criterion 2 does eliminate case (iii)$_2$ because we have $37 \equiv -1 \bmod 19$, $\tau_{6,8}(37) \equiv 4 \bmod 19$, and $\tau_{8,8}(37) \equiv 8 \bmod 19$. Let us state another criterion. This one is based on the following observation: for $\ell \leq 19$, if $\bar{\mathrm{r}}_{j,k;\ell}$ is not in case (iii)$_2$, then all its Jordan–Hölder factors are defined over $\mathbb{F}_\ell$. This follows from Proposition 10.4.9 and the fact that we have $\dim \mathrm{S}_\kappa(\mathrm{SL}_2(\mathbb{Z})) \leq 1$ for $\kappa \leq \ell + 1 \leq 20$.

Criterion 3. Suppose that we have $\ell \leq 19$ and that $\bar{\mathrm{r}}_{j,k;\ell}$ is not in case (iii)$_2$. *Suppose, moreover, that there exists a prime $p \neq \ell$ such that the polynomial $\mathrm{P}_p(t) := \det(t - \mathrm{r}_{j,k;\ell}(\mathrm{Frob}_p))$ in $\mathbb{Z}[t]$ is irreducible modulo ℓ. Then the representation $\bar{\mathrm{r}}_{j,k;\ell}$ is irreducible.*

As we have already determined the $\tau_{j,k}(p)$ for $p \leq 113$ (Table C.3) and the $\tau_{j,k}(p^2)$ for $p \leq 29$ (Table C.4), formula (9.1.3) shows that we have polynomials $\mathrm{P}_p(t)$ at our disposal for $p \leq 29$. In the case $\ell = 19$, the criterion above holds for $(j,k) = (6,8)$ and $p = 3$, and for $(j,k) = (8,8)$ and $p = 13$.

It remains to justify the last assertion of Proposition 10.4.10 concerning the $\mathrm{m}(\bar{\mathrm{r}}_{j,k;\ell})$. We use the following observation: let $p \neq \ell$ be a prime with $\tau_{j,k}(p) \not\equiv 0 \bmod \ell$; then $\mathrm{m}(\bar{\mathrm{r}}_{j,k;\ell})$ is divisible by the order of p in $(\mathbb{Z}/\ell)^\times$.

For example, the prime 3 generates $(\mathbb{Z}/7)^\times$, and we have the congruences $3\tau_{6,8}(3) \equiv \tau_{4,10}(3) \equiv \tau_{8,8}(3) \equiv 4 \bmod 7$. This shows $\mathrm{m}(\bar{\mathrm{r}}_{6,8;7}) = \mathrm{m}(\bar{\mathrm{r}}_{8,8;7}) = \mathrm{m}(\bar{\mathrm{r}}_{4,10;7}) = 6$. The other cases are similar. For $\ell = 13$, we use that the prime 2 generates $(\mathbb{Z}/\ell)^\times$, and the congruence $\tau_{4,10}(2) \equiv 10 \bmod 13$, so that we have $\mathrm{m}(\bar{\mathrm{r}}_{4,10;13}) = 12$. On the other hand, we have $\tau_{6,8}(2) \equiv 0 \bmod 13$. We show $\mathrm{m}(\bar{\mathrm{r}}_{6,8;13}) \equiv 0 \bmod 6$ using the congruence $\tau_{6,8}(17) \equiv 7 \bmod 13$. □

It is clear that the ad hoc methods used above are rather coarse, and that it is possible to study the potential decompositions of the representations $\bar{\mathrm{r}}_{j,k;\ell}$ for characteristics $\ell > 19$. We postpone this study, as well as the more interesting matter of determining the images of the $\bar{\mathrm{r}}_{j,k;\ell}$, to a later work. To self-congratulate ourselves, we note that the triples $(j,k;\ell)$ that appear in the statement of Proposition 10.4.10 are exactly those that do not occur in the statement of Theorem 10.4.4 when we have $\ell \leq 19$.

Remark 10.4.11. Suppose that $\bar{\mathrm{r}}_{j,k;\ell}$ is irreducible over $\overline{\mathbb{F}}_\ell$ (and therefore, in particular, that we have $\ell > 5$). The following assertions are equivalent:

(i) For every prime p that is not a square modulo ℓ, we have $\tau_{j,k}(p) \equiv 0 \bmod \ell$.
(ii) The integer $\mathrm{m}(\bar{\mathrm{r}}_{j,k;\ell})$ divides $(\ell - 1)/2$.
(iii) The residual representation $\bar{\mathrm{r}}_{j,k;\ell}$ is induced by an irreducible representation of dimension 2, with coefficients in $\mathbb{F}_{\ell^2}$, of the absolute Galois group of $\mathbb{Q}(\sqrt{\ell^*})$ with $\ell^* = (-1)^{(\ell-1)/2}\ell$.

In the cases $(j,k) = (6,8)$ and $\ell = 13$, an examination of Table C.3 shows that the congruence of assertion (i) holds for every prime $p \leq 113$. It is tempting to conjecture that it always holds, in other words, that we have $\mathrm{m}(\bar{\mathrm{r}}_{6,8;13}) = 6$.

Appendix A
The Barnes–Wall Lattice and the Siegel Theta Series of Even Unimodular Lattices of Dimension 16

(Following Martin Kneser [124])

We describe in this appendix the elegant and elementary proof given by M. Kneser of the fact that the Siegel theta series with genus ≤ 3 of the two even unimodular lattices of dimension 16 coincide.

The Barnes-Wall lattice is a remarkable lattice of dimension 16 discovered by Barnes and Wall in 1959 [19]. It appears repeatedly in [68], where it is denoted by Λ_{16} or BW_{16} (Conway and Sloane mischievously mention that it has been rediscovered by many authors). In [19], Barnes and Wall in fact define a sequence of lattices $(\Lambda_{2^n})_{n\in\mathbb{N}-\{0\}}$, where Λ_{2^n} has dimension 2^n; the first three lattices in this sequence are isomorphic to, respectively, I_2, D_4, and E_8. For a simple and elegant introduction to the Barnes–Wall lattices, we recommend [154] and [155]; one can also consult [41]. The lattice U that appears in the reference [124] is an avatar of Λ_{16}. The definition of Λ_{16} we give below suffices for us.

Let I be a Lagrangian of the q-vector space $\mathbb{F}_2\otimes_{\mathbb{Z}}\mathrm{E}_8$ (we therefore have $\mathrm{q}(I)=0$ and $\dim_{\mathbb{F}_2} I = 4$). We denote by Λ_{16} the submodule of $\mathrm{E}_8\oplus\mathrm{E}_8$ consisting of the ordered pairs (x_1, x_2) of elements of E_8 whose reductions $\bar{x}_1$ and $\bar{x}_2$ modulo 2 satisfy $\bar{x}_1+\bar{x}_2\in I$; Λ_{16} can be viewed as an integral lattice (in the quadratic sense) in $\mathbb{Q}\otimes_{\mathbb{Z}}(\mathrm{E}_8\oplus\mathrm{E}_8)$.

Proposition A.1. *The lattice Λ_{16} has the following properties:*

(a) We have $\mathrm{q}(x)\geq 2$ (or, equivalently, $x.x\geq 4$) for every x in $\Lambda_{16}-\{0\}$.
(b) We have $\xi.\xi\in\mathbb{Z}$ for every ξ in the dual lattice $\Lambda_{16}^{\sharp}$, and $\Lambda_{16}^{\sharp}$ endowed with the quadratic form $\xi\mapsto\xi.\xi$ is isomorphic to Λ_{16} (as a $\widetilde{\mathrm{q}}$-module).
(c) The module underlying the qe-module $\mathrm{res}\,\Lambda_{16}$ *is annihilated by* 2, *and the* qe-module $\mathrm{res}\,\Lambda_{16}$ *is isomorphic to the hyperbolic* $\mathbb{F}_2$-q-*vector space* $\mathrm{H}(I)$ *via the canonical embedding of* $\mathbb{F}_2$ *in* $\mathbb{Q}/\mathbb{Z}$.

Proof of Part (a). Let $x=(x_1,x_2)$ be an element of $\Lambda_{16}-\{0\}$. If x_1 and x_2 are nonzero, then we have $\mathrm{q}(x)=\mathrm{q}(x_1)+\mathrm{q}(x_2)\geq 1+1$. If x_i is zero for some i, then x_{3-i} belongs to I (and is nonzero) and $\mathrm{q}(x_{3-i})$ is even. □

G. Chenevier, J. Lannes, *Automorphic Forms and Even Unimodular Lattices*, Ergebnisse der Mathematik und ihrer Grenzgebiete. 3. Folge / A Series of Modern Surveys in Mathematics 69, https://doi.org/10.1007/978-3-319-95891-0

Proof of Part (b). The lattice $\Lambda_{16}^{\sharp}$ is the submodule of $\mathbb{Q} \otimes_{\mathbb{Z}} (\mathrm{E}_8 \oplus \mathrm{E}_8)$ consisting of the elements $\xi = (\xi_1, \xi_2)$ such that $(\xi_1 + \xi_2, \xi_1 - \xi_2)$ belongs to Λ_{16}. Since we have the identity $\mathrm{q}(\xi_1 + \xi_2) + \mathrm{q}(\xi_1 - \xi_2) = 2(\mathrm{q}(\xi_1) + \mathrm{q}(\xi_2))$, this concludes the proof of property (b). □

Proof of Part (c). Let ΔI be the linear subspace of $\mathbb{F}_2 \otimes_{\mathbb{Z}} (\mathrm{E}_8 \oplus \mathrm{E}_8) = (\mathbb{F}_2 \otimes_{\mathbb{Z}} \mathrm{E}_8) \oplus (\mathbb{F}_2 \otimes_{\mathbb{Z}} \mathrm{E}_8)$ that is the diagonal image of I. Note that Λ_{16} is the submodule of $\mathrm{E}_8 \oplus \mathrm{E}_8$ obtained by taking the inverse image of $(\Delta I)^{\perp}$ under the homomorphism $\mathrm{E}_8 \oplus \mathrm{E}_8 \to \mathbb{F}_2 \otimes_{\mathbb{Z}} (\mathrm{E}_8 \oplus \mathrm{E}_8)$. Having made this observation, we see that property (c) is a manifestation of the general phenomenon described below.

Let L be a q-module over $\mathbb{Z}$. Let p be a prime and J a linear subspace of $\mathbb{F}_p \otimes_{\mathbb{Z}} L$ with $\mathrm{q}(J) = 0$. Let M be the submodule of L consisting of the elements whose reduction modulo p is orthogonal to J. Then the abelian group $\mathrm{res}\, M$ is annihilated by p and the qe-module $\mathrm{res}\, M$ is isomorphic to the hyperbolic $\mathbb{F}_p$-q-vector space $\mathrm{H}(J)$ $(\cong \mathrm{H}(J^*) \cong \mathrm{H}(L/M))$, via the canonical embedding of $\mathbb{F}_p$ in $\mathbb{Q}/\mathbb{Z}$. □

Corollary A.2. *We have $\xi.\xi \geq 2$ for every ξ in $\Lambda_{16}^{\sharp} - \{0\}$.*

Let us now explain how to use the properties of the lattice Λ_{16}, following Kneser's strategy, to deduce the equality of theta series

$$\vartheta^{(g)}_{\mathrm{E}_8 \oplus \mathrm{E}_8} = \vartheta^{(g)}_{\mathrm{E}_{16}}$$

for $g \leq 3$ (a result due to Witt for $g \leq 2$ [213]).

This equality can be reformulated in terms of representations of integral quadratic forms by $\mathrm{E}_8 \oplus \mathrm{E}_8$ and E_{16}. Let us explain the terminology. Let L be an even unimodular lattice and G a free, finite-dimensional $\mathbb{Z}$-module endowed with a quadratic form with integral values (in view of what follows, we may assume that these values are nonnegative); a *representation* of G by L is a homomorphism $f\colon G \to L$ with $\mathrm{q}(f(x)) = \mathrm{q}(x)$ for every x in G. We denote by $\mathrm{Rep}(G, L)$ the set of representations of G by L; this set is clearly finite. Here is the reformulation we announced:

Theorem A.3. *Let G be a free, finite-dimensional $\mathbb{Z}$-module endowed with a quadratic form with integral values. If we have $\dim G \leq 3$, then the sets $\mathrm{Rep}(G, \mathrm{E}_8 \oplus \mathrm{E}_8)$ and $\mathrm{Rep}(G, \mathrm{E}_{16})$ have the same cardinality.*

Proof. The key point is the following observation.

Let $\gamma\colon \Lambda_{16}^{\sharp} \to \mathrm{res}\, \Lambda_{16}$ be the passage to the quotient. We denote by $\mathfrak{I}$ the (finite) set of Lagrangians of $\mathrm{res}\, \Lambda_{16}$ and by $\mathfrak{L}$ the set of even unimodular lattices in $\mathbb{Q} \otimes_{\mathbb{Z}} \Lambda_{16}$ containing Λ_{16} (and thus contained in $\Lambda_{16}^{\sharp}$). Recall that the map $\mathfrak{I} \to \mathfrak{L}$ defined by $I \mapsto \gamma^{-1}(I)$ is a bijection (compatible with the inclusion relations).

Let J be a submodule of $\mathrm{res}\, \Lambda_{16}$ with $\mathrm{q}(J) = 0$ and $\dim_{\mathbb{F}_2} J = 3$. Let M be the lattice $\gamma^{-1}(J)$; we have $\mathrm{res}\, M \cong J^{\perp}/J \simeq \mathrm{H}(\mathbb{Z}/2)$ (see Proposition 2.1.1). Let B be the lattice that is the inverse image under the homomorphism $M^{\sharp} \to \mathrm{res}\, M$ of the nonisotropic "line" (nonisotropic in the quadratic sense but isotropic in the bilinear sense); B is an odd unimodular lattice (see the discussion "2-Neighbors, The Point of View of Borcherds" after Proposition 3.1.9). Since B is contained in

$\Lambda_{16}^{\sharp}$, Corollary A.2 implies that we have $x.x \geq 2$ for every x in $B - \{0\}$. Finally, Scholium-Definition 3.3.3.2 shows that B is isomorphic to the lattice Bor_{16} defined there and that the two even unimodular lattices that are the inverse images of the two isotropic "lines" of res M are nonisomorphic (note, incidentally, that this observation also shows that the isomorphism class of M is independent of the choice of J).

The observation above leads to the following lemma. We also use the term representation of G by $\Lambda_{16}^{\sharp}$ for a homomorphism $f\colon G \to \Lambda_{16}^{\sharp}$ with $\mathrm{q}(f(x)) = \mathrm{q}(x)$ for every x in G.

Lemma A.4. *Let f be a representation of G by $\Lambda_{16}^{\sharp}$ with $\dim G \leq 3$. Let $\mathcal{L}(f)$ be the subset of $\mathcal{L}$ consisting of the lattices L containing $f(G)$. Let $\mathcal{L}_1(f)$ and $\mathcal{L}_2(f)$ be the subsets of $\mathcal{L}(f)$ consisting of the lattices L isomorphic to, respectively, $\mathrm{E}_8 \oplus \mathrm{E}_8$ and E_{16}. Then $\mathcal{L}_1(f)$ and $\mathcal{L}_2(f)$ have the same cardinality.*

Proof. Let the subsets $\mathcal{I}(f)$, $\mathcal{I}_1(f)$, and $\mathcal{I}_2(f)$ of $\mathcal{I}$ be the respective inverse images of the subsets $\mathcal{L}(f)$, $\mathcal{L}_1(f)$ and $\mathcal{L}_2(f)$ of $\mathcal{L}$ under the inverse of the bijection $I \mapsto \gamma^{-1}(I)$. Let $\mathfrak{J}$ be the set of the submodules J of res Λ_{16} considered above; let $\mathfrak{J}(f)$ be the subset of $\mathfrak{J}$ consisting of the J with $J \supset (\gamma \circ f)(G)$. Let $\mathcal{K}(f) \subset \mathcal{I}(f) \times \mathfrak{J}(f)$ and $\mathcal{K}_i(f) \subset \mathcal{I}_i(f) \times \mathfrak{J}(f)$ for $i = 1, 2$ be the subsets consisting of the pairs (I, J) with $I \supset J$. Finally, let $\pi_{\mathfrak{J}} : \mathcal{K}(f) \to \mathfrak{J}(f)$ and $\pi_{\mathcal{I}} : \mathcal{K}(f) \to \mathcal{I}(f)$ be the maps $(I, J) \mapsto J$ and $(I, J) \mapsto I$. It is clear that $\pi_{\mathfrak{J}}$ is surjective and that its fibers all have two elements. Likewise, $\pi_{\mathcal{I}}$ is surjective and its fibers all have $2^{(4-\delta(f))} - 1$ elements, where $\delta(f) \leq 3$ is the dimension of the $\mathbb{F}_2$-vector space $(\gamma \circ f)(G)$. In view of the above, the maps $\mathcal{K}_1(f) \to \mathfrak{J}(f)$ and $\mathcal{K}_2(f) \to \mathfrak{J}(f)$ induced by $\pi_{\mathfrak{J}}$ are still surjective. Since we have $|\mathcal{K}_1(f)| + |\mathcal{K}_2(f)| = |\mathcal{K}(f)| = 2|\mathfrak{J}(f)|$ (where $| - |$ denotes the cardinality of a finite set), we see that we have $|\mathcal{K}_1(f)| = |\mathfrak{J}(f)|$ and $|\mathcal{K}_2(f)| = |\mathfrak{J}(f)|$. By definition, we have $\mathcal{K}_1(f) = \pi_{\mathcal{I}}^{-1}(\mathcal{I}_1(f))$ and $\mathcal{K}_2(f) = \pi_{\mathcal{I}}^{-1}(\mathcal{I}_2(f))$, giving the equality $|\mathcal{I}_1(f)| = |\mathcal{I}_2(f)|$. □

Proof of Theorem A.3, Continued.

Let $\mathcal{L}_1$ and $\mathcal{L}_2$ be the subsets of $\mathcal{L}$ consisting of the lattices isomorphic to, respectively, $\mathrm{E}_8 \oplus \mathrm{E}_8$ and E_{16}. Note that we have $\mathcal{L}_1 = \mathcal{L}_1(0)$ and $\mathcal{L}_2 = \mathcal{L}_2(0)$ (where 0 denotes the unique representation of 0 by $\Lambda_{16}^{\sharp}$) and therefore $|\mathcal{L}_1| = |\mathcal{L}_2|$. Denote by $\mathrm{Rep}(G, \Lambda_{16}^{\sharp})$ the (finite) set of representations of G by $\Lambda_{16}^{\sharp}$, set $\mathrm{r}_1(G) = |\mathrm{Rep}(G, \mathrm{E}_8 \oplus \mathrm{E}_8)|$ and $\mathrm{r}_2(G) = |\mathrm{Rep}(G, \mathrm{E}_{16})|$, and consider the subsets $\mathcal{R}_i(G)$ $(i = 1, 2)$ of $\mathcal{L}_i \times \mathrm{Rep}(G, \Lambda_{16}^{\sharp})$ consisting of the pairs (L, f) with $L \supset f(G)$. By projecting onto each of the two factors of the product $\mathcal{L}_i \times \mathrm{Rep}(G, \Lambda_{16}^{\sharp})$, we see that we have

$$|\mathcal{R}_i(G)| = |\mathcal{L}_i|\, \mathrm{r}_i(G) \quad \text{and} \quad |\mathcal{R}_i(G)| = \sum_{f \in \mathrm{Rep}(G, \Lambda_{16}^{\sharp})} |\mathcal{L}_i(f)|\,.$$

Lemma A.4 now implies the equality $\mathrm{r}_1(G) = \mathrm{r}_2(G)$. □

Appendix B
Quadratic Forms and Neighbors in Odd Dimension

In this appendix, we present the "odd-dimensional" counterpart of some of the theory we developed "in even dimension" in Chaps. 2 and 3.

B.1 Basic Concepts in the Theory of Quadratic Forms on a Projective Module of Odd Constant Rank

Let A be a commutative ring with unit. In Chap. 2, we defined a q-module over A to be a projective A-module L of finite type endowed with a quadratic form $\mathrm{q}\colon L \to A$ such that the associated symmetric bilinear form is nondegenerate. If 2 is not invertible in A and L has constant rank, then the nondegeneracy forces this rank to be even (consider a homomorphism from A to a field k of characteristic 2 and note that the symmetric bilinear form associated with the quadratic form on $k \otimes_A L$ is alternating). If L has odd constant rank, then, classically, q is called nondegenerate if the degeneracy of the associated bilinear form is "minimal." We will make this definition more precise further on. Our presentation emphasizes the notion of half-determinant (see, for example, [125]); for a more sophisticated presentation, see [71, Exp. XII].

Let $k \geq 1$ be an integer; we denote by

$$\psi_L^k\colon \Lambda^k L \to \Lambda^{k-1} L \otimes L \quad \text{and} \quad \phi_L^k\colon \Lambda^{k-1} L \otimes L \to \Lambda^k L\,,$$

respectively, the "coproduct" and "product" homomorphisms induced by the Hopf algebra structure of the exterior algebra ΛL (the tensor products and the exterior algebra are over A).

G. Chenevier, J. Lannes, *Automorphic Forms and Even Unimodular Lattices*, Ergebnisse der Mathematik und ihrer Grenzgebiete. 3. Folge / A Series of Modern Surveys in Mathematics 69, https://doi.org/10.1007/978-3-319-95891-0

Lemma B.1.1. *Let L be an A-module endowed with a quadratic form $q\colon L \to A$, and let $b\colon L \times L \to A$ be the associated symmetric bilinear form. The symmetric bilinear form associated with the quadratic form $(\Lambda^{k-1}b \otimes q) \circ \psi_L$ is $k\,\Lambda^k b$.*

Proof. This follows from the fact that the composed homomorphism $\phi_L^k \circ \psi_L^k$ is k times the identity on $\Lambda^k L$. □

Let β be a symmetric bilinear form. We denote the quadratic form $x \mapsto \beta(x,x)$ by $\mathrm{qd}(\beta)$; the associated bilinear form is $2\,\beta$.

The proofs of statements B.1.2–B.1.4 below are immediate.

Proposition-Definition B.1.2 (Odd Exterior Powers of a Quadratic Form). *Let L be an A-module endowed with a quadratic form q and k an odd integer. Set*

$$\Lambda^k q = (\Lambda^{k-1} b \otimes q) \circ \psi_L^k - \frac{k-1}{2}\,\mathrm{qd}\,(\Lambda^k b)\ ,$$

where b is the symmetric bilinear form associated with q ($\Lambda^k q$ is therefore a quadratic form on $\Lambda^k L$).

*The symmetric bilinear form associated with $\Lambda^k q$ is $\Lambda^k b$. The form $\Lambda^k q$ is called the k*th exterior power *of q.*

Proposition B.1.3. *Let L be a projective A-module of finite type; we denote the A-module consisting of the symmetric bilinear forms $L \times L \to A$ (resp. the quadratic forms $L \to A$) by $\mathcal{B}(L)$ (resp. $\mathcal{Q}(L)$). If L has rank 1, then the homomorphism of A-modules $\mathrm{qd}\colon \mathcal{B}(L) \to \mathcal{Q}(L)$ is an isomorphism.*

Proposition-Definition B.1.4. *Let L be a projective A-module of finite type of odd constant rank n, endowed with a quadratic form q; denote the symmetric bilinear form associated with q by b. The two symmetric bilinear forms $\Lambda^n b$ and $\mathrm{qd}^{-1}(\Lambda^n q)$ with which the projective A-module of rank 1 $\Lambda^n L$ is endowed are related by the equality*

$$\Lambda^n b \;=\; 2\,\mathrm{qd}^{-1}(\Lambda^n q)\ .$$

The symmetric bilinear form $\mathrm{qd}^{-1}(\Lambda^n q)$ (or the projective A-module $\Lambda^n L$ endowed with this form) is called the half-determinant *of q (recall that the symmetric bilinear form $\Lambda^n b$ is called the determinant of b; see Sect. 2.1). The half-determinant of q is denoted by $\frac{1}{2}\text{-}\det q$ or $\frac{1}{2}\text{-}\det L$.*

(The terminology and notation are of course justified by the fact that we have $\det L = 2\,(\frac{1}{2}\text{-}\det L)$.)

Example. Let a be an element of A. Recall that $\langle a\rangle$ is the A-module A endowed with the symmetric bilinear form $(x,y) \mapsto a\,xy$; we therefore denote by $\mathrm{qd}(\langle a\rangle)$ the A-module A endowed with the quadratic form $x \mapsto a\,x^2$. We did the necessary to ensure that the half-determinant of $\mathrm{qd}(\langle a\rangle)$ is $\langle a\rangle$.

Proposition B.1.5. *Let P and L be two projective A-modules of finite type of even and odd constant rank, respectively, endowed with a quadratic form. We have a* *canonical isomorphism of projective A-modules of rank* 1 *endowed with a symmetric bilinear form:*

$$\tfrac{1}{2}\text{-}\det(P \oplus L) \;\cong\; \det P \otimes \tfrac{1}{2}\text{-}\det L\,.$$

Proof. Let q_P, b_P, q_L, b_L, respectively, be the quadratic and bilinear forms on P and L; let m be the rank of P and n that of L. We must show that we have $\Lambda^{m+n}(q_P \oplus q_L) = \Lambda^m b_P \otimes \Lambda^n q_L$. By using the naturality of exterior powers of bilinear forms (resp. odd exterior powers of quadratic forms), we reduce to the "universal case." In this case, the ring A is a polynomial ring with coefficients in $\mathbb{Z}$, in $m(m+1)/2 + n(n+1)/2$ variables, and 2 is not a zero divisor. Since we have $\det(P \oplus L) \cong \det P \otimes \det L$, this suffices to conclude. □

Definition B.1.6. Let L be a projective A-module of finite type and odd constant rank endowed with a quadratic form q. We say that q is *minimally degenerate* if the symmetric bilinear form $\frac{1}{2}\text{-}\det q$ is nondegenerate. For short, we call a projective A-module of finite type and odd constant rank endowed with a minimally degenerate quadratic form a q-i-*module* over A.

Example-Remark. Let A be a ring, P a q-module over A of even constant rank, and u an element of $A^\times$. Proposition B.1.5 shows that the orthogonal sum $P \oplus \mathrm{qd}(\langle u\rangle)$ is a q-i-module over A. Proposition 1.2 of [71, Exp. XII] says that, locally for the étale topology, every q-i-module is of this type with, moreover, P hyperbolic.

Classical Groups (Continued)

Let L be a q-i-module over A of rank n; as in Chap. 2, an endomorphism α of the A-module underlying L is called *orthogonal* if it preserves the quadratic form. The naturality of exterior powers of quadratic forms shows that $\Lambda^n\alpha$ is an orthogonal endomorphism of the b-module $\frac{1}{2}\text{-}\det L$ and therefore that the endomorphisms $\Lambda^n\alpha$ and α are automorphisms. The orthogonal endomorphisms form a group for the composition, which is called the *orthogonal group* of L and denoted by $\mathrm{O}(L)$. The functor $R \mapsto \mathrm{O}(R \otimes_A L)$, defined on the category of commutative A-algebras and with values in the category of groups, is an A-group scheme that we denote by O_L. In view of the above, the composition $\mathrm{O}_L \to \mathrm{GL}_L \overset{\det}{\to} \mathbb{G}_\mathrm{m}$ induces a homomorphism $\det\colon \mathrm{O}_L \to \mu_2$ (note that $\mathrm{O}_{\frac{1}{2}\text{-}\det L}$ can be identified with μ_2). We denote the kernel of the induced homomorphism by SO_L (again, the group $\mathrm{SO}_L(A)$ is simply denoted by $\mathrm{SO}(L)$).

If L has rank 1, then the group O_L can again be identified with μ_2, so that we may not expect O_L to be smooth over A in all generality. It does, however, hold for SO_L.

Proposition B.1.7. *For every* q-i-*module L over a commutative ring A with unit, the A-group scheme* SO_L *is smooth over A.*

Proof. Since the property we wish to verify is local for the étale topology, we may assume, by [71, Exp. XII, Proposition 1.2], that we have $L = \mathrm{H}(A^n) \oplus \mathrm{qd}(\langle u\rangle)$ with

u in $A^\times$. Since the two q-modules $\mathrm{H}(A^n)$ and $\langle u\rangle \otimes \mathrm{H}(A^n)$ are isomorphic, we may, moreover, assume $u = 1$ and therefore, ultimately, $A = \mathbb{Z}$ and $L = \mathrm{H}(\mathbb{Z}^n) \oplus \mathrm{A}_1$; the group SO_L is then the group denoted by $\mathrm{SO}_{n+1,n}$ in Sect. 8.1. The fact that $\mathrm{SO}_{n+1,n}$ is smooth over $\mathbb{Z}$ is well known. Below, we show that this property can be seen as a consequence of Proposition 2.1.5; this (very indirect!) proof is in the spirit of Sect. B.2.

Set $P = \mathrm{H}(\mathbb{Z}^n) \oplus \mathrm{H}(\mathbb{Z})$. Let (e_1, e_2) be the canonical basis of the factor $\mathrm{H}(\mathbb{Z})$, and set $e = e_1 + e_2$ and $f = e_1 - e_2$; we therefore have $\mathrm{q}(e) = 1$, $\mathrm{q}(f) = -1$, and $e.f = 0$. We easily see that L can be identified (with its quadratic form) with the orthogonal complement of f.

Let $\mathcal{C}$ be the affine quadric with equation $\mathrm{q} = -1$ (q being here the quadratic form P is endowed with); this $\mathbb{Z}$-scheme is smooth over $\mathbb{Z}$. Let $\mathrm{O}_{P,f}$ be the subgroup of the group O_P (which is smooth over $\mathbb{Z}$ by Proposition 2.1.5) defined as the stabilizer of f for the obvious action of O_P on $\mathcal{C}$; "differential calculus" shows that the group $\mathrm{O}_{P,f}$ is smooth over $\mathbb{Z}$. The equality $L = f^\perp$ provides a homomorphism of group schemes $\omega\colon \mathrm{O}_{P,f} \to \mathrm{O}_L$. Statements B.1.8 and B.1.9 below concern this homomorphism; the second implies Proposition B.1.7. The proof of the first is left to the reader. □

Proposition B.1.8. *The diagram*

$$\begin{array}{ccc} \mathrm{O}_{P,f} & \xrightarrow{\ \omega\ } & \mathrm{O}_L \\ {\scriptstyle\widetilde{\det}}\big\downarrow & & {\scriptstyle\det}\big\downarrow \\ \mathbb{Z}/2 & \longrightarrow & \mu_2\,, \end{array}$$

in which the arrow denoted by $\widetilde{\det}$ *is the restriction of the homomorphism* $\widetilde{\det}\colon \mathrm{O}_P \to \mathbb{Z}/2$, *is commutative.*

Denote the kernel of the homomorphism $\widetilde{\det}\colon \mathrm{O}_{P,f} \to \mathbb{Z}/2$ by $\mathrm{SO}_{P,f}$. Proposition B.1.8 shows that the homomorphism ω induces a homomorphism from $\mathrm{SO}_{P,f}$ to SO_L, which we denote by ω_{S}.

Proposition B.1.9. *The homomorphism of* $\mathbb{Z}$*-group schemes*

$$\omega_{\mathrm{S}}\colon \mathrm{SO}_{P,f} \longrightarrow \mathrm{SO}_L$$

is an isomorphism.

Proof. Let A be a commutative ring with unit.

We first show the injectivity of the homomorphism $\mathrm{SO}_{P,f}(A) \to \mathrm{SO}_L(A)$. For this, consider the commutative diagram of $\mathbb{Z}$-group schemes

$$\begin{array}{ccccc} \mathrm{O}_{\mathrm{H}(\mathbb{Z}),f} & \longrightarrow & \mathrm{O}_{P,f} & \xrightarrow{\ \widetilde{\det}\ } & \mathbb{Z}/2 \\ {\scriptstyle\omega}\big\downarrow & & {\scriptstyle\omega}\big\downarrow & & \big\downarrow \\ \mathrm{O}_{\mathrm{A}_1} & \longrightarrow & \mathrm{O}_L & \xrightarrow{\ \det\ } & \mu_2\,, \end{array}$$

in which the two horizontal arrows on the left are the obvious inclusions. Since an element of $\mathrm{O}_P(A)$ whose restriction to $A \otimes_{\mathbb{Z}} \mathrm{H}(\mathbb{Z}^n)$ is the identity can be identified with an element of $\mathrm{O}_{\mathrm{H}(\mathbb{Z})}(A)$, it follows that the kernels of $\mathrm{O}_{P,f}(A) \to \mathrm{O}_L(A)$ and $\mathrm{O}_{\mathrm{H}(\mathbb{Z}),f}(A) \to \mathrm{O}_{\mathrm{A}_1}(A)$ coincide. We conclude by noting that the composition of the two upper horizontal arrows is an isomorphism.

Next, we show that the homomorphism $\mathrm{SO}_{P,f}(A) \to \mathrm{SO}_L(A)$ is surjective. Let α be an element of $\mathrm{O}_L(A)$. Write $A \otimes_{\mathbb{Z}} L = \mathrm{H}(A^n) \oplus Ae$ and set $M = \alpha(\mathrm{H}(A^n))$. Let $M^{\perp}$ be the orthogonal complement of M in $A \otimes_{\mathbb{Z}} P$, viewed as a submodule of $A \otimes_{\mathbb{Z}} P$; this module $M^{\perp}$ has the following properties:

- The A-module $M^{\perp}$ is projective of rank 2.
- The restriction of the quadratic form of $A \otimes_{\mathbb{Z}} P$ to $M^{\perp}$ is nondegenerate.
- The q-module $A \otimes_{\mathbb{Z}} P$ is isomorphic to the orthogonal sum $M \oplus M^{\perp}$.
- The discriminant $\Delta(M^{\perp})$ is trivial (see [71, Exp. XII, 1.11]).
- The elements $\alpha(e)$ and f of $A \otimes_{\mathbb{Z}} P$ belong to $M^{\perp}$.

The following proposition, whose proof is left to the reader, shows that the q-module $M^{\perp}$ is isomorphic to $\mathrm{H}(A)$ (note that we have $\mathrm{q}(\alpha(e)) = 1$).

Proposition B.1.10. *Let N be a* q*-module over A of rank* 2*. The following properties are equivalent:*

(i) *The module N is isomorphic to the hyperbolic* q*-module* $\mathrm{H}(A)$.
(ii) *The discriminant $\Delta(N)$ is trivial, and there exists an e in N with* $\mathrm{q}(e) = 1$.

Let γ be an automorphism of the q-module $A \otimes_{\mathbb{Z}} P = \mathrm{H}(A^n) \oplus \mathrm{H}(A)$ induced by the isomorphism $\alpha \colon \mathrm{H}(A^n) \to M$, an isomorphism $\beta \colon \mathrm{H}(A) \to M^{\perp}$, and the isomorphism $M \oplus M^{\perp} \to P$. Since the orthogonal group $\mathrm{O}_{\mathrm{H}(\mathbb{Z})}(A)$ acts transitively on the set of x with $\mathrm{q}(x) = 1$, we may assume $\beta(f) = f$ and therefore $\gamma \in \mathrm{O}_{P,f}(A)$. Since the homomorphism $\widetilde{\det} \colon \mathrm{O}_{\mathrm{H}(\mathbb{Z}),f} \to \mathbb{Z}/2$ is an isomorphism, we may, moreover, assume $\gamma \in \mathrm{SO}_{P,f}(A)$. By construction, $\alpha^{-1} \circ \omega_{\mathrm{S}}(\gamma)$ is then an element of $\mathrm{SO}_L(A)$ that is the identity on $\mathrm{H}(A^n)$; but such an element is the identity. □

Remark. The homomorphism $\mathrm{O}_{\mathrm{H}(\mathbb{Z}),f} \to \mathrm{O}_{P,f}$ can be identified with a homomorphism $\mathbb{Z}/2 \to \mathrm{O}_{P,f}$ that is a "central" section of the homomorphism $\widetilde{\det} \colon \mathrm{O}_{P,f} \to \mathbb{Z}/2$; it follows that the group $\mathrm{O}_{P,f}$ is canonically isomorphic to the product $\mathrm{SO}_{P,f} \times \mathbb{Z}/2$. Likewise, the group O_L is canonically isomorphic to the product $\mathrm{SO}_L \times \mu_2$, and the homomorphism $\omega \colon \mathrm{O}_{P,f} \to \mathrm{O}_L$ can be identified with the product of the isomorphism ω_S and the canonical homomorphism $\mathbb{Z}/2 \to \mu_2$.

We conclude this subsection with the following statement; the reader will have no trouble deciphering the notation.

Scholium B.1.11. *Let A be a commutative ring with unit. Let P be a* q*-module over A of even constant rank, endowed with an element e with* $\mathrm{q}(e) \in A^{\times}$*. Let L be the* *orthogonal complement of* e *in* P. *Then*

- L *is a* q-i-*module over* A;
- *the* A-*group* SO_L *can be identified with the* A-*group* $\mathrm{SO}_{P,\mathrm{e}}$;
- *the* A-*groups* $\mathrm{O}_{P,\mathrm{e}}$ *and* O_L *can be identified, respectively, with the products* $\mathrm{SO}_L \times \mathbb{Z}/2$ *and* $\mathrm{SO}_L \times \mu_2$;
- *the canonical homomorphism* $\mathrm{O}_{P,\mathrm{e}} \to \mathrm{O}_L$ *can be identified with the homomorphism induced by the homomorphism* $\mathbb{Z}/2 \to \mu_2$.

B.2 On the q-i-Modules over $\mathbb{Z}$

A q-i-module over $\mathbb{Z}$ is nothing but a free $\mathbb{Z}$-module L, of odd finite dimension, endowed with a symmetric bilinear form that is even (that is, such that $x.x$ is even for every x in L) with $|\det L| = 2$. In fact, part (a) of the following proposition shows that the properties of the bilinear form induce that of the dimension.

Proposition B.2.1 (Classification of the $\mathbb{Z}$-q-i-**modules).**

(a) *Let* L *be a free* $\mathbb{Z}$-*module of finite dimension, endowed with an even symmetric bilinear form, with* $|\det L| = 2$. *Then there exists an element* ϵ *of* $\{\pm 1\}$, *uniquely determined in terms of* L, *such that the signature of* L *satisfies the congruence* $\tau(L) \equiv \epsilon \pmod 8$ (*so that the dimension of* L *is odd*).

(b) *Let* P *be a free* $\mathbb{Z}$-*module of finite dimension, endowed with an even symmetric bilinear form, with* $|\det P| = 1$, *and an element* e *with* $|\mathrm{e}.\mathrm{e}| = 2$. *Let* L *be the orthogonal complement of* e, *endowed with the even symmetric bilinear form that is the restriction of that of* P. *Then* L *is a free* $\mathbb{Z}$-*module of finite dimension* (*since* P/L *is free of dimension* 1) *with* $\det L = (\mathrm{e}.\mathrm{e}) \det P$.

(c) *Let* (n, ϵ) *be an element of* $\mathbb{N} \times \{\pm 1\}$ *with* n *odd; let* $\mathrm{QI}_{n,\epsilon}$ *be the set of isomorphism classes of* $\mathbb{Z}$-q-i-*modules* L *with* $\dim L = n$ *and* $\tau(L) \equiv \epsilon \pmod 8$. *Let* (n, ϵ) *be an element of* $\mathbb{N} \times \{\pm 1\}$ *with* n *even; let* $\mathrm{QR}_{n,\epsilon}$ *be the set of isomorphism classes of* $\mathbb{Z}$-q-*modules endowed with an element* e *with* $\mathrm{e}.\mathrm{e} = 2\epsilon$. *Then the map*

$$(P, \mathrm{e}) \;\mapsto\; \mathrm{e}^{\perp}$$

induces a bijection from $\mathrm{QR}_{n,\epsilon}$ *to* $\mathrm{QI}_{n-1,-\epsilon}$.

Proof of Part (a). We view L as a $\widetilde{\mathrm{q}}$-module. The group underlying its residue $\operatorname{res} L$ can be identified with $\mathbb{Z}/2$, and its quadratic linking form satisfies $\mathrm{q}(\bar{1}) = \epsilon/4$ with $\epsilon = \pm 1$. We are therefore led to introduce the orthogonal sum of $\widetilde{\mathrm{q}}$-modules $L \oplus \langle -\epsilon \rangle \otimes \mathrm{A}_1$ (note that $\mathrm{A}_1 = \mathrm{Q}(\mathbf{A}_1)$ is nothing but the $\mathbb{Z}$-module $\mathbb{Z}$ endowed with the quadratic form $x \mapsto x^2$) whose residue can be identified with the orthogonal sum of qe-modules $\operatorname{res} L \oplus \langle -1 \rangle \otimes \operatorname{res} L$. This qe-module has a unique Lagrangian, namely the diagonal; we denote by P the q-module corresponding to this Lagrangian via Proposition 2.1.1. Since the signature of P is divisible by 8, we indeed have the congruence $\tau(L) \equiv \epsilon \pmod 8$. □

Proof of Part (b). This follows, for example, from assertion (c) of the following statement, which will be useful to have in this book. The verification of this statement is left to the reader. □

Proposition B.2.2. *Let A be a Dedekind domain. Let L be a* q*-module over A, and let M be a submodule and $M^\perp$ its orthogonal complement. Suppose that M is a direct factor in L (in other words, that the quotient L/M has no torsion; note that $M^\perp$ is a direct factor in L for every M).*

(a) We have $(M^\perp)^\perp = M$.

Suppose, moreover, that the restriction of the bilinear form of L to M is nonsingular, in other words, that the induced homomorphism $M \to \mathrm{Hom}_A(M, A)$ is injective; M is therefore a $\widetilde{\mathrm{q}}$-module over A. Then the following hold:

(b) The restriction of this bilinear form to $M^\perp$ is also nonsingular.
(c) The canonical homomorphism of A-modules $M \oplus M^\perp \to L$ is injective, and we have an exact sequence of A-modules

$$0 \longrightarrow M \oplus M^\perp \longrightarrow L \longrightarrow \operatorname{res} M \longrightarrow 0\,,$$

in which the homomorphism $L \to \operatorname{res} M$ is the composition of the isomorphism $L \to \mathrm{Hom}_A(L, A)$ induced by the bilinear form of L and the canonical homomorphisms $\mathrm{Hom}_A(L, A) \to \mathrm{Hom}_A(M, A)$ and $\mathrm{Hom}_A(M, A) \to \operatorname{res} M$.
*(d) The isomorphisms of A-modules $\operatorname{res} M \cong L/(M \oplus M^\perp)$ and $\operatorname{res} M^\perp \cong L/(M \oplus M^\perp)$ (note that M and $M^\perp$ play symmetric roles) induce an isomorphism of A-modules $\varphi\colon \operatorname{res} M \to \operatorname{res} M^\perp$ such that we have $\mathrm{q}(\varphi(\xi)) = -\mathrm{q}(\xi)$ for every ξ in $\operatorname{res} M$. In other words, we have a canonical isomorphism of A-*qe*-modules*

$$\operatorname{res} M^\perp \cong \langle -1 \rangle \otimes \operatorname{res} M\ .$$

(e) Via Proposition 2.1.1 *and the isomorphism of* qe*-modules $\operatorname{res}(M \oplus M^\perp) \cong \operatorname{res} M \oplus \operatorname{res} M^\perp$, the* q*-module L corresponds to the the graph of φ, which is the Lagrangian of $\operatorname{res} M \oplus \operatorname{res} M^\perp$.*

Remarks.

- There exists a "bilinear version" of Proposition B.2.2 in which the q-modules (resp. $\widetilde{\mathrm{q}}$-modules, resp. qe-modules) are replaced by b-modules (resp. $\widetilde{\mathrm{b}}$-modules, resp. e-modules).
- There also exists, in a particular case, a "bilinear-quadratic version" of Proposition B.2.2. Let us be more precise. Let L be an odd b-module over $\mathbb{Z}$; let u be a Wu vector of L that we assume to be indivisible and nonisotropic. By construction, the orthogonal complement $u^\perp$ of u in L is an even $\widetilde{\mathrm{b}}$-module, in other words, a $\widetilde{\mathrm{q}}$-module. Proposition B.2.2 says that as an e-module, $\operatorname{res} u^\perp$ is isomorphic to $\mathbb{Z}/u.u$ endowed with the linking form defined by $\bar{1}.\bar{1} = -1/u.u$. For its part, the quadratic linking form is defined by

$$\mathrm{q}(\bar{1}) \;=\; \frac{1}{2}\left(1 - \frac{1}{u.u}\right).$$

Proof of Part (c) of Proposition B.2.1.

Denote by $\omega_{n,\epsilon}\colon \mathrm{QR}_{n,\epsilon} \to \mathrm{QI}_{n-1,-\epsilon}$ the map induced by $(P, e) \mapsto e^{\perp}$. Let (n, ϵ) be an element of $\mathbb{N} \times \{\pm 1\}$ with n odd. By construction, the q-module P that appears in the proof of part (a) of Proposition B.2.1 is endowed with an element e with $e.e = -2\epsilon$. The uniqueness of the Lagrangian invoked in this construction shows that the map $L \mapsto (P, e)$ induces a map $\mathrm{QI}_{n,\epsilon} \to \mathrm{QR}_{n+1,-\epsilon}$, which we denote by $\pi_{n,\epsilon}$. The two maps $\omega_{n,\epsilon}$ and $\pi_{n-1,-\epsilon}$ are each other's inverses. □

Scholium B.2.3. *Let L be a* q-i-*module over $\mathbb{Z}$, and let ϵ be the element of $\{\pm 1\}$ defined by $\tau(L) \equiv \epsilon \pmod 8$. Then the* qe-*module* $\mathrm{res}\, L$ *is isomorphic to $\mathbb{Z}/2$ endowed with the quadratic linking form defined by* $\mathrm{q}(\bar{1}) = \epsilon/4$.

Proposition B.2.1 and Scholium 2.2.1 also lead to the following statement.

Scholium B.2.4. *Let L_1 and L_2 be two* q-i-*modules over $\mathbb{Z}$. The following two conditions are equivalent:*

(i) *The two* b-*vector spaces $\mathbb{Q} \otimes_{\mathbb{Z}} L_1$ and $\mathbb{Q} \otimes_{\mathbb{Z}} L_2$ over $\mathbb{Q}$ are isomorphic.*
(ii) *The two* b-*vector spaces $\mathbb{R} \otimes_{\mathbb{Z}} L_1$ and $\mathbb{R} \otimes_{\mathbb{Z}} L_2$ over $\mathbb{R}$ are isomorphic.*

Genus of a q-i-Module over $\mathbb{Z}$

The method used to prove Proposition B.2.1 provides the following statement (the heading refers to part (b)).

Proposition B.2.5. *Let L be a* q-i-*module over $\mathbb{Z}$ of dimension $2n+1$ and determinant 2ϵ with $\epsilon = \pm 1$; let p be a prime.*

(a) The q-i-*vector space $\mathbb{F}_p \otimes_{\mathbb{Z}} L$ is isomorphic to* $\mathrm{H}(\mathbb{F}_p^n) \oplus \mathrm{qd}(\langle (-1)^n \epsilon \rangle)$.
(b) The q-i-*module $\mathbb{Z}_p \otimes_{\mathbb{Z}} L$ is isomorphic to* $\mathrm{H}(\mathbb{Z}_p^n) \oplus \mathrm{qd}(\langle (-1)^n \epsilon \rangle)$.

Remark. For $p \neq 2$, a q-i-vector space over $\mathbb{F}_p$ (resp. a q-i-module over $\mathbb{Z}_p$) is nothing but a b-vector space (resp. b-module) of odd dimension.

The Positive Definite Case

Let L be a q-i-module over $\mathbb{Z}$ with $\mathbb{R} \otimes_{\mathbb{Z}} L$ positive definite. In view of the above, such an L is nothing but an (integral) even lattice of determinant 2. From here on, we abandon the term "positive definite q-i-module over $\mathbb{Z}$" (which is far from being classical!) for the term "even lattice of determinant 2."

Let L be an even lattice of determinant 2; part (a) of Proposition B.2.1 shows that we have the congruence $\dim L \equiv \pm 1 \pmod 8$.

We first study the case $\dim L \equiv -1 \pmod 8$. Part (c) of Proposition B.2.1 specializes as follows.

Proposition B.2.6. *For an integer $n > 0$ with $n \equiv -1 \pmod 8$, let X_n be the set of isomorphism classes of even lattices L with $\dim L = n$ and $\det L = 2$. For an even integer $n > 0$ with $n \equiv 0 \pmod 8$, let $\mathrm{X}_n^{\mathrm{A}_1}$ be the set of isomorphism classes of even unimodular lattices P of dimension n endowed with an element e with $e.e = 2$ (in other words, a root). Then the map*

$$(P; e) \;\mapsto\; e^{\perp}$$

induces a bijection from $\mathrm{X}_n^{\mathrm{A}_1}$ to X_{n-1}.

(We justify the notation $\mathrm{X}_n^{\mathrm{A}_1}$ as follows: giving a root of P is equivalent to giving a representation of A_1 by P. It is the counterpart of the notation $\mathrm{X}_n^{\mathrm{E}_7}$ introduced further on.)

Examples

Determination of X_7. Since X_8 has only one element, namely the class of E_8, and the Weyl group of $\mathbf{E}_8$ (which coincides with the orthogonal group of E_8) acts transitively on the set of roots, the set X_7 has only one element, namely the class of the orthogonal complement of a root in E_8. We use the notation E_7 for this orthogonal complement, which agrees with that adopted in Chap. 2: $\mathrm{E}_7 = \mathrm{Q}(\mathbf{E}_7)$.

Determination of X_{15}. Since the Weyl group of $\mathbf{D}_{16}$ (which coincides with the orthogonal group of E_{16}) acts transitively on the set of roots and the same holds for the orthogonal group of $\mathrm{E}_8 \oplus \mathrm{E}_8$, the set X_{15} has two elements:

- the class of the orthogonal complement of a root in E_{16}, say E_{15},
- the class of $\mathrm{E}_7 \oplus \mathrm{E}_8$.

Determination of X_{23}. In view of Proposition B.2.6 and Theorem 2.3.17, this determination is a consequence of the following observation.

Let L be an even unimodular lattice of dimension 24 with roots. Let e_1 and e_2 be two roots of L, and let R_i, for $i = 1, 2$, be the irreducible component of the root system $\mathrm{R}(L)$ to which e_i belongs. Then the following two conditions are equivalent:

(i) The two root systems R_1 and R_2 are isomorphic.
(ii) There exists an element α of the orthogonal group $\mathrm{O}(L)$ such that we have $\alpha(e_1) = e_2$.

The implication (ii)⇒(i) is obvious. The implication (i)⇒(ii) can be verified on a case-by-case basis. We give a few details on this verification below.

Consider the decomposition of the root system $\mathrm{R}(L)$ into irreducible components

$$\mathrm{R}(L) \;\simeq\; \coprod_{R \in \mathcal{R}} \mathrm{m}(R)\, R\,,$$

where $\mathcal{R}$ denotes the set of isomorphism classes of irreducible root systems of type ADE, $\mathrm{m}\colon \mathcal{R} \to \mathbb{N}$ is a map with finite support, and $\mathrm{m}(R)\,R$ is the disjoint union of

$\mathrm{m}(R)$ copies of R. Recall that we denote by $\mathrm{A}(\mathrm{R}(L))$ the orthogonal group of the lattice $\mathrm{Q}(\mathrm{R}(L))$ and that the Weyl group $\mathrm{W}(\mathrm{R}(L))$ is a normal subgroup of $\mathrm{A}(\mathrm{R}(L))$. Also recall that we denote by $\mathrm{G}(\mathrm{R}(L))$ the quotient group $\mathrm{A}(\mathrm{R}(L))/\mathrm{W}(\mathrm{R}(L))$ and that we have a group isomorphism (canonical in an obvious way)

$$\mathrm{G}(\mathrm{R}(L)) \cong \prod_{R\in\mathcal{R}} \left(\mathrm{G}(R)^{\mathrm{m}(R)} \rtimes \mathfrak{S}_{\mathrm{m}(R)}\right) = \left(\prod_{R\in\mathcal{R}} \mathrm{G}(R)^{\mathrm{m}(R)}\right) \rtimes \left(\prod_{R\in\mathcal{R}} \mathfrak{S}_{\mathrm{m}(R)}\right),$$

where $\mathfrak{S}_{\mathrm{m}(R)}$ is the symmetric group of order $\mathrm{m}(R)$ that has an obvious action on the group $\mathrm{G}(R)^{\mathrm{m}(R)}$. Finally, we paraphrase part (b) of Scholium 2.3.15: the orthogonal group $\mathrm{O}(L)$ is the subgroup of $\mathrm{A}(\mathrm{R}(L))$ defined as the inverse image under the homomorphism $\mathrm{A}(\mathrm{R}(L)) \to \mathrm{G}(\mathrm{R}(L))$ of the subgroup that stabilizes the Lagrangian $L/\mathrm{Q}(\mathrm{R}(L))$ of the qe-module res $\mathrm{Q}(\mathrm{R}(L))$. In [81], V.A. Erokhin specifies this stabilizer, which he denotes by $\mathrm{H}(L)$, case by case for the 23 isomorphism classes of even unimodular lattices of dimension 24 with roots. It is clear that we have a canonical exact sequence of groups

$$1 \to \mathrm{H}_1(L) \to \mathrm{H}(L) \to \mathrm{H}_2(L) \to 1\,,$$

where $\mathrm{H}_1(L)$ can be identified with a subgroup of the product $\prod_{R\in\mathcal{R}} \mathrm{G}(R)^{\mathrm{m}(R)}$ and $\mathrm{H}_2(L)$ with a subgroup of the product $\prod_{R\in\mathcal{R}} \mathfrak{S}_{\mathrm{m}(R)}$, namely the image of the restriction to $\mathrm{H}(L)$ of the canonical homomorphism $\mathrm{G}(\mathrm{R}(L)) \to \prod_{R\in\mathcal{R}} \mathfrak{S}_{\mathrm{m}(R)}$ (although this is not stated explicitly, the groups $\mathrm{H}_1(L)$ and $\mathrm{H}_2(L)$ are the groups G_1 and G_2 whose cardinalities appear in columns 5 and 6 of [68, Chap. 16, Table 16.1]).

The implication (i)⇒(ii) we are considering follows from the fact that for every R in $\mathcal{R}$, the image of $\mathrm{H}_2(L)$ in $\mathfrak{S}_{\mathrm{m}(R)}$ is a transitive subgroup, which we easily verify by going through the list in [81] (clearly, the only R we need to consider are those for which we have $\mathrm{m}(R) \geq 2$!). By way of example, let us give these transitive subgroups for the first five root systems of the list in question:

(1) $\mathrm{R}(L) = 24\mathbf{A}_1$
The image of $\mathrm{H}_2(L)$ in $\mathfrak{S}_{24}$ is the Mathieu group M_{24}.

(2) $\mathrm{R}(L) = 12\mathbf{A}_2$
The image of $\mathrm{H}_2(L)$ in $\mathfrak{S}_{12}$ is the Mathieu group M_{12}.

(3) $\mathrm{R}(L) = 8\mathbf{A}_3$
There exists a bijection from the set $\{1, 2, \ldots, 8\}$ to the set underlying $\mathbb{F}_2^3$, viewed as an affine space of dimension 3 over $\mathbb{F}_2$, which induces an isomorphism from $\mathrm{H}_2(L)$ to the subgroup of affine transformations.

(4) $\mathrm{R}(L) = 6\mathbf{A}_4$
There exists a bijection from the set $\{1, 2, \ldots, 6\}$ to the set underlying $\mathbf{P}^1(\mathbb{F}_5)$, which induces an isomorphism from $\mathrm{H}_2(L)$ to the set of projective transformations. (Take note: the list of generators for $\mathrm{H}(L)$ given in [81] is incomplete.)
More precisely, in this case the exact sequence $1 \to \mathrm{H}_1(L) \to \mathrm{H}(L) \to \mathrm{H}_2(L) \to 1$ is isomorphic to the exact sequence $1 \to \mathbb{F}_5^\times/\{\pm 1\} \to \mathrm{GL}_2(\mathbb{F}_5)/\{\pm\mathrm{Id}\} \to \mathrm{PGL}_2(\mathbb{F}_5) \to 1$.

(5) $\mathrm{R}(L) = 4\mathbf{A}_6$

The image of $\mathrm{H}_2(L)$ in $\mathfrak{S}_4$ is the alternating subgroup $\mathfrak{A}_4$.

So, in the end, we see that the set X_{23} can be identified with the subset of the product $\mathrm{X}_{24} \times \mathcal{R}$ consisting of the pairs (x, r) such that r is the isomorphism class of an irreducible component of the root system $\mathrm{R}(x)$ (the abuse of notation is venial). The cardinality of X_{23} is obtained by considering the second column of [68, Chap. 16, Table 16.1] (our Table 1.1):

$$\begin{aligned}|\mathrm{X}_{23}| &= 32\\ &= 1+2+1+1+1+2+2+2+1+1+3+1+2\\ &\quad +1+1+2+1+2+1+1+1+1+1+0\end{aligned}$$

(this cardinality should clearly be compared with the number of representations in Table C.7).

Let us now study the case $\dim L \equiv 1 \pmod 8$.

Proposition B.2.7. *For an integer $n > 0$ with $n \equiv 1 \pmod 8$, let X_n be the set of isomorphism classes of even lattices L with* $\dim L = n$ *and* $\det L = 2$. *For an even integer $n > 0$ with $n \equiv 0 \pmod 8$, let $\mathrm{X}_n^{\mathrm{E}_7}$ be the set of isomorphism classes of even unimodular lattices P of dimension n endowed with a homomorphism $f\colon \mathrm{E}_7 \to P$ with* $\mathrm{q}(f(x)) = \mathrm{q}(x)$ *for every x in E_7 (in other words, with a representation of* E_7 *by P). Then the map*

$$(P; f) \;\mapsto\; (f(\mathrm{E}_7))^{\perp}$$

induces a bijection from $\mathrm{X}_n^{\mathrm{E}_7}$ to X_{n-7}.

Proof. This is a variant of the proof of part (c) of Proposition B.2.1. This time, we consider the orthogonal sum $L \oplus \mathrm{E}_7$. By Scholium B.2.3, the residue of this $\widetilde{\mathrm{q}}$-module is again isomorphic to $\operatorname{res} L \oplus \langle -1 \rangle \otimes \operatorname{res} L$. We conclude mutatis mutandis. □

Notation-Remark. Let G and L be two integral lattices; recall that a representation of G by L is a homomorphism of $\mathbb{Z}$-module $f\colon G \to L$ with $f(x).f(y) = x.y$ for all x and y in G and that the (finite) set of these f is denoted by $\mathrm{Rep}(G, L)$. We denote by $\overline{\mathrm{Rep}}(G, L)$ the quotient $\mathrm{Rep}(G, L)/\mathrm{O}(G)$ of the right action of the orthogonal group $\mathrm{O}(G)$ on $\mathrm{Rep}(G, L)$ (this action is free); $\overline{\mathrm{Rep}}(G, L)$ can be viewed as the set of submodules of L isomorphic to G as integral lattices.

Examples

Determination of X_{17}. We check on a case-by-case basis that the only irreducible root systems of type ADE that contain $\mathbf{E}_7$ are $\mathbf{E}_7$ and $\mathbf{E}_8$ (use, for example, [39, Chap. VI, Sect. 1, Proposition 24]). A further examination then shows that the only root systems appearing in the classification of Niemeier that contain $\mathbf{E}_7$ are $\mathbf{R}_2 = \mathbf{D}_{16} \coprod \mathbf{E}_8$, $\mathbf{R}_3 = \mathbf{E}_8 \coprod \mathbf{E}_8 \coprod \mathbf{E}_8$, $\mathbf{R}_6 = \mathbf{A}_{17} \coprod \mathbf{E}_7$, and $\mathbf{R}_7 = \mathbf{D}_{10} \coprod \mathbf{E}_7 \coprod \mathbf{E}_7$. Let P_i, for $i \in \{2, 3, 6, 7\}$, be "the" even unimodular lattice $\mathrm{R}(\mathrm{P}_i) \approx \mathbf{R}_i$ (we apologize for deviating here from the notation introduced at the beginning of Sect. 10.2); we easily verify that the orthogonal group $\mathrm{O}(\mathrm{P}_i)$ acts transitively on the set $\overline{\mathrm{Rep}}(\mathrm{E}_7, \mathrm{P}_i)$ in all four cases. The verification is immediate for the first three; for the fourth, we use the observation made at the end of the fifth illustration we gave of Proposition 2.3.13. Proposition B.2.7 therefore shows that the set X_{17} has four elements. Let us be inordinately precise. We choose a sublattice of P_i isomorphic to E_7 and denote by L_i the orthogonal complement of this sublattice; we have $\mathrm{X}_{17} = \{[\mathrm{L}_2], [\mathrm{L}_3], [\mathrm{L}_6], [\mathrm{L}_7]\}$ (where $[L]$ denotes the isomorphism class of an even lattice L of dimension 17 and determinant 2).

It is not difficult to give a definition ab initio of the lattices L_i:

- $\mathrm{L}_2 = \mathrm{E}_{16} \oplus \mathrm{A}_1$,
- $\mathrm{L}_3 = \mathrm{E}_8 \oplus \mathrm{E}_8 \oplus \mathrm{A}_1$,
- $\mathrm{L}_6 = \mathrm{A}_{17}^+$ (the qe-module res A_{17} is isomorphic to $\mathbb{Z}/18$ with $\mathrm{q}(\bar{k}) = 17k^2/36$, A_{17}^+ is the even lattice corresponding to the submodule generated by $\bar{6}$ via part (b) of Proposition 2.1.1),
- $\mathrm{L}_7 = (\mathrm{D}_{10} \oplus \mathrm{E}_7)^+$ (the qe-module $\mathrm{res}(\mathrm{D}_{10} \oplus \mathrm{E}_7)$ contains two isotropic nontrivial submodules, $(\mathrm{D}_{10} \oplus \mathrm{E}_7)^+$ is the even lattice corresponding to either one of these submodules via part (b) of Proposition 2.1.1).

Determination of X_9. Proposition B.2.7 shows that X_9 has a single element, namely the class of the lattice $\mathrm{E}_8 \oplus \mathrm{A}_1$.

On the Determination of X_{25}. The isomorphism $\mathrm{X}_{25} \cong \mathrm{X}_{32}^{\mathrm{E}_7}$ of Proposition B.2.7 cannot be used to determine X_{25} because X_{32} has not been determined yet. On the other hand, part (c) of Proposition B.2.1 can be applied; it says that X_{25} is in bijection with the set of isomorphism classes of pairs $(P; e)$ with P a q-module over $\mathbb{Z}$ of dimension 26 with signature 24 and e an element of P with $e.e = -2$. But all P of this type are isomorphic, by Theorem 2.2.7. As a representative of this class, we can choose the q-module $\mathrm{II}_{25,1}$ (notation of [68], [29]), which can be viewed as the lattice of the q-vector space $\mathbb{Q}^{26}$ endowed with the quadratic form $\frac{1}{2}(\sum_{i=1}^{25} x_i^2 - x_{26}^2)$, generated by the submodule of $\mathbb{Z}^{26}$ consisting of the $(x_1, x_2, \ldots, x_{26})$ with $\sum_{i=1}^{26} x_i$ even and the vector $\frac{1}{2}(1, 1, \ldots, 1)$. Let Y be the set of elements e of $\mathrm{II}_{25,1}$ with $e.e = -2$; in [29], Borcherds describes an algorithm to determine the set $\mathrm{O}(\mathrm{II}_{25,1})\backslash\mathrm{Y}$, which, by the above, is in bijection with X_{25} (note that $|\mathrm{X}_{25}| = 121$).

B.3 The Theory of p-Neighbors for q-i-Modules over $\mathbb{Z}$

Here is the counterpart (at least for the Dedekind ring $\mathbb{Z}$) of part (a) of Proposition 3.1.1.

Proposition B.3.1. *Let V be a* q-*vector space over* $\mathbb{Q}$. *Let L_1 and L_2 be two integral lattices (in the quadratic sense) in V of index 2 in their duals (in particular, the lattices L_1 and L_2 are two* q-i-*modules and $L_1 \cap L_2$ is a* $\widetilde{\mathrm{q}}$-*module over* $\mathbb{Z}$).

Set $I_1 = L_1/(L_1 \cap L_2)$, $I_2 = L_2/(L_1 \cap L_2)$, *and* $R = (L_1^\sharp \cap L_2^\sharp)/(L_1 \cap L_2)$.

(*a*) *We have* $L_1 \cap L_2^\sharp = L_1 \cap L_2$ *and* $L_2 \cap L_1^\sharp = L_1 \cap L_2$.

(*b*) *The two canonical homomorphisms* $R \to L_i^\sharp/L_i$, *for* $i = 1, 2$, *are isomorphisms.*

(*c*) *The three inclusions of* L_1, L_2, *and* $L_1^\sharp \cap L_2^\sharp$ *in* $(L_1 \cap L_2)^\sharp$ *induce a canonical isomorphism of abelian groups*

$$I_1 \oplus I_2 \oplus R \cong \operatorname{res}(L_1 \cap L_2)$$

(that allows the identification of the source and target).

(*d*) *The pairing* $I_1 \times I_2 \to \mathbb{Q}/\mathbb{Z}$ *induced by the linking form of the residue* $\operatorname{res}(L_1 \cap L_2)$ *is nondegenerate. For this form, the two submodules* $I_1 \oplus I_2$ *and* R *are orthogonal and canonically isomorphic, as* qe-*modules, to* $\mathrm{H}(I_1)$ *and* $\operatorname{res} L_1$, *respectively, so that the* qe-*module* $\operatorname{res}(L_1 \cap L_2)$ *is canonically isomorphic to the orthogonal sum* $\mathrm{H}(I_1) \oplus \operatorname{res} L_1$.

Proof. We verify part (a) and the isomorphism of qe-modules $R \cong \operatorname{res} L_1$; the proof of the rest of the statement is left to the reader.

Let $\mathcal{L}(V)$ be the set of integral lattices of V ordered by inclusion; we see that an integral lattice in V of index 2 in its dual is a maximal element of $\mathcal{L}(V)$ (in fact, all maximal elements are of this type). Part (a) follows from this observation. Consider the lattice $(L_1 \cap L_2^\sharp) + L_2$. It belongs to $\mathcal{L}(V)$ and contains L_2; we therefore have the equality $(L_1 \cap L_2^\sharp) + L_2 = L_2$, which implies $L_1 \cap L_2^\sharp = L_1 \cap L_2$.

The submodule L_1 of $(L_1 \cap L_2)^\sharp$ corresponds, via part (b) of Proposition 2.1.1, to the isotropic submodule I_1 of $\operatorname{res}(L_1 \cap L_2)$. Part (c) of this same proposition shows that we have $\operatorname{res} L_1 \cong I_1^\perp/I_1$. Since $I_1^\perp/I_1$ can be identified with R as a qe-module, this suffices to conclude. □

The verification of the following statement is immediate.

Proposition-Definition B.3.2. *Let* V *be a* q-*vector space over* $\mathbb{Q}$; *let* L_1 *and* L_2 *be two integral lattices (in the quadratic sense) in* V *of index* 2 *in their duals.*

Let p *be a prime. The following conditions are equivalent:*

(i) *The intersection* $L_1 \cap L_2$ *has index* p *in* L_1.

(ii) *The intersection* $L_1 \cap L_2$ *has index* p *in* L_2.

If these conditions are satisfied, the lattices L_1 *and* L_2 *are called* p-neighbors (*or* L_2 *is called a* p-neighbor *of* L_1). *In this case, the quotients* $L_1/(L_1 \cap L_2)$ *and* $L_2/(L_1 \cap L_2)$ *are the only nontrivial isotropic submodules of* $\operatorname{res}(L_1 \cap L_2)$.

We now fix a q-i-module L over $\mathbb{Z}$ and analyze the set of p-neighbors of L in $\mathbb{Q} \otimes_{\mathbb{Z}} L$; in this context, a p-neighbor of L is an integral lattice L' in $\mathbb{Q} \otimes_{\mathbb{Z}} L$ with L' of index 2 in $L'^\sharp$ and $L \cap L'$ of index p in L. Let $M = L \cap L'$. The analysis in question proceeds as in Chap. 3 and gives the following results:

- The lattice pL' is contained in M.
- The image of the composition $pL' \subset M \subset L \to L/pL$ is an isotropic line c in L/pL endowed with its structure of q-i-vector space over $\mathbb{Z}/p$.

- The lattice M is the inverse image of $c^{\perp}$ under the homomorphism $L \to L/pL$; here, $c^{\perp}$ denotes the linear subspace of L/pL orthogonal to the line c.
- The lattice L' is the inverse image, under the homomorphism $M^{\sharp} \to \operatorname{res} M$, of the unique nontrivial submodule distinct from L/M that is isotropic for the quadratic linking form.

Conversely, we have the following result.

Proposition B.3.3. *Let c be an isotropic line in L/pL, and let M be the submodule of L defined as the inverse image of $c^{\perp}$ under the homomorphism $L \to L/pL$. Then:*

(a) The qe-*module* $\operatorname{res} M$ *is isomorphic to* $\mathrm{H}(\mathbb{Z}/p) \oplus \operatorname{res} L$.
(b) The inverse image, under the homomorphism $M^{\sharp} \to \operatorname{res} M$, of the unique nontrivial submodule distinct from L/M that is isotropic for the quadratic linking form is a p-neighbor L' of L with $L \cap L' = M$.

Proof of Part (a). This is very similar to the proof of Proposition 3.1.4.

Let u be an element of L whose class modulo p generates the line c; since this line is isotropic, we have $\mathrm{q}(u) \equiv 0 \bmod p$. Let b be the symmetric bilinear form associated with the quadratic form of L/pL. If p is not 2, then b is nondegenerate, so that there exists an element v of L with $u.v \equiv 1 \bmod p$. This is also the case for $p = 2$. To see this, it suffices to show that the class of u modulo 2 does not belong to $\ker b$. Since L is a q-i-module over $\mathbb{Z}$, the quotient $L/2L$ is naturally a q-i-vector space over $\mathbb{Z}/2$; part (a) of Proposition B.2.5 shows that $\ker b$ is of dimension 1 and that the restriction of the quadratic form to $\ker b$ is nontrivial (in fact, this is a general phenomenon for q-i-vector spaces over a field of characteristic 2; see, for example, the remark following Definition B.1.6).

We see that v and u/p belong to $M^{\sharp}$; set $w = u/p - (\mathrm{q}(u)/p)v$. In $\mathbb{Q}/\mathbb{Z}$, the equalities $\mathrm{q}(v) = 0$, $\mathrm{q}(w) = 0$, and $v.w = 1/p$ hold. Let H be the submodule of $\operatorname{res} M$ generated by the classes of v and w (or those of v and u/p); the previous observation shows that the restriction of the linking form to H is nondegenerate, that H is a $\mathbb{Z}/p$-vector space of dimension 2 with a basis consisting of the classes of v and w, that H is isomorphic to $\mathrm{H}(\mathbb{Z}/p)$ as a qe-module, and finally that the qe-module $\operatorname{res} M$ is isomorphic to the orthogonal sum $H \oplus H^{\perp}$. Let I be the submodule L/M of $\operatorname{res} M$ (the submodule generated by v); I is isotropic, and the qe-module $I^{\perp}/I$ can be identified with $H^{\perp}$. Part (c) of Proposition 2.1.1 therefore shows that $H^{\perp}$ can be identified with $\operatorname{res} L$. □

Ultimately, we see that Proposition 3.1.5 on the q-modules over $\mathbb{Z}$ remains true, word for word, for the q-i-modules over $\mathbb{Z}$. The statement below only involves p-neighbors for p prime; the reader will have no trouble verifying that it extends to d-neighbors for every $d \geq 1$. Let us make this more precise. Let L be a q-i-module over $\mathbb{Z}$. We denote the associated projective quadric by C_L; it is again smooth over $\mathbb{Z}$ (this is, by the way, the elegant criterion chosen by Deligne in [71, Exp. XII] to characterize the quadratic forms he qualifies as nondegenerate). Let c be a point of $\mathrm{C}_L(\mathbb{Z}/p)$; we denote by $\mathrm{vois}_p(L;c)$ the lattice L' in $\mathbb{Q} \otimes_{\mathbb{Z}} L$ associated with c by the process described in part (b) of Proposition B.3.3, and by $\mathrm{Vois}_p(L)$ the set of p-neighbors of L in $\mathbb{Q} \otimes_{\mathbb{Z}} L$.

Proposition B.3.4. *The map*

$$\mathrm{C}_L(\mathbb{Z}/p) \to \mathrm{Vois}_p(L)\ , \quad c \mapsto \mathrm{vois}_p(L; c)$$

is a bijection.

B.4 The Theory of p-Neighbors for Even Lattices of Determinant 2

Let E be one of the two lattices A_1 and E_7 (so E is even of determinant 2); in this subsection, E is fixed unless mentioned otherwise.

Let $n > 0$ be an integer with $n \equiv -\dim \mathrm{E} \pmod 8$. Denote by U_n the $\mathbb{Q}$-vector space $\mathbb{Q}^n$ endowed with the quadratic form

$$(x_1, x_2, \ldots, x_n) \ \mapsto \ \frac{1}{2} \sum_{i=1}^{n-1} x_i^2 \, + \, x_n^2 \ .$$

Set $\mathrm{V}_n = \mathrm{U}_n \oplus (\mathbb{Q} \otimes_{\mathbb{Z}} \mathrm{E})$ (V_n is isomorphic, as a q-vector space, to $\mathbb{Q}^{n+\dim \mathrm{E}}$ endowed with the quadratic form $(x_1, x_2, \ldots, x_{n+\dim \mathrm{E}}) \mapsto \frac{1}{2} \sum_i x_i^2$).

Let $\mathcal{X}(\mathrm{U}_n)$ (resp. $\mathcal{X}(\mathrm{V}_n)$) be the set of even lattices of determinant 2 in U_n (resp. of even unimodular lattices in V_n). We therefore have $\mathrm{X}_n = \mathrm{O}(\mathrm{U}_n)\backslash\mathcal{X}(\mathrm{U}_n) = \mathrm{SO}(\mathrm{U}_n)\backslash\mathcal{X}(\mathrm{U}_n)$ (resp. $\mathrm{X}_{n+\dim E} = \mathrm{O}(\mathrm{V}_n)\backslash\mathcal{X}(\mathrm{V}_n)$). Let $\mathcal{X}(\mathrm{V}_n; \mathrm{E})$ be the subset of $\mathcal{X}(\mathrm{V}_n)$ consisting of the P containing E, and let $\mathrm{O}(\mathrm{V}_n; \mathrm{E})$ be the subgroup of $\mathrm{O}(\mathrm{V}_n)$ consisting of the elements that are the identity on E. With this notation, we can paraphrase Propositions B.2.6 and B.2.7 as follows.

Proposition B.4.1. *The map*

$$\mathcal{X}(\mathrm{V}_n; \mathrm{E}) \to \mathcal{X}(\mathrm{U}_n)\ , \quad P \mapsto P \cap \mathrm{U}_n$$

is an equivariant bijection with respect to the group isomorphism

$$\mathrm{O}(\mathrm{V}_n; \mathrm{E}) \xrightarrow{\cong} \mathrm{O}(\mathrm{U}_n)\ .$$

The specialization of Sect. B.3 to the positive definite case, in turn, leads to the following statement.

Proposition B.4.2. *Let P_1 and P_2 be two even unimodular lattices in V_n containing* E. *Let L_1 and L_2, respectively, be the two even lattices in U_n of determinant* 2, *namely $P_1 \cap \mathrm{U}_n$ and $P_2 \cap \mathrm{U}_n$. Let p be a prime. The following two conditions are equivalent:*

(i) *P_1 and P_2 are p-neighbors.*
(ii) *L_1 and L_2 are p-neighbors.*

Proof of (i)⇒ (ii). By its very definition, for $i = 1, 2$, the canonical homomorphism $L_i/(L_1 \cap L_2) \to P_i/(P_1 \cap P_2)$ is injective. By Proposition-Definition B.3.2, we have the following alternative: either $L_1 \cap L_2$ is of index p in L_1 and L_2, or $L_1 = L_2$. But the map $P \mapsto P \cap \mathrm{U}_n$ is injective. □

Proof of (ii)⇒ (i). Consider the following lattice N in V_n:

$$N \ := \ (L_1 \oplus \mathrm{E}) \cap (L_2 \oplus \mathrm{E}) \ = \ (L_1 \cap L_2) \oplus \mathrm{E}\,.$$

By part (d) of Proposition B.3.1 and part (d) of Proposition B.2.2, the qe-module $\operatorname{res} N$ is canonically isomorphic to the orthogonal sum

$$\mathrm{H}(L_1/(L_1 \cap L_2)) \ \oplus \ \langle -1 \rangle \otimes \operatorname{res} \mathrm{E} \ \oplus \ \operatorname{res} \mathrm{E}\,,$$

and we see that $P_1 \cap P_2$ is the submodule of $N^\sharp$ corresponding, via part (b) of Proposition 2.1.1, to the "diagonal" of the factor $\langle -1 \rangle \otimes \operatorname{res} \mathrm{E} \oplus \operatorname{res} \mathrm{E}$. Part (c) of the same proposition shows that we have $\operatorname{res}(P_1 \cap P_2) \cong \mathrm{H}(L_1/(L_1 \cap L_2))$. We conclude by applying part (a) of Proposition 3.1.1. □

As above, let $n > 0$ be an integer with $n \equiv -\dim \mathrm{E} \pmod 8$; let p be a prime. The Hecke operators $\mathrm{T}_p \colon \mathbb{Z}[\mathrm{X}_n] \to \mathbb{Z}[\mathrm{X}_n]$ are defined as in the case $n \equiv 0 \pmod 8$:

$$\mathrm{T}_p\,[L] \ := \ \sum_{L' \in \mathrm{Vois}_p(L)} [L']$$

for every even lattice L of determinant 2 and dimension n. Let L and L' be two even lattices of determinant 2 and dimension n; again as in the case $n \equiv 0 \pmod 8$, the integer $\mathrm{N}_p(L, L')$ is defined as the $[L']$-coordinate of $\mathrm{T}_p\,[L]$.

Part (a) of Proposition B.2.5 shows that all quadrics $\mathrm{C}_L(\mathbb{Z}/p)$ have the same cardinality, which we again denote by $\mathrm{c}_n(p)$; this time, we have

$$\mathrm{c}_n(p) = \sum_{m=0}^{n-2} p^m\,.$$

Proposition 3.2.2 still holds, word for word, giving the following statement.

Proposition B.4.3. *Let p be a prime. We have*

$$\sum_{y \in \mathrm{X}_n} \mathrm{N}_p(x, y) \ = \ \mathrm{c}_n(p)$$

for every x in X_n.

To state the analog of Proposition 3.1.10, we need to introduce some additional notation.

Denote by $\mathrm{B}_n(p)$ the set of isomorphism classes of $\widetilde{\mathrm{q}}$-modules M over $\mathbb{Z}$ with $\dim M = n$, $\mathbb{R} \otimes_{\mathbb{Z}} M > 0$, and $\operatorname{res} M \simeq \mathrm{H}(\mathbb{Z}/p) \oplus \langle -1 \rangle \otimes \operatorname{res} \mathrm{E}$. Note that every automorphism of the qe-module $\mathrm{H}(\mathbb{Z}/p) \oplus \langle -1 \rangle \otimes \operatorname{res} \mathrm{E}$ is the identity on the factor

$\langle -1\rangle \otimes \mathrm{res}\,\mathrm{E}$, so that the qe-module is canonically endowed with a direct factor $\langle -1\rangle \otimes \mathrm{res}\,\mathrm{E}$.

Denote by $\widetilde{\mathrm{B}}_n(p)$ the set of isomorphism classes of pairs $(M;\omega)$ with M as before and ω a bijection from the set of nontrivial isotropic submodules of $\mathrm{res}\,M$ to the set $\{1,2\}$. By definition, $\widetilde{\mathrm{B}}_n(p)$ is endowed with a left action of the symmetric group $\mathfrak{S}_2$ and the quotient set $\mathfrak{S}_2\backslash\widetilde{\mathrm{B}}_n(p)$ can be identified with $\mathrm{B}_n(p)$.

Let $(M;\omega)$ be as above. Denote by $\mathrm{d}_i(M;\omega)$, for $i=1,2$, the inverse image of $\omega^{-1}(i)$ under the surjection $M^\sharp \to \mathrm{res}\,M$; $\mathrm{d}_1(M;\omega)$ and $\mathrm{d}_2(M;\omega)$ are two even lattices of determinant 2 (and dimension n) that are p-neighbors in $\mathbb{Q}\otimes_{\mathbb{Z}} M$. By passing to isomorphism classes, we obtain two maps from $\widetilde{\mathrm{B}}_n(p)$ to X_n that we also denote by d_1 and d_2.

We can now state the analog of Proposition 3.1.10.

Proposition B.4.4. *Let p be a prime, and let x_1 and x_2 be two elements of X_n. We have*

$$\mathrm{N}_p(x_1,x_2) \;=\; \sum_{\beta\,\in\,\mathrm{d}_1^{-1}(x_1)\cap\mathrm{d}_2^{-1}(x_2)} \frac{|\mathrm{O}(x_1)|}{|\mathrm{O}(\beta)|}$$

with $|\mathrm{O}(\beta)| = |\mathrm{Aut}(M;\omega)|$, where $(M;\omega)$ represents β.

Let us now reformulate, at least in a special case, the statement above using the even unimodular lattices associated with even lattices of determinant 2.

Let L be an even lattice of determinant 2 and dimension n. As before, we set $P=(L\oplus \mathrm{E})^+$. By construction, P is canonically endowed with a representation $\mathrm{i}\colon \mathrm{E}\to P$. Denote by $\rho(L)$ (resp. $\bar\rho(L)$) the cardinality of the orbit of i (resp. $\mathrm{i}(\mathrm{E})$) under the action of the group $\mathrm{O}(P)$. As already mentioned (at least in the case $\mathrm{E}=\mathrm{E}_7$), we have $\rho(L)=\bar\rho(L)\,|\mathrm{O}(\mathrm{E})|$. We therefore have two functions, ρ and $\bar\rho$, from X_n to $\mathbb{N}-\{0\}$ (it will, in fact, be easier to work with ρ in the case $\mathrm{E}=\mathrm{A}_1$ and with $\bar\rho$ in the case $\mathrm{E}=\mathrm{E}_7$).

Let M be a $\tilde{\mathrm{q}}$-module over $\mathbb{Z}$, with $\dim M=n$, $\mathbb{R}\otimes_{\mathbb{Z}} M>0$, and $\mathrm{res}\,M\simeq \mathrm{H}(\mathbb{Z}/p)\oplus\langle -1\rangle\otimes\mathrm{res}\,\mathrm{E}$. Set $R=(M\oplus\mathrm{E})^+$. Let us decipher this. It follows from the observation made above that the qe-module $\mathrm{res}(M\oplus\mathrm{E})$ is canonically endowed with a direct factor $\langle -1\rangle\otimes\mathrm{res}\,\mathrm{E}\oplus\mathrm{res}\,\mathrm{E}$; R is the lattice corresponding, via part (b) of Proposition 2.1.1, to the "diagonal" of the direct factor, and part (c) of the same proposition shows that we have $\mathrm{res}\,R\simeq\mathrm{H}(\mathbb{Z}/p)$. Again, R is canonically endowed with a representation $\mathrm{i}\colon\mathrm{E}\to R$, and we again denote by $\rho(M)$ (resp. $\bar\rho(M)$) the cardinality of the orbit of i (resp. $\mathrm{i}(\mathrm{E})$) under the action of the group $\mathrm{O}(R)$. We therefore also have two functions ρ and $\bar\rho$ from $\mathrm{B}_n(p)$ to $\mathbb{N}-\{0\}$ with $\rho(-)=\bar\rho(-)\,|\mathrm{O}(\mathrm{E})|$.

The above shows that we have natural maps from X_n (resp. $\mathrm{B}_n(p)$) to $\mathrm{X}_{n+\dim\mathrm{E}}$ (resp. $\mathrm{B}_{n+\dim\mathrm{E}}(p)$); we denote these maps by π.

Finally, let $\mathrm{B}^0_n(p)$ be the subset of $\mathrm{B}_n(p)$ consisting of the isomorphism classes of the M as above such that the two even lattices L_1 and L_2 of determinant 2 associated with M are nonisomorphic. Denote by e the map from $\mathrm{B}^0_n(p)$ to the set of even elements of X_n that sends $[M]$ to the set $\{[L_1],[L_2]\}$.

Proposition B.4.5. *Let x_1 and x_2 be two distinct elements of* X_n. *We have*

$$\mathrm{N}_p(x_1, x_2) = \sum_{b \,\in\, \mathrm{e}^{-1}(\{x_1,x_2\})} \frac{|\mathrm{O}(\pi(x_1))|}{|\mathrm{O}(\pi(b))|} \frac{\rho(b)}{\rho(x_1)} = \sum_{b \,\in\, \mathrm{e}^{-1}(\{x_1,x_2\})} \frac{|\mathrm{O}(\pi(x_1))|}{|\mathrm{O}(\pi(b))|} \frac{\bar{\rho}(b)}{\bar{\rho}(x_1)} .$$

Proof. We see that the assumption $x_1 \neq x_2$ allows us to transform Proposition B.4.4 into Proposition B.4.5. Represent the element β of $\widetilde{\mathrm{B}}_n(p)$ in the former by a pair $(M;\omega)$, and set $L_1 = \mathrm{d}_1(M;\omega)$, $P_1 = (L_1 \oplus \mathrm{E})^+$, and $R = (M \oplus \mathrm{E})^+$. We have

$$\frac{|\mathrm{O}(L_1)|}{|\mathrm{O}(M;\omega)|} = [\mathrm{O}(M) : \mathrm{O}(M;\omega)]\, \frac{|\mathrm{O}(L_1)|}{|\mathrm{O}(M)|} = [\mathrm{O}(M) : \mathrm{O}(M;\omega)]\, \frac{|\mathrm{O}(P_1)|}{|\mathrm{O}(R)|} \frac{\rho(M)}{\rho(L_1)} .$$

The assumption $x_1 \neq x_2$ implies $\mathrm{O}(M;\omega) = \mathrm{O}(M)$; in fact, if this equality does not hold, there exists an element of $\mathrm{O}(M)$ that interchanges the two nontrivial isotropic submodules of res M, in which case $x_1 = x_2$. □

2-Neighbors, the Point of View of Borcherds (Continued)

Propositions B.2.6 and B.2.7 say that the map $\pi\colon \mathrm{X}_n \to \mathrm{X}_{n+\dim \mathrm{E}}$ considered above is the composition of a bijection $\mathrm{X}_n \cong \mathrm{X}^{\mathrm{E}}_{n+\dim \mathrm{E}}$ and a "forgetful map" $\mathrm{X}^{\mathrm{E}}_{n+\dim \mathrm{E}} \to \mathrm{X}_{n+\dim \mathrm{E}}$. Likewise, the map $\pi\colon \mathrm{B}_n(p) \to \mathrm{B}_{n+\dim \mathrm{E}}(p)$ is the composition of a bijection $\mathrm{B}_n(p) \cong \mathrm{B}^{\mathrm{E}}_{n+\dim \mathrm{E}}(p)$ and a forgetful map $\mathrm{B}^{\mathrm{E}}_{n+\dim \mathrm{E}}(p) \to \mathrm{B}_{n+\dim \mathrm{E}}(p)$; the definition of the set $\mathrm{B}^{\mathrm{E}}_{n+\dim \mathrm{E}}(p)$ holds no surprises: $\mathrm{B}^{\mathrm{E}}_{n+\dim \mathrm{E}}(p)$ is the set of isomorphism classes of pairs $(R; f)$, where R and f denote, respectively, an even lattice with $\operatorname{res} R \simeq \mathrm{H}(\mathbb{Z}/p)$ and a representation of E in R.

We saw in Chap. 3 (following Borcherds) that the set $\mathrm{B}_{n+\dim \mathrm{E}}(2)$ can be identified with the set, denoted by $\mathrm{B}_{n+\dim \mathrm{E}}$, of isomorphism classes of odd unimodular lattices of dimension $n + \dim \mathrm{E}$. Likewise, the set $\mathrm{B}^{\mathrm{E}}_{n+\dim \mathrm{E}}(2)$ can be identified with the set of isomorphism classes of odd unimodular lattices of dimension $n+\dim \mathrm{E}$ endowed with a representation of E, a set that we will denote by $\mathrm{B}^{\mathrm{E}}_{n+\dim \mathrm{E}}$. We also have an identification in the sense of Borcherds, $\mathrm{B}_n(2) \cong \mathrm{B}_n$, where B_n denotes the set of odd lattices of determinant 2: with an even lattice M with $\operatorname{res} M \simeq \mathrm{H}(\mathbb{Z}/2) \oplus \langle -1 \rangle \otimes \operatorname{res} \mathrm{E}$, we associate the lattice corresponding, via Proposition 2.1.1 (bilinear version), to the unique nontrivial submodule of $\operatorname{res} M$ that is isotropic in the bilinear sense but not in the quadratic sense.

Let $(Q; f)$ be an odd unimodular lattice of dimension $n + \dim \mathrm{E}$ endowed with a representation $f\colon \mathrm{E} \to Q$. The reader can verify that the correspondence $Q \mapsto M$ can be described in the following two ways:

- Consider the submodule R of index 2 of Q consisting of the x with $x.x$ even. We have $f(\mathrm{E}) \subset R$; M is the orthogonal complement of $f(\mathrm{E})$ in R.
- Consider the orthogonal complement Λ of $f(\mathrm{E})$ in Q. We see that Λ is odd (if this were not the case, we would have $R = \Lambda \oplus f(\mathrm{E})$, an equality that may not hold because the bilinear residue of the right-hand side is not hyperbolic). The module M is the submodule of Λ consisting of the x with $x.x$ even.

Finally, the map $\rho\colon \mathrm{B}^{\mathrm{E}}_{n+\dim \mathrm{E}} \to \mathbb{N} - \{0\}$ (resp. $\bar{\rho}\colon \mathrm{B}^{\mathrm{E}}_{n+\dim \mathrm{E}} \to \mathbb{N} - \{0\}$) induced by the identification $\mathrm{B}^{\mathrm{E}}_{n+\dim \mathrm{E}}(2) \cong \mathrm{B}^{\mathrm{E}}_{n+\dim \mathrm{E}}$ sends the class of the pair $(Q; f)$ to the cardinality of the orbit of f (resp. $\mathrm{f}(L)$) under the action of the group $\mathrm{O}(Q)$.

In Chap. 3, we noted, following Nebe and Venkov, that if an odd unimodular lattice L of dimension divisible by 8 represents 1, then the two even unimodular lattices that are 2-neighbors of L are isomorphic (Corollary 3.1.16). We conclude this section with the following technical statement that can be seen as the counterpart of this observation in the present context.

Proposition B.4.6. *Let $n > 0$ be an integer with $n \equiv -\dim \mathrm{E} \pmod 8$. Let L_1 and L_2 be two even lattices of determinant 2 in a* q*-vector space U of dimension n; set $V = U \oplus (\mathbb{Q} \otimes_{\mathbb{Z}} \mathrm{E})$. Suppose that L_1 and L_2 are 2-neighbors in U and nonisomorphic. Let P_i, for $i = 1, 2$, be the even unimodular lattice in V containing* E *associated with L_i, and let Q be the odd unimodular lattice in V containing* E *whose even 2-neighbors are P_1 and P_2.*

(a) In the case $\mathrm{E} = \mathrm{E}_7$, the lattice Q does not represent 1.
(b) In the case $\mathrm{E} = \mathrm{A}_1$, we have the following alternative:

(b.1) The lattice Q does not represent 1.
(b.2) The lattices P_1 and P_2 are isomorphic, and there exists an isomorphism of lattices $\phi\colon Q \cong \mathrm{I}_2 \oplus Q'$ with I_2 containing $\phi(\mathrm{A}_1)$ and Q' not representing 1.

Proof. Denote by Q^1 the submodule of Q generated by the elements x with $x.x = 1$ and by Q' its orthogonal complement. We have a decomposition $Q = Q^1 \oplus Q'$ as an orthogonal sum; the lattice Q^1 is isomorphic to I_m (with $m = \dim_{\mathbb{Z}} Q^1$), and the set $\mathrm{R}(Q)$ of roots of Q is isomorphic to the disjoint union $\mathrm{R}(\mathrm{I}_m) \coprod \mathrm{R}(Q') = \mathrm{R}(\mathrm{D}_m) \coprod \mathrm{R}(Q')$.

The Case $\mathrm{E} = \mathrm{E}_7$. By the above, E_7 is contained in Q', which implies $m = 0$. Indeed, if there exists an e in Q with $e.e = 1$, then the orthogonal symmetry s_e is the identity on E_7 and interchanges P_1 and P_2 (see Corollary 3.1.16), so that L_1 and L_2 are isomorphic.

The Case $\mathrm{E} = \mathrm{A}_1$ *and $m \neq 0$.* The lattices P_1 and P_2 are isomorphic by Corollary 3.1.16. Let α be a root of A_1, say "the" positive root. We necessarily have $\alpha \in \mathrm{Q}^1$ or $\alpha \in \mathrm{Q}'$. The same argument as before implies that we cannot have $\alpha \in \mathrm{Q}'$. We therefore have $\alpha \in \mathrm{Q}^1$, which implies $m \geq 2$. On the other hand, we have $m < 3$. Indeed, if we have $m \geq 3$, then there exists an element e in Q^1 with $e.e = 1$ and $e.\alpha = 0$ and we again have $\mathrm{s}_e(\alpha) = \alpha$ and $\mathrm{s}_e(P_1) = P_2$. □

B.5 Examples

B.5.1 Determination of T_2 *for* $n = 17$

The matrix of the Hecke operator $\mathrm{T}_2 : \mathbb{Z}[\mathrm{X}_{17}] \to \mathbb{Z}[\mathrm{X}_{17}]$ in the basis $(\mathrm{E}_{16} \oplus \mathrm{A}_1, \mathrm{E}_8 \oplus \mathrm{E}_8 \oplus \mathrm{A}_1, \mathrm{A}_{17}^+, (\mathrm{D}_{10} \oplus \mathrm{E}_7)^+)$ is the following (we also denote it by T_2) :

$$\mathrm{T}_2 = \begin{bmatrix} 20265 & 18225 & 153 & 63 \\ 12870 & 14910 & 0 & 90 \\ 16384 & 0 & 21624 & 18432 \\ 16016 & 32400 & 43758 & 46950 \end{bmatrix} .$$

Below, we explain how the theory of Sect. B.4 leads to this equality. We use the notation introduced at the end of Sect. B.2.

Let i and j be two elements of the set $\{2, 3, 6, 7\}$ with $i \neq j$. Borcherds' table [68, Chap. 17] provides the list of isomorphism classes of odd unimodular lattices of dimension 24 whose even unimodular 2-neighbors are isomorphic to P_i and P_j:

- The list is empty for $\{i, j\} = \{3, 6\}$.
- The list has only one element, which we denote by $[\mathrm{Bor}_{i,j}]$, for $\{i, j\} \neq \{3, 6\}$.

Borcherds' table also shows that in the case $\{i, j\} \neq \{3, 6\}$, the orthogonal group $\mathrm{O}(\mathrm{Bor}_{i,j})$ acts transitively on the set $\overline{\mathrm{Rep}}(\mathrm{E}_7, \mathrm{Bor}_{i,j})$.

Recall that the group $\mathrm{O}(\mathrm{P}_i)$ acts transitively on the set $\overline{\mathrm{Rep}}(\mathrm{E}_7, \mathrm{P}_i)$ for $i = 2, 3, 6, 7$.

Proposition 3.3.3.1 gives

$$\mathrm{N}_2(\mathrm{P}_i, \mathrm{P}_j) = 0 \text{ for } \{i, j\} = \{3, 6\} ,$$

$$\mathrm{N}_2(\mathrm{P}_i, \mathrm{P}_j) = \frac{|\mathrm{O}(\mathrm{P}_i)|}{|\mathrm{O}(\mathrm{Bor}_{i,j})|} \text{ for } \{i, j\} \neq \{3, 6\} .$$

Proposition B.4.5 and the discussion following it therefore give

$$\mathrm{N}_2(\mathrm{L}_i, \mathrm{L}_j) = 0 \text{ for } \{i, j\} = \{3, 6\} ,$$

$$\mathrm{N}_2(\mathrm{L}_i, \mathrm{L}_j) = \frac{|\overline{\mathrm{Rep}}(\mathrm{E}_7, \mathrm{Bor}_{i,j})|}{|\overline{\mathrm{Rep}}(\mathrm{E}_7, \mathrm{P}_i)|} \mathrm{N}_2(\mathrm{P}_i, \mathrm{P}_j) \text{ for } \{i, j\} \neq \{3, 6\} .$$

Consider, for example, the case $\{i, j\} = \{7, 2\}$. Borcherds' table mentioned above shows that we are in the case $|\overline{\mathrm{Rep}}(\mathrm{E}_7, \mathrm{Bor}_{7,2})| = 1$; in fact, Borcherds tells us that the set of roots of $\mathrm{Bor}_{7,2}$ (which has number 150 in the table) is isomorphic to $\mathbf{D}_{10} \coprod \mathbf{E}_7 \coprod \mathbf{D}_6 \coprod \mathbf{A}_1$. On the other hand, we have $|\overline{\mathrm{Rep}}(\mathrm{E}_7, \mathrm{P}_7)| = 2$ because the set of roots of P_7 is isomorphic to $\mathbf{D}_{10} \coprod \mathbf{E}_7 \coprod \mathbf{E}_7$. It follows that we have

$$\mathrm{N}_2(\mathrm{L}_7, \mathrm{L}_2) = \frac{1}{2} \mathrm{N}_2(\mathrm{P}_7, \mathrm{P}_2) = 63 .$$

Remark. The reader will note that we have $\mathrm{N}_2(\mathrm{E}_{16} \oplus \mathrm{A}_1, \mathrm{E}_8 \oplus \mathrm{E}_8 \oplus \mathrm{A}_1) = \mathrm{N}_2(\mathrm{E}_{16}, \mathrm{E}_8 \oplus \mathrm{E}_8)$ and $\mathrm{N}_2(\mathrm{E}_8 \oplus \mathrm{E}_8 \oplus \mathrm{A}_1, \mathrm{E}_{16} \oplus \mathrm{A}_1) = \mathrm{N}_2(\mathrm{E}_8 \oplus \mathrm{E}_8, \mathrm{E}_{16})$... and will have no trouble finding an explanation for this phenomenon.

As in the case $n = 24$, we see that the eigenvalues of the Hecke operator $\mathrm{T}_2 \colon \mathbb{Z}[\mathrm{X}_{17}] \to \mathbb{Z}[\mathrm{X}_{17}]$ are integral and simple. Hence, assuming Conjecture 8.1.2, part (ii) of Theorem 9.7.2 determines the Hecke operators $\mathrm{T}_p \colon \mathbb{Z}[\mathrm{X}_{17}] \to \mathbb{Z}[\mathrm{X}_{17}]$ for every prime p. For example, we obtain the following formula:

$$\frac{7}{286}\,\mathrm{N}_p(\mathrm{E}_{16} \oplus \mathrm{A}_1, \mathrm{E}_8 \oplus \mathrm{E}_8 \oplus \mathrm{A}_1) = (5p^4 + 7p^3 + 7p^2 + 7p + 5)\,\frac{p^{11} - \tau(p) + 1}{691} - 26\,\frac{p^{15} - \tau_{16}(p) + 1}{3617},$$

where $\tau_{16}(p)$ denotes the pth Fourier coefficient of the normalized modular cusp form (for $\mathrm{SL}_2(\mathbb{Z})$) of weight 16.

B.5.2 Determination of T_2 *for* $n = 15$

Part (b) of Proposition B.4.6 and the same arguments as before (rather, a simpler version of them) give

$$\frac{\mathrm{N}_2(\mathrm{E}_{15}, \mathrm{E}_7 \oplus \mathrm{E}_8)}{\mathrm{N}_2(\mathrm{E}_{16}, \mathrm{E}_8 \oplus \mathrm{E}_8)} = \frac{|\mathrm{R}(\mathrm{Bor}_{16})|}{|\mathrm{R}(\mathrm{E}_{16})|}, \quad \frac{\mathrm{N}_2(\mathrm{E}_7 \oplus \mathrm{E}_8, \mathrm{E}_{15})}{\mathrm{N}_2(\mathrm{E}_8 \oplus \mathrm{E}_8, \mathrm{E}_{16})} = \frac{|\mathrm{R}(\mathrm{Bor}_{16})|}{|\mathrm{R}(\mathrm{E}_8 \oplus \mathrm{E}_8)|},$$

where Bor_{16} is the odd unimodular lattice of dimension 16 introduced in Scholium-Definition 3.3.3.2, which "makes the jump" between E_{16} and $\mathrm{E}_8 \oplus \mathrm{E}_8$. Since we have $\mathrm{R}(\mathrm{Bor}_{16}) = \mathbf{D}_8 \coprod \mathbf{D}_8$ (Scholium-Definition 3.3.3.2), we find

$$\frac{\mathrm{N}_2(\mathrm{E}_{15}, \mathrm{E}_7 \oplus \mathrm{E}_8)}{\mathrm{N}_2(\mathrm{E}_{16}, \mathrm{E}_8 \oplus \mathrm{E}_8)} = \frac{7}{15}, \quad \frac{\mathrm{N}_2(\mathrm{E}_7 \oplus \mathrm{E}_8, \mathrm{E}_{15})}{\mathrm{N}_2(\mathrm{E}_8 \oplus \mathrm{E}_8, \mathrm{E}_{16})} = \frac{7}{15}. \tag{*}$$

It follows that the matrix of the Hecke operator $\mathrm{T}_2 : \mathbb{Z}[\mathrm{X}_{15}] \to \mathbb{Z}[\mathrm{X}_{15}]$ in the basis $(\mathrm{E}_{15}, \mathrm{E}_7 \oplus \mathrm{E}_8)$ is the following (we also denote it by T_2):

$$\mathrm{T}_2 = \begin{bmatrix} 10377 & 8505 \\ 6006 & 7878 \end{bmatrix}.$$

As before, assuming Conjecture 8.1.2, part (i) of Theorem 9.7.2 shows that the equalities (*) above generalize to every prime p:

$$\frac{\mathrm{N}_p(\mathrm{E}_{15}, \mathrm{E}_7 \oplus \mathrm{E}_8)}{\mathrm{N}_p(\mathrm{E}_{16}, \mathrm{E}_8 \oplus \mathrm{E}_8)} = \frac{p^3 - 1}{p^4 - 1}, \quad \frac{\mathrm{N}_p(\mathrm{E}_7 \oplus \mathrm{E}_8, \mathrm{E}_{15})}{\mathrm{N}_p(\mathrm{E}_8 \oplus \mathrm{E}_8, \mathrm{E}_{16})} = \frac{p^3 - 1}{p^4 - 1}.$$

B.5.3 On the Determination of T_2 for $n = 23$

The map $\pi\colon \mathrm{X}_{15} \to \mathrm{X}_{16}$ is injective; this implies that case (b.2) of Proposition B.4.6 does not occur for $n = 15$ (this argument has in fact been used above to determine T_2 for $n = 15$). On the other hand, the map $\pi\colon \mathrm{X}_{23} \to \mathrm{X}_{24}$ is not injective, and we will see that case (b.2) of Proposition B.4.6 does occur for $n = 23$.

Let us, for example, specify $\pi^{-1}([\mathrm{E}_{16} \oplus \mathrm{E}_8])$. The quotient $\mathrm{O}(\mathrm{E}_{16} \oplus \mathrm{E}_8)\backslash \mathrm{R}(\mathrm{E}_{16} \oplus \mathrm{E}_8)$ can be identified with the disjoint union $\mathrm{O}(\mathrm{E}_{16})\backslash \mathrm{R}(\mathrm{E}_{16}) \coprod \mathrm{O}(\mathrm{E}_8)\backslash \mathrm{R}(\mathrm{E}_8)$ (and is therefore a set with two elements). By Proposition B.2.6, it follows that we have $\pi^{-1}([\mathrm{E}_{16} \oplus \mathrm{E}_8]) = \{[\mathrm{E}_{15} \oplus \mathrm{E}_8], [\mathrm{E}_{16} \oplus \mathrm{E}_7]\}$.

Below, we show that the two elements $[\mathrm{E}_{15} \oplus \mathrm{E}_8]$ and $[\mathrm{E}_{16} \oplus \mathrm{E}_7]$ of X_{23} are 2-neighbors and that the set $\mathrm{e}^{-1}(\{[\mathrm{E}_{15} \oplus \mathrm{E}_8], [\mathrm{E}_{16} \oplus \mathrm{E}_7]\})$ (notation of Proposition B.4.5), viewed as a subset of $\mathrm{B}_{24}^{\mathrm{A}_1}$ (notation introduced in the discussion following Proposition B.4.5), is the singleton $\{[(\mathrm{I}_2 \oplus \Lambda; \iota)]\}$, where Λ denotes the unimodular lattice of dimension 22 corresponding to the obvious Lagrangian of the e-module $\mathrm{res}(\mathrm{E}_{15} \oplus \mathrm{E}_7)$ (we could also denote Λ by $(\mathrm{E}_{15} \oplus \mathrm{E}_7)^+$) and $\iota\colon \mathrm{A}_1 \to \mathrm{I}_2 \oplus \Lambda$ is the representation induced by the canonical representation $\mathrm{A}_1 \to \mathrm{I}_2$.

Set $S = \mathrm{A}_1 \oplus \mathrm{A}_1 \oplus \mathrm{E}_{15} \oplus \mathrm{E}_7$; denote by σ the obvious element of $\mathrm{O}(S)$ that interchanges the two factors A_1. Denote by ϖ_i the generator of the residue of the ith factor of S. The qe-module $\mathrm{res}\, S$ is therefore a $\mathbb{Z}/2$-vector space of dimension 4, with basis $\{\varpi_1, \varpi_2, \varpi_3, \varpi_4\}$, where the quadratic linking form is defined by $\mathrm{q}(\varpi_i) = 1/4$ for $i = 1, 2$, $\mathrm{q}(\varpi_i) = -1/4$ for $i = 3, 4$, and $\varpi_i.\varpi_j = 0$ for $i \neq j$. The structure of a Venkov qe-module is, for its part, determined by $\mathrm{qm}(\varpi_i) = 1/4$ for $i = 1, 2$ and $\mathrm{qm}(\varpi_i) = 3/4$ for $i = 3, 4$. The qe-module $\mathrm{res}\, S$ has two Lagrangians:

- the subspace J_1 generated by $\varpi_1 + \varpi_3$ and $\varpi_2 + \varpi_4$,
- the subspace J_2 generated by $\varpi_1 + \varpi_4$ and $\varpi_2 + \varpi_3$.

Note that these two Lagrangians are the graphs of the two isomorphisms of qe-modules from $\mathrm{res}(\mathrm{A}_1 \oplus \mathrm{A}_1)$ to $\langle -1 \rangle \otimes \mathrm{res}(\mathrm{E}_{15} \oplus \mathrm{E}_7)$ and that they are interchanged by σ.

For $k = 1, 2$, denote by P_k the even unimodular lattice with $S \subset P_k \subset S^\sharp$ and $P_k/S = J_k$. It is clear that P_1 and P_2 are both isomorphic to $\mathrm{E}_{16} \oplus \mathrm{E}_8$ and that they are interchanged by $\mathbb{Q} \otimes_{\mathbb{Z}} \sigma$.

Set $K = J_1 \cap J_2$; this is the subspace of $\mathrm{res}\, S$ of dimension 1 generated by $\varpi_1 + \varpi_2 + \varpi_3 + \varpi_4$. Denote by R the even lattice with $S \subset R \subset S^\sharp$ and $R/S = K$; we clearly have $R = P_1 \cap P_2$. We identify the qe-module $\mathrm{res}\, R$ with $K^\perp/K$ (Proposition 2.1.1); note that $\mathrm{res}\, R$ is generated by the classes of $\varpi_1 + \varpi_3$ and $\varpi_1 + \varpi_4$. This observation allows us to deduce that $\mathrm{res}\, R$ is isomorphic to $\mathrm{H}(\mathbb{Z}/2)$. Consequently, we see that P_1 and P_2 are 2-neighbors, say in $\mathbb{Q} \otimes_{\mathbb{Z}} S$.

Let L_k, for $k = 1, 2$, be the orthogonal complement of the first factor A_1 of S in P_k. We have $L_1 \simeq \mathrm{E}_{15} \oplus \mathrm{E}_8$ and $L_2 \simeq \mathrm{E}_{16} \oplus \mathrm{E}_7$. Proposition B.4.2 indeed shows that L_1 and L_2 are 2-neighbors (say in the orthogonal complement) of the first factor A_1 of S in $\mathbb{Q} \otimes_{\mathbb{Z}} S$.

Let J_3 be the linear subspace of $\mathrm{res}\, S$ generated by $\varpi_1 + \varpi_2$ and $\varpi_3 + \varpi_4$; then J_3 is a "Lagrangian of $\mathrm{res}\, S$ in the bilinear sense." The e-module $\mathrm{res}(\mathrm{A}_1 \oplus \mathrm{A}_1)$

(resp. $\operatorname{res}(\mathrm{E}_{15} \oplus \mathrm{E}_7)$) has a unique Lagrangian, which we denote by J_4 (resp. J_5); the subspace J_3 is the orthogonal sum of J_4 and J_5. The unimodular lattice associated with the pair $(\mathrm{A}_1 \oplus \mathrm{A}_1; J_4)$ is isomorphic to I_2. As stated before, we denote by Λ the unimodular lattice of dimension 22 associated with the pair $(\mathrm{E}_{15} \oplus \mathrm{E}_7; J_5)$; it is the first lattice of the table of Conway and Sloane [68, Chap. 16, Table 16.7, dim $= 22$]. We see that Λ does not represent 1 by invoking the structure of a Venkov qe-module of $\operatorname{res}(\mathrm{E}_{15} \oplus \mathrm{E}_7)$. The unimodular lattice associated with the pair $(S; J_3)$ is therefore isomorphic to $\mathrm{I}_2 \oplus \Lambda$. We have $K \subset J_3$; consequently, the Lagrangian of $\operatorname{res} R$ in the bilinear sense is J_3/K and the odd unimodular lattice whose even 2-neighbors are P_1 and P_2 coincides with the unimodular lattice associated with the pair $(S; J_3)$.

Finally, we deduce the following equality in $\mathrm{B}_{24}^{\mathrm{A}_1}$:

$$\mathrm{e}^{-1}(\{[\mathrm{E}_{15} \oplus \mathrm{E}_8], [\mathrm{E}_{16} \oplus \mathrm{E}_7]\}) \;=\; \{[(\mathrm{I}_2 \oplus \Lambda; \iota)]\}$$

by considering the first column of the table of Conway and Sloane mentioned above.

In view of this equality, we can determine $\mathrm{N}_2(\mathrm{E}_{15} \oplus \mathrm{E}_8, \mathrm{E}_{16} \oplus \mathrm{E}_7)$ using, for example, Proposition B.4.4:

$$\mathrm{N}_2(\mathrm{E}_{15} \oplus \mathrm{E}_8, \mathrm{E}_{16} \oplus \mathrm{E}_7) \;=\; \frac{|\mathrm{O}(\mathrm{E}_{15} \oplus \mathrm{E}_8)|}{|\mathrm{O}(\mathrm{A}_1 \oplus \mathrm{E}_{15} \oplus \mathrm{E}_7)|} \;=\; \frac{|\mathrm{O}(\mathrm{E}_8)|}{|\mathrm{O}(\mathrm{A}_1 \oplus \mathrm{E}_7)|} \;=\; 120\,.$$

To conclude this subsection, let us consider for $n = 23$ the observation made by Nebe and Venkov for $n = 24$: since the sum $\sum_{y \in \mathrm{X}_{23}} \mathrm{N}_2(x, y)$ is known for every x (Proposition B.4.3), to determine the Hecke operator $\mathrm{T}_2 : \mathbb{Z}[\mathrm{X}_{23}] \to \mathbb{Z}[\mathrm{X}_{23}]$, it suffices to compute $\mathrm{N}_2(x, y)$ for $x \neq y$. This observation and part (b) of Proposition B.4.6 (and the previous example) lead us to pose the following question:

Can we determine T_2 for $n = 23$, in the manner of Nebe–Venkov, by simply considering Borcherds' table [68, Chap. 17] and that of Conway and Sloane [68, Chap. 16, Table 16.7, dim $= 22$]?

Appendix C
Tables

In this appendix, we gather together the main tables of this book.

Table C.1 Standard parameters $\psi(\pi, \mathrm{St})$ of the representations π in $\Pi_{\mathrm{cusp}}(\mathrm{Sp}_{2g})$ generated by a Siegel modular form of weight 12 for $\mathrm{Sp}_{2g}(\mathbb{Z})$, in genus $g \leq 12$

g	$\psi(\pi, \mathrm{St})$	g	$\psi(\pi, \mathrm{St})$
		7	$\mathrm{Sym}^2\Delta_{11} \oplus \Delta_{17}[4] \oplus \Delta_{11}[2]$
1	$\mathrm{Sym}^2\Delta_{11}$	7	$\mathrm{Sym}^2\Delta_{11} \oplus \Delta_{15}[6]$
2	$\Delta_{21}[2] \oplus [1]$	8	$\Delta_{15}[8] \oplus [1]$
3	$\mathrm{Sym}^2\Delta_{11} \oplus \Delta_{19}[2]$	8	$\Delta_{21}[2] \oplus \Delta_{17}[2] \oplus \Delta_{11}[4] \oplus [1]$
4	$\Delta_{21}[2] \oplus \Delta_{17}[2] \oplus [1]$	8	$\Delta_{19}[4] \oplus \Delta_{11}[4] \oplus [1]$
4	$\Delta_{19}[4] \oplus [1]$	8	$\Delta_{21,9}[2] \oplus \Delta_{15}[4] \oplus [1]$
5	$\mathrm{Sym}^2\Delta_{11} \oplus \Delta_{19}[2] \oplus \Delta_{15}[2]$	9	$\mathrm{Sym}^2\Delta_{11} \oplus \Delta_{19}[2] \oplus \Delta_{11}[6]$
5	$\mathrm{Sym}^2\Delta_{11} \oplus \Delta_{17}[4]$	9	$\mathrm{Sym}^2\Delta_{11} \oplus \Delta_{19,7}[2] \oplus \Delta_{15}[2] \oplus \Delta_{11}[2]$
6	$\Delta_{17}[6] \oplus [1]$	10	$\Delta_{21}[2] \oplus \Delta_{11}[8] \oplus [1]$
6	$\Delta_{21}[2] \oplus \Delta_{15}[4] \oplus [1]$	10	$\Delta_{21,5}[2] \oplus \Delta_{17}[2] \oplus \Delta_{11}[4] \oplus [1]$
6	$\Delta_{21,13}[2] \oplus \Delta_{17}[2] \oplus [1]$	11	$\mathrm{Sym}^2\Delta_{11} \oplus \Delta_{11}[10]$
7	$\mathrm{Sym}^2\Delta_{11} \oplus \Delta_{19}[2] \oplus \Delta_{15}[2] \oplus \Delta_{11}[2]$	12	$\Delta_{11}[12] \oplus [1]$

G. Chenevier, J. Lannes, *Automorphic Forms and Even Unimodular Lattices*, Ergebnisse der Mathematik und ihrer Grenzgebiete. 3. Folge / A Series of Modern Surveys in Mathematics 69, https://doi.org/10.1007/978-3-319-95891-0

Table C.2 Fourier coefficients in $\frac{1}{2}$Gram of a well-chosen generator of $\mathrm{S}_{j,k}$

Gram \ (j,k)	$(0,10)$	$(6,8)$	$(0,12)$	$(8,8)$	$(12,6)$	$(4,10)$
$\begin{bmatrix} 2 & -1 \\ -1 & 2 \end{bmatrix}$	1	$Y^2X^2(X-Y)^2$	1	$Y^2X^2(X-Y)^2(X^2-YX+Y^2)$	$Y^4X^4(X-Y)^4$	$(X^2-YX+Y^2)^2$
$\begin{bmatrix} 2 & 0 \\ 0 & 2 \end{bmatrix}$	-2	$-2\,Y^2X^2(X^2+Y^2)$	10	$-2\,Y^2X^2(X^4-5\,Y^2X^2+Y^4)$	$-2\,Y^4X^4(X-Y)^2(X+Y)^2$	$-2\,(X^4-9\,Y^2X^2+Y^4)$
$\begin{bmatrix} 2 & -1 \\ -1 & 4 \end{bmatrix}$	-16	$-8\,Y^2(X-2\,Y)(X+Y)$ $\times(2\,X^2-2\,YX+Y^2)$	-88	$-8\,Y^2(2\,X^6-6\,YX^5+14\,Y^2X^4$ $-18\,Y^3X^3+14\,Y^4X^2-6\,Y^5X+3\,Y^6)$	$-16\,Y^4X^2(X-2\,Y)(X-Y)^2$ $\times(X+Y)(X^2-YX+Y^2)$	$-8(2X^4-4\,YX^3+21\,Y^2X^2$ $-19\,Y^3X-14\,Y^4)$
$\begin{bmatrix} 2 & 0 \\ 0 & 4 \end{bmatrix}$	36	$12\,Y^2(X^2+Y^2)$ $\times(3\,X^2-2\,Y^2)$	-132	$12\,Y^2(X^2-3\,Y^2)$ $\times(3\,X^4-5\,Y^2X^2-Y^4)$	$12\,Y^4X^2(3\,X^6-10\,Y^2X^4$ $+3\,Y^4X^2-4\,Y^4)$	$36\,(X^4-7\,Y^2X^2-7Y^4)$
$\begin{bmatrix} 2 & -1 \\ -1 & 6 \end{bmatrix}$	99	$3\,Y^2(33\,X^4-66\,YX^3$ $-91\,Y^2X^2+124\,Y^3X$ $-44\,Y^4)$	1275	$3\,Y^2(33\,X^6-99\,YX^5$ $+410\,Y^2X^4-655\,Y^3X^3$ $+343\,Y^4X^2-32\,Y^5X-132\,Y^6)$	$3\,Y^4(33\,X^8-132\,YX^7+142\,Y^2X^6$ $+36\,Y^3X^5-207\,Y^4X^4+200\,Y^5X^3$ $+88\,Y^6\,X^2-160\,Y^7X+80\,Y^8)$	$99\,(X^4-2\,YX^3+23\,Y^2X^2$ $-22\,Y^3X-7\,Y^4)$
$\begin{bmatrix} 4 & -2 \\ -2 & 4 \end{bmatrix}$	240	0	2784	$1344\,Y^2X^2(X-Y)^2(X^2-YX+Y^2)$	$-240\,Y^4X^4(X-Y)^4$	$-1680\,(X^2-YX+Y^2)^2$
$\begin{bmatrix} 2 & 0 \\ 0 & 6 \end{bmatrix}$	-272	$-16\,Y^2\,(X-Y)(X+Y)$ $\times(17\,X^2+13\,Y^2)$	736	$-16\,Y^2(X-Y)(X+Y)$ $\times(17\,X^4-62\,Y^2X^2+39\,Y^4)$	$-16\,Y^4(X-Y)(X+Y)$ $\times(17\,X^6-63\,Y^2X^4-6\,Y^4X^2-20\,Y^6)$	$-16\,(17\,X^4-96\,Y^2X^2$ $-144\,Y^4)$

Table C.3 Eigenvalues of Hecke operators in genus 2: the integer $\tau_{j,k}(p)$ for p prime and $p \leq 113$

p	$\tau_{6,8}(p)$	$\tau_{8,8}(p)$
2	0	1344
3	−27000	−6408
5	2843100	−30774900
7	−107822000	451366384
11	3760397784	13030789224
13	9952079500	−328006712228
17	243132070500	5520456217764
19	595569231400	−28220918878760
23	−6848349930000	79689608755152
29	53451678149100	−1105748270340
31	234734887975744	1851264166857664
37	448712646713500	22115741387845324
41	−1267141915544076	−29442241674311916
43	−1828093644641000	308109789751260712
47	−6797312934516000	43932618784857504
53	30226618925077500	−1178253142902441108
59	−51143734375273800	−3366234739477561080
61	7626516406720684	−8962102322409921476
67	−12252758021387000	14381861853876396664
71	−225641741059730736	40475791736823448944
73	486083162996216500	−11604559187113183148
79	1424574980940205600	14996327278915320160
83	−1351980902639367000	−154502893221792192168
89	−1127953215815294700	−49999331367987019020
97	−2710671093611565500	765838865005585444804
101	14595359522423307804	−1274759541025862678196
103	18796572299556586000	1130145111856472690992
107	−23385476046562641000	542230976527798722984
109	36219247764172458700	−884687494456719863780
113	−53733316769620465500	705599831303150185572

Table C.3 (continued)

p	$\tau_{12,6}(p)$	$\tau_{4,10}(p)$
2	−240	−1680
3	68040	55080
5	14765100	−7338900
7	−334972400	609422800
11	3580209624	25358200824
13	91151149180	−263384451140
17	−11025016477020	−2146704955740
19	−22060913325080	43021727413960
23	195863810691120	−233610984201360
29	−1743496339579620	−545371828324260
31	1979302106496064	830680103136064
37	−3685951226317460	11555498201265580
41	106065086529460884	−56208480716702316
43	74859021001125400	160336767963955000
47	156108802652634720	−116311331328502560
53	−1224706812408694260	−1944489787072554420
59	6289866383536712760	1843701997761637080
61	4857626575164933724	2376385974282228124
67	10336923176891703880	487223803841627560
71	−39237199980379430256	18272191888645387344
73	9078939377243940820	26899631446378070740
79	71873557961577515680	−80184572998399700960
83	94316650925918995560	157078549808482338120
89	115915137334350529140	228736927498417432220
97	894968190691418183620	−219326787347594393660
101	757457498875570442204	867394381514415093804
103	229164380766640031440	−657903326636255684720
107	−3571178446181577738600	−395867979731685155400
109	−2024515635534667135940	30287010492785677180
113	−4230007868022803115420	1657202008073896578660

Table C.4 Eigenvalues of Hecke operators in genus 2: the integer $\tau_{j,k}(p^2)$ for p prime and $p \leq 29$

p	$\tau_{6,8}(p^2)$	$\tau_{8,8}(p^2)$
2	409600	348160
3	333371700	748312020
5	−15923680827500	−395299890927500
7	−253514141409500	−155544419215478300
11	−75764187476725473836	19641545832571328136244
13	−4843967045593944889100	−596184280686941758305260
17	101161485715920379759300	−208424259842935445790738620
19	2430966330762186234484084	−1388004707990982166729991276
23	−129889399810754988793919900	−36435169742921431436190920540
29	−7216762572241226809807993676	−18636070203076686393140997747116

p	$\tau_{12,6}(p^2)$	$\tau_{4,10}(p^2)$
2	4276480	−700160
3	−8215290540	1854007380
5	722477627072500	−904546757727500
7	−1126868422025500700	−391120313742441500
11	−2263452414601610414156	−18738678558496864257356
13	−299941151717771094659180	323494600665947822387860
17	−94260803115254202283241660	70477693184423227137834820
19	−475514565037103383307581676	−1048771276144665792567133676
23	−505868492227965057753270620	−93299515424177439346879450460
29	−11097155072276494608459664937516	−2847689414234249875206600521516

Table C.5 Standard parameters $\psi(\pi, \mathrm{St})$ of the 24 representations π in $\Pi_{\mathrm{disc}}(\mathrm{O}_{24})$ such that π_∞ is trivial, ordered by increasing degree

i	ψ_i	λ_i	g_i
1	$[23] \oplus [1]$	8390655	0
2	$\mathrm{Sym}^2\Delta_{11} \oplus [21]$	4192830	1
3	$\Delta_{21}[2] \oplus [1] \oplus [19]$	2098332	2
4	$\mathrm{Sym}^2\Delta_{11} \oplus \Delta_{19}[2] \oplus [17]$	1049832	3
5	$\Delta_{19}[4] \oplus [1] \oplus [15]$	533160	4
6	$\Delta_{21}[2] \oplus \Delta_{17}[2] \oplus [1] \oplus [15]$	519120	4
7	$\mathrm{Sym}^2\Delta_{11} \oplus \Delta_{19}[2] \oplus \Delta_{15}[2] \oplus [13]$	268560	5
8	$\mathrm{Sym}^2\Delta_{11} \oplus \Delta_{17}[4] \oplus [13]$	244800	5
9	$\Delta_{21}[2] \oplus \Delta_{15}[4] \oplus [1] \oplus [11]$	145152	6
10	$\Delta_{21,13}[2] \oplus \Delta_{17}[2] \oplus [1] \oplus [11]$	126000	6
11	$\Delta_{17}[6] \oplus [1] \oplus [11]$	99792	6
12	$\mathrm{Sym}^2\Delta_{11} \oplus \Delta_{15}[6] \oplus [9]$	91152	7
14	$\mathrm{Sym}^2\Delta_{11} \oplus \Delta_{19}[2] \oplus \Delta_{15}[2] \oplus \Delta_{11}[2] \oplus [9]$	69552	7
16	$\mathrm{Sym}^2\Delta_{11} \oplus \Delta_{17}[4] \oplus \Delta_{11}[2] \oplus [9]$	45792	7
13	$\Delta_{15}[8] \oplus [1] \oplus [7]$	89640	8
15	$\Delta_{21,9}[2] \oplus \Delta_{15}[4] \oplus [1] \oplus [7]$	51552	8
17	$\Delta_{19}[4] \oplus \Delta_{11}[4] \oplus [1] \oplus [7]$	35640	8
18	$\Delta_{21}[2] \oplus \Delta_{17}[2] \oplus \Delta_{11}[4] \oplus [1] \oplus [7]$	21600	8
19	$\mathrm{Sym}^2\Delta_{11} \oplus \Delta_{19,7}[2] \oplus \Delta_{15}[2] \oplus \Delta_{11}[2] \oplus [5]$	17280	9
20	$\mathrm{Sym}^2\Delta_{11} \oplus \Delta_{19}[2] \oplus \Delta_{11}[6] \oplus [5]$	5040	9
21	$\Delta_{21,5}[2] \oplus \Delta_{17}[2] \oplus \Delta_{11}[4] \oplus [1] \oplus [3]$	−7920	10
22	$\Delta_{21}[2] \oplus \Delta_{11}[8] \oplus [1] \oplus [3]$	−16128	10
23	$\mathrm{Sym}^2\Delta_{11} \oplus \Delta_{11}[10] \oplus [1]$	−48528	11
24	$\Delta_{11}[12]$	−98280	12

Table C.6 The 8-tuples of integers $(m_1, \ldots, m_8)$ satisfying $4 \geq m_1 \geq m_2 \geq \cdots \geq m_8 \geq 0$ with $V_\lambda^\Gamma \neq 0$, where V_λ is the irreducible representation of $\mathrm{SO}(\mathbb{R}^{16})$ of highest weight $\lambda = \sum_{i=1}^{8} m_i \varepsilon_i$ and $\Gamma = \mathrm{SO}(\mathrm{E}_8 \oplus \mathrm{E}_8)$

$(m_1, \ldots, m_8)$	$\dim V_\lambda^\Gamma$	$(m_1, \ldots, m_8)$	$\dim V_\lambda^\Gamma$	$(m_1, \ldots, m_8)$	$\dim V_\lambda^\Gamma$
$(0,0,0,0,0,0,0,0)$	1	$(4,4,0,0,0,0,0,0)$	1	$(4,4,4,4,2,2,0,0)$	1
$(2,2,0,0,0,0,0,0)$	1	$(4,4,2,2,0,0,0,0)$	1	$(4,4,4,4,2,2,2,2)$	1
$(2,2,2,2,0,0,0,0)$	1	$(4,4,2,2,2,2,0,0)$	1	$(4,4,4,4,4,0,0,0)$	1
$(2,2,2,2,2,2,0,0)$	1	$(4,4,2,2,2,2,2,2)$	1	$(4,4,4,4,4,2,2,0)$	1
$(2,2,2,2,2,2,2,2)$	1	$(4,4,4,0,0,0,0,0)$	1	$(4,4,4,4,4,4,0,0)$	1
$(4,0,0,0,0,0,0,0)$	1	$(4,4,4,2,2,0,0,0)$	1	$(4,4,4,4,4,4,2,2)$	1
$(4,2,2,0,0,0,0,0)$	1	$(4,4,4,2,2,2,2,0)$	1	$(4,4,4,4,4,4,4,0)$	1
$(4,2,2,2,2,0,0,0)$	1	$(4,4,4,4,0,0,0,0)$	2	$(4,4,4,4,4,4,4,4)$	2
$(4,2,2,2,2,2,2,0)$	1				

Table C.7 Standard parameters of the 32 representations π in $\Pi_{\mathrm{cusp}}(\mathrm{SO}_{23})$ with $\pi_\infty = 1$, assuming Conjecture 8.4.25

$[22]$	$\Delta_{11}[11]$
$\Delta_{15}[7] \oplus [8]$	$\Delta_{17}[5] \oplus [12]$
$\Delta_{19}[3] \oplus [16]$	$\Delta_{21} \oplus [20]$
$\Delta_{17}[5] \oplus \Delta_{11} \oplus [10]$	$\Delta_{19}[3] \oplus \Delta_{11}[5] \oplus [6]$
$\Delta_{19}[3] \oplus \Delta_{15} \oplus [14]$	$\Delta_{21} \oplus \Delta_{11}[9] \oplus [2]$
$\Delta_{21} \oplus \Delta_{15}[5] \oplus [10]$	$\Delta_{21} \oplus \Delta_{17}[3] \oplus [14]$
$\Delta_{21} \oplus \Delta_{19} \oplus [18]$	$\Delta_{21,9} \oplus \Delta_{15}[5] \oplus [8]$
$\Delta_{21,13} \oplus \Delta_{17}[3] \oplus [12]$	$\Delta_{19}[3] \oplus \Delta_{15} \oplus \Delta_{11}[3] \oplus [8]$
$\Delta_{21} \oplus \Delta_{17}[3] \oplus \Delta_{11}[3] \oplus [8]$	$\Delta_{21} \oplus \Delta_{19} \oplus \Delta_{11}[7] \oplus [4]$
$\Delta_{21} \oplus \Delta_{19} \oplus \Delta_{15}[3] \oplus [12]$	$\Delta_{21} \oplus \Delta_{19} \oplus \Delta_{17} \oplus [16]$
$\Delta_{21,13} \oplus \Delta_{17}[3] \oplus \Delta_{11} \oplus [10]$	$\Delta_{21} \oplus \Delta_{19} \oplus \Delta_{15}[3] \oplus \Delta_{11} \oplus [10]$
$\Delta_{21} \oplus \Delta_{19} \oplus \Delta_{17} \oplus \Delta_{11}[5] \oplus [6]$	$\Delta_{21} \oplus \Delta_{19} \oplus \Delta_{17} \oplus \Delta_{15} \oplus [14]$
$\Delta_{21,5} \oplus \Delta_{19} \oplus \Delta_{17} \oplus \Delta_{11}[5] \oplus [4]$	$\Delta_{21,9} \oplus \Delta_{19} \oplus \Delta_{15}[3] \oplus \Delta_{11} \oplus [8]$
$\Delta_{21,9} \oplus \Delta_{19,7} \oplus \Delta_{15}[3] \oplus \Delta_{11} \oplus [6]$	$\Delta_{21,13} \oplus \Delta_{19} \oplus \Delta_{17} \oplus \Delta_{15} \oplus [12]$
$\Delta_{21} \oplus \Delta_{19} \oplus \Delta_{17} \oplus \Delta_{15} \oplus \Delta_{11}[3] \oplus [8]$	$\Delta_{21} \oplus \Delta_{19,7} \oplus \Delta_{17} \oplus \Delta_{15} \oplus \Delta_{11}[3] \oplus [6]$
$\Delta_{21,5} \oplus \Delta_{19,7} \oplus \Delta_{17} \oplus \Delta_{15} \oplus \Delta_{11}[3] \oplus [4]$	$\Delta_{21,13} \oplus \Delta_{19} \oplus \Delta_{17} \oplus \Delta_{15} \oplus \Delta_{11} \oplus [10]$

References

1. J. Adams, *Discrete series and characters of the component group*, in *Stabilization of the Trace Formula, Shimura Varieties, and Arithmetic Applications*, ed. by L. Clozel, M. Harris, J.-P. Labesse, and B.-C. Ngô (International Press, 2011), Chap. 4.2.
2. J. Adams, D. Barbasch, D. Vogan, *The Langlands classification and irreducible characters for real reductive groups*, Progr. Math., vol. 104 (Birkhaüser, Boston-Basel-Berlin, 1992).
3. J. Adams, J. Johnson, *Endoscopic groups and packets of non-tempered representations*, Compos. Math. **64** (1987), pp. 271–309.
4. A. N. Andrianov, *Euler products corresponding to Siegel modular forms of genus* 2, Uspekhi Mat. Naut **29** (1974), pp. 43–110.
5. A. N. Andrianov, *Quadratic forms and Hecke operators*, Grundlehren math. Wiss., vol. 286 (Springer-Verlag, 1987).
6. A. N. Andrianov, V. L. Kalinin, *On the analytic properties of standard zeta functions of Siegel modular forms*, Math. USSR, Sb. 35 (1979), pp. 1–17.
7. T. Arakawa, *Vector valued Siegel modular forms of degree two and the associated Andrianov* L*-function*, Manuscripta Math. **44** (1983), pp. 155–185.
8. N. Arancibia, *Paquets d'Arthur des représentations cohomologiques*, Ph. D. thesis, Univ. Paris 6 (2015).
9. J. Arthur, *Unipotent automorphic representations: conjectures*, in *Orbites unipotentes et représentations II : groupes p-adiques et réels*, Astérisque 171–172 (Soc. Math. France, Paris, 1989), pp. 13–71.
10. J. Arthur, *On the transfer of distributions: weighted orbital integrals*, Duke Math. J. **99**, pp. 209–283 (1999).
11. J. Arthur, *A note on the automorphic Langlands group*, Canad. Math. Bull. **45** (2002), pp. 466–482.
12. J. Arthur, *Automorphic representations of* $\mathrm{GSp}(4)$, in *Contributions to automorphic forms, geometry, and number theory* (Johns Hopkins Univ. Press, Baltimore, 2004), pp. 65–81.

G. Chenevier, J. Lannes, *Automorphic Forms and Even Unimodular Lattices*,
Ergebnisse der Mathematik und ihrer Grenzgebiete. 3. Folge / A Series of Modern
Surveys in Mathematics 69, https://doi.org/10.1007/978-3-319-95891-0

13. J. Arthur, *The endoscopic classification of representations: orthogonal and symplectic groups*, Colloquium Publ., vol. 61 (Amer. Math. Soc., 2013).
14. M. Asgari, R. Schmidt, *Siegel modular forms and representations*, Manuscripta Math. **104** (2001), pp. 173–200.
15. M.F. Atiyah, R. Bott, A. Shapiro, *Clifford modules*, Topology **3**, suppl. 1 (1964), pp. 3–38.
16. J. Barge, J. Lannes, *Suites de Sturm, indice de Maslov et périodicité de Bott*, Progress in Math., vol. 267 (Birkhäuser Verlag, Basel, 2008).
17. J. Barge, J. Lannes, F. Latour, P. Vogel, *Λ-sphères*, Ann. Sci. Éc. Norm. Sup. **7** (1974), pp. 463–505.
18. V. Bargmann, *Irreducible unitary representations of the Lorentz group*, Annals of Math. **48** (1947), pp. 568–640.
19. E.S. Barnes, G.E. Wall, *Some extreme forms defined in terms of abelian groups*, J. Austral. Math. Soc. **1** (1959), pp. 47–63.
20. J. Bellaïche, G. Chenevier, *The sign of Galois representations attached to automorphic forms for unitary groups*, Compos. Math. **147** (2011), pp. 1337–1352.
21. J. Bergström, C. Faber, G. van der Geer, *Siegel modular forms of degree three and the cohomology of local systems*, Selecta Math. **20** (2014), pp. 83–124.
22. F. van der Blij, T.A. Springer, *The arithmetic of octaves and the groups* G_2, Proc. Kon. Ak. Amsterdam **62** (= Ind. Math. **21**) (1959), pp. 406–418.
23. F. van der Blij, T.A. Springer, *Octaves and triality*, Nieuw Arch. Wiskd. **8** (1960), pp. 158–169.
24. S. Böcherer, *Über die Fourier-Jacobi-Entwicklung Siegelscher Eisensteinreihen*, Math. Z. **183** (1983), pp. 21–46.
25. S. Böcherer, *Über die Fourier-Jacobi-Entwicklung Siegelscher Eisensteinreihen II*, Math. Z. **189** (1985), pp. 81–110.
26. S. Böcherer, *Über die Funktionalgleichung automorpher* L*-Funktionen zur Siegelschen Modulgruppe*, J. reine angew. Math. **362** (1985), pp. 146–168.
27. S. Böcherer, *Siegel modular forms and theta series*, Proc. Symp. in Pure Math., vol. 49 (Bowdoin 1987, Amer. Math. Soc., 1989), pp. 3–17.
28. S. Böcherer, *Über den kern der thetaliftung*, Abh. Math. Sem. Univ. Hamburg **60** (1990), pp. 209–223.
29. R. Borcherds, *The Leech lattice and other lattices*, Ph. D. dissertation, Univ. of Cambridge (1984).
30. R. Borcherds, *The Leech lattice*, Proc. R. Soc. Lond. A **398** (1985), pp. 365–376.
31. R. Borcherds, E. Freitag, R. Weissauer, *A Siegel cusp form of degree 12 and weight 12*, J. reine angew. Math. **494** (1998), pp. 141–153.
32. A. Borel, *Some finiteness theorems for adele groups over number fields*, Publ. Math. de l'I.H.É.S. **16** (1963), pp. 101–126.
33. A. Borel, *Automorphic L-functions*, in *Automorphic forms, representations and L-functions, II* (Oregon State Univ., Corvallis, Ore.), Proc. Symp. in Pure Math. XXXIII (Amer. Math. Soc., Providence, RI, 1979), pp. 27–61.

34. A. Borel, *Linear algebraic groups, 2nd ed.*, Grad. Texts in Math., vol. 126 (Springer Verlag, 1991).
35. A. Borel, *Automorphic forms on* $\mathrm{SL}_2(\mathbb{R})$, Cambridge Tracts in Math., vol. 130 (Cambridge Univ. Press, 1997).
36. A. Borel, H. Jacquet, *Automorphic forms and automorphic representation* (Oregon State Univ., Corvallis, Ore.), in *Automorphic forms, representations and L-functions, II*, Proc. Symp. in Pure Math. XXXIII (Amer. Math. Soc., Providence, RI, 1979), pp. 189–203.
37. M. Borovoi, *Galois cohomology of reductive algebraic groups over the field of real numbers*, preprint available at http://arxiv.org/abs/1401.5913.
38. N. Bourbaki, *Éléments de mathématique, Algèbre, Chapitre 9* (Masson, Paris, 1981).
39. N. Bourbaki, *Éléments de mathématique, Groupes et algèbres de Lie, Chapitres 4, 5 et 6* (Masson, Paris, 1981).
40. S. Breulmann, M. Kuss, *On a conjecture of Duke-Imamoğlu*, Proc. Amer. Math. Soc. **128** (2000), pp. 1595–1604.
41. M. Broué, M. Enguehard, *Une famille infinie de formes quadratiques entières ; leurs groupes d'automorphismes*, Ann. Sci. Éc. Norm. Sup. **6** (1973), pp. 17–51.
42. K. Buzzard, *Notes on Siegel modular forms*, available at http://wwwf.imperial.ac.uk/~buzzard/maths/research/notes/siegel_modular_forms_notes.pdf.
43. K. Buzzard & T. Gee, *The conjectural connections between automorphic representations and Galois representations*, Automorphic forms and Galois representations Vol. 1, London Math. Soc. Lecture Note Ser. **414**, Cambridge Univ. Press, Cambridge (2014), pp 135–187.
44. A. Caraiani, *Local-global compatibility and the action of monodromy on nearby cycles*, Duke Math. J. **161** (2010), pp. 2311–2413.
45. *Variétés analytiques complexes et fonctions automorphes*, Séminaire H. Cartan, tome 6 (Éc. Norm. Sup. Paris, 1953/54).
46. *Fonctions automorphes*, Séminaire H. Cartan, tome 10 (Éc. Norm. Sup. Paris, 1957/58).
47. R. W. Carter, *Conjugacy classes in the Weyl group*, Compos. Math. **25** (1972), pp. 1–59.
48. P. Cartier, *Representations of p-adic groups: a survey*, in *Automorphic forms, representations and L-functions, I* (Oregon State Univ., Corvallis, Ore.), Proc. Symp. in Pure Math. XXXIII, (Amer. Math. Soc., Providence, RI, 1979), pp. 111–157.
49. P.-H. Chaudouard, G. Laumon, *Le lemme fondamental pondéré. I. Constructions géométriques*, Compos. Math. **146** (2010), pp. 1416–1506.
50. P.-H. Chaudouard, G. Laumon, *Le lemme fondamental pondéré. II. Énoncés cohomologiques*, Annals of Math. **176** (2012), pp. 1647–1781.
51. G. Chenevier, *Représentations galoisiennes automorphes et conséquences arithmétiques des conjectures de Langlands et Arthur*, mémoire d'habilitation à diriger des recherches, université Paris XI (2013).

52. G. Chenevier, M. Harris, *Construction of automorphic Galois representations II*, Cambridge Math. J. **1** (2013), pp. 53–73.
53. G. Chenevier, J. Lannes, *Kneser neighbours and orthogonal Galois representations in dimensions* 16 *and* 24, Algebraic Number Theory, Oberwolfach Rep. 31 (2011).
54. G. Chenevier, J. Lannes, PARI source code available at http://gaetan.chenevier.perso.math.cnrs.fr/articles/CL_Programmes.txt.
55. G. Chenevier, D. Renard, *Level one algebraic cusp form of classical groups of small rank*, Mem. Amer. Math. Soc., vol. 1121 (Amer. Math. Soc., Providence, RI, 2015).
56. C. Chevalley, *The algebraic theory of spinors and Clifford algebras*, oeuvres complètes vol. 2 (Springer-Verlag, Berlin, 1997).
57. C. L. Chai, G. Faltings, *Degeneration of abelian varieties*, with an appendix by D. Mumford, Ergeb. Math. Grenzgeb., vol. 22 (Springer-Verlag, 1990).
58. F. Cléry, G. van der Geer, *Constructing vector-valued Siegel modular forms from scalar valued Siegel mofular forms*, preprint available at http://arxiv.org/pdf/1409.7176v2.pdf (2014).
59. L. Clozel, *Motifs et formes automorphes : applications du principe de fonctorialité*, in *Automorphic forms, Shimura varieties, and L-functions, Vol. I* (Ann Arbor, MI., 1988), Perspectives in Math. vol. 10 (Academic Press, Boston, MA, 1990).
60. L. Clozel, *Purity Reigns Supreme*, Int. Math. Res. Not., I.M.R.N., vol. 2013 (2013), pp. 328–346.
61. L. Clozel, M. Harris, J.-P. Labesse, *Construction of automorphic Galois representations I*, in *Stabilization of the Trace Formula, Shimura Varieties, and Arithmetic Applications*, ed. by L. Clozel, M. Harris, J.-P. Labesse, B.-C. Ngô (International Press, 2011).
62. J. Cogdell, *Lectures on* L*-functions, converse theorems, and functoriality for* GL_n, in *Lectures on automorphic* L*-functions*, Fields Inst. Monogr. (Amer. Math. Soc., Providence, RI, 2004).
63. B. Conrad, *Reductive group schemes*, Notes of the Summer School SGA3 (Luminy, 2011), available at http://math.stanford.edu/~conrad/papers/.
64. J. B. Conrey, D. W. Farmer, *Hecke operators and the non-vanishing of L-functions*, in *Topics in number theory* (University Park, PA, 1997), Math. Appl., vol. 467 (Kluwer Acad. Publ., Dordrecht, 1999), pp. 143–150.
65. J. H. Conway, *A group of order 8,315,553,613,086,720,000*, Bull. London Math. Soc. **1** (1969), pp. 79–88.
66. J. H. Conway, *A characterisation of Leech's lattice*, Invent. Math. **7** (1969), pp. 137–142.
67. J. H. Conway, N. J. A. Sloane, *Twenty-three constructions for the Leech lattice*, Proc. Roy. Soc. London **381** (1982), pp. 275–283.
68. J. H. Conway, N. J. A. Sloane, *Sphere packings, lattices and groups*, Grundlehren math. Wiss., vol. 290 (Springer-Verlag, New York, 1999).
69. H. M. S. Coxeter, *Integral Cayley Numbers*, Duke Math. J. **13** (1946), pp. 567–578.

70. P. Deligne, *Formes modulaires et représentations ℓ-adiques*, Séminaire Bourbaki, no. 355 (1968/1969), pp. 139–172.
71. P. Deligne, N. Katz, *Groupes de monodromie en géométrie algébrique*, SGA7, Tome II, Séminaire de géométrie algébrique du Bois Marie 1967–69, Lecture Notes in Math., vol. 340 (Springer Verlag, 1973).
72. M. Demazure, P. Gabriel, *Groupes algébriques, Tome 1*, (Masson, Paris, 1970).
73. M. Demazure, A. Grothendieck, *Schémas en groupes*, SGA 3, Tome III, *Structure des schémas en groupes réductifs*, Séminaire de Géométrie Algébrique du Bois Marie 1962–1964, Doc. Math., vol. 8 (Soc. Math. France, 2011).
74. J. Dieudonné, *Pseudo-discriminant and Dickson invariant*, Pacific. J. Math. **5** (1955), pp. 907–910.
75. J. Dieudonné, *Sur les groupes classiques* (Hermann, Paris, 1997).
76. J. Dixmier, *Algèbres enveloppantes* (Gauthier-Villars Éd., 1974).
77. W. Duke, Ö. Imamoḡlu, *Siegel modular forms of small weight*, Math. Ann. **310** (1998), pp. 73–82.
78. M. Eichler, *Quadratischeformen und orthogonal gruppen*, Grundlehren math. Wiss. (Springer Verlag, 1952).
79. M. Eichler, D. Zagier, *The theory of Jacobi forms*, Progr. Math., vol. 55 (Birkhäuser, Boston, 1985).
80. V. Erokhin, *Theta series of even unimodular 24-dimensional lattices* (in Russian), Zap. Naučn. Sem. Leningrad Otdel. Mat. Inst. Steklov. (LOMI) **86** (1979), pp. 82–93.; English translation in J. Soviet Math. **17** (1981), pp. 1999–2008.
81. V. A. Erokhin, *Groups of automorphisms of 24-dimensional even unimodular lattices.* (in Russian), Zap. Naučn. Sem. Leningrad Otdel. Mat. Inst. Steklov. (LOMI) **116** (1982), pp. 68–73, 162.
82. C. Faber, *Modular forms and the cohomology of moduli spaces*, in *Automorphic forms and automorphic* L*-functions*, Surikaisekikenkyusho Kokyuroku **1826** (2013), pp. 129–137.
83. C. Faber, G. van der Geer, *Sur la cohomologie des systèmes locaux sur les espaces de modules des courbes de genre 2 et des surfaces abéliennes, I et II*, C. R. Acad. Sci. Paris, Sér. I **338** (2004), pp. 381–384.
84. S. Fermigier, *Annulation de la cohomologie cuspidale de sous-groupes de congruence de* $\mathrm{GL}_n(\mathbb{Z})$, Math. Ann. **306** (1996), pp. 247–256.
85. U. Fincke, M. Pohst, *Improved methods for calculating vectors of short length in a lattice, including a complexity analysis*, Math. Comp. **44** (1985), pp. 463–471.
86. E. Freitag, *Thetareihen mit harmonischen Koeffizienten zur Siegelschen Modulgruppe*, Math. Ann. **254** (1980), pp. 27–50.
87. E. Freitag, *Die Wirkung von Heckeoperatoren auf Thetareihen mit harmonischen Koeffizienten*, Math. Ann. **258** (1982), pp. 419–440.
88. E. Freitag, *Siegelsche Modulfunktionen*, Grundlehren der math. Wiss., vol. 254 (Springer Verlag, 1983).

89. G. van der Geer, *Siegel modular forms and their applications*, in *The 1–2–3 of modular forms*, ed. by J. H. Bruinier, G. van der Geer, G. Harder, D. Zagier, Universitext (Springer Verlag, Berlin, 2008), pp. 181–245.
90. S. Gelbart, H. Jacquet, *A relation between automorphic representations of* $\mathrm{GL}(2)$ *and* $\mathrm{GL}(3)$, Ann. Sci. Éc. Norm. Sup. **11** (1978), pp. 1–73.
91. S. Gelbart, F. Shahidi, *Boundedness of automorphic L-functions in vertical strips*, J. Amer. Math. Soc. **14** (2001), pp. 79–107.
92. I. M. Gel'fand, M. I. Graev, I. I. Pyatetskii-Shapiro, *Representation theory and automorphic functions* (Academic Press, 1990).
93. D. Ginzburg, S. Rallis, D. Soudry, *The descent map from automorphic representations of* $\mathrm{GL}(n)$ *to classical groups* (World Sci. Publishing, 2011).
94. W. Goldring, *Galois representations associated to holomorphic limits of discrete series I: unitary groups*, Compos. Math. **150** (2014), pp. 191–228.
95. R. Goodman, N. Wallach, *Representations and invariants of the classical groups* (Cambridge Univ. Press, 1998).
96. B. Gross, *Reductive groups over* $\mathbb{Z}$, Invent. Math. **124** (1996), pp. 263–279.
97. B. Gross, *On the Satake isomorphism*, in *Galois representations in arithmetic algebraic geometry*, ed. by A. Scholl, R. Taylor (Cambridge Univ. Press, 1998).
98. B. Gross, *Algebraic modular forms*, Israel J. Math. **113** (1999), pp. 61–93.
99. B. Gross, *On minuscule representations and the principal* SL_2, Electronic J. of Repr. Theory **4** (2000), pp. 225–244.
100. G. Harder, *A congruence between a Siegel and an elliptic modular form*, in *The 1–2–3 of modular forms*, ed. by J. H. Bruinier, G. van der Geer, G. Harder, D. Zagier (Universitext, Springer Verlag, Berlin, 2008), pp. 247–260.
101. Harish-Chandra, *Automorphic forms on semisimple Lie groups*, Lecture Notes in Math., vol. 62 (Springer Verlag, 1968).
102. J. Humphreys, *Introduction to Lie algebras and representation theory*, Grad. Texts in Math., vol. 9 (Springer Verlag, 1972).
103. J. Humphreys, *Linear algebraic groups*, Grad. Texts in Math., vol. 21 (Springer Verlag, 1975).
104. J. Humphreys, *Reflection groups and Coxeter groups*, Cambridge Stud. Adv. Math., vol. 29 (Cambridge Univ. Press, 1990).
105. J. Igusa, *On Siegel modular forms of genus two*, Amer. J. of Math. **84** (1962), pp. 175–200.
106. J. Igusa, *Modular forms and projective invariants*, Amer. J. of Math. **89** (1967), pp. 817–855.
107. J. Igusa, *Schottky's invariant and quadratic forms*, E. B. Christoffel (Aachen/Monschau, 1979) (Birkhäuser, Basel-Boston Mass., 1981), pp. 352–362.
108. T. Ikeda, *On the lifting of elliptic cusp forms to Siegel cusp forms of degree 2n*, Ann. of Math. **154** (2001), pp. 641–681.
109. T. Ikeda, *Pullback of the lifting of elliptic cusp forms and Miyawaki's conjecture*, Duke Math. J. **131** (2006), pp. 469–497.

110. T. Ikeda, *On the lifting of automorphic representations from* $\mathrm{PGL}_2(\mathbb{A})$ *to* $\mathrm{Sp}_{2n}(\mathbb{A})$ *or* $\mathrm{Sp}_{2n+1}(\mathbb{A})$ *over a totally real field*, available at https://www.math.kyoto-u.ac.jp/~ikeda/.
111. H. Jacquet, J. Shalika, *A non-vanishing theorem for zeta-functions of* GL_n, Invent. Math. **38** (1976), pp. 1–16.
112. H. Jacquet, J. Shalika, *On Euler products and the classification of automorphic representations I*, Amer. J. of Math. **103** (1981), pp. 499–558.
113. H. Jacquet, J. Shalika, *On Euler products and the classification of automorphic representations II*, Amer. J. of Math. **103** (1981), pp. 777–815.
114. J. Jantzen, *Representations of algebraic groups*, 2nd edn., Math. Survey Monogr., vol. 107 (Amer. Math. Soc., 2007).
115. T. Kaletha, *Rigid inner forms of real and* p*-adic groups*, preprint available at http://arxiv.org/abs/1304.3292.
116. M. Kashiwara, M. Vergne, *On the Segal-Shale-Weil representations and harmonic polynomials*, Invent. Math. **44** (1978), pp. 1–47.
117. C. Khare, *Serre's modularity conjecture : the level one case*, Duke Math. J. **134** (2006), pp. 557–589.
118. O. King, *A mass formula for unimodular lattices with no roots*, Math. Comp. **72**, no. 242 (2003), pp. 839–863.
119. A.W. Knapp, *Representation theory of semisimple groups* (Princeton Univ. Press, 1986).
120. A. Knapp, *Local Langlands correspondence: the archimedean case*, in *Motives* vol. II, Proc. Sympos. Pure Math., vol. 55 (Amer. Math. Soc., Providence, RI, 1994), pp. 393–410.
121. A. Knapp & G. Zuckerman, *Normalizing factors, tempered representations and L-groups*, in *Automorphic forms, representations and L-functions, I* (Oregon State Univ., Corvallis, Ore.), Proc. Symp. in Pure Math. XXXIII (Amer. Math. Soc., Providence, RI, 1979), pp. 93–105.
122. M. Kneser, *Klassenzahlen definiter quadratischer formen*, Archiv der Math. **8** (1957), pp. 241–250.
123. M. Kneser, *Strong approximation, algebraic groups and discontinuous subgroups*, Proc. Sympos. Pure Math., vol. 9 (Amer. Math. Soc., Boulder, 1966), pp. 187–196.
124. M. Kneser, *Lineare relationen zwischen darstellungsanzahlen quadratischer formen*, Math. Ann. **168** (1967), pp. 31–39.
125. M.-A. Knus, *Quadratic forms, Clifford algebras and spinors*, Sem. de Mat., vol. 1 (Campinas, 1988).
126. M.-A. Knus, R. Parimala, R. Sridharan, *On compositions and triality*, J. reine angew. Math. **457** (1994), pp. 45–70.
127. H. Koch, B. Venkov, *Über ganzzahlige unimodulare euklidische Gitter*, J. reine angew. Math. **398** (1989), pp. 144–168.
128. B. Kostant, *The principal three-dimensional subgroup and the Betti numbers of a complex simple Lie group*, Amer. J. Math. **81** (1959), pp. 973–1032.
129. R. Kottwitz, *Stable trace formula: cuspidal tempered terms*, Duke Math. J. **51** (1984), pp. 611–650.

130. R. Kottwitz, *Shimura varieties and λ-adic representations*, in *Automorphic forms, Shimura varieties, and L-functions* vol. I (Ann Arbor, MI, 1998), Perspect. Math., vol. 10 (Academic Press, 1990), pp. 161–209.
131. M. Kuga, G. Shimura, *On the zêta function of a fibre variety whose fibres are abelian varieties*, Annals of Math. **82** (1965), pp. 478–539.
132. N. Kurokawa, *Examples of eigenvalues of Hecke operators on Siegel cusp forms of degree two*, Invent. Math. **49** (1978), pp. 149–165.
133. J.-P. Labesse, R. Langlands, *L-indistinguishability for* SL(2), Canadian J. Math. **31** (1979), pp. 726–785.
134. J.-P. Labesse, J.-L. Waldspurger, *La formule des traces tordue d'après le Friday Morning Seminar*, CRM Monogr. Ser., vol 31 (Amer. Math. Soc., Providence, RI, 2013).
135. R. Langlands, *Problems in the theory of automorphic forms*, Lecture Notes in Math., vol. 170 (Springer Verlag, 1970).
136. R. Langlands, *Euler products*, Yale Math. Monogr., vol. 1 (Yale Univ. Press, 1971).
137. R. Langlands, *The classification of representations of real reductive groups*, Math. Surveys Monogr., vol. 31 (Amer. Math. Soc., Provindence, RI, 1973).
138. R. Langlands, *On the functional equation satisfied by Eisenstein series*, Lecture Notes in Math., vol. 544 (Springer Verlag, 1976).
139. R. Langlands, *Automorphic representations, Shimura varieties, and motives (Ein Märchen)*, in *Automorphic forms, representations and L-functions, II* (Oregon State Univ., Corvallis, Ore.), Proc. Symp. in Pure Math. XXXIII (Amer. Math. Soc., Providence, RI, 1979), pp. 205–246.
140. R. Langlands, D. Shelstad, *On the definition of transfer factors*, Math. Ann. **278** (1987), pp. 219–271.
141. J. Leech, *Some sphere packings in higher space*, Canad. J. Math. **16** (1964), pp. 657–682.
142. G. Lusztig, *Singularities, character formulas, and a q-analog of weight multiplicities*, Astérisque, vol. 101 (Soc. Math. France, Paris, 1983), pp. 208–227.
143. T. Mégarbané, work in progress.
144. J.-F. Mestre, *Formules explicites et minorations de conducteurs des variétés algébriques*, Compos. Math. **58** (1986), pp. 209–232.
145. P. Mezo, *Character identities in the twisted endoscopy of real reductive groups*, Mem. Amer. Math. Soc., vol. 222 (Amer. Math. Soc. Provindence, RI, 2013).
146. P. Mezo, *Tempered spectral transfer in the twisted endoscopy of real groups*, preprint available at http://people.math.carleton.ca/~mezo/ (2013).
147. S. Miller, *The highest-lowest zero and other applications of positivity*, Duke Math. J. **112** (2002), pp. 83–116.
148. J. Milnor, D. Husemoller, *Symmetric bilinear forms*, Ergebn. Math. Grenzgeb., vol. 73 (Springer Verlag, 1973).
149. I. Miyawaki, *Numerical examples of Siegel cusp forms of degree 3 and their zeta-functions*, Mem. of the Faculty of Sci., Kyushu University Ser. A **46** (1992), pp. 307–339.

150. S.-I. Mizumoto, *Poles and residues of standard L-functions attached to Siegel modular forms*, Math. Ann. **289** (1991), pp. 589–612.
151. C. Moeglin, J.-L. Waldspurger, *Le spectre résiduel de* $\mathrm{GL}(n)$, Ann. Sci. Éc. Norm. Sup. **22** (1989), pp. 605–674.
152. C. Moeglin, J.-L. Waldspurger, *Stabilisation de la formule des traces tordue VI & X*, preprint available at http://webusers.imj-prg.fr/~jean-loup.waldspurger/ (2014).
153. *Motives*, ed. by U. Jannsen, S. Kleiman, J.-P. Serre, Proc. Symp. in Pure Math., vol. 55 (Amer. Math. Soc., 1991).
154. G. Nebe, E. M. Rains, N. J. A. Sloane, *The invariants of the Clifford groups*, Des. Codes Cryptogr. **24** (2001), pp. 99–121.
155. G. Nebe, E. M. Rains, N. J. A. Sloane, *A simple construction for the Barnes-Wall lattices*, in *Forney Festschrift* (2002).
156. G. Nebe, B. Venkov, *On Siegel modular forms of weight 12*, J. reine angew. Math. **351** (2001), pp. 49–60.
157. B. C. Ngô, *Le lemme fondamental pour les algèbres de Lie*, Publ. Math. Inst. Hautes Études Sci. **111** (2009), pp. 1–169.
158. H.-V. Niemeier, *Definite quadratische Formen der Dimension* 24 *und Diskriminante* 1, J. Number Theory **5** (1973), pp. 142–178.
159. O. T. O'Meara, *Introduction to quadratic forms*, Classics in Math., vol. 117 (Springer Verlag, 1973).
160. The PARI Group, PARI/GP version `2.5.0`, Bordeaux, 2014, http://pari.math.u-bordeaux.fr/.
161. I. Piatetski-Shapiro, S. Rallis, *L-functions for the classical groups*, Lecture Notes in Math., vol. 1254 (Springer Verlag, 1987), pp. 1–52.
162. V. Platonov, A. Rapinchuk, *Algebraic groups and number theory*, Pure Appl. Math., vol 139 (1994).
163. V. Pless, *On the uniqueness of the Golay codes*, J. Combin. Theory **5** (1968), pp. 215–228.
164. V. Pless, N. J. A. Sloane, *Binary self-dual codes of length* 24, Bull. Amer. Math. Soc. **80** (1974), pp. 1173–1178.
165. G. Poitou, *Sur les petits discriminants*, Séminaire Delange-Pisot-Poitou, Théorie des nombres, tome 18, n. 1 exp. 6, 1–17 (1976).
166. G. Poitou, *Minorations de discriminants (d'après A. M. Odlyzko)*, Sém. Bourbaki no. 479 (1975/1976).
167. C. Poor, D. S. Yuen, *Dimensions of spaces of Siegel modular forms of low weight in degree four*, Bull. Austral. Math. Soc. **54**, no. 2 (1996), pp. 309–315.
168. C. Poor, D. S. Yuen, *Computations of spaces of Siegel modular cusp forms*, J. Math. Soc. Japan **59** (2007), pp. 185–222.
169. S. Rallis, *The Eichler commutation relation and the continuous spectrum of the Weil representation*, in *Non-Commutative Harmonic Analysis*, Lecture Notes in Math., vol. 728 (Springer Verlag, 1979), pp. 211–244.
170. S. Rallis, *Langlands' functoriality and the Weil representation*, Amer. J. Math. **104** (1982), pp. 469–515.

171. H. Resnikoff, R. Saldaña, *Some properties of Fourier coefficients of Eisenstein series of degree* 2, J. reine angew. Math. **265** (1974), pp. 90–109.
172. W. Rudin, *Real and complex analysis*, 3rd edn. (McGraw-Hill Int. ed.,1987).
173. Z. Rudnick, P. Sarnak, *Zeros of principal* L*-functions and random matrix theory*, Duke Math. J. **81** (1996), pp. 251–494.
174. I. Satake, *Theory of spherical functions on reductive algebraic groups over p-adic fields*, Publ. Math. Inst. Hautes Études Sci. **18** (1963), pp. 5–69.
175. R. Schulze-Pillot, *Siegel modular forms having the same L-functions*, J. Math. Sci. Univ. Tokyo **6** (1999), pp. 217–227.
176. J.-P. Serre, *Cohomologie galoisienne*, 5th edn., Lecture Notes in Math., vol. 5 (Springer Verlag, 1994).
177. J.-P. Serre, *Cours d'arithmétique* (Publ. Univ. France, Paris, 1970).
178. F. Shahidi, *Eisenstein series and automorphic L-functions*, Colloquium Publ., vol. 58 (Amer. Math. Soc., 2010).
179. D. Shelstad, *Orbital integrals and a family of groups attached to a real reductive group*, Ann. Sci. Éc. Norm. Sup. **12** (1979), pp. 1–31.
180. D. Shelstad, *L-indistinguishability for real groups*, Math. Ann. **259** (1982), pp. 385–430.
181. D. Shelstad, *Tempered endoscopy for real groups III : inversion of transfer and L-packet structure*, Represent. Theory **12** (2008), pp. 369–402.
182. D. Shelstad, *Examples in endoscopy for real groups*, notes from a talk at the conference *Stable trace formula* (Banff, 2008), available at http://andromeda.rutgers.edu/~shelstad/.
183. D. Shelstad, *Some results on endoscopic transfer*, notes from a talk at the conference *L-packets* (Banff, 2011), available at http://andromeda.rutgers.edu/~shelstad/.
184. D. Shelstad, *On geometric transfer in real twisted endoscopy*, Ann. of Math. **176** (2012), pp. 1919–1985.
185. D. Shelstad, *On the structure of endoscopic transfer factors*, preprint available at http://andromeda.rutgers.edu/~shelstad/.
186. D. Shelstad, *On spectral transfer factors in real twisted endoscopy*, preprint available at http://andromeda.rutgers.edu/~shelstad/.
187. G. Shimura, *On the modular correspondences of* $\mathrm{Sp}(n, \mathbb{Z})$ *and their congruence relation*, Proc. Natl. Aca. Sci. **49** (1963), pp. 824–828.
188. G. Shimura, *Introduction to the arithmetic theory of automorphic functions* (Princeton Univ. Press, 1971).
189. S.-W. Shin, *Galois representations arising from some compact Shimura varieties*, Ann. of Math. **173** (2011), pp. 1645–1741.
190. J. L. Spouge, *Computation of the Gamma, Digamma, and Trigamma functions*, SIAM J. Numer. Anal. **31** (1994), pp. 931–944.
191. T. Springer, *Linear algebraic groups*, 2nd edn., Progress in Math., vol. 9 (Birkhäuser, 1998).
192. T. Springer, *Reductive groups*, in *Automorphic forms, representations and L-functions, I* (Oregon State Univ., Corvallis, Ore.), Proc. Symp. in Pure Math. XXXIII (Amer. Math. Soc., Providence, RI, 1979), pp. 3–27.

193. J. R. Stembridge, *The partial order of dominant weights*, Adv. in Math. **136** (1998), pp. 340–364.
194. H. P. F. Swinnerton-Dyer, *On ℓ-adic representations and congruences for coefficients of modular forms*, in *Modular fonctions in one variable III*, Intl. Summer School on Modular fonctions (Antwerp, 1972).
195. O. Taïbi, *Dimensions of spaces of level one automorphic forms for split classical groups using the trace formula*, preprint available at http://arxiv.org/abs/1406.4247 (2014).
196. J. Tate, *Number theoretic background*, in *Automorphic forms, representations and L-functions, II* (Oregon State Univ., Corvallis, Ore.), Proc. Symp. in Pure Math. XXXIII (Amer. Math. Soc., Providence, RI, 1979), pp. 3–26.
197. R. Taylor, *Siegel modular forms of low weight*, Duke Math. J. **63** (1991), pp. 281–332.
198. J. Tits, *Reductive groups over local fields*, in *Automorphic forms, representations and L-functions, I* (Oregon State Univ., Corvallis, Ore.), Proc. Symp. in Pure Math. XXXIII (Amer. Math. Soc., Providence, RI, 1979), pp. 29–69.
199. R. Tsushima, *An explicit dimension formula for the spaces of generalized automorphic forms with respect to* $\mathrm{Sp}(2, \mathbb{Z})$, Proc. Japan Acad., vol. 59 (1983).
200. S. Tsuyumine, *Siegel modular forms of degree* 3, Amer. J. Math. **108** (1986), pp. 755–862.
201. B. Venkov, *On the classification of integral even unimodular* 24*-dimensional quadratic forms*, Chapter 18 in J. H. Conway, N. J. A. Sloane, *Sphere packings, lattices and groups*, Grundlehren math. Wiss., vol. 290 (Springer-Verlag, New York, 1999).
202. D. Vogan, *Representations of real reductive Lie groups*, Progr. Math., vol. 15 (1981).
203. D. Vogan, *On the unitarizability of certain series of representations*, Annals of Math. **120** (1984), pp. 141–187.
204. D. Vogan, G. Zuckerman, *Unitary representations with non-zero cohomology*, Compos. Math. **53** (1984), pp. 51–90.
205. J.-L. Waldspurger, *Engendrement par des séries thêta de certains espaces de formes modulaires*, Invent. Math. **50** (1979), pp. 135–168.
206. J.-L. Waldspurger, *Endoscopie et changement de caractéristique*, J. Inst. Math. Jussieu **5** (2006), pp. 423–525.
207. J.-L. Waldspurger, *Stabilisation de la formule des traces tordues I, II, III, IV, V, VII, VIII, IX*, preprints available at http://webusers.imj-prg.fr/~jean-loup.waldspurger/ (2014).
208. N. Wallach, *On the constant term of a square integrable automorphic form*, Operator algebras and group representations, Vol. II (Neptun, 1980), Monogr. Stud. Math., vol. 18 (Pitman, Boston, MA, 1984), pp. 227–237.
209. N. Wallach, *Real reductive groups I*, Pure and Applied math., vol. 132 (Acad. Press, 1988).
210. L. Walling, *Action of Hecke operators on Siegel theta series I*, Int. J. Number Theory **2** (2006), pp. 169–186.

211. A. Weil, *L'intégration dans les groupes topologiques et ses applications*, 2nd edn. (Hermann, 1940).
212. R. Weissauer, *Four dimensional Galois representations*, in *Formes automorphes (II), le cas du groupe* $\mathrm{GSp}(4)$, ed. by H. Carayol, M. Harris, J. Tilouine, M.-F. Vignéras, Astérisque, vol. 302 (Soc. Math. France, Paris, 2005).
213. E. Witt, *Eine Identität zwischen Modulformen zweiten Grades*, Abh. Math. Sem. Hansischen Univ. **14** (1941), pp. 323–337.
214. W. Wu, *Classes caractéristiques et i-carrés d'une variété*, C. R. Acad. Sci. (Paris, 1950), pp. 508–511.
215. H. Yoshida, *The action of Hecke operators on theta series* (Kinosaki, 1984), Algebraic and Topological Theories (Kinokuniya, Tokyo, 1986), pp. 197–238.
216. D. Zagier, *Sur la conjecture de Saito-Kurokawa (d'après H. Maass)*, Séminaire de Théorie des Nombres, Paris 1979–1980 (Sém. Delange-Pisot-Poitou), Progr. in Math., vol. 12 (Birkhäuser-Verlag, Boston-Basel-Stuttgart, 1980), pp. 372–394.

Postface

In this postface, added in 2018, we mention some results which are related to this book, but which have been proved after the end of its writing (June 2015).

(a) Conjecture 8.4.22 has been proved by Arancibia, Moeglin, and Renard in their work *Paquets d'Arthur des groupes classiques et unitaires*, to appear in *Annales de la faculté des sciences de Toulouse*. Moreover, conjectures 8.1.2 and 8.4.25 have also been proved by Taïbi, in his work *Arthur's multiplicity formula for certain inner forms of special orthogonal and symplectic groups*, to appear in *Journal of the European Mathematical Society*. As a consequence, our Theorems 8.5.2, 8.5.6, 8.5.8, 9.7.1, and 9.7.2, as well as the proofs in Sects. 9.2.10 and 9.2.11, are now unconditional. Moreover, by the discussion at the end of Sect. 8.5.7, this fully justifies all the information in Table 7.1, and by Theorem 9.5.9, this also shows the vanishing $\mathrm{S}_{11}(\mathrm{Sp}_{12}(\mathbb{Z})) = 0$ mentioned in Theorem G of the introduction.

(b) The methods of this book have been used by Mégarbané to study X_n for $n = 23$ and $n = 25$ in his work *Calcul des opérateurs de Hecke sur les classes d'isomorphisme de réseaux pairs de déterminant* 2 *en dimensions* 23 *et* 25, Journal of Number Theory **186**, pp. 370–416 (2018). He computes a matrix of T_2 acting on $\mathbb{Z}[\mathrm{X}_n]$ and studies the corresponding graphs $\mathrm{K}_n(p)$ for those two values of n. This allows him to refine some of the congruences of Theorem 10.4.4. For instance, he shows $\tau_{4,10}(p) \equiv \tau_{22}(p) + p^8 + p^{13} \bmod 9840$ for every prime p, which is an optimal strengthening of Harder's original congruence.

(c) The study of the modular Galois representations $\bar{\mathrm{r}}_{j,k;\ell}$ started in Sect. 10.4 has been pursued by Tayou in his work *Images de représentations galoisiennes associées à certaines formes modulaires de Siegel de genre* 2, International Journal of Number Theory **13**, 1129 (2017). Tayou substantially refines the statement of Proposition 10.4.10 and shows that for any prime $\ell > 19$, and any of the four relevant pairs (j, k), the representation $\bar{\mathrm{r}}_{j,k;\ell}$ is irreducible, except of

G. Chenevier, J. Lannes, *Automorphic Forms and Even Unimodular Lattices*, Ergebnisse der Mathematik und ihrer Grenzgebiete. 3. Folge / A Series of Modern Surveys in Mathematics 69, https://doi.org/10.1007/978-3-319-95891-0

course for $(j,k;\ell) = (8,8;23)$ and $(j,k;\ell) = (4,10;41)$. He also shows that when $\bar{\mathrm{r}}_{j,k;\ell}$ is irreducible, its image contains $\mathrm{Sp}_4(\mathbb{F}_\ell)$, except perhaps if we have $(j,k;\ell) = (6,8;13)$ or $(j,k;\ell) = (4,10;17)$ (the first of these two exceptions is compatible with our last Remark 10.4.11). It would be interesting to determine the image in these remaining two cases.

Notation Index

G. Chenevier, J. Lannes, *Automorphic Forms and Even Unimodular Lattices*,
Ergebnisse der Mathematik und ihrer Grenzgebiete. 3. Folge / A Series of Modern
Surveys in Mathematics 69, https://doi.org/10.1007/978-3-319-95891-0

Terminology Index

G. Chenevier, J. Lannes, *Automorphic Forms and Even Unimodular Lattices*, Ergebnisse der Mathematik und ihrer Grenzgebiete. 3. Folge / A Series of Modern Surveys in Mathematics 69, https://doi.org/10.1007/978-3-319-95891-0

H

I

L

M

N

O

P

Q

GPSR Compliance
The European Union's (EU) General Product Safety Regulation (GPSR) is a set of rules that requires consumer products to be safe and our obligations to ensure this.

If you have any concerns about our products, you can contact us on

ProductSafety@springernature.com

In case Publisher is established outside the EU, the EU authorized representative is:

Springer Nature Customer Service Center GmbH
Europaplatz 3
69115 Heidelberg, Germany

www.ingramcontent.com/pod-product-compliance
Ingram Content Group UK Ltd.
Pitfield, Milton Keynes, MK11 3LW, UK
UKHW021008290726
14059UKWH00001BA/25

* 9 7 8 3 3 1 9 9 5 8 9 0 3 *